OXFORD–WARBURG STUDIES

General Editors

ANN BLAIR *and* BILL SHERMAN

Oxford–Warburg Studies comprise works of original research on the intellectual and cultural history of Europe, with particular reference to the transmission and reception of ideas and images originating in the ancient world. The emphasis of the series is on elite rather than popular culture, and the underlying aim is to foster an interdisciplinary approach based on primary sources, which may be visual as well as written, and may extend to materials in a wide range of vernaculars and ancient languages. The authors of the series have addressed in particular the relationship between classical scholarship and the Christian tradition, the influence of modes of transmission on the uptake of ideas, the contributions of great scholars to the learning of their day, and the study of the Italian and Northern manifestations of humanism and their aftermath.

Former Series Editors

IAN MACLEAN
CHARLES HOPE
T. S. R. BOASE
J. B. TRAPP
DENYS HAY

OXFORD–WARBURG STUDIES

European Physico-theology (1650–c.1760) in Context: Celebrating Nature and Creation
KASPAR VON GREYERZ

The Reformation of Common Learning: Post-Ramist Method and the Reception of the New Philosophy, 1618–c. 1670
HOWARD HOTSON

The Invention of Papal History: Onofrio Panvinio between Renaissance and Catholic Reform
STEFAN BAUER

Making Mathematical Culture: University and Print in the Circle of Lefevre d'Étaples
RICHARD J. OOSTERHOFF

Criticism and Confession: The Bible in the Seventeenth Century Republic of Letters
NICHOLAS HARDY

Johann Heinrich Hottinger: Arabic and Islamic Studies in the Seventeenth Century
JAN LOOP

The Experience of Beauty in the Middle Ages
MARY CARRUTHERS

A History of Renaissance Rhetoric 1380–1620
PETER MACK

John Selden: A Life in Scholarship
G. J. TOOMER

The Church of England and Christian Antiquity: The Construction of a Confessional Identity in the 17th Century
JEAN-LOUIS QUANTIN

Latin Scientific Literature, 1450–1850

MARTIN KORENJAK

OXFORD
UNIVERSITY PRESS

OXFORD
UNIVERSITY PRESS

Great Clarendon Street, Oxford, OX2 6DP,
United Kingdom

Oxford University Press is a department of the University of Oxford.
It furthers the University's objective of excellence in research, scholarship,
and education by publishing worldwide. Oxford is a registered trade mark of
Oxford University Press in the UK and in certain other countries

Published in the United States of America by Oxford University Press
198 Madison Avenue, New York, NY 10016, United States of America

British Library Cataloguing in Publication Data
Data available

Library of Congress Control Number: 2023936285

ISBN 978–0–19–886605–3

DOI: 10.1093/oso/9780198866053.001.0001

Printed and bound by
CPI Group (UK) Ltd, Croydon, CR0 4YY

For Yin, Iris, and Flora

Preface

The present book results from the project *Nova Scientia: Early Modern Scientific Literature and Latin* (*NOSCEMUS*, 2017–23). A large part of its material basis comes from a database elaborated within this project which presents a hopefully representative selection of about one thousand early modern scientific writings in Latin (https://wiki.uibk.ac.at/noscemus/Main_Page). I draw on information from this database throughout without indicating it. When quoting from early modern Latin texts, I tacitly normalize spelling and interpunction, expand abbreviations, and ignore different fonts, unless I have specific reason not to. Trivial typographical errors are likewise corrected without comment, although accidentally omitted letters are added in angle brackets. The usually extensive titles of early modern works are abbreviated. The reference edition is always the *editio princeps*, unless indicated otherwise. Citations are given in the most reader-friendly way, whether this means a quire signature, a folio, page, column, chapter, or section number, or the title of a short paratext. Latin quotations in the main text (although not in the footnotes) are always translated or paraphrased, if their meaning is not evident from the context. All translations are my own. Personal names are given in the form in which they appear in the *Gemeinsame Normdatei*, except where another form is more current in English.

In applying for and conducting the aforementioned project, and even more in writing the book itself, I have incurred massive debts of gratitude towards several institutions and many people (none of whom is, of course, in any way responsible for the shortcomings of the final product). The members of a Scientific Advisory Board covering the main disciplines of early modern science, namely Kurt Girstmair (mathematics), Dag Nikolaus Hasse (astronomy/astrology), Robert Iliffe (physics), Florian Mittenhuber (geography), Brian W. Ogilvie (natural history), and Michael Stolberg (medicine), have been generous with their time and expertise. Günter Mühlberger from the department for Digitization and Electronic Archiving at Innsbruck was a highly competent and helpful partner concerning the digital side. The members of my core team, Ovanes Akopyan, Dominik Berrens, Johanna Luggin, and Irina Tautschnig, made the project an enjoyable experience, discussed many chapters with me at weekly meetings, and kindly allowed me to rely on database entries elaborated by them. Stefan Zathammer built the project's digital framework and worked out smart solutions to IT problems of all kinds far beyond his contractual duties. Our extraordinary and temporary team members William Barton, Florian Baumgartner, Manuel Huth, Urban Kirchler, Mārtiņš Laizāns, Sarah Lang, and Grigory Vorobyev provided stimulating information, comments, and criticism. In particular, I owe much mathematical advice and Figure 1 to Florian, information on Act and Tripos verses, snow dissertations, alchemy, and zoology to William, Urban, Sarah, and Grigory respectively. Single chapters were diligently read and commented upon by Tommi Alho, Sheldon Brammal, Oliver Budey, Julia Heideklang, Simon Smets, Pablo Toribio, and the members of the informal 'Leibnizkreis' that first gathered thirty years ago around Glenn W. Most in Heidelberg. Gábor Almási pointed out pertinent articles. Martin Bauer and Martin Lang provided information about medieval and ancient eastern matters. Bernhard Schirg introduced me to Johann Daniel Major. Some parts of the argument were presented to friendly and responsive audiences at Vienna, Berlin, Oxford, Budapest, Tübingen, Bochum, Bucharest, Albacete, Birmingham, Toronto, Prague, Pilsen, Leuven, and Bratislava. My colleagues at Innsbruck supplied fresh ideas, good spirits, and the most pleasant and stimulating working environment one could wish for. Notably, I could always rely on the scholarship and friendship of Wolfgang Kofler, Florian Schaffenrath, and Lav Šubarić. Ann Blair and Bill Sherman welcomed the book into the Oxford-Warburg Studies. The anonymous reader contributed valuable criticism and suggestions. My warmest thanks go to Ian Maclean, who had the kindness to preside over the aforementioned Advisory Board, suggested that literary genres could be understood as prototypical categories, encouraged me to send a proposal to Oxford University Press, and took it upon himself to see the book through the press after his official resignation as series editor.

In more practical terms, the European Research Council of the European Union provided generous funding for my project—basic research without any economical application—in a time of crisis and strained finances. During the application process, the background information given by Ylva Huber from the Austrian Research Promotion Agency was vital. The University of Innsbruck hosted the project and granted me an extraordinary sabbatical in 2021/22. The dedicated and reliable team of the university's Project Service Office helped with all kinds of organizational problems. Further administrative assistance and cartloads of books were contributed by Stefanie Lechner. Several libraries and archives gave me permission to use images from their collections; the Münchener Digitalisierungszentrum even did so for free through its generous download service 'Daten für die Forschung'. Andrea Obernauer diligently assisted me in composing the bibliography. The collaboration with the staff of the Press in general and my editor Cathryn Steele in particular was a pleasure.

I have always found it difficult to explain my work to those closest to me, and my wife and daughters have wondered for a long time what I am actually doing all day long. This book may not be much of an explanation, but I do not have a better one. It is dedicated to them with love and gratitude.

Contents

List of figures xi

Introduction 1
Subject and state of research 1
Aim and coverage 4
Synopsis 5
A note on terminology 10

PART I CONTEXTS

1. Science 13
An era of change 13
People and institutions 16
Intellectual context 20
Disciplines 23

2. Latin 34
Historical overview 34
Latin as the language of early modern science 36
Latin and other languages of science 39
Demise and persistence 41
The lexical peculiarity of scientific Latin 44

3. Literature 49
'Literature' in premodern times 49
The place of scientific texts 52
The literary form of scientific texts 53

4. Print 57
Early modern print 57
The printed scientific book 64

PART II TEXTS

5. Making sources accessible 77
Translations 77
Commentaries 103
Technical dictionaries 132
Bibliographies 146
Reviews 154

6. Presenting facts 163
Letters 163
Reports 190
Journal articles 203
'Historiae' 218
Table works 240

7. Arguing 251
Dialogues 251
Dissertations 265
Monographs 290

8. Summarizing knowledge 315
Pandects 315
Compendia 339
Textbooks 354

9. Publicizing science 382
Didactic poems 382
Panegyrical poems 395
Biographies 410
Orations 419

Conclusion 432
Towards modern scientific literature 432
What next? 435

Bibliography 437
Index of Names 507
Subject Index 517

List of figures

1. The disciplinary development of early modern science 25
Florian Baumgartner

2a–d. Four types of commentary layout 117
a: Bayerische Staatsbibliothek München, A.lat.a. 106#Beibd.1, p. 6, urn:nbn:de:bvb:12-bsb10170956-0; b: Bayerische Staatsbibliothek München, 4 Inc.c.a. 1236, sig. [e4^v], urn:nbn:de:bvb:12-bsb00058851-2; c: Bayerische Staatsbibliothek München, Res/2 Liturg. 322#Beibd.1, p. 2, urn:nbn:de:bvb:12-bsb10869723-2; d: Universitätsbibliothek Erlangen-Nürnberg, H00/4 PHS-V 2 o

3. A spread (fols. 2^v–3^r) from Gessner's *Catalogus plantarum*: plant names in Latin, Greek, German, and French 142
Bayerische Staatsbibliothek München, 4 A.lat.b. 11#Beibd., fols. 2^v–3^r, urn:nbn:de:bvb:12-bsb10151328-2

4. Title page of the first volume of the *Commentarii de rebus in scientia naturali et medicina gestis* 162
Bayerische Staatsbibliothek München, Med.g. 103-1, title page, urn:nbn:de:bvb:12-bsb10084150-0

5. The macrostructure of Jussieu's *Genera plantarum* in the form of a Ramist table (p. lxxi) 229
Staatsbibliothek Bamberg, Bot.o.24-a, p. lxxi, urn:nbn:de:bvb:12-bsb11437447-2

6. A broadsheet announcing a disputation at Bologna: the *Dubium astrologicum* 271
Archivio di Stato di Bologna, Riformatori dello Studio, 'Dispute e ripetizioni di scolari per ottenere letture d'università 1487–1515'

7. Title page of the *Consideratio physico-mathematica hiemis proxime praeterlapsae* 285
Universitätsbibliothek Augsburg, 02_VIII.2.4.14r, title page

8. Fermat's '2. figura': a set of straight lines (AQ, PV, TZ, YR, NM, OI), the sum of whose lengths is greater than the length of the curve APTYNOH (Tannery et al. 1891–1922, vol. 1, 216) 310

9. Frontispiece of Riccioli's *Almagestum novum* 328
Deutsche Meteorologische Bibliothek, Altbestand (1651), frontispiece

10. *Pharmacopoea Austriaco-provincialis*, p. 73: recipes presented in standardized layout 349
Bayerische Staatsbibliothek München, M.med. 807, p. 73, urn:nbn:de:bvb:12-bsb10288002-9

11. A typical page from Borrhaus's *Μαθημάτων στοιχεῖα*: one sentence per paragraph 368
Bayerische Staatsbibliothek München, Astr.u. 45, p. 19, urn:nbn:de:bvb:12-bsb10998850-4

12. The sophisticated layout of Biwald's *Physica generalis* 370
Bayerische Staatsbibliothek München, Phys.g. 45-1, p. 35, urn:nbn:de:bvb:12-bsb10130795-4

13. Presentation of the 'perca', a kind of bass, in the style of an emblem in Boussuet's *De natura aquatilium carmen* 387
Hathi Trust, public domain

14. Diagram illustrating the occurrence of a lunar eclipse from Mizauld's *De mundi sphaera* 391
Staats- und Stadtbibliothek Augsburg, Math 239#(Beibd.), p. 17, urn:nbn:de:bvb:12-bsb11267460-0

15. Portrait of Aldrovandi with epigram: 'This is not your image, Aristotle, but the one of Ulisse. / Your faces are different but your genius is equal.' 407
Staats- und Stadtbibliothek Augsburg, 2 Nat 1a-1, before the 'prolegomena', urn:nbn:de:bvb:12-bsb11200210-9

16. Grammatica's *Cometa poeticus* 409
Universitäts- und Landesbibliothek Tirol, 30.503, before p. 1

17. Herman Boerhaave delivering his speech on the lack of certainty in the natural sciences (frontispiece of *Sermo academicus de comparando certo in physicis*, 1715). Credit: Universitätsbibliothek Erlangen-Nürnberg, 4 TREW.R 341/342 425

Introduction

Subject and state of research

The early modern period produced a body of literature about the natural world that was more massive than anything written before and more diverse than anything written before or since on this theme. Inspired by the humanistic mindset of the contemporary intelligentsia and fuelled by the arrival of print, this literature reflected and promoted an intensifying and changing engagement with nature. In combination with sociological, economic, and institutional factors, it was instrumental in fostering modern science. Latin was the most important language of this literature for the better part of the period and remained one to be reckoned with until the nineteenth century. As the vernaculars made progress over the centuries, scientific texts written in these languages drew on Latin models with regard to literary form and terminology. This is still apparent in the predominantly Latin or Greco-Latin vocabulary of most modern disciplines.

The extent and importance of early modern scientific writings in Latin are well known in principle. 'Yet', as Anthony Grafton puts it in another context, 'the "translatio studii" from uncontroversial propositions to specific analysis is not easy.'[1] In fact, this body of texts has never been studied as a subject in its own right. Most historians of science have steered clear of it for linguistic reasons, most Latinists for reasons of content, and until recently, both groups because of problems of accessibility.

This is not to say that no work has been done in the field—far from it. First of all, many scholars not interested in Latin texts per se nonetheless were and are aware of their value as sources of information. It is true that Latin works were exploited more routinely in earlier days, when knowledge of the language was more widespread. An outstanding example of this approach is Lynn Thorndike's *History of Magic and Experimental Science*, whose later volumes list long stretches of pertinent books from the early modern period and describe their content.[2] On a smaller scale, however, much recent research still bases itself on information gained from Latin sources.

Apart from its use to other ends, Latin scientific literature in itself is to some extent documented in numerous bibliographies, which have been composed since early modern times. Robert Mortimer Gascoigne's catalogue of important

[1] Grafton 2008, 12. [2] Thorndike 1929–58, vols 4–8.

Latin Scientific Literature, 1450–1850. Martin Korenjak, Oxford University Press. © Martin Korenjak 2023.
DOI: 10.1093/oso/9780198866053.003.0001

scientific books contains a rich selection of Latin works in its early modern section.[3] The same is true of *Thornton's (and Tully's) Medical/Scientific Books, Libraries and Collectors*.[4] These selective overviews are complemented by others aiming at completeness but restricted in terms of chronology, geography, discipline, specific subject, literary genre, or combinations of these. Incunabula, for instance, received particular attention early on.[5] A monumental bibliography of the sciences in the Low Countries 1470–1700 has recently appeared.[6] Bibliographies also exist on subjects such as astronomy, Renaissance arithmetic, important mathematical works 1640–1940, botanical literature after 1753, or comets.[7] Dissertations have been bibliographized in the eighteenth and late twentieth centuries.[8] Some secondary literature contains separate bibliographical sections listing works from specific fields as well.[9] In addition to scholarly bibliographies, auction catalogues and catalogues of intact scientific libraries contain much of interest.[10] While most of the lists just cited include books irrespective of their language, some digital catalogues, *USTC* in particular, now enable users to apply this criterion and to focus their search on Latin works.

Important authors, for example Tycho Brahe, Nicolas Steno, Christiaan Huygens, or Euler, have been made accessible through 'opera omnia' editions.[11] Outstanding single works such as Vesalius's *Fabrica* (1543) or Newton's *Principia* (1687) have profited from similar initiatives.[12] Some such editions are equipped with commentaries: this is true, for instance, of Kepler's 'opera omnia' or Copernicus's *De revolutionibus* (1543).[13] For *De revolutionibus* and the *Fabrica*, censuses of all extant copies of their early modern editions have even been produced.[14]

This documenting work is supplemented by a great number of analytical studies on single disciplines, genres, authors, and texts.[15] Recently, short introductions to the field in its entirety have been provided in some handbooks of Neo-Latin, as the Latin of the modern period is commonly called.[16]

[3] Gascoigne 1984, 71–222.
[4] Besson 1990a; Hunter 2000.
[5] Klebs 1938; Sarton 1938.
[6] Hoogendoorn 2018.
[7] Astronomy: Houzeau and Lancaster 1882–7; Renaissance arithmetic: Smith 1908; important mathematical works: Grattan-Guinness 2005; botanical literature: Stafleu et al. 1976–; comets: Brüning 2000.
[8] Johann Carl Heffter, *Museum disputatorium physico-medicum* (1756–64); Marti 1982 (only partly on scientific topics).
[9] See, e.g., Paulus 2005, 405–21 (cosmology, astronomy, and meteorology from the Italian Renaissance); Korenjak 2019, 167–75 (didactic poetry).
[10] Many of these are cited in Besson 1990b, Mandelbrote 2000, and above all Overmier 2000.
[11] Brahe: Dreyer 1913–29; Steno: Maar 1910; Huygens: Société Hollandaise des Sciences 1888–1950; Euler: *LEOO*.
[12] *Fabrica*: Garrison and Hast 2014 (in English translation); *Principia*: Koyré et al. 1972.
[13] Kepler: *KGW*; *De revolutionibus*: Lerner et al. 2015.
[14] *De revolutionibus*: Gingerich 2002; *Fabrica:* Margósčy et al. 2018.
[15] Exemplary studies include Siraisi 1987 on Avicenna's *Canon* in the Renaissance, Friedrich 1995 on Conrad Gessner's *Historia animalium*, Pomata 1996, 2010, and 2011a on the medical 'observatio', Gärtner 2000 on German 'Rechenbücher' and their Latin pendants, and Habermann 2001 on German and Latin herbals and handbooks of surgery and distillation.
[16] *CNLS*, vol. 2, 324–61; *BENLW* 667–717; *OHNL* 263–77.

Despite all these valuable efforts, however, it is fair to say that the field as a whole is poorly known at present. There are still large blank areas on the map, and it is even difficult to locate them, because a systematic overview is lacking. As a consequence, Latinists and historians of science miss interesting aspects of their respective disciplines, run the risk of committing factual errors because they operate from a deficient source basis, and may even distort their subjects as a whole. Latinists' image of Neo-Latin literature tends to be overly dominated by belles-lettres and the precursors of the humanities, since the vast corpus of scientific and technical texts often remains outside their visual field. Historians of science incline to neglect aspects dominated by Latin, from dissertations and academic literature in general to the botanical tradition. On a more fundamental level, they sometimes underestimate the linguistic, philological, and literary component of early modern science and the extent to which it was integrated into the humanistic world of learning in general. In recent decades, these tendencies have been magnified by a general demise of the knowledge of Latin.[17]

In two respects, conditions currently appear favourable for changing this situation for the better. First, while scientific writing was hardly conceived as an appropriate object of philological study before the second half of the preceding century, interest in it has grown over recent decades (partly, it seems, due to the need to teach non-anglophone scientists scientific composition in English). Three main strands of this interest can be distinguished, although they show much overlap. One is linguistic and forms part of a more general upsurge in research in technical languages. Another focuses on stylistics, argumentation, and persuasion and is commonly referred to as 'rhetoric of science'. While these strands primarily target modern and contemporary science, a substantial number of publications nonetheless deal with early modern texts, even if these are mostly vernacular ones.[18] The third strand is historical from the outset and manifests itself in overviews of the scientific literature of particular periods. After Leonardo Olschki's study on the vernacular literature of the Italian Renaissance, published a full century ago and much ahead of its time,[19] recent decades have seen several such books on ancient literature[20] and certain kinds of early modern texts.[21]

[17] *CNLS*, vol. 2, 259.

[18] See, for instance, Jones 1966 (English scientific poetry of the eighteenth century); Vickers 1983 and Hallyn 1999, 620–7 (Galileo); Gross 1996 (including chapters on Copernicus, Newton, and Darwin); Fahnestock 1999 (rhetorical figures in early modern scientific texts); Gross et al. 2002 (English, French, and German scientific articles 1665 to present); Banks 2008 (English scientific texts from Chaucer to present). A number of seminal articles are collected in Harris 2018, a volume which also sketches the history and provides an overview of the field in its introduction (pp. 1–42).

[19] Olschki 1919–27. Olschki sees the use of Latin in the sciences as something historically doomed but discusses some of its aspects in vol. 2, 64–170.

[20] Kullmann et al. 1998; Asper 2007; Doody et al. 2012; Asper 2013, esp. 331–472; Taub 2017.

[21] Tebeaux 1997 discusses English technical texts, Paulus 2005 astronomical, meteorological, and cosmological texts of the Italian Renaissance in Latin and the vernacular. See also note 18 above.

In addition to this growing interest in scientific texts as literature, a non-academic development has made the present study timely—in fact, possible at all: the Digital Revolution. Until recently, a thorough account of early modern scientific literature in Latin would have been impossible to realize for practical reasons. More than a lifetime would have been needed to find the necessary materials in libraries throughout and beyond Europe, and the costs would have been prohibitive. This has now changed through the mass digitization of early modern books by initiatives and institutions such as Google Books, Internet Archive, Gallica, Münchener Digitalisierungszentrum, e-rara, and many others which has made an enormous and constantly growing number of Latin texts available online. The focus of each of the initiatives differs sufficiently to guarantee that the picture collectively provided by them is broadly representative. It literally has become possible to gain an overview of the field from one's desk.

Aim and coverage

To provide such a first overview in written form is the purpose of the present book. It aims at representativeness in regard to chronological and geographical spread, disciplines, and above all literary genres (that is, groups of texts sharing formal and content-related traits that guide authors in composing and readers in understanding them). The perspective from which this book is written is philological: works dealing with the natural world are described in terms of their social context, function, content, structure, and style, using the toolkit of philology and literary criticism. To curb different expectations, it is worthwhile to set this perspective apart from other possible approaches. I do not approach scientific texts in search of interesting positions or thought forms, as a historian of science would, and I have not chosen my examples for their scientific importance but as representatives of certain literary forms. While I hope that the book's results will be useful to historians of science, I am not qualified to make a contribution to this discipline myself. Likewise, despite the importance of book history for both literary history and the history of science, my book has no book historical aspirations. In the discussion of a scientific work, its physical form as a book is regularly taken into account, but more in order to elucidate points such as target readership, intended use, success, or the lack thereof than for its own sake. Finally, the present study is not a linguistic one. Although lexical, grammatical, and stylistic questions are addressed, the main objective is not to analyse Latin as a language of science. Necessary background information on the subjects just mentioned is provided in Part I of the book, as will become clear shortly.

In terms of chronology, texts are discussed that were produced between the invention of print around 1450 and the mid-nineteenth century, when Latin ceased to be an important language of science, with a particular focus on the sixteenth to eighteenth centuries. In the course of the four centuries covered, science and scientific literature underwent fundamental changes, and so treating them as a single period is open to objection. However, one can also point out a number of significant continuities beyond the persistent use of Latin. The university of 1800 was still medieval in many respects. Books were basically printed in a uniform manner until the introduction of the machine press. The history of many genres stretches from one end of the early modern era to the other. Reading habits apparently did not change greatly either. Of course, none of this would justify treating the period as a time-free zone, and I will draw attention to historical developments wherever appropriate.

In accordance with the present understanding of early modern science, I discuss works focusing on any aspect of the natural world, including disciplines such as astrology that are rated as occult or pseudo-scientific today.[22] On the other hand, works will be excluded in which such contents play a minor role and do not figure for their own sake but in the interest of other goals, for instance the development of a philosophical theory or a narrative plot.

Finally, I focus on printed books (and occasionally broadsheets) to the detriment of manuscripts. While the interaction of manuscript and printing culture was intense and manuscripts continued to play an important role throughout the entire early modern era, printed material had a larger readership and greater influence. Moreover, as already mentioned, a representative cross-section of the latter is currently available online, while the systematic inclusion of manuscripts would pose insurmountable problems in terms of heuristics, time, and finances. This said, side glances at manuscript material are provided where the occasion presents itself, and one very important genre that has for the better part remained in the sphere of handwriting, the letter, is discussed at length.

Synopsis

The book falls into two parts, the first of which may be called an extended introduction. It presents background information that enables or facilitates the understanding of what is said in the second part. Four aspects pertinent to Latin scientific literature are discussed in as many chapters: the early modern development of the disciplines concerned with the natural world; the role of Latin as the international language of the learned; the premodern idea of literature, which differed markedly from its modern counterpart; and the rise of print and its impact on scientific writing. Part I is entirely derivative. The reader may skip those chapters with whose subjects he or she is already familiar.

[22] Cf., for instance, the range of disciplines discussed in *CHS* 3 and 4.

The second and main part provides a survey of the major forms of early modern scientific literature. Its structure and contents need to be sketched in a bit more detail, starting with some conceptual problems. Generally speaking, the classification of scientific literature is a difficult task. To be satisfactory, a classification should in principle fully cover the field, make sense of its structure as a means to satisfy the needs of scientific teaching and practice, and provide a frame of reference within which any given text becomes understandable. Attempts in this direction have been made over the last decades for ancient Greek as well as for contemporary scientific (and technological) writing.[23] For the early modern period, nothing comparable exists, despite some rambling lists in contemporary bibliographies[24] and a few more systematic modern enumerations of genres in use in particular disciplines.[25] Regarding this era, the task of classification presents itself as especially challenging in two respects: first, early modern scientific literature is vast in terms of numbers and features a bewildering multitude of literary and (as Gianna Pomata terms them) epistemic genres.[26] Second, in the absence of a comprehensive theory or binding tradition of scientific writing, many genres bear no canonical name, show great internal variation, and have blurred boundaries. Authors enjoyed great freedom to create texts which belonged to several genres at the same time or were situated somewhere between them; occasionally, they found it impossible to classify their own production.[27] Whoever wants to impose some order on this vast and variegated landscape is hard pressed to decide which genres to postulate and how to define them.

To meet the first problem, the field has been prestructured by means of five basic communicative functions:[28] making written sources accessible (Chapter 5); presentation of facts and data (Chapter 6); argumentation (Chapter 7);

[23] See Asper 2007 and Göpferich 1995 (for a summary, Göpferich 1998–9) respectively.

[24] In the letter of dedication to his *Biblia iatrica* (1609), sigs.):(2^{v}–):(3^{r}, Johann Georg Schenck speaks of 'tabulis, lexicis, locis communibus, indicibus, aphorismis, dialogis, epistolis, quaestionibus, erotematibus, problematibus, isagogis, methodis, epitomis, diexodis, catalogis, consultationibus, responsis, contradictionibus, apologiis, paradoxis, disputationibus, depulsionibus et controversiis'. No system is discernible behind this, and many categories are difficult to grasp.

[25] See Reeds 1991, 136 for botany, Maclean 2002, 55–8 and Siraisi 2007, 65–9 for medicine. The categorization of the botanical books of Martin Fogel (1634–75) in Piepenbring-Thomas 2016, 256–7 intermingles chronological, formal, and content-related criteria.

[26] Pomata 2014, 2–3. Pomata distinguishes epistemic genres (examples named by her are treatise, commentary, textbook, encyclopedia, aphorism, dialogue, essay, medical recipe, case history, experimental article, atlas, and letter) from literary genres as 'that specific kind of genres whose primary goal is not the production of *meaning* but the production of *knowledge*' (her italics). Although her dichotomy formulates an important insight, it seems to me that in practice most genres discussed in this book aim at producing meaning as well as knowledge, which keeps me from adopting her terminology in what follows.

[27] In his autobibliographies, Girolamo Cardano resorted to default categories like 'extra ordinem' or 'varii incertique generis' (Maclean 2004, 16).

[28] Rough classification by communicative function has already been used regarding the scientific and technological literature of antiquity (Asper 2007, esp. 21: 'Wissenserschließung, Wissensvermittlung und Wissensthesaurierung') and the present (Göpferich 1995, esp. 123–8, 201–5: 'juristisch-normativ / fortschrittsorientiert-aktualisierend / didaktisch-instruktiv / wissenszusammenstellend').

summarizing knowledge (Chapter 8); and publicizing science (Chapter 9). These functions roughly correspond to five consecutive work steps in the natural sciences and thereby highlight the fundamental role of Latin literature in this context. To each function a number of major genres is assigned. (To discuss each and every minor genre has proved unfeasible; some of the omissions will be indicated in the Conclusion). Although the allocation of genres to functions is not always self-evident, as many genres fulfil two or more functions, one of these can usually be identified as pivotal.

The second problem is tackled by adopting a pragmatic combination of genres recognized as such in some way or other in their own time (the majority) and a few others recognizable in hindsight (for instance, reports, natural histories, textbooks), which, taken together, divide the field into meaningful and coherent units as far as possible. To account for the internal variation and fuzzy boundaries of many genres, genre is understood not as an Aristotelian but as a prototypical category. Such categories have no hard and fast criteria for inclusion and hence no clearly defined borders; rather, their members can be called ideal-typical, less perfect, or dubious depending on how many and how important criteria they meet associated with their particular category.[29] For instance, a perfect textbook is written for classroom use, covers a clearly delimited field corresponding to a certain part of a curriculum, is of modest size, clearly structured, and uses easy, pupil-friendly language. However, while a couple of early modern works perfectly fit this bill, the exact place of others in institutional teaching may remain unclear and they may be bulky, rambling, or hard to understand. My description of each genre will primarily be based on its ideal-typical representatives, but frequently occurring deviations will be taken into account as well.

The resulting model of Latin scientific literature, as any model, is intended to render an overly complex reality graspable by way of simplification. It hopefully provides a map orienting the reader in an alien territory from which it nonetheless remains categorially different. In what follows, a preliminary sketch of the map will be given.

The first communicative function, making written sources accessible, at first glance seems to be at odds with the main thrust of the Scientific Revolution which supposedly turned its back on canonical texts and oriented itself towards experience. However, the opposition of text and experience is misleading: with more empirical data available, science became not less, but more and more text-dependent—a process continuing to today, as shown by an ever-growing volume of publication. Even before this development gained momentum, one of its very preconditions was the widespread accessibility of the knowledge assembled in antiquity and the Middle Ages by Greek and Arabic authors. The first three genres

[29] On prototype theory in general, see Taylor 2003; on its application to text genres, Tophinke 1997.

to be discussed were instrumental to this end: translations of the authorities into Latin, commentaries elucidating them, and dictionaries making their terminology accessible. All three formats continued to play a role in later parts of the early modern period, when they were applied to contemporary scientific works. As books piled up ever higher, bibliographies became vital tools that helped scholars have an overview and decide what they had to read. From the later seventeenth century, reviews in journals collectively provided a kind of topical bibliography. Through their often extended summaries, they functioned as not only guides to but also convenient substitutes for books, if these were unavailable or their perusal too time-consuming.

The empiricism of the new science, its systematic efforts to collect fresh data and facts, found expression in a number of genres imparting this information to readers, typically as unencumbered as possible by any theory and in compartmentalized form. Most of the genres were 'discrete' texts, that is, strings of coordinate, formally unconnected units,[30] whether the result of many short texts being collected or one long one being subdivided into short sections. Letters were a foundational genre in this respect. Their exchange between scientists intensified over the early modern era, creating a continuous stream of information across the continent that provided raw material for new insights and theories. Letters were often published in collections and contributed to the emergence of two other genres, namely reports and journal articles. The former, first-person narratives of a particular experience, provided readers with a written equivalent of the personal observation of a natural phenomenon, disease process, section, or experiment. The latter—not strictly speaking a literary genre, but a 'publication genre', as many different kinds of texts may be turned into articles through publication in a journal—often originated from letters to a journal's editor and were written in report style. 'Historiae' (as I have chosen to call them) contained descriptive accounts of stones, plants, or animals, which were serialized and presented according to a uniform scheme. Collections of tables and charts constituted a particularly efficient way of presenting data in mathematics and strongly mathematized disciplines, astronomy in particular. As their use required considerable expertise, they were often supplied with detailed introductions.

The argumentative advocacy of certain positions follows as a logical third step after the uptake of pre-existing knowledge from books and the collection of fresh information from observation. An ancient genre into whose very texture this function was engrained was the dialogue. As the discussion between participants with different opinions was a vivid and elegant way of weighing the pros and cons of a given position, the genre was eagerly adopted to elucidate questions about the natural world. Dissertations were intimately connected to a ritualized form of

[30] The notion has been introduced by Asper 2007, 57–212, esp. 57–8.

discussion, the academic 'disputatio'. Although this practice was originally meant to enable students to find their way around the predefined truths of the curriculum, the printed texts were also used to present novel ideas. Ultimately, they evolved into the proof of capacity for original research that the dissertation is supposed to be today, while the 'disputatio' was abolished. A counterpart to the dissertation in terms of form and content was the monograph. In contrast to the former, however, it was independent of institutional structures and occasions. The focused, thoroughly argued treatment of a single, clearly circumscribed topic, whose author openly committed himself to a certain position, rose to new prominence in early modern science.

To the extent that a temporary consensus was reached by the exchange of arguments in a given field, the task of providing an overview of the agreed-on knowledge presented itself. In a time when knowledge accumulated and changed ever more rapidly, new works fulfilling this task were in continuous demand. The genres aspired to comprehensiveness to different extents and targeted different social settings, private ones as well as institutions of higher education. Pandects aimed at covering a certain discipline in its entirety and in detail. While these were expensive volumes in large formats, to be used on one's desk, the demand for portable summaries to be taken along on fieldtrips, sickbed visits, and similar occasions was satisfied by selective compendia. Finally, textbooks came to be written specifically for formal teaching, replacing commentaries on authoritative texts as standard teaching tools in the measure to which ancient and medieval wisdom was questioned and superseded.

Part II's final chapter discusses genres publicizing science, by which I mean two different but often combined activities: presenting and explaining scientific content in an easily understandable, enjoyable way, and praising science, scientists, and scientific results. Of course, science was often publicized in the vernaculars, but by no means exclusively so. Just as much popular science is in English today, in which form it reaches only a minority of the general public but does so on an international scale, Latin was an important language of popular science in early modern times. The addressees were the members of the 'respublica litteraria' as well as of the nascent scientific community itself. Publicizing science secured a minimum of common understanding in a time of increasing specialization, countered widespread hostility and ridicule, and projected a reassuring image of unanimity and selfless collaboration in the face of a reality often marked by dissent, factionalism, and personal vanity. In addition, it also responded to the economic necessity of book promotion. Didactic poetry, which had been an important genre for the presentation of scientific content in appealing form from ancient times, flourished in the early modern period. Poems of praise frequently introduced editions of scientific works, which they promoted by extolling the particular author's achievements and demonstrating the standing he enjoyed among his peers. Biographies of men of science, which formed a subgenre of biographies of intellectuals in general, were likewise used to introduce the works of their subjects but were also collected in self-contained volumes. Orations about scientific subjects constituted a subfield of epideictic oratory, which was vastly popular at the time, and graced occasions as diverse as academic festivities and funerals.

The individual sections of the chapters of Part II typically start with a literary contextualization of the genre in question, both diachronically in terms of ancient and medieval models and synchronically in terms of the contemporary literary environment (including non-scientific and vernacular texts, since few genres were exclusive to science and to Latin). The genre's social and institutional setting and its readership are then described, and its historical development is sketched. Finally, it is characterized in terms of content (scientific disciplines, level of technicality) and form (length, structure, verse or prose, style, vocabulary). However, these generic features are freely adapted to the peculiarities of the individual genres. The analysis is always illustrated by a number of typical works and text samples.

A note on terminology

The terminological dilemma in speaking about premodern 'science' is well known. Natural science did not exist as a unified field before the nineteenth century; yet no better collective designation for the individual disciplines than 'science' has been found so far. Following common practice, I therefore speak of early modern 'science' (and 'scientific' subjects, 'scientists', etc.) wherever convenient.[31] Similar, if slightly different problems occur and similar solutions are adopted regarding designations of disciplines like 'mathematics' or 'meteorology' whose early modern meaning differs from their present one. For the pre-eighteenth-century discipline comprising what is distinguished as 'alchemy' and 'chemistry' later, the term 'chymistry' has been proposed,[32] which I adopt.

[31] Cf., for instance, Blair 1996, 22–3; Shapiro 2000, 6 ('I occasionally use the terms "science", "scientist", and "scientific fact", but these terms are anachronistic').

[32] Newman and Principe 1998, 64–5.

PART I
CONTEXTS

1
Science

The basic information provided in this chapter falls into four parts. Initially, the most salient feature of early modern science, its dynamic evolution, is highlighted by contrasting its appearance at the beginning and end of the period. Its sociological and institutional background and its intellectual context are sketched in the two following sections. The chapter concludes with an overview of the internal structure of the field, its single disciplines, and their development.

An era of change

Natural science in the current sense of the word largely evolved in the time between the invention of print in the mid-fifteenth century and the decades around 1800. Whether this evolution, or part of it, should be called a 'Scientific Revolution' is debated and ultimately depends on the meaning given to the term 'revolution'.[1] In any case, the understanding of the natural world in Europe changed fundamentally within a few centuries. Human civilization and the surface of the earth began to look very different from how they had looked before on account of these changes.

Towards the end of the Middle Ages, the knowledges and practices partly corresponding to today's natural science did not constitute a unified, autonomous domain but formed a heterogeneous group of disciplines, as already transpires from their lack of a common name.[2] However, a precursor of science, in the sense of an intellectual field whose representatives claimed to describe and explain the natural world in its entirety, may be said to have existed in the shape of Aristotelian natural philosophy, which formed part of the academic curriculum and with which most educated people were familiar to some degree. As Aristotle explained in the *Posterior Analytics*, 'science' (ἐπιστήμη, comprising most of his philosophy) was deductive reasoning from certain and general first principles. Each subject

[1] The opening sentence of Stephen Shapin's *The Scientific Revolution* (Shapin 1996, 1) comes close to summarizing the currently dominant opinion: 'There was no such thing as the Scientific Revolution, and this is a book about it.' The opposing standpoint is advocated by Wootton 2015. For a notion of the Scientific Revolution not restricted to the seventeenth century but basically extending over the whole early modern period, see Knight 2014, 5.

[2] On medieval science, see *CHS* 2. Help with orientation within the extensive scientific literature of the era is provided by Thorndike 1929–58, vols. 1–4 and Thorndike and Kibre 1963.

Latin Scientific Literature, 1450–1850. Martin Korenjak, Oxford University Press. © Martin Korenjak 2023.
DOI: 10.1093/oso/9780198866053.003.0002

should be treated according to those distinct methodological premises that were appropriate to it; the transfer of methods from one discipline to another (*μετάβασις*) was illegitimate. Regarding nature, the goal was to understand how it worked always or for the most part. To this end, Aristotle made use of a number of pivotal notions and tenets: nature as the realm of things originating by themselves; the four causes contributing to a thing's existence—material, formal, efficient, and final; things as compounds of matter and form; the principle of plenitude, implying the impossibility of a vacuum; the four elements, their qualities (hot/cold, dry/wet), and their transmutability; and the dichotomy between the sublunary and superlunary spheres, the former the realm of the four elements, change, and rectilinear motion, the latter that of the fifth element, eternal being, and circular motion.

Besides natural philosophy, there were other disciplines concerned in some way or other with various aspects of nature, later versions of which came to form part of natural science in modern times. Despite the non-Aristotelian origin of these disciplines, Aristotelian dogmas played a fundamental role in a number of them (while they were less important in others such as chymistry). Astronomers, for instance, adopted the way things worked beyond the moon, physicians the elements and their qualities. Aristotle's epistemological ideas, in particular regarding the deductive character of science and the illegitimacy of *μετάβασις*, were also taken over to some degree. Because no self-evident first principles were as a rule available in the realm of nature, the tenets of authorities were often taken as starting points instead, which resulted in a marked traditionalism. Like natural philosophers, astronomers and doctors saw themselves as bound by tradition as it manifested itself in authoritative texts and aimed at the unadulterated transmission of their contents. Their deductive mindset made them wary of personal observation, because it yielded information about particulars, which could be transformed into general knowledge only by way of induction. All of this created some common ground, but due to the ban on *μετάβασις* as well as to their different histories and social contexts, the single disciplines nonetheless differed widely in their aims, methods, and modes of presentation and for the time being did not coalesce into one field governed by common rules.

Around 1800, by contrast, natural science was recognizably similar to what is understood by that term today. The main tenets of the Aristotelian view of nature had been abolished: the principle of plenitude had succumbed to the vacuum pump, the four elements to chemistry, the categorial difference between things sub- and superlunary to the universal validity of Newtonian mechanics. In terms of methodology and epistemology, the four causes as well as matter and form had become immaterial, as had the prohibition against *μετάβασις*. Deduction had been replaced by inductive reasoning from 'facts' (a notion unknown outside law to previous eras). The role of classical authors as providers of starting points for deduction had accordingly vanished, while gaining new facts by systematic observation and experiment—a contrivance to produce a certain effect in order to make an argument—had become pivotal. So had quantification, as mathematics had developed astonishingly and become a powerful tool for describing natural phenomena. It was in mathematical language that 'laws of nature' were now formulated. These laws were comparable to first principles in their generality but did not constitute starting points but endpoints of reasoning. Nature was understood as a homogeneous realm in which the same laws applied everywhere.[3]

Along with the basic tenets of the scientific disciplines, their self-conception and status had fundamentally changed as well. Science was now conceived as one coherent field. Its proponents no longer followed principles set by other disciplines but rather claimed intellectual leadership for themselves, contending to provide the best possible description and explanation of the world. Concomitantly, they had reoriented themselves in time. Since old texts no longer played a role as a source of reliable knowledge, there remained no reason to turn to the past for instruction. Scientists now looked towards an ever brighter future. The notion of scientific progress had become their guiding idea and driving force.[4]

How had all of this come about? Over the last century, the story has been told countless times and has become staggeringly complex as well as heavily contested in the process.[5] In fact, in accordance with the initial diversity of the disciplines involved, their integration into modern science—or, in some cases, their exclusion from it—followed widely diverging paths, and there is a different story to be told for each of them. However, at least a couple of agents of change that are commonly recognized as being of cross-disciplinary relevance can be named. The emergence and proliferation of the universities from the Late Middle Ages created a forum of intense engagement with pre-existing knowledge about the natural world from Greek, Arabic, and Latin sources. The available amount of such knowledge was greatly augmented by a fuller recovery of Greek science in the course of the humanist movement. At the same time, the discovery of the New World and other previously unknown regions exposed the limits of ancient wisdom, made early modern naturalists more self-confident, and offered a wealth of new objects and phenomena to be described, analysed, and debated. The invention and rapid spread of print halted rampant corruption in scientific texts and released energy for reflection on and discussion of their contents. Print also made it possible to reach larger groups of readers, to spread knowledge over long

[3] On the origins of facts and laws of nature, see note 28 below; on the experiment, see Dear 2000; *CHS* 3, 124–6, 130–1, 157–62, 302–4, 647–8; Wootton 2015, 310–60.

[4] Wootton 2015, 36–41, 61–2, 84, 341–6.

[5] The most authoritative encyclopedic overview to date is provided by *CHS* 3 and 4. Among recent monographs, Wootton 2015 represents a minority position, but he is clear and comprehensive and spells out his disagreements with his predecessors in detail. Henry 2002 and Knight 2014 provide more concise introductions for non-specialists.

distances, and to initiate collaboration *in absentia*.[6] New cooperations between scholars and artisans allowed tackling problems from theoretical and practical angles alike and developing more comprehensive approaches to nature.[7]

People and institutions

As already implied by the last point, people concerned with the natural world in early modern times did not form a group as homogeneous as modern scientists but recruited themselves from different echelons of society. The ways in which they dealt with nature and their motives for doing so differed accordingly.

A large and at the same time the most conspicuous proportion of early modern naturalists consisted of academically trained men. The most important professional group among them were physicians. They not only practised their proper craft in towns and at courts or taught it at universities but also saw themselves as experts in things natural in general and became responsible for a substantial part of the relevant research and publications. Moreover, they formed a strong international network early on—a predecessor of the later republic of letters and the scientific community.[8] However, as most university students became to some degree acquainted with the quadrivium and natural philosophy in the arts faculty, most academics were potential investigators of nature, and many realized this potential. Theologians were very active in the field. Besides the secular clergy, some orders—above all, the Jesuits[9]—developed a strong tradition of scientific teaching and research. A fair number of jurists engaged in the study of nature as well.[10]

Thus, only some of the academics who interested themselves in the sciences did so for professional reasons. There was a large segment of learned amateurs for whom the study of nature was a leisure activity. The gentlemen researchers of the Royal Society are perhaps the most famous of these, but they constitute only the tip of the iceberg. Across Europe, numerous clergymen, nobles, courtiers, and well-off citizens delighted in elucidating the secrets of the natural world. Nor were all of these academically trained. Many, for instance the cloth merchant Antonie van Leeuwenhoek (1632–1723), who stunned the learned world with his microscopic observations, had never seen the inside of a university. The same is almost by definition true of women, who were generally excluded from the academic

[6] Eisenstein 1979, 453–682. For a case study of the massive problems caused by textual corruption in the manuscript age, see Roelli 2023.

[7] Zilsel 1942 provides the classic formulation of this idea. For a recent overview, see Long 2011.

[8] On the network of physicians, see Maclean 2008; on the republic of letters, Waquet 1989.

[9] Harris 1989; Feingold 2003; Hellyer 2005; Udías 2015, 1–131; Waddell 2015.

[10] On jurists active in natural history, see *CHS* 3, 461. With regard to other disciplines such as mathematics and astronomy, one may think of names like Pierre de Fermat or Johannes Hevelius.

world but were often trained by and worked as assistants to their fathers and husbands.[11] Not only in their case but in early modern science in general, autodidactism and informal teaching, greatly facilitated by the spread of print, played a much more important role than today.

As already mentioned, there were also those involved in or connected to the study of nature in their capacity as artisans.[12] The list of the various professions includes, but is by no means restricted to, medical specialists from midwives and surgeons to bonesetters, cataract couchers, and apothecaries, surveyors, architects, engineers, fortress builders, artillery experts, miners, shipwrights, navigators, brewers, distillers, dyers, stonecutters, glassmakers, goldsmiths, instrument makers (who provided scientists with indispensable equipment), painters, engravers, and printers. In contrast to physicians, natural philosophers, and representatives of the quadrivium, craftsmen dealt with the ways of nature but incidentally in their professional lives. In contrast to amateurs, they had to make a living from their activities. However, when they collaborated with their betters, the social and economic divide was not impermeable: artisans might teach scholars their skills and acquire some learning in return. As a result, some of them became authors in their own right.

Of course, many of these players were not active in scientific establishments. In fact, the institutional background of early modern science was poorly developed by modern standards, although it grew stronger over time.[13] Moreover, it was weaker for research than for teaching, and institutions where people were paid for the practical application of scientific knowledge were rarer still.

The most important institutions were the universities, whose number greatly increased in the early modern period.[14] In older research, universities are often described as backward-looking and contenting themselves with the teaching of outdated dogmas. This is an exaggeration, but it is true that the *raison d'être* of the premodern university was teaching (in the two main forms of the lecture and the disputation) and that this teaching was as a rule conservative.Students were informed about nature in the arts and medical faculties. Medicine was exceptional among all scientific disciplines in having a faculty of its own. The subjects taught in the arts faculty included the quadrivium and natural philosophy,

[11] *CHS* 3, 192–205; 4, 184–210.

[12] The literature is vast but scattered. See note 7 above and, for some individual professions, see *CHS* 3, 272–319 (laboratory workers, war engineers); *CHS* 4, 511–35 (instrument makers); Gaida 2016, 278–9 (mathematical practitioners).

[13] On physical and institutional environments and contexts, see *CHS* 3, 206–362; 4, 44–128; Feingold and Giannini 2019.

[14] Some wider context for the paragraphs that follow is provided by the discussion of institutions of knowledge in general in Burke 2000, 32–52. Standard works on the early modern university include Rüegg 1993–2010, vol. 2; Grendler 2001 (Italian universities 1400–1600); Füssel 2017. On the universities and science, see *CHS* 4, 44–86.

as already mentioned.[15] Besides curricular teaching, professors also gave other classes which were more diverse and whose content could be more progressive. From the sixteenth century, the curricula underwent major changes. Classical texts were replaced by new textbooks. The medical curriculum was thoroughly revised, including new formats like dissections, botanical excursions, and later bedside teaching, as well as new subjects such as anatomy, surgery, and chymistry. In the arts faculties, Aristotelian natural philosophy was replaced by 'philosophia experimentalis' under the labels of Cartesianism and Newtonianism. New disciplines like geography appeared as well.[16] Conversely, astrology, which had been taught together with astronomy for centuries, was removed from the curricula in the eighteenth century.[17]

Research was not on the universities' official agenda but it was conducted de facto in various forms.[18] The questions arising from the perusal of and commenting on classical texts in lectures lent themselves to extensive discussions which are preserved in the rich academic commentary literature. The disputation was even more conducive to the critical consideration of difficult points from different perspectives. In addition, much discussion and reflection must have gone on in informal settings, and many professors took pen in hand to record the results of such musings. Professors were well represented among scientific authors, and their output was not restricted to commentaries and textbooks. Publications enhanced their reputation and became more important in this function over time. By the eighteenth century, innovative researchers brought their universities renown, attracted students from far away, and taught the best of them to do research on their own.[19]

At a level below that of the university, basic knowledge about the natural world was imparted at a number of types of school. In fact, there was often little difference between the level of education in an arts faculty and an advanced school. From the early sixteenth century, Protestants began to found several kinds of higher schools, in whose upper classes some natural philosophy, mathematics, and even medicine could be taught.[20] In the Catholic world, this model was imitated and standardized by the Jesuits. Teaching in their gymnasia was codified in the *Ratio studiorum*, the definitive version of which appeared in 1599[21] and was

15. For an overview of the situation in the Late Middle Ages and a case study (Ingolstadt, fifteenth and sixteenth centuries), see Schöner 1994; for natural philosophy, medicine, and mathematics in Italy, Grendler 2001, 267–352, 408–29.

16. Rüegg 1993–2010, vol. 2, 462–74, 486–91; on geography in England, see Cormack 1997.

17. Stuckrad 2007, 275.

18. See Feingold 2019, who takes Oxford and Cambridge as an example.

19. See Chang 2010, 351–5 on Albrecht von Haller.

20. For Germany, see Paulsen 1919–21, vol. 1, 276–387 (many scattered remarks;, see esp. 344), 489–90, 564–6, 572–4, 593–8, 602.

21. Edition: Lukács 1986. German translation with introduction: Duhr 1896. See also Paulsen 1919–21, vol. 1, 387–443.

implemented in hundreds of gymnasia in Europe and even in the New World, which were open to future Jesuits and others alike.[22] From the later sixteenth century, knight academies conveyed some natural knowledge to young nobles.[23] In the eighteenth century, specific schools were founded for the training of civil servants and bureaucrats in the state sciences, a multiform field which usually included several natural scientific disciplines and for which chairs were established at universities as well.[24]

Besides universities and schools, other institutional contexts offered work and, in some cases, research opportunities to those concerned with the natural world as well. Cities and towns not only were the obvious working environment for practising doctors, but they also hired municipal physicians to take care of public health issues and to supervise non-academic practitioners like surgeons, pharmacists, and midwives.[25] Princely courts availed themselves of the services of personal doctors, astrologers, alchemists, and a broad range of other specialists from architects and engineers to military experts.[26] In some cases, princes extended their practices of cultural patronage to scientific research in the proper sense out of personal interest or for reasons of prestige, providing time, money, and infrastructure. The funding of Tycho Brahe's observatories on the island of Hven by Frederick II of Denmark is a famous, if exceptional example of such patronage.

Academies were the last of the major institutions fostering early modern science to appear.[27] Although such institutions were founded from the end of the fifteenth century in imitation of the Platonic Academy, the first academies dedicated specifically to the natural world emerged only in the second half of the seventeenth century. They were linked both to cities and to courts, as their members often recruited themselves from the local intelligentsia and enjoyed princely protection. Apart from this, scientific academies were quite heterogeneous, their lowest common denominator being little more than the desire of their members to discuss questions about the natural world. Famous examples like the Royal Society should not be seen as prototypical. The importance of academies lay in their capacity to pool scattered scientific interest and expertise and to give it purpose and direction. Following the example of the French Académie des Sciences, some academies became fully funded state institutions in the eighteenth century. They could afford to publish the research of their members in their own journals and in a few cases even hosted large collaborative projects; the Societas

[22] Feldhay 2011. See also Feingold 2003 on the interplay of Jesuit scientific teaching, dissemination, and research.

[23] For Germany, see Paulsen 1919–21, vol. 1, 521–3.

[24] On the early history of the state sciences, particularly in Germany, see Lindenfeld 1997, 11–45, esp. 35 on the natural scientific disciplines represented in a number of curricula.

[25] *CHS* 3, 416–20; *EEMHO*, s.v. 'Physicus'.

[26] *CHS* 3, 251–67.

[27] *CHS* 3, 267–71; Hirschi 2017.

meteorologica Palatina at Mannheim, for instance, issued the first international meteorological ephemerides. Such organizations were the direct antecedents of modern national academies.

Intellectual context

The original lack of conceptual unity among the disciplines that were to coalesce into modern science corresponded with an openness towards and a dependence on fields and discourses which are clearly separated from science today. In some cases, the interaction focused on quite specific points. Jurisprudence, for instance, contributed to key scientific notions such as 'fact' or 'law of nature',[28] while it conversely was indebted to its academic sister discipline medicine for the concept of 'dogma'.[29] The relevance of several other fields, by contrast, was more general and pervasive. In what follows, I will briefly discuss the three important cases of philology, theology, and history. Their influences on the study of nature showed a large overlap and came together in exemplary fashion in biblical science, the discussion of which will conclude this section.

In stark contrast to present-day conditions, the early modern natural sciences were not strictly separated from the text-centred world of humanism.[30] Most naturalists had a humanist education, and philology, the methodical study of texts, constituted an important part of their toolkit. In the fifteenth and sixteenth centuries, the increased accessibility of Greek classics like Ptolemy, Galen, or Dioscorides through textual criticism, translation, and commentary created necessary preconditions for the further study of these fields, and their correct understanding remained vital to many disciplines for a long time. Until the seventeenth century, the dominant mode of teaching in higher education, the lecture, consisted of reading out a text to a class and providing a detailed interpretation. The corresponding written format, the commentary, was a major genre of scientific literature.

In some disciplines, the philological slant was particularly obvious. The embedding of medicine in humanist culture and the importance accorded to philological research on medical texts have been described in terms such as 'medical humanism' and 'iatrophilologia'.[31] In natural history, much time and energy was invested in terminological questions, since identifying the modern plant

[28] On the much-discussed origins of the fact, see, among others, Shapiro 2000 (who derives the notion from English common law) and Wootton 2015, 251–309, esp. 284–6 (who points to Italian and French sources). On laws of nature, see *CHS* 3, 70–105.

[29] Herberger 1981, 211–344. On the relationship between medicine and law in general, see Maclean 2002, 84–7.

[30] See in general Schmitz and Krafft 1980. Grafton 1991 is a treasure trove of telling examples.

[31] Schmitz and Keil 1984; Jaumann 2001; Maclean 2002, 18–22; Siraisi 2004, esp. 192 and 204; Siraisi 2007, 4–8.

designated by a Greek, Latin, or Arabic name and vice versa and matching up synonyms was pivotal in avoiding Babylonian confusion. As the number of known rocks, plants, and animals kept growing, the development of a unitary naming system became a major challenge. Naturalists time and again highlighted the importance of philological expertise for their discipline.[32] Astronomy could not dispense with old textual sources because of its concern with *longue durée* phenomena such as the precession of the equinoxes. Some astronomers became skilled critics and interpreters of ancient texts in order to make the best of their data.[33] In all disciplines, techniques of observing phenomena in texts and nature were remarkably similar. Careful observation and empiricism had always formed part of the philological 'Denkstil', and this may have contributed to the development of the corresponding scientific virtues.[34]

Although the relation between early modern religion and science is dominated by antagonism and censorship in the popular perception, the influence of religious thinking on science actually ran much deeper than that. While philology impacted the way in which the study of nature was approached, theology vitally influenced how nature was seen. The idea that the Scientific Revolution was conditioned by religious and, in particular, Protestant motives has been proposed in many variants for almost a century.[35] However this may be, Catholics and Protestants clearly agreed in their fundamental understanding of nature. In contrast to ancient and modern views of the natural world as the realm of things not made but emerging, nature was conceived as created by God, who had purposefully and artfully designed its every detail so that it served the needs of man as efficiently as possible. The early modern conception of nature was to a very great extent teleological. The idea that meticulous study of the natural world would lead to a better understanding of God's plans and actions constituted an important motivation for naturalists of all confessions and disciplines. As late as the eighteenth century, this thought attained new prominence and was applied to numerous specialized fields in the movement known as physico-theology.[36]

Naturalists also partook in the early modern interest in history and in the rise of a historical mindset. This was only logical in an era in which the history of nature and humankind appeared to be coextensive, as both had begun with the Creation some 6,000 years before. A bridge between the study of the human past and that of the natural world was already provided by the Latin word 'historia' itself, which rose to prominence in the sixteenth century and encompassed a number of meanings seamlessly blending into one another: the study of the past,

[32] Reeds 1991, 26–8, 135–65; Friedrich 1995, 91–106 (on Gessner's *Historia animalium*); Habermann 2001, 212 (on sixteenth-century herbals).

[33] Grafton 1991, 178–203 (on Kepler); Mosley 2000, 118, 128.

[34] Miert 2013b.

[35] Harrison 1998 (who provides an overview of the idea's variants at pp. 5–8).

[36] Michel 2008; Steinmann 2008 (on the rhetorical structure of physico-theological discourse); Blair and von Greyerz 2020.

the study of natural and cultural objects, the facts determined by such study, and the written record of these facts.[37] Several scientific disciplines traced their own history as an object of study.[38] More importantly, accounts of past events were seen as important sources and were systematically collected and assessed for that reason. This is particularly true of medicine, where a thriving literature of case histories and related forms developed,[39] but also of other disciplines such as astronomy and astrology, whose practitioners tried to provide it with an empirical basis in the form of collections of genitures.[40] In some areas, researchers systematically combined information gathered from historical sources and nature. Starting with Joseph Scaliger's *De emendatione temporum* (1583), chronographers learned to interlink calculations of astronomical events such as eclipses with the interpretation of historical documents in order to obtain a precise understanding of the chronology of the past. Conversely, chronology was understood as an ancillary science to history, as one of its two 'eyes' together with geography.[41] Finally, many scholars indiscriminately studied remnants of cultural and natural history. English antiquarians, for instance, happily unearthed ancient inscriptions along with fossilized fish.[42]

Philological methods, theological background, and historical thinking came together in scientific approaches to the Bible, the text of which rose to new prominence in the age of humanism and the Reformation. A look at the ways in which men of science dealt with the biblical text illustrates several of the remarks so far and provides a graphic impression of some major differences between early modern science and its modern counterpart. Two main approaches may be distinguished: scientific commentary on phenomena mentioned in the Bible, on the one hand, and instrumentalization of the Bible in the context of debated questions, on the other.

Regarding the first point, a whole branch of medicine and natural philosophy dealt with pertinent phenomena in the Bible. Numerous works on 'sacred medicine' and 'sacred' (or 'Mosaic') 'physics' were published between the sixteenth and the eighteenth century.[43] Georg Wolfgang Wedel, professor of medicine in Jena, for instance, issued 130 short contributions, collected into 13 booklets titled *Exercitationes medico-philologicae*,[44] in which he discussed questions such as how the stabbing of Christ on the cross described in John 19:34 took place exactly ('De latere Christi aperto', 1687). Johann Jakob Scheuchzer's four monumental, lavishly illustrated tomes of *Physica sacra* (1731–5) provided a complete natural

[37] On 'historia' as a bridge between the study of nature and culture in the early modern period, see Pomata and Siraisi 2005.

[38] On astrology and esp. medicine, see Siraisi 2007, 14, 106–33. [39] Siraisi 2007.

[40] Grafton 2000, esp. 64. [41] Grafton 1991, 104–44, esp. 104–5, 139–40.

[42] Poole 2010, 85–93.

[43] Blair 2000a; Touber 2021. Roling 2013 splendidly illustrates the continuities between the Middle Ages and early modern times in this respect.

[44] Maclean 2019, 87.

historical commentary on the Bible, in which each and every pertinent passage was discussed.[45] The original version, published simultaneously in Latin and German, was so successful that the work subsequently appeared in French and Dutch as well.

Concomitantly, the wording of the biblical text was used in deciding contested issues. The cosmological implications of Joshua's command to the sun—not the earth—to stand still during his battle against the Amorites (Joshua 10:12–13) were heatedly debated between Copernicans and their adversaries.[46] In his detailed reconstruction of Noah's Ark (*Arca Noe*, 1675), Athanasius Kircher demonstrated that it had been spacious enough to hold all sexually reproducing species known in his own day, thereby confirming the idea of the immutability of species.[47] How the changes undergone by the earth in former times had led to its present shape was discussed at great length in the light of the tales of the Creation and the Flood. A detailed, spectacular, and influential account of the respective events was given by Thomas Burnet in his *Telluris theoria sacra* (1681, definitive edition 1689).[48]

Disciplines

As already mentioned, the disciplines involved in the early modern investigation of nature differed greatly with regard to their subjects, presuppositions, aims, and methods.[49] However, their diversity did not keep them from overlapping and interacting in manifold ways. Their interplay was stimulated by the fact that the degree of specialization was much lower than today and that many scholars were active in several sectors.[50] The resulting conglomerate of disciplines, subdisciplines, and interdisciplinary areas cannot be mapped onto the internal structure of modern science. In fact, it already defied contemporary attempts at categorization, resulting in a plethora of different taxonomies.[51] Moreover, the structure of the field was in constant flux and evolved over time. At the beginning of the early modern era, it was characterized, in the absence of an overarching notion of science, by a number of smaller clusters comprising several disciplines each. At its end, such an overarching notion had formed and most disciplines encompassed by it had become recognizably similar to their present appearance.

[45] Müsch 2000. [46] Roling 2013, 180–204.
[47] Browne 1983, 3–8; Asúa and French 2005, 171–4; Enenkel 2014a, 104–10.
[48] Poole 2010, 55–63.
[49] The information provided in this section is largely based on *CHS* 3, 365–723; 4, 241–507.
[50] Pomata 2011b, 64.
[51] For the eighteenth century, this has been illustrated by the various classificatory schemes deployed in contemporary encyclopedias (*CHS* 4, 241–66). On early modern problems regarding the classification of knowledge in general, see Burke 2000, 81–115.

To outline this complex and elusive state of affairs in the space available here is only possible at the price of radical simplification. I make two attempts, looking at the situation from two mutually complementary angles in a manner that is somewhat similar to my approach in the first section of this chapter. First, the major clusters typical of the earlier part of the era are presented, including the disciplines comprised by them, which are, however, not described in any detail. After that, I take the single disciplines as we know them today as vantage points and look back from them to their early modern predecessors, which I characterize and whose development over time I sketch. While the perspective on early modern science is thus late medieval/early modern in the first case and modern in the second, the order of presentation of the various fields of knowledge is in both cases chronological, starting with those inherited from antiquity and the Middle Ages and progressing to those emerging at later moments (see Figure 1 for a visual summary).

Among the disciplinary clusters into which the study of nature was organized during the first centuries of the early modern period, the first, chronologically as well as in terms of significance, has already been mentioned at the beginning of this chapter: Aristotelian natural philosophy.[52] Aristotle's studies of the natural world—which had eclipsed their Presocratic predecessors early on and established themselves as the main fountainhead of the subsequent tradition—aimed at providing causal explanations of the processes going on in the material world. They were sometimes ordered in a sequence starting from fundamentals, descending from the heavens to the sublunary realm, and then progressing from inanimate matter to living beings and their principle of life, the soul. Each of these fields, embodied in one or more Aristotelian works, formed the subject of a specific discipline. How the physical world worked in principle was set out in the *Physics*. Cosmology, based on *De caelo*, explained the structure of the cosmos in general and in particular that of the unchanging world beyond the sphere of the moon.[53] Meteorology, embodied in the eponymous work, reviewed the sublunary world of change, dealing with atmospheric and aquatic phenomena as well as with those originating underground, such as earthquakes and volcanic eruptions.[54] The monumental *Historia animalium* and a number of shorter works focusing on specific questions provided the basis of biology. That of psychology, the doctrine of the soul and its faculties, was *De anima*.

Another cluster with ancient roots was the quadrivium. Following upon the trivium, the three verbal arts of grammar, logic, and rhetoric, it formed the second part of the seven liberal arts, which as a whole constituted a propaedeutics

[52] On its early modern tradition, see Leijenhorst et al. 2002, esp. the last eight contributions to that volume; *CHS* 3, 365–406.

[53] On the early modern history of the term 'cosmologia' and its application to various more or less Aristotelian conceptions of the field, see Tessicini 2022.

[54] Martin 2011.

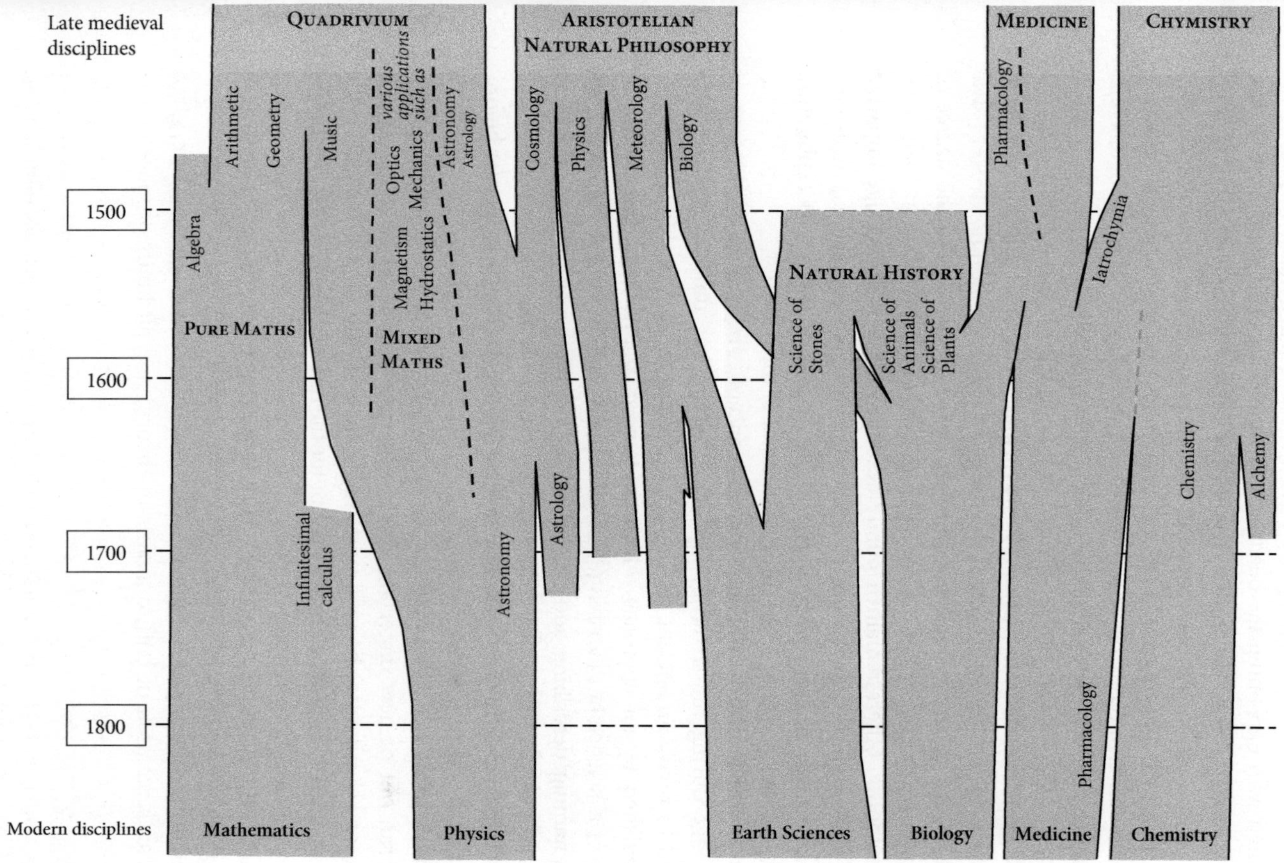

Figure 1 The disciplinary development of early modern science. Small capitals indicate (clusters of) disciplines inherited from earlier times or forming in the first half of the early modern period, designations in bold standard typeface, modern disciplines. Dotted lines denote fluid disciplinary boundaries. Natural philosophy and its subdisciplines are marked with the epithet 'Aristotelian' to distinguish them from their modern homonyms. Most minor fields are omitted.

to philosophy. The disciplines comprised by the quadrivium—arithmetic, geometry, music (that is, acoustics), and astronomy, often collectively called 'mathematics'—were strictly separated from natural philosophy: they were not concerned with explaining the behaviour of matter but with quantity in and of itself. According to a schematism going back to Nicomachus of Gerasa (first/second century AD), they were ordered according to the oppositions of 'discrete vs continuous' and 'unmoved vs moved': arithmetic was the science of discrete, unmoved quantity, geometry of continuous, unmoved quantity, music of discrete, moved quantity, astronomy of continuous, moved quantity.[55]

In contrast to natural philosophy and the quadrivium, cosmography emerged only during the sixteenth century. It was a fuzzy and internally diverse field, amalgamating cosmological, astronomical, and geographical knowledge.[56] The weight accorded to the single components varied greatly, as did the degree of mathematization: while some works presupposed or taught much mathematical understanding, others remained purely descriptive. Much knowledge imparted by cosmographical texts was practically useful for engineers, sailors, surveyors, mapmakers, and instrument makers. The new needs of the age of exploration provided a strong incentive for the development of the field.

Like cosmography, natural history was conceptualized as a distinct area of knowledge during the sixteenth century. Invented by humanist scholars who claimed to revive an ancient model that had never existed in this form, it was distinguished by an approach which renounced analysis and explanation in favour of a 'science of describing'[57] or, to expand on this formula, of collecting, listing, naming, describing, and classifying. This approach was applied to the three realms of nature—first plants, next animals, and to a lesser degree minerals—all of which had primarily been studied for their pharmacological properties before. This medical concern was now gradually complemented by an interest in numerous other properties. Given the rapid expansion of the known world, natural history never ran out of new material, which over time became so substantial that naming and classification became most urgent concerns. The search for a universally applicable method came to an end (except for the mineral world) with the acceptance of the Linnaean system.[58]

Finally, the quadrivium as well as cosmography overlapped with 'mixed mathematics', a term which became common around 1600 for the application of mathematical techniques to the material world (often seen as problematic, given the strict separation of both domains in theory) and all kinds of more or less

[55] Roelli 2021, 199.

[56] See Lestringant 1994, *CHS* 3, 469–96, and the articles collected in Dupré and Hallyn 2009, esp. Mosley 2009. For a concise overview, see Gaida 2016, 282–7.

[57] This is the title of Ogilvie 2006, the standard monograph on the subject.

[58] On the development of taxonomy in natural history, see Atran 1990, esp. 123–81; for a more theoretical approach, Pavlinov 2021.

practical questions.[59] Over time, the realm of mixed mathematics expanded to the extent that the approach to nature as a whole was mathematized. By the early eighteenth century, the field comprised not only venerable disciplines like astronomy and music but also many others from mechanics, (hydro-)statics, and optics to chronology and demography—indeed, a substantial part of science in general.

In what follows, I will start with the major modern disciplines and briefly trace their early modern prehistory. Medicine held an exceptional position due to its institutionalized education, its strong practical dimension, its identity as a recognized profession, and its long tradition dating back to the times of Hippocrates. In the course of the early modern period, the discipline changed profoundly.[60] At its beginning, it was conceived not only as the art of healing but also as a science of nature (φύσις) in general, on which the comprehension of the human body and the ability to cure it were based. The word 'physician' still preserves the memory of this conception. At many universities, the study of medicine smoothly took over from the study of natural philosophy in the arts faculty:[61] the first of the five subjects ('institutiones') comprised by the canonical curriculum was physiology, which dealt with nature as it presented itself in the healthy body (to be followed by pathology, semiotics (the interpretation of symptoms), dietetics, and therapeutics).[62] The equally common subdivision of medicine into 'theoria' and 'practica' likewise embedded the discipline in a natural philosophical framework: 'theoria' meant knowledge of nature in general; 'practica' was an application of this knowledge to the human body (though it still remained on a theoretical level). In the course of the seventeenth and eighteenth centuries, however, the link that bound medicine to natural philosophy weakened with the demise of the latter and was finally severed, a development epitomized in a semantic shift in both terms: 'theoria' now denoted the field originally covered by 'practica', while the new 'practica' was clinical practice, which had formerly been marginal in medical education.

In terms of medical authorities, Arab influence had been dominant in the Middle Ages, and Avicenna's *Canon* remained a pivotal text in the medical faculties until the sixteenth century in Northern Europe and until the seventeenth in Italy.[63] From the late fifteenth century, many Greek sources were rediscovered, translated, and introduced into the curriculum. Avicenna encountered competition and was eventually superseded by Galen, a process eased by the fact that the latter was Avicenna's own main authority. In the seventeenth century,

[59] See *CHS* 3, 4, 119–21, 156, 696, 744–5, and the contributions collected in the special issue 7/1 (2018) of the *Journal of Early Modern Studies*.

[60] *CHS* 3, 407–34; 4, 463–84.

[61] For Italy, see Grendler 2001, 268, 409.

[62] An important part of therapeutics was pharmacology, based on substances derived from plants, animals, and minerals and harking back to ancient and medieval texts like Dioscorides' *Materia medica*, Galen's *De simplicibus*, his two treatises *De compositione medicamentorum*, and books 2 and 5 of Avicenna's *Canon*. On the early modern history of pharmacology, see Schmitz et al. 1998–2005, vol. 2, 1–415.

[63] Siraisi 1987, 77–124.

the systematic Galen was in turn displaced by Hippocrates, who was understood as a paragon of the undogmatic empiricism that was held in great esteem at the time.[64] Concomitantly, however, medicine's foundation in classical authority in general eroded, the system based on the latter disintegrated, and its attitudes and tenets were abandoned. Humorism was laid to rest along with deductive reasoning from natural philosophical principles and causal explanation by means of innate faculties of organs. The place of these notions was taken by observational knowledge of the structures and functions of body parts. Numerous specific positions, for instance parts of Galen's anatomy and his view of the movement and function of blood, were refuted. Full-blown alternatives to the Galenic system appeared in the form of iatrochemistry and iatromechanics, both of which replaced the natural philosophical basis of medicine with a modern discipline. In the eighteenth century, iatromechanists, who tended towards a machine-like understanding of the body, were in turn opposed by vitalists, who held that living beings did not behave according to the same rules as dead matter. All the while, medical knowledge advanced on all fronts, as new anatomical structures, physiological processes, diseases, and remedies became known. The curricular reforms mentioned earlier contributed to these developments, as did new instruments like the microscope. By 1800, medicine admittedly was not much more successful in curing patients than a couple of centuries before, but its knowledge had greatly expanded and its agenda thoroughly changed.

Astronomy[65] had been the most highly developed natural science in antiquity alongside medicine, as well as the only one based on mathematical foundations back then: Ptolemy's *Almagest* already applies a mature spherical trigonometry. Much energy had to be invested to recover this level at the beginning of the modern era. At that time, the discipline was commonly subdivided into 'astronomia sphaerica' and 'astronomia theorica'. The former, which dealt with the structure of the cosmos in general, was also part of Aristotelian physics, while the latter, which observed and calculated the movements of the planets, was understood as a purely mathematical discipline.[66] After the question of the correct world system became dominant with the (initially hesitant) reception of Copernicus's *De revolutionibus* (1543), this distinction was abolished, and the whole discipline was reimagined. Henceforth, mathematical models were supposed not just to accord with the celestial phenomena and enable predictions but to represent physical reality. From Kepler, who went one step further and hypothesized about the forces that kept the planets in their orbits, astronomy also tried to explain why things happened in the skies the way they did and developed towards a celestial mechanics, a trend which reached its first culmination in Newton's *Principia* (1687).

[64] Hippocrates was commonly accepted as the author of the entire corpus named after him (see Chapter 5, 'Translations').

[65] *CHS* 3, 562–95; 4, 328–53.

[66] Voelkel 2001, 5.

With this work, the classical onion skin model of the cosmos was definitively replaced by an immense space full of sun-like stars (an idea speculatively advocated already by Giordano Bruno). All the time, observational astronomy, boosted by the invention of the telescope in the early seventeenth century and practised in an increasing number of observatories, kept pace and interacted with an astronomical theory based on a mathematics of ever-growing sophistication.

Along with calendar reckoning, astrology (a discipline fitting uneasily into the overall structure of this section, as it does not count as a science today) was the most important practical application of astronomy.[67] Its conceptual basis was the Aristotelian tenet that the superlunary world influenced the sublunary, which appeared reasonable in view of common phenomena such as day and night, the seasons, and the tides. Like the *Almagest*, the principal text of ancient astrology, the *Tetrabiblos*, was produced by Ptolemy. The system presented therein was much expanded by Muslim scholars, whose contributions continued to influence early modern astrology. Important subdisciplines were natal, electional, horary, and mundane astrology, which predicted the character and fate of individuals, determined the best moment for starting an enterprise, answered specific questions posed to the astrologer, and forecast major developments in the human and natural world respectively. Their common tool was the horoscope, a sketch of the heavens as seen at a particular time and from a particular place on earth. Already in the Middle Ages, the application of astrology to the weather had spawned astrometeorology, an early form of weather forecasting whose popularity reached its apogee in the early modern period.[68] (Aristotelian meteorology, by contrast, was not interested in prognosis.[69]) In iatromathematics, doctors used astrology to adjust their treatments to stellar influence.[70] Although astrology was frequently attacked on technical, epistemological, and theological grounds (most influentially in Pico della Mirandola's *Disputationes adversus astrologiam divinatricem*, 1496)[71] and came under pressure from the post-Tridentine Church's hostility towards divination in general, it became not less but rather more popular during the sixteenth and seventeenth centuries. Only in the eighteenth century was the discipline banned from academic teaching, as already mentioned, and downgraded to a pseudo-science.

Geography fell into two subdisciplines: mathematical geography (including cartography), which built on geometrical and astronomical foundations, and descriptive geography (including its small-scale variety, chorography), which was

[67] *CHS* 3, 541–61; Stuckrad 2007; Ernst and Giglioni 2012; Dooley 2014.

[68] Lawrence-Mathers 2020, esp. 173–202. This tradition overlaps to some degree with the lore of weather signs first codified in the pseudo-Theophrastean *De signis*; however, that is primarily based on terrestrial signs (Martin 2011, 11–13).

[69] Martin 2011, 5.

[70] Sudhoff 1902; Müller-Jahncke 1981. On theoretical speculations about analogies and connections between the human body and the skies, see Hirai 2012 and 2014.

[71] Modern edition: Garin 1946–52. See also Pompeo Faracovi 2012b; Akopyan 2021.

non-mathematical and interested not only in natural but also in cultural and historical aspects.[72] Both versions had their ancient classics: Ptolemy (once more), on the one hand, Strabo and Pomponius Mela, on the other. However, while this subdivision by and large remained in place in the early modern period, much information provided by the ancient authorities had already become obsolete in the sixteenth century, as the world expanded far beyond what was known to the Greeks and Romans during the age of discovery.[73] Besides making many new countries accessible, the voyages of the discoverers also refuted fundamental ancient and medieval tenets about planet earth. In particular, the equatorial zone, thought uninhabitable due to its heat, proved well populated. The notion that water, as the lighter element, should in principle cover the dry land which only emerged in one specific region of the globe due to some anomaly quietly gave way to the terraqueous globe as we know it today, a structure whose surface is covered with dry land and water in an irregular pattern which can only be ascertained empirically.[74] New factual and procedural knowledge—for instance mathematically based techniques of surveying and mapping such as the Mercator projection—accumulated rapidly. Geography and cartography rose to pivotal economic and political importance, a development indicated by a new kind of book devoted primarily to these fields, the atlas.

As for mathematics in the modern sense of the word, its two basic kinds, geometry and arithmetic, mostly had developed along different lines in antiquity and the Middle Ages.[75] For a long time, the former was more advanced than the latter. The Hindu-Arabic numeral system, known in Europe from around 1200 and adopted in practice from the fifteenth century, in the long run enabled a higher evolution of arithmetic and its extension, algebra, which provided general rules for the solution of equations. Building on this development, the most momentous innovations in early modern mathematics may be said to be two: first, the emergence of an increasingly sophisticated mathematical notation, which allowed the easy formulation of ideas that would have been forbiddingly unwieldy in natural language;[76] second, the development of analytical geometry, calculus, and analysis, which fused geometric and arithmetic thought and resulted in modern mathematics as a coherent field. The discipline, which had been strictly separated from natural philosophy and subordinated to it by Aristotle for its failure to provide causal explanations, rose in status. The certainty of its results was emphasized, and its representatives began to venture into the territory of natural

[72] On early modern geography, see Broc 1986; Cormack 1997 (who focuses on England); for a short overview, see *EEMHO*, s.v. 'Geography'. The aforementioned dichotomy ultimately goes back to Ptolemy's *Geography* (1.1.1). On cartography, see Woodward 2007; Edney and Pedley 2019; *EEMHO*, s.v. 'Cartography'.

[73] Randles 2000.

[74] Wootton 2015, 110–37.

[75] On early modern mathematics, see *CHS* 3, 696–723; 4, 305–27; Merzbach and Boyer 2011, 245–463.

[76] Cajori 1928–9.

philosophy, until both fields merged under the label of 'physico-mathematics'.[77] In the title of his magnum opus, Newton calmly presupposed that 'philosophia naturalis' had 'principia mathematica', something hardly imaginable a century earlier. Henceforth, mathematics developed without hindrance into a more and more powerful tool for the quantitative description and prediction of natural phenomena.

Early modern chemistry and alchemy are difficult to tell apart, and their exact relation and its development are hard to grasp.[78] Until the late seventeenth century, the two terms (and their orthographic variants) were used indiscriminately, and it has therefore been proposed to subsume the entire field up to that date under a common term, namely 'chymistry'. Chymistry comprised a range of activities concerned with the manipulation of substances that ranged from charlatanism and various crafts to scientific research and attempts at spiritual improvement. It produced an extensive literature along with a fantastic imagery.[79] Experimentation played an important role, but the interpretation of texts did as well. These texts, often of Greek and Arab origin or inspiration, were notoriously obscure, since chymists jealously guarded their secrets, which hindered the development of a common technical language. For these traits, as well as for preying on credulous souls, the discipline had been reviled since the Middle Ages but flourished nonetheless. The efforts of Paracelsus (1493–1541) to put pharmacology on a chymical basis engendered a field called 'chymiatria' or 'iatrochymia'.[80] From around 1600, this brand of chymistry began to be taught in medical faculties, and a textbook tradition evolved. Towards the end of the seventeenth century, people started to distinguish alchemy from chemistry. The former term could designate a particularly noble brand of chymistry but was soon restricted to attempts at making gold and denigrated as utterly fraudulent. Chemists, on the other hand, shed the secretive habits of their predecessors. They performed their experiments in public, freely exchanged information, and formed a scientific community in the modern sense of the word. Conceptually, corpuscularianism, which understood matter as consisting of invisibly small particles of an indefinite number of kinds, allowed chemists to go beyond the theories of matter of Aristotle and Paracelsus, who had envisaged just four (earth, water, air, fire) and three elements (sulphur, mercury, salt) respectively. With the establishment of the modern notion of 'element' by Antoine de Lavoisier in the late eighteenth century, chemistry took on a recognizably modern look.

[77] Dear 1995, esp. 151–79. See also Burke 2000, 203–4 on seventeenth-century enthusiasm for the 'geometrical method'.

[78] Newman and Principe 1998; Levere 2001, 1–79; *CHS* 3, 497–517; 4, 375–96; Wootton 2015, 353–60.

[79] Völlnagel 2012.

[80] Schmitz et al. 1998–2005, vol. 1, 267–343.

The part of Aristotelian natural philosophy codified in Aristotle's works on animals, some of his *Parva naturalia*, and *De anima*, and Theophrastus's works on plants can justifiably be said to constitute an ancient form of biology, a real science of life. However, no direct path leads from there to modern biology, although much biologically relevant knowledge accumulated in the course of the early modern period in anatomy and physiology as well as in natural history. The botanical branch of natural history in particular, which was also an auxiliary science of medicine because of its pharmacological relevance, grew fast. It became more empirical through botanical gardens and field trips, and the number of known plants expanded exponentially.[81] Moreover, from the late sixteenth century, attempts were made to emancipate botany from medicine and to constitute it as an independent science.[82] However, because of the concomitant demise of Aristotelianism and the increasing specialization conditioned by the incessant accumulation of new material, the aforementioned disciplines did not coalesce into one unified field for some time. It was only in the eighteenth century that a new interest in the phenomenon of life as such made itself felt.[83] The questions of what actually constituted life, how it emerged, and what were the differences, if any, between dead and living matter were intensely discussed. Around 1800, the word 'biology' began to be used in order to distinguish the science of life from the disciplines dealing with the non-living world.

The case of physics was similar: only antecedents of the modern discipline existed in the early modern period.[84] The term 'physica' meant 'natural philosophy'. Although many later subjects of physics were studied—for instance mechanical, optical, and acoustic phenomena, free fall, fluids, electricity, and magnetism—they originally were not understood as parts of one comprehensive discipline. In the seventeenth century, however, some of these fields were brought together under the corpuscular and mechanistic paradigm of Cartesian physics, which superseded its Aristotelian predecessor. They became more experimental and quantitative, a trend partly instigated by newly invented instruments like the barometer and the thermometer. An additional incentive for the progress of later physical subdisciplines was provided by the evolution of astronomy towards a celestial mechanics. Newton was the first to bring a substantial body of natural phenomena under mathematically formulated laws of universal validity in his *Principia*, and Newtonianism subsequently ousted Cartesianism as the leading paradigm in the field. The eighteenth century saw further significant attempts at describing natural phenomena mathematically, on the one hand, and making them accessible through experiments, on the other, although these two strands by and large developed in isolation from each other, as was already epitomized by

[81] Morton 1981, 115–361; Reeds 1991; Pavord 2005, 125–403; Ogilvie 2006. On medical botany in Italy, see Grendler 2001, 342–51.

[82] Jensen 2001, esp. 194, 204–6.

[83] *CHS* 4, 397–416.

[84] *CHS* 4, 354–74.

the contrast between Newton's *Principia* and *Opticks* (1704) in this respect. In the wake of the former work, classical mechanics was the only one of the three major parts of classical physics to take shape in the early modern period. Electrodynamics and thermodynamics were children of the nineteenth century, which also saw the definitive merging of mathematization and experimentalism.

The modern earth sciences, devoted to all aspects of planet earth from its core to the atmosphere, were even slower to develop and consolidate. This is somewhat paradoxical, since they (like biology) had a well-developed Aristotelian predecessor in the form of meteorology. However, when Aristotle's world view disintegrated, his meteorology was not replaced by a unitary new paradigm (modern meteorology took little from it but its name).[85] As in the case of physics, one rather finds a number of quite diverse subjects, tackled with varying degrees of empiricism, experimentalism, and mathematization. Mineralogical knowledge was first accumulated in connection with mining, as exemplified by the *oeuvre* of Georg Agricola (1494–1555). A theory of earth magnetism founded on experiment was propounded by William Gilbert (*De magnete*, 1600). The tides were much discussed until Newton proposed a conceptually correct, mathematically formulated explanation in the *Principia*.[86] From the mid-seventeenth century, systematic weather observations (and occasionally forecasts) were made with the help of new instruments, and progress made in fluid mechanics was later used to understand the behaviour of the atmosphere in general.[87] Much work on the history of the earth remained tied to biblical exegesis, as already mentioned. Nicolas Steno had formulated basic principles of stratigraphy as early as 1669 in his *De solido intra solidum naturaliter contento*, but geology did not emerge as a discipline in its own right before the early nineteenth century.[88] Many other branches of the earth sciences from geophysics to glaciology took even longer to come into their own. The internal coherence of the entire field remains weak to the present day.

[85] Martin 2011, 5. [86] Bonelli and Russo 1996; Omodeo 2017.
[87] Frisinger 1977, 47–143; Cassidy 1985; Feldman 1990. [88] Porter 1977.

2
Latin

Historical overview

Latin originated as the regional dialect of Latium in Central Italy.[1] Like most other languages in use in pre-Roman Italy, it was a Indo-European inflected language with a rich morphology. As the Roman sphere of influence expanded, it gradually became the common language of Italy, the western Mediterranean, and a number of adjacent countries. (The eastern part of the Roman Empire, by contrast, spoke Greek.) Everyday Latin had never been the same as the literary Latin of Cicero and other late republican and early imperial authors, which we know as the classical form of the language. It increasingly diverged from this standard and, moreover, formed regional varieties in the course of the imperial era. After the end of the Western Roman Empire, this development led to the emergence of the Romance languages. Latin itself became a 'dead language', as the technical term goes, that is, a language without any living native speakers, in the early Middle Ages. Nonetheless, it continued to be used actively and passively, in speech and writing, prose and poetry, and remained the common language of the learned in a Western and Central Europe[2] that was otherwise not only politically but also linguistically fragmented, comprising speakers of Romance, German, Slavic, Celtic, Finno-Ugric, and isolated languages. From a global perspective, Latin therefore belongs to a group of dead and grammatically fixed languages continuing to serve as common languages in formal contexts, which also includes languages such as Sumerian, Sanskrit, classical Chinese, and classical Arabic.[3]

As a dead language, Latin depended for its survival on the institutions in which it was learned and used: the schools, exclusively run by the Church initially; the Church itself; the institutions of higher education, among which the universities gained prominence from the later Middle Ages; the law; and the courts with their chanceries. In these contexts, Latin served as the language of education (as its medium as well as its main subject, since already learning the alphabet mostly took place in Latin), religion, higher learning, law, political correspondence,

[1] A number of comprehensive histories of Latin have appeared in recent decades: Vossen 1979; Farrell 2001; Janson 2004; Ostler 2007; Stroh 2007; Leonhardt 2013. The present section summarizes some of their insights.

[2] The Byzantine (later Ottoman) and Russian spheres did not adopt Latin as a common language.

[3] Leonhardt 2013, 17–40; on 'dead' and 'fixed', see also Roelli 2021, 374–7.

Latin Scientific Literature, 1450–1850. Martin Korenjak, Oxford University Press.
DOI: 10.1093/oso/9780198866053.003.0003

and diplomacy. Moreover, the learned also used it for their individual purposes, as the language of personal correspondence and literature.

Linguistically, medieval Latin retained a core of classical grammar and vocabulary but also developed numerous subforms for particular uses, from Church Latin, based on the Vulgate's literal translation of the Hebrew and Greek of the Old and New Testament's, to scholastic Latin, whose roots lay in Latin translations of Aristotle, and the Latin of legal instruments. Each of these exhibited a distinctive set of non-classical features.[4] In addition, in a number of grammatical aspects most medieval varieties diverged from classical Latin: for instance, reflexive and non-reflexive third-person pronouns were used indiscriminately; the typical accusative and infinitive (that is, constructions of the type 'I believe this to be correct') was replaced with 'quod' phrases ('quod' meaning 'that': 'I believe that this is correct'); tenses and moods could be played with freely. Moreover, much new vocabulary was coined or adopted from learned languages like Arabic and the vernaculars. Pronunciation changed as well, finding expression in orthographical peculiarities such as the replacement of classical 'ae' and 'oe' by simple 'e', with resultant forms such as 'celum' instead of 'caelum'.

With the beginning of humanism in fourteenth-century Italy, some forms of medieval Latin, particularly scholastic Latin, were denounced as barbaric, while the ancient classics, especially Cicero, were held up as models of good Latin. The perusal and imitation of them gained an important place in the curriculum. Soon, print made it comparatively easy to spread the new standards throughout Europe in the form of editions, programmatic statements, grammars, and textbooks. By the mid-sixteenth century, Latin grammar and idiomatics had mostly been restored to classical standards as far as this was possible without 'hard' philological research into classical Latin of the type conducted from the nineteenth century.[5] However, 'unclassical' varieties remained in use to some degree in certain environments such as the Church and the universities (and certainly in oral communication in general).[6] On the other hand, the question how far one should model one's Latin exclusively on Cicero gave rise to a heated debate, the 'Ciceronian controversy', in the fifteenth and sixteenth centuries, and similar discussions continued as long as Latin remained in active use.[7] Thus, the customary designation for the Latin of the modern period, 'Neo-Latin', is an umbrella term covering very different forms of the language.[8]

[4] Stotz 1996–2004 provides a comprehensive overview of medieval Latin, highlighting its internal diversity.

[5] On this approach to Latin, see Leonhardt 2013, 272–5.

[6] On oral Latin, see Waquet 2001, 152–71.

[7] Fumaroli 1980, 35–230; Greene 1982, 171–86; McLaughlin 1996, esp. 185–274.

[8] For linguistic overviews of Neo-Latin, see *CNLS* 2, 377–433; *BENLW* 5–54; *OHNL* 13–26. Overviews of Neo-Latin and Neo-Latin literature in general include, besides the works just cited, Moul 2017a. For a sociological approach, see Burke 1993; Waquet 2001.

In addition to the earlier institutions that used it, early modern Latinity gained a new, supra-institutional basis in the printed book, a much more powerful medium than the handwritten codex. The rise of print generated an explosive increase in Latin works.[9] Exact estimates are impossible, but the number of early modern Latin publications is in the seven-digit range, and over 95 per cent of all Latin preserved was produced after the invention of print, dwarfing the extant medieval and still more the ancient output. Latin also scores impressively in terms of printed works when compared with the European vernaculars. Initially it was absolutely dominant, as 70 per cent of all incunabula were in Latin. In the course of the early modern era, this dominant position was lost at different moments in different countries. In Italy, France, and England, the respective vernaculars overtook Latin in the sixteenth century. In Germany, by contrast, Latin maintained its leading role until the late seventeenth century: the first year in which more German than Latin books were advertised at the Frankfurt Book Fair was 1683. In some areas of Central Europe, in Eastern Europe, and in Scandinavia, this point was reached only in the eighteenth century. If one looks beyond the borders of the emerging national states and weighs the Latin publications of all countries against those in the individual vernaculars, this, of course, results in an even more impressive picture of the prominence of Latin in print.

Latin as the language of early modern science

In antiquity, the language of natural science was Greek, not Latin. Little written in Latin survives from this field, and most of it is translations and adaptations from Greek.[10] In the Middle Ages, Latin continued to serve as an intermediary for Greek and especially Arabic science, but it did so on a much larger scale, as Latin translations were systematically used in teaching. Moreover, they were increasingly accompanied by original work in the language of the Romans during the last centuries of the period. At the beginning of the early modern era, Latin was thus long established as the leading language in the study of nature in the West. It was to retain this role for several centuries.[11]

There were a number of reasons for this beyond Latin's traditional function as the language of learning in itself. To meet the challenges that this function entailed, Latin had developed an array of helpful lexical and syntactical devices,

[9] The following publication statistics are based on Waquet 2001, 81–4, Korenjak 2016, 18–19 and 21 (both with further literature), and Korenjak 2018b, 57–8. Leonhardt 2013, 2–4 takes the manuscript record into account as well. Raw data for the Frankfurt Book Fair are provided by Schwetschke 1850–77.

[10] Poccetti et al. 2005, 358–76.

[11] For a short overview of Latin as the language of science, see Gordin 2015, 23–49; on medieval Latin in particular, see Roelli 2021, 197–289.

as will be discussed in the last section of this chapter. By virtue of these, it was ahead of the vernaculars, most of which only began to consolidate as unified, regulated languages at the outset of the modern era. Moreover, while they, as living languages, were subject to change over time, Latin was, in important respects, fixed once and for all, which made it a guarantor of intellectual continuity: it was, in Leibniz's words, 'Europe's universal language, durable so as to reach posterity' ('lingua Europaea universalis et durabilis ad posteritatem').[12] Sociologically, most people who studied the natural world worked at universities or were academically trained, and many were clergy. They thus came from two institutions that were major strongholds of Latinity. In addition, science was an international enterprise, and the small communities of experts in the various disciplines were spread over a linguistically diverse area, which made the use of a common language mandatory.[13] This need for mutual understanding acquired a financial edge when books became a commodity in the age of print: for marketing reasons, publishers had an interest in ensuring that their products could be read in the whole of Europe. To all these practical reasons, one should add the influence of Latin on the way in which the sciences were perceived. As many results of the study of nature came to sound increasingly revolutionary, counter-intuitive, and disturbing, Latin proved useful for keeping dangerous ideas away from the undiscriminating masses, thereby protecting their authors from allegations of fuelling unrest and upheaval. At the same time, the age-old language of the learned, along with traditional literary forms, presumably had a soothing effect on those fluent in it. It lent respectability to potentially offending content by giving it a traditional, familiar look and made the new science appear as just another area of traditional learning.

As a result of all these factors, many people automatically thought in the language of the Romans when it came to science. Private documents show that this still applied at the very end of the early modern era: the many thousand notepads[14] left by Albrecht von Haller (1708–77) are exclusively in Latin, as is Gauss's mathematical diary for the years 1796–1814.[15] Men of science also regularly talked to each other in Latin, above all in the universities, where lectures, disputations, and orations were in that language and everyday conversation was at least supposed to be,[16] but also when they met colleagues from other countries. Quite naturally, conversation at a distance, that is, learned correspondence, frequently took place in Latin as well. People wrote in Latin even to compatriots and could switch to Latin in a vernacular letter as soon as the subject became scientific.

[12] Quoted in Schmied-Kowarzik 1916, p. xxii; on this aspect, see also Roelli 2021, 378–9.

[13] A particularly bold expression of this idea can be found in Linnaeus 1737a, p. 48.

[14] On the rise of the notepad as a tool for managing one's information from the eighteenth century, see Zedelmaier 2015, 63–74.

[15] Krebs 2005, 358–9. Klein 1903; Biermann et al. 2005.

[16] Rüegg 1993–2010, vol. 2, 60–1, 227, 267, 271, 448, 455–8; for Northern Italy in particular, see Grendler 2001, 151–2.

It therefore comes as no surprise that most publications in the natural sciences were in Latin as well. Truly representative statistics are not available, but some figures will at least give an impression of the situation. For the first half of the early modern period, *USTC*, which covers the years 1450–1650, is quite helpful. Its list of subjects includes science under the headings 'Medical Texts', 'Science', 'Astrology and Cosmography', and 'Mathematics'. In these four fields, some 15,700 out of 25,200, 3,300 out of 5,400, 2,300 out of 5,100, and 1,200 out of 2,700 books listed, that is, 62, 61, 45, and 44 per cent respectively, are in Latin. Taking all four categories together, one finds that a total of 22,500 out of 38,400, that is, over 58 per cent, are Latin books (while the most important of over a dozen vernaculars, German, reaches about 16 per cent). However, this figure does not include the numerous works concerned with the natural world among the 'University Publications' (63,500 items, 99 per cent in Latin). For later periods, one may note that 40–50 per cent of all chymical books printed up to 1690 and 15–20 per cent of all such works printed from 1691 to 1783 are in Latin.[17] Dissertations continue to be almost exclusively in that language until the early nineteenth century. Some additional figures for Germany will be given in a moment. Broadly speaking, Latin was the most important language of scientific publishing up to 1700, one of the most important during the eighteenth century, and still not negligible in the early nineteenth.

This numerical supremacy quite naturally translated into prestige and influence. Authors as a rule turned to Latin when they felt they had something important to say. Among forty-four works presented in an anthology of zoological works 1450–1700, thirty-two are written in Latin.[18] Of seventy-seven 'landmark writings' in mathematics, the first twenty-three, covering the years 1649–1809, include twelve Latin ones.[19] In all disciplines, one encounters a long series of Latin milestones, from Copernicus's *De revolutionibus* (1543) to Gauss's *Disquisitiones arithmeticae* (1801). In terms of literary form, numerous genres, from the monograph and the textbook to the dialogue and the didactic poem, originated in Latin and were imitated in the vernaculars later.[20] Innovations in layout and printmaking, such as source citations in marginalia or colour print, made their first appearance in Latin books as well. Bibliographers reacted to the situation by showing a strong bias towards Latin works. Small wonder that many authors Latinized their own names and equipped even vernacular books with flashy Latin titles.

[17] Brüning 2004–7, vols 1–2. [18] Pinon 1995.

[19] Grattan-Guinness 2005. Carl Gustav Jacob Jacobi's only Latin work after 1809 is his *Fundamenta nova theoriae functionum ellipticarum* (1829).

[20] With regard to learned literature in general, this phenomenon is already noted by Kristeller 1975, 226.

Latin and other languages of science

As already mentioned, Latin somewhat paradoxically owed its importance to the linguistic fragmentation of Europe—the fact that there were so many other languages around. A closer look at the linguistic context of Latin and its interaction with it is, therefore, in order. The languages of science other than Latin basically fall into two groups: other classical, learned languages, on the one hand, and the contemporary European vernaculars, on the other.

As far as science was concerned, the most important learned languages were Greek and Arabic, while others like Hebrew, Syriac, or Persian played minor roles. Knowledge of Greek was most widespread by far, but that does not mean much.[21] Although learning it was advocated by the humanists who wanted to gain direct access to ancient Greek literature, and although instruction in Greek was institutionalized in many places (it was prescribed, for instance, by the Jesuit *Ratio studiorum*), few attained a good command of the language. That some scientific writings, particularly in astronomy and medicine, are introduced by Greek poems of praise[22] and interspersed with Greek quotations and other phrases[23] does not contradict this statement. While the former were meant to be admired as virtuoso pieces, the latter were usually translated into Latin or paraphrased. The authors of these works liked to flaunt their linguistic expertise but did not expect the same level from their readers. The situation of Arabic was more precarious still. It did not command the same respect as Greek, a language of admired classical antiquity, but rather faced widespread hostility as a language of infidels whose achievements were seldom recognized. Despite the important spur given to Western science by Arab philosophers, astronomers, and physicians, these intellectuals were often declared to have found out nothing new by themselves and to have corrupted what they had taken from the Greeks. Formal teaching of Arabic developed only slowly and mainly in theological contexts. Only a tiny minority of scholars ever mastered the language.[24]

All in all, then, medieval conditions persisted to a remarkable degree: Greek and Arabic science was printed and became influential mostly in Latin translations, which were often equated unthinkingly with the originals by their readers. Both languages exercised a strong lexical influence on scientific Latin. However, this influence developed in opposite ways: Arabic terms, which had been taken over into medieval scientific Latin in large numbers, were ousted from most

[21] For a summary with further literature, see *DNP* suppl. 9, cols. 410–13.

[22] For overviews of early modern poetry written in classical Greek and of early modern literature in this language in general, see Pontani and Weise 2022 and Weise 2016 respectively.

[23] For a case study, see Wenskus 2023 on Kepler's Greek.

[24] On attacks on and defences of Arab science in the Renaissance, Hasse 2016, esp. pp. xi, 180–9, 250–4. On teaching and knowledge of Arabic in early modern times, see Fück 1955; *EEMHO*, s.v. 'Arabic studies'.

disciplines in early modern times. Greek, which had played a lesser role in the Middle Ages, suffered no such suppression and became an ever richer source of new terminology throughout the early modern era and beyond.

Latin's interaction with the numerous vernaculars was more complex.[25] Many early modern scientists will have discussed their work orally with others in Latin as well as one or more vernaculars, depending on their interlocutors and the specific occasion. Little evidence of such conversations survives, but analogous phenomena can be noticed in correspondence, which was often carried out in Latin and several vernaculars; code-switching within one and the same letter occurred as well, as already noted. Similar bilingual and multilingual phenomena are also in evidence in printed works. Translations in both directions were common, not least for economic reasons, as involving more languages meant addressing more potential buyers and readers. Latin versions could make vernacular works accessible to a Europe-wide educated readership. On the other hand, texts were also translated from Latin into the vernaculars for the benefit of regional literate but non-Latinate audiences, including practitioners, courtiers, townsmen, and women.[26] Some readers who were not themselves scholars were undeterred even by massive and learned works, as shown by the example of Athanasius Kircher's *Mundus subterraneus* (1665, 2nd edn 1678), translated into Dutch in 1682.[27] In certain traditions, Latin and vernacular strands were closely intertwined. The so-called 'Gart tradition', for instance, a series of botanico-pharmaceutical works starting with the German *Gart der Gesundheit* (1485) and continuing to 1783, involves numerous translations and adaptations in both directions.[28] There were also simultaneous editions in two languages, bilingual editions, and books containing components in different languages such as Peter Crüger's *Uranodromus cometicus: Ein außführlicher Tractat vom grossen Cometen deß 1618 Jahrs* (1619), whose technical parts were in Latin, the rest in German.

The centuries of Latin-vernacular multilingualism involved much exchange in terms of pronunciation, lexicon, idiomatics, and syntax, which left its mark on the languages. At least as far as science was concerned, however, the relationship between Latin and the vernaculars was asymmetrical: by and large, the former was the giving, the latter the receiving partners. This imbalance is particularly evident in the lexicon: little was borrowed by Latin from the vernaculars,[29] while these are saturated with Latin technical terms to the present day.

[25] There is a substantial body of literature on the interplay between Latin and the vernaculars in the early modern era. Short overviews: *CNLS* 2, 420–2; *BENLW* 275–86; *OHNL* 27–40; Deneire 2017. In-depth treatments: Tournoy and Tunberg 1996; Burke 2004; Deneire 2014; Bloemendal 2015; Winkler and Schaffenrath 2018. Specifically on the situation in science: Pantin 1996; Blair 1996 (focusing on France); Maclean 2002, 39–41 (on medicine); Cohen 2015; several contributions to Prinz and Schiewe 2018 (for Germany).

[26] See Blair 1996, 24–5, 28 (scientific works in Latin and their vernacular popularizations), 40–1 (courtiers and women).

[27] Nummedal and Findlen 2000, 186–7.

[28] Habermann 2001, 245–73.

[29] The reluctance to adopt vernacular expressions was not restricted to science (Tournoy and Tunberg 1996). See also the last section of this chapter.

Demise and persistence

From the beginning, the use of the vernaculars in scientific works was not restricted to translations for a wider readership. As early as the incunabula era, original works were composed in these languages by and for specialists who had little or no Latin, such as engineers, pharmacists, surgeons, midwives, and distillers.[30] The relationship between Latin and vernacular publications was marked by peaceful coexistence and productive exchange rather than fierce competition, as both parts of the scientific book market complemented each other.[31] Over the centuries, however, the balance shifted in favour of the vernaculars. The reasons behind this process are hard to grasp (what could induce the present scientific community to forsake English as its common language?) but certainly include general as well as science-specific ones.[32] Among the former, one may count a growing readership of groups without any higher education such as artisans, merchants, and women, politically motivated linguistic patriotism, and the Enlightenment notion that knowledge should be accessible to everyone. Regarding science in particular, the aforementioned non-Latinate practitioners came to play a more important role over time. Moreover, the vernaculars became more and more like Latin themselves, adopting its technical terminology, integrating the same mathematical symbolism, and following the same protocols of reasoning. A disciplinary 'Sprachbund'[33] emerged whose single members evolved into fully fledged languages of science in their own right and shared enough traits to enable at least some mutual understanding between scientists with different mother tongues. Under these circumstances, it appeared feasible to many to forgo one common language.[34]

Whatever the exact reasons, the replacement of Latin by the vernaculars took longer in science than in most other fields, because science remained a domain of the learned and a supranational enterprise. Moreover, the process proceeded differently in different countries, institutions, and disciplines.[35] Geographically,

[30] The intended audience is sometimes indicated in paratexts. In the Latin dedication to the second edition of his French *Géométrie practique* (1551), Charles de Bouelles writes that he has been asked to compose a vernacular introduction to geometry for craftsmen ('auturgis manuve operariis', sig. Aii^r) in need of some geometrical knowledge.

[31] Much of interest on this aspect can be found in Habermann 2001. See also the concise remarks in Nummedal and Findlen 2000, 176.

[32] Gordin 2015, 41–2; Prinz and Schiewe 2018, 4; Roelli 2021, 338–57.

[33] In transferring this notion, originally applied to languages in their entirety by Nikolai Trubetzkoy, to technical languages, I modify it in a hopefully acceptable way.

[34] Two additional reasons are often adduced, namely the alleged incapability of Latin to deal with the realities of the modern world and the rise of learned societies, which had a penchant for the vernaculars. The first idea is easily refuted (Stroh 2007, 237–8). The authority of the learned societies may have been influential, but since their members were as a rule perfectly Latinate, their preference for their respective vernaculars must be seen as an expression of the more widespread linguistic patriotism mentioned in the main text.

[35] Such differences were sometimes noted by contemporaries. In his *Assertio pro libello suo Germanico* (1585), a companion piece to a German plague ordinance of his, Johannes Crato highlights that the vernacular is held in lower esteem in Germany than in Italy and France: 'Neque enim, ut apud Italos atque Gallos, magni fit vulgaris linguae cultura' (p. 6).

the language used in scientific publications followed the logic of the book market in general: in big, politically unified territories which had developed a common vernacular standard early on, this standard was applied more readily to science as well as to other subjects than could be done in small, politically and linguistically fragmented countries.[36] The former group is best represented by Spain, France,[37] England, and (despite its political fragmentation) Italy. Their authors and publishers could rely on a solid market for books in their respective national languages, took pride in them, and could even expect some foreigners to master them as well (there was little formal teaching of modern vernaculars, but the practices of the 'peregrinatio academica' and the Grand Tour will have helped).[38] Prominent members of the latter group are to be found in German-speaking countries in the widest sense (including Austria, Switzerland, and the Low Countries), Scandinavia (especially the sparsely populated but multilingual Swedish Empire), and Eastern Europe. For inhabitants of such regions, where a standardized vernacular was slow to develop or books written in it were not marketable because of the small number of prospective buyers, participation in the international learned community was only possible through Latin. For Germany, some figures are available to document the tenacity of Latin in science. The catalogues of the Leipzig Book Fair for the years 1740, 1770, and 1800 show the numbers of scientific publications in Latin dropping from 66 per cent to 30 per cent to 15 per cent respectively (to be compared with 28 per cent, 14 per cent, and 4 per cent for all publications). However, their absolute numbers were still on the rise (eight, fourteen, twenty-one) due to the expansion of the book market in general. Although the representativeness of these statistics can be questioned, since only a small part of the scientific production made it to the fair, the long life of Latin in German science is confirmed by the *Wolfenbütteler Verzeichnis medizinischer und naturwissenschaftlicher Drucke, 1472–1831*: there, German briefly overtakes Latin as early as 1520 and again around 1680, but otherwise, the latter dominates until the 1770s.[39]

In terms of institutions, the universities clearly remained the main stronghold of scientific Latin, even though there is insufficient information on how long the language was used orally in class and daily dealings. Geographical and disciplinary differences must have been considerable in this respect. The first substantial inroads of the vernaculars into Latin territory seem to have been made in the

[36] In spite of the dominant role of Latin in the Catholic Church, no clear correlation emerges between a country's religious allegiance and its readiness to turn to its vernacular in science.

[37] On France, see Blair 1996, 26, 29.

[38] On instruction in the vernaculars between 1500 and 1800, see Schröder 1992. Around 1600, some North Italian universities established (or tried to establish) chairs of Italian to attract German students (Grendler 2001, 152).

[39] Pörksen 1986, 49–56. For countries from the former group, the beginnings of the decline of Latin in science can now be traced in a similar way through combined searches in *USTC*. The representativity of the data thus gained, however, is once again hard to assess.

seventeenth century, but their victory was definitive only two centuries later.[40] On official occasions and in writing, Latin remained in use far into the nineteenth century, as testified by academic orations, textbooks, and above all dissertations. The academies and learned societies showed themselves much more open to the vernaculars. However, there were noteworthy exceptions, such as the German Academia Naturae Curiosorum,[41] and many journals issued by academies remained bilingual until around 1800. Princely courts funding research, by contrast, were by and large vernacular venues.[42]

Regarding disciplines, nascent ones and those which had many practitioners outside academia (for instance, the earth sciences, chymistry, and anatomy) tended to adopt the vernacular more readily than those with a venerable tradition, a solid academic standing, and an established Latin vocabulary (mathematics, astronomy, medicine, botany). However, this rule of thumb was easily broken when circumstances demanded it. Meteorology, for example, originally the natural philosophy of the sublunary realm but reinvented as a descriptive and predictive science based on quantitative measurements in the early modern period, developed an efficient technical language in German during the eighteenth century.[43] Nonetheless, the first Europe-wide meteorological 'ephemerides', based in Mannheim, were still published in Latin as late as 1783–95 because of the need for international cooperation. The two disciplines with the longest-lived Latin traditions were pure mathematics and botany. In the former, where a universally understandable symbolism seems to have made the language question less pressing, the young Gauss wrote his aforementioned diary entirely in Latin and continued to publish mainly in that language until the 1830s. János Bolyai's outline of a non-Euclidean geometry, the 'Appendix scientiam spatii absolute veram exhibens', dates from 1832, Giuseppe Peano's axiomatization of arithmetic, the *Arithmetices principia*, from 1889. In descriptive and systematic botany and zoology, Latin became particularly entrenched, not least through the success of Linnaean nomenclature.[44] Latin descriptions of new plant species were made mandatory as late as 1906, a rule abolished only in 2011, and scientific plant and animal names still have to be in Latin today.

Contemporary scientists were divided about the demise of Latin and the rise of the various vernaculars. While the development was welcomed by many out of patriotism or an Enlightenment attitude, others lamented the linguistic fragmentation of the scientific community. Having to learn six or eight different languages in order to participate in scientific discourse was a terrible burden, Albrecht von

[40] For the German-speaking world, where this process seems to be documented best, half a dozen case studies are contained in Prinz and Schiewe 2018.

[41] Pörksen 1986, 56–63.

[42] On academies and courts in general, see *CHS* 3, 251–71.

[43] Klein 1999, 148–96.

[44] Plant physiology adopted the vernacular earlier than plant description and systematics (Krook 2022, 44).

Haller complained, and similar statements were commonplace in the later eighteenth century.[45] In fact, vernacularization proved a mixed blessing. It certainly opened or facilitated the way into science for many talented minds, but it did so at a considerable price. Despite the linguistic convergence of the major vernaculars, the situation before the establishment of English as the one dominant language of science in the later twentieth century has been characterized as a 'scientific Babel'. Numerous attempts to counteract it, including the idea of introducing a common Latin-based artificial language, proved abortive.[46]

The lexical peculiarity of scientific Latin

I have reserved my remarks on the distinctive traits of scientific Latin to this final section for two reasons: the topic is intertwined with Latin's previously discussed multilingual context, and it will be taken up again, from a different angle, in Chapter 3.

At the beginning of the early modern era, Latin was already deeply marked by its long career as a language of learned discourse in general, having served the presentation and analysis of complex fields from law to history and from philosophy to architecture for almost two millennia. In addition to its inherited rich range of inflexional forms, it had developed a sophisticated syntax, including an involved hypotaxis[47] capable of expressing various perspectives, shades of reality and possibility, and all kinds of logical relations between different statements. Its vocabulary had also expanded with every new field covered, and helpful types of word formation had been conventionalized: abstract nouns in '-io' like 'gravitatio' (literally 'the fact of being heavy' or 'of causing heaviness'), adjectives in '-bilis' designating a possibility ('concoctibilis', 'digestible'), verbs in '-scere' indicating a process or a lower degree of something ('caerulescere', 'to become bluish'), and many more.

Scientific Latin did not stand out from this background in terms of morphology and syntax. In these respects, it by and large shared the classicism that was typical of Neo-Latin in general.[48] By contrast, it developed a massive technical lexicon of its own. Although medieval Latin was already lexically productive in the realm of science, and although the Neo-Latin lexicon as a whole evolved greatly over the centuries (it contains many more new lexemes than words attested from

[45] Krebs 2005, 368. For similar complaints by Jean-Baptiste Le Rond d'Alembert (1751) as well as authors in Uppsala (1752) and Greifswald (1794), see Roelli 2021, 343–5 and Alvermann 2018, 25 and 30–1 respectively.

[46] On the 'scientific Babel', see Gordin 2015; on artificial languages derived from Latin, see also Waquet 2001, 268–70.

[47] A hypotactic sentence is a combination of main and dependent clauses.

[48] For medical Latin, this is shown by Nutton 1995.

antiquity),[49] both tendencies were magnified in scientific Neo-Latin: there, the vocabulary simply exploded.[50]

Up to and including the sixteenth century, this development was somewhat restricted by Ciceronianism and classicism in general, the proponents of which tried to limit the Latin lexicon to expressions used in ancient texts, resorting to circumlocutions in the case of new objects or phenomena if possible.[51] The tension between classicism and the requirements of science surfaced in excuses for unclassical terminology that can be found in introductory paratexts. Oddo degli Oddi, for instance, justified his use of 'barbaric' ('barbara') expressions in his commentary on the first *fen* of Avicenna's *Canon* (1575) with their long-standing currency ('per multum temporis cum medicis consueverunt') and the fact that medicine is there for the sake of health, not linguistic correctness ('non ut nomina propria instituamus, sed ut sanitatem tueamur'; pp. 2–3). As time went on, however, authors and readers grew tired of the topos. Excuses became less frequent and the practical necessity of using new words for novel contents was tacitly accepted. New expressions were introduced either by a stipulative definition—that is, a statement that X will henceforth be used in the meaning Y[52]—or even without comment, if they were transparent in themselves. When the author of an eighteenth-century textbook of mineralogy wrote 'Lapis est terra semivitrificata' (Benedikt Stattler, *Mineralogiae principia physica*, 1765, vol. 2, thesis XI appended to the work), he saw no need to explain that 'semivitrificatus' meant 'half turned into glass': the prefix 'semi-' ('half'), the stem 'vitr-' ('glass'), and the suffix '-ficatus' ('made or turned into') are immediately understandable to anyone who knows Latin. Less self-evident neologisms were sometimes collected in technical dictionaries, which document part of the terminological upsurge.

In terms of word class, most new words are substantives, which is in keeping with a general tendency of scientific language towards nominalization.[53] Adjectives and participles were created as well, as were verbs, albeit in smaller numbers.[54] Typologically, neologisms of sense, the use of pre-existing words in a new meaning, can be distinguished from neologisms of form, the creation of new words (like 'semivitrificatus'). The overwhelming majority of neologisms of form, for their part, are made up of pre-existing components by derivation with the

[49] No comprehensive Neo-Latin dictionary exists today, but see Johann Ramminger's *Neulateinische Wortliste* (http://nlw.renaessancestudier.org, accessed 16 February 2023), Hoven and Renard-Jadoul 2006, and Pitkäranta 2018. Overviews of the subject are given in *CNLS* 2, 382–404 and *BENLW* 37–54.

[50] Much of the following is taken from *BENLW* 51–4 and Berrens 2022; see also Waquet 2001, 91–5. To keep things simple, I restrict myself to single words and do not discuss terms consisting of two or more words, which are very common in fields like botany, zoology, or anatomy.

[51] French 1985, 71.

[52] Robinson 1954, 59–92.

[53] See, for instance, Roelli 2021, 389–90, 392–3, 397.

[54] A rich source of adjectives is Stearn's overview of descriptive terminology in botany (Stearn 1992, 229–51, 303–44).

help of prefixes and suffixes or by word composition.[55] The source materials for such neologisms may be taken from Latin itself (again, 'semivitrificatus' is a good example), the vernaculars, Greek, or even combinations of these: hybrid formations are no rarity. Vernacular expressions make up a much smaller part than Greek ones and are seldom used in compounds. They are usually either Latinized by means of a fitting ending, like 'harengus' for 'herring' (presumably from the French or Dutch form, 'hareng'/'haring') or subject to loan translation: the German name for the wolverine, 'Vielfraß', which means 'glutton', was rendered with the late antique Latin synonym 'gulo'. The majority of foreign-language neologisms, however, involve Greek components. Greek qualified as a donor of technical terms because of the rich scientific lexicon it had developed in antiquity, its nearly unlimited capability in word composition, and its high prestige. Greek neologisms typically are either simple words which have been semantically reinterpreted or compounds which do not exist as such in ancient Greek. Both classes are Latinized in terms of script, orthography, and endings, as in the case of 'scotophorus' (*σκοτοφόρος*, 'darkness-bringer') for silver nitrate, which darkens under the influence of sunlight, which was used as the title of a pertinent report by Johann Heinrich Schulze (1721).

The terminological histories of the single disciplines differ considerably. Each of them developed its own technical jargon, encountered its own problems, and found its own solutions. Mathematics and most of the strongly mathematized disciplines (astronomy and astrology, mathematical geography, physics) featured a comparatively restricted vocabulary, some of which had already been created in antiquity[56] and the Middle Ages on the basis of Greek and Arabic. In astronomy and astrology in particular, and to a lesser degree also in arithmetic and algebra, terms of Arabic origin like 'zenith' or 'ciphra' remained in use throughout the modern period. Fixed stars were designated with various circumlocutions until Johann Bayer in his *Uranometria* (1603) introduced the system still in use today, assigning each star a Greek letter followed by its constellation in the genitive (for instance, 'λ Geminorum'); however, more conspicuous stars bear additional Greek, Latin, and Arabic proper names to this day.[57] With the development of a sophisticated symbolic language in mathematics, some previous coinages became superfluous. It was no longer necessary to call a fifth power 'surdesolidus' with Michael Stifel (*Arithmetica integra*, 1544) when one could simply write 'a^5'.[58] Similar, if less pervasive translingual symbolisms were used in astronomy, astrology,

[55] Completely new words are hardly ever created. Seeming exceptions may turn out to be malapropisms, as in the case of Jan Baptist van Helmont's 'gas', where the orthography and pronunciation of 'chaos' have been changed on purpose (Loewe 1936).

[56] On ancient astronomical and astrological Latin, see Poccetti et al. 2005, 362–6.

[57] Hamel 2010, 24–6, 61–3; on the history of the Arabic star names (most of them loan translations of circumlocutions used by Ptolemy) in Europe, see Kunitzsch 1959.

[58] On Stifel's terminology for powers, see Glei 2017; on the development of mathematical notation, see Cajori 1928–9.

chymistry, and pharmaceutics. Like the terminology of the mathematical disciplines, the medical one had ancient roots[59] and was partly based on Greek and Arabic, with the addition of some Hebrew, Syriac, and Persian elements. Andreas Vesalius banned all 'barbaric' and even most Greek terms from anatomy in an attempt at creating a purely Latin nomenclature, but much Greek was later reintroduced. In the terminology of other medical fields, for instance pathology, Greek remained an important or even dominant element throughout.[60] In the earth sciences, Georg Agricola found an already well-developed mining terminology in German and tried to create a Latin equivalent.[61] Two special cases were chymistry and, as already mentioned, natural history. For a long time, chymists were notorious for their cryptic expressions and terminological confusion, since they did not manage to agree on a commonly intelligible nomenclature due to the esoteric nature of their craft.[62] The discipline turned into a science in the modern sense—a collective endeavour based on shared knowledge—so late that the chemical terminology in use today, which marked a sharp break with the past, was already founded in a vernacular, namely French, by Antoine Lavoisier.[63] Early modern naturalists likewise found it difficult to establish commonly accepted names for the steadily increasing number of plants and animals they were confronted with. However, their problems were ultimately resolved in Latin—in a way reminiscent of Bayer's star nomenclature but on a larger scale—by the probably most efficient naming system ever created, the Linnaean. In addition to names for species and other taxa, the extremely precise descriptions produced especially in botany brought forth a correspondingly subtle lexicon covering all aspects of a plant from anatomical structures to finest shades of colours.[64]

The Latin scientific lexicon expanded as long as Latin was used as a language of science (and even beyond that point: it still keeps growing in botany and zoology). In the later stages of this process, Latin and vernacular word coinage more and more often went hand in hand: the vernaculars took over the respective mechanisms of word formation, and new expressions in one language immediately found parallels in others, tightening the aforementioned 'Sprachbund'. A short example from an article on arsine by Friedrich Stromeyer ('De hydrogenii arseniati natura', 1808, p. 153) may illustrate the situation towards the end of the Latin era.

[59] Poccetti et al. 2005, 367–71.

[60] Much of interest on the history of anatomical terminology and medical terminology in general is contained in Hyrtl 1880 and Dirckx 1983 respectively.

[61] Berrens 2020.

[62] On the secretiveness of early chymistry, see Wootton 2015, 353–60; on early modern chymical lexica seeking in vain to change this, see Chapter 5, 'Technical dictionaries'. A helpful modern dictionary is Abraham 1998.

[63] A history of chemical terminology is provided by Crosland 1962.

[64] Stearn 1992, 303–44; for colours in particular, see pp. 229–51, esp. 234–5.

> Itaque 200 gas hydrogenii arseniati partes adhibitae 111.5 gas oxygenii partes combustione absorpserunt atque 56.45 milligrammata acidi arsenicosi produxerunt.
>
> So the 200 parts adhibited of the gas arsenic hydrogen [that is, arsine] absorbed 111.5 parts of the gas oxygen through combustion and produced 56.45 milligrams of arsenic acid.

Of the technical expressions in this extract, all but 'combustio', 'absorbeo', 'acidum', and 'produco', which are genuinely Latin, stem from Greek roots. 'Gas', 'arseniatum', and 'arsenicosi' were derived directly from Greek. By contrast, 'hydrogenium', 'oxygenium', and 'milligramma' originated as 'hydrogène', 'oxygène', and 'milligramme' in French and were taken over into Latin from there.[65] Regardless of their origin, all technical terms (with the exception of 'hydrogenium arseniatum', which has been simplified to 'arsine') as well as most others find an exact etymological equivalent in modern English. The only English words without any counterpart in the Latin—those underlined in my translation—are conjunctions, prepositions, and articles. Even without a translation, a chemist should still be able to guess the meaning of the Latin text today. At the same time, it is understandable that many of Stromeyer's contemporaries did not see much sense any longer in learning complex Latin grammar in order to say in Latin what they could express in their own vernaculars with the very same words.

When all is said and done, it is thus logical that what remains of Latin in the sciences today is basically words without grammar, whether in their original form (as in biology, anatomy, and pharmacy) or with vernacular endings. What once was a language lives on as a terminology.[66]

[65] The adoption of Greek in modern languages via Latin as well as the other way round is not restricted to science (Adrados 2001, 261–83). On the (Greco-)Latin lexicon of the modern Western languages in general see Munske and Kirkness 1996 and any dictionary of foreign words.

[66] This state of things is documented by practical introductions to scientific Latin such as Nybakken 1959 and Helander 2021 (for science in general); Stearn 1992 (botany); Ahrens 1988 (medicine and related fields); Bracke and Deumens 2000 (medicine).

3
Literature

In the preface to his *Griechische Wissenschaftstexte*, Markus Asper announces that he will analyse from a literary angle 'what never seemed "literature" strictly speaking to most readers'.[1] Asper's remark can easily be transferred to the modern perspective on early modern scientific texts. In order to make it understandable why these can be discussed as literature at all, some insight into the early modern notion of 'literature' in general will be provided, along with a discussion of how far texts about the natural world were included in this notion. This will be followed by some remarks on the actual character of early modern scientific literature and on how it compares with what we are used to in modern texts of this kind.

'Literature' in premodern times

To a large extent, the early modern concept of literature resembles the ancient and medieval one and markedly contrasts with the current notion that developed in the decades around 1800.[2] In premodern times, there existed no word for the class of texts that forms the core subject of literary studies today—fictional works like novels, poems, and dramas. The etymological equivalents of 'literature'—'litterae', 'litteratura', 'res litteraria'—first and foremost meant learning in the broadest sense.[3] As far as they were applied to texts, they designated texts in which any kind of learning was put into writing. 'Litterae' in this sense were supposed to be inherently informative and didactic.

Within this field, a smaller class of texts received special attention. No specific designation was in use for this class, but its notional existence can be deduced from its subdivision into genres and the theoretical reflections devoted to these.[4] In his influential *Institutio oratoria* (10.46–131), the early imperial teacher of

[1] 'Mit anderen Worten, hier soll literaturwissenschaftlich betrachtet werden, was den meisten Lesern nie "Literatur" im engeren Sinne zu sein schien' (Asper 2007, 9).

[2] On the history of the notion of 'literature' in general, see *HWRh* 5, cols. 276–306, s.v. 'Literatur'; on the transition from premodern to modern concepts, see Abrams 1953 and Rosenberg 1990; and on the early modern era, see Fumaroli 1980, 17–34. The following section highlights only a handful of aspects of a much more complex story.

[3] 'Litterae' could also denote the letters of the alphabet as well as an epistle.

[4] In early modern times, this field overlapped but was not identical with the 'studia humanitatis', which comprised grammar, rhetoric, history, poetry, and moral philosophy and were the domain of the humanists (Buck 1984).

Latin Scientific Literature, 1450–1850. Martin Korenjak, Oxford University Press. © Martin Korenjak 2023.
DOI: 10.1093/oso/9780198866053.003.0004

rhetoric Quintilian provides a catalogue of excellent authors and works which a budding speaker should study to improve his style. Compiled by a highly cultivated, well-read Roman who had given the subject serious thought, his list points to this more circumscribed class of texts, the closest premodern counterpart to our present-day notion of literature. (I will refer to it as 'literature' in inverted commas throughout this chapter.) Besides all major kinds of poetry, Quintilian's catalogue also contains works by historiographers, orators, and philosophers (but not, for instance, novels). In its purpose as well as its content, it highlights that there still remains a major difference between premodern and modern ideas on the subject. Quintilian's decisive criterion is linguistic form: a text qualifies for inclusion if it lives up to certain formal standards. Absent, by contrast, as shown by the prose genres included, is the criterion of fictionality, a notion only partly developed and sporadically applied to 'literature' before the eighteenth century.[5] Capable language use, not the ability to invent a counterfactual story, was what made a good writer.

This difference was closely linked to another: the ability to write well was seen less as a result of individual talent, in particular powers of imagination, than of teaching and training. And taught it was: a great part of the total energy invested in education went into turning pupils into capable speakers and writers—first and foremost in Latin, the language of education par excellence. The education system as a whole was heavily text-based and oriented towards the production of texts. Together with wisdom and religious sentiment, eloquence was one of its supreme goals, as epitomized in the formula 'sapiens et eloquens pietas'.[6] Three aspects of this preoccupation with texts are worth highlighting because of their particular impact on early modern writing, including scientific writing: philology, imitation, and rhetoric. The theories, practices, and techniques subsumed under these keywords were of ancient origin but were revived in Renaissance humanism.

Philology, the systematic study of texts, comprised a large number of approaches, such as textual criticism, grammatical analysis, paraphrase, glossography, commentary, allegorical interpretation, generic classification, and canon formation. Originating in Hellenistic Alexandria and first applied to Homer and other Greek classics, these approaches were later transferred to Latin authors and to the Bible and became ubiquitous in the ancient, medieval, and early modern classroom.[7] Pupils undergoing philological training for years developed a sensitive attitude towards texts, especially towards their formal features.

This way of reading was transformed into a way of writing through the theory of 'imitatio'.[8] The basic idea was that excellence as a writer could be attained by

[5] *HWRh* 3, cols. 342–7, s.v. 'Fiktion'. For the early modern period in particular, Trappen 1998 is illuminating; see also *EEMHO*, s.v. 'Fictionality'.
[6] Sohm 1912, 31–123; Rüegg 1993–2010, vol. 2, 42, 108–9.
[7] Pfeiffer 1968–76.
[8] *HWRh* 4, cols. 235–303, s.v. 'Imitatio'.

mimicking classical models recognized as excellent. Sometimes, one single author was held up as the only or most important model. Alternatively, a mix of particularly felicitous traits of several authors was recommended. In any case, it was commonly accepted that the model(s) to be imitated depended on the genre in which one was writing. Imitation was not supposed to be slavish copying. The classical authors rather provided raw material to be skilfully adapted to one's own needs. To many writers, this way of composing became second nature: classical words, phrases, and structural forms automatically came to their mind as soon as they took pen in hand. In particular, Cicero was widely recognized and imitated as the dominant model of prose style and the Ciceronian period—the syntactically complex, extended, yet harmoniously balanced sentence—became the hallmark of good Latin.[9]

This comparatively simple and intuitive theory of text production was supplemented by a much more complex one: rhetoric. In fact, classical rhetoric probably remains the most elaborate system of organizing human speech the world has ever seen.[10] First sketched in classical Athens but developed and expanded throughout antiquity, this system was founded on a number of powerful basic categories. Among the most important of these were the notion of persuasion as the overarching goal of speech; the three dimensions of speech—subject, speaker, audience—and the three corresponding ways of persuasion—rational argument, self-presentation, and emotional arousal; the five tasks of the speaker: finding what to say, ordering the material, formulating the speech, learning it by heart, delivering it; the three basic genres of speech, judicial, political, and epideictic, dealing with the past, the future, and the present respectively; the structure of a judicial speech (adaptable to other genres as well), consisting of proem, narration, argumentation, refutation of the adversary's arguments, and conclusion; the four virtues of elocution, correctness, brevity, adornment, and situational adequacy, the last of which also provided a general guideline for the speaker's behaviour in general; and the three types of style, the lowly ('genus humile'), the middle ('genus medium'), and the grand ('genus grande').

On this basis, a finely ramified and extremely sophisticated theory—inter alia, it comprised hundreds of different tropes and figures—was developed over the centuries. Some parts of this theory, which was laid out in numerous Greek and Latin handbooks, remained known in the Middle Ages and evolved in new directions, resulting in textbooks of preaching and letter-writing. From the fourteenth century, ancient rhetoric was rediscovered on a broader scale and became more dominant than ever.[11] It was extensively taught at school, and a vast production of

[9] On the notion of the rhetorical period, see *HWRh* 6, cols. 750–65, s.v. 'Periode'.

[10] On the system, see *HlRh*; on its history up to the present day, see Vickers 1988; *HWRh*; MacDonald 2017.

[11] On Renaissance and early modern rhetoric, see Fumaroli 1980 (focused on France) and 1999; Mack 2011; *OHNL* 119–34; Poel 2017; and the contributions in MacDonald 2017, 375–568.

didactic literature began. Over 3,800 titles in over 12,000 editions, most of them in Latin, have been listed for the period 1460–1700.[12] Far beyond its application in oral speech,[13] rhetoric exerted a pervasive influence on most written texts, in which all of its tools were freely used. The literature of the early modern period was deeply impregnated with rhetoric.[14]

The place of scientific texts

What does all of this imply for the place of scientific texts within the field? They formed part of 'litterae' as a matter of course. Regarding 'literature', the situation was more complicated. In some respects, texts about the natural world could be regarded as falling into that category as well. The great majority of early modern men of science had gone through an education in which the perusal and imitation of the classics as well as rhetorical composition held a prominent place, and they had learned to see these exercises as the natural basis of their own writing, regardless of its subject. Factuality and informativeness certainly did not exclude science from 'literature'. Its Latinity worked positively towards its inclusion, since Latin was the very language in which the verbal arts were taught and learned.

Nonetheless, the position of scientific writing, along with other kinds of technical writing, was marginal within 'literature' from the beginning. The problem was that the content of crafts, arts, and sciences was usually thought to be too unwieldy to allow for a formally appealing presentation. Ancient authors of technical works often apologized for their off-putting style and unclassical vocabulary or declared that style was immaterial and only content mattered in their case.[15] Accordingly, technical writing was largely disregarded in literary theory. It is no coincidence that (with one single exception, to be mentioned in the note 16) such works do not figure in Quintilian's catalogue. No attempts were made to describe and characterize specifically technical genres such as the monograph or the textbook. Occasional praise for single authors like Hippocrates or Celsus[16] remained just that.

This situation did not change fundamentally in early modern times. As far as theoretical reflections were undertaken at all (often in dedications and prefaces), they were mostly limited to the association of technical writing with the 'genus

[12] Green and Murphy 2006. [13] See Poel 2007 and, for a case study, O'Malley 1979.

[14] See, for instance, *OHNL* 129. Peter Mack's guide to rhetorically informed reading (Mack 2017) can be applied profitably to early modern texts.

[15] Janson 1964, 98–100, 127, 133–4. A good example of such a statement is Manilius's characterization of his astronomical/astrological subject, 'ornari res ipsa negat contenta doceri' (3.39), much quoted in early modern times.

[16] Hippocrates was famous for his terse and pithy style in antiquity and early modern times alike (Rütten and Rütten 1998, 27 and 43), as Celsus, the 'medical Cicero', was for his polish and elegance (Quintilian, *Institutio oratoria* 10.1.124; Habermann 2001, 379–80).

humile', the lowest of the three styles of classical rhetoric.[17] An attempt at giving such texts a firmer place in the rhetorical system was made by Melanchthon, who in his influential *Elementa rhetorices* (1531) supplemented the three traditional genres of speech—judicial, political, and epideictic—with a fourth, called 'didascalicum' ('suited for teaching'), which was taken over by a number of later textbooks.[18] Conversely, the problematic status of scientific writing was exacerbated in the seventeenth century by a growing hostility towards rhetoric, and even towards the trivium in general, among proponents of the new science. Authors like Galilei, Descartes, Pierre Gassendi, and a number of members of the Royal Society conjured up an antagonism between the deceptive and fruitless verbal arts and science, devaluing the verbal aspect of the latter in the process.[19] Its association with the 'genus humile' was radicalized by Thomas Sprat, who advocated a kind of zero-degree style, a non-rhetorical, purely fact- and truth-oriented kind of writing, in his *History of the Royal Society* (1667, pp. 111–15).[20] Such ideas fell on fertile ground in the eighteenth century, when rhetoric in general came more and more to be seen as an art of specious embellishment. Revaluations of the literary character of scientific texts were attempted even at that late stage, but they were few and far between and of restricted scope and little influence. When Christoph Stay, for instance, declared scientific didactic poetry to be the greatest poetic genre of all ('De poesi didascalica dialogus', 1792, esp. pp. xxi–xxiv), this was an idiosyncratic position motivated by the activity of the author's brother Benedict Stay in this field, and one to which nobody seems to have paid attention.[21]

All in all, then, as far as literary theory and criticism were concerned, the place of scientific texts within the field of 'literature' remained hazy and ill-defined. They were not categorically excluded from the realm of aesthetically interesting texts, but they were relegated to its little-observed fringes.

The literary form of scientific texts

So much for theory. How are the respective attitudes and opinions reflected in practice, that is, in the structure and style of early modern scientific texts themselves? Generally speaking, their theoretical marginalization results in a notable degree of freedom and internal variety but not in a fundamental difference from

[17] Habermann 2001, 209–10, 216, 219–20, 222, 224–32, 243–4 (esp. regarding herbals).
[18] Wels 2001, 40–59; Mack 2011, 104, 114, 197, 308.
[19] On the relationship between early modern science and rhetoric, see Pera and Shea 1991; Hallyn 1999; Moss 2017; Walmsley 2017.
[20] Sprat's remarks have been much discussed. See *CHS* 3, 153–4 with further literature there.
[21] Schindler 2018.

writings from other thematic fields.[22] In fact, the formal distance between texts concerned with nature and others in early modern times is much smaller than the one between early modern and modern scientific literature. It is pertinent to name briefly some of the latter's characteristics as points of comparison.[23]

Modern scientific literature features a restricted number of genres, the most important of which are the article and the textbook. The former can be called the scientific format par excellence.[24] It conforms to a set of strict rules, part of which it shares with other scientific text types. At the same time, these rules sharply distinguish articles from most non-scientific kinds of writing. As will be described in a little more detail later (Chapter 6, 'Journal articles'), modern articles are characterized by their uniform structure, their syntactical peculiarities, and above all their impersonal mode of presentation in which both author and reader seem to vanish in order to let the subject speak for itself. A scientific article masks the fact that it is actually an act of communication. Because of the high degree of standardization underlying these characteristics, writing is seen not as a creative but as a mechanical activity by most scientists—the fitting of novel content into a predefined form.

Early modern texts concerned with the natural world provide an entirely different image. Their composers saw themselves very much as authors, and for good reasons. This is not the place to pre-empt what will be said in the main part of this book regarding different genres and texts, but a few general points are worth anticipating.

As already suggested, there was no sharp formal divide between scientific works and other kinds of writing. Scientific authors used devices like technical terminology, artful layout, and illustrations more freely than others, but apart from this, they basically availed themselves of the same toolkit as non-scientific writers. Opening an early modern book on the natural world, one will usually not find it specifically 'scientific' in its structure and style. In most cases, the way in which such a work is written has close parallels in other types of learned 'literature'. Few literary genres are restricted to works of natural science, and most of these are not typical of science as such but of this or that particular discipline: 'consilia' belong to medicine, 'historiae' mainly to botany and zoology, Euclid's *Elements* provide a blueprint for geometric works, and so forth. Most of the genres to be discussed in this book, by contrast, are indiscriminately applied to scientific and non-scientific subjects.

This leads straight to my next point, which has already been addressed in the Introduction: unlike its modern counterpart, early modern scientific writing

[22] Briefly noted for the eighteenth century by Gunnarsson 2011, 3.
[23] A concise yet thorough overview is provided by Göpferich 1998–9.
[24] Frasca-Spada and Jardine 2000, 9; Johns 2000, 408; Gross et al. 2002, p. vii.

shows a bewildering generic richness.[25] From the foregoing, the reasons for this multitude of genres and the fluent boundaries between them are evident. Not only could scientific authors draw on the stock of literary genres in general. In the absence of regulation by literary theory (and publishers' rules), they were free to adapt and blend them and to invent new ones *ad libitum* in order to fit any specific combination of subject, audience, and communicative situation.

In contrast to a modern article, an early modern scientific text does not hide its communicative nature but rather highlights it. Its author and reader are much more than an implied source and recipient of information respectively: they are emphatically present in the text. This has to do with the fact that the basic speech act of a scientific text tended to be understood not just as information but as persuasion. When Blaise Pascal (1623–62) reflected on the transmission of scientific knowledge in his *De l'esprit géométrique*, it appeared self-evident to him that this transmission was a persuasive act.[26] Accordingly, the impact of rhetoric with its focus on persuasion and its precise ideas about the roles of speaker and audience is clearly visible throughout scientific literary production. So, even more fundamentally, is language itself, whose role goes far beyond that of a simple carrier of information. Far from 'absolutely disappearing into the thought' (as a memorable phrase by Nathaniel Hawthorne has it),[27] it claims a part of the reader's attention and appreciation in its own right.

This fact finds expression in a last but perhaps most conspicuous trait of early modern scientific writing, namely its impressive stylistic diversity, which once more starkly contrasts with present-day conditions. To begin with, not only prose but also poetry could be used to convey scientific content. Within prose, the whole stylistic range from the simple factual 'genus humile' to the richly ornamented emotional 'genus grande' is present, often within one and the same work. It is true that the former is the dominant register, as it is in technical literature in general, not least because authors and publishers were aware that an overly complex style might deter potential buyers.[28] The contrast between dedications and prefaces written in elaborate periods and the syntactically simpler, lexically freer main text is often striking.[29] Sometimes, the syntax and word order of the latter, while grammatically impeccable, are more reminiscent of the vernaculars than of Ciceronian Latin, making the text more intuitively comprehensible to the early modern (and also to the present-day) reader.[30] However, such simplicity was seen less as the linguistic equivalent of plain facts, as it is in the modern article, than,

[25] This is briefly highlighted by Frasca-Spada and Jardine 2000, 6.

[26] Mesnard 1991, 360–429; Scholar 2006, 2–5.

[27] Quoted in Heard 2014, 64.

[28] For the case of a historical work in which this actually happened, see Maclean 2012, 179.

[29] This contrast once more goes back to antiquity: Janson 1964, 127; Nutton 1995, 193–4 (on Dioscorides). Habermann 2001, 231–2 observes it in early modern herbals.

[30] See Leonhardt 2013, 239–41 with an example from Newton. Cf. the 'Sprachbund' mentioned in Chapter 2, 'Demise and persistence'.

in accordance with Pascal's reflections, as a technical device to convince the audience of one's claims. As noted already by Melanchthon in his *Elementa rhetorices*, the plain style 'ambushes the listener by the very semblance of simplicity' ('ipsa simplicitatis specie insidiatur auditori').[31] Not all authors liked it that simple, though. Many flaunted their skills of expression by producing elegant periods, and some even aspired to the most elevated register. Nature was often exploited as a source of greatness and emotion in scientific writing. Such an approach had already been advocated in the first centuries AD by the author of *De sublimitate*, which became a very influential text in the second half of the early modern era,[32] and it suggested itself all the more in view of the physico-theological background of much scientific writing, by which nature was intimately associated with the Creator himself. Between the two poles of the 'genus humile' and the 'genus grande', one finds all shades of stylistic plainness and audacity.

Nor is this all. In addition, one encounters a great number of genre-specific styles that find no place within the narrow framework of the theory of the three styles. To give a handful of examples, commentaries offer a quasi-scholastic Latin that reflects medieval teaching traditions. Many reports are characterized by lively narrative. The reader of 'historiae plantarum' is presented with descriptions so full of detail that they can provide a substitute for an image of a plant or a plant itself ('vicarious descriptions'). Dialogues display a studied colloquiality, aphorisms an oracular terseness. In different ways, textbooks and didactic poems take the reader by the hand, so to speak, and guide him through their respective subjects; in doing so, the latter strongly draw on the ancient models of Lucretius and Virgil. In the wake of Euclid, the language and train of thought of geometric works is often starkly formulaic; they are perhaps the only premodern text type that is comparable to the modern article in terms of standardization. In panegyrical poetry, hyperbole is de rigueur. Moreover, many authors show a strong stylistic individuality: Pietro Bembo displays his Ciceronian purism, Kepler is as precise as he is inventive and artfully interweaves his Latin with Greek expressions,[33] Athanasius Kircher indulges in exuberant periods and heavy doses of pathos, Linnaeus, an excellent Latinist, tosses even basic linguistic correctness overboard when it comes to plant and animal names, and so on.

To ascertain any overarching tendencies in this diversity is difficult. Two may tentatively be formulated by way of conclusion: first, texts for a wider audience, such as didactic and panegyrical poems, orations, and biographies, are as a rule less technical, stylistically more elevated, and more strongly oriented towards classical models than those written exclusively for the author's peers. Second, the writing becomes more technical and drier towards the end of the early modern period.

[31] Wels 2001, 318; Habermann 2001, 226.

[32] *HWRh* 2, cols. 1357–89, s.v. 'Erhabene, das', esp. cols. 1364–77.

[33] Wenskus 2023.

4
Print

Although early modern authors sometimes expressed anxiety about the possible profanation of their knowledge in print and often frustration about the practicalities of the new medium, most were nonetheless convinced that, in principle, its invention was an epoch-making event and all kinds of learning were receiving a huge boost as a result of it.[1] Modern scholarship agrees regarding the first part of this conviction. As for the second, and the subject of the present book, print was not indispensable for the implementation of a standardized, classicizing Latin as the common language of the learned but surely facilitated it.[2] That it constituted a 'conditio sine qua non' of the Scientific Revolution has famously been asserted by Elisabeth Eisenstein. Her claims have provoked much criticism, but the question need not concern us here.[3] No matter how big or small the impact of print on the development of science, that it decisively shaped scientific *literature* is obvious. In the first section of this chapter, I will provide some basic information on early modern print in general. In the second, the influence of print on the shape of the scientific book will be delineated.

Early modern print

Although Johannes Gutenberg could build on many pre-existing technologies when he invented print with moveable type, he had to invest years of patient experimentation to adapt them to his needs and to combine them in order to realize an overall vision entirely his own.[4] By the standards of his time, the result was high-tech, and it presupposed an expensive, proto-industrial infrastructure.[5] All paraphernalia involved, from the paper, the alloy used for the type, and the printing ink to the manual casting instrument and the press itself, had to be

[1] Reservations: Howard 2022. Enthusiasm: Giesecke 1991, 146–67.

[2] On Neo-Latin and print, see *BENLW* 171–271. That the standardization and fixation of a language of learning is in principle possible without print is shown by examples such as Sumerian, classical Greek, and classical Arabic (Leonhardt 2013, 17–40).

[3] Eisenstein 1979, 453–682; on the widespread critique of her thesis, see Frasca-Spada and Jardine 2000, 3–4.

[4] This section is largely based on a number of classic accounts of early modern print: Febvre and Martin 1976; Eisenstein 1979; Giesecke 1991; Martin 1994, 182–330; Pettegree 2010. Many aspects of printing culture in general are concisely explained in Rautenberg 2015.

[5] Febvre and Martin 1976, 45–76; Giesecke 1991, 67–123; Gaskell 1995, 5–185.

Latin Scientific Literature, 1450–1850. Martin Korenjak, Oxford University Press. © Martin Korenjak 2023.
DOI: 10.1093/oso/9780198866053.003.0005

produced with a high degree of precision and standardization in order for their interplay to result in a legible printed page. The logistics involved, outside and inside the printing shop, were correspondingly complex. Representatives of various crafts were needed to fabricate the necessary raw materials and instruments: papermakers, calligraphers, medallists, type founders, producers of ink, and many more. Within the shop, compositors, inkers, pressmen, correctors, and rubricators collaborated according to sophisticated organizational schemes to make sure that the printing process proceeded swiftly, efficiently, and without major mishaps.[6] It is all the more astonishing that the entire technology and logistics were already mature when Gutenberg presented the world with his first results. His Bible, which he finished printing in 1456, is not a beginner's but a master's work. In spite of numerous improvements and refinements, the fundamentals of Gutenberg's technique remained unchanged for three and a half centuries. It was only around 1800, when the manual press was superseded by the machine press, that a new era of printing dawned.[7]

Originally, print emerged in the context of the professionalized scribal culture of the Late Middle Ages and its mass-manufactured books (more manuscripts are extant from the fifteenth than from any earlier century).[8] The core idea behind the new technology was to solve the problems incurred by scribes to meet an ever-growing demand for written material by producing greater amounts of more standardized text in shorter time. For quite some time, commercial text production by manual copying and by print coexisted and interacted, and many works were simultaneously transmitted in both media.[9] Even after this state of things came to an end around 1500, the interplay between manuscript and printing culture remained intense: not only were prints based on manuscripts, but printed books also continued to be privately copied by hand. Readers collected the fruits of their reading into handwritten commonplace books, which gave birth to a genre of printed books in turn.[10] Scribal publication remained an important phenomenon for the better part of the early modern era.[11]

In line with this interpenetration of manuscript and printing culture, the appearance of the first prints mimicked the manuscripts of the time. Printing fonts copied current handwriting styles, including their numerous abbreviations. The same goes for the mise-en-page, which appears underdeveloped to a modern reader. Paragraphs were mostly indicated by pilcrows (¶) within the line and/or rubrication. Line breaks were used sparingly, if at all. Like manuscripts, early

[6] Febvre and Martin 1976, 129–36; Giesecke 1991, 86–123. [7] Gaskell 1995, 189–310.

[8] Febvre and Martin 1976, 15–28 (an addition by Marcel Thomas); *BENLW* 265–71.

[9] Schmitz 2018, 11–42. For a case study (commentaries on Hippocrates's *Aphorisms*), see Pesenti 2000.

[10] On commonplace books and their wider context of information management, see Moss 1996; Blair 2010, esp. 62–116 and her index, s.v. 'commonplace books'.

[11] See, for instance, Maclean 2009, 414–17 on scribal publication in seventeenth-century Paris.

books had no title page but gave bibliographical data (date, place, author, printer, some sort of title) in a colophon at the end. The distinctive character of fifteenth-century prints is recognized by their modern designation as 'incunables' ('books in their cradle').[12] From the end of the century, print was increasingly understood as a medium in its own right, which resulted in substantial changes on the levels of typography and layout, some of which will be mentioned in the second half of this chapter. In many respects, the resulting look of the early modern book, somewhere in between that of a manuscript and a modern book but closer to the latter, remained fairly constant over the next few hundred years, despite a host of minor trends and local styles and a new tendency towards modern plainness and neatness in the eighteenth century.[13]

If the history of printing up to 1800 was not distinguished by qualitative change, it was undoubtedly characterized by quantitative expansion. Despite permanent minor crises of the book market and a profound one around the mid-seventeenth century,[14] the number of printing shops grew from century to century, especially in the German-speaking world, the Low Countries, Italy, France, and later England, and so did the number of books printed. In the second half of the fifteenth century, *c.*27,000 editions were produced in Europe as a whole. The corresponding figures for the German-speaking world alone may provide an impression of the development in the course of the sixteenth, seventeenth, and eighteenth centuries. In the first of the three national bibliographies, the *Verzeichnis der im deutschen Sprachbereich erschienenen Drucke des 16. Jahrhunderts*, 106,000 editions have been catalogued. For the two ongoing follow-up projects, the estimates are 300,000 and 600,000 respectively. Still, it has been argued that these figures only represent one half to two thirds of the real production.[15]

In the course of its development, print also greatly expanded the literary milieu in the widest sense of the word—the part of the population concerned in some way or other with the production, distribution, and consumption of texts. It also changed the roles of all participants. The numerous artisans taking part in the printing process have already been mentioned. Apart from them, the most important groups concerned with the printed word were authors, printers, publishers, booksellers, readers, and collectors.

For authors, print provided a new incentive to produce texts: as they were distributed much faster and farther than manuscripts, printed works promised to be a source of self-esteem, prestige,[16] and other social benefits such as the

[12] An introduction to the field is provided by Schmitz 2018. The leading catalogue is the *Incunabula Short Title Catalogue* (http://www.bl.uk/catalogues/istc/, accessed 16 February 2023).

[13] On the last point, see Barker 1981; Martin 1994, 317–20.

[14] Martin 1994, 250–1, 256; Maclean 2012, 211–34.

[15] Beyer 2011. On other modern national bibliographies, see Pettegree 2010, 353–6.

[16] On scientific publications as a source of prestige, see Siraisi 1987, 5; Grendler 2001, 160.

strengthening of personal networks and the acquisition of patrons. On the other hand, the effort to be invested to this end increased considerably. As a result, publication became a more conscious and deliberate step than it had been before. Nonetheless, an author's control over the printing process and the final product was limited, if he did not oversee the work in person, which happened only rarely. Often, he showed little concern about the exact wording of his text from the outset and tacitly accepted or even welcomed interventions by the printer.[17] On the other hand, complaints about the carelessness of printers and the number of misprints produced by them were commonplace. In response to these problems, some authors set up their own presses in order to realize their publication projects in the envisaged form. Famous examples from science include Regiomontanus (1436–76), Gemma Frisius (1508–55), and Tycho Brahe (1546–1601).[18] Others were helped by friends and colleagues who took on the role of 'editors' (as Nicole Howard calls them) and, besides proofreading and overseeing the printing process in general, might provide a number of additional services such as correcting and translating manuscripts, encouraging publication, finding a publisher, and taking over the costs of printing.[19]

The printer, the publisher, and the bookseller[20] originally were just one person, since the earliest printers invested their own money in their projects and also acted as their own salesmen. Soon, however, some division of labour became common. Since printing a substantial work often meant a higher investment than a printer himself could afford, a publisher could step in, investing the money, taking the risk, raking in the profits, and paying the printer for his efforts. Itinerant booksellers and those with a permanent place of work emerged who took care of the problems of distribution. However, the combination of two or all three of these roles remained an option. The printer-publisher who owned his own type and presses and invested his own money in his books remained a common phenomenon. Also, most publishers owned bookshops as well.

By far the biggest group of those affected by print were readers. Developed in response to a growing demand for reading material, print created additional demand in turn.[21] Literacy rates rose after its introduction, and the ability to read extended beyond the clergy and the learned to groups such as nobles, townspeople, and women. As books became more reader-friendly, the act of reading itself changed as well. While people had often, although by no means exclusively, read aloud in the manuscript age, since articulation provided help towards

[17] Lowry 1979, 224–8.

[18] Febvre and Martin 1976, 163 (self-publishing in France); Lowood and Rider 2000, 13–15; Mosley 2000 (Brahe); Howard 2022, 117–39 (including authors developing alternative reproduction techniques for text and authors printing their own illustrations).

[19] Howard 2022, 140–72.

[20] Febvre and Martin 1976, 136–59; Maclean 2012, 101–4; for the incunable era, see Schmitz 2018, 175–86.

[21] Schmitz 2018, 26–7.

understanding, silent reading gradually became standard, and reading speed increased.[22] Some characteristic habits of early modern readers of learned texts are worth noting: like their predecessors in the manuscript age, they freely annotated their books—a practice recommended and even taught—and the notes they left reveal much about their attitudes and interests;[23] they often did not read cover to cover but selectively; and they 'read to reap', to extract nuggets of information for their aforementioned commonplace books.[24]

Over time, many well-off people got into the habit of assembling collections of books, not necessarily with the intention of reading everything they bought. The trend towards building libraries was indicated and furthered by how-to books like Gabriel Naudé's *Advis pour dresser une bibliothèque* (1627). From the mid-sixteenth century, book auctions were set up in the Netherlands, from the seventeenth, they became popular in England.[25] Collectors did much to keep certain segments of the learned book market alive.[26] Many collections included or even focused on scientific books.[27] So did the libraries of universities and other learned institutions, which kept being founded and growing.[28]

Considered from an economic point of view, print fundamentally transformed the book market. Commercial scribes had mostly worked to meet a local demand. Now books, especially learned books in Latin, were bought and sold on an international scale and became a much more important commodity than they had been before.[29] However, the workings of the early modern book market differed from those of its modern equivalent in several respects. Not only was trading in books fraught with logistical problems and uncertainty to a much higher degree than it is today, as market participants had to cope with risks arising from lack of information, transport hazards, censorship, piracy, and many other factors, but the roles of the single players involved were unlike those of today as well.

This is most obvious in the case of authors.[30] While they could gain symbolic capital from publications, they were usually not paid by publishers for letting

[22] Giesecke 1991, 89–90. A concise overview of the history of reading in general from the Middle Ages to modern times is provided by Zedelmaier 2015, 5–16.

[23] An introduction to the study of annotations is provided by Jackson 2001, who focuses on the period 1700–2000 but provides a broader historical overview on pp. 44–80. Two major pertinent case studies are Gingerich 2002 and Margócsy et al. 2018, 56–120, who discuss the annotations of all extant copies of the first two editions of Copernicus's *De revolutionibus* (1543, 1566) and Vesalius's *Fabrica* (1543, 1555) respectively.

[24] The phrase is taken from Grafton 2008, 27. On the various practices of excerption, see Zedelmaier 2015, 45–61. On early modern reading habits and their relative uniformity in general, see also Burke 2000, 179–82 and, for a case study, see Margócsy et al. 2018, 11, 131–2.

[25] Myers et al. 2001.

[26] See Maclean 2016, 307–8, 312, 315, 317, 328 on their role in the eighteenth-century law book market.

[27] Besson 1990b; Mandelbrote 2000; Overmier 2000.

[28] Pettegree 2010, 319–32.

[29] See Febvre and Martin 1976, 109–27, 216–47; Burke 2000, 159–73 (embedded in a discussion of knowledge in general as a commodity); Pettegree 2010, 65–90, 249–70. On the market for learned books in particular, see Maclean 2009 and 2012.

[30] Febvre and Martin 1976, 159–66.

them print and sell their works. In fact, the contrary applied in many cases: even famous authors such as Kepler had to subsidize their publications. If they were unable to do so, they might find no publisher willing to take the risk, especially in the case of large, complex works.[31] In addition, copyright was unknown, so an author forsook any legal right in his work once he handed it over to a publisher. It was only from the eighteenth century that the author's financial and legal situation slowly changed for the better. Publishers began to pay for manuscripts on a regular basis. A modern notion of authorship, including intellectual ownership, legal responsibility, and control over one's text, emerged from around 1700 in England and spread on the Continent over the subsequent period.[32]

The situation of the publisher (who was often also a printer and bookseller, as already mentioned) was not easy either. He could rarely survive from publishing substantial books alone. The backbone of his business was, rather, cheap broadsheets and pamphlets of all kinds, from indulgences to proclamations of rulers and thesis prints, which he produced on request and which guaranteed a sure profit.[33] When he was offered a bulky manuscript, by contrast, the burden of uncertainty weighed heavily on him, and he had to rely on educated guesses to make a decision that might result in his prosperity or ruin. His initial investments in infrastructure, in particular in type and presses, were in any event considerable, and so were those to be made for every single book. For these, paper was the largest cost, followed by the workforce (and the expense would rise in the case of complex layout and illustrations), while everything else played a minor role.[34] The high paper costs go some way to explaining the popularity of the paper-saving octavo and smaller formats and make it understandable that publishers seldom risked high print runs even in the case of a promising work, preferring to set it afresh if it proved successful. The usual number of copies was a few hundred in the incunable age, rose to about 1,000–1,500 from the sixteenth century, and more or less remained at that until the eighteenth.[35] A publisher often tried to protect for a number of years the profit he expected in exchange for his investment and risk by a privilege against the rampant phenomenon of unauthorized reprinting. However, privileges only covered the dominion of the issuing authority and were often disregarded even there.[36] From the sixteenth century, the big fairs (Frankfurt and Leipzig in particular) provided important meeting places for publishers and booksellers. Initially, they engaged in barter there and on other

[31] In the later seventeenth century, the subscription system was introduced to finance such publications (Burke 2000, 167, 172, 177; Nummedal and Findlen 2000, 190–1).

[32] Burke 2000, 162, 165–6; Lowood and Rider 2000, 7; Jardine 2000, 397 (all with further literature).

[33] Martin 1994, 295–9; Schmitz 2018, 142–4, 191–2.

[34] Febvre and Martin 1976, 109–15.

[35] Schmitz 2018, 186–92 (incunables); Febvre and Martin 1976, 216–22 (later times).

[36] On privileges and their functions, see Mosley 2000, 123–4; Rautenberg 2015, 133–4. On the ubiquity of counterfeit editions, see Moureau 1988.

occasions to enlarge their retain stock of books, but this system collapsed in the time of the Thirty Years War.[37]

What readers actually had to pay for their books is a thorny question, partly because prices were often adapted to the needs of the situation, partly because our data are incomplete and partly because the complexities of the premodern monetary and accounting systems render their evaluation tricky: amounts given in account books in particular may be based either on real currency or on a fictitious currency used for accounting purposes only.[38] Under these conditions, the general statements one can make are few and mostly commonsensical: printed books were cheaper than comparable manuscripts from the beginning; prices dropped substantially during the first decades of printing history; nonetheless, books remained much more expensive than today; the price of a book increased with its size, quality, and additional features.

From the incunable era, the wider spread of information through print raised concerns among ecclesiastic and secular authorities alike. They responded to the new situation by implementing a host of institutions and procedures of censorship.[39] The Catholic Church was the most important organization in this respect, and the *Index librorum prohibitorum*, whose first edition appeared in 1564, the most famous instrument, but the phenomenon was ubiquitous. At the Frankfurt Fair, for instance, an imperial book commission examined books from the late 1560s.[40] Censorship could take place before publication, namely through an obligatory imprimatur, or after it, through lists of forbidden books like the *Index*, corrected editions, and sanctions against authors, printers, publishers, sellers, and owners. The reasons for censorship were mostly religious, but dissident political opinions, betrayal of state secrets, libel, and violation of public morals could lead to the banning of a book as well. In most cases, however, the aim was not the annihilation of the work in question but the deflection of dangers from readers and authorities, which could also be achieved through the blackening of problematic passages, a purged reissue, or restriction of access. As a rule, censorship was only moderately efficient,[41] not least because in a politically and religiously fragmented Europe, it could be evaded by printing in a different region. Regarding scientific books in particular, the first thing that springs to mind is the case of Galilei's *Dialogo* (1632), but this affair was not only unusually high-profile but also exceptional in a number of other ways. Since it was acknowledged that

[37] Maclean 2012, 181–5.

[38] For a sample of incunable prices, see Schmitz 2018, 209–18. A database of European book prices 1540–1630 has been produced by the project EMoBookTrade (https://emobooktrade.unimi.it/, accessed 16 February 2023). On the problems with interpreting early modern prices, see Ammannati 2018.

[39] Giesecke 1991, 441–5, 462–70; Martin 1994, 266–82; *EEMHO*, s.v. 'Censorship'; Schmitz 2018, 197–201.

[40] Maclean 2009, 13–14 with further literature.

[41] On the modest impact of censorship of Vesalius's *Fabrica*, see Margócsy et al. 2018, 121–30.

forbidden books about the natural world might still contain useful knowledge, people in need of access to such information for professional reasons could apply for personal permits to read them.[42]

The printed scientific book

The early modern book market fell into two major parts, one comprising works for scholars written predominantly in Latin, the other, vernacular ones for a broader readership. Learned books represented nearly one half of the total production,[43] and a substantial section of them was concerned with the natural world.[44] In this field, as in others, the first publications were texts of proven popularity in the manuscript age, that is, classics from antiquity, the Arab world, and the Latin Middle Ages.[45] The great majority of Sir William Osler's over 200 *Incunabula medica* printed between 1467 and 1480, for instance, belong to this category.[46] Within a few decades, however, there was an upsurge in contemporary texts. Cheap broadsheets featuring almanacs, 'regimina pestis', and recipes[47] led the way, but books followed suit. The non-anonymous scientific works printed over the entire incunable era were composed by *c.*300 contemporary alongside 200 ancient and medieval authors, although between them the latter were responsible for the majority of the works.[48] Thereafter, the number of new works grew from century to century. The medical titles available, for example, seem to have increased by over a hundredfold between 1500 and 1600.[49] For later periods, contemporary bibliographies provide a rough impression of the expansion of output. Johann Carl Heffter's *Museum disputatorium physico-medicum* (1756–64), for instance, lists 18,500 German dissertations on medicine and the natural world in general from the preceding one hundred years. Georg Rudolf Boehmer's bilingual *Handbuch der Naturgeschichte/Bibliotheca scriptorum historiae naturalis* (1785–9) comprises 65,000 titles.[50]

The large-scale book market brought about by print put pressure on authors, publishers, and anyone else in the business to present their products in most

[42] See Marcus 2020, who primarily deals with Catholic censorship of medical books in sixteenth- and seventeenth-century Italy but also discusses the Galilei affair on pp. 226–36.

[43] See the statistics for the period 1450–1600 in Pettegree 2010, 357.

[44] Of *USTC*'s currently 841,000 titles from the period 1450–1650, around 60,000 (*c.*7 per cent) may be assigned to the natural sciences, although the mixture of content-related and formal categories in this database makes estimates precarious. On early modern books and science, see Besson 1990a; Hunter 2000; Frasca-Spada and Jardine 2000; Pettegree 2010, 273–318.

[45] An overview of the learned medieval works (including those concerned with the natural world) printed in early modern times can be gained from Schönberger et al. 2011.

[46] Osler 1923; see also Arrizabalaga 1998, 4. On medical incunables in general, Rhodes 1990.

[47] Schmitz 2018, 142.

[48] Klebs 1938; Sarton 1939, esp. 198–226.

[49] Jones 2000, 131.

[50] Brock 2000, 323; see also pp. 307–29 in general for more early modern bibliographies with numbers of entries.

saleable shape. As a result, books were moulded into certain forms and developed distinctive features they had not possessed in the manuscript age. Books about the natural world participated in this general trend but also developed a number of particular traits, as some of the challenges they had to meet were specific to science (or at least to learning in general). In what follows, an overview of the most salient among the respective features will be provided. I will start with some remarks on the generic landscape and will then discuss the typographical structure of scientific books, their paratextual apparatus, and the inclusion of non-text.

Comparing the kinds of scientific texts and books that existed in the early modern period with those from earlier times, one notices two opposed trends: generic creativity on the one hand, and a recycling mentality, on the other. Print engendered many literary genres and publication formats which had no or only embryonic precedents in the manuscript era. With regard to genres, one encounters technical dictionaries, bibliographies, reports, dissertations, and textbooks in the table of contents of the present book alone, a list which can be supplemented by less widespread or thematically restricted genres such as medical 'observationes' or atlases. The most important novel publication format in science by far was the journal, with its two main components, the review and the article, the latter destined to become *the* scientific genre *tout court*. The credit for new text types such as these is shared by inventive authors and enterprising printer-publishers.

The same is true of the concomitant trend towards derivative forms, with the difference that the authors of original works were often replaced by later editors and translators. Derivative forms were nothing new in themselves. Already in the manuscript era most works were 'fluid texts' which could appear in a wide spectrum of related, yet different versions.[51] However, the practice of reworking and recycling texts became more targeted and purposeful in early modern times: selling one text in different versions was not only used to keep track of the development of science but also to target new groups of readers, prevent market saturation, and generate additional profit.[52] Starting with the simple, perhaps slightly corrected reprint, there were manifold kinds of adaptations: texts were set in new type, layout, and format, expanded, epitomized, changed, censored, moved to another genre, versified, translated, annotated and commented, illustrated, equipped with new paratexts, and brought together into thematic collections and 'opera omnia'. In some cases, a work changed beyond recognition in the process. Pietro Andrea Mattioli's Dioscorides commentary (1st edn 1544), for

[51] The term 'fluid text' was coined by Bryant 2002, who is primarily (although not exclusively) concerned with the effects of authorial revision before publication. A more precise term for the mutability of texts already in circulation would be the expression 'unfester Text' ('unfixed text') commonly used in German medieval studies.

[52] For an overview of derivative forms in early modern scholarly publishing in general, see Maclean 2012, 82–91.

one, not only grew massively over its publication history, but was also translated from Italian into Latin and thence into French, Czech, and German, and finally epitomized again and republished as a compendium. Linnaeus's *Systema naturae* grew from a dozen folio pages in its first edition (1735) to four volumes and nearly 2,300 octavo pages in its twelfth (1766–8), not to mention the replacement of introductory paratexts, the coming and going of illustrations, translations into several vernaculars, and the consistent adoption of the binomial nomenclature from the tenth edition.

A few decades after the invention of print, the typographical structure of books, which originally imitated that of manuscripts, began to be adjusted to the new medium. With due respect for the achievements of late medieval scribes,[53] it seems safe to say that printed books became more intuitive and reader-friendly than manuscripts had ever been. Not all aspects of this complex and finely ramified development are of concern in the present context. Two issues of particular importance regarding the scientific book, whose argumentative complexity holds specific challenges for the reader, are traits facilitating perusal and elements providing orientation within and beyond the book.

Reading was made more convenient by a number of devices. Type fonts (originally of the Gothic, later predominantly of the roman type) departed from their handwritten models and unashamedly presented themselves as print.[54] In learned Latin books in particular, a combination of clear-cut roman fonts for normal text and capital scripts for highlighted parts established itself. This ultimately led to the habit of beginning certain kinds of words but especially sentences in upper case, which together with the gradual establishment of punctuation resulted in a clearer expression of a text's syntax. The abbreviations typical of handwritten texts were substantially reduced, also because they cluttered the typecase with too much special type. Orthography was slowly being standardized, something easier to achieve in Latin, which had been regulated by grammarians for over a millennium, than in the vernaculars.[55] Width was increased and words were more clearly separated from one another. Moreover, lines were spaced more generously, line breaks used systematically to set text units apart, and blanks discovered as an efficient way of structuring an account.[56] As a result, texts were read with greater ease and faster than before, and silent reading became the rule, as already mentioned.

Orientation within the printed book in general was provided by an explicit hierarchical structure, which tended to become more fine-grained and rigid over time: a larger work might not only fall into several books and its books into chapters (as already in antiquity and the Middle Ages), but chapters were also divided

[53] Schmitz 2018, 14–15, 129.
[54] Schmitz 2018, 253–79.
[55] On French orthography, see Martin 1994, 305–8.
[56] Schmitz 2018, 133–4.

into sections, subsections, and so forth. On each level, contents and their order of presentation could be standardized to varying degrees.[57] All or some of the resulting units featured more and more often in an opening table of contents and were marked in the text itself by subtitles in different fonts and sizes, as well as by page, double-page, and column titles. Even more detailed orientation could be provided by marginalia. (Colour, by contrast, which had been important in manuscripts, largely disappeared as a structuring device.[58]) Counting systems became more pervasive and efficient. Not only were the aforementioned units, that is, parts of the respective *work*, numbered, but the *book* was equipped with a counting system as well. Folio, page, and column numbers made precise cross-referencing and indexing possible.[59] Moreover, they enabled other authors to refer to a book thus equipped more accurately. Initially, such references were integrated into the running text itself. Soon, marginalia came to be used for this purpose as well, followed by footnotes from the seventeenth century.[60] The former were sometimes, the latter, which could also serve as a selective or running commentary, always keyed to the main text by figures, letters, or symbols. Both devices embedded books more firmly in an overarching discourse and created a new, more precise, and unambiguous kind of learned intertextuality.

Devices like those just discussed form part of a book's paratextual apparatus, the richness and scope of which constitute one of the most conspicuous characteristics of the early modern scientific book.[61] In most publications, a considerable number of paratexts, composed by authors, editors, publishers, printers, friends, colleagues, public authorities, and others, supplement the main text. Some of these text types predated the advent of print and even rose to considerable importance in the manuscript era,[62] but their pluralism was a result of the printing revolution. So was their organization into a highly flexible, but nonetheless easily recognizable paratextual system, within which the single components were assigned specific, mutually complementary functions.

[57] See Tebeaux 1997, 35–89 on English technical literature; Habermann 2001, 274–93 on Latin and German herbals.

[58] Martin 1994, 231.

[59] Folio numbers first appeared in fourteenth-century manuscripts, page numbers sporadically from the 1470s (Schmitz 2018, 103–4).

[60] I have encountered marginalia referencing other authors from *c.*1500 (Giovanni Manardi, *De erroribus Symonis Pistoris*), marginalia referencing works by book and chapter from the 1530s (Johannes Lonicer, *Nicandri Theriaca et Alexipharmaca*, 1531). On precise marginal references in commentaries on Avicenna's *Canon* from *c.*1550, see Siraisi 1987, 170–1; on frequent references in Dillingen dissertations from *c.*1600, see Leinsle 2006, 44. The earliest footnote references known to me are in the 1642 edition of Pascasius Justus Turcq's *Alea*, pp. 49 and 61. In general, see also Grafton 1997.

[61] On the notion of the paratext, see Genette 1997; for a recent discussion with respect to Neo-Latin works, see Enenkel 2015, 1–53. On the development of paratexts in incunables, see Schmitz 2018, 219–51; on paratexts to editions of Avicenna's *Canon* and to herbals, see Siraisi 1987, 161–74 and Habermann 2001, 98–223 respectively.

[62] See Enenkel 2015, who studies dedications, prefaces, author portraits, and dedication images from 1350 to 1650.

In many cases, the dividing line between main text and paratexts cannot be drawn with confidence, as some paratexts may be indispensable to the understanding of the main text and introductory pieces may lead the reader step by step into the work itself.[63] The classification of paratexts proves difficult as well. For a start, one may distinguish two groups: paratexts accompanying the main text like the ones previously mentioned (subtitles, marginalia, footnotes, page numbers), on the one hand, and those introducing and concluding it, on the other. The latter quite ample group comprises inter alia (in their approximate order of appearance within the book, although this order in fact varies considerably) the frontispiece,[64] the author portrait, the title page, poems of praise, the dedication, the preface to the reader, the imprimatur of the board of censors, the privilege, the table of contents, the biography of the author, the glossary,[65] the index, the errata, and the colophon;[66] in a wider sense, shorter works appended to the main work can be said to belong to this category as well. While no early modern book contains all of these types of paratext, all contain some.

Unsurprisingly, most functions of the paratextual apparatus were related to print in one way or another. A few paratexts, such as the sheet title or the imprimatur and the privilege, were linked to its technical and legal aspects respectively. Most, however, served an economic purpose in the context of the competitive early modern book market: they tried to give readers good reasons to buy this specific book instead of another among the many vying for their attention and money. This could be done by way of direct advertisement (for example in the form of a laudatory poem), by making the book more usable (table of contents, glossary, index, errata), or by a combination of both, as in the case of the title page, the dedication, or the preface to the reader. In what follows, these last three items as well as the index will be briefly discussed, since they are very common, relevant to scientific works, and hence figure prominently in the second part of the present book.

Supplanting the colophon of the manuscript era, the title page developed from the later fifteenth century and had acquired its early modern standard form by the first quarter of the sixteenth.[67] Providing first information about as well as some promotion of the book in question, it combined functions of the modern title page, the imprint, and the blurb. Title pages often contained much more text

[63] Genette 1997, 1–2; Enenkel 2015, 8–10.

[64] On frontispieces in seventeenth-century scientific (broadly speaking, mathematical) books, see Ashworth 1991; Remmert 2006 and 2011. Remmert 2006 distinguishes four functions of frontispieces: courting patrons, taking sides in scientific debates, legitimizing fields of study, and advertising books.

[65] For instance, Georg Agricola, *Bermannus* (1530), pp. 131–5; William Gilbert, *De magnete* (1600), after the letter of dedication. On glossaries to Avicenna's *Canon*, see Siraisi 1987, 165.

[66] Schmitz 2018, 229–35.

[67] On the development of title pages in general, see de Vinne 1902, 1–191. On their early history, see Smith 2000; Rautenberg 2008; Schmitz 2018, 220–8, esp. 226. On the colophon and title page in herbals, see Habermann 2001, 109–32.

than today, as the full title could be long and syntactically elaborate, the principal phrase being supplemented by relative clauses, non-finite clauses, and appositions. Topics to be mentioned comprised but were not restricted to the book's rich contents, its careful elaboration and general excellence, its novelty (indicative of a strong belief in progress in early modern science),[68] the profession and high status of the author, the quality of the edition, which was often announced as having been 'cleansed of innumerous misprints' ('ab innumeris mendis purgata') or the like, and the targeted readership. The other bibliographical data (printer, publisher, bookseller, place, date, privilege) usually followed at the bottom of the page. Title pages were eye-catchers, displaying studied layout, a hierarchy of different fonts and type sizes, and pictorial elements such as the printer's or publisher's mark, a title woodcut, or a title copper. They even could be executed as fully engraved copper titles.[69] One long-term effect of the title page was the promotion of the title itself, a previously rather neglected, fluid entity, to a prominent and permanent part of the work, which made the book easier to reference by bibliographers and other authors.

Dedicating one's work to some particular person was an ancient custom, but in printed books, the dedication became ubiquitous.[70] This development was noticed by contemporaries and judged problematic because of the self-interested motives thought to underlie it, as shown by its discussion in Giovanni Fratta's *Della dedicatione de' libri, con la correttion dell' abuso* (1590). Dedicatees were usually friends, political authorities, or patrons, but others could be cast in this role as well, including abstract entities and even the Almighty himself, as in Jan Baptist van Helmont's *Ortus medicinae* (1648). In the first two cases, a desire to strengthen personal relationships or one's position in society may be alleged. A dedication to a patron could be a token of gratitude for his financing of the print, an attempt to attract some future compensation, or an effort to secure one's work a friendlier reception or better sales by investing it with additional prestige. The dedication customarily comes as an open letter, although the opening and concluding formulas typical of this text type can be handled loosely. Often, the letter falls into two parts: the first extols the importance and value of the work's subject, perhaps in combination with a more or less modest presentation of the work itself; the second explains in flattering terms why the addressee is the worthiest dedicatee imaginable. However, both aspects may also be mixed together,

[68] Thorndike 1951.

[69] On the development of the printer's or publisher's mark in incunables, see Schmitz 2018, 236–42. On the engraved title page ('Kupfertitel'), the title woodcut ('Titelholzschnitt'), and the title copper ('Titelkupfer'), see Rautenberg 2015, 244–5, 386. All three devices could also be called 'frontispiece' in early modern times (Remmert 2006, 239) but should nonetheless be distinguished from the real frontispiece.

[70] On ancient dedications, see Janson 1964; on late medieval ones, Enenkel 2015, esp. 38 and 47; on the incunable era, Schmitz 2018, 242–5; on the sixteenth century, Schottenloher 1953. For further literature, see Enenkel 2015, 22–46.

and, in general, many authors did their best to circumvent the hackneyed conventions of the text type or to vary them as inventively as possible. Other forms than letters were used as well, in particular so-called 'ellogia', that is, texts imitating the stately language and layout of an inscription in print. One also finds combinations of different types, as in Thomas Ittig's *Lucubrationes academicae de montium incendiis* (1671) which feature an 'ellogium' and a letter of dedication to John George II, Elector of Saxony.

In contrast to the dedication apostrophizing a real person, the preface addressed to the abstract general reader hardly had any direct manuscript predecessors[71] but was a child of print. Some preforms, such as short addresses in the colophon or opening epigrams to the reader, already existed in the incunable era. The classic preface at the beginning of a work, often directly following upon or (more rarely) followed by the dedication, appeared in the first decades of the sixteenth century and became common around the middle of it.[72] It responded to the fact that in the age of print, authors, editors, and publishers were facing a physically distant, anonymous, and unknown readership which should nonetheless be given some information about the book in question in advance. To this end, the multitude of readers was linguistically reduced to one ideal reader who was kindly apostrophized as 'friendly' or 'favourable reader' ('amice lector', 'benevole lector', and the like; cf. note 71) and whose expectations and questions were addressed and answered. In the best case, every real reader would thus be provided with all necessary previous knowledge. In addition, he would gain the reassuring impression of entering into a personal relationship with the author of the book.

In some regards, the preface resembled the dedication. Both were styled as letters—usually by the author, more rarely by the editor or publisher—and, like the dedication, the preface often exhibited a bipartite structure: in its case, a commendation of the subject in general was regularly followed by topics specifically pertaining to the book. At the same time, both text types were supposed to be mutually complementary, one supplementing what was absent in the other, as their vicinity would render any redundancies glaringly obvious. Although many aspects could in principle be addressed in the dedication and the preface alike, it was usually the latter that offered more book-specific information. In many cases,

[71] However, Ovid's opening and concluding addresses to the general reader (epigram at the start of the *Amores*; *Tristia* 3.1.1–2: 'lector amice'; 4.1.1–2: 'lector'; 4.10.131–2: 'candide lector') were presumably an inspiration. The general reader is also addressed briefly at the beginning of Apuleius' *Metamorphoses* (1.1.1: 'lector, intende: laetaberis').

[72] Concluding notes to the reader: see, for instance, the colophon to the Latin *editio princeps* (1475) of Ptolemy's *Geography* (*Cosmographia*, 1475); initial epigram: Georg von Peuerbach and Johannes Regiomontanus, *Epitoma in Almagestum Ptolemaei* (1496), sig. a3^r; early example of a fully fledged preface: Johann Winter, *Pauli Aeginetae Opus de re medica* (1532), verso of the title page.

the preface is our best source about a work's genesis, purpose, and target readership, as well as about the author's hopes and fears regarding its reception.[73]

Indexes by and large originated in the Late Middle Ages, when theological, philosophical, medical, and encyclopedic texts began to be made accessible in this way.[74] Medieval indexes mostly referred to books and chapters only, and their alphabetical order did not extend beyond the first two or three letters of the individual entries. Printed indexes appeared from the 1480s and became more and more frequent in the course of the following century, as well as more precise by referring to folios, pages, and columns, fully alphabetized (following a trend also evident in lexicons and other kinds of lists), and better organized through subheadings and cross-references.[75] Like a table of contents—which could likewise be called 'index' in Latin—an index enabled readers to find contents of interest to them quickly. However, it was usually more detailed as well as searchable by terms, while a table had to be searched thematically.[76] Both devices made the use of books faster and more efficient, as linear perusal could be supplemented or replaced by selective consultation—a huge advantage in an era when information overload became rampant and scholars ran short of time. Indexes differed greatly in length, thoroughness, and content. Encyclopedic works in particular featured impressive examples. An all-time high point may have been reached with Antonio Musa Brasavola's *Index refertissimus* (1586) to the Giuntine 'opera omnia' edition of Galen, a separate folio of some 1,100 pages and 150,000 entries, each of which not only indicated but also summarized a passage and resulted in an alphabetized epitome of Galen's *oeuvre*. The importance of indexes was widely recognized by contemporaries. When a book had one, this was often advertised on its title page, a fact which is confirmed by remarks that books with indexes sold better.[77] Occasionally, owners added indexes by hand to books lacking them.[78] Theoretical reflections on indexing can also be found. The main part of Conrad Gessner's *Pandectae* (1548–9), for instance, which comprises the first nineteen of twenty-one projected books, contains a short treatise 'De indicibus librorum' on fols. 19^{v}–22^{v}.

The use of non-text in early modern books is a wide and diverse field on which much research has been done in recent decades. Only some very general remarks can be given here. The borders between text and non-text are fluid. To name but

[73] On scientific prefaces, see Howard 2022, 49–89 (including a provisional typology at p. 56 and numerous examples); on prefaces to herbals of the fifteenth and sixteenth centuries, see Habermann 2001, 168–223.

[74] On the history of the index in general, see Duncan 2021; on its medieval origins, see Witty 1973; Meyer 1991.

[75] Blair 2000b, 75–85 (on indexes in scientific works); Schmitz 2018, 245–9 (in incunables).

[76] On both devices and their interplay in early modern herbals, see Habermann 2001, 133–44.

[77] Blair 2000b, 78; Schmitz 2018, 246.

[78] With regard to incunables, see Schmitz 2018, 248–9. The practice persisted for a long time. A much later example is provided by a certain Lucas Schröck's copy (Staats- und Stadtbibliothek Augsburg, Med 4039) of Samuel Schönborn's *Manuale medicinae practicae Galenico-chymicae* (1637). See also Duncan 2021, 264–6.

one important question: does mathematical notation—not natural language but more or less readable like normal text—belong to the one or the other category? I adopt an inclusive notion of non-text embracing everything markedly different from normal text (which is why I speak of 'non-text' instead of 'illustrations' in the first place). It is likewise difficult to establish a typology of non-text.[79] In terms of function, one may roughly distinguish ornamental and advertising elements—for instance, the aforementioned paratextual illustrations—from informative, explanatory, and argumentative ones, which are dominant in scientific works. Semiotically, non-text can be said to be representational (illustrations), symbolic (expressing logical structures: tables and charts, tree diagrams, Ramist tables, mathematical notation), or in between (schematic representations, maps, geometric diagrams).[80]

In the manuscript era, non-text had been highly liable to degradation through repeated copying by hand. It therefore profited to a particular degree from the introduction of print, which halted or at least slowed down corruption through exact reproduction in great numbers. Moreover, while some kinds of non-text such as mathematical notation presented great typographical challenges,[81] in other cases, the print-related techniques of the woodcut and the copper engraving allowed for new levels of fineness and precision. Occasionally, for instance in Erhard Ratdolt's *editio princeps* of Euclid's *Elements* (1482), novel, highly sophisticated techniques were adopted and perfected to obtain optimal results.[82] As many issues discussed in science are highly susceptible of visualization, it is quite logical that non-text experienced a boom in early modern scientific books, where it appeared in great quantities and many forms. Numerous kinds predating print, from botanical illustrations to mathematical notation, developed dynamically, while others, such as Ramist tables, were now invented. Astonishing craftmanship and inventiveness are evident in some devices, for instance in paper instruments to be cut out and reassembled by the reader or in the so-called 'volvelles', movable illustrations visualizing subjects as diverse as time-reckoning, cosmological models, and human anatomy.[83] Non-text can be found in most types of early modern scientific works and in all disciplines.

While non-textual elements were helpful in many respects, their use was not without its problems. Techniques were costly, and good artisans could be hard to find. Even if these obstacles were overcome, authors often found it difficult to explain how exactly they wanted their illustrations and diagrams to be executed,

[79] On the difficulties besetting the demarcation and classification of non-text, see Lüthy and Smets 2009.

[80] This trichotomy is based on a dichotomy proposed by Maclean 2002, 60–3, who divides the field of medical non-text into 'representational items' (images) and 'pedagogical tools' (tree diagrams, Ramist charts).

[81] Topham 2000, 327–9 shows that this was the case as late as the nineteenth century.

[82] Baldasso 2008, 72–84.

[83] Gaida 2016; Crupi 2019.

were unable to control what was delivered and printed, or did not care in the first place.[84] One result of all these circumstances was the custom of reusing old woodblocks or copperplates wherever possible. For instance, the majority of the numerous herbals printed until the early seventeenth century drew on just half a dozen sets of woodcuts.[85] Generally speaking, illustrations were often poor, and text and non-text might have little to do with each other.[86] Some kinds of books—dissertations, for instance—largely renounced non-text for financial reasons. In addition, certain authors objected to illustrations on principle: they felt, for instance, that an image could only portray the accidents but not the essence of a plant, that its two-dimensionality did not do justice to the physical reality of the human body, or that it was incapable of showing the internal variation within a certain class of objects.[87] Moreover, humanistically educated readers found it easier to understand even complex and difficult texts than to 'read' novel kinds of non-text. More attention was therefore devoted to the text of Vesalius's *De humani corporis fabrica* (1543) than to the woodcuts on which the work's fame rests today. However, authors' and readers' attitudes and abilities evolved over the centuries. Ultimately, a high amount of non-text became a hallmark of scientific literature.

[84] Grafton 1997, 173; Kusukawa 2000a, 97.

[85] Lowood and Rider 2000, 16–17.

[86] Kusukawa 2000a, 99–105. The problems of identification caused or exasperated by discrepancies between text and image in botany are illustrated by Reeds 1991, 145–62.

[87] Kusukawa 2000a, 107–8. The second objection is raised by Jacques Dubois, *Ordo et ordinis ratio in legendis Hippocratis et Galeni libris* ([1535]), sigs. b2^v–3^r. See also Andrea Cesalpino, *De plantis* (1583), sig. [a4^r]: 'Non enim omnes differentias pictura exprimit, ut oratio' and Jensen 2001, 194.

PART II
TEXTS

5
Making sources accessible

Translations

Historical overview

Translation is a category of literary activities and texts so fundamental that it becomes prone to being overlooked. Since the invention of script, translation of written texts was an essential dimension of cultural transfer, as it is still today. Like today, as many or more books were read in translation than in the original, and like today, most readers presumably were not interested in the fact and little aware of it. When perusing a work in translation, they normally relied entirely on the translation and more or less equated it with the work itself.[1] The long-standing status of the Vulgate as the Bible *tout court* in the Latin West and the common practice of quoting texts in translation without further ado are just two testimonies to this simple view of things. Bilingual editions, which could have encouraged people to challenge it, remained an exception.[2] Nor were the importance of and the usually blind trust in translations shaken by the humanist 'ad fontes' movement. In practice, 'ad fontes' more often than not meant 'ad translationes'. The paradoxical combination of fundamental importance and common neglect repeats itself in modern scholarship. Although translation studies are flourishing, the existence and reliability of translations is still tacitly taken for granted in the study of many fields of literary and cultural history, including early modern science.[3]

The language of ancient Rome first became a literary language in translations from Greek.[4] Among the first literary works composed in Latin was Livius Andronicus's version of Homer's *Odyssey*, the *Odusia*. It became the starting point for a long series of translations from Greek (and occasionally other languages like

[1] Rener 1989, 22.

[2] In early modern science, such exceptions include the Greek-Latin *Opera omnia* of Archimedes (1544) as well as bilingual editions of natural historical works and of didactic poetry.

[3] A number of partial bibliographical overviews and analytical studies, mostly undertaken by philological specialists, will be cited in the course of this chapter. The general lack of interest in translation on the part of historians of early modern science is epitomized in the lack of an entry 'translation' among the *c.*2,000 headings of the index of *CHS* 3. Students of medieval science seem to be more philologically minded: *CHS* 2 has a long entry comprising two dozen subheadings.

[4] On Roman translation and its theory, see Copeland 1991, 9–36; Seele 1995; Bettini 2012; McElduff 2013.

Latin Scientific Literature, 1450–1850. Martin Korenjak, Oxford University Press.
DOI: 10.1093/oso/9780198866053.003.0006

Punic or Hebrew). Cicero proved an influential figure in this field, as in others. We have no complete translation from his pen, but he Latinized Greek philosophical terminology and sketched a theory of translation which proved helpful in some respects but is deficient in others, as will become clear later on. Outlined very briefly in chapter 14 of his *De optimo genere oratorum*, taken up and expanded in Jerome's *Epistle* 57, it focuses on the relation between original and translation. The goal of correspondence in terms of content, which is taken for granted, can be reached by translating either verbatim ('ad verbum', 'ad litteram') or analogously ('ad sensum'). The first possibility means sticking to the original words and word order, the second allows for greater freedom in the interest of fluent, idiomatic Latin. When knowledge of Greek was vanishing in late antiquity, increased efforts were made to save important texts for the Latin-speaking West. Besides the Vulgate, Boethius's translation of the Aristotelian *Organon* was particularly influential, as it became a point of reference for later translators and a starting point for scholastic Latin. Boethius and a few other authors, for instance the doctor Caelius Aurelianus (fifth century AD), also translated or paraphrased a couple of Greek scientific texts such as Nicomachus of Gerasa's works on musicology and arithmetic or the medical writings of Soranus.

The Middle Ages were a period of extensive translation from and into Latin as well as between various vernaculars.[5] The translation of scientific texts came to a halt for some centuries in the first half of the era, but from the late tenth century, Greek works on the natural world were again Latinized, either from the original or via Arabic (and sometimes Hebrew) intermediaries. So were Arabic commentaries and independent works.[6] To name some of the most important specimens: from the *Corpus Aristotelicum*, the books on cosmology, meteorology, and zoology (collectively known as *De animalibus*), psychology (*De anima*), and theory of science (*Analytica posteriora*) were translated. In medicine, over fifty Galenic works including the *Ars parva*, Rhazes's encyclopedic *Liber continens*, the Salernitan collection *Articella*, Haly Abbas's *Pantegni*, Avicenna's *Canon*, and Averroës's *Colliget* should be mentioned. In the realm of mathematics and physics, Euclid's *Elements*, a large part of the works of Archimedes,[7] and Alhazen's *Book of Optics* became available. The consummation of ancient astronomy, Ptolemy's *Almagest*, received a Latin translation as well, as did Alhazen's *On the Configuration of the World* and a number of Arabic astronomical tables (*zīj*), first and foremost the *Toledan Tables*. Many of these translations followed the source

[5] *LMA* 8, cols. 1147–63 and 1163–6, s.vv. 'Übersetzer, Übersetzungen', and 'Übersetzung, allgemeine Voraussetzungen und theoretische Grundlagen'; Copeland 1991.

[6] Part of the following is taken from the more complete overview in *CHS* 2, 341–6. See also Livesey 2000, 74–6, and, for translations of scientific texts into both the vernaculars and Latin, Goyens et al. 2008.

[7] Clagett 1964–84.

text closely, sometimes at the cost of idiomatically correct Latin and intelligibility. Many others, by contrast, were easy to understand.[8]

In the early modern period, the number of translations in general grew exponentially.[9] Print accelerated book production and created an incentive to have texts translated for profit. In its capacity as the pan-European idiom of the learned, Latin played a crucial role in this context.[10] The host of works written in other learned tongues, from Greek, Arabic, and Hebrew to exotic languages such as Chinese that became known at the time, mostly reached their European readers in Latin garb. Moreover, at a time when more and more non-academic readers perused books in their respective vernaculars, every work that targeted both them and the academy had to undergo translation from or into Latin: successful Latin books were vernacularized to reach national audiences; the opposite procedure turned promising vernacular works into bestsellers for the Continent's educated elite.

As a result, Latin translations from the early modern period form an impressive body of texts.[11] The significance of sheer numbers is enhanced by the fact that translations were mostly made of initially successful texts and therefore usually proved influential as well. They were often reprinted several times and served as source texts for further renderings. In some cases, a Latin version constituted the final edition of a work, as the occasion was seized to incorporate corrections and additions.

Developing in tandem with practice, translation theory expanded far beyond Cicero's and Jerome's remarks in treatises like Leonardo Bruni's *De interpretatione recta* (probably 1420s), Martin Luther's *Sendbrief vom Dolmetschen* (1530), Lawrence Humphrey's *Interpretatio linguarum* (1559), and Pierre-Daniel Huet's *De interpretatione* (1661).[12] Because the Ciceronian controversy impacted on translation theory from the mid-sixteenth century,[13] Ciceronianism had substantial

[8] On the literalism of medieval translations, see *LMA* 8, s.v. 'Übersetzer, Übersetzungen', col. 1148, and s.v. 'Übersetzung, allgemeine Voraussetzungen und theoretische Grundlagen', col. 1165; Grabmann 1939, 9–10 (translations from Aristotle); Roelli 2021, 219–23 (Boethius as a translator). However, the comparison of medieval and early modern translations of the same text shows that differences are rather gradual than categorial. This can be seen, for instance, from the beginnings of five medieval and eight Renaissance translations of book 4 of Aristotle's *Meteorology* reproduced in Martin 2008, 176–80 (*pace* Martin's own remarks on p. 161).

[9] Burke and Hsia 2007; Sumillera et al. 2020.

[10] On early modern translations into Latin, see *CTC*; *CNLS* 2, 488–501; *DNP* 15/3, cols. 726–39; *BENLW* 127–39; Grant 1954; Burke 2007; Cortesi 2007; Korenjak 2016, 142–54; Fery-Hue and Zinelli 2018.

[11] A catalogue of early modern translations into Latin has been compiled by Peter Burke, who summarizes and discusses his insights in Burke 2007. Although his list does not aspire to completeness, and despite its restriction to vernacular source texts and the additional exclusion of short, unpublished, and late (after 1800) translations, it still contains 1,140 items.

[12] On early modern translation theory, see Norton 1984; Rener 1989; Botley 2004 (pp. 5–62 on Bruni). On Humphrey in particular, see Brammall 2018.

[13] Brammall 2018, esp. 58.

implications not only for technical texts originally composed in Latin but also for translations.

Early modern translations of scientific texts

Science was no exception to the general popularity of translations, which played a particularly important role in this field: bridging linguistic gaps between authors and readers and making important works accessible to all those interested in the natural world, translations disseminated new results widely, helped the development of the individual disciplines gain momentum, and contributed to turning scientific progress into a self-propelling process. Indeed, early modern science owed its ability to penetrate into most European countries and different social strata to a large extent to translations in all directions: from Greek, Arabic, and Hebrew into Latin, from these languages into various vernaculars, from the vernaculars into Latin, and from non-Western languages into Latin and the vernaculars.[14]

Translations into Latin in particular[15] proved vital to the development of early modern science in a society where few people could read Greek, almost no one Arabic, and the teaching of the vernaculars was in an embryonic state.[16] Young medical students would have been as incapable of absorbing their Galen or Avicenna in Greek and Arabic as the Continental intelligentsia were of reading about Newton's optical discoveries in English. To overcome such linguistic problems, Latin was often the best option. In paratexts to translations from the vernaculars in particular, the power of Latin to internationalize science is a much-used commonplace.

In terms of subjects, medicine was the most-translated discipline on account of its academic standing. Since Latin was spoken in universities and academically trained doctors formed a large body of readers, Latin versions of medical texts, starting with such classics as Hippocrates, Galen, and Avicenna, were in continuous demand. Next came botany, which established itself in universities as a handmaiden to medicine from the sixteenth century. Numerous botanical works originally

[14] A number of recent collections of essays on scientific translation are dedicated to or include (part of) the early modern era: Goyens et al. 2008; Duris 2008; Cook and Dupré 2012; Dietz 2016; Fransen et al. 2017; Dupré et al. 2018; Manning and Owen 2018. Not too many of the case studies in these volumes focus on translations from Latin and only a handful focus on translations into Latin (see note 15).

[15] For some general reflections, see *OHNL* 270–3. *CTC* contains entries on the following scientific authors: Aristarchus Samius, Autolycus, Hypsicles (vol. 1), Strabo (vol. 2), Dionysius Periegetes, Thessalus astrologus (vol. 3), Dioscorides, Paul of Aegina (vol. 4), Nemesius Emesenus (vol. 6), Cleomedes (vol. 7), Geminus of Rhodes/pseudo-Proclus, Hanno (vol. 8). Case studies include Beullens 2008; Ventura 2008; Martin 2008; Selosse 2012; van Leeuwen 2017.

[16] Schröder 1992.

written in the vernaculars for apothecaries and village doctors turned out to be of interest to academic physicians as well. Due to its place in the arts faculty, the quadrivium, notably geometry, arithmetic, and astronomy (including astrology), was also well-represented. Other disciplines such as zoology, geography and cartography, the earth sciences, chymistry, or modern physics, which developed and joined academia later, followed at a certain distance.

Regarding translation technique, strict literalism of the type found in some medieval translations was an option only taken if there were specific reasons for doing so: in his commentary on Galen's *De ossibus ad tirones* (1603), for example, Giovanni Filippo Ingrassia translated each section to be commented upon more or less word for word, without any regard for stylistic elegance, in order to convey the most precise understanding possible of the Greek original.[17] The anonymous translator of 'De pulsibus libri quattuor e Sinico translati' (1682) adopted a similar strategy to cope with an overwhelmingly difficult original, as will be discussed later in this section.[18] In most cases, by contrast, a middle course was steered. Formulas like 'nunc ad verbum, nunc ad sensum'[19] often occur in dedications and prefaces. The question how Ciceronian a translation should be mostly boiled down to the more or less extensive use of neologisms and unclassical vocabulary in general.[20] As time went on, the lexical aspect of Ciceronianism came to be silently ignored, and translators contented themselves with adhering to classical standards on the grammatical level.

Regarding the source language, there were some translations from non-Western scientific traditions and languages such as Chinese, but these played a minor role. The great majority of texts turned into Latin belonged to two groups: on the one hand, ancient and medieval works in Greek and the languages by whose mediation Greek science came to Europe, namely Arabic (which also

[17] See his first comment (p. 1): 'Tametsi nostri Siculi Balamii versionem commendamus, clarioris tamen ac tutioris doctrinae gratia (quoad possibile sine praeiudicio fuerit) fere verbum verbo reddere conabimur Latino interim candidiori faciliorem apertioremque ac tutiorem sermonem praeferentes.'

[18] In other cases, the purpose of the method is less clear. In the *Geographia Nubiensis* (1619), a Latin translation of Muḥammad Ibn-Muḥammad al-Idrīsī's twelfth-century world geography by the two Maronites Gabriel Sionita and Johannes Hesronita, the paratactic syntax of the Arabic original and other Arabisms are preserved to the detriment of syntactic correctness. This seems to be done on purpose, as the contrast with the Ciceronian periods of the paratexts suggests, but no reason is given. See, for instance, p. 53 (my italics): '*Cum* terra Mahra iunguntur a septentrione regiones Oman, *et de* regionibus Oman sunt duae urbes Tsur et Cqelhat, *atque* haec sita est ad oras maris salsi Persici, distant*que* abinvicem hae duae urbes longa statione itinere terrestri, maritimo vero minus.' ('*With* the country of Mahra, the regions of Oman border at the north side, *and from* the regions of Oman, there are the two cities Tsur and Cqelhat, *and* the latter is situated at the coast of the salty Persian Gulf, *and* these two cities are separated from each other by a long day's journey by land, but by sea, the distance is less.') The model of the Vulgate may have played some role.

[19] Johann Winter, *Galeni De anatomicis administrationibus libri novem* (1531), sig. $\alpha 2^v$. See also Rener 1989, 126–7, 289.

[20] In a passage to be quoted more extensively and translated later in this section, Francesco Barozzi captures this focus on terminology when he ridicules people who 'Ciceroniana lingua scientiarum vocabula (quod fieri non potest) exprimere voluerint'.

contributed a fair number of original works[21]), and, to a lesser degree, Hebrew and Syriac; on the other, mostly contemporary works in various European vernaculars. Because translations were first needed to channel Greek and Arab knowledge into Europe, whereas, later on, they kept an intra-European exchange of information and ideas going, these two groups formed two waves of translation which followed one upon the other, albeit with considerable chronological overlap. The first wave, which continued and completed late antique and medieval attempts at appropriating Greek and Arab science, began to grow in the mid-fifteenth century and peaked in the mid-sixteenth before abating in the seventeenth. Translations from the vernaculars became common from the sixteenth century,[22] reached their greatest popularity around 1700, and became obsolete only in the nineteenth century, when Latin had lost its dominant position in science for good. In what follows, each group will be reviewed briefly.[23]

As already noted, Greek and Arabic works were Latinized in considerable numbers during the High and Late Middle Ages. However, access to texts in these source languages remained restricted, the circle of scholars able to do the job was small, and a translation in manuscript was always prone to be forgotten or lost. In the early modern era, conditions improved: texts unknown before streamed into the West and were actively hunted by humanists, knowledge of Greek became more widespread (but never so widespread as to restrict the demand for translations), and print ensured that translations would find a readership and not get lost again.

In many respects, continuities between the medieval and the early modern translation movement were strong. Despite humanist polemics against the 'barbarian' versions of earlier centuries,[24] the earliest translations printed were often medieval ones, and some of them remained in use for a long time. The Aristotelian *Meteorology*, for instance, was frequently read and taught in Wilhelm of Moerbeke's version during the Renaissance, although the vocabulary was sometimes updated. The situation was similar with the philosopher's other works.[25] Dioscorides's *Materia medica* first appeared in 1478 in an anonymous late antique or early medieval version. The *editio princeps* of Euclid's *Elements*

[21] See Hasse 2016, 69–133 (analysis of fifteenth- and sixteenth-century translations from Arabic) and 318–407 (a helpful list titled 'The Availability of Arabic Authors in Latin Editions/Translations of the Renaissance').

[22] Of the 115 translations of scientific texts included in the library of Columbus's second son Ferdinand (†1539), a full one hundred are into Latin, but there is not a single translation into the vernacular from Latin (Sumillera 2020, 19–20).

[23] The field is similarly subdivided in *OHNL* 270–3.

[24] An important starting point for these polemics was Leonardo Bruni's remarks about Aristotle's medieval translators (Botley 2004, 51–2). See also Rener 1989, 285, on some pertinent remarks by Giannozzo Manetti. Some aspects of humanist criticism were anticipated in the Late Middle Ages themselves by authors like Roger Bacon, who found Bartholomew of Messina's version of the pseudo-Aristotelian *Mechanics* incomprehensible (Ventura 2008, 129–30).

[25] On the *Meteorology*, see Martin 2008, 155, 158; in general, see Kusukawa 2000b, 122.

(1482) featured the Latin text of the expanded and annotated version by Campanus of Novara (thirteenth century), who had himself drawn on a number of medieval predecessors.[26] The first printed text of Ptolemy's *Geography* (1475) was a translation by Manuel Chrysoloras and Jacopo d'Angelo finished in 1406;[27] the first complete edition of the *Almagest* (1515) featured the Latin version by Gerard of Cremona (twelfth century). Avicenna's *Canon* was printed in (corrected versions of) the same scholar's rendering as long as the book played a role in medical teaching, that is, until the mid-seventeenth century.[28] Sometimes, medieval translations were secretly exploited. Plagiarists revised them a bit or not at all and passed the result off as their own work. Such practices were decried and exposed by contemporaries but persisted nonetheless.[29]

However, Greek and Arabic works already translated in the manuscript age were also Latinized afresh from the beginning of the age of print. Theodorus Gaza's rendering of Aristotle's animal treatises into Ciceronian Latin (1476) superseded the medieval translations by Michael Scot and William of Moerbeke (as well as the contemporary one by George of Trebizond, which remained in manuscript) and became the only one to be reprinted for centuries.[30] In 1516, not one but two new translations of Dioscorides, by Ermolao Barbaro and Jean Ruel, appeared at the same time to replace the version printed in 1478.[31]

It was soon felt that retranslation was not enough. As scholars gained a better overview of Greek and Arab science, they began to realize how much untranslated material of interest there was around. In an academic oration given in 1464 at Padua (but printed only in 1537), Johannes Regiomontanus peppered a thematic overview of the mathematical disciplines with hints at works still awaiting a Latin version (sigs. β^{v}, $\beta 2^{r}$):

> Diophanti autem tredecim libros subtilissimos nemo usque hac ex Graecis Latinos fecit, in quibus flos ipse totius arithmeticae latet...Succedit Theon Alexandrinus commentator Ptolemaei clarissimus, cuius adhuc scripta nemo docte satis traduxisse videtur.
>
> But so far, no one has translated the thirteen extremely sophisticated books by Diophantus, wherein the very best of arithmetic at large lies hidden, from Greek into Latin...Ptolemy was followed by his illustrious commentator Theon of Alexandria, whose writings no one seems to have translated yet with sufficient expertise.

[26] Modern edition: Busard 2005. For Campanus's reworking of the complex medieval tradition, see vol. 1, 32–40. On the diagrams in particular, see Baldasso 2008.

[27] Codazzi 1950, 22–9. [28] Siraisi 1987, 361–6.

[29] Siraisi 1987, 127–74, esp. 149 and 155 n. 83. An example will be discussed at the end of this chapter.

[30] Perfetti 1995; Monfasani 1999; Perfetti 2000, 11–28, esp. 21–5; Beullens and Gotthelf 2007; Beullens 2008, 112–15.

[31] Touwaide 2012, esp. 374–5.

Latin versions of texts never translated before began to appear as early as 1469, when Strabo's *Geography* was published in a rendering by Guarino Veronese and Gregory Tifernas. In medicine, Alban Thorer published the first complete translation of the medical compendium in seven books by Paul of Aegina in 1532, four years after the Greek *editio princeps*. In mathematics, the first four books (of eight) of Apollonius of Perge's *Conica* appeared in 1566 in Federico Commandino's translation; a fairly complete Latin version was published by Edmund Halley in 1710 (see note 52). Diophantus' *Arithmetica* appeared in Latin in 1575 thanks to Wilhelm Xylander, while Pappus' *Mathematicae collectiones* (1588) were once more due to Commandino. 'Opera omnia' editions of famous authors often contained previously untranslated material as well. The first edition of Galen aspiring to completeness was published in 1490.[32] The Hippocratic Corpus first appeared in Latin in 1525 (in Greek the year after).[33] A complete Greek-Latin Archimedes came out in 1544.

Translators, publishers, and readers did not content themselves with first-rank works. Their appetite was omnivorous and indiscriminate. Many texts little known today became popular through translations. Thomas Linacre's 1499 version of the pseudo-Proclan *Sphaera* turned this apocryphal work, in fact an extract of some chapters from the astronomical *Isagoge* by Geminus of Rhodes, into an elementary astronomical textbook which attracted numerous commentaries.[34] The publication of a Byzantine compilation of works on horse medicine under the title *Veterinaria medicina* (1530) by Jean Ruel proved successful enough to prompt three more translations into other languages in the decades that followed.[35] The *Definitions of the Terms Used in Geometry* by Heron of Alexandria were published by Konrad Dasypodius in 1570 and, in a fuller version, in 1579.

Translators usually made little effort to distinguish genuine from spurious material. Publishers were interested in having works printed under great names, because they sold better that way.[36] In medicine, the whole Hippocratic Corpus, none of which can be attributed to the physician from Cos with certainty, was in practice treated as composed by him in spite of its subdivision into four classes according to the degree of (in)authenticity by Girolamo Mercuriale (*Censura de Hippocratis operibus*, 1583).[37] Galen's works also bristled with pseudepigrapha

[32] Fortuna 2012, 394–5.

[33] Maloney and Savoie 1982, nos. 92 and 102. Maloney and Savoie's bibliography also provides an overview of Latin translations of Hippocratic works in general, which amount to a good part of all books concerned with Hippocrates until the early seventeenth century.

[34] Todd 1993; *CTC* 8, 7–48. [35] McCabe 2007, 49–50.

[36] In the preface to his translation of *De anatomicis administrationibus* (sig. α2^{v}), Winter mentions an older translation of the same (genuine) Galenic work, 'quam typographi Bononienses viro cuidam magni nominis vita defuncto, ut esset vendibilior, sicut conicio, adscripserunt'. Although he does not speak of a famous classical author but of a renowned early modern scholar, the principle is the same.

[37] The de facto acceptance of the entire corpus as genuine continued medieval practice, documented in Kibre 1985. On Mercuriale's classification, see Nutton 1989, 423.

like the popular treatise *De urinis*, first translated by Józef Struś in 1535. About 30 per cent of the works included in Galen's first printed 'opera omnia' (1490) are not by him.[38] The one hundred astrological aphorisms known as *Centiloquium*, a work of presumably Arab origin but commonly ascribed to Ptolemy, remained as popular as they had been in medieval times.[39] Even self-conscious early modern forgeries were passed off as translations of ancient works. Giovanni Battista Rasario, for one, faked several attested but lost commentaries by Galen, partly to provide added value to a new edition of the Galen's complete works produced by Vincenzo Valgrisi in 1562–3.[40] Gilles Personne de Roberval playfully composed an *Aristarchi Samii De mundi systemate* (1644)—allegedly a polished version of a barbarous Latin translation of an Arabic rendering of the Greek original—in order to proclaim his allegiance to the Copernican system behind the mask of the ancient father of heliocentrism.[41]

The effort to have as many works as possible translated was accompanied by a marked tendency towards multiple translation. No doubt partly fuelled by economic reasons, as is still the case today, this trend went far beyond the perceived necessity of superseding medieval translations. That a work was turned into Latin several times was the rule rather than the exception. For instance, no fewer than ten translations of the *Centiloquium*, six medieval and four early modern ones, are extant.[42] Five different versions of Aristotle's *Parva naturalia* appeared in print within four years in the early sixteenth century (1520–3).[43] Not only the one complete translation already mentioned, but three of them (1532, 1532, 1556) followed the Greek *editio princeps* of Paul of Aegina. In addition, half a dozen partial translations are on record.[44] Two toxicological poems by the Hellenistic poet Nicander (second century BC), the *Theriaca* and the *Alexipharmaca*, inspired five renderings into Latin (one of them via a French translation) in the sixteenth century.[45] Even a minor work such as Theophrastus's *De lapidibus* had two Latin versions produced in 1578 and 1647. Turning from there to Galen's massive *oeuvre*, one finds no fewer than twenty-two Latin 'opera omnia' editions (often, it is true, recycling the same translations) between 1490 and 1625.[46]

That only two editions of Galen's Greek 'opera omnia' appeared during the same time span is indicative of the fact that Latin translations overshadowed Greek and Arabic originals in science as much as in other fields.[47] Not only were

[38] Fortuna 2019, 438. See also the index of Bouras-Vallianatos and Zipser 2019, s.v. 'pseudo-Galenic'.
[39] Lemay 1978; Rinaldi 1999.
[40] Savino 2019. On Valgrisi's edition, see Fortuna 2012, 405–6.
[41] Babeş 2022.
[42] See *PAL*, B.1.1–10.
[43] Perfetti 1999, 444.
[44] *CTC* 4, 145–91, esp. 147, 150–76.
[45] Santamaría Hernández 2004; 2018 (an edition of Pedro Jaime Esteve's Latin version of the *Theriaca*, 1552).
[46] Fortuna 2012. See also Durling 1961, whose census of Galen editions also includes single works and collections which are not 'opera omnia'.
[47] Cf., for instance, Schmitt 1983, 36–9, 44, on the situation concerning the *Corpus Aristotelicum*: he estimates that under 10 per cent of all editions of Aristotelian works contain a Greek text.

more Latin versions published than originals, but they also proved more popular and successful. Gaza's translation of Aristotle's animal books (1476), for one, went through forty-two complete or partial editions,[48] the two most successful Renaissance versions of the *Meteorology*, through thirty-six and twenty-six.[49] Moreover, translations tended to precede originals in terms of chronology: most *editiones principes* of Greek and Arabic classics were Latin translations.[50] As already remarked, the first major work about any aspect of the natural world to appear in print, Strabo's *Geography*, was published in Latin in 1469. In Greek, it appeared half a century later, in 1516. Some Latin versions even held the field alone until modern philology took care of the matter. For the first comprehensive Greek edition of Pappus' *Συναγωγή*,[51] one had to wait nearly 300 years after Commandino's translation (1588). During the first centuries of the modern era, the West by and large came to know Greek and Arab science in Latin garb.

By 1600, most extant Greek as well as numerous Arabic works of scientific content had become available in Latin editions.[52] At the same time, modern science began to overtake its predecessors, the study of whom gradually became less important, and the vernaculars were already gaining ground in science. This development led not only to vernacular versions of Latin works for readers without an academic education[53] but also to Latin translations of vernacular texts of interest to an international learned readership. As the vernacular output increased and scientific progress emerged as a reality of which one had to keep abreast, men of science increasingly relied on vernacular-Latin translations.

Translators themselves were not shy of promoting their Latin versions of vernacular originals as a service to the republic of letters. Joachim Camerarius concludes the preface to the first volume (1532) of his translation of Dürer's *Vier Bücher von menschlicher Proportion* (1528) with a proclamation of his wish 'to disseminate what seems to be written for one nation only to all' ('illa quae uni dumtaxat nationi scripta videntur propagandi in omnes', sig. Aiiii^{v}). When Matthias Bernegger turned Galilei's *Dialogo sopra i due massimi sistemi del mondo* (1632) into a *Systema cosmicum* (1635), he did so 'in order to share the book, intelligible to few while speaking in the Tuscan tongue, with the entire cultivated and educated world in Latin translation' ('ut, quem Hetrusce loquentem pauci

[48] Beullens and Gotthelf 2007, 508–13. [49] Martin 2008, 157.

[50] Again, this is true beyond science. See the data gathered in *DNP* suppl. 2.

[51] Hultsch 1876–8.

[52] In some complicated cases, the process took more time. Books 1–4 of Apollonius' *Conica* are preserved in Greek, books 5–7, in various Arabic versions, while book 8 seems to have been lost for good. The first half of the work was published in a faulty and a good Latin translation in 1537 and 1566 respectively. An abridged Arabic version of books 5–7 came out in Latin in 1661 on the initiative of Ferdinando II of Tuscany. In 1710, Halley combined translations of books 1–4 and of a fuller Arabic version of books 5–7 with his own reconstruction of book 8.

[53] One area in which this happened particularly often was botany. For example, Otto Brunfels's illustrated herbal, the *Herbarum vivae eicones* (1530–6), was turned into German as *Contrafayt Kreüterbuch* (1532–7).

intelligerent, Latine conversum cum orbe cultiori doctiorique universo communicarem'; first page of the address to the reader).[54] That such statements were no empty bragging is confirmed by Kepler's angry reaction to Galilei's insistence on publishing in his native Italian, which he could only read with some difficulty.[55]

Accordingly, a considerable amount of vernacular science was Latinized from the sixteenth to the eighteenth centuries. In the fields in which vernacular-Latin translations were important, science ranked third in terms of numbers after religion and history.[56] Major early examples include Camerarius's aforementioned rendering of Dürer's book on human proportions (1532–4), two Latin editions—besides others in French, German, and Czech—of Pietro Andrea Mattioli's successful Italian commentary on Dioscorides (1554, 1565), and Sebastian Münster's German *Cosmographia* (1544), whose similarly multilingual translation history also included a Latin edition (1552). In the seventeenth century, the number of translations increased. Francesco Redi's refutation of the notion of spontaneous generation in his *Esperienze intorno alla generazione degl'insetti* (1668), Latinized by Andreas Frisius as *Experimenta circa generationem insectorum* (1671), is but one arbitrarily chosen example among many. At turn of the eighteenth century, Johann Jakob Scheuchzer's *Specimen geographiae physicae* (1704) made John Woodward's spectacular reconstruction of the Flood and its geological effects, the *Essay towards a Natural History of the Earth* (1695),[57] known in Continental Europe. Newton's pupil Samuel Clarke produced a Latin version (1697) of Jacques Rohault's *Traité de physique* (1671), a popular scientific exposition of Cartesian physics. He also Latinized his master's second magnum opus, the *Opticks* (1704), as *Optice* (1706).

Beyond single works, a number of first-rank authors depended on translation into Latin for their Europe-wide influence. Among the earliest to do so was Paracelsus, whose idiosyncratic language, a variant of Swiss German interspersed with neologisms, rendered his works barely accessible to most readers even within the German-speaking world. His collected works were published in Latin in 1575 (Basel, two volumes; an abortive 'opera omnia' edition), 1603–5 (Frankfurt, twelve volumes), and 1658 (Geneva, three volumes).[58] No match to Paracelsus's influence, but still an important figure was the Portuguese mathematician and astronomer Pedro Nuñez (one of the fathers of the Vernier scale, called 'nonius' in several languages after his Latin name). His collected works (1566),

[54] On the genesis and early reception of Bernegger's translation, Kremer 2014, 115–18. For yet another case in point, see Francesco Calzolari's proclamation regarding the Latin version of his Italian description of Monte Baldo near Verona ('Iter Baldi', 1571) at the beginning of his introductory letter to Pietro Andrea Mattioli: 'Communi hac lingua nunc conscriptum omnibus in Europa nationibus…facile innotescet.' Examples could be multiplied.

[55] Olschki 1919–27, vol. 2, 71; Kleinert 2021 (who shows that the expression 'crimen laesae humanitatis', which is often quoted in this context, is Olschki's, not Kepler's).

[56] Burke 2007, 73. [57] Poole 2010, 64–7.

[58] Sudhoff 1894, nos. 165–6, 258–63, 269–74, 381–3, and pp. 692–5.

which combined Latin originals and works translated from Portuguese, were key to his reception outside his small homeland.[59] Later on, Galilei, Descartes, and Bacon partly owed it to Latin translations that they were read across the Continent.[60] Robert Boyle, who did not publish a line in a language other than English himself, entirely depended on translators in this respect. Over thirty Latin editions of single works, 'opera varia', and 'opera omnia' by him are on record.[61]

In addition to collections of translations defined by author, one also finds thematic collections, many of which unite texts originally composed in Latin and translations from one or more other source languages. To this class belongs a series of collections of chymical works starting with Johannes Petreius's *De alchemia* (1541) and culminating in the six tomes of Lazarus Zetzner's *Theatrum chemicum* (1602–61), which contains texts originally written in Latin, Greek, Arabic, and half a dozen vernaculars. For the first systematic presentation of Europe-wide weather observations, the *Ephemerides* published by the Societas meteorologica Palatina from 1783 to 1795, contributions from collaborators from the whole continent had to be Latinized, which caused considerable delay, as explained in the first volume (sig.)(3^{r-v}).

As may be guessed from the examples adduced so far, the most common source languages were French, English, and Italian. These nations not only found themselves at the forefront of scientific progress and had a strong publishing industry (although it developed later in England than in the other two countries), but they were also particularly proud of their own vernaculars and reluctant to write in Latin. Germans, by contrast, stuck to the language of the Romans for a longer time to the degree that, even for dissemination within Germany, Latin was sometimes preferred to the mother tongue. On the third page of the 'introductio' to his *Amphitheatrum zootomicum* (1720), an expanded translation, published in Frankfurt and Giessen, of the *Memoir's for a Natural History of Animals* (1688),[62] Michael Bernhard Valentini presents himself musing that a Latin version of the English work would be welcome to German scholars:

> cogitare olim coepi, annon quis gratiam penes eruditos in Germania inire possit, si librum hunc nitidissimum et ob reflexiones eruditas interspersas utilissimum Latio donaret.

[59] For an example of this, Nuñez's reception by Simon Stevin, see Randles 2000, ch. 14, p. 88.

[60] Galilean and Cartesian examples are discussed in the present chapter. Important translations of English works by Bacon ('New Atlantis', *Naturall Historie*) are 'Nova Atlantis' (1638, a self-translation: Licht 2006, esp. 118) and *Sylva sylvarum* (1648, by Jacob Gruter).

[61] For full bibliographical information, see Fulton 1961; a concise list has been put online by the Centre for Neo-Latin Studies at Cork (https://www.ucc.ie/en/cnls/resources/findinglist/robertboyle/, accessed 16 February 2023). See also Burke 2007, 74.

[62] This publication is itself an English translation, made on the initiative of the Royal Society, of a French work issued by the Académie des Sciences that was apparently not distributed through the book trade.

some time ago, I began to ponder whether one could not merit the thanks of the learned in Germany, if one were to present this most splendid and, through the learned considerations scattered in it, most useful book to Latium.

The idea of a German translation apparently did not occur to him.[63]

Regardless of whether an ancient, medieval, or contemporary source text was translated, modalities of production and publication could differ in several respects. Presumably, the original initiative often came from publishers, especially in the case of big, complex enterprises like the 'opera omnia' editions of classical authors. Sometimes, a publisher even took the translation upon himself, as did Henri Estienne with the Greek extracts of which his *Dictionarium medicum* (1564) is composed or Frisius with Redi's *Esperienze intorno alla generazione degl'insetti*.[64] In the case of contemporary works, self-translation occurred: Jacques Peletier brought out his *Algèbre* (1554), a treatise on advanced arithmetic, in Latin garb in 1560 as *De occulta parte numerorum*, Johannes de Laet his *Nieuwe Wereldt* (1625), the fullest description of the Americas of its time, as *Novus Orbis* (1633). More often, however, the translator was someone else. This is true, for instance, of Galilei's *Dialogo* (1632), of Newton's *Opticks* (the Latin version of which was, however, supervised or even corrected by the master),[65] and of Jan Swammerdam's *Bybel der Natuure/Biblia Naturae* (1737), the author of which had died as early as 1680. Newton's equally posthumous *Method of Fluxions and Infinite Series* (1736) presents a special case. 'Translated from the author's Latin original' for publication in London, as the title page states, it had to be retranslated into Latin for the benefit of Continental readers to form part of a collection of Newton's opuscula printed in Lausanne and Geneva (1744). The translator's work itself was sometimes eased by the possibility of dictating to a scribe, as Gaza did with the pseudo-Aristotelian *Problemata*.[66]

As is evident from most of the above examples, even in the case of contemporary works a few years or even decades usually elapsed between the appearance of the original and the Latin version. In some cases, by contrast, authors or

[63] Forty years later, August Johann Rösel von Rosenhof's *Historia naturalis ranarum nostratium* (Nuremberg, 1758) exclusively dealt with German frogs and presumably catered to a German rather than an international readership. That it was nonetheless published in Latin and German implies that part of the public still preferred the former. The situation was different in the realm of belles-lettres, where countless translations from French and English into German were produced during the eighteenth century (Knufmann 1967, esp. 2677).

[64] Maclean 2021b, 280–345 gives a full overview of Frisius's career as a scholar and publisher, including a few more translation-editions of works by Redi and Antonio Neri.

[65] In his preface, Clarke asserts that even minor changes in the interest of clarity were introduced only according to Newton's command or with his permission ('vel iussu auctoris vel eiusdem permissu').

[66] The testimony of his secretary Nicolaus Gupalatinus is contained in the preface to the second of two incunable editions (1473, 1475; reported in Beullens and Gotthelf 2007, 473). On Gaza as a translator of the *Problemata* in general, see Ventura 2008, 143–4.

well-organized publishers managed to issue both versions simultaneously. A work on embryology and obstetrics by Jacob Ruf was printed by Froschauer at Zurich as *Trostbüchle* and as *De conceptu et generatione hominis* in 1554. The Dutch and Latin editions of Maria Sibylla Merian's *Metamorphoses of the Insects of Surinam* (1705) present a similar case.

Bilingual editions constituted a logical next step. The possibility of presenting source text and Latin translation in one volume was realized already for some Greek classics, the Archimedes of 1544 being a prime example. An early contemporary work to be published in this way was Maria Cunitz's *Urania propitia* (1650). For some reason, bilingual editions were particularly well liked in eighteenth-century didactic poetry and in the natural history of the same period, where a format presenting Latin (to the left) and the vernacular (to the right) facing each other in two columns on the same page became popular. Swammerdam's *Biblia Naturae*, a Latin-French edition of Merian's work on the insects of Surinam (1726),[67] Georg Eberhard Rumpf's *Herbarium Amboinense/Het amboinsch kruidboek* (1741–50) on Indonesian plants and animals, and August Johann Rösel von Rosenhof's Latin-German history of frogs (1758) were laid out in that manner. In most such books, as already in the case of simultaneous Latin and vernacular editions, it remains unclear which version was translated from which.

Occasionally, only the Latin version of a work appeared in book format. The case of Pappus has already been mentioned. The Dutch originals of Leeuwenhoek's letters about his microscopic discoveries remained in manuscript during the early modern era, and those he sent to the Royal Society were usually read to the fellows and printed in the *Philosophical Transactions* in English.[68] By contrast, the first collections published for the benefit of a wider readership, namely the *Anatomia et contemplatio nonnullorum secretorum* (1685), the *Anatomia seu interiora rerum detecta* (1687), the massive *Arcana naturae detecta* (1695), its *Continuatio* (1697), and the four volumes of Leeuwenhoek's *Opera omnia* (1715–22), all came out in Latin.

What was a translation? A broad notion and its varieties

So far, I have tried to provide a reasoned overview of a field which is large but does not seem to hold great surprises at first sight. However, a closer look at the subject disproves this first impression. A number of seemingly self-evident modern assumptions about translations do not hold for the early modern era, and the

67 In the cases of Cunitz and Merian, Latin–vernacular bilingualism seems to correspond to the fact that these women remained on the fringes of the academic Latin world despite their learning and expertise.

68 Henderson 2012. On several letters sent to the fellows in Latin translation, see there, pp. 245–6, 255.

field bristles with unexpected complications. In the rest of this section, I will discuss two pertinent questions: What was actually covered by the notion of translation? And what specific difficulties did translators have to face?

As long as one considers questions of literalness, freedom, and terminology, the choices of an early modern translator seem to be rather limited. By contrast, as soon as one ventures beyond the level of word and sentence, one is astonished at the sheer variety of the activities and texts that could be termed 'translation'. This is already suggested by the terminology. Latin words for 'to translate' were numerous, including 'aemulari' ('to contend'), 'explicare' ('to explain'), 'exprimere' ('to express'), 'interpretari' ('to construe'), 'mutare' ('to change'), 'transferre' ('to carry across'), 'traducere' ('to lead across'), 'reddere' ('to render'), '(con)vertere' ('to transmute'), and their respective substantives, plus some additional terms such as 'commentarius' or 'paraphrasis'. No single term prevailed over all others during the era.[69] The varied and open nomenclature fitted a notion which was less sharply defined and covered a broader field in practice than it does today. Early modern translators took much greater liberties with their texts than their modern colleagues: in a work designated as a translation on its title page, the original could be subject to a wide range of changes and manipulations.[70]

To begin with, translators could explicitly or tacitly correct the text of their model, if they judged it to have been corrupted in the process of transmission. A splendid example of this practice is provided by Francesco Barozzi's translation of Proclus' commentary on Euclid's first book (1560). The translator's attention to textual matters is highlighted on the title page, where the four books of the commentary are said to have been 'cleansed from all blemishes with the utmost labour, care, and diligence' ('summa opera, cura ac diligentia cunctis mendis expurgati'). Barozzi uses no fewer than five different Greek manuscripts, which he enumerates in his letter of dedication (fourth- to second-last page), and highlights variant readings by marginalia marked with a dagger. His philological expertise and acumen even secured his work a place as a textual witness in the only modern edition of the Greek original.[71] Gaza's interventions in the text of Aristotle's *Historia animalium* are on a larger scale and more problematic. He not only tacitly emended the Greek text in translating it, as Barozzi had done, but also

[69] For terms already in use in ancient Rome, see McElduff 2013, 189–96. On 'commentarius', see note 87 below. A rendering titled 'paraphrasis' is Andreas Vesalius's translation (1537) of book 9 of Rhazes's *Liber ad Almansorem*. One interesting use of premodern translation terminology not discussed in what follows was intra-linguistic: it could be applied to a revision or a digest of an earlier Latin translation without reference to the source text (Grabmann 1939, 190; Hasse 2016, 122–3).

[70] For some remarks on the licence taken by premodern translators, see Rener 1989, 283, 285–7, 329–30; on ancient and medieval translations, see McElduff 2013 and *LMA* 8, s.v. 'Übersetzung, allgemeine Voraussetzungen und theoretische Grundlagen', cols. 1163–4.

[71] Friedlein 1873, esp. p. vi.

excluded book 10 as spurious, rearranged large blocks of text within book 8, and placed book 9 before book 7.[72]

Next, source texts might be expanded or shortened. The former could happen not only because of a translator's wish to clarify and complete the account of his template (a peculiar example of this will be discussed shortly) but also on the initiative of the original author, who took the occasion of a Latin translation to supplement his work with additional material. The length of the third and last book of Newton's *Opticks* tripled in Clark's version. A good example of an abbreviating translation is provided by Miguel Jerónimo Ledesma, who substantially shortened the first *fen* of Avicenna's *Canon* in his fluent and readable Latin version (1547).[73] In some cases, pruning could be so severe as to result in anthologization. Thomas Burnet's *Hippocrates contractus* (1685) presents a slim, handy anthology of the Hippocratic corpus ordered by topics and titles. Born from Burnet's personal 'easily portable Hippocrates' ('Hippocrates facile portabilis', preface), conceived as a help for doctors too poor to buy the whole corpus, and as an enticement for others to read the original, the work was quite a success. New editions appeared until 1765, a German translation as late as 1791.[74] At times, expansion or shortening was accompanied by rearrangement, as in the case of Pierre Gilles's translation (1562) of Aelian's *On the Characteristics of Animals*, in which an attempt was made to bring some order into the deliberate confusion of the Greek work.[75]

Occasionally, texts from various sources were compiled to form a new whole. Valentini combined his translation of the *Memoir's for a Natural History of Animals* with sections from Latin works by the Italian doctor Marco Aurelio Severino and the Dutch doctor Gerard Blasius as well as extracts from unspecified, presumably vernacular scientific journals. This procedure made his *Amphitheatrum zootomicum* into a veritable encyclopedia of animal anatomy (see his 'introductio', third to fourth page).

A less drastic but more common intervention was stylistic revision in an attempt to adapt a work to the needs of a new audience. This happened particularly often in translations from the vernaculars, which were intended for a less practically minded but more educated readership than their respective originals.

[72] He was right with book 10. His treatment of books 8 and 9 dictated their presentation even in editions of the Greek original until the nineteenth and late twentieth century respectively but is now regarded as unjustified (Beullens and Gotthelf 2007, 471–7, 504). Three more examples of translation-cum-emendation are provided by Niccolò Leoniceno's translation of the pseudo-Aristotelian *Mechanics* (1525; Leeuwen 2017, 186–7), Commandino's Latin version of Pappus' *Συναγωγή*, and Kepler's rendering of Plutarch's *De facie in orbe lunae* (1634), in which Kepler filled lacunae with his own supplements and noted conjectures in the margins.

[73] Hasse 2016, 109–14.

[74] For another, peculiar type of abbreviating translations, 'review translations', see the section 'Reviews' ot this chapter.

[75] Berrens 2023.

In its German version, Rufs *Trostbüchle*, an innovative handbook of obstetrics, gynaecology, and embryology, was mainly directed at the mothers, midwives, and municipal authorities of the city of Zurich. For this audience, Ruf chose a style abounding in strings of colourful adjectives, moralistic reasoning, personal interventions, and apostrophes. In the simultaneous Latin translation by Wolfgang Haller, all these traits are much reduced, and the style becomes less personal and more sober. This tendency is in evidence already in the title, the generalizing singulars and abstractions of which have a less alluring and dramatic but more scientific ring than the original one: *Ein schön lustig Trostbüchle von den empfengknussen und geburten der menschen unnd jren vilfaltigen zuofälen und verhindernussen* ('A Pretty, Pleasurable Booklet of Comfort about the Conceptions and Births of Men, their Numerous Emergencies and Impediments') gives way to *De conceptu et generatione hominis et iis, quae circa haec potissimum considerantur* ('On the Conception and Procreation of Man and on the Things to be Observed in This Respect in the First Place').[76]

A special variety of stylistic revision, and one particularly disconcerting to the modern reader, is versification. For instance, some thirty renderings of Hippocrates' *Oath*, *Prognostics*, and especially his *Aphorisms* into Latin hexameters, elegiacs, and iambics are known.[77] Most of these were didactically motivated, a phenomenon in line with the premodern tendency to regard poetry as more pleasurable and easier to memorize than prose and to use it as a vehicle of instruction. As the title of Reginald Sturm's hexameter translation of the *Aphorisms* (1582) announces, the book is 'extremely helpful to memorize them' ('ad illorum memoriam summe utilis'). Another example is the elegiac *Aphorismi Hippocratis versibus redditi* (1599) by the Veronese doctor Paullus Dionysius. Their editor claims in his letter of dedication that the translator has not only versified but also clarified Hippocrates' compressed prose, 'so that every reasonably educated person can very easily understand its meaning' ('ut quilibet vel mediocriter eruditus ipsum [*sc.* sensum] facillime percipere possit'). This aim is mostly achieved through the expansion of elliptic formulations and the addition of explanations. Moreover, Hippocrates' matter-of-fact statements are sometimes rendered more appealing by emotionally charged expressions. Aphorism 3.10 dryly states: *τὸ φθινόπωρον τοῖσι φθίνουσι κακόν*, 'Autumn: bad for the consumptive'. Dionysius (p. 22) expands this to 'His nocet autumnus, tristi qui tabe laborant / ulcere pulmones depopulante fero', 'Autumn is harmful to those who suffer from grim consumption, because an ulcer devastates their lungs'. In doing so, he directly or indirectly draws on the commentary on the *Aphorisms* by Galen, who discusses whether Hippocrates specifically refers to tuberculosis.[78]

[76] Keller and Steinke 2008, 324–8. [77] Seidenschnur 1847, 780–5.
[78] Kühn 1821–33, vol. 17.2, 577.

Originals could be improved not only regarding their textual foundation and their style, but also in terms of content. This was usually done in order to bring a work in line with the current state of knowledge.[79] An instructive example of an updated translation is provided by Georg von Peuerbach's and Regiomontanus's *Epitoma in Almagestum Ptolemaei* (1496). In addition to the abridgement indicated by the title *Epitoma*, the *Almagest* is modernized through a whole series of changes and additions: the sine and cosine functions and their respective laws are introduced, the determination of the sun's apogee is improved, Ptolemy's epicycle theory of the moon is refuted (because of its consequence that the moon's apparent size would change by a factor of four), and epicycle and eccentre models are demonstrated to be geometrically equivalent for all planets, not only the superior ones, as Ptolemy believed.[80] Such upgrades could also include the addition of illustrations. To his translation of the pseudo-Aristotelian *Mechanics* (1525), Niccolò Leoniceno added the diagrams mentioned in the text but missing in his manuscript,[81] as did Barozzi in his version of Proclus' commentary on Euclid.

Last but not least, translations were embedded in paratexts such as dedications, letters to the reader, subtitles, marginalia, notes, or commentaries,[82] all of which influenced the way in which they were read and understood. This conforms to early modern publication practice in general. What is astonishing at times, however, is the sheer mass of paratexts by which a translation could be overgrown. In the second and definitive two-volume edition of Frans van Schooten's Latin version of Descartes's *Géometrie* (1637), the *Geometria a Renato des Cartes edita* (1659–61), the one hundred pages of the original French edition became close to one thousand through the addition of introductory paratexts, two commentaries, ten monograph treatises relating to particular aspects of Cartesian geometry, and even a dialogue in Greek elegiacs in praise of the recently deceased Schooten. All of this was seen as added value: the title page proudly proclaims that the new edition is 'equipped with richer commentaries and adorned with many superb additions, which contribute to the fuller clarification of this *Geometry* as well as to the enhancement of its magnificence' ('locupletioribus commentariis instructa multisque egregiis accessionibus tam ad uberiorem explicationem quam ad ampliandam huius Geometriae excellentiam facientibus exornata'). By embedding the *Géometrie* in thick layers of adjuncts, Schooten turned a novel, pithy, and

[79] Ideologically motivated corrections seem to be comparatively rare. A case in point is provided by Johannes Florianus, who added some religious polemics to his Latin version (1556) of Leo Africanus's 'Descrittione dell'Africa' (1550; Black 2002, 268–70).

[80] See in particular book 3, proposition 14, book 5, proposition 22, and the beginning of book 12. Cf. Hartner 1980, 124; *CHS* 2, 472.

[81] Leeuwen 2017, 190–204.

[82] On the close connection of translations and commentaries, see the section 'Commentaries' of this chapter. The conception of the *Catalogus Translationum et Commentariorum* does justice to this fact.

difficult book for mathematical experts with good French into a kind of encyclopedia of Cartesian mathematics for the whole of educated Europe. The edition was reprinted in 1683 and did much to acquaint readers across the Continent with Descartes's fusion of geometry and algebra.[83]

In sum, early modern 'translationes' often amounted to heavily edited versions of their originals, and many of them overstretched the modern notion of a translation. This state of affairs reflects the fact that in the early book market an author, whether dead or alive, usually had little influence on the fate of his work once it was published. The case of Newton and his obedient disciple Clarke is exceptional. As a rule, publishers and translators did not bother to ask about an author's wishes, nor did they feel obliged to produce a philologically accurate reproduction of his work in a different language. By contrast, they were very much interested in the wants and needs of their prospective readers.[84] They aimed at providing not the most authentic, but the most useful and saleable version of a given text. The motives behind the composition and publication of early modern translations were not primarily philological but practical.[85]

A difficult job

The Ciceronian tradition of translation theory is problematic as a guide to its subject not only because of its focus on small text units but also because it idealizes translation by tacitly assuming that it takes place under ideal conditions. In fact, translators often worked under circumstances which presented daunting difficulties, and their results accordingly left much to be desired. In what follows, I will try to convey an impression of the obstacles with which they had to wrestle by lending an ear to their remarks in dedications and prefaces and having a look at some problem-ridden specimens of actual translations.

Translators were fond of highlighting the great obstacles they had to overcome, as well as of pointing out the shortcomings of their colleagues, in introductory paratexts. Of course, they partly did so in order to enhance the value of their efforts, but if this implies that their assertions are selective, it does not follow that they are outright fictions. Two pertinent statements may suffice to illustrate the range of topics brought up on such occasions.

[83] On the *Géometrie*, see Serfati 2005; Merzbach and Boyer 2011, 309–20. On Schooten's role as a translator and editor, see Howard 2022, 153–4.

[84] On the needs of the reader as a yardstick justifying all manipulations of the original, see Rener 1989, 257.

[85] A similar point is made for editions and translations of medical texts in particular by Nutton 1989, 429–30.

In the letter of dedication to the French king Francis I which Johann Winter prefixed to his translation of Galen's *De anatomicis administrationibus* (1531),[86] one reads the following (sig. $\alpha 2^{v}$):

> Vix enim tuti erimus a rabiosulis quibusdam et male feriatis hominibus, qui ingenuos cuiusque conatus elevare, non meliora facere didicerunt. Quippe hi omnem orationem ut putidam respuunt et quodcumque verbum allatrant lautioris orationis institores opinanturque ceteros quasi in Tullianorum officina non perpolitos nequiquam ad palmae speciem aspirare aut secundae aut tertiae. Quantum sane ad nostras partes attinet, laboratum est in hoc commentario haud mediocriter, primum quod fere nulli inter emunctos Latinae linguae scriptores habeantur, quos in anatome explicanda aemulari potuerimus, deinde quod Graeci in ea potissimum medicinae parte multo Latinis sint locupletiores et in effingendis rerum vocabulis multo liberaliore natura et beatiore, demum quod non liceat nobis esse tam disertis, qui severiores Musas colimus. Porro auxit laborem exemplar ab Asulani officina profectum tot scatens mendis, ut in suum id nitorem restituere immensi cuiusdam operis esse videatur.
>
> For I will hardly be safe from some foamy-mouthed manikins with too much time who have learned to belittle the noble efforts of all others but not to make it better themselves. For these people reject all speech as rotten and bark against every word whatsoever in their role of wardens of refined speech. They think that the others, who have not, so to say, got their fine sanding in the Ciceronians' workshop, are striving in vain for the victory palm or for second or third prize. As for myself, I have worked really hard at this translation,[87] first because there are as good as no cultivated authors writing in Latin whom I could have tried to equal in the explanation of anatomy, secondly because the Greeks are, particularly in this part of medicine, much richer and much more freely and happily coin words for the phenomena, finally because as a worshipper of the more austere Muses, I am unable to be so eloquent. Moreover, my toil was increased by the text [of the Greek original] issued from the printing shop of [Andrea] Asolano, as it is so full of errors that is seems to require immeasurable work to restore it to its original splendour.

Barozzi, for his part, was less concerned with his own work than with the incompetence of his colleagues in the preface to his translation of Proclus' Euclid commentary. His is one of the most pronounced statements of a feeling that translators were generally ill-equipped for their work and did a poor job ('Ad lectorem', first to second page):

[86] On Winter's translation and preface, see Berrens (forthcoming).

[87] In the Latin, the translation is called a 'commentarius', perhaps in order to emphasize its explanatory function.

neque eos imitatus sum, qui in vertendis libris non pauca de suo adiciunt, permulta praetermittunt aut seriem auctorum atque ordinem perturbantes commutant... varias siquidem multorum vidi conversiones, quae certe ab omnibus sunt deridendae. Nam alii, ut iam dixi, nescio cuius rei causa multa addunt, omittunt atque permutant, alii vero pulcherrima auctorum et lucidissima sensa obscurissima falsaque reddunt, aut quia Graecum sermonem perfecte non callent aut quia scientias atque artes ignorant, de quibus auctores illi pertractant, aut demum quia cum Ciceroniana lingua scientiarum vocabula (quod fieri non potest) exprimere voluerint, inextricabiles labyrinthos ingressi eos etiam secum una pessum trahunt, qui eorum scripta legunt. Alii autem barbariem passim adamantes ita libros e Graeco sermone in Latinum convertunt, ut in quamlibet potius aliam linguam quam in Latinam conversi dici possint.... Alii denique nec linguas nec scientias possidentes, dum paedagogorum more Graecas dictiones Latinis et Graecis characteribus conscribunt, egregie hallucinantur.

and I did not follow those who in translating books add much of their own, leave out even more, or change the sequence and order created by the original authors... since I have seen various translations by many which all people must no doubt laugh at. For some, as I already said, add, leave out, and change much for I do not know what reason; others render the very beautiful and entirely clear meanings expressed by the authors in a totally cloudy and wrong way, either because their Greek is less than perfect or because they are ignorant of the disciplines and crafts discussed by those authors or finally because they want to express the technical words of the disciplines in the language of Cicero, which is impossible: thus, they enter labyrinths from which there is no escape and draw their readers to the depths with themselves. Others again are lovers of barbarism throughout and translate books from Greek into Latin in such a way that they may be said to have been translated into any other language rather than Latin.[88] ... Lastly, some have no command either of the languages or of the disciplines: writing down Greek expressions in Latin and Greek letters, as schoolteachers do, they produce exquisite nonsense.

Despite their polemic stance, these two statements cover a number of real major obstacles with which translators frequently had to struggle: they had to work from badly corrupt manuscripts or prints based on such. The contents they dealt with were forbiddingly technical and complex. Their mastery of the source language might leave something to be desired, as they had often learned it by

[88] For a similar criticism of excessive literalism, see the dedication of Winter's translation of Galen's *De anatomicis administrationibus*, where he chastises an earlier translator of the same work (sig. $\alpha 2^v$): 'tam obscure vertit, ut a nemine queat intelligi, dum videlicet schemata Graeca pugnante Latinorum idiomate reddere totidem verbis nititur.'

themselves and with inadequate tools. Terminological troubles arose from a dearth of classical Latin authors in many fields and the resulting lack of Latin technical terms; they were exacerbated by the restricted possibilities of word composition in Latin (especially compared with Greek), the lack of dictionaries, and the constraints of Ciceronianism, which deterred translators from resorting to non-classical expressions.[89] A look at a few specimen translations will flesh out Winter's and Barozzi's remarks.

Theodorus Gaza's translation of Theophrastus's *Historia plantarum*, published posthumously in 1483, is the first Latin version of this work ever produced and constitutes an early milestone in the appropriation of Greek science in print. As a Greek, Gaza should have had little problems with his source text, but as he remarks in his 'praefatio' (sigs. Aiii^v^–Aiiii^r^), two aspects made him sweat: Theophrastus's botanical terminology and the textual corruption reigning in the only manuscript at his disposal.[90] Both problems surface in almost any given passage. In the first lines of book 6 (6.1–2 in modern editions), where Theophrastus announces his ensuing treatment of undershrubs and herbs, they can be easily identified. Gaza is confronted with a string of four plant names (*ἕρπυλλος, σισύμβριον, ἑλένιον, ἀβρότονον*) for which he has no Latin counterparts. Resorting to the common strategy decried by Barozzi, he simply transliterates these terms, changing their endings to Latin ones, as *serpyllum, sisibrium, beleium,* and *abrotanum*.[91] The apparent inaccuracy of all four transliterations suggests that plant names were consistently misspelled in Gaza's Greek manuscript, but this is still a comparatively harmless type of corruption. Having declared that he will first discuss wild undershrubs, Theophrastus then comes up with a statement printed as follows in modern editions:

[89] It is telling that stylistic difficulties in the narrow sense of the word do not figure in this list. Under early modern conditions, they were luxury problems, and pertinent remarks are rare. Hieronymus David Gaub's reflections on the difficulty of reproducing Swammerdam's terse and pointed Dutch style in Latin (*Biblia Naturae*, 1737, between Boerhaave's biography and the beginning of Swammerdam's work) are an exception.

[90] As another case of grave problems caused by a faulty source text, the *Geographia Nubiensis* may be mentioned. As the translators explain in their letter to the reader (sig. B^r^), their Arabic manuscript and the print based on it were corrupt in many places. Where emendation and translation proved impossible, they retained the original expression, which is why their Latin text is sometimes interspersed with Arabic script.

[91] On the popularity of transliteration in such cases, see Rener 1989, 99–103; Beullens 2008, 108, 110–11 (fish names). Martin 2008, esp. 160, points out the limited success of humanist attempts to oust transliteration in the case of the fourth book of Aristotle's *Meteorology*. Sometimes, the reader is helped by the addition of a Latin paraphrase. In Struś's version of the pseudo-Galenic *De urinis*, for instance, *ὑπόστασις πιτυροειδής* ('bran-like deposit') is rendered as 'hypostasis pityroeides, id est, quae habet quiddam simile furfuri' (fourth-last, unnumbered page). Difficult to understand is Tomeo's decision to replace his Latin translations of technical terms in the *Mechanics* with transliterations at a later stage. The explanation given in Leeuwen 2017, 188–9, that he thus conformed to the expectations of a humanist-minded readership is unconvincing. The ultimate result of the practice of transliteration is the modern Greco-Latin terminology of science and scholarship at large.

Ἔστιν αὐτῶν εἴδη καὶ μέρη πλείω, ἃ δεῖ διαιρεῖν καὶ τοῖς καθ' ἕκαστον γένος καὶ τοῖς ὅλοις εἴδεσι.

The Loeb translation[92] renders this as:

> Of these are several classes and sub-divisions, which we must distinguish by the characteristics of each sub-division as well as by those of each class taken as a whole.

Such sense as this translation seems to make rests on the impossible assumptions that the 'sub-divisions' are initially called *μέρη* but *γένος* a few words later, that *εἶδος* changes its meaning from 'class' to 'characteristic', and that *ὅλος* can signify 'valid for a whole class'. Nor does any other modern translation come up with a better solution. In all likelihood, the transmission is corrupt. How does Gaza react to this conundrum? He renders the Greek as:

> Horum igitur species partesque plures cum singulis generibus tum totis speciebus existunt; quas utique discernamus oportet.

This is every bit as hard to understand as the original. The implication that a 'species' should be composed of several 'genera' seems bizarre. *γένος* (accusative singular) becomes 'generibus' (ablative plural; perhaps Gaza's text once more differed from modern editions at this point). The attempt to take the relative clause *ἃ δεῖ διαιρεῖν* ('quas utique discernamus oportet') as independent from the following is misguided. Presumably, Gaza failed to see any sense in the original and tried to reproduce at least all the words and the grammatical structure in the hope that some readers would be able to make more of the result than himself. In this, as in many other cases, the difficulties he faced were simply insurmountable. They often are today too, and translators still react by pretending that they do not exist.

Textual corruption and plant names apart, Gaza was well able to cope with the content of the work he translated. Other translators were not. Some telling examples may be found in the preface to Clarke's rendering of Rohault's *Traité de physique*.[93] As Clarke had been preceded as a translator of this text by Théophile Bonet (1674), he felt the need to justify his own work and did so by pointing out a number of glaring mistranslations in his predecessor. In the following example, Bonet fails to understand the difference in Rohault between the terms 'vitesse' (speed in our sense) and 'mouvement' (volume of a body in motion times

[92] Hort 1916.

[93] Another example of severe criticism of a translation of seventeenth-century science is provided by Henry Oldenburg's judgements about Christoph Sand's Latin version of the *Philosophical Translactions* in the 1670s. Sand himself complained that the various novel contents of the *Transactions* would require a translator who was 'one and a half men' ('sesquivir'; Toribio 2020, 131–8, esp. 137).

distance covered; not in use as a physical quantity today).[94] In his translation, this leads to the bizarre result that of two equally fast bodies of different size, the one eight times bigger is also eight times faster than the other. In Clarke's words (4th edn 1718, sig. A4^{r}):

> Ad priorem versionem quod attinet, quam male multaverit auctorem nostrum interpretis haud satis felicis inscitia, uno aut altero allato exemplo satis patebit. Parte I., capitulo 11., articulo 9. 'Proinde', inquit interpres, 'si ea corpora aequali celeritate moveantur, corpus A debet octuplo celerius moveri corpore B.' Ridiculum! Verba auctoris sunt haec: 'Par consequent, si ces corps se meuvent d'egale vitesse, le corps A doit avoir huit fois autant de mouvement que le corps B', id est 'Proinde si ista corpora pari celeritate motae fuerint, corpus A octies tantum motus habere debebit quam quantum corpus B'. Quam longe diverso sensu, quis non videt?
>
> Regarding the earlier translation, what a raw deal the ignorance of the unlucky translator gave to our author will become clear enough by one or two examples. In part I, chapter 11, paragraph 9, the translator says: 'If, therefore, these bodies move with equal speed, body A must move eight times as speedily as body B.' Ridiculous! The author's words are as follows: 'Par consequent, si ces corps se meuvent d'egale vitesse, le corps A doit avoir huit fois autant de mouvement que le corps B', that is 'If, therefore, these bodies move with the same speed, body A must have eight times as great an amount of movement as body B.' Who fails to see how much the meaning is different?

Rohault's translators dealt with a text from their own culture written in a major European language. The anonymous intellectual who undertook one of the first attempts to Latinize a work of Chinese medicine, a sphygmological handbook ascribed to the famous physician Wang Shuhe (*c.* AD 180–270), found himself in a less comfortable position.[95] On the initiative of the doctor and merchant Andreas Cleyer, who resided in Jakarta as an employee of the Dutch East India Company, the translation was printed as the first of six thematically related texts in a collection titled *Specimen medicinae Sinicae* in 1682 at Frankfurt. In this case, the translator's

[94] Predictably, Bonet says nothing of his own lack of factual understanding. Most translators are silent in this respect, but there are exceptions. In the preface to the first two books of his Latin version of Dürer's *Vier Bücher von menschlicher Proportion* (1532, sig. Aiii^{v}), Camerarius openly confesses: 'Equidem non ignorabam quantum operis susciperem in tractanda re non satis mihi perspecta.'

[95] The Chinese author is given as 'Vám Xó Hó' by the translator, which corresponds well to 'Wang Shuhe' in view of his habits of transcription. However, the work in question is not, as one would expect, the sphygmological classic Maijing ('Canon of Pulse'; modern English translation: Yang 1997) for which Wang Shuhe is famous, and Wang Shuhe, therefore, may not be the real author of the work under discussion. The translator has been assumed to be the Jesuit Michał Piotr Boym (*Neue Deutsche Biographie*, vol. 3, 290; Walravens 2014, 297–8, 302–3), but this attribution is based on a confusion with the *Clavis medica ad Chinarum doctrinam de pulsibus* (1686), which is a work by Boym himself comprising only some translated extracts from Chinese pulse books.

problems were not restricted to identifying the meaning of isolated expressions and finding Latin equivalents for them. Pivotal notions of Chinese medicine such as 'qì' or 'yáng' make sense within their original conceptual framework only and cannot be extricated from it. This framework presumably was little known to the translator and certainly unknown to his readers. Moreover, whether due to the sorry state of his Chinese template, lack of linguistic competence, or both, the translator often failed to grasp the syntax of the original. In his attempts to cope with these daunting obstacles, he combined strategies already noted in previous examples: he provided transliterations of key concepts in combination with rough Latin equivalents and for the rest remained as literal as possible. In the latter respect, he had to go so far that the result simply became incomprehensible over long stretches, as in the following section (p. 34):

> De novem viarum pulsibus
>
> Primus pulsus cham seu longus est naturae yam seu caloris primigenii, digitus infra perquirens reperit in tribus locis instar hastae, cui quis innititur, cum elevatur est nimius; vocatur longus, transiliendo enim propriam sedem etiam appellatur longus; dominatur perturbationi corporis, est calidus; noctu aeger dormiens non quiescit.
>
> On the pulses of the nine ways
>
> The first pulse, cháng or long, is of the nature of yáng or the primordial heat, the finger which looks for it below finds it in three places, similar to a lance on which one steadies oneself, when it rises, it is too much; it is called long, for by overleaping its own place, it is also called long; it rules over the confusion of the body, it is warm; at night, the sick, while sleeping, finds no rest.

None of the above examples is meant to ridicule translators or to suggest that their efforts were futile. Given all the difficulties, it rather is remarkable that many decent and some excellent translations were produced after all. Moreover, even a translation full of serious mistakes might still prove useful in some respect. By way of conclusion, I would like to demonstrate this last point by means of an extreme example, which combines more or less all the types of problem discussed so far with blatant deception.

The first work by Archimedes ever to appear in print was one of his most difficult and sophisticated. In 1503, the astronomer, astrologer, and mathematician Luca Gaurico issued a small collection of mathematical writings which included a translation of the *Quadratura parabolae*. In this treatise, the Syracusan mathematician gives two ingenious but involved proofs for the fact that the area of a parabolic segment always equals four thirds of the area of the biggest triangle that can be inscribed into it. In the first proof, he uses an imaginary balance and the law of the lever; in the second, a series of triangles whose areas form a geometric progression is inscribed into the segment.

Gaurico's translation is characterized by incomprehension on all possible levels. To begin with, he is often unable to handle the Greek of Archimedes, who writes in a Doric dialect which significantly diverges from the standards of Attic Greek. In the opening sentence of his letter of dedication to a certain Dositheus, the author mourns the death of his friend and mathematical correspondent Conon and announces that, in future, he will send his findings to his addressee, another friend of the deceased:

> Ἀκούσας Κόνωνα μὲν τετελευτηκέναι, ὃς ἦν ἔτι βλέπων ἡμῖν ἐν φιλίᾳ, τὶν δὲ Κόνωνος γνώριμον γεγενῆσθαι καὶ γεωμετρίας οἰκεῖον εἶμεν τοῦ μὲν τετελευτηκότος εἵνεκεν ἐλυπήθημες ὡς καὶ φίλου τοῦ ἀνδρὸς γεναμένου καὶ ἐν τοῖς μαθημάτεσσι θαυμαστοῦ τινος, ἐπροχειριξάμεθα δὲ ἀποστεῖλαι τοι...

In Gaurico's translation, γεωμετρίας οἰκεῖον ('familiar with geometry') becomes 'geometriae domesticum' ('from the house of geometry'), τὶν ('you') becomes 'quendam' ('a certain person'), ἐπροχειριξάμεθα ('I decided') becomes 'praeconati sumus' ('I proclaimed'), and the whole sentence becomes meaningless gibberish. When it comes to mathematics, important terms are mistranslated: παρὰ ('along') is rendered as 'apud' or 'penes' ('by'), χωρίον ('area') as 'spatium' ('distance'). Others are simply transliterated, but how this is done can be telling as well. The crucial term ἐπίτριτος ('by one third'), which indicates the difference between the areas of the parabolic segment and the inscribed triangle, is consistently corrupted to 'epitricus' or 'epytricus', which shows that Gaurico had no idea what the word meant: he therefore remained in the dark about Archimedes's goal of proof. Archimedes's expression for a parabola, ὀρθογωνίου κώνου τομά ('section through a right-angled cone'), is faithfully Latinized as 'sectio rectanguli coni', but this does not imply that its meaning was clear to the translator. The volume's title page announces a collection of treatises on the squaring of the circle ('circuli quadratura'), and Archimedes's work is in fact accompanied by two others on this very subject. Moreover, the diagrams with which the present translation is equipped do not show parabolas, but semicircles. In sum, Gaurico seems to have believed that he translated a text discussing an indeterminate ratio between the areas of semicircles and inscribed triangles. In fact, however, he translated nothing at all, but simply plagiarized a medieval predecessor: he made a sloppy copy of a bad manuscript of a translation done by Moerbeke in 1269 and passed the result off as his own.[96]

If ever a translation was completely useless, one is inclined to think, it must have been this one. But that is not the case. In 1543, Gaurico's text was republished on folios 19^{v}–29^{r} of his *Opera Archimedis* by Niccolò Tartaglia, who had little humanist training and certainly not enough Greek to consult the original, but was one of the leading mathematicians of his day. Tartaglia left the text

[96] See Clagett 1964–84, vol. 2.1, 72, and, for an edition of Moerbeke's text, 141–56.

basically untouched, presumably because he did not trust his own Latin enough to intervene. Nonetheless, the circles of Gaurico's diagrams miraculously metamorphosed into true parabolas: apparently, Tartaglia managed to achieve a correct understanding of Archimedes's treatise by applying his mathematical acumen to a deeply uncomprehending translation.[97] Only one year later, it is true, Moerbeke's and Gaurico's version was laid to rest for good by the epoch-making Basel edition of Archimedes's *Opera omnia*, which ignored it and based its Latin translation on a far superior mid-fifteenth-century rendering by Iacopo da Cremona instead, which had, moreover, been revised by Regiomontanus.[98] Still, the impeccable diagrams of the Basel edition may build on Tartaglia's.[99]

Commentaries

The commentary in premodern times

A commentary is a text whose stated aim it is to elucidate another text.[100] In the Western tradition, designations for such texts were manifold. In Greek, commentaries were often called *ὑπομνήματα* or simply titled *Εἰς* ('On, About') followed by the title of the primary text. Latin expressions include 'enarratio', 'expositio', 'glossa', and 'commentarius', a word which also occurs in the (neuter) plural ('commentarii', 'commentaria') and in the short form 'commenta'. There are also other plural expressions such as *σχόλια*/'scholia', 'notae', or 'annotationes', which can but need not indicate lighter, less coherent annotation.[101] The situation is complicated by the fact that not everything denoted by the 'commentari-' words which have come to designate the genre in most vernaculars today was in any sense a commentary. In some cases, the works were, rather, sketches, unpretentious narratives (like Caesar's *Commentarii de bello Gallico*), studies, or monographs.[102]

[97] On Tartaglia's way of dealing with Moerbeke's translations of Archimedes in general, see Clagett 1964–84, vol. 3.3, 538–607; see esp. p. 566 for Tartaglia's replacement of the semicircles with parabolas.

[98] On Iacopo da Cremona's rendering, see d'Alessandro and Napolitani 2012, esp. 281–327 (Latin text and Italian translation of the *Quadratura parabolae*).

[99] Although Regiomontanus's corrections also pertained to the diagrams, he left the semicircles untouched (d'Alessandro and Napolitani 2012, 139–45, 187–9, 248–50). An example comparable to the present one is provided by Winter's predecessor as a translator of Galen's *De anatomicis administrationibus*, mentioned in note 88 above: his close adherence to the Greek original made his translation incomprehensible in and of itself but occasionally ('in locis aliquot') helped Winter to guess the original wording and meaning of his own, heavily corrupt Greek manuscript.

[100] For this opening section in general, see Most 1999; Goulet-Cazé et al. 2000; Gibson and Kraus 2002; Geerlings and Schulze 2002–4; Boer 2018.

[101] Cf. Paulus 2005, 93 (additional terms), 154 (terminological trends over time), 230 (Latin and vernacular terms).

[102] Neglect of this fact (as in Häfner and Völkel 2006, 1–3 and Henderson 2012, esp. Céard 2012) results in an overextension of the genre. The polysemy of the respective expressions was noted in early modern times. As Céard 2012, 4–5, points out, Juan Luis Vives (*De ratione dicendi*, 1533, pp. 230–2), distinguishes commentaries 'in aliud' (commentaries in the present sense of the word) from 'simplices' (others).

The diverse terminology fitted the openness of the genre. In spite of occasional reflections,[103] no binding theory of the commentary existed. As will become clear later, the elucidation of another text was not necessarily a commentary's only or most important goal. Moreover, commentaries could merge fluently into neighbouring forms such as discussions of recently published works in review or book format, texts taking their cue from one or a few classical quotations, 'animadversiones', that is, series of text-critical and exegetical remarks, as well as revisions and expansions of originals.[104] Full-blown commentaries and light annotation were just two extremes with a continuum in between.

While commentaries and translations can be separated neatly, their relationship was, nonetheless, close. They were similar and were perceived as such in their exegetical nature.[105] Their respective spheres of influence could be disputed: explanatory additions in translations were criticized with the argument that this task should be reserved for the commentary.[106] Authors used commentaries to explain their own translations and to compare and criticize those of others.[107] On a practical level, a translation of a text and a commentary on it were often combined in one book.

From its first attested specimens in the Ancient Near East,[108] the commentary was intimately linked to teaching. Most written commentaries originated in lectures and recorded their contents for later consultation, including their use in new lectures. Teaching by commentary instead of textbook, a key feature of premodern education, reflected the fact that instruction was sought first and foremost in classical authors and works. As most of these had not been conceived for pupils in the first place and moreover came from chronologically and geographically distant cultures, they had to be explained to serve their new didactic purpose.

Many ancient and medieval commentaries preserve traces of their origin in the classroom. The exposition often starts with a preliminary discussion of general topics, the so-called *προλεγόμενα* or 'accessus', which provided students with a

[103] Besides Vives, Cardano, for instance, gave the genre some thought (Maclean 2004, 30–4).

[104] To illustrate these text types by examples from early modern science: for reviews, see the section below. For a book-length discussion of a recent work, see Kepler's *Dissertatio cum Nuntio sidereo* (1610; Rosen 1965). A text starting with some classical quotations is Rudolf Jacob Camerarius's academic oration 'De generatione ex primaevo semine' (1694), which uses a discussion of three (pseudo-) Aristotelian statements (*De plantis* 2; *De generatione animalium* 2.1 and 2.5) to propound the theory that every existing sperm was created at the beginning of the world. For the serialization of this formula in 'animadversiones' and similar works, see Filippo Beroaldo's 'Annotationes Galeni' in his *Opusculum de terrae motu et pestilentia* (1505), sigs. Eiiii^v–Gii^v, as well as the works on 'sacred medicine' and 'sacred physics'. Revisions and expansions may be represented by Georg von Peuerbach's and Johannes Regiomontanus's *Epitoma in Almagestum Ptolemaei* (1496) and Franco Burgersdijck's schoolbook edition of Sacrobosco's *Sphaera* (1626), in which he distinguishes his additions by italics but otherwise integrates them into the original text. For some additional academic genres from Renaissance Aristotelianism bordering on the commentary, see Lohr 1988, pp. xiv–xv.

[105] Copeland 1991, 63–86, esp. 86.

[106] Martin 2008, 173 n. 66.

[107] Martin 2008, 166, 168, 171.

[108] See the *c.*900 texts presented in the *Cuneiform Commentaries Project* (https://ccp.yale.edu/, accessed 16 February 2023).

framework for understanding the text to be discussed in class and the teacher's explanations.[109] The syllogistic reasoning of medieval commentaries reflects the academic thought style and method of explanation of the time, as does the frequent division of the commenting procedure into four steps: 'littera' (the teacher read the text), 'divisio' (he analysed its structure), 'expositio' (he explained the meaning of one of the resulting sections), and 'disputatio' (he discussed specific questions).[110] The two main varieties of the genre correspond to the two most important didactic formats of the medieval university: the exegetic commentary covering the text in full, the standard structure of which has just been detailed, matches the lecture, the 'quaestiones' commentary focusing on selected problems, the 'disputatio'.[111]

However, although most commentaries were somehow connected to formal teaching, not all remained exclusively restricted to that sphere. A commentary could also be used by its author as a means of presenting his own concerns and be perused privately by scholarly readers. The traditionalist habitus of premodern learning made it an attractive option to present one's opinions as the explication of something implicit in earlier authorities. It was first and foremost in commentaries, for instance, that the philosophical traditions invoking Plato and Aristotle evolved far beyond the thinking of their founding fathers in antiquity, the Muslim world, and the medieval West.

Over time, a massive body of commentaries accumulated in belles-lettres, philosophy, theology, law, and other fields of technical writing. The natural world was one such field, although its commentary tradition could not compare with the aforementioned in terms of numbers.[112] The most substantial ancient body of texts is due to Aristotle's Greek commentators, beginning with Alexander of Aphrodisias, who left nearly twenty expositions of the *Physics*, *De caelo*, *Meteorology*, *De anima*, and the minor biological works.[113] In medicine, ancient commentaries on Hippocratic writings stretch from the first century BC to the seventh century AD, culminating in the twelve extant commentaries (out of seventeen attested) forming part of the Galenic corpus, which was commented upon in turn in late antiquity.[114] Expositions of introductory as well as advanced mathematical works by authors such as Euclid, Archimedes, Apollonius of Perge,

[109] See Mansfeld 1994 for antiquity, Quain 1945 and Copeland 1991, 65–76 for the Middle Ages.
[110] Paulus 2005, 94–5.
[111] Paulus 2005, 93; with respect to Aristotle in the Middle Ages, see Grabmann 1939, esp. 34–6, 48, 50. Predominantly exegetic commentaries often contained some 'quaestiones' as well (Perfetti 2000, 4–5).
[112] A good overview is provided by Von Staden 2002; see esp. 124–5, 130 n. 65.
[113] *Commentaria in Aristotelem Graeca*, vols. 3.1–2, 5.2–4, 5.6, 7, 9, 10, 11, 12.1–2, 14.1–3, 15, 16, 17, 22.1–2, 23.1. Most of these are now accessible in English in the volumes published by the project *Ancient Commentators on Aristotle* founded by Richard Sorabji. On their early modern printing history and influence, see the concise remarks in Schmitt 1983, 49.
[114] On the commentaries by an elusive figure called John of Alexandria, see Nutton 1991.

and Nicomachus of Gerasa, some of which formed part of the Neoplatonic curriculum,[115] survive from around AD 300 to the end of antiquity. Within the science of the stars, Ptolemy's dominant position is reflected by fourth-century and later commentaries on the *Almagest*, the *Handy Tables*, and the *Tetrabiblos*.[116] By contrast, no ancient expositions are extant in areas like geography, botany, or chymistry.

The medieval material is more copious, because the teaching of pertinent subjects was institutionalized on an unprecedented scale at the universities: natural philosophy and the quadrivium were taught in the arts faculty, medicine in its own. In these disciplines, a canon of works formed which were routinely lectured on and discussed and attracted a massive body of commentary.[117] Its core included Aristotle's 'libri naturales',[118] Euclid's *Elements*, Sacrobosco's *Sphaera*, Hippocrates' *Aphorisms*, Galen's *Ars medica*[119] (both of which also formed part of the collection known as *Articella*), and Avicenna's *Canon*, even though the curriculum as a whole was considerably richer.

Commenting on science in the early modern era

The early modern age is commonly regarded as the era in which the commentary lost its dominant place in teaching to the textbook. This is true of the second half of the period. Until the seventeenth century, by contrast, commentaries by and large remained the most important teaching tools. The first two centuries after the invention of print were marked not by a decline but by an efflorescence of the format.[120] The rise of universities, the rediscovery of Greek and Arab classics across a broad front, the relative ease of turning lecture notes into books, and the education-driven demand for such material made the commentary a highly

[115] On Proclus's commentary on the first book of Euclid's *Elements*, see Morrow 1970, esp. pp. xxiv–xxv.

[116] On Pappus' *Almagest* commentary, see *CDSB* 10, 298–9; on Theon's expositions of the *Almagest* and the *Handy Tables*, see *CDSB* 13, 321–2; on an anonymous *Tetrabiblos* commentary, see *PAL*, C.2.11.

[117] Grendler 2001, 267–8, 318–24, 410–11 (for Italy).

[118] See Grabmann 1939 and, for the most complete bibliography, Lohr 2010–13 (within an overview of medieval Aristotle commentaries at large in both cases). On *De generatione et corruptione* in particular, see the contributions in Thijssen and Braakhuis 1999, 69–205; on the pseudo-Aristotelian *Problemata*, see Ventura 2008, 131–41. Livesey 2000, 78–9 knows of 2,500 medieval Aristotle commentaries (by 850 commentators) but does not indicate how many are concerned with the 'libri naturales'.

[119] On the medieval translations and commentaries of the *Aphorisms*, see Kibre 1985, 29–90; on Sacrobosco, see Thorndike 1949; on the *Ars medica*, see Niedling 1924, 9–17, 19–21.

[120] On Renaissance and early modern commentaries in general, see Buck and Herding 1975; Pade 2005; Häfner and Völkel 2006; Henderson 2012; Enenkel and Nellen 2013, esp. the editors' introduction (pp. 1–76); Enenkel 2014b (esp. 327–410 on commentaries on botanical, zoological, and geographical texts); Boer 2018; Suter-Meyer 2020, 25–42. Much pertinent material, including commentaries on scientific authors, is found in *CTC*.

popular genre. In keeping with manuscript precedents, classical authors were regularly printed together with their commentators—Virgil with Servius, Aristotle with Averroës,[121] and so on. Moreover, the growing role of books in general as a means of communication within disciplinary communities led to the wider use of commentaries as outlets for new results, to be studied autodidactically outside a formal teaching context.

The spread of commentaries was such as to provoke stark criticism on occasion. Montaigne, for one, complained (*Essays*, book 3, ch. 13): 'Il y a plus affaire à interpréter les interprétations qu'à interpréter les choses, et plus livres sur les livres que sur autre sujet: nous ne faisons que nous entregloser. Tout fourmille de commentaires; d'auteurs, il en est grand cherté.' But in the natural disciplines at least, commentators found it easy to counter such laments: expositions were indispensable, they asserted, because the primary texts simply were forbiddingly difficult. Conrad Heingarten, author of an unpublished commentary on the *Tetrabiblos*, remarked on that book:

> Cuius tanta est obscuritas tum propter succinctam sui brevitatem, tum propter verborum ponderosam gravitatem, ut vix a quoquam intelligi potuerit.[122]
>
> Because of its concise brevity as well as the weighty pithiness of its words, it is so opaque that hardly anyone has been able to understand it.

Francesco Barozzi had similar things to say about the *Almagest* in the preface to his *Cosmographia in quattuor libros distributa* (1585, sig. [b5]^{v}):

> Quod propter auctoris gravitatem et rerum astrologicarum maximam difficultatem demonstrationumque arithmeticarum et geometricarum (praesertim hisce nostris temporibus) obscuritatem ac etiam Graeci exemplaris depravationem et malam Latinam versionem non ab omnibus et praecipue a tirunculis intelligi facile potest..., sed locupletissimis indiget commentariis.
>
> Because of the author's profound mind, the extreme difficulty of the subjects of astronomy, and the obscurity of the arithmetical and geometrical proofs, particularly for these our times,[123] moreover because of the corruptness of the Greek manuscript and the bad Latin translation, it cannot be easily understood by all, particularly not by inexperienced beginners..., but is in need of most extensive commentaries.

[121] Two monumental Latin Aristotle-Averroës editions came out in Venice in 1472–4 (7 vols) and 1550–2 (11 vols); see Grendler 2001, 274, 277, and, specifically on the second one, Burnett 2013.

[122] Quoted in Grafton 1999, 256 n. 36.

[123] The underlying idea—eloquently spelled out, for instance, in Konrad Dasypodius's *Oratio de disciplinis mathematicis* (1579)—is that the excellence of ancient mathematics has been lost since the end of antiquity.

Because commenting was a craft as demanding as it was indispensable, it carried considerable scholarly prestige. The reputation of a professor largely depended on his ability to expound the classic texts of his discipline in his lectures,[124] many of which circulated as manuscripts. Their publication could further enhance the fame gained from this source.[125] Many teacher, therefore, took the trouble of publishing their commentaries themselves. In other cases, eager disciples helped out. After the death of the eminent Paduan professor Giovanni Battista da Monte, his pupil Walenty z Lublina issued a number of his medical lectures, including a commentary on the ninth book of Rhazes's *Liber ad Almansorem* comprising over 700 quarto pages (1554). Such a task required considerable dedication, because the publication process was laborious. As professors usually did not write down their lectures in full (it would have been too time-consuming, and it could even be seen as shameful to depend on a preformulated manuscript),[126] the text of an edition had to be reconstructed from the professor's notes, students' transcripts ('reportationes', 'recollectiones'),[127] or personal memories.

Most readers of commentaries were professors and students as well. The former used them to prepare their own classes, the latter found them a welcome help in following lectures but also in their private reading.[128] That this last practice was widespread is confirmed by criticism of it. In a 'study guide' talk delivered around 1570 to first-year medical students in Padua, Girolamo Mercuriale advised his listeners to read the classics without commentaries, 'which tend to confound the judgement and render the mind lazier in tracing truth' ('qui et confundere iudicia et ingenia ad veritatem indagandam pigriora reddere soleant'). However, he had to make exceptions for difficult passages as well as for Hippocrates and Avicenna *in toto*.[129]

As with other genres, the introduction of print did not make itself felt as a sharp break in the case of commentaries. In terms of disciplines, the continuities with the medieval tradition are immediately evident. (That the picture is also similar to the one obtained when looking at translations confirms the close links between both genres.) Most commentaries continued to be written and printed in those fields which formed part of the academic curriculum: medicine, together with its ancillary discipline, pharmacological botany,[130] on the one hand, and the

[124] Grendler 2001, 159–60; cf. also Paolo Giovio's praise of Pietro Pomponazzi's brilliance in expounding Aristotle (*Elogia clarorum virorum*, 1546, fol. 44^r).

[125] Siraisi 1987, 5, 63–4. [126] Grendler 2001, 159.

[127] That Pomponazzi spoke so slowly and clearly that his pupils were able to take down his sentences in full ('explicatas sententias notis exciperent') must have been exceptional, since the fact is singled out for special praise by Giovio (see note 124 above). Giovio's words are confirmed by the full and correct 'recollectio' of a Pomponazzi course on *De partibus animalium* 1–2 published by Perfetti 2004.

[128] Pesenti 2000, 74 and 78, respectively.

[129] *De modo studendi*, ed. Durling 1990, esp. lines 154–6, 162–70.

[130] On medical commentaries, see Maclean 2002, 206–33; on botanical commentaries in Italy around 1500, see Dilg 1975.

disciplines treated in the arts faculty, namely the quadrivium, in particular astronomy and cosmology, and natural philosophy, on the other.[131] In all these areas, much effort focused on comparatively few texts or even single books or chapters. To the medical favourites belonged Hippocrates's *Aphorisms* and *Prognostics*, Galen's *Ars medica*, and Dioscorides's *Materia medica*. Nearly 140 commentaries on the *Aphorisms* were on record by 1679. About forty were written on Dioscorides. Some Arabic works were much commented upon as well, for instance the ninth book of Rhazes's *Liber ad Almansorem*, which discusses over ninety diseases from head to toe, and above all several *fen* (1.1, 1.4, 4.1) of Avicenna's *Canon*. It attracted over sixty manuscript and printed expositions from *c.*1500 to the late seventeenth century.[132] For Sacrobosco's *Sphaera*, some 170 editions of commentaries published between 1472 and 1650 have been described.[133] The body of expositions devoted to the pertinent parts of the Aristotelian encyclopedia is far larger still: some 1,700 items are on record for the period 1500–1650.[134] Much attention was devoted, for instance, to Aristotle's works on meteorology and zoology, the master texts of their respective disciplines.[135] At the same time, much less prominent works from medicine and the quadrivium were commented upon as well, repeatedly so in some cases. The half-dozen sixteenth-century translations of Nicander's *Theriaca* and *Alexipharmaca* featured commentaries partly based on the ancient scholia accompanying the poems in the manuscripts. Jacopo Berengario da Carpi wrote a monumental commentary (1521) on a late medieval manual of dissection, the *Anatomia* by Mondino de Luzzi.[136] Even a decidedly scholastic work from the same period, such as Albert of Saxony's *Tractatus de proportionibus* on mathematical proportions and their application to the theory of motion, could be explained by a humanist doctor of arts and medicine, Benedetto Vettori (1506). By contrast, works from disciplines without any academic standing, such as higher mathematics or chymistry, trailed far behind. If commentaries on such texts were written at all, they were not meant for formal

[131] A range of astronomical, cosmological, and meteorological commentaries from Italy is scrutinized by Paulus 2005, 91–233.

[132] On *Aphorisms*, *Prognostics*, *Ars medica*, and the *Canon* as the core texts of sixteenth-century teaching, see Maclean 2002, 29. On the *Aphorisms* in particular, see Maclean 2021a, 222–3; on the *Ars*, see Niedling 1924, 9–19, 21–5; on Dioscorides, see *CTC* 4, 44–116; on the *Canon*, see Siraisi 1987, 175–217, 367–76.

[133] The census of these, along with an analysis of the whole tradition in terms of the sociology of knowledge, is due to the project *The Sphere* conducted at the Max Planck Institute for the History of Science by Matteo Valleriani (https://sphaera.mpiwg-berlin.mpg.de/, accessed 16 February 2023). The main results of the project are summarized in Valleriani 2017 and 2020; see also Valleriani and Ottone 2022.

[134] See the pertinent material in Lohr 1988. For an analytical overview, see Schmitt 1983, 39–52; Lohr 2002.

[135] The bibliography of Martin 2011, which does not aspire to completeness, contains fourteen commentaries on the *Meteorology*; for commentaries on the zoological works, see Perfetti 2000 and 2004. Cf. also Lo Presti 2021 and Ventura 2008, 142–52 for expositions of the *Parva naturalia* and the *Problemata* respectively.

[136] Lind 1975, 159–65; French 1985.

teaching but for private study. The same is true of scientific expositions of non-scientific texts, the Bible in particular (see Chapter 4).

The transition to the age of print appears smooth not only with regard to the texts commented upon but also with regard to the expositions themselves: among the first results of the new technology was a wider circulation of prominent medieval commentaries. Whether they were of Arab or Western origin, they were already available in Latin and ready to be printed at the onset of the incunabula era. With regard to natural philosophy, this is most true of the *oeuvre* of Averroës (see note 121 above), as well as of many Western commentators on Aristotle.[137] In medicine, part of the ample body of medieval expositions of the Hippocratic *Aphorisms*, the Galenic *Ars*, the *Articella*, and the *Canon* became successful in print from the 1470s to the mid-sixteenth century.[138] In astrology, Ali ibn Ridwan's *Tetrabiblos* commentary was printed in 1493 and 1519.[139] Greek commentaries profited a bit later, as they had to be translated first. This applies, for instance, to Galenic commentaries on Hippocrates such as those on *De natura humana* and *De salubri victus ratione*, which were Latinized and published together with their primary texts by Hermann Cruser (1534).[140] Two Latin translations of an anonymous late antique exposition of the *Tetrabiblos* appeared in 1502 (by Giorgio Valla) and 1559.[141] When Federico Commandino translated books 1–4 of Apollonius' *Conica* (1566), one of the most advanced works of Greek geometry, he added the late antique commentary by Eutocius. In 1710, Edmund Halley still did so in his fuller edition of the same work.

Nonetheless, most commentaries published in the early modern period were contemporary. When an ancient work was rediscovered, one of the first things done was often to equip it with an exposition. In this way, Sacrobosco's *Sphaera* got company from a very elementary homonymous work, which was ascribed to Proclus but was in fact just an extract from the *Eisagoge* by Geminus of Rhodes (first century BC). After its first publication, in Latin, by Thomas Linacre in 1499, this text was frequently used in teaching and attracted at least ten commentaries, most of which were printed, in the sixteenth and early seventeenth centuries.[142] More difficult works were in even greater need of commentary. To his *editio princeps* of Pappus' *Mathematical Collections* (1588), for instance, Commandino added detailed discussions of the work's single 'theoremata', 'problemata', and 'propositiones'. That this was perceived as a standard approach transpires from a forgery: when Gilles Personne de Roberval released an alleged Latin translation of an ancient exposition of the heliocentric system by Aristarchus, he made

[137] Their early modern editions are noted in Lohr 2010–13.
[138] Pesenti 2000; Niedling 1924; French 1985, 49, and Arrizabalaga 1998; Siraisi 1987, 63–4.
[139] *PAL*, A.2.5, C.2.1.
[140] On the popularity and influence of Galen's Hippocrates commentaries, see Siraisi 1987, 177–8.
[141] *PAL*, C.2.12.1, C.2.12.2.
[142] Todd 1993; *CTC* 8, 7–48, esp. 38–48.

his concoction look more genuine by the addition of a number of longer and shorter 'notae'.[143]

Not only newly found, but also freshly translated or otherwise topical works often came with a new commentary. Both the first early modern translation and the Latin *editio princeps* of the astrological aphorisms known as *Centiloquium* and wrongly ascribed to Ptolemy, by George of Trebizond (1453/54, printed 1544) and Giovanni Pontano (1512) respectively, were equipped with substantial commentaries.[144] After the publication of the Greek *editio princeps* of Paul of Aegina in 1528, Johann Winter, the author of the second complete Latin translation to appear in print in 1532, premised some 'annotatiunculae' to the individual books, which he replaced with a fuller commentary in the second edition (1542). All in all, nine different translations and fourteen (albeit mostly partial) expositions of the work appeared in the course of the sixteenth century.[145] New interest in the Hippocratic *De vulneribus capitis*, kindled by disputes about the treatment of gunshot wounds, resulted in at least eight published commentaries between 1566 and 1616.[146]

Cases such as these are apt to exemplify the close relation of commentary and translation mentioned initially. Translator and commentator were often the same person. As a rule, the text commented was given in Latin translation only—something valid for most Latinized works in general—and no reference was made to the original, which was thereby pushed into the background and out of the reader's awareness. Sometimes, as in Oddo degli Oddi's *In primam fen primi libri Canonis Avicennae expositio* (1575), not only one translation, but the two most popular at the time were given.[147] That both a Greek text and a Latin translation were provided, as in Giovanni Filippo Ingrassia's commentary on Galen's *De ossibus ad tirones* (1603) or Johannes de Laet's on Theophrastus's *De lapidibus* (1647), became somewhat less unusual under the influence of humanist thinking but nonetheless remained the exception rather than the rule. So did criticism of a translation in the exposition based on it.[148]

Roberval's aforementioned annotation was added to a text which was not in fact a newly found classic but a contemporary work. This practice was not restricted to forgeries. From the end of the incunabula era, commentaries appeared on newly written texts which were perceived as important but difficult. Georg von

[143] In doing so, Roberval may also have followed the model of the famous forger Annio da Viterbo, on whose commentary of his *Antiquitates*, see Grafton 1997, 131.

[144] On the *Centiloquium*, its complicated tradition, and George of Trebizond's and Pontano's contributions, see *PAL*, B.1, B.1.7, B.1.9, Lemay 1978, and esp. Rinaldi 1999. Rinaldi 2013 analyses the reception of Pontano's commentary.

[145] *CTC* 4, 147, 151–89. [146] Nutton 1989, 432–3.

[147] On these two translations, see Hasse 2016, 98–115, 361–2. French 1985, 70, speaks of cases in which three translations were given but provides no examples.

[148] Some examples are provided by commentators on the Aristotelian *Problemata* who found fault with Theodorus Gaza's translation of this work (Ventura 2008, 154).

Peuerbach's *Theoricae novae planetarum* ([1472]), which ousted its predecessor, the twelfth-century *Theoricae planetarum*, and became the standard text of planetary astronomy in the curriculum, provides the earliest important example. Alongside an unspecified, but high number of manuscript commentaries, printed ones appeared by over a dozen professors like Albert Brudzewski (1495), Franciscus Capuanus (1495), Erasmus Reinhold (1542), Erasmus Oswald Schreckenfuchs (1556), or Pedro Nuñez (1566).[149] A few decades later, Gemma Frisius's introduction to arithmetic, the *Arithmeticae practicae methodus facilis* (1540), became a big success and was republished several dozen times until the early seventeenth century. From 1545, most editions featured 'annotationes' (as well as other additions) by Jacques Peletier. From the later seventeenth century, commentaries were written almost exclusively on early modern works, as will become clear shortly.

Roberval presents an eccentric case of another common phenomenon as well, namely self-commentary. In his *Curationum centuriae*, seven sets of which were published from 1551 to 1566, Amatus Lusitanus successfully introduced this feature into the genre of collections of medical case histories, the so-called 'observationes' or 'curationes'. His 'scholia' are separated from the case histories proper and comment upon them by giving reasons for what happens, generalizing observations, and adding further information.[150] Kepler's *Somnium* (1634), in which he clads his ideas about the moon in a fantastic, semi-autobiographical narrative, is followed by a series of 223 'notae' twice as long as the text itself. The notes not only explain factual detail but also invite the reader to have a look at the author's workshop.[151] Linnaeus's *Fundamenta botanica* (1735) was turned into a comprehensive *Philosophia botanica* (1751) by a self-commentary to each of the 365 aphorisms.[152] Self-commentaries were particularly often added to didactic poems, which catered to a broader educated audience but nonetheless presupposed considerable expertise in their respective subjects.[153] Ancient and early modern commentaries on classical authors of scientific didactic verse like Nicander, Lucretius, or Manilius provided models to follow. As early as 1556, Nicolas Biese equipped his Lucretian poem *De universitate* with extensive notes. Two centuries later, Roger Joseph Boscovich explained his *De solis ac lunae defectibus* (1760) in this way, György Alajos Szerdahélyi, his *Historia Uraniae Musae* (1787) about the discovery of Uranus by Wilhelm Herschel. The last two works are highly technical and would have been forbiddingly hard reading on their own. Like Lusitanus, these learned poets conceived their self-commentaries in tandem with their

[149] Aiton 1987, 7; Malpangotto 2020, 111–246, 694–707. Individual commentaries went through from one to nine editions.

[150] Pomata 2011b, 55–9.

[151] This apparently was a habit of Kepler's. See the self-commentary added to the second edition (1621) of his *Mysterium cosmographicum* (*KGW* 8, 6–128) and Grafton 1997, 28–9 (with some additional remarks on self-commentaries in general).

[152] English translation: Freer 2003.

[153] On commentaries on didactic poems in general, see Chapter 9, 'Didactic poems'.

primary texts, the actual shape of which was conditioned by the former. The possibility of explaining themselves in prose gave them the freedom to present complex facts and theories in compressed, allusive manner while at the same time indulging in the ornate Latin and far-fetched erudition sanctioned by tradition and expected by readers.

From the later sixteenth century, two interconnected developments made themselves felt more and more strongly: first, classical authors began to lose their status as authorities on the natural world; secondly, commentaries as teaching tools encountered competition from textbooks. Under these circumstances, writing and publishing new expositions of the classics slowly but surely began to look like a gratuitous enterprise. The reputation attached to the practice dwindled, and it fell out of fashion. Depending on the field, the production of new commentaries declined and finally ceased between the second half of the sixteenth and the second half of the seventeenth century. Subsequently, the use of commentaries in teaching declined as well. The genre lost the pivotal role it had held in the didactic sphere for millennia.[154]

As the link between commentary and classroom loosened, other functions which had been fulfilled by the format all along became more prominent. This trend was heralded by vernacular commentaries catering to a non-academic readership, on the one hand,[155] and Latin expositions for mostly expert readers, on the other.[156] Many of the aforementioned commentaries on newly discovered works, which for the most part did not enter the curriculum, belong to the latter class. The numerous attempts at explaining Paul of Aegina are a case in point. For later examples, one may turn to the exposition of Aristotle's *Historia animalium* by Julius Caesar Scaliger (posthumously 1619)[157] or to the first edition of books 5–7 of Apollonius' *Conica* (1661). Translated from the Arabic by Abraham Ecchelensis, they were provided with 'notae' more extensive than the text by Giovanni Alfonso Borelli. However, intellectuals also commented on works which did have a place in the curriculum in order to make them accessible to private readers, voice their own ideas, and gain fame: Cardano's commentary on Ptolemy's notoriously difficult *Tetrabiblos* (1554) is a prominent instance.[158]

[154] See Siraisi 1987, 5, and Perfetti 2000, 231–2, with regard to expositions of Avicenna's *Canon* and Aristotle's zoology respectively.

[155] Precursors date back to the Late Middle Ages. Of the seven astronomical and cosmological commentaries in the vernacular analysed by Paulus 2005, 157–230, one dates from the fourteenth century, one from 1498, and the rest from 1550–1604.

[156] For the typological division between didactic and scholarly commentaries, see Siraisi 1987, 178–9, who exemplifies it by expositions of Avicenna's *Canon*.

[157] Perfetti 2000, 155–81, esp. 180.

[158] Grafton 1999, 251–89; Pompeo Faracovi 2012a, esp. 134–7. That the *Tetrabiblos* was taught at the arts faculties is noted by Grafton 1999, 256–7, and Grendler 2001, 412, but Cardano himself apparently never lectured on the work.

In the seventeenth and eighteenth centuries, except for didactic poetry, most commentaries without a formal didactic background dealt with works by contemporary or near-contemporary celebrities. As mentioned earlier in this chapter, the final Latin version of Descartes's *Géométrie* (1637), the *Geometria a Renato des Cartes edita* (1659–61), included 'notae breves' by Florimond de Beaune as well as more extensive 'commentarii' by Frans van Schooten. Christiaan Huygens's 'De ratiociniis in ludo aleae' (1657) was equipped with a full commentary by Jacob Bernoulli when the latter appropriated this treatise, making it the first part of his *Ars coniectandi* (1713).[159] The most ambitious and advanced commentary on a work of early modern science unsurprisingly concerned Newton's *Principia*. Lengthy, elliptic, technical, and novel, this work provoked commentary projects almost from the moment of its appearance in 1687, but by the same qualities frustrated all of them for half a century. Finally, a three-volume commentary based on the third and definitive edition (1726) came out in 1739–42. Originally composed by the two Minim friars Thomas Le Seur and François Jacquier in Rome, supplemented and edited by the Genevan mathematician Jean Louis Calandrini, this effort built bridges between Catholics and Calvinists, Newtonians and Leibnizians, and provided help to generations of readers. It was reprinted, re-edited, reworked, and abridged until 1833 and occasionally is still of use for modern historians of science who need to understand details of Newton's text.[160]

Different kinds of commentaries

Turning from general observations to the commentaries themselves, one finds it difficult to bring some order to a bewildering variety of forms fluently merging into one another. If the genre still shows great internal diversity today, this was the case to an even higher degree in the early modern era. In what follows, some orientation will be provided by a number of parameters with respect to which commentaries differed from one another. Some of these are more or less self-explanatory and have already been mentioned in passing: discipline, level of technicality, length, and (non-)identity of commentator and author of the primary text. Other topics, to be discussed presently, include language and style, the relation of a printed exposition to its classroom origins, its presentation on the page, and its degree of selectiveness and problem orientation. A more detailed discussion and a separate section will be reserved for one of the most

[159] Jan Swammerdam deals in a similar way with the *Circa partes generationis observationum prodromus* (1668) by Johannes van Horne, from which he quotes passages and comments on them in the opening section (pp. 3–29) of his *Miraculum naturae* (1672).

[160] Guicciardini 2015.

striking aspects of early modern commentaries, their widely varying relationship to the text commented on.

The syllogistic reasoning typical of medieval expositions quickly fell into disuse, but apart from that, the commentary proved a difficult terrain for humanist classicism. Many commentators stuck to some degree to the syntactically simplified, lexically uninhibited Latin of their medieval predecessors. As is evident from the elaborate style of their dedications and prefaces, they did so not out of an inability to write Ciceronian periods: they simply found the scholastic idiom an efficient and reliable tool to explain things in class. On average, an early date, close ties to the classroom, and an Aristotelian primary text tended to correspond with a scholastic style, the opposite of these factors with linguistic classicism. Of course, individual preferences of commentators played a role as well, as did editorial choices.

As already mentioned, most commentaries originated in orally delivered lectures, and in many ancient and medieval examples, the written text retains traces of its origin. This did not change fundamentally with the advent of print. Of course, the necessary elaboration of professors' notes or students' 'reportationes' in terms of structure, language, and content could in principle result in a complete loss of the classroom character. However, it was not always deemed necessary or desirable to edit out all traces of the original viva voce performance. After all, most readers were familiar with a commentary's original setting. Some may even have appreciated hearing some echoes of the lecturer's voice in the written version. A frequently preserved structural trait of the lecture was the 'accessus'. In Syllanus de Nigris's *In Almansoris nonum collectiones* (time of origin unclear, printed 1476), for instance, four topics are clarified beforehand (first, unpaginated page): the efficient, material, formal, and final causes of the book under consideration, its title, the part of philosophy it belongs to, and its place in Rhazes's work as a whole. In 1495, Franciscus Capuanus opens his commentary on Georg von Peuerbach's *Theoricae novae planetarum* by discussing 'the aim…or topic of this work, its usefulness, subdivision, structure, and manner of teaching' ('intentionem…seu operis huius subiectum, utilitatem, divisionem, ordinem et viam doctrinae', sig. a2^{r}). In 1554, Giovanni Battista da Monte introduces another commentary on book 9 of the *Liber ad Almansorem* in a similar manner (fols. 1^{r}–6^{v}).[161] It was only towards the end of the sixteenth century that the 'accessus' fell into disuse. Other structural features reminding the reader of the classroom were the contrasting formats of the exegetic and the 'quaestiones' commentary and the occasional division of commentaries into single lessons ('lectiones').[162]

Also in terms of wording and style, editors might try to remain close to the version delivered by the professor. Some students almost saw this as a moral

[161] Maclean 2002, 210–12. [162] Siraisi 1987, 176.

obligation. Da Monte's exposition of Rhazes published by his admiring pupil Walenty z Lublina is a good example. When the title page announces that the commentary has been 'faithfully transmitted' ('fideliter communicata') to the readers, this may, of course, pertain less to linguistic form than to content. In his letter of dedication (sig. iii^{r}), however, Walenty promises to 'fulfil the duty of a true pupil' ('functurus officio veri discipuli') by publishing the lectures in a form 'most unscathed' ('integerrima'), which he contrasts with the practice of bad editors, who not only obscure the structure of the original but also 'fill it with solecisms' ('repleta soloecismis effingunt'). Language concerns obviously mattered to him. That this could imply reproducing the original lecture style up to a point becomes clear from the printed version itself. Walenty refrains from editing out inelegant word repetition, strings of paratactic phrases, subordinate clauses introduced by 'quod' (instead of a classical accusative with infinitive), and even muddled syntax. Da Monte's frequent addresses to his students are preserved as well. Another stylistic peculiarity presumably going back to oral delivery and occasionally preserved in print is the use of 'didactic formulae', as one might call them. In the edition of Oddo degli Oddi's commentary on the first *fen* of Avicenna's *Canon* by the commentator's son Marco, for example, the discussion of possible objections mostly follows the pattern 'you will say...—I thus take the view that it must be answered...' ('dices...—respondendum igitur censeo...'). Similar formulaic language is found in a manuscript commentary on *De partibus animalium* from the 1520s by Pietro Pomponazzi.[163]

In a general sense, the layout of a commentary could reflect its classroom origins as well. In early sixteenth-century Italy, for instance, folio format, two-column setting, and black-letter typeface marked a book as academic.[164] More importantly, however, the distribution of primary text and commentary on the printed page conveyed different ideas of their relationship and influenced the way in which both were read. A number of types can be distinguished (Figure 2a–d).[165] The lemmatized style common in modern philology, in which one or more lemma words are followed by the commentator's explanation, existed but had not yet become standard. It is perhaps no coincidence that the best example known to me, Joseph Scaliger's commentary on Manilius' *Astronomica* (1579), concentrates on philological aspects—a focus shared by comparatively few commentaries in the early modern era.[166] A number of other formats were more widespread: In incunabula and post-incunabula commentaries like Capuanus's on Peuerbach or Vettori's on Albert of Saxony, chunks of text and commentary simply alternate on the page—a reproduction of the way the teacher broke up the text in class,

[163] Perfetti 1999, 444; 2000, 40. [164] French 1985, 71.

[165] On different types of layout in early modern commentaries in general, see the illustrated overview in Enenkel and Nellen 2013, 39–54.

[166] Enenkel and Nellen 2013, 60–1.

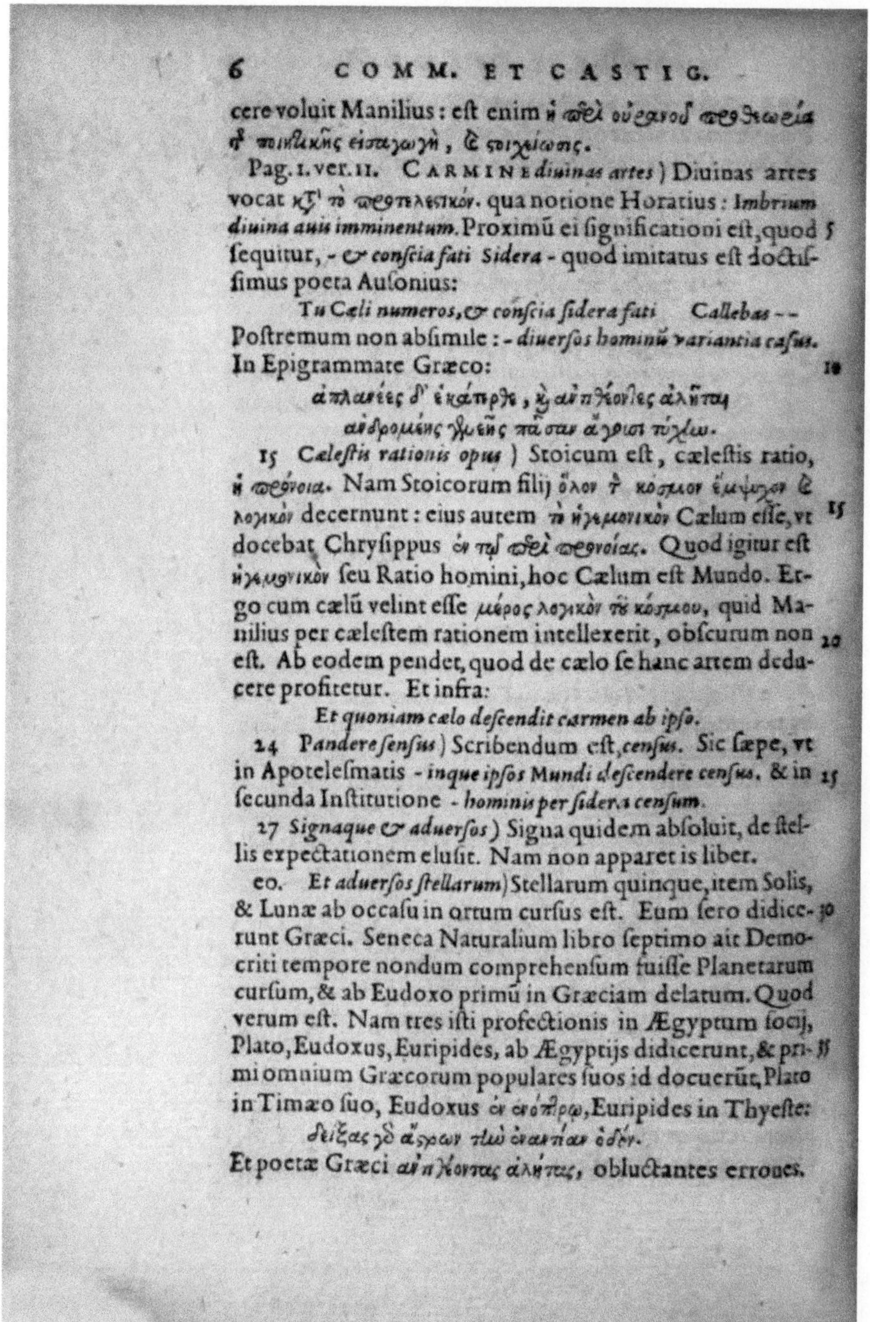

6 COMM. ET CASTIG.

cere voluit Manilius: eſt enim ἡ περὶ οὐρανοῦ προθεωρία τῆς ποικιλικῆς εἰσαγωγῆς, καὶ στοιχειώσεως.

Pag. 1. ver. 11. CARMINE *diuinas artes*) Diuinas artes vocat κτ̓ τὸ προτελεστικόν. qua notione Horatius: *Imbrium diuina auis imminentum*. Proximũ ei ſignificationi eſt, quod ſequitur, - *& conſcia fati Sidera* - quod imitatus eſt doctiſſimus poeta Auſonius:

Tu Cæli numeros, & conſcia ſidera fati Callebas --

Poſtremum non abſimile: - *diuerſos hominũ variantia caſus*. In Epigrammate Græco:

ἀπλανέες δ' ἑκάτερθε, καὶ αἰπήεντες ἀλῆται
αἰθρομένης γενεῆς πᾶσαν ἄγουσι τύχην.

15 *Cæleſtis rationis opus*) Stoicum eſt, cæleſtis ratio, ἡ πρόνοια. Nam Stoicorum filij ὅλον τὸν κόσμον ἔμψυχον καὶ λογικὸν decernunt: eius autem τὸ ἡγεμονικὸν Cælum eſſe, vt docebat Chryſippus ἐν τῷ περὶ προνοίας. Quod igitur eſt ἡγεμονικὸν ſeu Ratio homini, hoc Cælum eſt Mundo. Ergo cum cælũ velint eſſe μέρος λογικὸν τοῦ κόσμου, quid Manilius per cæleſtem rationem intellexerit, obſcurum non eſt. Ab eodem pendet, quod de cælo ſe hanc artem deducere profitetur. Et infra:

Et quoniam cælo deſcendit carmen ab ipſo.

24 *Pandere ſenſus*) Scribendum eſt, *cenſus*. Sic ſæpe, vt in Apoteleſmatis - *inque ipſos Mundi deſcendere cenſus*. & in ſecunda Inſtitutione - *hominis per ſidera cenſum*.

27 *Signaque & aduerſos*) Signa quidem abſoluit, de ſtellis expectationem eluſit. Nam non apparet is liber.

eo. *Et aduerſos ſtellarum*) Stellarum quinque, item Solis, & Lunæ ab occaſu in ortum curſus eſt. Eum ſero didicerunt Græci. Seneca Naturalium libro ſeptimo ait Democriti tempore nondum comprehenſum fuiſſe Planetarum curſum, & ab Eudoxo primũ in Græciam delatum. Quod verum eſt. Nam tres iſti profectionis in Ægyptum ſocij, Plato, Eudoxus, Euripides, ab Ægyptijs didicerunt, & primi omnium Græcorum populares ſuos id docuerũt, Plato in Timæo ſuo, Eudoxus ἐν ὀκτάπρῳ, Euripides in Thyeſte:

δείξας γὰρ ἄστρων τὴν ἐναντίαν ὁδόν.

Et poetæ Græci αἰπήεντας ἀλήτας, obluctantes errones.

Figure 2a–d Four types of commentary layout: lemmatized commentary (Scaliger on Manilius, p. 6), alternating chunks of text and commentary (Capuanus on Peuerbach, sig. [e4v]), text surrounded by commentary (Vadianus on Mela, p. 2), footnotes commentary (Biese on himself, p. 11).

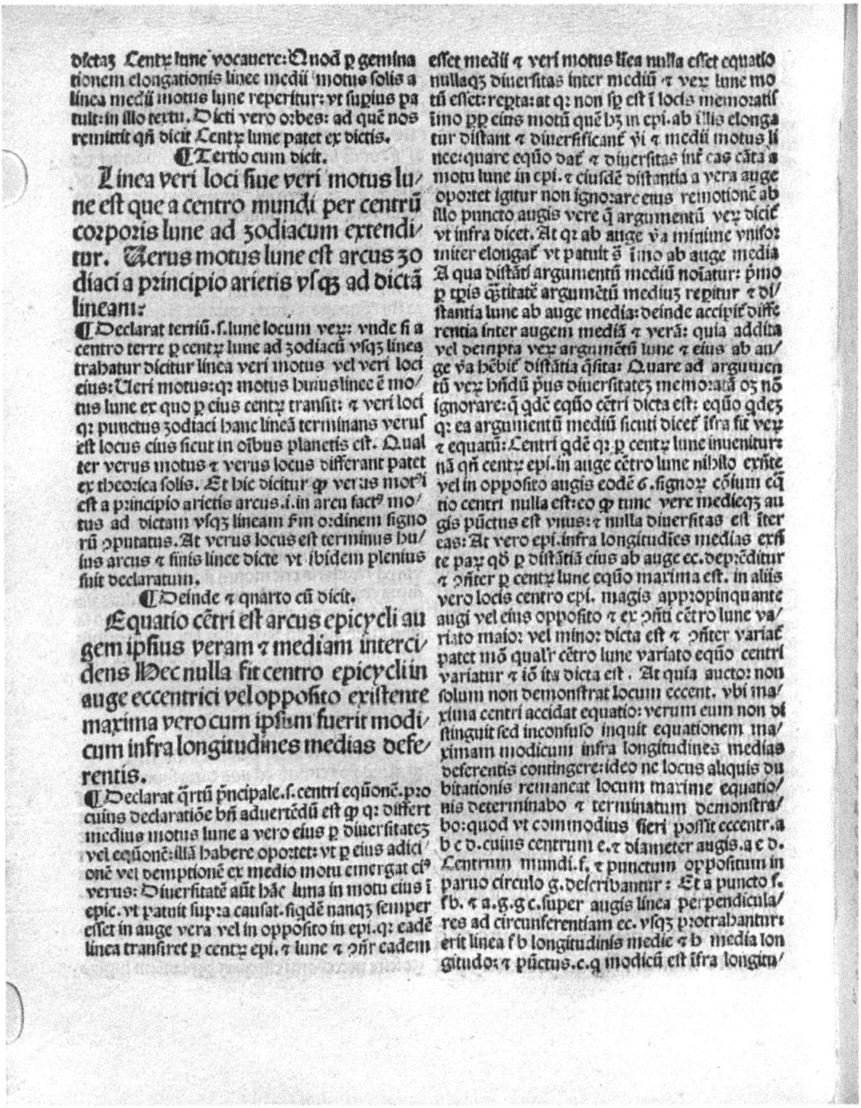
dictaʒ Centꝝ lune vocauere: Quod p gemina tionem elongationis linee mediĩ motus solis a linea mediĩ motus lune reperitur: vt supius pa tuit: in illo textu. Dicti vero orbes: ad quē nos remittit qñ dicit Centꝝ lune patet ex dictis.
Tertio cum dicit.
Linea veri loci siue veri motus lune est que a centro mundi per centrũ corporis lune ad ʒodiacum extenditur. Verus motus lune est arcus ʒodiaci a principio arietis vsqʒ ad dictã lineam:
Declarat tertiũ.s.lune locum veꝝ: vnde si a centro terre p centꝝ lune ad ʒodiacũ vsqʒ linea trahatur dicitur linea veri motus vel veri loci eius: Veri motus: q: motus huius linee ē motus lune ex quo p eius centꝝ transit: ꝛ veri loci q: punctus ʒodiaci hanc lineã terminans verus est locus eius sicut in oĩbus planetis est. Qualiter verus motus ꝛ verus locus differant patet ex theorica solis. Et hic dicitur ꝙ verus mot9 est a principio arietis arcus.i.in arcu fact9 motus ad dictam vsqʒ lineam sm ordinem signorũ ꝯputatus. At verus locus est terminus huius arcus ꝛ finis linee dicte vt ibidem plenius fuit declaratum.
Deinde ꝛ quarto cũ dicit.
Equatio cētri est arcus epicycli augem ipsius veram ꝛ mediam intercidens Hec nulla fit centro epicycli in auge eccentrici vel opposito existente maxima vero cum ipsum fuerit modicum infra longitudines medias deferentis.
Declarat q̃rtũ pncipale.s.centri eqũonē.p10 cuius declaratiōe hn̄ aduertēdũ est ꝙ q: differt medius motus lune a vero eius p diuersitateʒ vel eqũonē: illã habere oportet: vt p eius adictionē vel demptionē ex medio motu emergat ei9 verus: Diuersitatē aũt hāc luna in motu eius i epic. vt patuit sup:a causat. siquidē nanqʒ semper esset in auge vera vel in opposito in epi. q: eadē linea transiret p centꝝ epi. ꝛ lune ꝛ ꝯr̄ eadem esset mediĩ ꝛ veri motus linea nulla esset equatio nullaqʒ diuersitas inter mediũ ꝛ veꝝ lune motũ esset: repta: at q: non sp est i locis memoratis imo pp eius motũ quē hʒ in epi. ab illis elongatur distant ꝛ diuersificant vi ꝛ mediũ motus linee: quare eqũo dat ꝛ diuersitas int eas cãta a motu lune in epi. ꝛ eiusdē distantia a vera auge oportet igitur non ignorare eius remotionē ab illo puncto augis vere q̃ argumentũ veꝝ dicit vt infra dicet. At q: ab auge vã minime vniformiter elongat vt patuit s imo ab auge media ã qua distãtĩ argumentũ mediũ noiatur: pmo p tpis q̃titatē argumētũ mediuʒ repitur ꝛ distantia lune ab auge media: deinde accipit differentia inter augem mediã ꝛ verã: quia addita vel dempta veꝝ argumētũ lune ꝛ eius ab auge vã hēbit distãtia q̃sita: Quare ad argumentũ veꝝ hn̄dũ pus diuersitateʒ memoratã oʒ nõ ignorare: q̃ qdē eqũo cētri dicta est: eqũo qdeʒ q: ea argumentũ mediũ sicuti dicet ifra fit veꝝ ꝛ equatũ: Centri qdē q: p centꝝ lune inuenitur: nã qñ centꝝ epi. in auge cētro lune nihilo existe vel in opposito augis eodē 6. signoꝝ cõium eq̃ tio centri nulla est: eo ꝙ tunc vere mediqʒ augis pũctus est vnus: ꝛ nulla diuersitas est inter eas: At vero epi. infra longitudies medias exñte paꝝ qð p distãtiã eius ab auge ec. dep:ēditur ꝛ ꝯñter p centꝝ lune eqũo maxima est. in aliis vero locis centro epi. magis appropinquante augi vel eius opposito ꝛ ex ꝯñti cētro lune variato maior vel minor dicta est ꝛ ꝯñter variat patet mõ qualr cētro lune variato eqũo centri variatur ꝛ iõ ita dicta est. At quia auctor non solum non demonstrat locum eccent. vbi maxima centri accidat equatio: verum eum non distinguit sed inconfuso inquit equationem maximam modicum infra longitudines medias deferentis contingere: ideo ne locus aliquis dubitationis remaneat locum maxime equationis determinabo ꝛ terminatum demonstrabo: quod vt commodius fieri possit eccentr. a b c d. cuius centrum e. ꝛ diameter augis. a e d. Centrum mundi. f. ꝛ punctum oppositum in paruo circulo g. describantur: Et a puncto f. fb. ꝛ a. g. g c. super augis linea perpendiculares ad circunferentiam ec. vsqʒ protrahantur: erit linea f b longitudinis medie ꝛ b media longitudo: ꝛ pũctus. c. q modicũ est ifra longitu

Figure 2a–d *Continued*

reading one passage, then adding his commentary, before moving on to the next. But the format was also applied in the absence of an actual lecture background, presumably because it was easy to typeset. For instance, Cruser had his translation of two Galenic commentaries printed this way, although Galen wrote them at the request of friends who perused them privately. The 'frame-commentary', which surrounded the primary text on all or some sides, reflecting the way a student annotated it, had been the most common choice in the manuscript age.

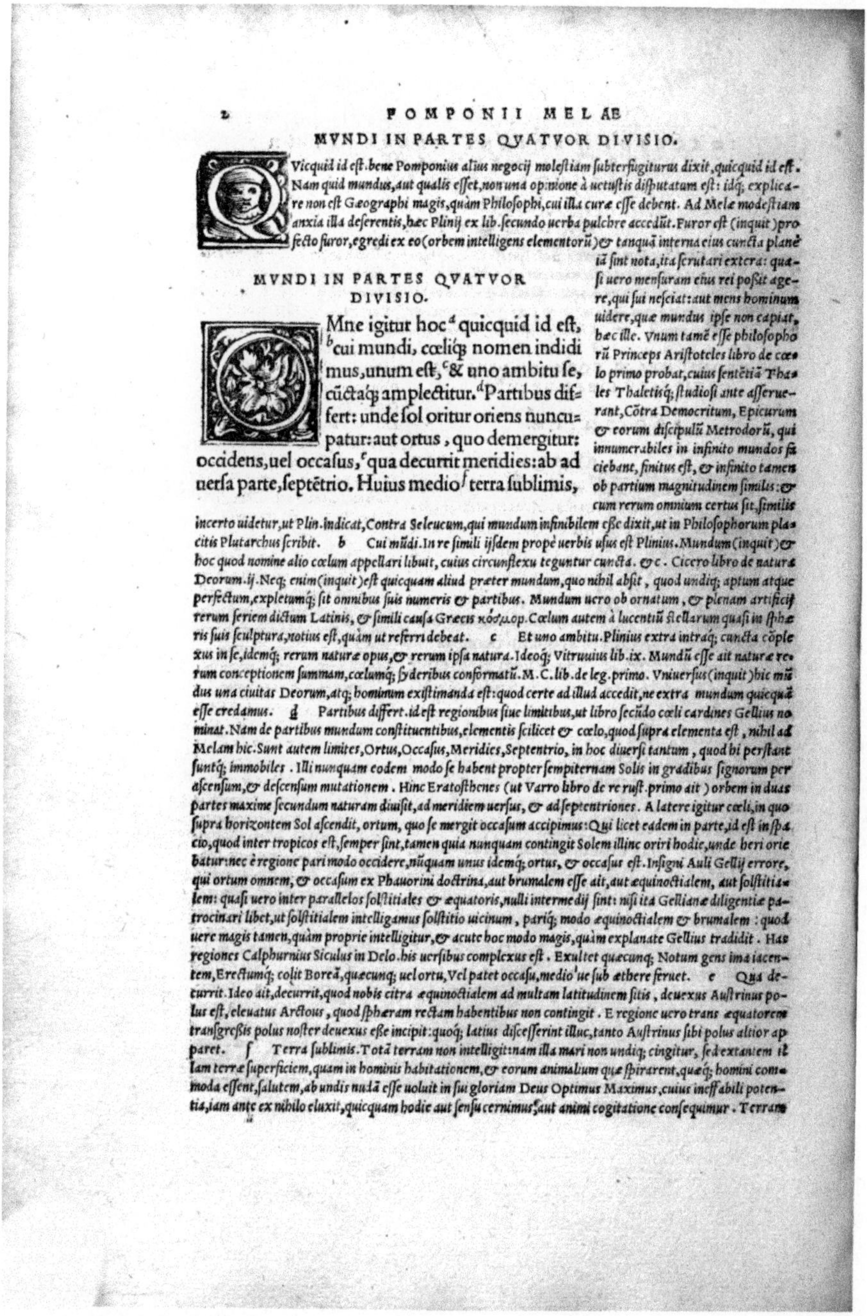

2 POMPONII MELAE

MVNDI IN PARTES QVATVOR DIVISIO.

Vicquid id est. bene Pomponius alius negocij molestiam subterfugiturus dixit, quicquid id est. Nam quid mundus, aut qualis esset, non una opinione à uetustis disputatum est: idq; explicare non est Geographi magis, quàm Philosophi, cui illa curæ esse debent. Ad Melæ modestiam anxia illa deferentis, hæc Plinij ex lib. secundo uerba pulchre accedũt. Furor est (inquit) profecto furor, egredi ex eo (orbem intelligens elementorũ) & tanquã interna eius cuncta planè iã sint nota, ita scrutari extera: quasi uero mensuram eius rei posit agere, qui sui nesciat: aut mens hominum uidere, quæ mundus ipse non capiat, hæc ille. Vnum tamẽ esse philosophorũ Princeps Aristoteles libro de cœlo primo probat, cuius sententiã Thales Thaletisq; studiosi ante asseruerant, Cõtra Democritum, Epicurum & eorum discipulũ Metrodorũ, qui innumerabiles in infinito mundos faciebant, finitus est, & infinito tamen ob partium magnitudinem similis: & cum rerum omnium certus sit, similis

MVNDI IN PARTES QVATVOR DIVISIO.

Omne igitur hoc[a] quicquid id est, [b]cui mundi, cœliq; nomen indidimus, unum est, [c]& uno ambitu se, cũctaq; amplectitur. [d]Partibus differt: unde sol oritur oriens nuncupatur: aut ortus, quo demergitur: occidens, uel occasus, [e]qua decurrit meridies: ab aduersa parte, septẽtrio. Huius medio[f] terra sublimis,

incerto uidetur, ut Plin. indicat, Contra Seleucum, qui mundum infinibilem esse dixit, ut in Philosophorum placitis Plutarchus scribit. b Cui mũdi. In re simili ijsdem propè uerbis usus est Plinius. Mundum (inquit) & hoc quod nomine alio cœlum appellari libuit, cuius circumflexu teguntur cuncta. &c. Cicero libro de naturæ Deorum. ij. Neq; enim (inquit) est quicquam aliud præter mundum, quo nihil absit, quod undiq; aptum atque perfectum, expletumq; sit omnibus suis numeris & partibus. Mundum uero ob ornatum, & plenam artificij rerum seriem dictum Latinis, & simili causa Græcis κόσμον. Cœlum autem à lucentiũ stellarum quasi in sphæris suis sculptura, notius est, quàm ut referri debeat. c Et uno ambitu. Plinius extra intraq; cuncta cõplexus in se, idemq; rerum naturæ opus, & rerum ipsa natura. Ideoq; Vitruuius lib. ix. Mundũ esse ait naturæ rerum conceptionem summam, cœlumq; syderibus conformatũ. M. C. lib. de leg. primo. Vniuersus (inquit) hic mũdus una ciuitas Deorum, atq; hominum existimanda est: quod certe ad illud accedit, ne extra mundum quicquã esse credamus. d Partibus differt. id est regionibus siue limitibus, ut libro secũdo cœli cardines Gellius nominat. Nam de partibus mundum constituentibus, elementis scilicet & cœlo, quod supra elementa est, nihil ad Melam hic. Sunt autem limites, Ortus, Occasus, Meridies, Septentrio, in hoc diuersi tantum, quod hi perstant suntq; immobiles. Illi nunquam eodem modo se habent propter sempiternam Solis in gradibus signorum per ascensum, & descensum mutationem. Hinc Eratosthenes (ut Varro libro de re rust. primo ait) orbem in duas partes maxime secundum naturam diuisit, ad meridiem uersus, & ad septentriones. A latere igitur cœli, in quo supra horizontem Sol ascendit, ortum, quo se mergit occasum accipimus: Qui licet eadem in parte, id est in spacio, quod inter tropicos est, semper sint, tamen quia nunquam contingit Solem illinc oriri hodie, unde heri oriebatur: nec è regione pari modo occidere, nũquam unus idemq; ortus, & occasus est. Insigni Auli Gellij errore, qui ortum omnem, & occasum ex Phauorini doctrina, aut brumalem esse ait, aut æquinoctialem, aut solstitialem: quasi uero inter parallelos solstitiales & æquatoris, nulli intermedij sint: nisi ita Gellianæ diligentiæ patrocinari libet, ut solstitialem intelligamus solstitio uicinum, pariq; modo æquinoctialem & brumalem: quod uere magis tamen, quàm proprie intelligitur, & acute hoc modo magis, quàm explanate Gellius tradidit. Has regiones Calphurnius Siculus in Delo. his uersibus complexus est. Exultet quæcunq; Notum gens ima iacentem, Erectumq; colit Boreã, quæcunq; uel ortu, Vel patet occasu, medio' ue sub æthere feruet. e Qua decurrit. Ideo ait, decurrit, quod nobis citra æquinoctialem ad multam latitudinem sitis, deuexus Austrinus polus est, eleuatus Arctous, quod sphæram rectam habentibus non contingit. E regione uero trans æquatorem transgressis polus noster deuexus esse incipit: quoq; latius discesserint illuc, tanto Austrinus sibi polus altior apparet. f Terra sublimis. Totã terram non intelligit: nam illa mari non undiq; cingitur, sed extantem illam terræ superficiem, quam in hominis habitationem, & eorum animalium quæ spirarent, quæq; homini commoda essent, salutem, ab undis nudã esse uoluit in sui gloriam Deus Optimus Maximus, cuius ineffabili potentia, iam ante ex nihilo eluxit, quicquam hodie aut sensu cernimus, aut animi cogitatione consequimur. Terram

Figure 2a–d *Continued*

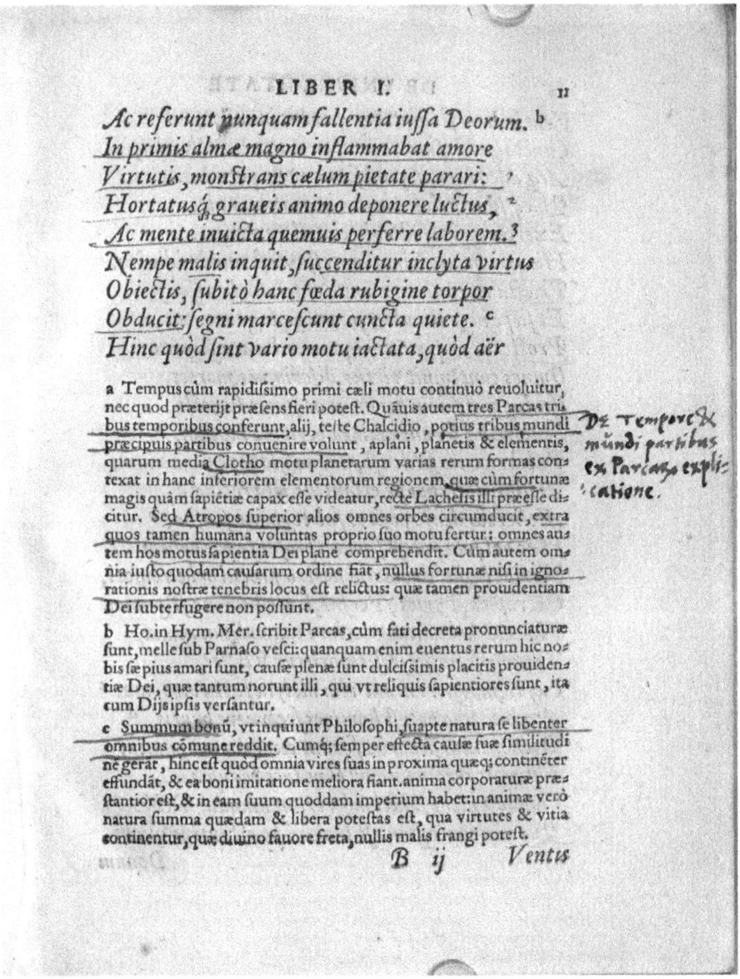

Figure 2a–d *Continued*

Joachim Vadianus's commentary on Pomponius Mela's *De situ orbis* is a lavish example from the era of print, especially in its definitive 1522 edition.[167] In the course of time, it became common to print commentaries as footnotes, often linked to theit respective text passages by letters, as in Biese's *De universitate.*[168] This was a typographically demanding arrangement, but it respected the unity of the text, emphasized its primacy over the commentary, and was reader-friendly, as readers could peruse the text in one go, checking the notes at the bottom only

[167] Suter-Meyer 2020, 84–6. The term 'frame-commentary' is taken from Enenkel and Nellen 2013, 50. On its very common use in Renaissance law, see Maclean 1992, 37–9, 41, 46.

[168] This type of footnote differs from the one studied by Grafton 1997, which relates to the emergence of critical historiography.

when they deemed it necessary. Finally, text and commentary could be separated more completely by printing the latter after the former as a series of endnotes, as in Peletier's edition of Gemma Frisius's arithmetic textbook, or even by printing only the latter, as in the case of da Monte's Rhazes commentary and many early Aristotle commentaries.[169]

In vast majority of cases, text and commentary were set apart typographically, so that the reader could distinguish them at a glance.[170] This differentiation could also be used to establish or underline a hierarchy between text and commentary: the latter was often set in smaller type (Figures 2b–d) or, more rarely, with less spacing, as in Wenceslaus Fabri's commentary on Sacrobosco's *Sphaera* ([*c.*1495]). Alternatively, a different font could be used, for instance italic instead of roman or vice versa.

The medieval distinction between exegetical and 'quaestiones' commentaries found its continuation in commentators' markedly different approaches to the problems presented by a primary text. Most aimed at uniform coverage of all difficulties, while others focused on a smaller number of specific problems, which were discussed in correspondingly greater detail.

The aspects brought up in the first type were so manifold as to defy classification, precisely because commentators tried to deal with every problem that came up in order to enable an understanding of the text as complete as possible. Of course, the aspects in need of clarification depended on the discipline and the overall make of the primary text. Instead of attempting a comprehensive overview, I try to provide an impression of their spread by means of two quite different examples. Pappus' *Mathematicae collectiones* (fourth century AD) is a collection of results from advanced Greek geometry, presented and proven in Euclidean manner, stylistically terse and elliptic in their reasoning.[171] The Greek text has been mutilated and heavily corrupted in the course of transmission. In annotating his 1588 translation, Commandino dedicates himself fully to explaining his emendations[172] and clarifying Pappus' argument. His commentary abounds in entries such as the following, in which the designation of points in a diagram is at issue: 'Graecus codex mendose habebat... πρὸς μθ, cum legendum sit πρὸς μα' (fol. 3ʳ; 'The Greek manuscript wrongly had... πρὸς μθ, when one must

[169] Schmitt 1983, 50.

[170] In the *editio princeps* of Euclid's *Elements* (1482), by contrast, the comments by Campanus of Novara are typographically indistinguishable from Euclid's proofs. This probably does not reflect a lack of skill on the part of the otherwise accomplished and innovative printer Erhard Ratdolt but the then common belief that the proofs were not the work of Euclid either but a kind of commentary by Euclid's late antique editor Theon of Alexandria (Verdonk 1968, 378–9). In fact, both the proofs and Campanus's commentary are printed in smaller type than the theorems.

[171] Greek text: Hultsch 1876–8. Characterization: *CDSB* 10, 294–8. The only book-length study on Pappus and his works is Cuomo 2000.

[172] On textual criticism in the early modern commentary in general, see Enenkel and Nellen 2013, 29–31.

read πρὸς μα'). The lengths Commandino has to go in spelling out all steps in the argument omitted by Pappus are evident on every page. Where Pappus immediately jumps to the conclusion 'ergo recta linea est, quae per FEG transit' ('so the line through F, E, and G is straight', book 7, theorem 58, proposition 61, fol. 196ᵛ), Commandino adds no fewer than ten intermediate steps:

> Iungatur FG secans rectam lineam AD in puncto M. Quoniam igitur anguli FMB, GMC ad verticem sunt aequales et triangulum MFB triangulo MBC simile, ergo ut FB ad BM, ita GC ad CM, et permutando ut FB ad GC, ita BM ad MC. Erat autem ut FB ad BM, ita BE ad EC. Quare et ut BE ad EC, ita est BM ad MC, et componendo ut BC ad CE, ita BC ad CM. Ergo CM ipsi CE est aequalis et punctum M idem quod E. Recta igitur linea est, quae per FEG transit.
>
> [1] Let [the line] FG, which cuts the straight line AD in the point M, be drawn. [2] So, since the angles FMB and GMC are equal at their vertex [3] and the triangle MFB is similar to the triangle MBC, [4] as FB [relates] to BM, so GC to CM, [5] and conversely, as FB [relates] to GC, so BM to MC. [6] However, as FB related to BM, so BE to EC. [7] Therefore [it is] also [the case that] as BE [relates] to EC, so BM relates to MC, [8] and putting this together [it results that] as BC [relates] to CE, so BC to CM. [9] Thus, CM is equal to CE itself [10] and the point M [is] the same as E. So the line through F, E, and G is straight.

Several times per page, moreover, the commentator supplies tacitly presupposed theorems from Euclid's *Elements*, other works such as Apollonius' *Conica*, and earlier parts of the *Collectiones* themselves. Finally, Commandino presumably spent much effort on correcting and supplementing the diagrams without which the written text would be unintelligible.[173]

In elucidating Theophrastus' mineralogical treatise *De lapidibus*, in which different stones and soils are presented and considerable attention is paid to their quarrying and uses,[174] de Laet has to grapple with textual issues as well, but otherwise his problems are of a different nature. Laconic statements and obscure expressions have to be expanded and clarified: 'Per *lapidibus praestantiora* intelligit gemmas, ut in progressu videbimus' (sig. [**1]ʳ; 'By *things more valuable than stones* he understands gems, as we will see in what follows'). As customary in natural historical texts, the identification of individual items presents puzzles. De Laet often has to discuss the identity of this or that stone, and many of his

[173] How much work exactly is unclear. Commandino's son-in-law and posthumous editor Valerio Spacciolo only says that the *Collectiones* have been 'nonnullis exornata figuris' by his father-in-law (letter of dedication, sig. †2ʳ). Hultsch is silent about the state of the diagrams in the manuscript transmission. In general, diagrams were peculiarly prone to corruption or omission. On the title page of his edition of Proclus' commentary on the first book of the *Elements* (1560), Francesco Barozzi claims to have equipped the work with 'figuris, quae in Graeco codice omnes desiderabantur'.

[174] Modern edition with translation and commentary: Eichholz 1965.

comments start with sentences like 'Quaeritur, quae sit haec hyaloeides, id est vitriformis' (sig. ***3^{v}; 'There is an open question, what [stone] this [stone called] "hyaloeides", that is "glassy", [actually] is'). Theophrastus' remarks about the economic and cultural history of minerals need to be contextualized with the help of antiquarian knowledge. In a discussion of quicklime (whose chemical reaction with water produces much heat), the philosopher mentions a ship transporting clothes which become wet for some reason and cause the whole vessel to burn down. This obscure anecdote is elucidated by de Laet as follows: 'Sed observandum est veteres consuevisse vestibus cretam, gypsum et calcem aspergere' (sig. [****8]^{v}; 'But it must be noted that the ancients used to dust clothes with chalk, gypsum, and lime'). To solve textual problems, identify stones, and clarify cultural historical issues, other ancient writers such as Aristotle, Pliny, or Solinus are routinely quoted, as well as much early modern scholarship.

Successors of the medieval 'quaestiones' commentaries were less widespread. That they were recognized as representing a specific type is sometimes indicated by their titles, which may contain formulas such as 'quaestiones super' ('questions about') or 'dubitationes in' ('doubts regarding').[175] Most such expositions were written on works which already had been treated in the 'quaestiones' format in the medieval university. In particular, Aristotle's natural philosophical treatises continued to be discussed in this manner by authors like Chrysostomus Javellus (*Super tres libros Aristotelis De anima quaestiones subtilissimae*, 1552), Domingo Báñez (*Commentaria et quaestiones in duos libros De generatione et corruptione*, 1585), or Pietro Pomponazzi. In his posthumous *Dubitationes in quartum Meteorologicorum Aristotelis librum* (1563), Pomponazzi discusses and resolves 128 questions arising from the fourth book of the *Meteorology*. The argument presented in the individual 'dubitationes' may comprise all or some of the following steps: the identification of the problem, the opinions of earlier commentators (above all Alexander of Aphrodisias, Themistius, Averroës, Albertus Magnus, and Thomas Aquinas), Pomponazzi's own opinion, arguments in its favour, and the refutation of counterarguments. The point under discussion may be the correct interpretation of Aristotle's text, the truth of the matter, or both. Occasionally, Pomponazzi exhorts his readers to reconsider the topic for themselves, as in the short 'dubitatio XXXIII' dedicated to differences between various cooking processes (*Meteorology* 4.2–3):

> Aliud dubium: Dicit Aristoteles, quod frixa sunt assata, quia attrahunt humidum et fiunt humidiora quam elixa, et per hoc conveniunt cum assis, sunt autem humidiora intus et per hoc differunt ab elixatis. Videtur, quod haec ratio non

175 However, such a title alone is no guarantee that a work belongs to this subgenre. For instance, Hartmann Beyer's successful *Quaestiones novae in libellum de sphaera* (1549) is not a problem-oriented commentary on Sacrobosco but a catechism.

valeat, quoniam assati humidum est alterius generis quam frixorum, cum assorum humiditas sit propria, frixorum autem aliena et extrinseca. Super hoc considerate.

Another doubt: Aristotle says that fried victuals are at the same time roasted, because they attract moisture and become moister than boiled victuals. And in this respect[, he says,] they are similar to roasted victuals, but in that they are moister inside, they differ from boiled victuals. It seems that this reasoning is not cogent, because the moisture of roasted victuals is of a kind that differs from the moisture of fried victuals, because the moisture of roasted victuals is their own, whereas the moisture of fried victuals is foreign and external. Think about this.[176]

That treatment by 'quaestiones' was not restricted to Aristotelian works is exemplified by Vettori's exposition of Albert of Saxony's *Tractatus proportionum*. Although an exegetical commentary of each section is also provided, the better part of the work is taken up by the discussion of thirty-three specific problems, highlighted by a 'Dubitatur' or the like, written in large print, in the text and listed in a 'table of doubts' ('tabula dubiorum') at the end. Normal exegesis and problems could also be separated more systematically, which resulted in a 'commentum cum quaestionibus',[177] as done by Niccolò Cabeo in the four tomes of his massive *Commentaria et quaestiones* on the *Meteorology* (1646). While Cabeo ascertains the meaning of Aristotle's text as precisely as possible in his 'commentaria', in the 'quaestiones' he discusses specific problems without feeling bound by the authority of the master.[178]

Beyond the primary text: 'free' commentaries

If the last examples might be perceived as somewhat irritating by the modern reader, because they seem to overstep the limits of a format characterized by its subservience to a primary text, this highlights a point mentioned at the beginning of this section but not pursued further so far: the genre of commentary was much less strictly defined in this respect in early modern times than it is today. This is not to say that there were no commentators whose efforts were first and foremost directed towards the explanation of their primary text. Indeed, many were quite exclusively dedicated to this goal, producing admirable examples of self-effacing

[176] Pomponazzi's efforts to instil scepticism and self-reliant thinking in his audience are evident also from an unpublished commentary on *De partibus animalium* (Perfetti 1999; 2000, 48–63).

[177] Discussed as a distinct type by Perfetti 1999, 444–5.

[178] Martin 2011, 109. For a medieval predecessor of this approach, Gentile da Foligno's *Ars medica* commentary, see Edwards 1976, 287.

explanation. When Capuanus comments on Peuerbach's *Theoricae novae planetarum*, Oddo degli Oddi on the first *fen* of Avicenna's *Canon*, or Commandino and de Laet on Pappus and Theophrastus, their own convictions and preferences vanish, and all their energy is channelled into the elucidation of the text. De Laet even programmatically proclaims this attitude in the first entry of his exposition. Commenting on Theophrastus' initial assertion that metals are formed from the element of water, stones, from earth, he remarks: 'Non est autem nobis propositum disquirere, an haec illius sententia vera sit necne; maioris enim hoc operae foret; tantum ipsius mentem explicamus' (sig. [**1]^{r}; 'But it is not my purpose to examine whether this opinion of his is true or not, for this would require more work: I only explain his thinking.')

However, such a statement only made sense if alternatives existed. Indeed, their hard and diligent work in the service of a primary text did not keep many commentators from pursuing their own agenda as well. While elucidating the text's meaning, they also tried to shed new light on the subject under discussion.[179] In addition to its role as 'explanandum', the work commented upon became a starting point for the commentator's own musings. At least one prominent contemporary theorist of commentary, Girolamo Cardano, found that a good commentator was not only free but even obliged to make a contribution of his own, when he determined his four duties as follows (*De libris propriis*, 1562, p. 94; Maclean 2004, 352):

> auctoris mentem declarare, demonstrare quod ita sit quod dicitur, amplificare illius doctrinam et tueri ipsum a calumniatoribus.
>
> to explain the opinion of the author, to prove that the things said [by him] are indeed so, to add to his learning, and to defend him against detractors.

All but the first of these tasks regard the factual content of the work commented upon and the truth of the matter rather than the meaning of what is said.[180] Cabeo provides a textbook example of this approach, but the phenomenon was by no means restricted to 'quaestiones' commentaries. In fact, many commentators were more radical than Cabeo, minimizing or even ignoring the explanatory function of their work. As noted initially, commentaries had been a customary medium for discussing problems and developing positions already in antiquity

[179] The distinction between text- and subject-centred commentary is due to Paulus 2005, 139. On the emancipation of early modern commentaries from their primary texts in general, see Enenkel and Nellen 2013, 8–12.

[180] It is true that not all early modern critics agreed with Cardano. Juan Luis Vives, for one, had mixed feelings in this respect. He noted as a matter of fact that those who compose longer commentaries may count as authors in their own right ('non tam alieni operis expositores dici debeant quam auctores proprii') but also condemned long excursuses and admonished commentators: 'Dum enim in alium scribis, illi est ubique subserviendum' (*De ratione dicendi*, 1533, pp. 231–2). Erasmus found fault with Guillaume Budé's overly digressive practice of commentary (Céard 2012, 11).

and the Middle Ages. To cite an early mathematical example, Proclus' commentary on the first book of the *Elements*, translated by Francesco Barozzi in 1560, is selective in commenting on theorems and proofs, but rich in general reflections, not only on Euclid's manner of thinking and on mathematical terminology but also on the philosophy of mathematics, an aspect about which Euclid is altogether silent.[181] In early modern times, self-assertive expositions of scientific texts multiplied. As the evolution of scientific disciplines accelerated, such commentaries became a medium for coping with and furthering developments.

The trend towards the 'free' commentary, as one might call it, in science made itself felt from the sixteenth century, as a couple of instances may illustrate.[182] Da Monte's massive commentary on the ninth book of Rhazes's *Liber ad Almansorem* (1554) picks out only nine of ninety-five diseases treated in the book. While discussion of them—first practice-oriented, following the Empiric school, and then more theoretical according to Dogmatic and Methodic principles—is extensive, Rhazes's text is not given and mostly not even referenced by means of lemmata. In fact, da Monte does not mention the Persian physician at all for long stretches. His work reads like a series of independent monographs inspired by the topics of some of Rhazes's chapters. Pedro Nuñez's remarks on Peuerbach's *Theoricae novae planetarum* (1566) also proceed without their primary text, which is at best paraphrased. A series of extensive 'annotationes', 'propositiones', 'lemmata', and 'corollaria' focus on the application of Peuerbach's planetary theory in—often quite advanced—astronomical calculations. In Christopher Clavius's highly successful edition of the *Sphaera* (1570), which went through more than twenty editions in the decades that followed, Sacrobosco's booklet of a few dozen pages gains encyclopedic proportions due to the commentator's contributions.[183] While Clavius does explain Sacrobosco, he invests much more energy into bringing his treatment up to the present state of knowledge. In this way, his commentary and the massive early modern tradition of Sacrobosco commentaries as a whole, made it possible to keep the medieval work as an academic textbook when it had long become glaringly outdated in a period of rapid astronomical progress. Ingrassia's commentary on the Galenic *De ossibus ad tirones* proves Galen wrong and sometimes right on numerous points. His index contains about one hundred entries titled 'Galeni error' ('Galen's mistake') or similar, but also 'Galeni defensio' ('vindication of Galen') or 'Galeni detractorum error' ('mistake of Galen's detractors'). In addition, Ingrassia presents several findings of his own, most prominently the smallest bone in human anatomy, the stirrup bone in the ear.[184]

[181] Morrow 1970, pp. xxiv–xliii.

[182] In addition, see Siraisi 1987, 178–9 on *Canon* commentaries.

[183] The editions are listed in the database 'Sphaera CorpusTracer' produced in the project *The Sphere* (see note 133 above). On the commentary's printing history, see Maclean 2022, 187–90, 198–210, 216–17, 219–20; on its use in Jesuit mathematical teaching, see Grendler 2022.

[184] Dispenza et al. 2013. As early as 1521, da Carpi had used his commentary on Mondino's *Anatomia* as an outlet for his own anatomical discoveries (French 1985, esp. 51).

As may be guessed from the works just mentioned, 'free' commentaries could use their freedom in various ways. Once more, classification varies from difficult to pointless, but some frequently taken directions can be indicated. In what follows, I illustrate three of them, marked by digression, supplementation, and criticism, by means of one typical example for each.

The concise world geography by Pomponius Mela, who wrote around AD 40, was popular in early modern times. In 1518, the Swiss humanist Joachim Vadianus, who had lectured on the work in Vienna, published the first commented edition of it. A much expanded second and definitive version (from which I quote in what follows) came out in 1522.[185] As indicated by four more editions up to 1564, the book enjoyed some success. Of the 220 densely printed folio pages filled with primary text and exposition in the second edition, about one tenth is taken up by the former, the rest by Vadianus's comments surrounding it in a smaller font (Figure 2c). Of these, only a minor part—for instance, explanations of geographical terminology—is needed to clarify Mela's remarks, which are mostly quite clear in themselves. Another part adds geographical information left out by Mela, from individual cities or rivers to the existence of the New World. Last but not least, a very substantial portion of the commentary is made up of digressions on a broad variety of topics—from the salt mines of Bochnía and Wielíczka in Poland and the Antipodes to the style of Solinus, the self-blinding of Democritus (to escape distraction from his philosophical musings), and the psychology of dreaming.[186] This exuberant encyclopedism is self-conscious: when Vadianus in a concluding address to the 'candidus lector'[187] counters the possible reproach that he has composed 'a wordy commentary' ('verbosa commentaria'), he makes no attempt at styling Mela as a difficult author in need of long explanations, but simply invokes the 'practically endless immensity' ('immensitate quasi intermina') of all that there is to know in the world.

An example may illustrate how Vadianus constructs his excursuses. On p. 34 of the second edition, he narrates an ascent of Mount Pilatus near Lucerne undertaken by him in the summer of 1518. The point of departure of this digression is Mela 1.34 (p. 33), where a cliff in the Cyrenaica is described which is sacred to Auster (the South Wind) and causes sandstorms when touched by a human. After discussing the power of sandstorms in general, Vadianus turns to the question whether such freak weather can really be caused by human transgressions. Among the examples leading him to answer in the affirmative, there is the

[185] Vadianus's commentary is presented and analysed in exemplary fashion by Suter-Meyer 2020. See in particular pp. 74–82 on his digressions in general and pp. 204–20 on the example discussed below.

[186] Polish salt mines: pp. 180–3; Antipodes: pp. 7–8; Solinus: p. 74; Democritus: p. 105; dreaming: pp. 37–44.

[187] The numerous paratexts, which fill some 130 pages, are divided between the first and last pages of the book in a quite unusual way. The one-page address to the reader follows immediately after the commentary.

so-called 'lacus Pilati' situated on the northern slopes of the homonymous mountain. It is said to house the corpse and ghost of Pontius Pilate, who reacts with terrible thunderstorms when frivolously disturbed by some object thrown into the lake. The wish to see this miraculous lake with his own eyes and put the stories about it to the test motivates Vadianus's climb up the mountain, which is narrated in rich and captivating detail. On his way to the top, he indeed passes the lake, which is unspectacular in terms of size and depth ('palus verius quam lacus', 'swamp rather than lake') but nonetheless evokes Gothic awe by its 'black, underworldly water' ('aqua nigricans, Phlegetontaea'), the 'bristly wood' ('horrida silva') surrounding it, and the 'deep silence' ('altum silentium') reigning. Impressive as all of this is, the crucial experiment of throwing something into the lake cannot be conducted, because the shepherd guiding Vadianus and his friends panics and implores them not to put his life at risk. Thus, the ascent of Mount Pilatus is no help in deciding whether Mela's account of the North African cliff is true or not. Nonetheless, Vadianus obviously found his adventure too good to go unrecorded, and his readers agreed. His hiking tour was cited several times by other Swiss proto-alpinists from the sixteenth century and earned him some fame as a mountaineering pioneer in recent times.[188]

In natural history, which grew rapidly by the accretion of new material, commentaries added to primary texts not so much through excursuses as through the description of further plants, animals, or minerals.[189] Because of the additive nature of his work, Dioscorides was an author particularly well suited to this approach. Among the numerous commentaries on him, the most successful by far came from the pen of the Venetian doctor, botanist, and personal imperial physician Pietro Andrea Mattioli. Its first version (1544) was in Italian, but the 160 pages of this edition were superseded by a completely rewritten Latin version more than four times as voluminous in 1554. This was in turn greatly expanded in 1565, resulting in a definitive edition comprising 1,222 folio pages of text and commentary, not counting over 150 pages of introductory material and some 70 pages of text and commentary on the spurious sixth book of the *Materia medica*. Between them, the two Latin versions went through eleven editions and provided the basis of sixteen more in German, French, and Czech.[190]

In the final Latin version, Dioscorides's text (taken from Jean Ruel's 1516 translation but revised in some places in the light of new textual findings)[191] accounts for under 10 per cent of the volume's bulk. The rest is filled with Mattioli's

[188] See, for instance, Conrad Gessner, *Descriptio Montis Fracti* (1555), pp. 55–7; Coolidge 1904, pp. xxxviii, 23**–25**.

[189] A thorough analysis of botanical commentaries on Pliny and Dioscorides in the Renaissance is provided by Beer 2014; see esp. pp. 342–51 on Mattioli's Dioscorides commentary.

[190] *CTC* 4, 92–6; Stannard 1969; Bäumer 1991–6, vol. 1, 182–5.

[191] *CTC* 4, 18.

commentary and woodcut illustrations.[192] The commentary often contains an independent description of the substance in question or expands on Dioscorides', adds medical uses and other cultural historical information, adduces ancient parallels (in particular from Pliny and Galen), summarizes and comments on early modern discussions, and lists each item's names in Greek, Latin, Arabic, and the modern vernaculars. Above all, however, it adds new items related to the one under consideration or bearing a similar name, characterizing them, in the manner of Dioscorides himself, in terms of habitat, appearance, medical powers, and usage. How much is actually subjoined to Dioscorides's *c.*800 vegetable, 100 animal, and 100 mineral items is difficult to tell. At the end of the introductory paratexts, Mattioli provides a 'catalogus imaginum plantarum et animalium' ('list of the images of plants and animals') comprising some 1,100 entries, but since not every plant or animal described gets a woodcut and minerals never do, this list is far from complete. Moreover, it is often unclear whether an item added by Mattioli is a new species, a subspecies, or just a phenotypic variety, particularly in those frequent cases in which a Dioscoridean plant, animal, or mineral is subdivided: for instance, while Dioscorides knows of only one kind of valerian, alum, and narcissus, Mattioli has three ('phu magnum', 'parvum', and 'minimum'), six, and eight respectively (pp. 39–42, 616, 1240–7). Overall, one gets the impression that the number of simples described by Dioscorides has at least doubled.

Expositions criticizing and correcting their primary texts can be found in considerable numbers from the sixteenth century. Avicenna and Sacrobosco, for instance, were carped at by their Renaissance interpreters.[193] In the decades around 1700, Descartes suffered a similar fate at the hands of Samuel Clarke, a pupil and translator of Newton, although the work censured was not by the French philosopher himself. Rather, Clarke found fault with Jacques Rohault's *Traité de physique* (1671), a popular introduction to Cartesian physics translated by him. The translation went through four editions between 1697 and 1718, was reprinted on the Continent until 1740, turned into English in 1723, and became 'the outstanding scientific textbook in England (and America) in the first half of the eighteenth century'.[194] From the start, Clarke added 'adnotationes' to his Latin version, and this commentary grew together with the work's success. The title page of the fourth and definitive Latin edition explains that the translator 'has extended and adorned' the text 'with notes drawn for the most part from the philosophy of the highly famous Isaac Newton' ('adnotationibus ex illustrissimi Isaaci

[192] The practice of adding fresh illustrations to a commentary is not common, but examples can be found in other disciplines besides botany, for instance in da Carpi's anatomical commentary on Mondino (French 1985, 61–2) and perhaps also in Commandino's on Pappus.

[193] For Avicenna, see Siraisi 1987, 189–91 and 200–2, who speaks of 'the use of the commentary as a vehicle of attack on the commented-upon author' (p. 189). For Sacrobosco, see the text before note 183 above on Clavius's commentary.

[194] Sarton 1948, 141–5 (quotation on p. 144). On Rohault and his *Traité* in general, see *CDSB* 11, 506–8; Clair 1978, 5–8 (bibliography of editions), 17–99 (biography).

Newtoni philosophia maximam partem haustis amplificavit et ornavit'), which are now 'expanded by half' ('dimidia parte auctiores'). The argument and tone of these notes may be exemplified on the basis of a fundamental tenet of Cartesian physics stated early on by Rohault. At the beginning of chapter 8 of part 1 (p. 41), he states: 'De ce que nous venons d'establir touchant l'essence de la matiere, nous conclurons premierement que le vuide des Philosophes est impossible', which is faithfully translated by Clarke (4th edn, p. 24): 'Ex his, quae de natura materiae posuimus, colligere licet primo inane quod vocant philosophi nullum esse posse.' However, since Newton holds the exact opposite opinion, Clarke immediately corrects Descartes's and Rohault's error in a long footnote running over to p. 25 of which it takes up three-quarters:[195]

> 1. Inane nullum esse posse] Consentaneum hoc quidem ei dicere, qui essentiam esse materiae extensionem dixerit. Verum ex gravitatis natura, quae infra breviter exponetur, constat iam omnino aliquod inane, et multo id quidem maximum, in rebus esse. Praeterea demonstratur inane, ut iam dixi, ex cometarum motibus.... Idem quoque constat ex corporum pendulorum vibrationibus.... Quod hic commentus est Cartesius materiae suae subtilis tenuitatem in causa esse, ne eius resistentia sensibilis sit..., ineptissimum est et tum rationi tum experientiae contrarium. Demonstravit enim illustrissimus Newtonus mediorum fluidorum densitatem proportione respondere resistentiae ipsorum quam proxime (Optic. p. 311)... Quare si nullum esset spatium inane, sequeretur corpus proiectum in aëre vel in loco, unde exhaustus sit aër, non minus difficile moveri debere quam in argento vivo.... Cum igitur materiae essentia non in extensione, sed in soliditate impenetrabili posita sit, dicendum est mundum universum constare ex corporibus solidis, quae moveantur in inani.
>
> 1. That there can be no void] It is consistent to say so for him [Descartes] who said that the essence of matter is extension. But from the nature of gravity, which will briefly be explained below, it is completely certain that there is some void in the world and that it even takes up the biggest part of it by far. Moreover, the existence of void is proven, as I already said [p. 22 n. 2], from the movement of the comets.... The same is also certain from the oscillations of pendula [which continue indefinitely in an artificial vacuum].... What Descartes has fabricated in this respect, namely that the tenuousness of his fine matter should be the reason why its resistance cannot be perceived..., is most stupid and contradicts reason as well as experience. For the veryy famous Newton has shown that the density of liquid media is most closely proportionate to their resistance (*Optice*, p. 311)... If no empty space existed, it therefore would follow that a body set in

[195] For an even more extended example, see the note on pp. 312–15 explaining that the vortex theory is false and must be replaced by Newtonian gravitation.

> motion would have to move with no less difficulty in air or a place from which the air has been sucked out than in quicksilver.... So since the essence of matter does not consist in extension but in impervious solidity, one has to say that the entire world is made up of solid bodies moving in the void.

Clarke did not write like this out of malice. Nor was he motivated by a perverse desire to keep a worthless book in print: quite the contrary: he wanted to save a work which he thought still valuable despite its partly misguided basic assumptions. In view of the rise of Newtonianism, which established itself as a powerful competitor to Cartesian physics in the first decades of the eighteenth century, Rohault's prolonged success was at least partly due to Clarke's commentary, which kept his work up to date—a small-scale analogue to the aforementioned salvation of Sacrobosco by Clavius and others.[196]

'Free' commentaries such as these refused to subordinate themselves to the primary text. They formed a new, symbiotic unity with it or even overgrew it, using it as a convenient framework within which to pursue their own ends. No longer meant to be read in the first place for the light they shed on the primary text but for their own content, they in some sense became works in their own right, as may be seen from various indications. Sometimes it suffices to open a commentary to get an impression of the relative importance of primary author and commentator. On the title page of the *Commentarii in sex libros Dioscoridis*, Mattioli, placed on top in monumental capitals, eclipses the Greek pharmacologist below, whose name is set in a smaller font and normal script. Poems of praise in commented Sacrobosco editions likewise tend to assign the books to their respective commentators rather than to the medieval author.[197] The addition of an index is a sign that a commentary follows its own encyclopedic aspirations and caters to readers looking for a work of reference. Da Monte's commentary on Rhazes and Vadianus's on Mela feature indexes of *c*.300 and 2,000 entries respectively. Both are dwarfed by Mattioli's elaborate index, which contains some 5,000 entries and carefully distinguishes Mattioli's contributions from those of Dioscorides.[198] The publication history of many 'free' commentaries reminds one of encyclopedic works as well. Their growth from edition to edition was not due to efforts to explain the primary text more and more precisely but to the conviction that the more information the better. Some of them became objects of supercommentaries, although this happened less often in science than in other fields like Aristotelian

[196] A counterexample is provided by Melchior de Polignac's didactic poem *Anti-Lucretius*, the Cartesian content of which was generally perceived as outdated at its belated and uncommented edition in 1747 (Fleischmann 1965, 61–3).

[197] I owe this information to Irina Tautschnig (personal communication).

[198] On the encyclopedic character of early modern commentaries in general and the use of indexes, see Enenkel and Nellen 2013, 23–9, 54–7; on Vadianus's index, see Suter-Meyer 2020, 61–6.

philosophy or rabbinic scholarship.[199] The Roman physician Bartolomeo degli Emanuelli commented upon the commentary on *Canon* 4.1 by his famous late medieval colleague Gentile da Foligno (1524). Barozzi added 'scholia' (in the form of marginalia) to his edition of Proclus' Euclid commentary, as announced on the title page. Vadianus's commentary is followed by a 'locorum aliquot ex suis in Melam commentariis explanatio', an apologetic explication of fifteen passages from it, which runs to sixty pages. Finally, it could happen that a commentary simply shed its original format and crossed over into another genre. A shortened version (*c.*900 quarto pages) of Mattioli's commentary came out under the title *Compendium de plantis omnibus* in 1571[200] and was republished in 1586 as *De plantis epitome utilissima*. In both editions, Dioscorides's text and any trace of the original commentary character have vanished, and the work has become a classic handbook.

Technical dictionaries

Questions of definition

This section is about Latin dictionaries, but both parts of this designation are in need of clarification. Dictionaries (or lexicons) are even more difficult to distinguish from related forms—such as simple glossaries, catalogues, encyclopedias, natural histories, or textbooks—than many other genres,[201] and while 'Latin' is mostly an uncontroversial characterization, this is not the case here. Some preliminary definitions are thus in order. An ideal-typical dictionary may be defined as (i) a self-contained work (ii) structured by lemmata defined by words which are (iii) arranged in alphabetical order[202] and (iv) contain not only factual but also lexical information.[203] A dictionary typically consists of three types of text modules: the usual paratexts, lemma words, and their description in the entries.

[199] Commenting on Aristotle often implied commenting on Averroës's commentaries as well (Hasse 2016, index, s.v. 'supercommentaries'; Sahr 2023, 88–9, 95). The Tanakh and Talmud commentaries by the French rabbi Rashi (1040–1105) attracted, and continue to attract, hundreds of supercommentaries (and super-supercommentaries); see Lawee 2006.

[200] Stannard 1969, 72.

[201] On the theory and practice of lexicography in general, see Zgusta et al. 1971; Hausmann et al. 1989–91. For a history (and some attempts at a definition) of the dictionary, see Green 1996, 20–4. On the difficulties of distinguishing dictionaries from neighbouring genres, see Klein 1998–9, 1910–11. These difficulties become evident, for example, in the attempts of Tonelli 1971 (esp. 1–2) to distinguish between real dictionaries and 'pseudo-dictionaries, dictionaries not ordered alphabetically, or of minor importance'.

[202] On the early history of alphabetical arrangement, see Daly 1967.

[203] The contrast between language dictionaries and subject dictionaries (encyclopedias) and the problems attached to it are discussed by Hupka 1989. Most works discussed in the following combine both aspects, but one like Pietro Bongo's encylopedia of numerology, *Numerorum mysteria* (1599), pays no attention to the numeral words and can thus be excluded from this chapter.

To call a dictionary a Latin book, lemma words in that language are neither a necessary nor a sufficient condition.[204] By contrast, Latin as the sole or main language of the paratexts and the descriptions in the entries is.

Historical overview

Dictionaries are among the oldest text types preserved in writing.[205] Lexical lists were repositories of knowledge and didactic tools in Mesopotamia from the late fourth millennium BC. In classical antiquity, the earliest lexicons presumably originated in Hellenistic Alexandria. They first recorded glosses, that is, difficult words from important authors and their explanations. In ancient and medieval manuscript culture, lexicography and glossography maintained close links, as dictionaries collecting glosses were used to gloss texts in turn. Early Greek lexicons often dealt with specific authors. Under the Roman Empire, the elevation of the Attic dialect of the fifth and fourth centuries BC to good Greek *tout court* gave rise to a number of Atticistic lexicons from which this linguistic norm could be learned. The major Latin work in the field was *De verborum significatu* by the Augustan grammarian Verrius Flaccus, which survives in digests only. In the Byzantine era, the results of ancient Greek lexicography were collected into monumental dictionaries, the best-known of which, the *Suda* from the tenth century, lists some 30,000 entries. In the West as well, glossographical and lexicographical activities continued through the Middle Ages, culminating in massive dictionaries like Papias's *Elementarium doctrinae rudimentum* (mid-eleventh century) and John of Genoa's *Catholicon* (1286). The latter became one of the first books ever printed around 1460, followed after some decades by Ambrogio Calepino's *Dictionarium* (1502), the most influential lexicon of the early modern era.

A fraction of this production was specifically concerned with aspects of the natural world. The grammarian Erotianus (first century AD) wrote a *Collection of Words in Hippocrates*, Galen, an *Explanation of Hippocrates' Difficult Terms*. Heron of Alexandria (first century AD), better known for his descriptions of technical playthings, authored a terminologically oriented introduction to Euclidean geometry. In the Byzantine Empire, medical and botanical glossaries were compiled. In the West, a late medieval chymical dictionary went by the name of the

[204] The latter is obvious from the numerous dictionaries with Latin lemma words which are still published today.

[205] For the following, see in general Green 1996, 39–54; Klein 1998–9. For Mesopotamian word lists in particular, see Bienkowski and Millard 2000, 181, s.v. 'Lexical texts'. On ancient and medieval glossography and lexicography, see *DNP* 4, cols. 1097–1101 and 7, cols. 123–6, s.vv. 'Glossographie' and 'Lexikographie'; *LMA* 4, cols. 1507–15 and 5, cols. 1933–5 s.vv. 'Glossen, Glossare' and 'Lexikon'; *CHS* 2, 359 n. 74 (astronomical, astrological, medical glossaries). *DNP* 15/1, cols. 126–49, s.v. 'Lexikographie' informs about the reception of ancient lexicography in medieval and modern times.

grammarian and poet John of Garland. Above all, however, there were a considerable number of alphabetical collections of medical and pharmaceutical terms, some of which appeared in print in incunabula and sixteenth-century editions.[206] An impressive example is the *Pandectae medicinae* (*c.*1317) by the physician and botanist Matthaeus Silvaticus, a huge pharmacopoeia and medical dictionary. Because of his predominantly Arabic terminology, Silvaticus was scolded by some of his early modern successors such as Otto Brunfels (*Ὀνομαστικὸν medicinae*, 1534, letter to the reader):

> In quo tertia fere semper dictio corrupta vitiataque aut certe barbara, obsoleta vel exotica est. Nec paulo diversa omnia ab magorum et necromanticorum praestigiosis nomenclaturis, ut nescias rerumne an malorum potius daemoniorum sint nomina.
>
> In his work, more or less every third word is corrupt, perverted, or in any case barbarian, outdated, or outlandish. The whole does not differ, even a little, from the deceptive jargon of wizards and necromancers, so that one is unsure whether [the lemma words] are designations of [medical] substances or rather of evil demons.[207]

The humanists' interest in language in general and their intent to purify technical terminology of Arabic and other 'barbarian' accretions in particular were one reason for the proliferation of early modern lexicography in the natural sciences.[208] However, they were not the most important one, nor were they long-lived. The impetus given to them seems to have slowed in the sixteenth century. More concrete needs and possibilities played a greater part in the development. One had to come to terms with the language of newly discovered Greek authors. Contemporary writers began to coin a plethora of neologisms for new things and notions. Print spread the results of lexicographical efforts and provided generations that followed with a strong basis on which to continue. (Like most 'discrete' texts, lexicons are seldom written from scratch; most are corrected and expanded versions of or compilations from earlier lexicons.[209]) Printing technology also developed typographical and layout techniques such as markup of lemma words, insertion of paragraphs, indentation, or headlines, which made dictionaries more user-friendly and efficient. In combination with an upsurge in interest in the

[206] For a short overview of the field, see Osler 1923, 32–6. There were also collections of natural historical vocabulary from Aristotle ('tabulae'; Grabmann 1939, 124–55), but these were very full indexes rather than lexicons.

[207] Some more criticism of Silvaticus can be found in Osler 1923, 35.

[208] A bibliographical overview of sixteenth- to eighteenth-century Latin and vernacular dictionaries from the natural sciences and other fields can be found in Tonelli 1971.

[209] Green 1996, 27–30. On the moral and legal implications of lexicographical copying, see Zgusta 1988.

natural world, all of this led to unprecedented lexicographical productivity in the various scientific disciplines, which only continued to grow over the centuries.

The development is all the more remarkable because dictionaries were a genre intensely loathed by their authors. Compiling a lexicon was—and even in digital times still is—dull, tiresome work which, nonetheless, demanded high focus. In his aforementioned letter to the reader, Brunfels describes lexicography as the most tedious kind of writing: 'Inter tam multa scribendi argumenta et genera…nescio an ullum molestius extet quam lexicorum commentarios scribere.' That this is more than just self-fashioning as a martyr to science is indicated by Joseph Scaliger's even more drastic words (*Poemata omnia ex museio Petri Scriverii*, 1615, p. 35):

> Si quem dura manet sententia iudicis olim,
> damnatum aerumnis suppliciisque caput,
> hunc neque fabrili lassent ergastula massa,
> nec rigidas vexent fossa metalla manus:
> lexica contexat, nam cetera quid moror? Omnes
> poenarum facies hic labor unus habet.
>
> When someone condemned to tribulations and punishments once awaits the judge's harsh sentence, may no workhouse wear him out with the blacksmith's material nor digging for ore torture his stiff hands. May he compile dictionaries! What do I care about the rest? All facets of punishment are united in this one toil.[210]

That dictionaries were, nonetheless, written in great numbers highlights their indispensability.

Contents: early modern scientific dictionaries from different disciplines

Lexicons appeared in most major fields of the study of nature,[211] but due to its institutional background, medicine came first and developed the strongest tradition by far. In the wake of late medieval dictionaries, some new (that is, freshly compiled) lexicographical works were printed in the late fifteenth century. Saladinus de Asculo's *Aromatariorum compendium* (1488), for instance, contains a word list of the technical terms in the *Antidotarium Nicolai*, an influential twelfth-century pharmacopoeia (sigs. b2^{v}–b5^{r}). Fully fledged lexicons followed a

[210] That the poem was popular is indicated by a German adaptation on the second-last page of the introduction to Johann Christoph Sommerhoff's *Lexicon pharmaceutico-chymicum* (1701).

[211] A first impression of the disciplines covered by dictionaries from the invention of print is conveyed by Tonelli 1971, 39–41.

little later,[212] starting with Symphorien Champier's *Vocabulorum medicinalium epitoma* (1508). Terminological difficulties with the ancient authors encouraged efforts in the realm of iatrophilology. Henri Estienne's *Dictionarium medicum* (1564) represented an important step towards mastering the Greek lexicon of the discipline.[213] The success of such works contributed to a prevalence of Greek vocabulary in medicine to the present day. As a subject of author-specific dictionaries, Hippocrates remained popular and found a sixteenth- and seventeenth-century successor in Paracelsus, whose readers were much in need of help because of his wanton creation of neologisms.[214] Gerhard Dorn's *Dictionarium Theophrasti Paracelsi* (1573; second, expanded edition 1583) is one example among half a dozen. Medical lexicons not restricted to one particular author were far more numerous. Some of them, like Philipp Andreas Nemnich's *Lexicon nosologicum polyglotton* (1801), focused on certain aspects such as diseases. Others even encompassed auxiliary and neighbouring disciplines, in particular pharmaceutics, botany, and chymistry. In the full title of his *'Ονομαστικὸν medicinae*, Brunfels boasts that the work covers forty areas from these disciplines as well as from metrology, natural philosophy, magic, and astronomy/astrology. Johann Christoph Sommerhoff's *Lexicon pharmaceutico-chymicum* (1701) includes medical, chymical, botanical, zoological, and mineralogical objects and notions. A contributor of a laudatory poem to this dictionary apparently found it easy to come up with the following list of medieval and early modern medical lexicographers (third-last page of the introductory paratexts):

> Scripserunt equidem medicorum lexica quondam
> Toxites, Stephanus, Brunsfellus atque Rulandus,
> Gorraeus, Thurneisserusque, Foësius, Alstedt,
> Johnsonus, Lycius, Sylvaticus et Dieterici,
> Mullerus, Blancard, Castellus, Bruno: quis omnes
> enumeret?
>
> In fact, Toxites, Estienne, Brunfels and Ruland, Gorris, Thurneysser, Foësius, Alsted, Jonson, Lycius, Silvaticus, Dietrich, Müller, Blankaart, Castelli, Bruno once wrote dictionaries for physicians: who could rehearse them all?

212 Some inspiration may have come from the academic sister disciplines of theology and law, whose first real lexicons were published as early as the 1470s and 1480s (Tonelli 1971, 39). Five Latin dictionaries are included in Michael Stolberg's microfiche edition of early modern medical dictionaries and encyclopedias (Stolberg 1993–2006).

213 Its medical material was taken over into Estienne's monumental *Thesaurus Graecae linguae* (1572), on which, see Pfeiffer 1968–76, vol. 2, 110.

214 Klein 1998–9, 1918; Müller 2001, 490–9. For a list of Hippocratean and Paracelsian dictionaries, see Tonelli 1971, 41. (For reasons unknown, no specifically Galenic dictionaries seem to be on record. The two items listed by Tonelli under Galen's name are of a more general nature.)

In addition to the sheer number of authors and works, many medical dictionaries were hugely successful. To stay with the authors just enumerated, Bartolomeo Castelli's *Lexicon medicum Graecolatinum* (1598) and Steven Blankaart's *Lexicon medicum Graeco-Latinum* (1679) were reissued for nearly two centuries, until 1795 and 1832 respectively. Their editions constantly expanded, which testifies to the proliferation of the medical vocabulary in the early modern period. Blankaart's dictionary in particular grew from a moderate volume of some 300 pages to two bulky tomes comprising between them over 1,800 pages in much denser print.[215]

Among the remaining disciplines, chymistry was tied to medicine via the iatrochemical movement represented by Paracelsus and his followers, but its traditionally vague and ambiguous terminology made it a special case. Starting with the work ascribed to John of Garland, first printed in 1560, and with the Paracelsian dictionaries, many attempts were made to bring order into the Babylonian confusion of chymical nomenclature. Martin Ruland the Younger's *Lexicon Alchemiae* (1612) is a prominent example, but it was no more successful than any other. A few decades later, an anonymous author was able to publish a polemical dictionary of alchemical nonsense terms, *Non-entia chymica* (1645), which was reprinted twice and was read and used by Robert Boyle.[216]

While some aspects of natural history were included in medical dictionaries, only a handful of lexicons specifically concerned with this field seem to exist. The confusion created by the many different names for animal and especially plant species was not much less grave than in chymistry, but before a commonly accepted solution was found by Linnaeus, the problem was often tackled with the help of synonym lists in the discipline's massive 'historiae'. However, a number of pertinent dictionaries, for example Conrad Gessner's quadrilingual *Catalogus plantarum* (1542)[217] or Caspar Bauhin's monumental dictionary of botanical synonyms, the *Πίναξ theatri botanici* (1623), became famous and influential.[218]

Names for rocks and minerals were partly accommodated in pharmaceutical and chymical dictionaries. Works like John Jonston's *Notitia regni mineralis* (1661), which dealt with the subject from a natural historical perspective, existed as well. Something like a mineralogical and geological nomenclature in Latin, corresponding to the terminology of mining which already existed in German and other vernaculars, was coined by Georg Agricola, who concluded some of his works with substantial glossaries. The 'Interpretatio Germanica vocum rei metallicae' styled as a letter to Agricola's friend Wolfgang Meurer, which rounds off the

[215] On later editions and reworkings of Castelli, see Tonelli 1971, 11–12, and the preface to the tenth issue of Stolberg 1993–2006 (http://www.haraldfischerverlag.de/hfv/reihen/AEL/ael_3-10_einleitung.php, accessed 16 February 2023).

[216] Thorndike 1929–58, vol. 7, 196–7 (including a discussion of the authorship); Principe 1998, 55.

[217] Müller 2001, 483–5.

[218] Slaughter 1982, 57–8; Reeds 1991, 125–7.

great collection of his writings published in 1546 in Basel (pp. 469–87), is the most important of these.[219]

The areas covered by dictionaries titled 'mathematical' mostly comprise several parts of the quadrivium. Konrad Dasypodius's *Λεξικὸν seu dictionarium mathematicum* (1573) provides the basic terminology of all four mathematical disciplines plus the practical art of surveying ('geodaesia'). Accounting ('logistica') is announced on the title page as well but actually gets no proper section. Even more untrue to its title is the *Lexicon mathematicum, astronomicum, geometricum* (1668) by Girolamo Vitali, which deals primarily with astronomy and astrology, then with music, geography, and astrological medicine, but hardly at all with geometry and arithmetic.

Forms: the genre and its fringes

At the beginning of this section, an ideal-typical dictionary was characterized as (i) self-contained, (ii) lemmatized, (iii) alphabetized, and (iv) oriented towards words, not just things. One need not doubt that a genre broadly conforming to this definition existed in early modern times. Many books can intuitively be recognized as dictionaries from a glance at their layout and the perusal of a couple of lemmata. Early modern authors themselves were conscious of the genre. They identified such books as particularly tedious to compile, as already mentioned, and they gave genre-specific titles to the results of their drudgery. Roughly one half of the works discussed here are titled 'lexicon' (from λέξις, 'word'), 'dictionarium' (from 'dictio'; perhaps a loan translation of 'lexicon'), 'onomasticon' (from ὄνομα, 'name'), or 'nomenclator' ('name-caller'). In addition, some individually chosen titles such as Ole Borch's *Lingua pharmacopoeorum* (*Language of Pharmacists*, 1670) indicate the work's generic affiliation as well.[220]

Nonetheless, the number of borderline cases failing to conform to one or more of the above criteria or being exceptional in some other way is striking. Many scientific lexicons were not published as individual works, but appeared either appended to a more substantial work or in combination with other works. The former applies to Agricola's glossaries, for instance, the latter to Champier's *Vocabulorum medicinalium epitoma* (1508), which is buried in the midst of a confused miscellany of works by Champier and others. In Josse de Harchies's *Enchiridion medicum* (1573), a lexicon of simples is combined with a piecemeal didactic poem on the same topic.[221] Some works which qualify as dictionaries with regard to their intention to familiarize people with vocabulary from a certain field, for instance Joachim Camerarius's anatomical and physiological *Commentarii*

[219] Naumann 2007, 85; Berrens 2020, 236 n. 5.
[220] For some other typical titles, see Green 1996, 20.
[221] Müller 2001, 488–90.

utriusque linguae (1551),[222] lack the typical lemmatic structure, although most of these, like Franz Joseph Märter's terminological introduction to Linnaean botany titled *Fundamenta et termini botanici* (1789), are constructed of some kind of small, typographically separated units. A fortiori, such texts are not alphabetized but arranged thematically (Camerarius, for one, proceeds 'a capite ad calcem'). The same is also true of some works which do consist of lemmata. Bauhin, for instance, opts for thematical instead of alphabetical order in his *Πίναξ theatri botanici*. He can afford to do so, because readers can find items by name with the help of an index. By contrast, the content of Johannes Kentmann's Latin-German 'Catalogus rerum fossilium' (1565), which is also thematically structured, is made accessible by an equally thematical table of contents only. Although it looks like an ordinary dictionary at first sight and may have been used as such, this text in fact originated as a catalogue of the author's personal collection of 'fossilia'. Gessner's *Nomenclator aquatilium animantium* (1560)[223] belongs to a group of dictionaries which explain the phenomena denoted by their lemma words in such detail that one feels tempted to assign them to some more fact-oriented genre. Lavishly illustrated, discussing over 700 water animals in systematic order, and including much factual information, while keeping its focus on their designations, this work looks like a 'historia animalium' in spite of its typical lexicon title and is in fact a combination of both genres. Sometimes one and the same work appears to be a simple wordbook and an encyclopedia depending on the entry one looks at. In Ruland's *Lexicon Alchemiae*, one finds one-word German translations, short essays, and anything in between.

Even while remaining more or less within the limits of the ideal type, lexicographers proved inventive and created idiosyncratic forms. The Greek lemma words in Henri Estienne's *Dictionarium medicum*, for instance, are not explained or translated in isolation. Rather, their meaning is made clear by their use in extended quotations from medical authors, given in the Greek original and in Latin translation but without any further comment. Claude Saumaise's *De homonymis hyles iatricae* (1689) is a collection of concise, extremely learned essays on pharmaceutical homonyms, which exhibits all characteristics of a dictionary except alphabetization but at the same time clearly belongs to the humanist genre of the philological 'miscellanea' initiated by Poliziano.[224]

Languages, users, and uses

In most of the works discussed so far, the interplay of different languages is worthy of attention.[225] Although most scientific lexicons of the period qualify as

222 Baier 2017. 223 Pinon 1995, 98–9.
224 Pfeiffer 1968–76, vol. 2, 44–5.
225 On Latin and German in early modern dictionaries, see Klein 1998–9, 1917, 1920–3. On bilingual and multilingual dictionaries in general, see Hausmann et al. 1989–91, vol. 3, 2711–3119.

'Latin' in the sense outlined at the beginning, complete Latin monolingualism is the exception rather than the rule. Forms of multilingualism are manifold, as they involve not only lemma words but paratexts and entries as well. From the possible combinations of the linguistic treatment of these parts, a broad range of varieties results.

The languages involved alongside Latin fall into two groups: the classical languages (ancient Greek having pride of place), on the one hand, and the modern vernaculars on the other. Most of the linguistic range can be exemplified by one early and one late work, Gessner's *Nomenclator aquatilium animantium* (1560) and Nemnich's *Lexicon nosologicum polyglotton* (1801). The index to Gessner's work comprises the following sections: Latin, Hebrew and Arabic, Greek, Italian, Spanish, French, German, English, Hungarian, 'Illyrian' (all Slavic tongues understood as one single language with different dialects), American and Canary languages. In Nemnich's dictionary of diseases, German, Dutch, Danish, Swedish, English, French, Italian, Spanish, and Portuguese are represented alongside Latin.

Greek obviously is a common ingredient in lexicons dealing with specific Greek authors but also with disciplines having a strong Greek background. Regarding the former, Anutius Foësius's influential *Oeconomia Hippocratis* (1588) may be mentioned beside Estienne's dictionary. It has Latin paratexts, Greek lemmata, and Latin entries interspersed with numerous Greek references from the Hippocratic Corpus and other authors such as Erotianus and Galen. In his *Λεξικὸν seu dictionarium mathematicum*, Dasypodius made a radical choice to do justice to the overwhelming importance of Greek mathematics. As already suggested by its bilingual title, this non-lemmatized dictionary comprises two sections which are precise translations of each other. The first gives and explains the mathematical terminology exclusively in Latin, the second, in Greek. Despite the fundamental importance of the Arabic tradition, no scientific lexicons featuring comparable combinations of Latin and Arabic (or Hebrew, for that matter) exist. Humanist prejudices against 'barbarian' terminology and the difficulty of printing Arabic presumably combined to prevent such works.[226]

Combinations of Latin and the vernaculars appeared at an early date, as shown by Gessner's lexicography, and they became more common and varied as time went on. The simplest way to add vernaculars to a Latin dictionary was to enrich only the entries with information on vernacular designations while keeping the whole rest of the book in Latin, as in Gessner's work on water animals. Vernacular-Latin lexicons could be created by simply translating the lemma words of a purely Latin or Greco-Latin lexicon. Blankaart's *Lexicon medicum Graeco-Latinum*,

[226] Arabic designations are discussed in a few works like Saumaise's *De homonymis hyles iatricae* (where they are also printed in Arabic script).

for instance, was apparently published in a German-Latin version in 1705.[227] At about the same time, bidirectional lexicons, which combined a Latin-vernacular and a vernacular-Latin section, became popular. The bidirectional character could be highlighted in the title, as in Sommerhoff's *Lexicon pharmaceutico-chymicum Latino-Germanicum et Germanico-Latinum*. More complex, because more polyglot, is the structure of Nemnich's *Lexicon nosologicum polyglotton*, which falls into ten sections. In the first, Latin lemma words are provided with equivalents in nine vernaculars, in the remaining nine, terms from these vernaculars are rendered into Latin. Neither in Sommerhoff nor in Nemnich is the relationship between Latin and the vernaculars symmetrical. In Sommerhoff, many Latin lemmata in the so-called Latin-German section in fact lack a German translation, and the German-Latin section is much shorter than its counterpart. In Nemnich, the prevalence of Latin is evident from its function in the lead section and its presence in all ten sections. More egalitarian is the presentation of Latin, Greek, German, and French plant names in four columns, spread over two facing pages, in Gessner's *Catalogus plantarum*. Still, the alphabetical order of the words is determined by Latin. Moreover, additional information is given in the Latin and German columns, while many plant names are lacking in the French column (Figure 3).

Bilingualism or multilingualism in the paratexts is a rarer phenomenon, but it occurs. Sommerhoff mixes German and Latin in the introductory section of his lexicon. A Latin title and dedication are followed by a German address to the 'dear, well-disposed reader' ('Lieber Günstiger Leser') and poems of praise in both languages. Part two features a German title page of its own. The Viennese pharmacist Friedrich Müller von Löwenstein and his publishers, the Endter brothers in Frankfurt, simultaneously brought out his *Lexicon medico-galeno-chymico-pharmaceuticum* (1661) in two editions featuring a Latin and a German title page and letter to the reader respectively,[228] but being completely identical otherwise. The two versions primarily cater to academic physicians and pharmacists respectively, as is clearly indicated by the wording of the full titles. The Latin version promises to be useful 'to those learning medicine as well as to those who confer fame to the art of Paion [a Greek god of healing]' ('tum medicinam addiscentibus, tum Paeoniae arti nomen dantibus'), the German one, 'particularly to pharmacists' ('absonderlich denen Apotheckern').[229]

[227] Klein 1998–9, 1920 (however, I have been unable to find this version in the common online bibliographies). Conversely, the lemma words were kept in Greek and Latin with all the rest translated into the vernacular in an English version (1693).

[228] In the German version only, a dedication to the Estates of Lower Austria is included as well.

[229] The different ways in which some notions are phrased in the addresses to the reader is revealing as well. 'There are few true pharmacists' is expressed with a learned quotation from Plato's *Phaedo* (69c) in the Latin version: 'pauci Bacchi, sed multi thyrsigeri'. In the German edition, this becomes the more down-to-earth 'nicht alle / die lange Messer tragen gute Köche seyn' (both quotations from the books' respective first pages).

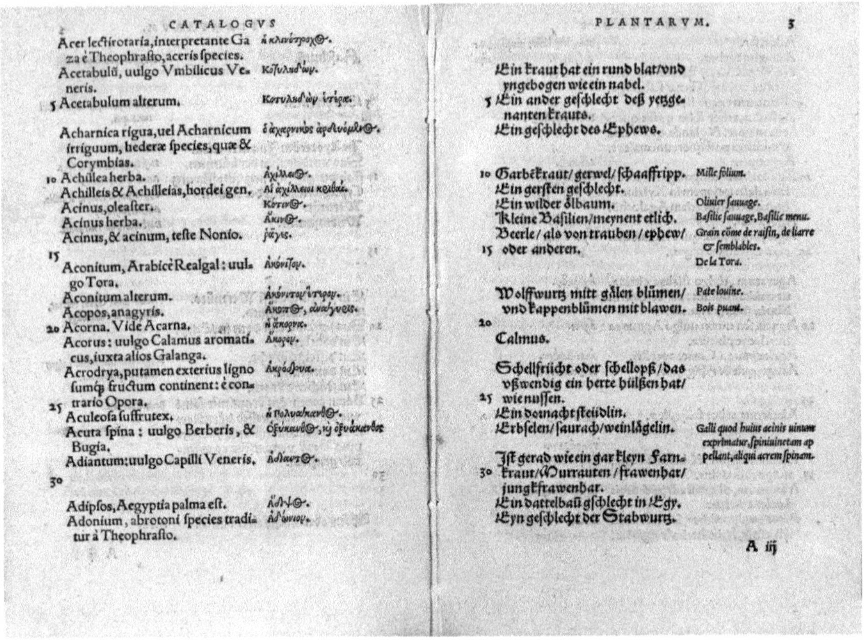
CATALOGVS

Acer lectirotaria, interpretante Gaza è Theophrasto, aceris species.
Acetabulũ, uulgo Vmbilicus Veneris. Κοτυληδών.
Acetabulum alterum. Κοτυληδὼν ἑτέρα.
Acharnica rigua, uel Acharnicum irriguum, hederæ species, quæ & Corymbias.
Achillea herba. Ἀχίλλειος.
Achilleis & Achilleias, hordei gen. Αἱ ἀχίλλειοι κριθαί.
Acinus, oleaster. Κότινος.
Acinus herba. Ἄκινος.
Acinus, & acinum, teste Nonio. ῥᾶγες.
Aconitum, Arabicè Realgal: uulgo Tora. Ἀκόνιτον.
Aconitum alterum. Ἀκόνιτον ἕτερον.
Acopos, anagyris.
Acorna. Vide Acarna.
Acorus: uulgo Calamus aromaticus, iuxta alios Galanga. Ἄκορον.
Acrodrya, putamen exterius ligno sumq; fructum continent: è contrario Opora. Ἀκρόδρυα.
Aculeosa suffrutex.
Acuta spina: uulgo Berberis, & Bugia.
Adiantum: uulgo Capilli Veneris. Ἀδίαντος.
Adipsos, Aegyptia palma est. Ἄδιψος.
Adonium, abrotoni species traditur à Theophrasto. Ἀδώνιον.

PLANTARVM. 5

Ein kraut hat ein rund blat/vnd yngebogen wie ein nabel.
Ein ander geschlecht deß yetzgenanten krauts.
Ein geschlecht des Ephews.
Garbekraut/gerwel/schaaffripp. Mille folium.
Ein gersten geschlecht.
Ein wilder ölbaum. Oliuier sauuage.
Kleine Basilien/meynent etlich. Basilic sauuage, Basilic menu.
Beerle/als von trauben/ephew/oder anderen. Grain cõme de raisin, de liarre & semblables.
De la Tora.
Wolffwurtz mitt gälen blůmen/vnd kappenblůmen mit blawen. Pate louine. Bois puant.
Calmus.
Schellfrücht oder schellopß/das vßwendig ein herte hülßen hat/wie nussen.
Ein dornacht steüdlin.
Erbselen/saurach/weinlägelin. Galli quod huius acinis uinum exprimatur, spiniuinetam appellant, aliqui acrem spinam.
Ist gerad wie ein gar kleyn Farnkraut/Murrauten/frawenhar/jungkfrawenhar.
Ein dattelbaũ gschlecht in Egy.
Eyn geschlecht der Stabwurtz.

A iij

Figure 3 A spread (fols. 2ᵛ–3ʳ) from Gessner's *Catalogus plantarum*: plant names in Latin, Greek, German, and French.

The obvious advantage of multilingual lexicons lay in their appeal to a broader audience comprising readers of a different mother tongue, linguistic capacity, and, in some cases, disciplinary expertise. Authors were conscious of this. Gessner explains his decision to present German and French plant names in the *Catalogus plantarum* with his wish 'that the book may be useful in many ways, bringing benefit not to one people only and not to the learned alone either, but also to beginners' ('ut πολύχρηστος esset liber, non solum uni genti, neque doctis tantum, sed tironibus quoque usui futurus'); Greek, on the other hand, was included 'to please the learned' ('in eruditorum gratiam'; letter of dedication, sig. αiiᵛ). Michael Toxites's and/or Johann Fischart's multilingual 'Onomasticon Theophrasti Paracelsi' (1574)[230] claims on its title page to be helpful to students of Paracelsian medicine 'of any nation whatsoever' ('cuiuscumque nationis sint').

Together with other factors such as organization, layout, and format, the roles of Latin and other languages not only impacted on who used a dictionary, but also on how they did this. Of course, one and the same work could be intended to serve different ends, and readers will have put books to uses unforeseen by their

[230] Tonelli 1971, 32, attributes the Paracelsian lexicon to Fischart alone, but the terms in which both Fischart and Toxites are speaking in their initial addresses to the dedicatee and the reader respectively are too vague to permit any certain conclusions about this.

authors.[231] Nonetheless, two core functions of early modern scientific dictionaries can be determined with reasonable certainty: first, help with reading pertinent literature (the function responsible for their inclusion in the present section) and, secondly, help with the composition of scientific texts.

Greek-Latin, Latin-Latin, and Latin-vernacular dictionaries will have served first and foremost as reading aids. They helped readers to cope with the unfamiliar terminology of rediscovered ancient authors as well as that of contemporary writers, from whose pens a torrent of new words to express new notions was gushing forth. Girolamo Mercuriale advised students of medicine to equip themselves with lexicons before embarking on their private reading programme.[232] In this respect, scientific lexicons were no different from other dictionaries, which were also philological tools in the first place. It thus makes sense that such works were sometimes written not by representatives of their respective disciplines, but by Greek and Latin scholars, as in the case of Estienne's *Dictionarium medicum*. However, men of science knew about the need for lexical tools as well, as evidenced by Heron's terminological introduction to the perusal of Euclid and its translation by Dasypodius. As Dasypodius explains in the preface to his 1579 version, Heron wrote the work 'so that we may be equipped with a plentiful command of things and words, the tools of insight, so to speak, and gain an easier understanding of what we read in books of this kind [i.e. such as Euclid's *Elements*]' ('ut copia et facultate rerum et verborum instrumentis quasi cognitionis instructi facilius quae legimus in eiusmodi voluminibus intelligamus'). Müller von Löwenstein promotes his lexicon as 'very useful for an easier grasp of the more recondite writings of Galenic as well as chymical inspiration' ('ad scripta tam Galenica quam chymica obscuriora facilius capessenda perquam utile') on the title page of the Latin version.

In works tailored to the terminology of one specific author, the philological orientation is particularly evident. Dorn stresses this aspect in the first sentence of his note to the reader (*Dictionarium Theophrasti Paracelsi*, p. 9):

> Ad Paracelsi quoscumque libros intelligendum ex eius mente genuina in primis ac prae ceteris maxime tibi necessarium erit singula, quibus ludere suo more delectatur suis cum discipulis interdum, vocabula peregrina probe et ad unguem cognita habeas.
>
> In order to understand any book by Paracelsus whatsoever according to his true intention, you will first and foremost have a great need to gain a sound and precise knowledge of each and every unusual word with which he takes delight in playing together with his pupils, as is his manner.

[231] Even today it is difficult to ascertain by whom, how, and to what ends dictionaries are used. For an overview of the many possibilities, see Kühn 1989.

[232] *De modo studendi* (*c.*1570), ed. Durling 1990, lines 143–51.

The *Onomasticon Theophrasti Paracelsi* is advertised in the German version of its title as leading 'to…a better understanding and useful perusal of Theophrastus's books as well as others' ('zu…richtigerem verstand und förderlicher lesung der Theophrastischen und sonst bucher'). In the same way, an author like Georg Agricola could help readers to a better understanding of his own works. In all these cases, the lexicon or glossary became a kind of lexical (self-)commentary.

Clear indications are lacking whether and to what degree Latin-vernacular dictionaries were also used for the composition of scientific texts in the vernaculars, but the idea does not seem unreasonable in itself. After all, a scientific nomenclature had to be developed in each vernacular before that language could be used as a linguistic medium for science. It appears natural to assume that dictionaries played an important role in this process. In later times, vernacular-Latin lexicons very probably fulfilled a double function, working as reading tools for texts in foreign vernaculars and providing help with the composition of scientific texts in Latin. Their emergence in the latter part of the early modern era makes good sense in terms of the linguistic developments in science. By that time, the vernaculars had massively intruded into scientific writing, and scientists were complaining about the difficulty of mastering several vernaculars in order to read the literature. At the same time, much of the steadily increasing mass of new scientific publications still had to be in—increasingly specialized—Latin for commercial reasons, institutional constraints, or prestige purposes. Moreover, the scientific community began to grow apart from the 'respublica litteraria' and accommodated more and more members capable in their field but in need of help with Latin.

Apart from reading and writing, were scientific dictionaries also intended to help people with conversation (be it in Latin or in the vernaculars)? Lack of fluency with technical terms in general was sometimes perceived as a problem. As Hadrianus Junius expounds in the preface (sig. †3r) to his *Nomenclator omnium rerum* (1567), an influential dictionary of such terminology from all fields:

> Etenim locuturi de re quapiam etiam illi, qui magnam in litteris sui opinionem excitarunt, ad generalia communiaque confugiunt vocabula vel barbaris et peregrina e lingua petitis utuntur vocibus vel gravem sibi ipsi turpitudinem infligunt, dum tamquam in salebris haerentes caput scabiunt, unguem arrodunt neque se explicare valent.
>
> For when they are supposed to speak about some topic, even those who have made people think highly of their learning take refuge in generic, common expressions, use barbaric words taken from some foreign language, or bring heavy shame upon themselves, being stuck in a pothole, so to speak, scratching their head, biting their nails, and finding no escape.

That people might have used scientific dictionaries to spare themselves such embarrassment is conceivable in principle. If this did indeed happen, however, it was usually passed over in silence. At least one prominent physician and chymist, however, found it worth his while to write a dictionary which exclusively aimed at a specific aspect of the correct use of terminology in speaking. Ole Borch's *Lingua pharmacopoeorum* teaches pharmacists not the meaning, but the pronunciation of some 400 difficult expressions for medications and their ingredients. Pharmacists were situated on the fringes of academia, often lacked a proper education in Latin, and were held in disrespect by many academic doctors. The ability to pronounce his drugs' names correctly mighy well have enhanced a pharmacist's social standing.

Finally, it bears mentioning that a number of scientific dictionaries were conceived as introductory textbooks. This is less astonishing than it sounds at first. Even today, the contents and terminology of a scientific field are learned together. In early modern times, the acquisition of higher learning was closely interlinked with that of the learned language, Latin. Schoolbooks like Johannes Honterus *Rudimenta cosmographica* (1542) or Comenius's famous *Orbis pictus* (1658) testify to the habit of conveying 'res' and 'verba' together. Accordingly, Dasypodius's *Λεξικὸν seu dictionarium mathematicum* provides a first overview of the mathematical disciplines at large. In its dedication to the young student Johann Baptist Hainzel (sig. iii^{r}), Dasypodius characterizes the booklet as a short propaedeutics to the field. Märter's *Fundamenta et termini botanici* was written for the author's students at Leuven as a companion piece to an introductory course. As the title intimates, the work simultaneously conveys the conceptual foundations and the basic terminology of Linnaean botany. While one may object that these works represent borderline, non-lemmatized forms, Jean Baptiste Callard de LaDucquerie's *Lexicon medicum etymologicum* (1692) is fully lemmatized and alphabetized. Nonetheless, as detailed on the first two pages of the note to the reader, the author read the manuscript to a class of first-year medical students at the University of Caen and subsequently published it at the request of a later generation of students. He still understands the printed book as an introduction to be read through rather than as a simple reference work: 'Qui igitur artis Apollineae orgiis audacter initiari flagrant, in huius libri lectionem serio incumbant.' ('So may those possessed by the ardent desire to be bravely initiated to the mysteries of Apollo's art dedicate themselves seriously to the perusal of this book!'[233])

[233] For a similar attitude in a preface to a thematically unspecific dictionary, see Basilius Faber's injunction to read his *Thesaurus eruditionis scholasticae* (1571) through several times from beginning to end: 'Proinde ad assiduam eius lectionem docentes et discentes etiam atque etiam cohortamur. Eorum enim solum studiis proderit hic Thesaurus, qui ab initio ad finem usque aliquoties attente et diligentissime integrum evolverint volumen' (fourth page of the preface to the reader; cf. *DNP* 15/1, cols. 134–5). For modern dictionaries meant to be read through and serving as textbooks, see Kühn 1989, 119–21.

Bibliographies

General remarks

A bibliography was usually called a 'bibliotheca', a virtual library, in early modern times.[234] Nonetheless, the word 'bibliographia', very rare and simply meaning 'writing books' in ancient Greek, got its modern meaning in this period (to be more precise, in mid-seventeenth-century France).[235] This is fitting, since the thing as it is known today originated in early modern times as well: a text, ideally a self-contained book, exclusively providing an alphabetically and/or thematically ordered list of books with their bibliographical data from a particular field. Earlier bibliographical writings, beginning with Callimachus' lost *Pinakes* (mid-third century BC), departed from this definition in several respects, of which the following appear crucial: first, they were biobibliographies, ordered by authors and including information on them. Secondly, just as the authors were usually presented in chronological order, chronology provided the guiding structural principle of the bibliographical parts as well. Thirdly, what was listed was works, not editions. Fourthly, thematically, there were only a few types of bibliographies: universal, Greek, Roman, and ecclesiastical. All of this changed during the early modern period.[236]

The catalyst was print. Producing ever vaster numbers of books faster and faster, print made it difficult and urgent to keep an overview of them. Scholars needed information on what to read in their fields of interest, and publishers needed scholars to be informed for business reasons.[237] Along with the modern bibliography in the strict sense of the word, a number of other forms both old and new flourished which fulfilled similar functions but should nonetheless be distinguished from it. In the interest of clarity, it is pertinent to enumerate the most important of these related forms. Biographies of learned men were still complemented by lists of their works. In the wake of Pliny the Elder's catalogues of sources for all thirty-seven books of his *Historia naturalis*, authors added similar lists to their books. Catalogues of libraries (public and private) had existed since antiquity. They were supplemented by topical catalogues listing the books for sale by specific publishing houses and at the Frankfurt and Leipzig Fairs.[238]

[234] Burke 2000, 93, 103. [235] Blum 1969, esp. cols. 1029–33; Maclean 2009, 429.

[236] On the history of bibliography from Callimachus to the sixteenth century, see Blum 1983. The above four points are distilled from his pp. 207–8.

[237] See Maclean 2009, 18–21 on the function of bibliographies as 'finding lists'.

[238] On publisher's catalogues, see Maclean 2012, 18, 177, 179–81. For a brief overview of the fair catalogues, see Blum 1983, cols. 205–6. There are some modern reprints, but the fullest collection is available (for a fee) at http://www.olmsonline.de/kollektionen/messkataloge/ (accessed 16 February 2023).

From the late seventeenth century, abstracts and reviews in journals helped people keeping track of recently produced works. For some journals, this aspect was their main *raison d'être*.[239] Many study guides contained reading lists.[240] The 'historia litteraria' of the seventeenth and eighteenth centuries, a genre conceived as a propaedeutics to academic studies, traced the development of the totality of learning and summarized the current state of knowledge mainly in the form of commented bibliographies.[241]

The modern bibliography, strictly speaking, originated in the late fifteenth century.[242] The first printed bibliography, one of ecclesiastical writers (1494), was the work of Johannes Trithemius. Conrad Gessner's alphabetically ordered *Bibliotheca universalis* (1545) and its complement, the thematically structured *Pandectae* (1548–9), raised the genre to a new level (the former, picking up the title of the *Βιβλιοθήκη* by Photius, a patriarch of Constantinople in the ninth century, also established its dominant early modern name).[243] Gessner's efforts served as a model for several successors who continued his universal aspirations.[244] An exceptionally comprehensive and thoroughly structured example is Georg Drauds *Bibliotheca classica* (1611). Both Gessner's and Draud's *Bibliotheca* knew massively expanded editions in 1578 and 1625 respectively—a treatment typical of an accumulative genre which turned bibliographies into repositories of old books *and* tools to keep track of recent publications at the same time.

Most early modern bibliographies were in Latin and focused on Latin books.[245] Although some examples written in the vernaculars and listing vernacular publications appeared from the mid-sixteenth century,[246] in most fields Latin bibliographies outshone vernacular ones in numbers and importance until the second half of the eighteenth. Originally, the dominance of Latin in bibliographies naturally followed from their learned nature and content. But even when more and more vernacular learning had to be recorded from the seventeenth century, the bibliography itself usually remained in Latin, since the genre was international,

[239] See Kronick 1976, 171–83 ('abstract journals').

[240] A nice example is the 'Manuductio ad medicinam' preceding Johannes Antonides van der Linden's *De scriptis medicis libri duo* (1637). See also Maclean 2002, 65.

[241] Grunert and Vollhardt 2007; Grunert and Syndikus 2015. A late example (uncharacteristically) restricted to one scientific discipline is Johann Friedrich Blumenbach's *Introductio in historiam medicinae litterariam* (1786).

[242] Most of the material discussed in what follows was discovered in Besterman 1955–6. Searching this astonishing bibliography of over 84,000 bibliographies by thematic keywords such as 'Academic writings', 'Chemistry', or 'Medicine', one canfind many more examples.

[243] Blum 1969, cols. 1024–5, 1027–9; 1983, cols. 189–96, 202–4; Friedrich 2002, 402–4; Zedelmaier 2015, 22–33 (who notes on pp. 28 and 31 that the *Pandectae* in and of themselves are not so much a bibliography as an encyclopedic collection of 'loci communes').

[244] Maclean 2012, 204–5.

[245] On the bias in favour of Latin works, see Pantin 1996, 56–7.

[246] Blum 1983, col. 205.

listing books from everywhere and targeting a pan-European readership. In his 'Preface on the usefulness of this book', Pierre Borel counted it as the eighth of ten assets of his *Bibliotheca chimica* (Paris, 1654) that it informed foreigners about French books and the French about foreign books. Not for nothing, Latin still held its place among the most important languages of bibliography as late as 1956. With 2,653 titles on record, it ranked fifth after English, German, French, and Italian, but ahead of Spanish, Dutch, and Russian.[247]

Looking for science among the subjects of bibliography, one finds that it played a substantial role but was in no way presented as a unity. In fact, bibliographies like Gessner's and Draud's illustrate how different the conceptualization and subdivision of the disciplines concerned with the natural world was from ours today. Within the projected twenty-one books of Gessner's *Pandectae*,[248] arithmetic, geometry (including optics and catoptrics), music, astronomy, and astrology fill books 5–9, the quadrivium forming part of the work's liberal arts section. Geography, in book 11, is a counterpart to history in book 12. 'Chymia' is to be found under titulus 9 of book 13, which is dedicated as a whole to 'various unlettered and mechanical arts as well as to others which are useful for the life of humankind' ('diversis artibus illiteratis, mechanicis et aliis humanae vitae utilibus'). Natural philosophy (book 14) includes cosmology, meteorology, mineralogy, and botany. Book 20 would have been dedicated to medicine. Draud subdivides his material into the seven categories of the contemporary catalogues of the Frankfurt Book Fair: theology; law; medicine; history, geography, and politics; philosophy; poetry; music.[249] Medicine is the only scientific discipline that constitutes a category of its own; it includes some botany and zoology, which contributed to 'materia medica'. Geography (mathematical and descriptive alike) is coupled with history and politics. Mathematics, physics, astronomy, astrology, cosmology, meteorology, earth sciences, chymistry, and some minor disciplines only form subsections of philosophy. Later, when Gessner's and Draud's universal approach was abandoned in favour of a focus on more specific fields, such fields could be constituted by single scientific disciplines or literary genres. Science as a single category, by contrast, was not envisaged before the nineteenth century.[250]

In the following two sections, I will review bibliographies defined by discipline and by genre respectively. The discussion will be concluded by a look at autobibliographies.

[247] Besterman 1955–6, preface to vol. 3.
[248] Book 21, devoted to theology, appeared one year after books 1–19, book 20 (medicine) not at all.
[249] Maclean 2012, 61, who notes the popularity of this subdivision of the field of learned books in general.
[250] In Brock 2000, the first part (pp. 298–307), dedicated to scientific bibliography in general, lists exclusively modern works. An early approximation of a bibliography of science is the *Scientia naturalis* section in Reuss's *Repertorium*, to be discussed shortly. For the rest, see Besterman 1955–6, vol. 3, cols. 3714–48.

Bibliographies of scientific disciplines

The number of bibliographies dedicated to most disciplines is not great.[251] Moreover, these disciplines may turn out to be very loosely conceived. Alongside mathematics strictly speaking, the *Bibliographia mathematica et artificiosa* (1688) by the Dutch bookseller and bibliographer Cornelis à Beughem not only covers fields such as geography, fortifications, and painting, in which mathematics plays a certain role, but even stenography and cookery (p. 249). More to the point is Johann Friedrich Weidler's *Bibliographia astronomica* (1755), the earliest in its discipline. As in the fair catalogues and in Draud's work, geography is usually linked to history and politics, for instance in Matthias Bartels's *Biblionomia historico-politico-geographica* (1682). The *Orbis terraqueus* (1670) compiled and printed by Simon Pauli at the instigation of Dutch mapmakers is exceptional as a bibliography not of books, but of maps, as well as for its primarily economic rather than scholarly motivation. In natural history, Scheuchzer's *Bibliotheca scriptorum historiae naturali omnium terrae regionum inservientium* (1716) restricts itself to books dedicated to specific countries. This peculiarity reflects the work's function as a 'forerunner' ('prodromus') of a national natural history by Scheuchzer himself, his vernacular *Natur-Historie des Schweitzerlandes*, published 1716–18. Botany, which was also covered in medical bibliographies, was first treated separately in Ovidio Montalbani's *Bibliotheca botanica* (1657) and last discussed in Latin in Georg August Pritzel's huge *Thesaurus literaturae botanicae*, the second and definitive edition (1872) of which was reprinted several times until 1995. As an example from chymistry, Borel's aforementioned *Bibliotheca chimica* (1654) can be named.[252]

Medicine, by contrast, has over forty bibliographical treatments dedicated to it according to Besterman, surpassing all other disciplines concerned with the natural world together.[253] Medical bibliographies are representative of early modern bibliographies in general in their great variety of form and content, which defies systematization. A few examples may provide an impression of this pluralism. The earliest works, Symphorien Champier's 'De medicinae claris scriptoribus' ([1506]) and Otto Brunfels's *Catalogus illustrium medicorum* (1530), are still conceived as biobibliographies, with the authors being at least as important as the works.[254] They also try to honour their discipline by including famous figures

[251] Some of the titles named in this section as well as a number of others are also cited in in Brock 2000, 307–8 (mathematics in the widest sense), 311 (astronomy), 315–16 (chemistry), 323 (natural history), and 326–7 (botany).

[252] On some bibliographical lists of chymical works in earlier publications, see Thorndike 1929–58, vol. 5, 548.

[253] On medical bibliographies, see Fulton 1951; Symons 1990, esp. 243–9.

[254] Blum 1983, col. 201.

who are not authors or doctors, such as Apollo and Aristotle (Brunfels, *Catalogus*, pp. 1 and 42–50).[255]

By around 1600, medical bibliography has matured. One finds a series of competent attempts at covering the entire discipline within a reasonable space such as Paschalis Gallus's *Bibliotheca medica* (1590), Johann Georg Schenck's *Biblia iatrica* (1609), or Johannes Antonides van der Linden's successful *De scriptis medicis libri duo* (1637), two more editions of which appeared in 1651 and 1662, and an updated, greatly expanded rearrangement, the *Lindenius renovatus*, was published in 1686 by Georg Abraham Mercklin.[256] Martin Lipenius's *Bibliotheca realis medica* (1679), a work alphabetically arranged by keywords and comprising over 500 quarto pages, should also be mentioned. A century later, such syntheses have grown to daunting proportions. Albrecht von Haller's plan of a multivolume *Bibliotheca medica*, which should have covered the most important subfields and auxiliary disciplines,[257] was never completed. Still, the parts that appeared—*Bibliotheca botanica* (1771–2), *Bibliotheca chirurgica* (1774–5), *Bibliotheca anatomica* (1774–7), and *Bibliotheca medicinae practicae* (1776–88)—comprise ten volumes and over 50,000 titles, chronologically ordered by authors, between them. Still, Haller's efforts are dwarfed by the 175,000 entries in Wilhelm Gottfried Ploucquet's thematically structured *Litteratura medica digesta* (1808–9), although this work has no botanical section. Works such as these were in need of and did indeed develop sophisticated arrangement systems, including huge indexes, to keep the mass of information manageable, something often announced in their titles. Carl Gottlob Kühn's *Bibliotheca medica continens scripta medicorum omnis aevi ordine methodico disposita* ('...structured in an orderly and methodical way', 1794) is a case in point.

Encyclopedism was complemented by selectivity. Students, in particular, were provided with select reading lists, presented in running text and embedded in a first overview of the discipline in general—small-scale equivalents of the encyclopedic 'historiae litterariae'. Pedro de Castro introduced his *Bibliotheca medici eruditi* (1654) with a letter to 'his most beloved students of practical medicine' ('suis dilectissimis praxeos studiosis') and only presented some 400 titles to them in a lively, if somewhat disorderly account, highlighting the importance of the fields touched upon and the merits of the books presented.[258] Other bibliographies are limited in terms of subdisciplines and genres (see the next subsection) or geographically, focusing on works from certain regions, countries, or universities.

[255] Similar things are still found in much later works. In his *Bibliotheca chimica* (1654, p. 2), Borel includes Abel because he is called a chymist in an obscure *Liber de gloria mundi*.

[256] Maclean 2002, 38–9; 2012, 61; Symons 1990, 244.

[257] See the preface to his *Bibliotheca botanica*.

[258] Two more examples, which also show the internal variety of the subgenre, are van der Linden's concise 'Manuductio ad medicinam' and Herman Boerhaave's extensive and systematic *Methodus discendi medicinam* (1726).

Towards the end of the Latin tradition, national bibliographies like Mathias Winther's *Bibliotheca Danorum medica* (1832) appeared. On the other hand, Kurt Sprengel acquainted his German colleagues with medical literature published since 1750 *outside* Germany in his *Literatura medica externa recentior* (1829). Finally, one finds bibliographies of rare works and editions which target collectors rather than doctors, such as the two parts of the *Bibliotheca librorum rariorum physico-medicorum* (1751–2) by Friedrich Börner.

Bibliographies of genres

Generically limited bibliographies seem to be a phenomenon of the eighteenth and nineteenth centuries. They focus on two kinds of writings, both of which were exceptionally productive at this late stage and therefore posed particular heuristic problems: academic dissertations and journal articles. In each case, one exceptionally comprehensive and informative specimen stands out.

Of bibliographies of dissertations, Besterman lists ten Latin items. Most of them are thematically unspecific, enumerating dissertations from various natural scientific disciplines together with others, but three focus on the sciences. Of these, two are concerned with medical subdisciplines, namely anatomy and physiology, on the one hand, and obstetrics, on the other.[259] The third and most important by far is Johann Carl Heffter's *Museum disputatorium physico-medicum tripartitum* (1756–64). Technically, this is a catalogue of the private collection of the author, a physician in the small Saxon town of Zittau and an avid dissertation hunter. Each of its two volumes, the first of which saw two editions in 1756 and 1763, falls into three parts. In the first and most extensive one, the individual items are given in alphabetical order (by the last name of the 'praeses') with their full bibliographical data, the second is an 'index rerum', and the last, an index of defendants. Between them, the two volumes comprise 18,500 dissertations, which are carefully and consistently numbered in all three parts. Most of them are medical, but many deal with topics from all other natural sciences, including fringe areas such as magic. In chronological terms, the great majority stem from the time between 1650 and the compilation of the work. Geographically, the focus is on Germany. The preface not only details how the work was compiled and structured but also provides important information on early modern dissertations in general, their production, sociological setting, circulation, and collection. The *Museum* as a whole allows one to estimate the magnitude of the overall production of dissertations in early modern times (a six-figure number at least). Moreover, it provides insight into the enormous thematic variety of the genre and

[259] Besterman 1955–6, vol. 1, cols. 6, 25, 30–1 (thematically indiscriminate); cols. 9, 18 (anatomy and physiology; obstetrics); col. 16 (*Museum disputatorium*).

the boundless curiosity of the era, listing subjects from snow containing insects and La Mettrie's *L'homme-machine* to the identity of the Behemoth and the therapeutic use of music.[260]

Information on journals in general could be found by an early modern reader in works such as Christian Juncker's *Schediasma historicum de ephemeridibus* (1692). He mainly learned about individual journal articles from other journals, where they were reprinted, translated, abstracted, excerpted, and reviewed. The first and only real bibliography of journal articles, Jeremias David Reuss's *Repertorium commentationum a societatibus litterariis editarum*, appeared as late as 1801–21.[261] It focused on articles published in the periodicals of academies and other learned societies, which usually contained more original and novel material than journals issued by commercial publishers.

The *Repertorium* comprises sixteen volumes, over 10,000 pages, and over 100,000 titles. Most of this material stems from the natural sciences. The first six volumes are collectively titled *Scientia naturalis* and deal with natural history, zoology, botany, chemistry, physics, astronomy, and 'oeconomia', that is, husbandry. These disciplines are followed by three separate volumes (vols 7–9) devoted to mathematics and its technical applications, history (including geography as an auxiliary science), and philology respectively. The second long part featuring a collective title, *Scientia et ars medica et chirurgica*, comprises vols 10–16 and deals with pharmaceutics, medicine proper, obstetrics, and veterinary medicine. The individual volumes fall into short thematic 'sectiones', whose order may represent the structure of the field in question or be alphabetical. Within the sections, articles are listed randomly. Each volume has an index of authors.

Of the articles listed, some 10–20 per cent are in Latin, but the proportion varies greatly depending on topic. That the work itself still appeared in Latin testifies to the enduring prestige of the old language in German academia and also to its international usefulness (copies are extant in all major European countries and beyond). However, the difference from a vernacular edition is not great, as the amount of Latin paratext is minimal compared with the immense lists of titles and bibliographical data, which are given without commentary. There is no preface to explain the purpose and rationale of the work. Only an opening dedication to the Königliche Gesellschaft der Wissenschaften zu Göttingen (which celebrated its fiftieth anniversary in 1801), the title pages, an overview of the whole project, as far as it has progressed, at the beginning of each volume, and the detailed tables of contents ('elenchi sectionum') are in Latin. Still, a reader interested in the measurement of the capacity of barrels had to understand the relevant entry, 'Mensura doliorum', in the 'elenchus sectionum' of volume 7 in order to find what he wanted.

[260] See vol. 1, part 2, s.vv. 'Behemoth', 'Homo machina', 'Insecta cum nive delapsa', 'Musica'.
[261] Kronick 1976, 273.

Autobibliographies

I would like to conclude with a bibliographical variety that looks exotic at first sight only: the self-bibliography. In fact, this subgenre goes back to the prolific last centuries of classical antiquity. Both Galen and Augustine left such writings, which were well known in early modern times. Galen's *De libris propriis* and *De ordine librorum suorum* informed readers how and in what order to read his works. In his *Retractationes*, Augustine critically re-examined his entire *oeuvre*. In the early modern period, autobibliography made an early and conspicuous start. In the 1520s and 1530s, Erasmus composed two such texts, the second of which defined the order of his works in his posthumous *Opera omnia* (1540).[262]

Among scientific writers, Cardano is easily the most famous and fertile self-bibliographer, as he is also the most famous self-biographer. Between 1538 and the year of his death, 1576, he produced and for the most part published no fewer than six autobibliographical texts, which differ greatly in their construction, length, and purpose. In doing so, he saw himself following in the footsteps of Galen and Erasmus. Provision for the publication of his manuscript legacy was a motive for Cardano, as it had been for Erasmus, but he, moreover, was personally obsessed with the cataloguing and classification of his many writings, an activity he seems to have seen as a way to self-understanding.[263]

Even apart from this spectacular but eccentric case, quite a few self-bibliographies by authors with scientific interests can be named. Another sixteenth-century example is Conrad Gessner's *De libris a se editis epistola*. Originally written at the request of the English divine and naturalist William Turner, who had asked for a list of Gessner's works, it was later appended to Josias Simmler's biography of his recently deceased friend (1566). One and a half centuries later, Gessner's compatriot Johann Jakob Scheuchzer produced an autobibliography under the exotic title 'Acarnania', derived from the name Acarnan ('The Laureate') which he had adopted as a member of the Academia Naturae Curiosorum. The text was first published in 1718 in the learned journal *Miscellanea Lipsiensia*. A second, updated and expanded version appeared in 1723 as an appendix to the author's *Οὐρεσιφοιτής Helveticus*. Presenting his readers with a chronological list of his printed works and manuscripts, Scheuchzer in many cases includes their time of writing and purpose as well. His literary production thus appears embedded in his professional biography, to which it contributes in turn. Speaking much in the first person, Scheuchzer sketches a flattering self-portrait of a tireless and heroic naturalist despite his initial protestation that his text 'does not smell encomiastic' ('non encomium spirat', 1723, p. 609). If the reader certainly

[262] Maclean 2004, 15.

[263] All six texts are published with introduction and notes in Maclean 2004, on which this paragraph is based.

receives some help in finding this or that work by Scheuchzer, he also learns to admire the author.

The pride taken in their literary production by authors like Cardano, Gessner, and Scheuchzer was noted and satirized already in John Barclay's famous and influential *Icon animorum* (1614), where one reads about the Germans: 'Plura, quam legerint, scribunt et suam famam ex voluminum, quae edunt, numero aut magnitudine aestimant.' ("They write more than they have read and believe their reputation to be proportional to the number or bulk of the tomes they publish.'[264]) The way from early modern self-bibliographies to the modern publication list which has become such an essential aspect of academic self-presentation may be long and winding. Nonetheless, the texts just discussed testify to the beginnings of the process by which the sum of a scholar's publications grew into the embodiment of his professional identity.

Reviews

Early modern scientific reviews in general

A review may be defined as a brief presentation and/or critical assessment of a learned work in a journal.[265] From the very beginning, the format was applied to scientific works alongside books from other fields such as theology, law, and history. The review had several early modern predecessors: bibliographies informed people of the existence of all kinds of books old and new,[266] publishers' and fair catalogues, in particular those from Frankfurt and Leipzig, of books recently published. Of course, news about books travelled by letter as well. Moreover, the sixteenth and seventeenth centuries saw the rise of the 'review monograph', the book-length comment on another book, which often pursued a strong personal agenda. Kepler's *Dissertatio cum Sidereo nuntio* (1610) was a friendly example, but Kepler nonetheless tried to fit Galilei's discoveries into his own cosmological system.[267] Other review monographs coolly exposed factual errors, as did Rasmus Bartholin's *Specimen recognitionis* ([1668]) with regard to Albert von Curtz's *Historia coelestis* (1666). Many were frankly polemical, which could lead to fierce exchanges.[268]

264 Chapter 5, quoted after p. 59 of the edition from Frankfurt, 1625.

265 A short overview, including a number of historical remarks, is provided by Fritz 2020.

266 On the relation between bibliographies and reviews, see Blum 1969, cols. 1045–7.

267 Latin text: *KGW* 4, 281–311. English version with introduction and commentary: Rosen 1965.

268 For a moderately polemical example, see Giovanni Antonio Magini's *Confutatio diatribae Scaligeri de aequinoctiorum praecessione* (1617), in which he refutes Joseph Scaliger's misguided ideas about the precession of the equinoxes (*De aequinoctiorum anticipatione diatriba*, 1613); for a fierce exchange of criticisms and countercriticisms, see Leonhart Fuchs's *Errata medicorum recentiorum* (1530), Sebastien de Monteux's *Annotatiunculae in Errata recentiorum medicorum* (1533), and Fuchs's *Paradoxorum medicinae libri III* (1535).

While the review may in a loose sense be called a successor to all these text types, it was not a direct descendant of any of them. Rather, it originated together with the scholarly journal (on which publication format, see the section on journal articles) in 1665. The two very first journals, the *Journal des sçavans* and the *Philosophical Transactions*, already contained reviews, which made up the majority of contributions in the *Journal*. Many periodicals followed suit, at least among those published on the initiative of commercial publishers, some of which were more or less exclusively composed of reviews.[269] The academies, by contrast, were mainly content to present the original research of their members in their journals—which became themselves objects of reviews alongside books. The new genre shared to a large extent the non-committed nature of the bibliography and the fair catalogue. Like the catalogue and the 'review monograph', it was topical, and like the latter, it focused on one book only. It was more informative than the first two genres and more reliable, because less biased, than the last. In sum, it proved ideal for a time in which keeping track of new publications had become a major challenge for scholars. Reviews soon became important distributors of novel ideas.[270] Their success led to their production in such numbers that they already repelled some readers in the mid-eighteenth century.[271]

The beginnings of the new form were as unspectacular as they were sudden. Reviews often appeared intermingled with original contributions and other kinds of news, such as appointments to professorships and funerals, and they were not distinguished as a separate category in tables of contents.[272] In fact, it seems that initially reviews were often not recognized as texts *sui generis* but rather understood as yet another ingredient in the miscellany that made up many early modern journals. For some decades, they even lacked a distinctive name. It was only in the eighteenth century that 'recensio' became common in review titles and opening statements of journals[273] (while less specific terms such as 'iudicium' or 'commentarius' continued in use as well). In such statements, but also in reviews of journals containing reviews and in answers to reviews, the new genre became the object of some explicit reflection. The German theologian Johann Christoph Greiling even published a kind of theory of the review ('Einige allgemeine Grundsätze zu einer Theorie der Recensionen', 1799).[274] Nonetheless, reviews

[269] Kronick 1976, 184–201. On Leibniz's abortive plans for a review journal titled *Nucleus librarius semestralis*, see Nummedal and Findlen 2000, 200 and 204.

[270] Fritz 2020, 302.

[271] Kronick 1976, 103.

[272] Such is the case, for instance, in Lorenz Christoph Mizler's *Acta litteraria regni Poloniae et magni ducatus Lithuaniae* (1755–7).

[273] See, e.g., Christian Maximilian Spener, 'Recensio Itineris Alpini Ioannis Iacobi Scheuchzeri', *Miscellanea Berolinensia*, 1 (1710), p. 139; 'Instituti ratio', in *Commentarii de rebus in medicina et scientia naturali gestis*, 1 (1752), pp. x, xiii. (The verb is used there as well: 'in libris recensendis' and 'recensitis', pp. x–xi). Mizler distinguishes 'recensiones' from 'dissertationes' in the opening letter to his *Acta litteraria*, 1 (1755), p. 11. 'Rezension' is still the German word for 'review' today.

[274] Fritz 2020, 302–3.

were apparently regarded as lowly work, which was probably one reason why most reviewers remained anonymous until the late eighteenth century (others being the wish to avoid personal animosities and a general tendency towards anonymity in early modern journals).[275]

In accordance with the slow recognition of them as a genre, early modern reviews were quite diverse in a number of respects. Their length could vary from a few lines to several dozen pages. Books chosen for review might be fresh off the presses or published many years ago. Reviewers mostly focused on the work at hand but sometimes allowed themselves astonishing digressions. In the first volume of his *Acta litteraria* (1755, pp. 62–7), Lorenz Christoph Mizler discusses a French epidemic treatise published seventeen years before and uses a quarter of his space to compare two Latin translations of Sirach 38, a passage dealing with the reputation of doctors in ancient Israel. Regarding their attitude to the publication in question, some reviews contextualized it with respect to the previous literature in the field or voiced criticism, but the most common ingredient was a more or less extended summary of the work's content.[276] Readers first of all wanted to learn what was being published about what topics, and many journal editors were averse to strong criticism. They knew all too well, not least from the aforementioned 'review monographs', how quickly argument could degenerate into polemics. Their journals were supposed to be calm meeting places of ideas, not intellectual battlefields.[277] Only from *c.*1750 onwards did analysis, evaluation, and even polemics become more prominent.[278] The descriptive character of most reviews entailed a sober language aiming first and foremost at precision and clarity. Later on, stylistic lucidity was sometimes aided by tables and mathematical formulae, as reviews could become more technical over time.[279]

Latin reviews

Because the learned journal, like the review with it, was a latecomer on the scene of scientific literature, the majority of reviews appeared in the vernaculars from the beginning, and the career of the Latin review was comparatively short-lived. Nonetheless, the number of reviews published in Latin until the early nineteenth century was not negligible.

[275] On anonymity in reviews, see Fritz 2020, 303. It was impossible in the case of one-man-journals such as Mizler's *Acta litteraria*. On anonymity in early modern journals in general, see Kronick 1976, 83–4, 101, 130–2, 281.

[276] Cf. Laeven 1990, 87–96 on the types of reviews in the *Acta eruditorum*. In the opening letter to his *Acta litteraria*, Mizler finds it worth emphasizing that he writes no 'naked reviews' ('nudae recensiones') but 'corrects and adds' ('corrigendo et addendo', p. 11).

[277] Laeven 1990, 77–83 (on the *Acta eruditorum*).

[278] Fritz 2020, 296, 302.

[279] For an example, see the review of Johann Heinrich Lambert's *Pyrometrie* in the April issue of the *Nova Acta Eruditorum* 1776, pp. 162–72.

This partly has to do with the fact that Latin had a special function as a review language, which paradoxically became not less but more important over time as the scientific world became more and more fragmented linguistically through the rise of the different vernaculars. Alongside and more than French, Latin was still the common international language of the learned world. As such, it was the perfect medium for distributing news about books written in languages as different as French, Italian, German, Dutch, English, or Swedish, many of which were not understood by the majority of potential readers. For a Dutch periodical like the *Verhandelingen uitgegeeven door de Hollandse Maatschappy der Weetenschappen te Harlem* (1754–93), reviews in a Latin journal like the *Commentarii de rebus in scientia naturali et medicina gestis* that was for sale not just in the Netherlands, but in Germany, England, France, and Italy were the best way to bring at least part of its contents to the attention of an international audience.[280] That editors of review journals were conscious of this is shown by Johann Gottlieb Krause's preface to the first issue of his Latin *Nova litteraria* (1718, sigs. A^{r}–$A2^{r}$). A major obstacle to keeping an overview of the numerous learned journals, he notes, is 'the various languages in which they are written and all of which not everyone is capable of knowing' ('linguae, quibus exarantur, variae et quas non cuiusvis sit omnes cognoscere'). Having published himself a learned journal in German, the *Neue Zeitungen von Gelehrten Sachen*, for some years, he found that 'several persons, particularly foreigners unfamiliar with our language... seemed to wish that these *News* of ours should be available in Latin' ('plures, imprimis exteri usu linguae nostrae destituti...optare viderentur, ut Latine prostarent haec *Nova* nostra'). In Latin 'review translations', a function of the review came to the fore which it accomplished in other respects as well (as will be seen shortly): that of a proxy for the original text.[281]

Moderate numbers of Latin reviews of scientific books and periodicals appeared in several journals devoted to medicine and studies of nature or to learning in general. I will give just a handful of examples. The five volumes of the *Acta medica et philosophica Hafniensia* (1671–9), edited by Thomas Bartholin, are divided into a first part comprising several hundred pages of 'observationes' and a shorter second part devoted to books recently published in Copenhagen or by authors from that city. In the first volume, this part contains thirty-four short reviews plus ten books named by title only. The two parts are somewhat intertwined, as the 'observationes' contain not only original work but extracts from the books reviewed as well. In the later volumes, however, the second part is reduced to a bare bibliography or omitted altogether. Among other kinds of news, Krause's

[280] Volumes 1, 3, and 5 of the *Verhandelingen* were reviewed in *Commentarii*, 6 (1757), pp. 309–23; 9 (1761), pp. 3–22; and 11 (1763), pp. 120–39.

[281] That is also the function envisaged by Leibniz for his *Nucleus* (see note 269 above): as its title intimates, its reviews should acquaint readers unable to procure a book for themselves with the core subject matter discussed in it.

aforementioned *Nova litteraria* (1718–23), conceived as a supplement to the *Acta eruditorum* and to his own *Neue Zeitungen*, contains numerous reviews of books published and previews of books about to be published. A fair portion of this material concerns work in the natural sciences. While the previews obviously constitute an advertisement that must have been welcome to authors and publishers, the reviews are in fact summaries of reviews from other journals, which compress the topic treated and the judgement passed ('operis argumentum censuramque ab auctore diarii de eo latam', 1718, sig. A^v) into a few lines. As explained a little earlier (sig. A^r), reading not only the many books themselves, but also the great number of reviews had already become cumbersome. In the *Acta litteraria regni Poloniae et magni ducatus Lithuaniae* (1755–7), one finds a limited number of longer reviews, a few of which deal with scientific books.

However, examples such as these are dwarfed by the massive number of reviews contained in two unusually successful and long-lived journals. The first of these is the famous *Acta eruditorum* (1682–1731) founded by the Leipzig professor of practical philosophy, morals, and politics Otto Mencke, together with its successor, the *Nova acta eruditorum* (1732–76), and their respective *Supplementa* (1692–1734, 1735–57). In the course of one hundred years, the *Acta* published some 3,000 reviews of natural scientific publications alongside an even greater number of non-scientific ones.[282] (The number of articles equalled that of reviews at the beginning but diminished over time.) The mostly descriptive and summary reviews comprised four to five pages on average. The second and even more remarkable major source of scientific reviews in Latin is the *Commentarii de rebus in scientia naturali et medicina gestis*. This periodical is a good place to turn for a closer look at the way early modern reviewing worked and the role of Latin in it.

The *Commentarii de rebus in scientia naturali et medicina gestis*

Unlike most other journals, the *Commentarii* focused nearly exclusively on reviews. (Some shorter original notes, news from the academic world, and obituaries can be found interspersed.) In fact, it was the first exclusive review journal of some importance in the history of the natural sciences.[283]

The *Commentarii* was founded in 1752 on the initiative of a group of doctors at Leipzig in tandem with the publishing house Gleditsch. Over the next half-century, until 1806, thirty-seven regular plus four supplementary volumes came

[282] This figure is based on the *Acta*'s contents as published at http://www.izwtalt.uni-wuppertal.de/Acta.html (accessed 16 February 2023). For a full list and classification of the reviews 1682–1706, see Laeven 1990, 299–360.

[283] Kronick 1976, 188.

out.[284] A normal volume comprised four trimestral issues and a total of 700–800 octavo pages. An issue contained between twenty and twenty-five reviews on average and ended with a news section ('Nova physico-medica') and a bibliography of works not reviewed. Every tome featured an extensive, carefully produced index. Over the journal's lifespan, a body of over 3,500 reviews filling some 30,000 pages was produced. The individual reviews were quite extensive. Most of them comprised between five and thirty pages, although shorter pieces of no more than one page can be found in later volumes. Not only books, but also single volumes of other scientific journals were reviewed. A substantial proportion of the works discussed was still in Latin, but the vernaculars gained ground as time progressed. While the ratio was 68:30 in favour of Latin in the first volume (1752), it had become (40:41) in volume 28 (1785). Although the *Commentarii* was printed in Leipzig, it was sold in various places outside Germany as well, as recorded on the title pages. Cities in whose bookshops the journal could be found included Amsterdam, Leiden, London, Edinburgh, Stockholm, Paris, Padua, Venice, Milan, Turin, and Rome, although not all of them at the same time.

The founders, who remain anonymous apart from the first editor Christoph Gottlieb Ludwig, open the first volume with an extended statement titled 'Instituti ratio' (pp. iii–xviii). This statement constitutes a revealing source for many aspects of reviewing and academic life in general in the mid-eighteenth century. Apart from the story of the journal's genesis and organizational questions, the following issues may be highlighted: the journal's *raison d'être* is the pressing need for students of nature to remain *au courant* regarding the newest literature. Among physicians in particular (a profession to which the founders themselves belong), the wish for specialized bibliographical aids is widespread. However, the periodical takes into account all the other natural sciences as well, because they are seen as auxiliary sciences to medicine (pp. iii–v). Atmospheric and ethical issues are a major concern: authors whose books are not reviewed should not be angry (pp. x–xi). No offence should be taken at criticisms (pp. xii–xiii). No polemics will be argued out in the *Commentarii*. The editors are incorruptible. They will publish no favourable reviews in return for gifts of books (p. xvii). The reviews themselves will cover not only major books, but also shorter works and issues of scientific journals (p. ix). Reviews will be written from personal reading of a work, never on the basis of other people's judgement (p. x; one is reminded of the secondhand reviews in the *Nova litteraria*). The diversity of the works considered does not allow for a uniform way of reviewing, but purpose, method, and novel aspects will always be indicated and a sincere judgement expressed ('nostram denique . . . sententiam de ipso, quod indicamus, opere libere profiteamur'). The judgements will

[284] For bibliographical details, see Rogg 1830, 287, and Bolton 1897, no. 1325 (with minor differences between them). Vols. 1–28 (1752–86) can be found online in the Biodiversity Library (https://www.biodiversitylibrary.org/, accessed 16 February 2023).

be neither adulatory nor hypercritical: 'Absint elogia vana, adulationis plenissima, absint econtra contumeliosae vituperationes' (pp. xi–xii). The language will be Latin, the common learned language of Europe, which is even understood by good surgeons and pharmacists (who obviously form part of the intended readership, p. xv). In terms of style, reviewers will aim at clarity ('perspicuitas'), not elegance (pp. xv–xvi).

The actual reviews broadly conform to these programmatic statements. They bear no formal title, but are headed by the title (plus its translation into Latin in the case of a vernacular work) and bibliographical data of the book in question. The review proper usually starts with a brief contextualization and characterization of the work in question and proceeds to a paraphrase of its reasoning and results, which is often quite detailed and includes page numbers in the form of marginalia.

The one major deviation from the announcements made in the 'Instituti ratio' is the fact that praise strongly predominates over criticism. In most reviews, the author receives complimentary epithets such as 'most famous' ('celeberrimus' or 'illustrissimus'), and in the introduction and conclusion to the review in particular, his professional experience and expertise, his diligence, and his learning are extolled in more specific terms, along with the usefulness of his work, its elevated intellectual level, and its good Latin style. In many cases, this technique of framing a descriptive review with a laudatory introduction and conclusion seems to be applied consciously and methodically.[285] Criticism is voiced much less openly. The reviewer of Pietro Moscati's *Discorso academico dei vantaggi della educazione filosofica nello studio della chimica* (1784)[286] was clearly not enthusiastic about it. But apart from some distancing formulas such as '[the author] convinces himself' ('persuadet sibi') and one or two mildly critical remarks, his dissatisfaction mainly transpires in the very shortness of his account and in his change, towards the end of his summary, to a telegraphic style, as if the book were not worth the effort of formulating full sentences.

All in all, however, not only criticism but also praise remain comparatively marginal. Pervaded by a sober and neutral atmosphere, the reviews are above all an exercise in precise paraphrase and explanation. The grammatically correct, syntactically simple, lexically mildly technical and nominalizing Latin used to this end may be exemplified by the following explanation, found in a review of a journal article on medical properties of air, of why wind feels cooler than calm air (vol. 9, 1761, p. 7):

> Est vero recreationis refrigerationisque, quae corpori humano ab aëre moto contingit, ratio haec: Cum calor in corpore naturalis 92 vel 94 gradus thermometri

[285] A good example is provided by the review of Johann Ernst Hebenstreit's *Anthropologia forensis* (1751) in vol. 1 (1752), pp. 365–78.

[286] Vol. 27 (1785), pp. 491–2.

> Fahrenheytiani cernatur, sequitur calorem in homine eum, qui in aëre insit, longe superare. Quare in illam aëris portionem, quae hominis superficiei proxima est, si in quiete sit constituta, caloris copia ingreditur et sic humanum corpus circumdatum est atmosphaera aliqua calida, cuius calorem ipse homo percipit. Iam spirante vento dispergitur haec atmosphaera atque homo hanc caloris privationem sentiens frigoris gradum percipit.
>
> The reason for the refreshment and cooling that the human body receives through moving air is the following: because the natural warmth in the body is seen to be 92 or 94 degrees on Fahrenheit's thermometer, it follows that the warmth in a person is far greater than that in the air. Therefore, an amount of warmth passes into that part of the air which is closest to the surface of the person, provided the air is at rest, and so the human body is surrounded by a certain warm atmosphere, whose warmth is perceived by the person himself. But when the wind blows, this atmosphere is dispersed and the person who feels this deprivation of warmth perceives a certain degree of coldness.

Reviews such as this one served the community of physicians and students of nature in two different ways. On the one hand, of course, they guided readers to books and journals of interest to them. On the other hand, however, many books described in the *Commentarii* will have proved unavailable, too costly, dauntingly long, or written in an unintelligible language. In such cases, the review provided a surrogate for the work itself. That the reviews in the *Commentarii* could be perused and even enjoyed for and in themselves is already implied by its title copper (Figure 4), which shows a beehive situated in a bucolic landscape, along with the motto 'SIBI ET ALIIS' ('FOR THEMSELVES AND OTHERS'), which seems to allude to the genesis of the journal.[287] While the beehive clearly symbolizes the contributors' collective effort, honey production is also a long-standing image for the creation of a new, autonomous work of literature on the basis of previous ones.[288] Here, this topos is applied to scientific texts: readers, it is suggested, can taste the honey of science in the *Commentarii* without visiting every single flower in person.

[287] The 'Instituti ratio' (vol. 1, p. vii) narrates how the founders initially wanted to provide something useful to themselves ('nobis ipsis') and their friends only. Soon, however, they felt the ambition of helping the learned world at large.

[288] The *locus classicus* is Seneca, *Epistulae morales* 84. On the ensuing tradition, see Stackelberg 1956.

COMMENTARII
DE REBVS
IN
SCIENTIA NATVRALI
ET
MEDICINA
GESTIS.

SIBI
ET
ALIIS

VOLVMINIS I. PARS I.

LIPSIAE
APVD IOH. FRIDERICVM GLEDITSCH.
MDCCLII.

Figure 4 Title page of the first volume of the *Commentarii de rebus in scientia naturali et medicina gestis*.

6
Presenting facts

Letters

From antiquity to the early modern period

Among the main advantages of writing is its ability to traverse long distances unchanged. Accordingly, letters became an early application of writing. Until the rise of modern communications technology, they remained the only means of long-distance communication apart from human messengers. It is this communicative function which defines a letter in the first place. Such formal features as are commonly regarded as characteristic of the genre are subservient to this purpose. Moreover, most of their traits, even the initial address and the valediction, are in fact less stable than one might be inclined to think. They can be standardized in different ways depending on time, place, and occasion, and may even be lacking altogether.

In Greek and Roman antiquity, letters (ἐπιστολαί/'epistolae' or 'litterae') were ubiquitous.[1] This is evident not only from letter corpora by celebrities like Cicero, Pliny the Younger, Paul, Ambrose, or Jerome, but also from scores of personal and business letters on papyrus preserved by the sands of Egypt. The art of letter writing was a highly regarded skill which was diligently taught and learned, as testified by a number of textbooks. Epistolary theory, which developed as an offshoot of classical rhetoric, included the influential ideas that correspondence was a written conversation among friends and that a single letter represented one half of a dialogue. While a letter was supposed to convey a true image of the writer's soul, it was at least as important that it fitted the addressee's position, needs, and interests. In this sense, it was supposed to be shaped as much by the recipient as by the sender. Friendship, the social basis of epistolary exchange, implied the obligation of reciprocity; each partner was supposed to answer a letter he had received in kind. Numerous subtypes of letters were distinguished according to the particular relationship between the correspondents and the occasion. In terms of form, brevity, a loose structure, and an informal style were recommended.

[1] On ancient letters in general, see Koskenniemi 1956; Thraede 1970; Stowers 1986; *DNP* 2, cols. 771–5, and 3, cols. 1161–6, cols. 1166–9, s.vv. 'Brief', 'Epistel', 'Epistolographie'. On epistolary theory, see Malherbe 1988.

Latin Scientific Literature, 1450–1850. Martin Korenjak, Oxford University Press. © Martin Korenjak 2023.
DOI: 10.1093/oso/9780198866053.003.0007

The theoretical attention dedicated to this partly reflected the fact that even the private letters dominating the scene in antiquity were more than just private documents. Recipients had considerable freedom to share them with whomever they saw fit. Often, letters were read to family, friends, and colleagues. The writer, in turn, instrumentalized them not only to foster pre-existing friendships but also to extend his networks, to project a certain image of himself, and ultimately to achieve his social and political goals. The common distinction between private real-life letters and literary letters conceived for publication from the outset, therefore, turns out to be problematic. The hastily scribbled note to a close friend and the polished treatise in epistolary guise mark only two ends of a continuum, and the place of a given epistle on this scale can be hard to pin down in hindsight. The semi-public status of letters persisted until new notions of privacy caught on in modern times.

In the Middle Ages, letters continued to be produced in great numbers. Unlike the situation in antiquity, however, the medieval record is dominated by official correspondence. Most letters transmitted from this era were written by secretaries of clerical or worldly rulers or of cities and addressed other such officials and their superiors. The Apostolic Chancery in particular took a leading role in the development of the public letter, and its missives served as models for those of many other institutions. Epistolary theory and didactics adapted to the new circumstances. In the 'artes dictandi' which once more applied the teachings of rhetoric to the art of letter writing and became dominant from the twelfth century, epistolography was turned into a highly formulaic affair. The form of address, for instance, was strictly regulated, as it had to correspond exactly to the rank and function of the addressee.[2]

From the mid-fourteenth century, Petrarch and his fellow humanists returned to the personal letter, taking its Roman representatives as models and ancient theory, which they developed further in the process, as guidance.[3] The rhetoric of friendship, the associative structure, and the relaxed style of the ancient letter were all revived, as were the classical lexicon, grammar, and turns of phrase. The latter often but not always included the characteristic third-person address of the type 'Gaius Marco s(alutem) d(icit)' ('Gaius wishes Marcus well') and the closure with a valediction and a Roman-style date like 'Vale! Romae, a(nte) d(iem) V kal(endas) Sept(embres)' ('Farewell! Rome, the 5th day before the Calends of September'). At the same time, formerly non-literate groups like craftsmen or

[2] On medieval letters in general, see Constable 1976; Høgel and Bartoli 2015. On the 'artes dictaminis', see Camargo 1991; Hartmann 2013.

[3] See Witt 1982, esp. 26–31, who also illuminates the social nexus between the 'ars dictaminis' and humanism. Overviews of early modern epistolography, especiallyin Latin are provided by Nellen 1993; *CNLS* 2, 218–28; Houdt et al. 2002; *BENLW* 335–49; *OHNL* 167–82; *EEMHO* s.v. 'Letter'. On the development of epistolary theory from the 'artes dictandi' to humanist guides to letter writing, see Worstbrock 1981.

merchants began to write letters as well, understandably preferring their vernacular for this. Overall, letter writing grew more common in response to factors such as population growth, higher literacy rates, greater mobility, the establishment of postal services,[4] and cheaper writing material (paper as opposed to parchment), until it became an essential part of general culture in the eighteenth century.

In the realm of learned discourse, intellectuals like Erasmus or Justus Lipsius had established sizeable correspondence networks by the sixteenth century, and such networks continued to expand in the course of the following centuries. As knowledge multiplied faster and faster, correspondence developed into an important tool for learned men in providing part of the information they needed to do their work. Writing letters became a part of their daily routine which was indispensable but often time-consuming and annoying too, as shown by an incessant stream of complaints about one's correspondents' silence and excuses for one's own. Letter networks were constitutive of the nascent 'respublica litteraria', which truly became a 'republic of letters' in both senses in accordance with the polysemy of the term 'litterae'. Although the better part of early modern learned correspondence is no doubt lost for good, enough remains to ascertain that the letter was among the most prolific text types of the era.

Learned correspondence increased in tandem with the expansion of print, as both letters and printed books profited from the general growth of distant communication made possible by the aforementioned factors. The relationship between both media was not competitive but complementary. Books aimed at presenting coherent information in a form palatable for large numbers of readers but could not take individual sensitivities into account. Letters were more informal and less constrained by the demands of completeness and coherence. They did not aspire to the thematic unity of a book but mostly carried single, apparently disparate pieces of information. However, this content could be adapted to the interests of a particular addressee. Moreover, a letter expressed and strengthened a unique relationship between writer and addressee, featuring an interpersonal dimension that was lacking in books despite attempts at creating an impression of it by dedicatory epistles and letters to the reader.

Such epistolary paratexts point to another essential aspect of the relationship between letter and book: letters not only ran in parallel to print as another strand of written communication, but also interacted with it and were themselves drawn into the sphere of the book. Letters often informed their addressees about new books, contained pertinent excerpts, or were accompanied by books. They also fed into books, as many works could not have been written without the information gathered 'per litteras'. Finally, single letters and letter collections were

[4] Broman 2000, 226–8.

subsequently printed as and in books or conceived for publication in print from the outset.

The importance of letters in the early modern period also extended to the natural sciences, the representatives of which were in particular need of information exchange. Scientific letters developed as a subspecies of learned letters, from the other varieties of which they were not sharply distinct. Formally, they resembled letters from fields like theology, law, philology, and antiquarianism, all of which followed a similar logic.[5] In terms of content, the multiple subjects addressed in most letters make a clear demarcation illusionary as well.

In what follows, I will first provide an overview of scientific correspondence conducted and preserved in manuscript. After that, the various ways in which letters were turned into printed books will be elucidated. Each section will be concluded by a short analysis of a pertinent example.

Scientific correspondence

In recent decades, the extent and impact of early modern letter exchanges has become more and more visible and attracted unprecedented attention. While the correspondence of major figures has long been published separately or as part of 'opera omnia' editions,[6] two interlinked developments have recently boosted interest in such material and raised the engagement with it to a new level: the possibility of publishing massive text corpora online[7] and the rise of network analysis. Correspondence has been recognized as granting privileged access to intellectual networks and thereby providing insight into the channels and mechanisms through which discoveries, inventions, techniques, and notions spread throughout Europe and beyond.

A good place to start an overview of early modern scientific correspondence is the digital platform *Early Modern Letters Online* (*EMLO*). Uniting metadata, and in some cases the texts themselves, of currently *c.*160 sixteenth-, seventeenth-, and

[5] This transpires from the contributions collected in Miert 2013a, which together provide a good introduction to the learned letter as a medium of information exchange in early modern times.

[6] Important scientific examples include the correspondence of Tycho Brahe (Dreyer 1913–29, vols 6–8, 14), Kepler (*KGW* 13–18), Marin Mersenne (Waard et al. 1932–88), Christiaan Huygens (Société Hollandaise des Sciences 1888–1950, vols 1–10 and 22), Henry Oldenburg (Hall and Hall 1965–86), Leibniz (series I–III of the 'Akademie-Ausgabe', 1923–), John Flamsteed (Forbes et al. 1995–2002), Albrecht von Haller (Boschung et al. 2002; a repertory, not an edition), and Leonhard Euler (*LEOO*, ser. IVA).

[7] There are major online editions of Carolus Clusius (https://clusiuscorrespondence.huygens.knaw.nl/, accessed 16 February 2023), the Bernoulli dynasty (https://ub.unibas.ch/bernoulli/index.php/Hauptseite, accessed 16 February 2023), Linnaeus (https://c18.net/linnaeus/index.php, accessed 16 February 2023), and medical letters 1500–1700 from the German-speaking world (http://www.aerztebriefe.de/, accessed 16 February 2023).

eighteenth-century letter exchanges, *Early Modern Letters Online* has established itself as the most important contact point in the field. The figures given on this website comprise numerous personalities who were active in the exploration of the natural world. Table 1 includes thematically pertinent correspondence which comprises one hundred letters or more in chronological order.[8]

Of course, this table cannot provide more than a first impression, which is in need of interpretation and supplementation. In what follows, some of its implications will be spelled out and discussed. First of all, numbers: between them, the 37 collections of correspondence included comprise some 48,000 letters. However, as *EMLO* does not aspire to completeness, this figure only represents a fraction of the material currently known. Many important letter writers, such as Euler, of whose correspondence some 3,000 items are preserved, are represented by part of the material only. Others are not included at all. This is true of authors like Conrad Gessner (*c*.500 letters to 255 correspondents preserved), Carolus Clusius (a correspondence of *c*.1,600 letters[9]), and the members of the Bernoulli dynasty (over ninety letter exchanges comprising one to several hundred letters), but it is also true of some of the most prolific epistolographers of all time: the combined output of Leibniz (*c*.20,000 letters), Linnaeus (*c*.6,300),[10] and Albrecht von Haller (*c*.17,000) alone nearly equals all the scientific letters presented by *EMLO*. Moreover, it is also enlightening to compare the results of an endeavour which does not focus on big names but tries to cover a particular disciplinary, regional, and chronological section as fully as possible: the project *Frühneuzeitliche Ärztebriefe des deutschsprachigen Raums* (1500–1700) (see note 7) has unearthed about 65,000 letters so far (52,000 of which are already filed in its database). Taking evidence such as this together and considering that large portions of the field have not been surveyed at all to date, one may reckon that all scientific letters preserved from the early modern era add up to a high six-figure or a seven- figure number, not to speculate about the much bigger proportion lost.

As is evident from the names mentioned so far, letters were exchanged by representatives of most scientific disciplines. The logical exception to this rule was secretive disciplines such as chymistry or magic, the exponents of which refused to participate in the exchange of knowledge which became a prerequisite of the emergence of modern science. Of the remaining disciplines, medicine in all probability was the most productive, as suggested by the *Ärztebriefe* project. This assumption also ties in with the fact that it was physicians who created

[8] For each correspondence, the total number of letters is not given according to the list of correspondence catalogues, because the numbers are often manifestly wrong (a problem recognized and explained in a 'known issues' section). Instead, the numbers of letters sent and received by each scholar have been determined in the 'search+' section and added.

[9] On Gessner and Clusius, see Delisle 2013 (esp. 29) and Egmond 2013.

[10] For an overview of Linnaeus's correspondence, see Jönsson 2011, esp. 171–5.

Table 1 Early modern scientific correspondence

Name	Dates of birth and death	Number of letters	Most important language*
Abraham Ortelius	1527–98	483	Latin (437)
Joseph Scaliger	1540–1609	1,716	Latin (1,103)
Tycho Brahe	1546–1601	526	Latin (49)
Giovanni Antonio Magini	1555–1617	118	Latin (67)
Johannes Kepler	1571–1630	958	Latin (872)
Johannes Isacius Pontanus	1571–1639	373	Latin (56)
Nicolas-Claude Fabri de Peiresc	1580–1637	2,570	French (2,276) (Latin: 150)
Matthias Bernegger	1582–1640	695	Latin (647)
Caspar Barlaeus	1584–1648	759	Latin (11)
Joachim Jungius	1587–1657	493	Latin (421)
Marin Mersenne	1588–1648	1,489	French (970) (Latin: 680)
John Winthrop the Elder and the Younger	1588–1649, 1606–76	1,696	English (1,667) (Latin: 43)
René Descartes	1596–1650	1,108	French (340) (Latin: 106)
Athanasius Kircher	1602–80	2,760	Latin (1,628)
Pierre de Fermat	1607–65	204	French (165) (Latin: 84)
Gaspar Schott	1608–66	227	Latin (184)
John Wallis	1616–1703	2,508	English (1,448) (Latin: 403)
Elias Ashmole	1617–92	769	English (637) (Latin: 18)
Henry Oldenburg	1618–77	5,987	English (2,676) (Latin: 1,451)
John Collins	1625–83	463	English (353) (Latin: 70)
John Aubrey	1626–97	1,870	English (784) (Latin: 6)
Philipp Jacob Sachs von Lewenheimb	1627–71	167	Latin (157)
Robert Boyle	1627–92	2,234	English (1,810) (Latin: 136)
Christiaan Huygens	1629–95	3,179	French (389) (Latin: 71)
Antonie van Leeuwenhoek	1632–1723	682	English (136) (Latin: 13)
Jan Swammerdam	1637–80	213	Latin (17)
Edward Bernard	1638–97	819	English (533) (Latin: 271)
Martin Lister	1639–1712	2,761	English (2,133) (Latin: 194)
Robert Plot	1640–96	189	English (178) (Latin: 2)
Isaac Newton	1643–1727	1,571	English (1,250) (Latin: 309)
Edmund Halley	1656–1742	454	English (221) (Latin: 47)
Edward Lhuyd	1660–1709	3,971	English (3,704) (Latin: 97)
Royal Society	1665–1800	751	English (372) (Latin: 99)
James Jurin	1684–1750	932	English (741) (Latin: 124)

Leonhard Euler	1707–83	776	French (343) (Latin: 316)
Thomas Pennant	1726–98	679	English (666) (Latin: 20)
Charles Hutton	1737–1823	138	English (136) (Latin: 0)

* The figure for the most-used language is given, plus the figure for Latin, if both are not identical. Letters can be written in more than one language, so the figures in this column may add up to a number higher than the total, as in the case of Mersenne. On the other hand, no language may be indicated for most letters of a correspondence, which leads to implausibly low numbers of letters written in its dominant language (e.g. in the case of Tycho Brahe).

Source: *Early Modern Letters Online*.

discipline-specific subforms of the letter in print, as will be discussed later. Natural history, represented by personalities such as Lhuyd or Swammerdam—many of them doctors of medicine as well—seems to come second. The sheer mass of hitherto unknown animals, plants, and minerals naturalists were confronted with made letters an indispensable tool for them to remain *au courant*. Linnaeus would have been unable to complete and update his classification of the three realms of nature in remote Uppsala without an intense 'commercium epistolicum'. Astronomers like Brahe, Magini, or Kepler needed one another's data to calibrate their own observations and to learn about events which had been unobservable to them due to bad weather or other reasons. They also discussed the meaning of their observations and the conclusions to be drawn from them.[11] Mathematical and descriptive geographers and cartographers like Ortelius relied on information obtained 'per litteras' because they could not travel the whole world for themselves. Physicists such as Huygens and Newton corresponded about experiments and their interpretation. Mathematicians, for instance Wallis and Euler, exchanged results with colleagues and discussed unsolved problems.[12] Moreover, one finds among early modern letter writers numerous individuals not readily assignable to any major scientific discipline: from natural philosophers (Descartes), polymaths (Kircher), and chronologists (Scaliger) to experimenters (Schott), microscopists (van Leeuwenhoek), and, last but not least, organizers and so-called intelligencers, that is, people with an extensive network which they used to spread information systematically among the members of the republic of letters (Peiresc, Mersenne, Oldenburg).[13] Moreover, many intellectuals were active in several fields, and the scientific content of their letters was correspondingly diverse.

The table also provides some hints about the chronological development of early modern scientific correspondence. Of course, many collections of correspondence

[11] Mosley 2002, esp. 468; 2013, esp. 116 and 134.

[12] An exemplary instance of an ongoing mathematical discussion is provided by the correspondence between Euler and Christian Goldbach (*LEOO*, ser. IVA, vol. 4).

[13] On their correspondence, and especially on of Ismaël Boulliau (1605–94), see Hatch 1998.

span parts of two centuries, but one may broadly assign four to the sixteenth, twenty-five to the seventeenth, and eight to the eighteenth century. As already mentioned, some extremely extensive collections of correspondence from the late seventeenth and eighteenth centuries are not covered by these figures. All in all, then, the data suggest that scientific correspondence (with the partial exception of medicine, some of the letter formats of which date back to the Late Middle Ages) developed out of humanist letter writing and learned correspondence in general with some delay and expanded over the centuries.[14] Indeed, while examples of collectionsof scientific correspondence from the fifteenth and early sixteenth centuries exist, their scope is quite restricted. One gets the impression that corresponding had not become part of the daily working routine yet. To take astronomy as an example, the numbers of letters sent and received by authors such as Regiomontanus, Copernicus, and Georg Joachim Rheticus are modest compared with the those of later luminaries like Brahe, Kepler, or John Flamsteed.[15] From the mid-sixteenth century, correspondence networks and letter corpora became more extensive and bulky by the decade, as the number of people interested in the natural world kept growing and material conditions improved, while scientific journals did not exist or had not yet reached their later dominant position as distributors of new knowledge. After this happened, in the decades around 1800, scientific letters continued to be written, sometimes in imposing numbers, as in the well-known case of Charles Darwin. However, they gradually lost their vital importance for scientists as the most important means of exchanging fresh information.

An idea of the geographical extension of letter networks, not displayed in the table, can be gained with the help of the 'places' categories in *EMLO*'s 'search+' section. Two things transpire: first, most letters sent and received by a given individual remained within his vicinity in the wider sense, that is, within a range of a dozen to a hundred kilometres, and within the same language community. Secondly, impressive distances were spanned beyond that core region, and these distances tended to increase with time. In the mid-sixteenth century, Gessner's network reached 'from Caen to Warsaw, from Utrecht to Rome'. In the mid- and late eighteenth century, Linnaeus's literally spread over the whole of Europe and beyond: he received letters from Russia, Turkey, Africa, and North America.[16] In some cases, the geographical development of colections of correspondence indicates important developments which affected the community of naturalists. After Peter the Great founded the Russian Academy of Sciences in 1724 and was

[14] Cf. Miert 2013a, 246–7, whose assertion that the apogee was reached in the seventeenth century seems to be in need of qualification.

[15] The comparison is taken from Mosley 2013, 134.

[16] Delisle 2013, 29 (see also Maclean 2008, 23–4 for sixteenth-century medical letters in general); Jönsson 2011, esp. 173.

able to attract many Western men of science, in particular from Germany, St Petersburg became an important hub of scientific correspondence.

Regarding language in general and the role of Latin in particular, *EMLO*'s corpus is not representative. Of a current total of about 192,000 letters, some 70,900 are in English, 37,800 in Latin, 21,800 in French, 11,900 in Dutch, 9,200 in German, 4,800 in Italian, 900 in Spanish, and the rest in several other vernaculars.[17] One recognizes at a glance, for instance, that English is grossly over-represented, that the figures for Dutch and German are out of proportion, and that German, Italian, and Spanish are all under-represented. Moreover, no language at all is indicated for a substantial number of letters. None of this inspires confidence that the role of Latin will be adequately represented either in general or with regard to scientific correspondence. Nonetheless, some generalizations can be made with due caution and the help of other sources: from the sixteenth century, Latin had to compete with French and other vernaculars as a language of scientific correspondence. In most large letter exchanges, it was used along with one or more vernaculars. Language distribution varied widely according to time, the area covered by a correspondent's network, and his and his partners' mother tongues, foreign language skills, and preferences. Speakers of major and fairly standardized vernaculars such as Italian, French, or English were more inclined to use them in their letters (as they also did otherwise) than authors whose mother tongues were minor and underdeveloped. The role of Latin increased with the internationality of a correspondence and decreased over time.

A few examples may illustrate these rules of thumb but also show that not everything can be explained with their help. Before and around 1600, Carolus Clusius, culturally Flemish but probably of French mother tongue and widely travelled, received letters in six languages—one third in Latin and French each, the rest in Italian, German, Spanish, and Dutch.[18] Only a few decades later, the Provençal Peiresc, most of whose correspondents were French, wrote and received letters overwhelmingly in the national idiom. By contrast, German doctors of the sixteenth and seventeenth centuries preferred to correspond with their compatriots and colleagues in Latin. Of the 52,000 letters filed in the database of the *Ärztebriefe*, over 40,000, more than three-quarters, are in Latin. This situation had not changed fundamentally by the mid-eighteenth century, as will soon become clear. The Swedes, on the other hand, preferred their vernacular among themselves but wrote to the international community in Latin, if the Linnaean correspondence is in any way representative: 45 per cent of its letters are in Swedish, 45 per cent in Latin, only 10 per cent in other vernaculars.[19]

[17] The comparatively high figures for ancient Greek and Hebrew (*c.*1,550 and 150 respectively) mainly result from quotations within letters in other languages.

[18] Miert 2013a, 226.

[19] Jönsson 2011, 171.

When one looks at individual collections of correspondence in greater detail, the mix of linguistic competence, practical motives, considerations of prestige, and personal taste leading to certain decisions can become nearly impossible to reconstruct. Some choices made by letter writers appear simply whimsical. For the Italian astronomer Magini, for instance, it made obvious sense to exchange letters with Brahe in Latin while using Italian for his noble compatriot Francesco Gonzaga, Duke of Mantua. But why Magini's Italian colleague Bartolomeo Cristini wrote to him in Latin while Magini himself addressed Galilei in Italian is hard to tell.[20]

Finally, it should be noted that the multilingualism which characterized collections of correspondence as a whole can be observed on a smaller scale as well. It is not to be taken for granted that two correspondents used the same language on every occasion or for the whole duration of their correspondence. Sometimes, translation had to intervene because the partners had no common language at all.[21] Even within a single letter, code-switching could be used not only on account of quotations, but also because of a change of topic. When entering into technical detail about combinatorics in a letter to Fermat, Pascal turned from French to Latin: 'Par exemple, et je vous le dirai en latin, car le français n'y vaut rien: Si quotlibet litterarum, verbi gratia octo...'[22]

Not too much needs to be said for the moment about the character of the single letter in general, as a representative example will soon be analysed, but a few general remarks are nonetheless in order. The most salient trait is the thematic structure, which was characterized by diversity and lack of coherence. A typical letter presented a series of pieces of information which were independent of one another—'atomistic knowledge', as it has been appositely termed.[23] These individual items could be medical case reports, information about new plants, astronomical data, mathematical findings, or anything at all. They did not have to relate to one overarching topic, a given discipline, or even science strictly speaking. In fact, information about the natural world was usually intermingled with news from academia (colleagues, controversies, jobs, publication problems), private matters (health, family, career prospects), and remarks on the correspondence itself (one's correspondent's or one's own prolonged silence,

[20] See Favaro 1886, nos. XII, XIV, XVI–XVIII, suppl. 4–7, 10–11 (Magini and Brahe), nos. LIX, suppl. 15–19, 21, 23 (Magini to Gonzaga), nos. XXIX, XXXV–XXXVIII (Cristini to Magini), nos. suppl. 27–9 (Magini to Galilei).

[21] An example of a correspondence fluctuating between Latin and French is provided by that between Albrecht von Haller and Leopoldo Marcantonio Caldani (Boschung et al. 2002, vol. 1, 84–8). For a once and for all switch from German to Latin and for translations, see later in this section .

[22] Pascal to Fermat, 29 July 1654; About 1983, 41. Numerous similar examples are contained in the correspondence between Euler and Christian Goldbach; see, for instance, Euler to Goldbach, 6 March 1742 (*LEOO*, ser. IVA, vol. 4, no. 47).

[23] Maclean 2008, 28 (see also 26).

difficulties—in the absence of national postal services—to find a messenger, letters delayed, damaged, or lost).[24]

This thematically discrete format was advantageous for the writer in several respects. It saved him time because it spared him structural considerations, it was flexible enough to accommodate everything he saw fit, and it allowed for any last-minute additions, if he thought it necessary (or just wanted to leave no blank space and waste no paper).[25] The recipient, for his part, was interested specifically in the single 'atoms of knowledge' and cared little about the coherence of the letter as a whole. The value attached to each piece of information could even induce him to cut letters up in order to file the interesting pieces in a personal card index or to glue them into a volume prepared for that purpose[26]—practices which came as a natural extension of established thinking and working habits: early modern scholars were used to paying little attention to large-scale textual structures and much more to useful individual facts and phrases, which they collected systematically (as noted in Chapter 4, 'Early modern print').

Not all relevant information was necessarily included in the letter itself. In many cases, its contents were supplemented by attachments on separate slips ('schedia'): notes, lists, tables, diagrams, images, and sensitive information (in view of the custom of sharing letters with acquaintances; alternatively, such matters could be entrusted to the bearer). In natural history in particular, the most precious kinds of information were usually specimens of certain plants, animals, or minerals themselves. In all disciplines, books belonged to the most common objects requested and sent.[27] Letters thus often came as packages.

Style could vary widely, but often, as little time and energy were devoted to this aspect as to structure. Many scientific letters starkly contrast with humanist epistles, the authors of which used the quality of their Latin as an intellectual business card. Over time, stylistic negligence seems to have become more and more acceptable, as correspondence grew and available time shrank. Sixteenth-century examples are usually plain and unadorned, but Gessner still edited his medical and natural historical observations for inclusion in his letters.[28] This seems to have changed by the eighteenth century, as suggested by the example to which we will turn presently.

[24] The compartmentalized and disparate character of the early modern scientific letter has been noted by Egmond 2013, 69, and Boran 2013, 181, for Clusius's and James Ussher's correspondence respectively. Further good examples are to be found in Mersenne's and Euler's: see, for instance, Isaac Beeckman to Mersenne, March 1629 (http://ckcc.huygens.knaw.nl/epistolarium/letter.html?id=beec002/0013, accessed 16 February 2023); Christian Goldbach to Euler, 20/31 July 1730 (*LEOO*, ser. IVA, vol. 4, no. 8).

[25] Euler to Goldbach, 4/15 June 1730 (*LEOO*, ser. IVA, vol. 4, no. 5, p. 114): 'Ut pagina haec impleatur, transcribam quadraturam quandam circuli.'

[26] Miert 2013a, 229; Delisle 2013, 38–42 (on Gessner's 'letterbook').

[27] Miert 2013a, 5–6; 2013a, 109–10; Delisle 2013, 27, 30. On attachments to botanical and astronomical letters in particular, see Egmond 2013, 50, 61; Mosley 2013, 120–1.

[28] Plain style: Maclean 2008, 21; Egmond 2013, 49–50 (Clusius). Gessner editing his observations: Delisle 2013, 32–3.

The correspondence of the Bernese physician, anatomist, and botanist Albrecht von Haller (1708–77), professor in Göttingen and one of the leading naturalists of the eighteenth century, has been thoroughly examined in a repertory and an analytical overview.[29] About 3,700 letters from Haller and 13,300 addressed to him have been preserved. Their length ranges from a few lines to several dozen pages. Haller exchanged them with some 1,150 correspondents, including 50 women, in 8 different languages. Overall, Latin ranks third in terms of quantity after German and French, followed by English; Italian, Dutch, Swedish, and Spanish come far behind. In total, about 3,600 letters or 21 per cent of the total preserved are in Latin. However, these statistics change decisively if one focuses on the strictly scientific part of the letter exchange. While the private, societal, and political parts of it (Haller held several offices in his home town) are overwhelmingly in German and French, Latin plays a far more prominent role in scientific topics. Moreover, Haller himself clearly thought such letters the most valuable part of his correspondence. His own edition of select letters to him started with those in Latin, and the six volumes of the *Epistolarum ad Albertum Hallerum pars I* (1773–5) were the only part of this project he managed to complete in his lifetime.

Unsurprisingly, Haller's scientific letters were exchanged above all with his peers: professors of medicine, anatomy, and botany as well as practising doctors sent and received the lion's share. In geographical terms, Haller's network extended as far as Spain, Scotland, Sweden, and Russia. However, about two-thirds of his Latin correspondence remained within the limits of the German-speaking world. This corresponds to the longevity of Latin in Germany, where a vernacular standard language was slow to develop, but also to Haller's own predilection for the old language. To him, Latin was more than just a convenient lingua franca. By virtue of its immutability and richly developed technical terminology, it constituted the scientific language par excellence. In view of Haller's standing, this preference may have exerted some pressure on his German correspondents to use Latin as well.[30] As a common third language, Latin served in Haller's dealings with speakers of vernaculars like Italian (which he understood but did not write himself) and Swedish (Linnaeus in particular). Thematically, the letters feature the aforementioned mix of scientific, professional, and private subjects. Along with discussions of topical research questions, the identity of plants, or the treatment of patients, the correspondents alert each other to new publications, enquire about colleagues, discuss appointments to professorships, form alliances,

[29] Boschung et al. 2002; Stuber et al. 2005. Within the latter publication, see esp. Krebs 2005, from whom much of the information given in what follows is taken. For a short overview, cf. also Korenjak 2016, 159–62.

[30] However, Latin correspondence between intellectuals of the same mother tongue was restricted neither to Haller and his correspondents nor to speakers of German. In England, one of the first countries in which the vernacular asserted itself, James Ussher (1581–1656) corresponded with colleagues like William Camden in Latin (Boran 2013, 182–3).

chat about their careers, families, and state of health and mind, and speculate about the whereabouts of earlier postal items. Sometimes, scientific documentation, for example in the form of plant lists, is added to a letter as an enclosure. Books and plant specimens are included as well.

To illustrate the character of this correspondence, a closer look at one of Haller's letters to Johann Georg Gmelin (1709–55) is helpful.[31] Gmelin and Haller became friends as teenagers, when they both studied medicine in Gmelin's home town of Tübingen between 1723 and 1725. In the years that followed, they conducted a lively correspondence, which continued after Gmelin moved to St Petersburg in 1728 to become a member of the newly founded Russian Academy of Sciences. In 1733, he joined the Great Northern Expedition and vanished in the vast expanses of Siberia, to reappear only in 1743, accompanied by a wealth of botanical material, the publication of which could only be completed posthumously by his nephew (*Flora Sibirica*, 1747–69). When Haller learned of Gmelin's return, he immediately wrote to him, greeted him enthusiastically, invited him to renew their friendship, and begged him to share his knowledge about Siberian flora. The ensuing letter exchange continued until Gmelin's untimely death. Tellingly, while the friends had corresponded in German as teenagers and young men, they wrote exclusively in Latin now that they were reputable members of the scientific community. The following letter was sent from Göttingen to St Petersburg a few months after Gmelin's return. Haller presumably expected it to be read not only by Gmelin but also by two mutual friends, the physicians António Nunes Ribeiro Sanches and Johann Friedrich Schreiber, who likewise held positions in St Petersburg.

Amicissimo Gmelino

s(alutem) p(lurimam) d(icit) A. Haller

Ecce iuramentum, quo se obligabit d(ominus) P(appelbaum), prius quam diploma accipiat. Id ergo sua manu signatum ad me mittat necesse est. Diplomata signata et disputatio impressa est, cuius errores emendavi. Seposui exemplaria et <curavi>, ne unum distribuetur ante medium mensem Martium, nihil ultra mihi licet contra statuta.

Ad d(ominum) Sanches nuper scripsi. Schreiberi epistolam avidus expecto, tum lapillum. Facile invenero editorem, sed placebit aeque, si cl(arissimus) Wagnerus curaverit.

Gerberiana auctor optimus manu sua mihi commendavit, quae apud me sunt. Si fieri potest, manibus amici satisfaciam. Ego totus sum in anatomicis. Arteriae

[31] The correspondence between Haller and Gmelin has been edited, translated, and commented by Krebs 2008. For the letter discussed in what follows and more detail about its subjects, see there no. 30 (pp. 103–5).

carotidis externae plenam iconem paravi sculpendam. Vasa ab vagina ad int(estinum) rectum manifeste demonstravi. Sextus tomus Boerh(aav)i ex prelo prodiit. Quintus nunc incipitur. Omnia vobis misi Lipsiam, mittam denuo, si viam novi<s>tis Hamburgi aut Lubecae.

Accepi Nortwyckii uterum gravidum. Albini Eust(achiu)m quotidie expecto, qui ex prelo prodiit nuper. In Germania nihil prodit alicuius momenti, et nescio quo modo retinentur omnium manus, ne maius aliquid edant; adeo inveniendi curam relinquunt exteris.

Vale. Saluta m(eo) n(omine) cl(arissimum) Schreiberum. Dedi raptim d(ie) 5. Dec(embris) 1743.

To his best friend Gmelin

A. Haller wishes all the best.

Here you have the oath by which Mr Pappelbaum has to bind himself before he can receive the diploma. So he must send it back to me signed by his own hand. The diplomas are signed and the disputation printed; its errors I have corrected. I have put aside the copies and have seen to it that not even one will be distributed before mid-March. I can do no more against the statutes.

To Mr Sanches I wrote a short while ago. Schreiber's letter I expect impatiently, and then the small stone. I will easily find a publisher but will be equally content if the most famous Wagner takes care of it. Gerber's writings the excellent author consigned to me with his own hand; they are now with me. If it can be done, I will satisfy the soul of my dead friend. I am completely absorbed by things anatomical. Of the external carotid artery, I have produced a full picture for engraving. The vessels from the vagina to the straight intestine I have shown plainly. Volume six of Boerhaave has been published. Volume five is begun now. All of this I have sent you in Leipzig. I will send something again, if you know a street in Hamburg or Lübeck.

I have received Noortwyck's pregnant womb. Albinus's Eustachi I expect daily. It has recently been published. In Germany, nothing of any importance appears, and for reasons I do not know everyone's hand is held back from publishing anything substantial; they leave the trouble to find out something completely to the foreigners.

Stay well. Pass on my greetings to the most famous Schreiber. Given in a hurry on 5 December 1743.

This letter is emblematic of eighteenth-century scientific correspondence in several respects. The first thing to note is its information density. Haller has a few friendly words for Gmelin, Sanches, and Schreiber, but he keeps these niceties brief. Instead, he shows his affection by squeezing into his letter as many chunks of information as possible. They regard (1) the current state of the affair of

Gmelin's friend Georg Christoph Pappelbaum, a physician in St Petersburg, who seeks a medical doctoral degree from Göttingen; (2) letters expected; (3) a manuscript by Schreiber perhaps to be published by Haller; (4) the written legacy of his deceased friend Traugott Gerber, which he will likewise try to publish; (5) Haller's ongoing anatomical work; (6) the progress of his edition of the *Praelectiones academicae* (1739–44) of his former teacher Herman Boerhaave; (7) his attempt to provide Gmelin and his friends with the above works; (8) books received and expected by Haller; and (9) the dearth of innovative work in Germany. All of this is presented without much order and coherence. Haller simply juxtaposes otherwise unconnected facts and data. Nor does the information strike one as being carefully selected to suit the needs and interests of Gmelin and his friends. The first three points and the seventh clearly are relevant to them, but for the rest, Haller seems to commit to paper anything crossing his mind which has some chance of being of some interest. The contrast with a humanist letter, unobtrusively yet carefully structured and fine-tuned to suit its addressee's character, situation, and needs, is once more noteworthy.

The same goes for language and style.[32] In addition to many intentional abbreviations, resolved in round brackets in the Latin text, Haller commits outright errors, leaving out letters and entire words ('novi<s>tis', '<curavi>'). Tenses and moods are treated with great negligence. An imperfect subjunctive is replaced by a simple future ('distribuetur' instead of 'distribueretur'), and a future by a future perfect ('invenero' instead of 'inveniam'). A Greek term may be used although a perfectly Latin equivalent would be available ('iconem' instead of 'imaginem'). An expected participle like 'ducentia' is omitted in 'Vasa ab vagina ad int(estinum) rectum'. In contrast to normal Latin usage, Haller constantly puts direct objects at the beginning of sentences to stress their importance ('Gerberiana...', 'Arteriae carotidis externae plenam iconem...', 'Vasa...'). Even considering that he can presuppose much shared knowledge from earlier letters, he is often elliptical in the extreme. For 'the manuscript of Schreiber's treatise on kidney stones', he simply writes 'lapillum', 'the small stone'; books by Noortwyck and Albinus are alluded to in similar manner. The same words and phrases are repeated a short distance apart ('ex prelo prodiit', 'ex prelo prodiit', 'nihil prodit'). Connecting particles may be omitted and short sentences follow upon one another in staccato style, as in the second paragraph.

That Haller's lexical, grammatical, and stylistic mastery of Latin was, in fact, beyond reproach is attested by thousands of printed pages. The reason for his casual letter style surfaces in the set phrase 'Dedi raptim' in the valediction: like many of his colleagues then and now, Haller was always short of time. Letter writing was an additional strain on his time, which is why he focused on getting the

[32] On the Latin of Haller's letters in general, see Krebs 2005, 362–8.

information across to the detriment of elegance. In this respect, his attitude to Latin was practical and down to earth. Although he was strongly convinced of Latin's superiority over the vernaculars as language of science and wanted to maintain it in this function, as already mentioned, he thought of it as an instrument, not an end in itself. Exaggerated care for linguistic form was useless in his eyes. When he once heard from Gmelin that Sanches felt embarrassed because of his shaky Latin, he told him to reassure their mutual friend: 'Antonio nostro nollem ullum ob dictionem scrupulum esse; viri enim sumus et seria et res ipsas tractamus.' ('I would not want our António to have any concerns regarding the wording: for we are grown men and deal with serious matters, with the things themselves.'[33])

Letters in print

From the incunabula era, letters were printed regularly. The appetite of the public for this kind of reading was great, particularly in the case of famous authors. Tycho Brahe could be assured of this when he published the first volume of his *Epistolae astronomicae* (1596). As he remarks in the preface:[34]

> epistolae multa et varia complectantur et ob id a quamplurimis avide leguntur, idque minori taedio quam alia eadem fere tractantia et copiose replicantia volumina.
>
> letters comprise many different things and are, therefore, avidly read by a lot of people; readers become less weary of them than of other volumes, which treat more or less the same topics, unfolding them at length.

Not much of an excuse was needed, therefore, when one set about satisfying the readership's desire for letters. When Caspar Bauhin edited a work on plant names by his brother Johann (*De plantis a divis sanctisve nomen habentibus*, 1591), he appended a number of hitherto unpublished letters by Conrad Gessner to the main text. As he explains (pp. 93–4), he did so because Johann's treatise was otherwise slim ('alias parvo'), because the letters contained much information on botany and pharmacology, and because he had noted that 'the great man's letters are very much asked for and yearned for' ('tanti viri litteras maximopere expeti et desiderari').

As noted earlier, letters were semi-public documents from the outset. Their authors might write them with an eye to publication in the first place and

[33] Haller to Gmelin, 18 February 1745, discussed in Krebs 2008, 53.

[34] Dreyer 1913–29, vol. 6, 22. On the *Epistolae astronomicae* in general, see Mosley 2002. For the emphasis on variety, see also Maclean 2008, 21.

distribute copies to people other than the primary addressee. Recipients often read them to or copied them for friends and colleagues, excerpted them in their own letters, or forwarded them to other correspondents.[35] Publication in print, therefore, did not contradict a letter's original nature in principle. Nonetheless, the transition from manuscript to printed page implied a quantum leap in terms of reach and impact and affected the documents in several ways.

To begin with, in print, letters were decontextualized and recontextualized. Entire correspondences, such as Brahe's aforementioned *Epistolae astronomicae*, which he exchanged with William IV, Landgrave of Hesse-Kassel, and his court astronomer Christoph Rothmann, were published only rarely,[36] as most people did not bother to keep copies of their own letters. (Brahe had secretaries to whom he dictated his letters[37] and who presumably also made copies for his personal archive; few others could count on such help.) Usually, the reader was faced with one side of a correspondence only or simply with individual letters written at various times by different authors or to different addressees. In most cases, he will have lacked the intimate biographical knowledge needed to grasp fully the original purpose of the documents in front of him. On the other hand, a published letter was placed in a new context of originally unconnected letters and of paratexts—not only the usual introductory matter but also aids to orientation in the form of summaries and indexes.[38] When Antonie van Leeuwenhoek's long letters to the Royal Society were collected and published (*Arcana naturae detecta*, 1695), they were prefaced by abstracts, and their various subjects were listed in an index. In short, the change of context in print amounted to emphasizing the factual content of letters at the expense of their communicative function in a specific situation.

In accordance with their new context, letters were often edited for publication in terms of content and form. Regarding the former, private matters could be excised completely or in part. In the preface to his *Epistolarum ad Albertum Hallerum pars I* (vol. 1, sig. *3^{r-v}), Haller explains:

> Earum etiam, quas admisi, plurima loca expunxi, quae inter amicos gratiam suam olim habuerunt, aliis lectoribus fastidio essent futura aut certe nihil haberent, quod scire cuperetur.

[35] With regard to early modern science, see Mosley 2002, esp. 465–8; Maclean 2008, 24; Delisle 2013, 36; Mosley 2013, 118–20.

[36] Some cases were prompted by extraordinary circumstances and purposes. The *Commercium epistolicum domini Iohannis Collins et aliorum de analysi promota* (2nd edition 1722) was published to bolster Newton's position in his priority dispute with Leibniz.

[37] Dreyer 1913–29, vol. 6, 23. The French intelligencer Ismaël Boulliau usually kept copies of his outgoing Latin letters; of those in other languages he only retained the ones he deemed particularly important (Hatch 1998, 71).

[38] Maclean 2008, 21.

> Even of those [letters] included [for many were left out altogether], I deleted a great number of passages which once had their charm among friends but would be tedious to other readers or in any case contain nothing one would want to know.[39]

Besides lack of general interest, the desire to avoid trouble through overly frank statements, be it harsh judgements about colleagues or heterodox opinions, provided an ulterior motive for leaving out certain passages. Sometimes, so much was cut out that only excerpts remained to be printed.[40] Titles emphasizing this operation, such as 'Excerpta ex epistolis ad Gulielmum Crabtrium' (in Jeremiah Horrocks's astronomical *Opera posthuma*, 1673, pp. 241–338) or *Excerpta ex litteris ad Iohannem Hevelium perscriptis* (1683), suggest that readers liked the idea of getting straight to the heart of the matter without losing time with frills.

The language was usually revised as well. The plain, casual style seen as genre-typical was as a rule preserved[41] and excessive attention to linguistic form rejected, as announced, for instance, by Giovanni Manardi in the opening letter of his *Epistolae medicinales* (1521): 'nec Latinam linguam discere sed medicinam cupientibus haec scripta sunt.' (fol. 2^v; 'the following has been written for those who want to learn, not Latin, but medicine'[42]). However, slips of the pen, grammatical errors, and stylistic infelicities were removed. Haller implies as much on the second-last page of his aforementioned preface when declaring that he departed from this practice in the case of Linnaeus's letters only ('immutata reliqui omnia neque vitia aliqua stili ex properatione nata emendavi').[43] Sometimes, publication presupposed translation, as in the case of Antonie van Leeuwenhoek, who wanted to share his microscopic observations with the Royal Society but whose letters would have been incomprehensible to the fellows and most other readers in the original, because he spoke only Dutch. Therefore, letters like the one containing the first description of spermatozoa came out in Latin in the *Philosophical Transactions* ('Observationes de natis e semine genitali animalculis', 1679), and later collections of them were published in that language as well (see Chapter 5, 'Translations'). When Johann Erich Olhoff published the *Excerpta* of letters by famous colleagues to Johannes Hevelius in 1683, he apparently found French examples acceptable for the reading public but translated Italian ones.

As is evident from the examples just discussed, the most important publication format was the collection, which had ancient (Cicero, Seneca, Pliny the Younger),

[39] For some examples of this operation, see Krebs 2005, 365–8.

[40] Maclean 2008, 21.

[41] Maclean 2008, 21.

[42] For another, quite extended example of this prefatory topos, see the preface to Brahe's *Epistolae astronomicae* (Dreyer 1913–29, vol. 6, 23).

[43] Because Linnaeus and Haller had clashed a number of times in the course of their careers, Haller wanted to avoid any suspicion that he could have tampered with the evidence in order to show himself in a better light (a point not made perfectly clear in Krebs 2005, 365).

late medieval (see note 45 below), and early modern models (Petrarch and other humanists). The practice of publishing letters in this form was transferred to print soon after its invention.[44] Criteria for inclusion varied, but in most cases, personal and thematic aspects were combined: letters were published together which had been written by the same sender, to the same addressee, formed a correspondence, or dealt with the same subjects. The collections by Leeuwenhoek and Haller, for instance, combined the criteria of author and addressee respectively with the one of content (microscopy in the first case, botany, medicine, and anatomy in the second). As shown by the former example or by the thirty-nine letter exchanges brought together in Stanisław Lubieniecki's *Theatrum cometicum* (1668), collections could focus on quite specific subjects. In most cases, however, the field covered was broader, comprising one or more disciplines. This is true in particular of medicine and related fields.

Medicine also stood out in its development of proper subgenres of letter collections. In fact, the earliest extant collections of medical letters already belong to a specific subtype: they are 'consilia', a form which originated in the thirteenth century in the orbit of the Northern Italian medical faculties.[45] A 'consilium' was usually composed by a renowned doctor as an answer to a request from a colleague facing a difficult case. Formally, the text could be addressed to the patient, his relatives, or his friends, but the real addressee was the treating physician, as medical laymen would have been unable to understand the technical Latin and the content alike. The basic structure of a 'consilium' was tripartite, comprising a description of the patient and his disease, dietary recomendations, and recomendationsfor the medical treatment strictly speaking, but there was much room for variation. More often than not, considerations restricted to the case at hand were combined with more general reflections. Authorities could be cited in great numbers. From the very beginning, 'consilia' were brought together in manuscript collections, which became larger and more consistently structured over time. A popular structure followed books 3 and 4 of Avicenna's *Canon*, which are dedicated to localized disorders, on the one hand, and generalized diseases and physical injuries, on the other. Such collections could assume encyclopedic dimensions, as in the case of the about 350 'consilia' by Bartholomaeus de Montagna, which fill over 650 two-column folio pages in their 1476 *editio princeps*. In the age of print, collected 'consilia' became popular as works of reference and media of theoretical discussion. The full title of Antonio Maria Venusti's *Consilia medica* (1571) proudly announces the thorough treatment of many difficult medical questions ('multae et arduae quaestiones medicae pertractantur'). In this context, the individual cases, which could be reported in great detail, served as medical 'exempla', demonstrating the practical implementation of a largely

[44] Clough 1976, esp. 41–58.
[45] The following paragraph summarizes results obtained by Agrimi and Crisciani 1994.

theoretical academic instruction. Collections of 'consilia' continued to be printed until the seventeenth century. While they had remained an Italian speciality in the manuscript era, print brought them to other countries as well, in particular Germany.

Collections titled *Epistolae medicinales* and the like constituted a more open variety, which could include 'consilia' but much else as well. They encompassed 'the humanist discussion of ancient sources and the referents for words found there; consilia; polemic with colleagues over remedies and the identification of illnesses and plants; engagement with the latest publications and latest ideas; accounts of rare and marvelous happenings; acts of eye-witnessing'.[46] Of this type of collection, which was inaugurated by Manardi's exceptionally successful *Epistolae medicinales*, at least twenty-one specimens are on record over the period 1521–1626. Many of them went through several editions.

Apart from collections, another way of publishing smaller numbers of letters or single items was to append them to major publications of a different kind as bonus material. Among the numerous and diverse pieces added by Gessner to his 1559 edition of the botanical works of the recently deceased Valerius Cordus, one finds a description of the flora of the Calanda, a mountain near Chur in Switzerland, which he had received 'per litteras' from Johannes Fabricius Montanus (fol. 236^{v}: 'Galandae montis stirpium enumeratio'[47]). Sometimes, such additions came in pairs whose second item answered the first. In Sebastien de Monteux's *Annotatiunculae in errata recentiorum medicorum* (1533), Bernhard Unger wrote to his teacher Symphorien Champier in defence of the use of laxatives in cases of fever caused by yellow bile (fols. xvii^{v}–xix^{v}). This 'Apologetica epistola pro defensione Arabum medicorum' was directly answered by an 'Apologetica epistola responsiva' (fols. xx^{r}–xxiii^{r}) in which Champier explained to Unger why he was wrong and warned him against the Arabs from whom he had derived such dangerous opinions.[48]

In some cases, letters were not added as an appendix but integrated into the body of a work and interwoven with its argument. Such insertions can be found, for instance, in Brahe's posthumous *Astronomiae instauratae progymnasmata* (1602), in which letters by William IV, Landgrave of Hesse-Kassel, Caspar Peucer, and others are reproduced and discussed,[49] or in Athanasius Kircher, who often draws on his extended correspondence. When he talks about mineral resources in his *Mundus subterraneus*, for instance, two letters inform the reader about conditions in the Tyrol and Burgundy respectively.[50]

[46] Maclean 2008, esp. 19 (quotation), 29–30 (bibliography of collections).
[47] For comparable examples from the works of Clusius, see Egmond 2013, 50.
[48] A more extended, but purely fictitious set of letters by a teacher to his pupil is appended to Thomas Moffett's dialogue *De iure et praestantia chymicorum medicamentorum* (1584).
[49] Mosley 2013, 119.
[50] They are found on pp. 246–8 of the 3rd edition (1678). The first letter is described in *TL* 518–19.

Individual letters could be published as self-contained items either as articles in journals or as books. The former option, which became available from the late seventeenth century only, will be discussed in the relevant section of this chapter. Here I just take the opportunity to draw attention to the fact that journals, and bodies and individuals connected to journals, invited such contributions by open letters to groups of people who had valuable information to share. These letters could take the form of printed questionnaires, adopting a format generally becoming popular at the time. This is true, for instance, of the 'inquiries of natural history' sent out by the Royal Society.[51] In an epistle published in the *Philosophical Transactions*, James Jurin invited other meteorologists to make a systematic record of and share information on their local weather, providing a blueprint of a meteorological diary at the end ('Invitatio ad observationes meteorologicas communi consilio instituendas', 1723). By contrast, Johann Michael Fehr's 'Epistola invitatoria ad celeberrimos Europae medicos', opening the first volume of the *Miscellanea curiosa* (1670), only constituted an invitation in general terms to contribute to the journal. The practice was sometimes taken up independently of a journal by enterprising individuals like Johann Jakob Scheuchzer. His *Charta invitatoria* ([1699]), which asked no fewer than 186 questions about Swiss natural history, was printed in two versions, German and Latin, and sent to correspondents all over Switzerland, most of whom sadly did not take the trouble to respond.[52]

That an individual letter which had originally been destined for a particular addressee was subsequently published in book format was not very common due to the difference in length between most letters and even a short book, but it was not excluded in principle, and some prominent examples can be cited. Columbus's letter about his first journey, originally written on his homeward voyage and addressed to the Spanish administrator of the royal privy purse, Luis de Santangél, spread the word of his discoveries throughout Europe as soon as its Latin translation appeared in Rome under the title *Epistola de insulis Indiae supra Gangem nuper inventis* (1493).[53] A decade later, news about the hitherto unknown lands in the west was disseminated even more efficiently by Amerigo Vespucci's *Mundus novus*, the Italian original of which to Lorenzo di Pierfrancesco de' Medici is lost, but its Latin publication became a bestseller responsible for both America's current name and its designation as the 'New World'.[54] In 1610, Kepler sent his *Dissertatio cum Sidereo nuntio* to Galilei, who had asked for his opinion

[51] Avramov 2013, 199–200. On the rise of the questionnaire in general, see Burke 2000, 121–2, 126–7, 130, 137.

[52] Boscani Leoni 2007, 204–7 (in particular on Scheuchzer's reception of English models); Cooper 2007, 133–7. Edition of the German version: Küster and Küster 1997, 14–31.

[53] Modern edition: Wallisch 2000.

[54] The *Mundus novus* was presumably first published in 1502/03, but the first dated edition appeared in 1504. Modern edition: Wallisch 2002.

about the *Sidereus nuntius*, from Prague in manuscript and rushed it into print a few weeks later.[55]

In the absence of contradicting evidence, however, the chances are that a longer letter making up an entire book or a substantial, self-contained part of one was meant for publication from the outset. In particular, letters were perceived as an attractive literary form in which to cast a short monograph.[56] Many works of this kind flaunted their epistolary character by titles such as 'dissertatio epistolica', 'dissertatio epistolaris', 'epistolaris disquisitio', or simply 'epistola'. Georg Caspar Kirchmaier, for example, informed his friend Christian Franz Paullini about a recent invasion of migratory locusts to Eastern and Central Europe in a *De locustis insolitis dissertatio epistolica* (1693). Friedrich Christian Lesser's physico-theological *De sapientia, omnipotentia et providentia divina ex partibus insectorum cognoscenda epistolaris disquisitio* (1735) is a special case in that its very publication served a private goal as well: it was conceived as a birthday gift to the addressee Albert Seba, who had recently turned 70.[57] Accordingly, the letter's six central paragraphs (§§ 2–7) are carefully framed by an opening and a close in which Lesser affectionately addresses his friend in person. Conversely, some published letters were polemical. An *Epistola admonitoria* (1662) of no fewer than 143 pages permitted Michael Bayer, a doctor of medicine who had recently graduated from Padua, to attack his colleague Francesco Cima on account of a disagreement regarding the correct treatment of a noblewoman and the comprehension of the term 'podagra' in what appears to have been a proxy war on behalf of the opponents' respective teachers.[58]

In contrast to the examples just mentioned, the letter format was eroded to the point where the generic affiliation of the text became unclear in other cases. Vincenzo Renieri's *De Etruscarum antiquitatum fragmentis disquisitio astronomica* (1638), an astronomical defence of some pseudo-Etruscan forgeries,[59] exemplifies this possibility. The work is addressed to the author's sovereign, Ferdinando II de' Medici, Grand Duke of Tuscany, on its title page without being expressly called a letter. The opening salutation is reduced to a 'Serenissime princeps' integrated into the first sentence, and the body of the text shows no epistolary features at all. At least, Renieri concludes with a formal valediction: 'Vale. Serenissimae Celsitudinis Tuae servus humillimus dominus Vincentius Renerius Olivetanus.' ('Farewell. Your most Serene Highness's most humble slave, Mr Vincenzo Renieri,

[55] An excellent translation and commentary are provided by Rosen 1965.

[56] Two typical examples that can be cited in addition to those discussed in the main text are Friedrich Lachmund's letter on the albatross to Johann Daniel Major (*De ave Diomedea*, 1674) and Johann Philipp Breyn's letter on petrified melons from Mount Carmel to Johann Anderson (*Epistola de melonibus petrefactis*, 1722).

[57] Trepp 2009, 394 and 411–12 (within an instructive chapter on Lesser and his writings in general, pp. 373–466).

[58] *TL* 519–20.

[59] Rowland 2004.

an Olivetan.') On the other hand, Simone Porzio's *De conflagratione agri Puteolani epistola* (1551) bears an epistolary title and features a formal address to the Neapolitan viceroy but lacks a valediction.

Independently of how clearly the epistolary character was formally articulated, styling a monograph as a letter brought several advantages. The brevity of the letter suited the concise treatment of a clear-cut question. Because a letter could easily be styled as forming part of an ongoing discussion, an author could present his results as preliminary and renounce the claim to have the last word on the subject. Moreover, the intimate character of the genre could make a work appear more alluring than an impersonal treatise. By dressing a publication as a whole in epistolary form, one was also able to dispense with a separate letter of dedication, which might have looked awkward in front of a very short text, while pleasing and obliging a friend and benefactor at least as much or even more than by the simple dedication of a non-epistolary publication. The addressee's interest in the matter under discussion and a request by him to be enlightened could serve as a justification for writing, as in a letter on fossils to John Ray appended to Edward Lhuyd's *Lithophylacii Britannici ichnographia* (1699), which starts with the words (p. 131):

> Rogas, vir insignissime, quid mihi tandem post aliquot annorum experientiam de marinorum corporum, quae e subterraneis eruimus, et mineralium foliorum ortu coniciendum videatur.
>
> You ask, most famous sir, what I, basing myself on several years of experience, think that one should assume concerning the origin of the bodies of sea animals which we dig out from the ground and of petrified leaves.

In some cases, an author presents himself as overcoming his timidity and fear of publication through the reassuring feeling of speaking to someone familiar and benevolent. Gabriele Falloppio claims to have surmounted his reluctance to publish anything on anatomy in view of Vesalius's overwhelming achievement partly by addressing his *Observationes anatomicae* (1562) to his friend and colleague Pietro Manna.[60] A famous addressee could provide respectability and credibility—certainly one of the reasons why the young Georg Joachim Rheticus published his *Narratio prima* (1540), the first printed report on the Copernican system, as a letter to the renowned mathematician and astronomer Johannes Schöner.[61]

Sometimes, the addressee even chose to speak himself. The result was a booklet consisting of a pair of letters, as in the case of Johann Daniel Major's *Dissertatio*

[60] The introductory address to the reader and the opening part of the *Observationes* themselves present a sustained reflection on this topic. See esp. fols. 5^{r-v}, 8^{r-v}.

[61] Introduction, text, French translation, commentary: Hugonnard-Roche et al. 1982. On Rheticus in general, see Danielson 2006.

epistolica de cancris et serpentibus petrefactis addressed to Philipp Jacob Sachs von Lewenheimb and answered by the latter with a *Responsoria dissertatio historico-medica* (1664). In such a dialogic arrangement, both contributors could not only present their results, but also show their friendship and harmonious collaboration. In fact, Major's and Lewenheimb's exchange formed part of the publication programme of the recently founded Leopoldina and was presumably meant to exemplify the new academy's programme of scientific cooperation.[62] Cases in which one letter booklet was answered by another can be seen in a similar light. In 1632, Pierre Gassendi published two letters to his friend Wilhelm Schickard in Tübingen, in which he reported, among other things, on his recent observation, the first ever, of a transit of Mercury.[63] The same year, Schickard responded in kind, discussing the information provided by his French colleague in detail. He found Gassendi's opinion that Mercury was much smaller than previously assumed unconvincing but nonetheless conceived his answer as a testimony to their friendship.

By way of conclusion, some characteristics of the literary letter and its potential as a medium of scientific communication will be exemplified by an elaborate item, Bartolomeo Eustachi's 'Epistula de auditus organis'. Published on pp. 148–64 of Eustachi's *Opuscula anatomica* (1564), this short work is one of three treatises on new anatomical discoveries in letter format contained in the book.[64] Such discoveries could be presented on a couple of pages and were attractive as 'gifts' to the addressee because of their very novelty.[65] The 'Letter on the Instruments of Hearing', which presents a number of findings regarding the structure of the human middle ear, constitutes a textbook example of how to put the letter format to good use in a scientific publication. In several respects, it strongly contrasts with Haller's correspondence, discussed earlier.

The text is addressed to the jurist Francesco Alciati, a younger relative of Andrea Alciati, who was bishop of San Severo in Apulia at the time, soon to be promoted to the dignity of a cardinal in 1565. In the 'valedictio', the epistle is dated Rome, 12 October 1562, and there is no need to doubt that it actually was completed and sent on that day. However, it is equally clear that it was from the outset meant to be published: Eustachi expressly states that, because Alciati

[62] Cf. Maclean 2019, 83. The format could be expanded by adding a response to the response, as in Constanzio Varolio's and Girolamo Mercuriale's exchange about the 'nervus opticus' and other anatomical structures of the human brain (1573).

[63] Joy 1987, 109–10. On the 1631 Mercury transit and the publications prompted by it in general, see Van Helden 1976.

[64] The other two are 'Examen ossium' and 'De motu capitis', combined into one single letter to Fabio Amicio (pp. 165–227, 227–60).

[65] As a seventeenth-century counterpart to Eustachi's letter, Nicolas Steno's 'De vitelli in intestina pulli transitu epistola' to Paul Barbette may be cited. The letter discusses the vitelline duct in the chicken embryo and was included as a companion piece to Steno's *De musculis et glandulis observationum specimen* (1664, pp. 76–90; Bäumer-Schleinkofer 1993, 125–6).

provided him with an occasion to commit his results to paper, 'all students of anatomy may confess to be deeply indebted to you' ('omnes anatomiae studiosi tibi se plurimum debere fateantur', p. 151.29–31).[66] Moreover, the sheer technicality of the content of the letter cannot have been fully understandable by Alciati, who apparently had no medical training. In fact, in the absence of illustrations, the technical passages must have been difficult to grasp even for an anatomist who had not already dissected a number of middle ears himself.

The letter starts with a formal epistolary opening and launches into fulsome praise of the addressee's numerous virtues. Among Alciati's merits as far as Eustachi is concerned, it is highlighted that he countered attempts of the latter's enemies to alienate him from his patron, Cardinal Carlo Borromeo. Because he turned a deaf ear to the allegations and because of his interest in anatomy, Eustachi finds it fitting to dedicate his new insights about the anatomy of the ear and the mechanisms of hearing to him (pp. 148.18–152.9). The work's technical part starts with a discussion of previous research on the ear, including the work of Vesalius, which Eustachi finds wanting in many respects (pp. 152.9–156.20). He promises to make good these deficits by meticulously elucidating the structure and function of the ear in the present work (pp. 156.20–157.30). His first discovery regards the musculus tensor tympani, one of only two muscles in the middle ear. Precise prescriptions are given to the dissector on how to find this tiny muscle (pp. 157.31–159.7). Next, the innervation of the middle ear is described in meticulous detail (pp. 159.7–160.15), and the structure of the cochlea is explained (pp. 160.15–161.16). The last finding presented regards the tube connecting the middle ear to the nasopharynx which still bears Eustachi's name (pp. 161.16–164.13). In the final lines, the author returns to Alciati and concludes with a formal 'valedictio', giving the place and date in Roman style ('Romae, quarto Idus Octobris. MDLXII'; p. 164.13–32).

The differences between Haller's hasty, informal letter and Eustachi's artful work strike the eye in several ways. First, as already suggested, Eustachi addresses two different (groups of) readers: while the initial and final parts are directed at Alciati, the author speaks to his fellow anatomists in the core part of the text. This becomes particularly evident when he gives prescriptions for dissection, as in the case of the musculus tensor tympani: 'Tu si illum videre cupis, aperta calvaria os incide, quod petram refert' (p. 158.4–5; 'If you want to see it for yourself, open the skull and make an incision into the bone which resembles a rock'). Quite obviously, it is not Alciati who is asked here to pick up the scalpel.

Secondly, while Haller transmits information, Eustachi displays and advertises his findings and himself. Time and again, he emphasizes that he has been the first to note this or that phenomenon. These statements are often backed up by

[66] Quite unusually, the *Opuscula anatomica* features not only a page but also a line count.

polemics against his predecessors, who failed to observe the structures or provided erroneous descriptions despite the great claims made in their writings. The importance of such novel insights is stressed repeatedly, for instance as follows (p. 162.29–32):

> Hoc callidissimum Naturae artificium a me inventum contemni, ut opinor, non debet, siquidem tum philosophis, tum medicis non parum utilitatis afferre potest.
>
> This extremely sophisticated craftwork of Nature which I have detected must not be despised, I think, since it can be of no small use to philosophers and physicians alike.

In many cases, Eustachi further enhances his findings by replacing the 'Natura artifex' invoked in the preceding passage by the Creator himself and by describing the structure of the middle ear in physico-theological terms. For instance, the technical discussion strictly speaking is introduced by this very trope (p. 157.25–8):

> Quis autem in horum ossiculorum articulis componendis praecipuus sit usus, ut opificis nostri artem ac potentiam, quae vel in minimis elucet, contemplari atque mecum admirari possis, iamiam aperio.
>
> But now I already start to show the particular usefulness lying in the connection of these small bones by joints, so that you can consider and admire with me the artfulness and power of our Creator, which lights up even in the tiniest things.

A third point of difference concerns the style. Not only in those parts directly addressed to Alciati and in framing and programmatic passages like the ones just quoted but also in technical descriptions, Eustachi writes a thoroughly classical, Ciceronian Latin. He avoids the careless shorthand Latin of the real-life letter writer, and his language is even a far cry from the studied casualness of classics of the literary letter like Pliny the Younger's. This applies not only to the syntax, which features long, impeccably constructed, and well-balanced periods, but, more strikingly, to the vocabulary as well: wherever possible, technical terms are avoided and replaced by circumlocutions. In the following example devoted to the description of the musculus tensor tympani (p. 158.10–22), the only technical expression is the Greco-Latin 'apophysis' ('process of a bone'), which is sanctioned by Hippocratic and Galenic precedent. By contrast, the bones called 'malleus', 'os temporale', and 'os sphenoidale' today are described, rather than named, as 'os, quod cuneum imitatur', 'temporis os', and 'ossiculum malleo comparatum'.

> Hoc sane experta manu ubi effeceris, statim musculus conspiciendum se exhibebit, qui etsi omnium minimus est, elegantia tamen et constructionis artificio nulli cedit. Oritur a substantia ligamentis simili, qua parte os, quod cuneum

imitatur, cum temporis osse committitur, indeque carneus evadens redditur sensim ad medium usque aliquantulum latior, deinde vero angustior effectus tendinem gracillimum producit, qui in maiorem apophysim ossiculi malleo comparati fere e regione minoris apophysis eiusdem inseritur et in canibus parum ante insertionem carneam rotundamque glandulam aliquantulum depressam habet.

After you have done so [that is, uncovered the middle ear] with an experienced hand, the muscle, which, although the smallest of all, is second to none in its elegance and artfully crafted structure, will forthwith present itself to your eyes. It originates from a matter similar to ligaments at the place where the bone which imitates a wedge meets the bone of the temple, and from there, turning fleshy, it gradually becomes a little broader up to its middle, but after that point, having become leaner again, it extends a most slender tendon, which, coming more or less from the direction of the minor process, merges into the major process of the small bone comparable to a hammer and shortly before the merging point has a round, but somewhat flattened, fleshy bulge in dogs.[67]

Finally, the carefully crafted structure of the letter as a whole is eye-catching. As already mentioned, Eustachi does not try to create the impression of a private letter that has later been published but openly acknowledges the public nature of his text. Nonetheless, he starts with effusive thanks for and praise of his addressee, to whom he also returns at the end. This results in a ring composition (reminiscent of Lesser's aforementioned *Epistolaris disquisitio*) in which a monographic core is embedded in an epistolary frame, creating an appealing symmetry. Most importantly, the combination of both parts is presented not as an arbitrary juxtaposition, but as a result of their correspondence in terms of content. This is hinted at already in the introduction, when Eustachi praises his addressee for 'having listened with your ears in such a way to the insults and slanders with which my enviers tried to harass me that they could do me no harm because of your patronage' (p. 151.19–21, 'auribus maledicta et calumnias, quibus me invidi persequi conabantur, ita audivisti, ut nihil mihi te patrono obesse potuerint'). At the end, the correlation is spelled out and at the same time artfully linked to the 'Natura artifex' motif (p. 164.19–31):

Satis mihi fuit in praesenti aliqua ex parte indicasse, qua sollertia auditus organa Natura construxerit atque munierit et qua conditione vocem per varias concavitates et spiras quasi per diversos gradus sensim elaboratam et quodam modo fractam in cerebrum ipsum ascendere voluerit: quasi omnes tacite admoneat,

[67] In passages such as this, Eustachi still does not go to the extremes of some other authors like Luigi Mondella, who is capable of writing in long, lofty periods even about topics such as frog's leg soup (*Epistolae medicinales*, 1543, p. 77).

> ne orationi, quae est animi nuntia, sine diligenti praemeditatione maturoque iudicio, praecipue quando honori et dignitati alicuius detrahitur, facile assentiendum sibi proponant. Te vero, qui hac ratione ea, quae ab invidis et malevolis audiveras, iudicasti, etiam atque etiam oro, ne me amare ac protegere in posterum quoque desistas.
>
> To me it was enough for the moment to have pointed out to some extent with what care Nature has constructed and protected the instruments of hearing and in what way she wanted the voice to ascend to the brain through various hollows and turns, as if through different steps, in a form gradually worked up and in a certain way refracted: as if she would silently admonish all people that they should intend not to approve easily of speech, the mind's messenger, without careful deliberation and well-thought-out judgement, in particular when someone's honour and dignity are impaired. But as for you, having judged in this way what you heard from enviers and malicious people, I implore you over and over again not to stop loving and protecting me in the future as well.

Thus, Eustachi treats the letter format neither as a mere convention nor as a simple vehicle to convey his anatomical findings. By presenting them in a literary form which is at the same time a form of social interaction and by making social interaction itself a topic of discussion in doing so, he highlights the importance of his discoveries, which extends beyond the realm of anatomy and medicine to human life in general.

Reports

An early modern genre

Unlike most genres discussed in this book, the report, the detailed narrative of a personal experience made at a particular date and place, is a quintessentially early modern genre as far as science is concerned. The reasons for its chronological restriction to the early modern period are epistemological. Before 1500, the dominance of the Aristotelian notion of 'science' (ἐπιστήμη) as deductive reasoning starting from commonly accepted truths left little room for the unique experiences of individuals. After 1800, a concept of science based on positivism, objectivity, and discipline-specific methodology shaped the article, the dominant form of modern scientific writing, into an impersonal text characterized, inter alia, by the so-called 'ego' and 'narrative taboos'.[68] In the intervening period, by contrast, the particular observation made by a trustworthy person—who could interfere

[68] Kretzenbacher 1994 and, more generally, Gross et al. 2002. See also the section 'Journal articles'.

with the course of events, as in an experiment, or not—came to be seen as an important source of knowledge. 'Observatio' or 'historia', as it was called, rose to a respected and influential category in scientific discourse. It promised fresh facts, independent of the bookish authorities whose status was waning and unencumbered by big theories and ideological concerns. Over time, a treasure of new knowledge would be amassed in this way and engender scientific progress. The report, which was often just titled 'observatio', emerged as a fitting literary medium for sharing new observations with one's fellow observers and became a popular format.[69]

This novel and in some respects peculiar form of scientific writing was noted and discussed as such already in its own time. Robert Boyle in particular produced a thoughtful theory of the report (which he called 'essay') in general and of his own experimental reports in particular in a 'Proëmial Essay' prepended to his *Certain Physiological Essays* (1661).[70] Partly taking their cue from this text, modern historians of science directed their attention towards the format from the 1980s, capitalizing on the way in which Boyle and other fellows of the early Royal Society communicated their experiments in it.[71] As noted by Peter Dear, the fellows' predilection for this untheoretical and fact-oriented form, which they used not only in books like Boyle's but also in the articles of the *Philosophical Transactions*, stemmed from their desire to realize Francis Bacon's ideal of science as a collaborative enterprise in the face of their dissent with regard to fundamental theoretical questions. However, as Dear also observed, the narrative presentation of scientific experience practised by the fellows was no isolated phenomenon, nor was it invented by them. Rather, it was an English variety of a pan-European—and, one can add, largely Latin—trend. Widening the focus to include the Continent and the early modern period as a whole, one finds a large number of similar texts from different areas. The roots of the genre and authors' motives for resorting to it turn out to be manifold. The report comprises several subtypes, and one encounters many borderline examples which strain the initially given definition.[72]

[69] On the history of the epistemological notion of 'observatio' (τήρησις) from antiquity to the eighteenth century, see Pomata 2011a; on its rise in the early modern period, particularly in medicine, and the accompanying emergence of a literary genre devoted to it, see Pomata 1996, 2010, 2011a.

[70] The text was Latinized in the year of its publication together with the rest of the *Physiological Essays*, which appeared as *Tentamina quaedam physiologica*.

[71] Shapin 1984 (partly reprinted in Shapin and Schaffer 1985, 22–79); Dear 1985; Bazerman 1988, 59–79 (reprinted with minor changes in Harris 2018, 263–79).

[72] In fact, plenty of such examples can also be found among the reports in the *Philosophical Transactions*. In Newton's 'New Theory about Light and Colours' (1672), for instance, most of the experiments described are left undated, and the last of them (pp. 3085–7) is formulated not as a report, but in recipe style. Moreover, the text includes a section presenting no experiments at all but the author's theory of light announced in the title (pp. 3081–5).

Overview

From the second half of the seventeenth century, many reports appeared in journals such as the *Philosophical Transactions*, the *Miscellanea curiosa*, the *Acta eruditorum*, and the serial publications of the academies, to whose editors they were often sent as letters.[73] Before that date, the comparatively short genre had to be accommodated somehow to the normal book format. Few reports were long and complex enough to make up a volume on their own. In many cases, authors resorted to serialization, producing report collections or continuous narratives comprising several reports. Often, reports did not make up the whole of a work but were inserted into an account of another kind.

Thematically, such texts were written in most disciplines except pure mathematics, whose deductive structure was incompatible with the genre. Apart from the ubiquitous medical and astronomical reports to be discussed shortly, one finds specimens dealing with physical (Rasmus Bartholin, *Experimenta crystalli Islandici disdiaclastici*, 1669), chemical (Johann Heinrich Schulze, 'Scotophorus', 1721), botanical (Georg Bernhard Bilfinger, 'De tracheis plantarum ex melone observatio', 1735), and meteorological phenomena (Anders Celsius, 'Observationes meteorologicae', 1744), as well as technical applications (Johann Jakob Hemmer, 'Conductorum fulmineorum vis egregia', 1790).

In the present subsection, I will leave aside reports in journals, as they will come up again in the next section, and focus on self-contained publications. To be more precise, I will discuss four subgroups which between them account for the majority of all reports published in book format. Two of them—the 'observationes' of physicians and astronomers—are defined by discipline, while the other two describe specific kinds of activities, namely scientific excursions and experiments.

The strongest tradition of reports developed in medicine,[74] a discipline dealing by definition with exceptional—hence rare or unique—phenomena. Doctors wrote reports to show how their largely theoretical training could be put into practice and at the same time to advertise their own skills. In doing so, they could build on several literary forerunners of the genre, the most important of which were the case history,[75] in particular the terse, precise accounts in books 1 and 3 of the Hippocratic *Epidemics*, which attracted numerous commentaries in early modern times, and the late medieval collections of 'consilia'. Both these kinds of writings share with the medical report a concern with an individual case and the practical approach to medicine, without being reports themselves. In the former,

[73] Reports were also styled as letters in other cases. Boyle, for one, did so on several occasions (Shapin 1984, 489–90). A Latin example is provided by Gabriele Falloppio's *Observationes anatomicae* (1562).

[74] The following paragraph provides a very short summary of Gianna Pomata's rich and insightful articles on the topic (Pomata 1996, 184–92; 2010; 2011a, 52–64).

[75] Hurwitz 2006.

the narrative is usually completely impersonal. In the latter, the speaker, who gives advice to a colleague, may not have seen the patient himself, and the disease in general is often more important than the case at hand, which mainly serves to illustrate the points made in the learned discussion of it. By contrast, in the 'observationes' (also titled 'curationes', 'casus', or 'historiae')[76] which began to compete with the 'consilia' from about 1550 and eventually superseded them, case histories are presented in the form of autonomous first-person accounts told by the attending physician, and such academic discussions as may still form part of them are neatly separated from these narratives. In Amatus Lusitanus's *Curationum medicinalium centuriae*, of which seven instalments came out from 1551 to 1566 and whose success did much to establish the new genre, theoretical considerations are relegated to sections titled 'scholia' and following the 'curationes' proper. Medical 'observationes' became highly popular: some seventy collections, many of them bulky, are on record for the period 1551–1676.[77] Moreover, the format inspired one of the earliest scientific journals, the *Miscellanea curiosa* (1670–), whose initial volumes consist of numerous 'observationes' reporting primarily medical cases but also other peculiarities.

Astronomy, the second discipline developing a distinctive report tradition, was well suited to the new genre too: date and place formed an integral part of any astronomical observation, as the position of a star was worthless without the exact time and position from which it had been recorded; details of circumstances like the weather conditions and the instruments used were almost as essential; and the person of the observer himself was of crucial importance, because only someone equipped with deep a theoretical understanding and great technical skill was able to produce reliable results.[78] However, while ancient and medieval astronomers invested immense energy in their observations, they usually recorded the results of their efforts in 'tabulae' whose figures did not represent the raw data gained at a given occasion but described the celestial movements in general terms. From the mid-sixteenth century, such table works were joined by others, titled (just as in medicine) 'observationes' or 'historiae', which documented positions observed at precisely defined moments and geographical locations. Concomitantly, as rare phenomena like novas, comets, and transits were turned into arguments in cosmological debates and the telescope broke down the limits of naked-eye astronomy, a tradition of running-text reports on such events and other new findings in the skies developed. The novas of 1572 and 1604 were made the subjects of reports by Tycho Brahe (*De nova stella*, 1573) and Kepler (*De stella*

[76] Pomata 2010, 199–226, distinguishes 'curationes' from 'observationes' on account of their focus on successful cures and the sharing of interesting information respectively, but as suggested by the frequent pairing of both terms in titles like *Sylloge curationum et observationum medicinalium* (Georg Hieronymus Welsch, 1667), the boundary between both subforms is not clear-cut.

[77] See the list in Pomata 2010, 232–6, who makes no claim to completeness.

[78] Pomata 2011b, 49–50.

nova in pede serpentarii, 1606) respectively. Brahe also narrated in detail his observations of the comet of 1577 (*De mundi aetherei recentioribus phaenomenis*, 1588). Transits of Mercury and Venus, which provided rare opportunities to determine the dimensions of the solar system, were observed and described by numerous seventeenth- and eighteenth-century astronomers from Gassendi (*Mercurius in sole visus*, 1632) to Maximilian Hell (*Observatio transitus Veneris*, [1761]). In his *Sidereus nuntius* (1610), Galilei narrated what he found out about the moon's surface, the nature of the Milky Way, and the moons of Jupiter with the help of his telescope. Only one year later and clearly inspired by his famous predecessor, Johann Fabricius gave a lively account of his discovery of sunspots (*De maculis in sole observatis*, 1611).[79]

Among subforms not restricted to a particular discipline, travelogues—often headed 'iter' or 'itinera'—had a long pedigree and thrived in the Age of Discovery. Early printed examples like Columbus's letter from his first journey or Vespucci's *Mundus Novus* contain some information about the exotic nature of newly discovered regions but are not informed by scientific interests. From the beginning of the seventeenth century, by contrast, botanical field trips, which had risen to popularity as a didactic means in the previous century, became a well-liked subject of travel reports. Typically, such works provide a first-person narrative of the trip, the route taken, and the experiences had, which are regularly interrupted by lists of plants found at particular places.[80] Thomas Johnson, for instance, a prominent member of the Apothecaries' Company at London and an enthusiastic botanist, published a series of field trip reports such as the *Iter in agrum Cantianum* (1629).[81] His lively narratives not only inform readers about the flora of the regions around London, such as Kent, but also commemorate the botanical achievements of the company's members, their adventures, and their friendship. Similar works, such as Albrecht von Haller's *Iter Helveticum* (1740), continued to appear in the eighteenth century. Less regularly, travels to see and study other aspects of the natural world were described as well. Pietro Bembo's *De Aetna* (1496), formally a dialogue between the author and his father, but de facto largely a narrative of the former's ascent of the active Mount Etna in 1493, provides an early example of a 'volcano report'. One and a half centuries later, Athanasius Kircher left an emotional account of his nightly climb of Mount Vesuvius in 1638 and his observation of the volcanic activity from the crater rim (*Mundus subterraneus*, 1665, preface, ch. 3).[82] Some excursionists showed an omnivorous

[79] Galilei is invoked as a model on sigs. B3^{v} and C2^{v}.

[80] Other works, such as Francesco Calzolari's 'Iter Baldi' (1571), are structurally similar and clearly based on their authors' personal experience as well, but take the form not of first-person reports, but of guidebooks (see note 96 below).

[81] Cited by the edition by Ralph 1847 in what follows. See also Pavord 2005, 344, 349.

[82] See also Luggin 2023, 132–9, on Kircher's account of his ascent of Mount Etna in book 4 of the same work.

appetite for all aspects of nature (and culture) they encountered. Johann Jakob Scheuchzer's narrative of nine journeys through the Swiss Alps in the summers of 1702–7 and 1709–11 (*Οὐρεσιφοιτής Helveticus*, 1723) contains lists of plants, barometric altitude measurements, descriptions of glaciers, geological reflections, thoughts on the water of health spas, information about the hunting of chamois, and much more.[83]

Although experimental reports are presented as the quintessential form of the genre in modern literature, they actually were latecomers compared with the varieties discussed to far, since the experiment established itself as a recognized method of doing science only from the middle of the seventeenth century. Moreover, experimental reports in book format (as opposed to journal articles) remained comparatively rare despite the popularity of Boyle's works.[84] Most reports by Boyle were quickly translated into Latin, and the same could happen to similar works by others, as shown by Andreas Frisius's version (1671) of Francesco Redi's *Esperienze intorno alla generazione degl'insetti* (1668). However, many authors writing in Latin seem to have been reluctant to compose experimental reports themselves, perhaps because the hands-on approach involved and the implied clash with Aristotelian epistemology sat uncomfortably with their academic training. Even in works discussing experiments at length, other formats were often preferred. While William Gilbert's *De magnete* (1600)[85] was seminal in its development of a new branch of natural philosophy with the help of systematic experimentation, the author does not present his numerous experiments as first-person narratives but either summarizes their results in the form of general statements or instructs his readers how to repeat them. In other publications, experimental reports are found intermingled with accounts of vaguely related topics such as technical toys or marvellous phenomena in general. In Gaspar Schott's *Technica curiosa* (1664), for instance, Otto von Guericke's and Boyle's vacuum experiments, which fill the first two of twelve books, are titled 'mirabilia' and followed by subjects such as 'mirabilia cyclometrica' and 'cabalistica' ('marvels of measuring [that is, squaring] the circle' and 'cabalistic marvels', books 8 and 12). Early experimental reports which were clearly meant and presented as such include Jean Pecquet's *Experimenta nova anatomica* (vivisections of dogs leading to the discovery of the cisterna chyli, 1651), the third book of Guericke's *Experimenta nova Magdeburgica de vacuo spatio* (1672), and Johann Sigismund Elsholtz's *De phosphoris quattuor observatio* (luminance experiments, 1676).

[83] Korenjak 2012, esp. 209–10.

[84] Even in a journal such as the *Philosophical Transactions*, experimental reports accounted for only around 5–20 per cent of all articles per volume up to 1800 (Bazerman 1988, 65, reprinted in Harris 2018, 267).

[85] Roller 1959; Kay 1981, 82–149.

Narrating science

The overall design of reports varies greatly in many respects. This applies not only to such obvious features as subject, length, and style, the last ranging from plain to convoluted, generally intelligible to technical, and dry to emotional.[86] It is also true in relation to the genre's initial definition as a 'detailed narrative of a personal experience made at a particular date and place'. In what follows, I will try to provide an impression of how the single aspects of this definition—narrativity, occasionality, detail, and personal character—can be highlighted or played down in literary practice.

The narrative aspect, while essential, is not always equally dominant. In contrast to the narrator's own actions, his observations are often described rather than recounted: it is not the operation of observing, but the facts observed which are foregrounded. In the reports of botanical field trips, for instance, this procedure leads to an alternation of narrative and descriptive parts, as already mentioned. A farther-reaching departure from pure narration is the addition of theoretical reflections, which presumably responded to the expectations of many readers who would have been dissatisfied with naked facts and no attempt to make sense of them. An author who offered no explanation at all of his observations could feel the need to apologize for this defect, as did Schulze in his 'Scotophorus', a short report on the darkening of silver nitrate under the influence of sunlight (§ XIII): 'Aetiologiam huius effectus adderem, si ipse mihi in eo satisfacerem.' ('I would add a causal explanation of this phenomenon, if I could do so to my own satisfaction.') Sometimes, theoretical considerations contextualize the narrative embedded in them, as in Guericke's *Experimenta nova Magdeburgica*: his 'experimenta' properly speaking, which make up the third book, are framed by reflections about the cosmos in general contained in the two preceding and four following books. In other cases, theory is rather offered as a supplement to the narration, for instance in many medical reports, as the practice of Amatus Lusitanus's *Curationum medicinalium centuriae* was widely imitated in later representatives of the genre. In texts of this kind, readers could profit from the facts presented but remained free to accept the author's interpretation of them or form their own opinions. Often, by contrast, the phenomena observed and the interpretation of them are found intermingled. Fabricius, for one, freely switches from one to the other and back again in his account of the discovery of sunspots.

In principle, a report is supposed to refer to one specific event and to make this clear in its linguistic form: time, place, and the original audience[87] should be

[86] *Pace* Shapin 1984, 495, Boyle ('A Proëmial Essay', pp. 304–5) allows for considerable stylistic variety in reports, including his own. For an example of a report partly written in the 'genus grande', see Korenjak 2012, 221–30, on Scheuchzer's *Οὐρεσιφοιτής Helveticus*.

[87] In the early Royal Society, those present at an experiment had to sign the report (Shapin 1984, 488). For an earlier example of the importance of witnesses from Pecquet's *Experimenta nova anatomica*, see below.

detailed to bolster the account's veracity. In practice, however, this ideal is not always fully realized, and the occasionality of the event itself as well as its verbal expression may leave something to be desired. An author may fail to indicate date and place, as in many medical reports (where sex, age, constitution, and social position of the patient are often given instead). He also may leave it open whether his narrative refers to an observation or experiment made only once or to one repeated on several occasions. Sometimes, a unique event is artificially created by conflating several observations made at different times in order to simplify things and correspond to readers' expectations. A well-known example of this approach is presented by Newton's report on his prism experiments: the arrangements and observations feeding into the report's first experiment were not made, as the author claims, on a particular day in early 1666, but in the course of several years of research.[88] While Newton tacitly fictionalized his actual work, at least one Latin author revealed his own literary technique. At the end of his *Experimenta nova anatomica*, in which he recounts a short series of three vivisections, Pecquet states (p. 17):

> Ita, mi lector, habes exactam lactearum venarum historiam. Intra triplicis dissectionis spatium assiduum ferme trium annorum laborem coarctavi, quia tantilli temporis dispendio potes ab erroribus desciscere. Trinum tibi, ut expono, canicidium dabit, quod mihi centena plus quam vivorum[89] animantium exenteratione vix tandem concessum est.
>
> Thus, my reader, you have my precise report on the milky vessels.[90] I have condensed into the space of three dissections the continuous work of approximately three years, because you can thus put away your errors investing an absolute minimum of time. The killing of three dogs, as I describe it, will give you an understanding which was hardly conceded to me in the end after I disembowelled more than one hundred living animals.

If Pecquet's confession that he had fictionalized his narrative in order to facilitate the replication of his experiments did not detract from his credibility, it probably came as no great surprise to his readers, at least to those experienced in such matters. There seems to have been a tacit understanding that techniques like Pecquet's were acceptable if applied for the sake of brevity and clarity.

When reports first attracted the interest of historians of science, they were intrigued by the amount of precise and graphic detail to be found in such texts. Boyle's reports in particular won fame for this feature. The numerous details contained in them were interpreted as an attempt to place the reader in the position

[88] 'New Theory about Light and Colours', pp. 3075–6; Dear 1985, 154–5.

[89] The print's 'vivarum' seems to be a typographical error. The dissected dog depicted at the end of the work has testicles.

[90] The thoracic duct and its tributaries, which carry chyle, a milky mix of lymph and emulsified fat.

of a witness to the experiment ('virtual witnessing') and to convince him in this way of Boyle's truthfulness.[91] This is plausible, but a few points should be added when speaking about the genre as a whole. First, and most obviously, not all reports contain the same amount of detail. Many are much sparer than Boyle's. Nor do all reports present the same kind of detail. In this respect, a distinction can be made between the observation or the experiment itself, on the one hand, and its circumstances—time, place, audience, and so on—on the other, even though this distinction is not always sharp, as noted in the previous subsection with regard to astronomical reports. Reports of botanical field trips, for instance, can be quite detailed on the circumstances—something partly explicable by the fact that a plant's habitat and time of flowering and fructification form an integral part of a botanical observation—but summarize the plants observed themselves in the form of simple lists. Authors of experimental reports, by contrast, may dwell lovingly on the particulars of their experiments while being reticent about the circumstances. Secondly, convincing readers by means of detail is not Boyle's invention. In fact, he simply makes good use of the techniques subsumed under the heading 'descriptio' in classical rhetoric. Ancient rhetoricians and their early modern followers agree that speech sounds truer and proves more efficient if it bristles with pertinent particulars, because the audience gets a quasi-sensual experience of what is told and unquestioningly accepts it as real. They also provide clear instructions how to achieve this goal.[92] Thirdly, the objective and effect of the technique is not just persuasion. It fulfils other functions as well, which are at least as important. The right kind of detail is vital for making the complex events described comprehensible in the first place. Precise information on particulars is also needed in order to replicate the observation or experiment, should the reader wish to do so. Aspects from which the author jumps to wrong conclusions or draws none at all may be put to better use by an alert reader.[93] Finally, an author may use detail to put himself in a good light, for example by underlining his care and circumspection, or to satisfy his literary ambitions by imbuing his narrative with drama. Several of these functions can be identified in the following (admittedly gruesome) extracts from the *Experimenta nova anatomica* (pp. 4–5, 15) in which Pecquet narrates the first steps of his vivisections and the moment of his discovery of the cisterna chyli.

[91] Shapin 1984, 490–4; Shapin and Schaffer 1985, 60–5; Dear 1985, 153. This technique may be compared with the 'vicarious description' practised by botanists (see the text before note 201 below).

[92] *HlRh*, §§ 810–19, 1133; *HWRh* 1, cols. 1495–510, s.v. 'Beschreibung'; 2, cols. 549–53, s.v. 'Descriptio'; 3, cols. 33–47, s.v. 'Evidentia, Evidenz'.

[93] That a comprehensive account of what has been done and observed enables the reader to correct the author's conclusions, if necessary, is repeatedly stated by contemporaries. See Gilbert, *De magnete*, preface to the reader, sig. *ii^{v}: 'Illis [my experiments and findings] gaudete et in meliores usus, si poteritis, fruimini'; Boyle, 'A Proëmial Essay', pp. 303–4, with Shapin 1984, 499.

Ergo diffisso viventis, quae media est, alvo molossi inchoo[94] extispicium. Nec mora: cor rescissis, quibus reliquo adhaeret corpori, vasculorum retinaculis avello. Tum exhausta, quae statim restagnaverat spectantisque confuderat obtutus, copiam cruoris albicantem subinde lactei liquoris nec certe parum fluidi scaturiginem intra venae cavae fistulam, circa dextri sedem ventriculi, miror effluere. Superstitis sanguinis residuae guttulae colorem infecerant, sic ut delitescentis intra thoracem forte saniem abscessus ex cruenti puris imagine suspicarer. At cum avulsum penitus a sede cor inque mensam proiectum nonaginta ferme systolas aequabili vehementia compressisset totidemque explicuisset diastolas et supremos tum demum efflavisset aliquanta palpitatione spiritus, prioris cogitationis puduit, nec ultra tantum morbum credidi tanto vigori conterminum....

Ergo lancinata ilico receptaculi tunica chylum effudit, et secutus per eiusdem vulneris rimam, quem turgidis mesenterii lacteis relaxato liberaveram ligamine, dubium omne revulsit scaturienti evidentia.

So I cut open the abdomen of a Molossian dog in the middle and start my examination of the viscera. Without any delay, I rip out the heart, cutting off the ties of the small vessels by which it clings to the rest of the body. Then, having skimmed off the abundant blood, which has instantly formed pools and is clouding my view as I am looking, I am astonished to see a rich white fountain of milky, highly fluid liquid flowing out forthwith at the inside of the tube of the vena cava, in the region of the right chamber of the heart. Some remaining drops of blood left over had changed its colour so that I suspected a festering abscess was hiding in the chest because of what looked like bloody pus. But when the heart, completely wrenched from its seat and thrown on the table, had contracted into ninety systoles with uniform power, expanded into as many diastoles, and finally blown out its last spirits with notable convulsion, I felt ashamed of my previous thought and did not believe any longer that such a great disease could be the neighbour of such strength...

So the wounded cover of the reservoir poured out its chyle, and as I had freed the latter from the swollen milky veins of the mesentery by loosening the strapping band, it followed through the fissure and uprooted all doubt with gushing obviousness.

Pecquet gives his readers little in terms of background. He says nothing about the date and place of his experiments, nor would this have been possible, as already noted. Witnesses appear in some other passages of his account (pp. 7–9, 13), and their names will be given at the end (p. 19) to 'crush the stubborn mind' ('obstinatum...animum obruere') of incredulous readers. Here, however, the report is

[94] Pecquet begins in the present tense to create an impression of vivacity and immediacy (cf. *HlRh*, § 814.2), a feature present in many other reports as well.

made understandable and convincing, at least to a knowledgeable audience, by its detailed and well-structured step-by-step exposition of what Pecquet does, sees, feels, and thinks. The narrator himself comes across as purposeful and determined, but not immune from error. At the same time, his initial misinterpretation of the bloody lymph serves to dramatize the events: the road to knowledge is presented as difficult and hazardous, and Pecquet seems always to be in danger of getting on the wrong track. The highly rhetorical climax, when lymph and understanding gush forth at the same moment from the cisterna chyli, is all the more triumphant and impressive.

Rhetorical resources were not always judged sufficient to conjure up all the details of an event in the reader's mind's eye. In such cases, non-text in the form of tables and illustrations could be of help. Galilei graced his *Sidereus nuntius* with images of the moon as it looked through his telescope (fols. 8^{r}, 9^{v}, 10^{r-v}), in addition to other, more diagrammatic illustrations. In later astronomical reports, such as Hell's *Observatio transitus Veneris*, whose exact numerical data were more important than the observed event in itself, positions of celestial bodies were tabulated. The engravings showing sturdy fellows and horses working in vain against the force of air pressure in the third book of Guericke's *Experimenta nova Magdeburgica* contributed much to the work's renown. Likewise, Scheuchzer's *Οὐρεσιφοιτὴς Helveticus* won fame for its graphic copper tables, whose printing was financed by the members of the Royal Society.[95] All in all, however, it is striking how many reports manage to do with few images or none at all. Pecquet finds one single table featuring an eviscerated dog and a diagram of the lymphatic system sufficient illustration for his *Experimenta nova anatomica*. Medical 'observationes' are bare text only, as are most botanical 'itinera'. Fabricius and Gassendi need nothing but words to explain how they discovered the sunspots and observed the transit of Mercury for the first time respectively.

In many reports, a feature as conspicuous as the richness in factual detail is the strong visibility of the narrator in his own narrative, a feature which goes far beyond the use of the first person singular. (Indeed, this grammatical marker may even be missing without great detriment to the narrative's personal flavour.[96]) The narrator often tells the reader far more about himself than would be necessary for an adequate understanding of his account—a surplus which starkly contrasts with the almost total reticence in modern scientific writing. Some of the most conspicuous forms in which this personal perspective expresses itself will be discussed briefly by way of conclusion.

[95] The engravings can be admired in the database Viatimages (https://www2.unil.ch/viatimages/, accessed 16 February 2023).

[96] Calzolari's 'Iter Baldi' is no first-person narrative but clearly presupposes the author's personal experience of Monte Baldo near Verona, an aspect underlined by its strong aesthetic and emotional response to the mountain (Korenjak 2013b, 204–5, 209–11, 213–14, 216–17).

An obvious way of personalizing one's account, and one that ties with the previous section, is the inclusion of personal detail unrelated to the scientific content of the report proper. Amatus Lusitanus, for instance, goes out of his way to record his gratitude for the 300 gold ducats bestowed on him by a generous patient (*Curationum medicinalium centuria prima*, p. 66). Johnson's *Iter in agrum Cantianum* bristles with lively details of this kind: a thunderstorm threatening the party on the Thames (described in mock-heroic fashion in the words of *Aeneid* 1.88–91; p. 1), the sight of the royal fleet near Gravesend (p. 4), a pastor's hospitality (p. 5), the splendours of a castle built by Edward III (p. 6), the torments of Tantalus suffered during a thirsty walk by the sea (p. 7), and an exquisitely prepared lamb (p. 10) all find a place in the narrative.

In other cases, the reader is invited into the head of the author, so to speak, who shares his reasoning with him. Here is Amatus Lusitanus again, doing a bit of detective work in his examination of a girl bitten by a poisonous snake (*Curationum medicinalium centuria prima*, p. 56):

> Quapropter cunctatius morsum locum animadvertens in pedis parte alta, tarso sive pedis raseta dicta, duo nigerrima signa deprehendo, ex quibus iudico morsum illum a masculo vipera illatum esse, cum duos tantum caninos vipera masculus habeat dentes, femina vero quattuor.
>
> Taking, therefore, a closer look at the place bitten, which was located in the higher part of the foot the called 'tarsus' or 'the raseta of the foot',[97] I notice two deep black marks. From them, I judge this bite to have been inflicted by a male viper, because the male viper has two fangs only, but the female, four.

While this zoologically unfounded conclusion is clearly meant as a specimen of Amatus's acumen, other authors do not shy away from reporting their errors as well. An example from Pecquet's *Experimenta nova anatomica* has already been quoted; another, from Luigi Galvani's 'De viribus electricitatis in motu musculari commentarius' (1791), will be discussed in the section on journal articles. In astronomy, unfounded disbelief is even used as a kind of formula to announce a great discovery: the narrator, unable to trust his own eyes, feels compelled to call other witnesses to verify what he is seeing. Brahe does this when observing the nova of 1572 and the comet of 1577,[98] and so does Fabricius when he first notes dark spots on the sun's surface (*De maculis in sole observatis*, sig. C3^{r}):

[97] Both expressions refer to the region of the four bones known today as os cuneiforme I–III and os cuboideum; see Andreas Vesalius, *De humani corporis fabrica* (1543), p. 168, c, c1,2,3, d, e, f, and g, plus the three preceding woodcuts of the human skeleton.

[98] *De nova stella*, sig. A^{r}; *De mundi aetherei recentioribus phaenomenis liber secundus*, sigs. [*^{v}]–*2^{r}.

> Nolui tamen mihi etiam in manifesto isto oculorum testimonio propter monstrositatem et inusitatam solis apparitionem assentiri. Vocavi ilico parentem, cui tum temporis ex Batavia reversus aderam, ut ad hanc observationem, non quidem sine metuendo oculorum damno habendam, accurreret.
>
> Nonetheless, I did not want to agree with myself despite the evident testimony of my own eyes, because the sun's appearance was so contrary to nature and anything but normal. At once I called my father, with whom I lived at the time after my return from the Netherlands, to hurry to make this observation as well, although it could not be made without fear of harming one's eyes.

In addition to their train of thought, some authors of reports grant their readers access to their emotional life as well. At the very beginning of Haller's *Iter Helveticum* (§ I), for instance, the narrator's desire to go out into the mountains and botanize becomes simply irresistible.

> Quando eo vere, quod proxime elapsum est, ad patriam urbem iter feci, nondum adultum ver erat et nimbosa hiems plantas, etiamsi floruissent, legi vetabat. Bernae me alia et quae mihi quidem maiori<s> momenti essent detinuerunt. Non potui tamen ferre aspectum eorum montium, qui patriam urbem coronant, absque dulci desiderio stirpium, quas in iis toties innocentissima voluptate perfusus iunior legeram. Eripui me quantum licuit urbanioribus negotiis et quam primum humiliores montes nix deseruit, ad eos raptim iter feci.
>
> When I travelled to my home town last spring, spring had not fully arrived there yet and cloudy winter would have forbidden collecting plants even if they had flowered at all. Other occupations which were quite important, to me at least, kept me in Bern. Nonetheless, I could not bear the sight of the mountains surrounding my home town without feeling a sweet yearning for the plants I had collected there so many times, filled with the most innocent pleasure, in my younger days. I shook off as far as possible my business in town and as soon as the snow had uncovered the lower mountains, I hiked towards them in a rush.

Scheuchzer's *Οὐρεσιφοιτὴς Helveticus* bristles with aesthetic delight (and sometimes horror) at rainbows, glaciers, gorges, and prospects from mountain tops.[99]

Finally, the narrator may not only open up to the reader but also try to establish direct contact with him by apostrophizing him directly. This comes quite naturally in epistolary reports such as Gabriele Falloppio's *Observationes anatomicae* (1562), where the readership as a whole is represented by the addressee, but it can be observed in other texts as well. In his *Experimenta nova anatomica*, Pecquet uses the technique in quite provocative fashion, as he

[99] Korenjak 2012, 218–21.

challenges the reader not to trust him blindly but to repeat his experiments and see for himself (p. 13):

> Nec, o lector, meis, quos in tanti ponderis argumento exertissimos defigebam, oculis confidito!... Ipse, ut irritavi pridem et ad laudabile facinus provocavi animum tuum, aude et in arenam descendere! Ignavus est, cuius laboris formidine tota in auribus fides, nec profecto quis sapere dignus, dum errorem sectatur, quem possit eluere.
>
> Reader, do not trust my eyes, although I had them fixed most intently on a topic of such weight!... Since I have already given you an incentive and called upon you to accomplish a praiseworthy deed, dare to step forward into the arena as well! He is a slacker who puts all his faith in his ears for fear of hard work. Indeed, nobody deserves to know when he follows an error he could do away with.

Like the detail of many reports, their conspicuous personal element seems to be due to more than one reason. The desire to win the reader's sympathy and admiration presumably constitutes a weighty motive, as does the wish to make the narrative more attractive and captivating. In any case, given that the report is an autobiographical format by definition, placing himself at the centre of his own narrative is a quite obvious option for any author.

Journal articles

Early modern learned journals

Like the report, the article was a latecomer among the types of scientific writing discussed in this book: it came into being only in the second half of the seventeenth century. Moreover, it was exceptional in originally constituting, not a literary genre, but rather a publication format embedded in a specific context. Before I turn to the format itself, it is, therefore, necessary to adumbrate this context: the early modern learned journal.[100]

[100] The paragraphs that follow mainly distil information from Kronick 1976 (see esp. pp. 16–18 for his definition of the journal), whose remarks can be supplemented by Broman 2000, Vittu 2005, and Pfeiffer and Vittu 2008. Bibliographies of early scientific journals and articles are provided by Reuss 1801–21 (articles in journals issued by academies only); Scudder 1879 (4,390 periodicals up to the author's own day); Bolton 1897 (*c.*8,600 periodicals up to 1895); Gascoigne 1985 (936 important early periodicals); Kronick 1991 (*c.*1,000 periodicals appearing under 1,858 titles, 1665–1800). The instructive account of the initial years of the *Acta eruditorum* by Laeven 1990 substantiates many of the general statements made in this section. On the development of medical journals, see Morton 1990, esp. 223–30.

An ideal journal may be defined as a periodical publication of indefinite duration overseen by one or several editors whose single items are uniform and contain a number of original, non-topical contributions by different authors, meeting certain thematic criteria and addressing a specific audience. Joining earlier periodical forms such as book fair catalogues, calendars, almanacs, and newspapers and often relying on personal correspondence for its content, the learned journal emerged in the 1660s. Within a few years, three archetypical representatives were launched, the *Journal des sçavans*, the *Philosophical Transactions* (both 1665), and the German *Miscellanea curiosa* (1670). In 1692, the new form was prominent enough to become the subject of a 300-page *Schediasma historicum de ephemeridibus* by Christian Juncker. By 1700 several dozen journals had sprung up, a few decades later, several hundred, a great part of them in Germany, which produced most periodicals by far.[101] Appearing in a rapid and steady rhythm and containing numerous short contributions for which no customized outlet had been available before, journals responded to the acceleration of intellectual production by accelerating dissemination. Articles in journals were free of charge for authors, who could thus circumvent the financial troubles bedevilling them in the publication of books. Edmund Halley, for one, published most of his original work in the form of eighty-one articles in the *Philosophical Transactions*, while Johann Bernoulli's mathematical *oeuvre* took the form of nearly two hundred contributions to the *Acta eruditorum* and other journals.[102]

Few journals issued before the French Revolution strictly conform to the ideal-typical definition given initially. Ranging from newspaper-like forms to publications resembling multivolume books, early modern journals show a 'bewildering array of formats'[103] and often defy modern expectations. Most strikingly, many were conceived less as collections of original research than as carriers of news. Fresh research results formed part of this category, but much else did as well. The opening declaration of the *Journal des sçavans* is revealing in this respect: as announced on the first page, the periodical will inform its readers, first, about the most important books recently printed; secondly, about the death of scholars, biobibliographies of whom will be provided; thirdly, about new results from physics, chymistry, astronomy, meteorology, and anatomy, and about new inventions; fourthly, about pertinent decisions of ecclesiastical and secular bodies; and in short, about everything 'digne de la curiosité des Gens de lettres'. In addition to the topics featured in this list, many journals also reprinted contributions

[101] Kronick 1976, 88–9, 121, 206. The popularity rapidly acquired by learned journals is underlined by a high number of counterfeit editions; see Vittu 1988 for the *Journal des sçavans* (pp. 313–15 for some further examples).

[102] Nummedal and Findlen 2000, 192; Pfeiffer and Vittu 2008, 289–90.

[103] Kronick 1976, 20.

published elsewhere or translations, extracts, and abridgements of such. Only about one half of all journals up to 1790 contained any original contributions at all.[104] Further potentially surprising and confusing features include a low degree of specialization, a fluid border between learned and popular journals, the anonymity of many contributions, frequent change of journal titles, and a usually short lifespan. The last point particularly applied to independent enterprises. Journals launched by academies and learned societies lasted longer, since they received public funding.[105] Moreover, because of the concentrated expertise of their publishing bodies, whose flagships they constituted, their scientific and scholarly level was on average higher than that of their independent counterparts.

Thematically, few journals confined themselves to the natural world from the outset. The *Philosophical Transactions* did so to a large degree, the *Miscellanea curiosa* exclusively, but most other early journals were, rather, dedicated to learning in general. The German *Acta eruditorum* (1682–1782), the most important independent periodical of its time, covered the following six fields according to its indexes: (i) theology and church history; (ii) law; (iii) medicine and 'physica'; (iv) mathematics; (v) history and geography; (vi) philosophy and philology.[106] Many learned periodicals contained little or nothing about the natural world.[107] Over time, however, the field became more prominent. For the period 1665–1800, about one thousand partly, predominantly, or exclusively natural scientific journals are on record.[108] Academy journals separated science from other areas by assigning contributions to different sections. The Societas regia scientiarum Gottingensis, for instance, divided its journal into three sections dedicated to the natural world, mathematics, and humanities respectively (*Commentationes physicae, mathematicae, historicae et philologicae*) from 1779.

Within science, the earliest area of specialization was medicine, to which already the *Miscellanea curiosa* was primarily devoted. It was joined by the laboratory sciences and natural history in the eighteenth century.[109] In astronomy, *Ephemerides* was published annually from the late seventeenth century, but this was a very special kind of journal, if it can be called one at all; however, some started publishing articles in addition to the usual tables in the eighteenth century.[110] In all disciplines, the steadily growing importance of journals can be

[104] Kronick 1976, 281. In the first twenty-five years of the *Acta eruditorum*, nearly half as many articles taken from other journals as original contributions were printed, and the ratio of reviews to original contributions was about 12:1 (Laeven 1990, 299–372).

[105] Cf. Broman 2000, 234.

[106] For details, see Laeven 1990, 52–5. The categories chosen seem to be loosely inspired by those of contemporary fair catalogues.

[107] For examples, see Gascoigne 1985, 96.

[108] Kronick 1991 (see note 100 above).

[109] Kronick 1976, 97–8; Gascoigne 1985, 120–1.

[110] Gascoigne 1985, 141–2.

read off the publication record of scientific authors. By classifying some 2,700 respected personalities by decade of birth, it has been found that only 17 per cent of those born between 1600 and 1609 ever published in periodicals. For the birth decades 1650–9, 1700–9, 1750–59, and 1770–9, this figure rises to 47, 65, 75, and 85 per cent respectively.[111]

The vernaculars and Latin

It is a widespread idea that the scientific journal was a quintessentially vernacular medium and that such traces of Latin as may have been present at its birth vanished quickly. An authoritative overview explicitly excludes Latin specimens because 'the decline of Latin in the periodical literature is too precipitous for us to make comparisons between the seventeenth century and later centuries'.[112] Such a point of view, which seems to reflect the assumption that the progressivity of the medium must have expressed itself also in its language, receives apparent confirmation from the two earliest and most famous journals, the French *Journal des sçavans* and the English *Philosophical Transactions*. In fact, however, the story was more complex and Latin less marginal than commonly supposed: it is present even in primarily vernacular journals, and numerous periodicals existed with contributions entirely or primarily in Latin.

The former point is well illustrated by the *Transactions*.[113] While Latin was absent from their very first volumes, Latin contributions appeared regularly from the 1670s to the 1770s. As a cursory survey of selected issues and (from the 1750s) volumes shows, some 10 per cent of the *Transactions*' content was in Latin up to 1750. In the following three decades, the figure remained at around 4 per cent. It dropped virtually to zero only after 1780. (By contrast, almost nothing was published in other major vernaculars such as German, Italian, or even French.) Most Latin papers were sent in by foreigners like Daniel Gabriel Fahrenheit ('Experimenta et observationes de congelatione aquae in vacuo factae', 1724) or Roger Joseph Boscovich ('De proximo Veneris sub sole transitu', 1759) or involved interactions with foreigners. When in 1672 the Frenchman Ignace Gaston Pardies voiced objections to Newton's new theory of light in a Latin letter to the *Transactions*' founder and first editor Henry Oldenburg, he was answered by Newton in the same language. Some Englishmen, for example Edmund Halley, also preferred to publish in Latin by themselves, presumably with an eye on an international readership. Finally, some contributions submitted in foreign vernaculars were translated into Latin for publication. This is true, for instance, of a

[111] Gascoigne 1985, 91–2. [112] Gross et al. 2002, 33.
[113] Their early volumes are available at https://royalsocietypublishing.org/loi/rstl (accessed 16 February 2023).

number of Antonie van Leeuwenhoek's Dutch letters about his microscopic discoveries of things like human semen (1677). Still, the *Transactions*' dominant language, English, the least known of Europe's major vernaculars at the time, caused Continental readers trouble, hindering for example the abstraction of the first volume in the *Journal des sçavans* before the its editor had 'enfin trouué vn interprete Anglois' to this end (vol. 1, 1665, p. 156). As in the case of several vernacular journals (not least the *Journal des sçavans* itself, which also appeared as *Ephemerides eruditorum* between 1665 and 1670), attempts were made to remedy the situation by means of Latin translations. During the 1670s, the first 68 issues, covering the years 1665–71 and filling some 2,700 duodecimo pages, were Latinized by two translators, John Sterpin and Christoph Sand. Both translations were harshly criticized by Oldenburg, but they seem to have sold. While Sterpin was expressly forbidden by Oldenburg to continue, Sand's efforts were brought to a halt only by his untimely death in 1680.[114] In addition, many individual articles were Latinized. Nearly ninety such translations feature in the first twenty-five volumes of the *Acta eruditorum*.[115]

More importantly, a substantial number of journals were originally published in Latin. In David A. Kronick's title index, there are some seventy Latin titles in about one thousand, which implies that 7 per cent of all scientific and technological journals published from 1665 to 1800 were in Latin. In particular, this is true of nine (22 per cent) out of the forty-one most important periodicals founded before 1790.[116] The monumental *Repertorium commentationum a societatibus litterariis editarum* by Jeremias David Reuss, published between 1801 and 1821 and exploiting periodicals from the one and a half preceding centuries, contains about 15 per cent of articles in Latin. Table 2 provides a selection of major journals (most of which appeared in several instalments) featuring substantial numbers of original scientific contributions and published entirely or mainly in Latin. Ten journals issued by academies and similar bodies come first, followed by three private enterprises. Within both categories, the periodicals are ordered chronologically.[117]

[114] On the Latin translations of the *Transactions*, see Toribio 2020. On the frequency of translation of entire journals in general (often coupled with abridgement), see Kronick 1991, p. xvii, and the pertinent remarks in his catalogue entries; Vittu 2005, 531–2. Of the Latin journals to be listed shortly, the *Miscellanea Berolinensia*, the *Acta literaria Sueciae*, the *Commentarii Academiae scientiarum imperialis Petropolitanae*, and the *De Bononiensi scientiarum et artium instituto atque academia commentarii* prompted vernacular translations.

[115] Laeven 1990, 367–72.

[116] Kronick 1991, 297–322; Gascoigne 1985, 93–4.

[117] The list is based on Gascoigne's catalogue of most influential journals (see note 116 above). The *Ephemerides* of the Societas meteorologica Palatina has been added for its importance and to represent the fringe form of the scientific journal. Two additional privately published periodicals supplement the *Acta eruditorum*. Names of academies and titles of journals are given according to Kronick 1991, whose entries may be consulted for additional information.

Table 2 Selection of journals featuring original scientific contributions entirely or mainly in Latin

Editing body or editor, place	Title (catalogue numbers in Kronick 1991)	Period of appearance in Latin	Remarks
Academia naturae curiosorum, Halle	*Miscellanea curiosa* (nos. 1150, 564, 57, 1334)	1670–1791	continued to the present day; still in Latin to a smaller extent in the first half of the nineteenth century
Academia scientiarum Berolinensis, Berlin	*Miscellanea Berolinensia* (no. 1147)	1710–43	
Societas scientiarum Upsaliensis, Uppsala	*Acta literaria Sueciae Upsaliae publicata* (nos. 50, 48, 59, 54)	1720–1850	some Latin contributions even in the subsequent series
Academia scientiarum imperialis Petropolitana, St Petersburg	*Commentarii Academiae scientiarum imperialis Petropolitanae* (nos. 475, 1336, 43, 1331)	1727–95	in fact continued to 1806; predominantly in Latin at the beginning, *c.*50% Latin, 50% French in the last volume
Reale Accademia delle scienze dell' istituto di Bologna, Bologna	*De Bononiensi scientiarum et artium instituto atque academia commentarii* (no. 408)	1731–91	continued in Latin as *Novi commentarii Academiae scientiarum Instituti Bononiensis* (10 vols, 1834–49; no separate entry in Kronick 1991)
Societas physico-medica Basiliensis, Basel	*Acta Helvetica physico-mathematico-botanico-medica* (nos. 47, 1333)	1751–87	
Societas regia scientiarum Gottingensis, Göttingen	*Commentarii Societatis regiae scientiarum Gottingensis* (nos. 478, 482, 1337, 483)	1752–1808	
Academia electoralis Moguntina scientiarum quae Erfurti est, Erfurt	*Acta Academiae electoralis Moguntinae scientiarum utilium quae Erfordiae est* (nos. 39, 42, 1330)	1757–1806	last volume with a substantial amount of Latin articles appeared in 1799
Academia electoralis scientiarum et elegantiorum literarum Palatina, Mannheim	*Historia et commentationes Academiae electoralis Palatinae* (no. 749)	1766–94	split into a science and a humanities subseries from vol. 3

Societas meteorologica Palatina, Mannheim	*Ephemerides Societatis meteorologicae Palatinae: Historia et observationes* (no. 559)	1783–95	contains meteorological observations conducted according to a uniform scheme
Thomas Bartholin, Copenhagen	*Acta medica et philosophica Hafniensia* (nos. 51, 465, 61, 58)	1673–1779	continuations appeared a century later (1774–9) than the original journal (1673–80)
Otto Mencke, Leipzig	*Acta eruditorum* (nos. 46, 1332)	1682–1782	various substantial supplements in Latin, one of which is no. 67
group of anonymous scholars from Lübeck	*Nova literaria maris Balthici et septentrionis* (no entry in Kronick 1991)	1698–1708	news from the learned world interspersed with original articles

All in all, then, Latin remained an important language of scientific journals throughout the eighteenth and sometimes even in the early nineteenth century, particularly in Germany and North Eastern Europe. The academies, in particular, in these areas stuck to it,[118] which qualifies the common notion that such bodies were always torchbearers of the vernaculars.

Latin scientific articles: archaic or modern?

As the dominant kind of publication in science today, the article belongs to the few scientific genres whose history has been explored in some detail. Alan G. Gross, Joseph E. Harmon, and Michael Reidy have traced its development from its seventeenth-century origins to its standardized form at the end of the twentieth century.[119] Their book is based on English, French, and German articles and excludes Latin ones, but it nonetheless seems useful to summarize their findings as a point of reference for the subsequent discussion of the Latin material.

The origins of the scientific article are closely connected to the letter. Many early articles originate as extracts from letters sent to journal editors (and

[118] The full text of most important German scientific journals *c.*1680–1810 (with the important exception of the *Miscellanea curiosa*) is presented online by the Interdisziplinäres Zentrum für Wissenschafts- und Technikforschung of the University of Wuppertal (http://www.izwtalt.uni-wuppertal.de, accessed 16 February 2023).

[119] Gross et al. 2002. Banks 2008, 79–177, who restricts himself to the *Philosophical Transactions*, by and large comes to similar conclusions. See also the even more specific study on the development of the experimental report in that journal by Bazerman 1988, 59–79 (Harris 2018, 263–79).

sometimes read out in front of the members of a learned society). Their text is short and shows no subdivisions. Nothing or little is given in terms of research context. The dominant mode is first-person narrative, used to report a specific observation or experiment made by the author. Illustrations and non-text in general are rare. Credibility is attained through personal authority and trustworthy witnesses. Authors are reluctant to venture explanations and generalizations. They show an epistemological scepticism typical of the seventeenth century, a wariness of large-scale hypotheses, and a hope of progress through the accumulation of small, ideologically neutral pieces of information.[120] Sentences are long and complex; technical terminology is used sparingly.

Over time, especially from the later eighteenth century, the genre develops in the direction of the modern scientific article. The text, which tends to become longer, is preprocessed for the reader by presentational features such as subheadings and footnotes. The latter, which contain references to previous literature, situate the paper in its research context; this is also an important function of an introduction and conclusion. The observations and experiments reported are no longer seen as interesting in themselves but as instances of general laws. Facts and data become subservient to explanations, which turn increasingly discipline-specific. Credibility is established through the display of professional methodology. Non-text becomes more important and reference to it is standardized. The language becomes more technical in terms of vocabulary and also of syntax, as sentences become shorter and less complex and colourful verbs disappear. The centre of the sentence is typically taken by a complex noun phrase in subject position. Actions of scientists are put into the passive voice. These features have the effect of letting nature itself take centre stage and, as it were, speak for itself. The writing becomes self-effacing, and an impression of total objectivity is created.

As already mentioned (see note 112), Gross, Harmon, and Reidy justify the exclusion of Latin papers from their overview by denying them any noteworthy development because Latin supposedly became extinct as a language of science too early for any evolution to happen. This amounts to assigning Latin items exclusively to the 'archaic' variety (as I will term it in what follows) of the scientific article. Is this characterization accurate? Yes and no. Unsurprisingly, truly modern-looking Latin papers are indeed hard to find. The structure known as IMRaD (Introduction, Methods, Results, and Discussion), which has become standard in modern scientific articles, is never fully realized in Latin ones, nor are the stylistic features singled out as quintessentially modern by Gross, Harmon, and Reidy, in particular, complex noun phrases as subjects and passivization. The same is true of some other characteristics not mentioned by them, notably the avoidance of the first person singular, narration, and metaphors.[121]

[120] On the epistemological volte-face behind this, see Pomata 1996.
[121] On these 'taboos', see Kretzenbacher 1994.

On the other hand, clear-cut examples of the archaic type turn out to be a rarity as well. In fact, Gross, Harmon, and Reidy's assessment of Latin papers seems to rest exclusively on the two earliest important Latin journals, the *Miscellanea curiosa* and the *Acta eruditorum*, the only ones mentioned by them,[122] while they omit the eighteenth-century academy periodicals whose articles have a more modern look to them. But even if one accepts this restriction, the archaic article in its pure form remains an elusive entity. One of the best examples I was able to find, an observation by the physician and polymath Johann Daniel Major printed in the third volume of the *Miscellanea curiosa* (1672, p. 11), may illustrate the point.

Observatio VI.

Doctoris Johannis Danielis Majoris

De pullo gallinaceo pullis incubante

Anno 1668 Lubecae in aedibus vicarii cuiusdam reverendi capituli ibidem ego et plures alii aliquotiens non sine voluptate vidimus pullum gallinaceum perdicis magnitudine nunc pipiendo praestare se pullis minoribus aliis aetate non multo grandiorem, nunc convocando eosdem, ducendo, cibando subque alas recipiendo ac fovendo defungi rectissime veteranae gallinae officio, memoratu digno specimine non in homine solum sagacitatem ingenii annorum mensuram quandoque antevertere, sed huius rei exemplum aliquod etiam in alacritate brutorum superesse, referendum in physicis ad sollertiam animalium, de qua, nisi fallor, Plutarchus aut Plato peculiarem sermonem scripsit.

[In litteris die 15. Junii anni 1671 datis ad doctorem Philippum Jacobum Sachs de Lewenhaimb]

Observation VI.

By Dr Johann Daniel Major

Regarding a Hen Chick Taking Care of Other Chicks

In 1668, at Lübeck, in the house of a certain vicar of the honorable chapter in this very place, I myself and several others saw a couple of times, not without pleasure, a chick the size of a partridge, which sometimes offered its help to the other, smaller chicks by peeping, although it was not much older itself, and sometimes fulfilled exactly the duty of an experienced hen by calling them together, leading them, giving them shelter, and keeping them warm under its wings, thus providing a memorable example that not only in humans the development of the mind sometimes hurries ahead of the measure of the years, but that some instance of this phenomenon exists also in the brightness of brutes,

[122] Gross et al. 2002, 33.

something to be correlated within the study of nature to the smartness of animals, about which topic, if I am not mistaken, Plutarch or Plato has written a separate treatise.

[In a letter given the 15 June 1671 to the doctor Philipp Jakob Sachs of Lewenheimb]

In a number of aspects, this short text conforms quite well to the archaic type. In an extract from one of his letters, Major narrates a single observation made by himself and backed up by some additional witnesses (although no details are given to enhance their credibility). He does so in the first person, foregrounding his own presence by emphasizing his pleasure at the sight of the precocious chick and his uncertainty about the author (Plutarch, in fact) of a thematically pertinent treatise. The whole incident is told in untechnical language and in one long, syntactically complex period. The recipient and editor, Sachs of Lewenheimb, contextualizes the observation by adding the author's name and a title at the beginning, as well as some information about the original letter at the end. Within the text, by contrast, one finds no structuring devices, neither textual nor in terms of layout. There is no non-text either. In one crucial aspect, however, Major deviates from the supposed standard of the archaic article: he does not let the incident stand on its own, but understands it as an instance of a general law—child prodigies exist among animals as well as among humans. His article is formally archaic but epistemologically modern.

A versatile format

As already intimated, most Latin articles contain more modern features than the one just discussed. Above all, however, most transcend both the archaic and the modern type alike in numerous ways. The more one reads, the less significant this dichotomy appears in itself. Ultimately, one is led to acknowledge that what was premodern about Latin journal articles was not the prevalence of a certain type but the very absence of clear-cut types. In many respects, they constituted a distinctly experimental format.[123] In what follows, I try to provide a first overview of this varied landscape, occasionally drawing on the useful descriptive categories introduced by Gross, Harmon, and Reidy but renouncing the temptation of grading articles according to their modernity or lack thereof, and to trace historical developments.

The first thing to note is great variety in terms of genre. As stated earlier, the journal paper started life not as a genre in the strict sense of the word, a kind of

[123] One suspects this to be true of their vernacular counterparts as well, but to corroborate or disprove this assumption is beyond the scope of this section.

text distinguished by specific conceptual features; it only grew into one over time. At its beginning, it was a novel publication format which drew on pre-existing genres and text types. The most fundamental of these was the letter, as many early articles simply were excerpts from letters sent to a journal editor. Complete letters could be printed as well. In fact, that practice survived for a long time. When Boscovich wanted to exhort those in the astronomical community to prepare themselves in advance for the observation of the 1761 Venus transit, he did so in a long open letter to the members of the Royal Society which was published in the 1759 volume of the *Philosophical Transactions*.

Among further genres feeding into the journal article, the observational or experimental report was suitable for this in its capacity as a short first-person narrative of a scientifically relevant experience. Fahrenheit's aforementioned 'Experimenta et observationes de congelatione aquae in vacuo factae' is a text-book example of a report turned into a paper. A later case in point is provided by Luigi Galvani's frog experiments, which led to the discovery of the role of electricity in muscular contraction ('De viribus electricitatis in motu musculari commentarius', 1791) and appeared in the *Commentarii* of the Reale Accademia delle scienze dell' istituto di Bologna. As it describes and discusses a single clear-cut phenomenon in considerable detail, this text of some fifty-five pages could also be termed a short monograph. For monographs, a journal might prove a more convenient and less costly outlet than an independent publication. Georg Christian Füchsel, for one, explains the appearance of his geological description of Thuringia in the *Acta Academiae electoralis Moguntinae* (1761) by the fact that he could find no publisher to bring it out as a book (p. 45).[124] As further examples of articles of monographic proportions, Johann Heinrich Hottinger's 'Montium glacialium Helveticorum descriptio', the first work ever entirely devoted to glaciers (*Miscellanea curiosa*, 1706), and Christian Maximilian Spener's 'De crocodilo in lapide scissili expresso', dedicated to a single fossil of a protorosaurus found in Thuringia (*Miscellanea Berolinensia*, 1710) may be cited. Papers published in academy journals frequently originated as 'orationes', speeches read out at the meetings of academies. After Georg Christoph Lichtenberg—better known for his aphorisms today, but also the holder of the first chair of experimental physics in Germany—reported to the Societas regia scientiarum Gottingensis on his determination of the coordinates of Hanover, Osnabrück, and Stade at their meeting on 14 December 1776, his talk appeared in the Society's *Commentarii* the following year ('Observationes astronomicae', 1777).[125] In fact, however, there is

[124] A journal could also be more advantageous than a book for the publication of a short article. Friedrich Zwinger first published his description of a rare fungus, to be mentioned shortly, in the third part (1749) of Daniel Bruckner's *Versuch einer Beschreibung historischer und natürlicher Merkwürdigkeiten der Landschaft Basel* but republished it in the *Acta Helvetica* two years after in the hope that this would be read more widely than a local history.

[125] Reprint with facing German translation: Hasse 1997, 8–53.

little oratorical about such texts. Their length, often some twenty to thirty pages, as well as the inclusion of calculations, charts, and footnotes, demonstrates that they constitute heavily edited versions of the original presentations.

While the genres mentioned so far were not restricted thematically to this or that discipline, others were discipline-specific. The case histories quite common in early journals were a medical genre that can be traced back to the Hippocratic *Epidemics* and became successful in book format from the mid-sixteenth century. Some of these 'curationes' also went by the more general title of 'observationes', which medical books shared with astronomical ones. The popularity of such collections was reflected in journals too: articles often reported observations and were titled accordingly.[126] The contributions to the medically oriented *Miscellanea curiosa* and their follow-ups were collectively given this title until the end of the eighteenth century, the single items being simply titled 'Observatio I., II.' etc. Natural history contributed descriptions of individual plants, animals, and stones, a text type that had been standardized and serialized since antiquity. In the first volume of the *Acta Helvetica* (1751), for instance, Friedrich Zwinger published a description of a hitherto unknown fungus he had found near Basel. In some cases, book-length series of descriptions could be printed. A late and elaborate example is provided by the ninety-seven genera of lichens from Scandinavia and Greenland described in Theodor Magnus Fries's 'Lichenes arctoi' (*Nova acta Regiae societatis scientiarum Upsaliensis*, 1861). The form also fitted the presentation of collections like Johann Friedrich Blumenbach's of human skulls, which he published in six instalments in the *Commentationes* of the Societas regia scientiarum Gottingensis (1791–1820). The theorem plus proof, a mathematical text type reaching back to Euclid, was ideally suited for journals because of its shortness and self-sufficient nature. Textbook examples are provided by Euler's numerous articles in the *Commentarii* of the Academia scientiarum imperialis Petropolitana, such as his 'Observationes circa bina biquadrata' (1772/73), in which he proves the existence of infinitely many numbers which can be represented as sums of two fourth powers in two different ways.

Along with pre-existing genres and text types, one encounters others which were born with the journal itself or whose rise to importance was due to the new publication format. This applies to 'methodus' pieces, as one might call them, articles presenting new techniques ('methodi') of solving theoretical questions or practical problems and announcing this by the respective keyword in their titles. As outstanding examples, Leibniz's first calculus paper, 'Nova methodus pro maximis et minimis' (*Acta eruditorum*, 1684),[127] and Edmund Halley's 'Methodus

[126] On the transfer of the 'observatio' from the book to the journal, see Pomata 2011b, 62 and 78 n. 98.

[127] A translation with introduction and commentary is provided by Struik 1969, 270–80. See also Merzbach and Boyer 2011, 386.

singularis' of using a Venus transit to calculate the distance from the Earth to the Sun (*Philosophical Transactions*, 1716) may be cited. Meteorological lists and tables for a particular place and time span, which were of only limited interest if taken individually, became much more widespread due to the possibility of publishing them in journals, which allowed for the cumulation of their data until general rules could be deduced from them. In the 1730s, for instance, the *Miscellanea Berolinensia ad incrementum scientiarum* published several tables of rainfall, temperature, air pressure, wind strength, and cloudage.[128] Initiatives to coordinate such local enterprises date back to the early eighteenth century, when James Jurin invited researchers to join forces in an 'Invitatio ad observationes meteorologicas communi consilio instituendas' (*Philosophical Transactions*, 1723). They first came to full fruition with the foundation of the Societas meteorologica Palatina in 1780, which issued twelve massive tomes of *Ephemerides* between 1783 and 1795.

In view of their generic range, it does not come as a surprise that journal articles proved malleable in many other respects as well. Although most comprised a couple of pages, their actual length could vary from a dozen lines, as in the example by Major quoted in full earlier, to monographic proportions: Füchsel's 'Historia terrae et maris' runs to 210 pages. Sometimes, the problem of excessive length was solved by serialization, as in the case of Blumenbach's skull papers. Moreover, long papers could also appear as self-contained books, as happened with Galvani's frog experiments, which came out as an offprint the same year and as a book in the following one.

Related to length are questions of structure and layout. The usually short articles of the earliest decades were subdivided by paragraphs at best, the only other structuring device being a title, which often served as a summary of sorts. Very short pieces even lacked paragraphs, as exemplified once more by Major's chick paper. Many items, especially those taken from letters, lacked an introduction and a conclusion as well: they went straight *in medias res* and ended abruptly. In the eighteenth century, when articles tended to grow longer, paragraphs became standard and numbering them became an option. This device could be used to establish a system of cross-references, as in the 278 paragraphs of Füchsel's 'Historia terrae et maris'. In addition, some papers featured subheadings: after an untitled introduction, Euler's 'Observationes circa bina biquadrata' fell into four sections—task ('Problema'), general solution ('Solutio'), and two numerical solutions ('Prima' and 'Secunda solutio specialis'). A more important innovation was the appearance of footnotes. In a piece like Johann Hofer's 'De possibilitate physica longaevitatis patriarcharum antediluvianorum' (*Acta Helvetica*, 1758), lettered

[128] See, for instance, vol. 4 (1734), pp. 349–76 (Augustin Grischow, pluviometric observations at Berlin); vol. 5 (1737), pp. 123–31 (anonymous, temperature in Pennsylvania); pp. 132–56 (Anders Celsius, air pressure, temperature, wind, cloudage, rainfall at Uppsala).

footnotes were systematically used to cite primary and secondary sources, while asterisk notes were reserved for personal asides.

Apart from providing general orientation, a clear and obvious structure could also serve more specifically to distinguish parts differing from one another in content and function. In particular, data and argumentation were often separated in this way in a partial anticipation of the later IMRaD style. Already to the epistolary reports in the first volumes of the *Miscellanea curiosa*, a 'scholion' comprising similar phenomena, explanations, or further considerations was frequently added by the recipient or a third party. While most of Hottinger's 'Montium glacialium Helveticorum descriptio' is indeed descriptive, chapter 7 (pp. 54–66) stands out as explaining the phenomena listed so far ('phaenomena recensita explicat'), as announced in its title. In Galvani's article on animal electricity, the experimental reports making up its main part are capped by 'some hypotheses and inferences' ('coniecturae et consectaria nonnulla') in the final section (pp. 394–415).

Although most authors were capable Latinists,[129] they did not agree on something like a common article style—quite the contrary. Some dispensed with technical vocabulary, which resulted in easily readable articles like Major's on the motherly chick. Other papers, by contrast, particularly later ones such as Friedrich Stromeyer's 'De hydrogenii arseniati natura' (*Commentationes Societatis regiae scientiarum Gottingensis*, 1808) or Fries's discussion of Arctic lichens, feature exceedingly technical language. When Stromeyer declares: 'Itaque 200 gas hydrogenii arseniati partes adhibitae 111.5 gas oxygenii partes combustione absorpserunt atque 56.45 milligrammata acidi arsenicosi produxerunt' (cf. Chapter 2, 'The lexical peculiarity of scientific Latin'), this is more readily understandable to a modern chemist than a Latinist. Syntactically, everything was possible from expansive, elaborate periods to keyword style. When the occasion was solemn and the author an accomplished writer, even a highly technical subject could be discussed in (apart from terminology) fully Ciceronian Latin. The Jesuit Georg Kraz published a prize essay ('Dissertatio pro praemio', 1767) on the weight and orbit of the moon in such a style in the journal of the Bavarian Academy.[130] On the other hand, even full sentences are a rarity in Anders Celsius's 'Observationes meteorologicae' from Uppsala (1744). In somewhat longer pieces, the introduction was often set apart from the main text not only typographically but also stylistically in accordance with a long-standing custom in technical writing: an opening in high style introduced a paper in technical prose. However, 'technical' did not necessarily imply 'dry', as lively and circumstantial description and first-person narrative remained widespread even in the late eighteenth century.

[129] If an author was not, this might prompt an apology. Daniel Bernoulli, editor of the *Nova Acta Helvetica*, begs the reader's pardon for the 'barbaram Latinitatem' of an anonymous contributor (vol. 1, 1787, p. 143).

[130] His essay stands out all the more because the Academy's *Abhandlungen* were otherwise published in German from the beginning.

Galvani, for one, describes his experimental arrangements and spectacular results (for instance, dead frogs jerking at the exact moment of a lightning strike, p. 375) graphically and minutely. Moreover, he intertwines his actions and perceptions with a detailed account of his train of thought, including mistakes and aberrations, observing and depicting his inner self no less carefully than his electric generators and frogs. In the following quotation, he shows himself tempted by the wrong idea that animals do not generate any electricity by themselves but only accumulate atmospheric electricity (pp. 377–8):

> Has tamen contractiones cum non nisi in libero aëre inspexissem, nondum enim rem aliis in locis temptaveram, parum abfuit, quin ab atmosphaerae electricitate in animal irrepta atque in eo cumulata et in unci cum cancellis ferreis contactu ab eodem rapide exeunte tales contractiones repeterem. Facile enim est in experiundo decipi et quod videre et invenire optamus et vidisse et invenisse arbitrari.
>
> But since I had observed these contractions only under the open sky—for I had not tried the thing in other places yet—I was not far from deducing such contractions from atmospheric electricity which had entered the animal and gathered in it and was swiftly escaping from it when the hook touched the iron grate. For it is easy to be deceived in experimenting and to believe that we have seen and found what we are hoping to see and find.

A last respect in which papers varied widely concerned non-text and its relation to the natural language text. While many articles contained text only, non-text fulfilled crucial functions in others, either complementing the text (something usually indicated by references within the latter) or replacing it. Complementary non-text was standard, for instance, in natural history, where verbal descriptions of plants, animals, and parts of these were routinely accompanied by lifelike illustrations. Newly invented machines and instruments were also depicted as well as described, and the same goes for experiments. In both cases, illustrations helped readers understand the text, while the latter translated immobile images into function and action. In an article on recently invented surgical instruments (*Acta eruditorum*, 1729), Johann Zacharias Platner detailed how the objects depicted in his engravings were to be used. Galvani's 1791 article set out the operating principles and rationale of the bizarre experimental designs presented on four copper plates at its end, as well as his own activities as an experimenter. Maps and sections played an obvious role in geographical and geological contributions respectively. Some of the earliest geological maps ever accompanied Füchsel's 'Historia terrae et maris'.[131]

[131] Oldroyd 1996, 77–8.

In other fields, information in natural language was substituted rather than complemented by non-text. In disciplines depending on serialized observations, such as meteorology or astronomy, tables and tabular lists were a crucial device. How seriously the question of presenting data in this form was taken is indicated by attempts at standardization such as the table of examples at the end of Jurin's open letter or the instructions for observers published by the Societas meteorologica Palatina in the first volume of their *Ephemerides* (pp. 8–17). In mathematics and disciplines increasingly dependent on it, such as astronomy and physics, mathematical notation developed and gained ground in addition to the usual diagrams dating back to Euclid. In Euler's article on sums of fourth powers, the equations are still fully embedded in running natural language text grammatically, but they are set apart typographically, and they account for no less than half of the paper's overall length.

'Historiae'

The genre and its origins

'Historia' is my designation, taken from the most common term to be found in the relevant titles, for a genre which is comparatively well defined in terms of form and content, widespread, and long-lived—Latin examples can be found until the twentieth century,[132] vernacular ones until today—but which has not been recognized as a genre so far,[133] and for which no comprehensive name has established itself. A typical 'historia' consists of a series of distinct entries, each of which presents one item from a clearly defined class (for instance, plants, water animals, or the minerals of Silesia) according to a uniform scheme; the text is often supplemented by illustrations. Among the genres discussed in this book, 'historiae' are special in their thematic restriction to one particular discipline: the items described are chosen from the subjects of natural history ('naturalis historia'), which deals with the objects and living beings of the natural world in a spirit not of analysis and explanation, but rather of description, cataloguing, and classification.[134] This approach is epitomized in the literary genre under discussion.

[132] See, for instance, Giacomo Bresadola et al., *Iconographia mycologica* (1917–60). Some additional examples can be found among the sources of the plant descriptions quoted in Stearn 1992, 154–93, at pp. 168, 172, 184.

[133] Reeds 1991, 136, lists 'local floras', 'garden catalogues', 'large general histories of plants', and 'illustrated herbals' as different botanical genres. All four will be discussed as varieties of the 'historia' in what follows.

[134] Some attempts to adapt the format to other disciplines such as astronomy (Johann Bayer, *Uranometria*, 1603) or medicine (François Boissier de la Croix de Sauvages, *Nosologia methodica*, 1763) remained isolated phenomena.

As natural history was an early modern invention, so was the 'historia'. However, the genre had close conceptual and structural models in antiquity and the Middle Ages, although these are not to be found where book titles would lead one to suppose. The origins of the 'historia' do not lie with Pliny's *Naturalis historia*, which amasses information about the natural world but is not structured by uniform entries. Nor can Aristotle's *Historia animalium* (united with *De partibus animalium* and *De generatione animalium* under the title *De animalibus* in the Latin tradition) or Theophrastus's *Historia plantarum* be considered its fountainheads, because they are primarily given to general reflections on their respective subjects. These are occasionally intermingled with discussions of specific kinds of animals and plants, but once more, these are not presented in a series of uniform entries.[135] By contrast, Dioscorides's influential *Materia medica* shows exactly this structure, each entry presenting and discussing one simple according to a standard procedure. The work differs from early modern 'historiae'only in the mixed content of its entries, most of which are dedicated to plants, but some to animals, specific parts of plants or animals, or minerals. No fewer than three different Latin translations of the *Materia medica* appeared in print as early as 1478 and 1516 (see Chapter 5, 'Translations'). Before that, Dioscorides already had numerous successors who imitated his basic formula in late antiquity and the Middle Ages. Many of these works were, moreover, uniform in terms of content, as they focused exclusively on plants. Alongside such herbals, one also finds bestiaries and lapidaries, either as independent works or as sections of encyclopedias like Vincent of Beauvais's *Speculum naturale*.[136] Medieval herbals provided crucial models for the first early modern 'historiae', although the humanistically inspired authors of the latter preferred to invoke Dioscorides as their most important predecessor.[137]

Early modern 'historiae': an overview

The 'historia' came into being in the first half of the sixteenth century when the structure common to 'materiae medicae', herbals, bestiaries, and lapidaries was

[135] Pliny's work is broken up into numbered and individually titled chapters in early modern editions, but these differ greatly from one another. General discussions like those in Aristotle and Theophrastus are found in several 'historiae' as well, but if so, they are separated as an introduction from the main part, which remains enumerative and descriptive.

[136] Much information on 'materiae medicae' and herbals from Dioscorides to the end of the Middle Ages is found in Pavord 2005, 71–160. A more analytical overview of their complex history is provided by Ventura 2017. On bestiaries, see *LMA* 1, 2072–80 and Hassig 1995, esp. 1–16; on lapidaries, see *LMA* 5, 1714–15 and Accordi 1981, 311–12. An overview of the encyclopedic predecessors of Conrad Gessner's *Historia animalium* is provided by Friedrich 1995, 15–24.

[137] Ogilvie 2006, 137–8, 269. Quite unusually, in his *De plantis* (1583), the Aristotelian Andrea Cesalpino explicitly rejects Dioscorides as a model because of his exclusive interest in medical properties (letter of dedication, sig. [A4^r]; Jensen 2001, 204).

put into the service of the emerging discipline of natural history. The 'historia' became the literary expression of this approach to nature par excellence and carried all its distinctive tasks: inventorying, naming, describing, and ordering nature's confusing wealth of living beings and dead matter. As mentioned earlier, however, the genre never got a name of its own despite its popularity, and its works carried all kinds of titles. 'Historia', often preceded or followed by a genitive or the adjective 'naturalis', seems to have been a default option.[138] However 'historia' titles were by no means restricted to the books discussed in this section, and for these, several other possibilities existed as well. 'Catalogus' and 'pinax', for instance, emphasized the serial structure, '(e)icones' and 'theatrum' the visual appeal of such books. In botany, 'Flora' was first used in Simon Paulli's *Flora Danica* (1648)[139] and soon became common. The analogous 'Fauna' was introduced as a title of zoological works by Linnaeus in order to emphasize that his *Fauna Suecica* (1746) was a companion piece to his *Flora Suecica* published the year before.

In terms of subject, 'historiae' could be written on each of the three 'regna naturae'—plants, animals, and minerals—as well as on any combination and part of them. Plant lore was dominant because it had been so already in the genre's ancient and medieval preforms and because this field evolved fastest in early modern times.[140] Animals came second, the first comprehensive accounts being Edward Wotton's *De differentiis animalium* (1552) and Conrad Gessner's *Historia animalium* (1551–8).[141] 'Historiae' of minerals, stones, and other constituents of the earth's crust, which were difficult to distinguish and classify before chemical analysis, developed least vigorously.

Rarely did a 'historia' cover all three kingdoms in their entirety. Even Linnaeus's *Systema naturae* (1735), which fostered this exceptional ambition, dropped minerals in its later, otherwise ever-expanding editions. Most books spanning all kingdoms restricted themselves in other ways, usually in terms of geographical extension. Even within the single realms, 'historiae' of particular kinds of plants and animals were common alongside more complete ones. Regarding animals, a rough subdivision into land animals, water animals, birds, and small animals (insects, worms), deeply entrenched in human perception and cognition,[142]

[138] Pinon 2005, 259 and 267, counts fourteen major sixteenth-century zoological works thus titled.

[139] Berrens 2019. Paulli's work is in Danish despite its title.

[140] Seminal works are discussed in Bäumer 1991–6, vol. 1, 185–217, 223–8, Pavord 2005, 161–403, and Ogilvie 2006.

[141] On zoological 'historiae', see Bäumer 1991–6, vol. 2, 20–155, esp. 32–73 on Wotton and Gessner; Bäumer-Schleinkofer 1998. A number of pertinent items are to be found in the anthology of early modern zoological books by Pinon 1995 (pp. 84–7 on Wotton's and Gessner's works). In-depth analyses of Gessner's *Historia animalium* (supplemented by two posthumous tomes in 1587 and 1634) are provided by Gmelig-Nijboer 1977, Friedrich 1995, who also discusses the work's vernacular translations and adaptations, and Pinon 2005.

[142] On such 'life forms' as commonsensical, ubiquitous subdivisions of the plant and animal kingdoms, see Atran 1990, 30–5.

was often taken as a basis for specialized 'historiae'.[143] On water animals, for instance, four substantial works appeared as early as the mid-sixteenth century: Pierre Belon's *De aquatilibus* (1553), Ippolito Salviani's *Aquatilium animalium historia* (1554), Guillaume Rondelet's *Aquatilium historia* (1554–5), and the fourth volume (1558) of Gessner's *Historia animalium*, which was subdivided more or less according to the aforementioned principle. Two hundred years later, similar works continued to be written, their scale reflecting the multiplication of known species in the meantime, as in the case of Mathurin-Jacques Brisson's Latin-French *Ornithologia/Ornithologie* (1760) which runs to six volumes and nearly 4,000 pages.[144] A corresponding basic subdivision of the plant kingdom had been codified in Theophrastus's four categories of trees, shrubs, subshrubs, and herbs.[145] While this classification was often used to order the contents of comprehensive 'historiae plantarum', only herbs were dominant in some works such as Otto Brunfels's *Herbarum vivae eicones* (1530–6), something due, as already in medieval herbals, to their pharmacological importance. By contrast, the progress of botanical research manifested itself in the comparatively early appearance of 'historiae' dedicated to more specific taxa, starting with Robert Morison's discussion of what he called 'umbrella-carrying' plants, a group broadly equivalent to the modern family of Apiaceae, in his *Plantarum umbelliferarum distributio nova* (1672). Such works are called biological monographs today, a designation which only appeared in the late eighteenth century in titles like Nikolaus Joseph von Jacquin's *Oxalis: Monographia* (1794).[146] At the same time, plants could also be selected for inclusion in a 'historia' according to wholly unscientific criteria: those discussed in Crispijn van de Passe's *Hortus floridus* (1614), for instance, are chosen for their attractiveness to amateur gardeners.

The recognition of Northern European botanists that the flora around them was not identical to that described by Dioscorides and the extension of geographical knowledge raised awareness of and kindled interest in the differences between the fauna, flora, and geology of different areas. From the second half of the sixteenth century, geographical criteria were often used to subdivide the field of natural history. In particular, a strongly perceived dichotomy between the indigenous and the exotic resulted in works focusing either on an author's homeland

[143] An overview of such works on birds, fish, and insects up to *c.*1650 is given by Bäumer 1991–6, vol. 2, 333–87; see esp. pp. 346–81 (and Gudger 1934) on ichthyological works. On Salviani's and Rondelet's works, see also Pinon 1995, 88–9 and 94–5.

[144] Bäumer 1991–6, vol. 3, 390–1.

[145] Pavord 2005, 22. Theophrastus also pondered the dichotomies of cultivated and wild, male and female, land and water, flowering and flowerless, fruiting and non-fruiting, evergreen and deciduous plants (Wöhrle 1985, 53–62, 98–110; Pavord 2005, 27–9, 359), to which he attached less importance and which became less influential.

[146] With few exceptions, such as John Caius's 'De canibus Britannicis' (1570), works dedicated to individual zoological taxa, which appeared from the early sixteenth century (Bäumer 1991–6, vol. 2, 194–220), were not 'historiae' but monographs in the modern sense of the word.

or on faraway regions.[147] Regional and, later, national natural histories tried to inventory the entire natural wealth of certain areas. Caspar Schwenckfeld, for instance, described the botanical, mineralogical, and zoological riches of Silesia in a work published in two instalments (*Stirpium et fossilium Silesiae catalogus*, 1600; *Theriotropheum Silesiae*, 1603),[148] while Johann Baptist Großinger's *Universa historia physica regni Hungariae* (five printed volumes plus four remaining in manuscript)[149] appeared as late as 1793–7. In botany, 'local flora' became an important subgenre. Such works, which listed plants growing in the near vicinity of a particular university town, served as guides for academic excursions, satisfied local pride, and advertised the town and its university abroad. An early example is provided by Ludwig Jungermann and Caspar Hofmann's *Catalogus plantarum circa Altorfium* (1615). The text type had a long life. Later examples could outgrow its original purpose and present themselves as serious works of science, as in the case of Alexander von Humboldt's *Florae Fribergensis specimen* (1793), which focuses on subterranean cryptogams to be found around the mining town of Freiberg in Saxony. Regional mineralogies (which, just like local floras, flourished particularly in Germany and were sometimes inspired by these) not only explored their subject for its own sake but also pursued an economic and political agenda: authors such as Johann Jakob Baier (*Ὀρυκτογραφία Norica*, 1708) discussed the natural resources of their homeland in order to alert their rulers to these untapped treasures. Regional or national faunas were rarer and came later, as indicated by the dates of Linnaeus's *Fauna Suecica* (1746) or Carl Clerck's bilingual *Svenska spindlar/Aranei Suecici* (1757), the first work to apply Linnaeus's binomial nomenclature to zoology.

Exotic counterparts of such works discussed the nature of non-European regions in general or focused on their flora. Georg Marggraf's *Historia naturalis Brasiliae* (1648) exemplifies the former variety. The botanical riches of foreign countries were treated in works as different as the *Flora Sinensis* (1656) by Michał Piotr Boym, who used the appeal of South East Asian vegetation to promote the Jesuit mission in China,[150] and Johann Georg Gmelin's four-volume *Flora Sibirica* (1747–69), which contained the results of his participation in the Great Northern Expedition. Sometimes, the fascination with the exotic inspired works whose content was not defined by a certain region, but simply by its unfamiliarity to a European readership. The exotism of books like Carolus Clusius's *Exoticorum libri decem* (1605) or Juan Eusebio Nieremberg's *Historia naturae, maxime peregrinae*

[147] On exoticism and the cult of the indigenous in early modern natural history, see Cooper 2007, esp. 116–51, 51–86, and 87–115 respectively for the regional natural histories, 'local floras', and regional mineralogies to be discussed in what follows.

[148] On the zoological part, see Bäumer 1991–6, vol. 2, 186–9.

[149] Stöger 1855, 109.

[150] Miazek-Męczyńska 2005 (text with English preface, Polish introduction and translation); Walravens 2014.

(1635) is already advertised in their titles.[151] In other cases, the related criterion of rarity was invoked, for instance by John Caius in his *De rariorum animalium atque stirpium historia* (1570)[152] or by Clusius in his *Rariorum stirpium per Pannoniam et Austriam observatarum historia* (1583). As university and private botanical gardens developed into repositories of rare and exotic plants from the sixteenth century, catalogues not only helped students and other visitors orient themselves in them but catered to the predilection for the strange as well.[153] Linnaeus's account of the garden of George Clifford in Hartekamp, the *Hortus Cliffortianus* (1737), is perhaps the best-known work of this kind. For similar reasons, cabinets of natural history were published on occasion.[154] The 1,766 minerals and fossils gathered by Edward Lhuyd, which he depicted and described in his *Lithophylacii Britannici ichnographia* (1699), illustrate the exotic appeal of such collections quite well, despite the fact that they all come from Britain, as proudly emphasized in the title.

In terms of functionality, books for the desk can be distinguished from books for the field. Many of the works mentioned so far, from Gessner's *Historia animalium* to Linnaeus's *Hortus Cliffortianus*, came as weighty tomes of many hundred pages, in folio format to accommodate large-size illustrations and equipped with extensive indexes emphasizing their encyclopedic aspirations. On the other hand, books to be taken on botanical field trips—a common practice at universities and elsewhere—were slender compendia in octavo or an even smaller format. Even before local floras came to be written specifically for academic excursionists, the two different purposes of 'historiae' were clearly perceived and consciously exploited by authors and publishers. The same work could be published in two versions, an *editio maior* to be consulted, worked through, and annotated at home and an *editio minor* to be taken out to the meadows and woods. After the success of his monumental *De historia stirpium* (1542), one of the most influential Renaissance plant histories,[155] Leonhart Fuchs produced a pocket version (1545) in which the 900 folio pages of the original are reduced to 500 octavo pages. On the title page, it is explained that this has been done 'so that all those possessed by a desire to become thoroughly familiar with botany can carry the book more conveniently in their pocket when going for a walk or on a journey and compare

[151] On the former work, see Bäumer 1991–6, vol. 2, 189–91; Pinon 1995, 108–9; Asúa and French 2005, 109–14; Egmond 2010, 191–207. On the latter, see Pinon 1995, 116–17; Enenkel 2014a, 99–104.

[152] Bäumer 1991–6, vol. 2, 181–2.

[153] On the catalogues of the botanical gardens of the universities of Leiden and Pisa, Pieter Pauw's *Hortus publicus Academiae Lugduno-Batavae* (1601) and Michaele Angelo Tilli's *Catalogus plantarum horti Pisani* (1723), see Grämiger 2016, 243–9 and Schmiedel 2016, 226–8 respectively.

[154] On such 'museum catalogues', as they are commonly called, and their close relationship to 'historiae', see Funder 2018, with previous literature.

[155] Meyer et al. 1999 (commentary volume and facsimile edition). Bäumer 1991–6, vol. 1, 194–202 (on the German version published in 1543); Pavord 2005, 174–204; Ogilvie 2006, 194–7 (on the illustrations).

it with the herbs growing on the spot' ('ut quicumque rei herbariae radicitus cognoscendae desiderio tenentur eas vel deambulantes vel peregrinantes in sinu commodius gestare adque nativas herbas conferre queant'). In general, the portability of small 'historiae' was frequently advertised in paratexts.[156]

The desire to maximize readership was also reflected in the different languages of 'historiae', which were among the most multilingual text types dealing with the natural world from the outset. Already medieval bestiaries and lapidaries were composed not only in Latin but in the vernaculars as well. More importantly, Latin herbals and 'materiae medicae' were printed in vernacular versions from the incunabula era in order to reach pharmacists and other medical practitioners without Latin besides academic doctors.[157] In the same spirit, 'historiae', whose content and often splendid presentation could in addition be of interest to gardeners, mining experts, rulers, courtiers, noblemen, and wealthy citizens, appeared not only in Latin but also in different vernaculars. Translations in both directions, often accompanied by revisions and adaptations, were common. Of the earliest plant histories from Germany, Brunfels's *Herbarum vivae eicones* (1530–6) was turned into German as *Contrafayt Kreüterbuch* (1532–7), while Hieronymus Bock's *New Kreütter Bůch* (1539) was Latinized in 1552 (*De stirpium commentarii*). German versions of four volumes of Gessner's *Historia animalium* were produced from 1563 to 1589.[158] In the same period, Spanish works on the natural history of the New World came to be known in Europe mainly through Latin translations produced by authors such as Clusius.[159] A complex case is presented by Belon's *De aquatilibus*, which elaborated on the author's *Histoire naturelle des estranges poissons marins* (1551) and was itself translated into French as *La nature et diversité des poissons* (1555).[160] In the Enlightenment, publishers often found it more convenient to produce bilingual 'historiae' from the outset. The six volumes of Georg Eberhard Rumpf's *Herbarium Amboinense* (1741–50) about the plants of the Indonesian island of Ambon, originally conceived in Latin but translated into Dutch for a non-academic East Indian readership (sig. ***r–v), were published together with a facing Latin retranslation by the Dutch physician and botanist Johannes Burman forty years after the author's death.[161]

In one specific section of 'historiae', namely in the lists of names which introduced the single entries, multilingualism was the rule even in otherwise monolingual works. Because names posed a protracted problem—it was often unclear to what plant or animal a Latin or Greek term referred, and the multiple vernaculars

156 For a late example, see Linnaeus, *Flora Suecica* (1745), p. viii: 'Brevitati operis consului, ne mole gravis ad herbarias excursiones evaderet liber, et hinc additarum observationum paucitas.'

157 For Germany, see Habermann 2001, esp. 98–167 (with a tabular overview at p. 99) and 245–73 (on the intertwined Latin and German traditions of the *Gart der Gesundheit*/*Hortus sanitatis*).

158 On the German versions of the three works, see Bäumer 1991–6, vol. 1, 186–94; vol. 1, 202–11; vol. 2, 65–8.

159 Ogilvie 2006, 24, 244–5.

160 Bäumer 1991–6, vol. 2, 351.

161 Sarton 1937, esp. 246–8.

and their dialects complicated things further—naturalists not only enumerated Latin designations given by different authors but also correlated names in different languages. Mostly, a name was at least given in Latin and the author's vernacular, but often, much more was done in this respect. In works on European plants, animals, and minerals, the old learned languages Greek, Hebrew, and Arabic as well as further vernaculars were often added in descriptions of non-European items, languages from the region in question. In Rumpf's *Herbarium Amboinense*, plants are named not only in Latin and Dutch, but also in Portuguese, Arabic, Chinese, and a number of Malayo-Polynesian languages.

Structures and content

Like a number of other genres discussed in this overview—notably dictionaries, bibliographies, and collections of tables—'historiae' are 'discrete' texts: they do not develop a coherent argument but consist of an asyndetic succession of small units of equal rank. To compensate for the lack of continuous reasoning and because they are designed to present as much material as possible as succinctly as possible, 'historiae' are frequently in need of genre-specific paratexts besides the standard ones such as title, dedication, preface, and index.

In some cases, the presentation of the individual items is introduced by a general 'philosophical' discussion of the subject. The opening books of Jean Ruel's *De natura stirpium* (1536), Andrea Cesalpino's *De plantis* (1583), and John Ray's *Historia plantarum* (1686–1704) are dedicated to this end.[162] The idea behind this approach may have been to produce a more orderly counterpart to Aristotle's and Theophrastus's *Historiae*, where general reflections and descriptions of particular organisms are mixed together. In works focusing on certain geographical areas, the enumeration and description of plants, animals, or minerals can be contextualized by an overview of the region. The 'praefatio' to Gmelin's *Flora Sibirica* (vol. 1, pp. iii–cxxx) is, in fact, a geographical sketch of Siberia along with an account of the Great Northern Expedition. The gradual development of a technical terminology, particularly in botany, could necessitate the elucidation of frequently used terms in the form of word lists. As early as 1542, Fuchs introduced *De historia stirpium* with an 'explanation of difficult words' ('vocum difficilium explicatio') covering about 130 expressions.[163] Rudimentary bibliographies in the form of lists of authors used, as already found in Pliny's first book and common in early modern times, were gradually replaced by more informative ones, giving

[162] On Cesalpino's work, see Slaughter 1982, 58–60; Atran 1990, 151–8; Bäumer 1991–6, vol. 1, 225–8; Jensen 2001; Pavord 2005, 237–41. On Ray's, see Bäumer 1991–6, vol. 3, 79–81; Pavord 2005, 386, 390–2.

[163] Pavord 2005, 189; *BENLW* 934.

titles too, as an extensive natural historical literature accumulated and authors tried to save space by replacing detailed descriptions with abbreviated references to earlier discussions. By Linnaeus's day, the inclusion of such bibliographies among the prefatory material had become routine. Finally, when taxonomy rose in importance[164] and many authors developed their own view on this topic, they often set out their classificatory system in advance. In Otto Frederik Müller's *Vermium terrestrium et fluviatilium historia* (1773–4), which discusses the taxonomically particularly tricky default category of 'vermes' ('worms', including micro-organisms), the structure of the field is tabulated on eleven pages before the account proper.

To move on from paratexts to the main text, the rise of taxonomy was also one of two developments which influenced the macrostructure of the latter, that is, the order of its individual entries. The other was the emancipation of natural history from older disciplines, in particular from medicine, and its attainment of autonomy. Although the latter trend logically and historically preceded the former, the chronological overlap was large, and both often made themselves felt in a given book in tandem.

The sequence of entries in the earliest 'historiae' still reflects natural history's original lack of independence as a discipline. Their authors made no effort to arrange the individual items according to an order perceived as inherent in the plant, animal, or mineral kingdom. Rather, entries were usually ordered according to the Greek or Roman alphabet in imitation of a structure already displayed in Galen's treatise on simples, *De simplicium medicamentorum temperamentis et facultatibus*, and late antique rearrangements of Dioscorides.[165] This even holds for otherwise seminal books such as Fuchs's *De historia stirpium* or Gessner's *Historia animalium*. So far as botanical and zoological categories played a role at all in the structuring of 'historiae', the aforementioned folk-biological life forms were used to this end, again folowing ancient predecessors. Theophrastus's subdivision of plants into trees, shrubs, subshrubs, and herbs was still applied by leading seventeenth-century figures such as Caspar Bauhin and John Ray.[166] Many sixteenth- and seventeenth-century works on animals followed and adapted the classifications of Genesis 1 (water animals, birds, four-footed animals, creeping animals) or Aristotle (animals with blood: mammals, birds, amphibia/reptiles, fish; without blood: molluscs, crustacea, insects, shellfish).[167] The six tomes of Gessner's *Historia animalium* discuss viviparous and oviparous quadrupeds, birds, water animals, snakes, and insects, while Edward Wotton (*De differentiis*

[164] On the history of taxonomy in natural history, see Slaughter 1982, 15–82, esp. 48–64, 75–82; Atran 1990; Bäumer 1991–6, vol. 2, 398–9, vol. 3, 73–98, and 257–83; Pavlinov 2021, 13–77, esp. 31–46. Pavord 2005 focuses on botanical works and also discusses the corresponding ordering of entries.

[165] Ventura 2017, 112; Pavord 2005, 78–81. Dioscorides himself had rejected alphabetical order.

[166] On Ray's Theophrastean subdivision, see Pavord 2005, 377.

[167] Atran 1990, 191.

animalium) and Ulisse Aldrovandi in his animal books (1599–1642) add a subdivision according to the presence or lack of blood at the uppermost level. In his *Icones animalium* (1553), Gessner begins by distinguishing between domestic and wild animals.[168]

Works renouncing structural principles foreign to natural history, such as alphabetical order or, in the case of plants, their medical properties, and moving beyond the intuitive categories of folk biology appeared from the latter half of the sixteenth century. In chapter XIIII of the 'praefatio' to his *De stirpium nomenclaturis propriisque differentiis commentarii* (1552), Hieronymus Bock announces:

> In his vero omnibus describendis eam fere rationem secutus sum, ut eas in tractatione coniungerem plantas, quas Natura formae similitudine coniunxisse videtur, ita tamen, ut singulas suis capitibus ab aliis secernerem. Neque volui litterarum ordinem sequi, ut veteres solent herbarii. Habet enim ea ratio multum incommoditatis, si quid iudico.
>
> In the description of all these plants, I basically followed the principle of uniting in my treatment those which Nature seems to have united by the similarity of their appearance, although I separated each plant from the others by assigning it a chapter of its own. I did not want to follow alphabetical order, as the old authors of herbals are used to. For this principle creates much inconvenience, if I have any judgement.[169]

Subsequently, questions of how to classify and arrange the contents of one's book became a major question in paratexts. In his large 'historia' of water animals, Rondelet dedicates a separate subchapter to the topic, because, as he remarks, 'I thought hard and long what order I should best follow' ('quem potissimum ordinem servarem, multum diuque cogitavi'; vol. 1, ch. 5.1, p. 113). Cesalpino lays out the order of his *De plantis* in the last two chapters of the work's first book (pp. 26–30).[170] Programmatic titles like *Historia generalis plantarum…per certas classes artificiose digesta* (Jacques Dalechamps, 1586) or *Plantarum umbelliferarum distributio nova* (Morison, 1672) became common as well, as did, somewhat

[168] See Bäumer 1991–6, vol. 2, 35, 78, and 59–60 respectively. Aldrovandi's 'encyclopedia' of animals is discussed on pp. 74–119 there.

[169] In the 'prolegomena' (pp. 7–8) to his *Ornithologia* (1599–1603), Aldrovandi also emphatically rejects alphabetical order as a structural principle but does not search for a natural order instead, as he rather starts from the eagle as the most dignified bird (Bäumer 1991–6, vol. 2, 80–1; see also note 177 below on axiological arrangements in general). Opposition to a medical order is expressed in Matthias de L'Obel's and Pierre Pena's *Stirpium adversaria nova* (1571), in whose 'argumentum operis' (sig. B^{r}) outer appearance is given precedence over medical powers, and in Cesalpino's *De plantis* (p. 26), where the classification of plants has to follow their 'substantia', not their 'accidentia': 'Idcirco neque ex facultatibus medicatis neque ex alia utendi ratione…genera eorum et species sunt constituendae.'

[170] The philosophical foundations of Cesalpino's system are explained by Jensen 2001. In addition, see also the opening 'ordinis ratio' (p. 1) in Clusius's *Exoticorum libri decem* (1605) and a similar passage in Aldrovandi's *De quadrupedibus digitatis* (1637), pp. 1–2.

later, the aforementioned tabulations of one's chosen taxonomy. A further indication of the widespread concern with classification and order was the use of Ramist charts, in paratexts, as in the body of a work, from the late sixteenth until at least the late eighteenth century. The device can be found as early as John Caius's *De canibus Britannicis* (1570)[171] or Matthias de L'Obel and Pierre Pena's *Stirpium adversaria nova* (1571) and as late as Antoine Laurent de Jussieu's *Genera plantarum* (1789; Figure 5).

Satisfying solutions proved more difficult than programmatic announcements—in fact, biological taxonomy still teems with unsolved problems today.[172] Everyone agreed that the goal was a 'methodus naturalis' of classification, and this became a much-used catchword, but naturalists disagreed heartily about its actual meaning. What criteria should be chosen, and how many? In retrospect, Linnaeus found all the grand claims of having discovered such a natural method in botany ridiculous,[173] but he nonetheless thought the disagreements important enough to develop a taxonomy of naturalists on their basis and to coin new words such as 'fructista', 'corollista', or 'calycista' ('advocate of the fruit, the corolla, the calyx') for colleagues championing different criteria.[174] In practice, many works present ad hoc solutions in which a number of aspects coexist uneasily. To cite just one zoological example, Belon divides his work on water animals into two books on the basis of the Aristotelian distinction between animals with and without blood. The members of each class are then arranged according to complexity, anatomical properties, propagation, habitat, and other criteria, sometimes with strange results: the ichneumon finds itself among the oviparous animals together with the crocodile, with which it was said to be locked in deadly enmity (pp. 40–5). Problems were even more daunting with respect to 'fossilia', where it was hard to find any clue to a meaningful taxonomy before the arrival of modern chemistry.[175] Many authors simply presented their own collections and transferred its order to the printed page. Johannes Kentmann's 'Catalogus rerum fossilium' (1565) even starts with a depiction of the cupboard ('arca') which apparently held his collection in its twenty-six shelves. The arrangement of Lhuyd's *Lithophylacii Britannici ichnographia* equally reflects the order of the author's collection, but because his 'fossilia' are mostly fossils in the modern sense, he can draw on botanical and

171 Page 16 in the edition by Roberts and Venn 1912. On the work, see Bäumer 1991–6, vol. 2, 194–7.

172 Pavlinov 2021, esp. 169–99.

173 *Genera plantarum* (1737), 'Ratio operis', § 9: 'rideo omnes methodos naturales hactenus exclamatas'. This did not deter later naturalists from making similar claims. In the preface to his *Stemmata avium/Geschlechtstafeln der Vögel* (1759, sig. *2^{r-v}), Jacob Theodor Klein takes pride in his 'methodus mere naturalis' of classifying birds. See also Bäumer 1991–6, vol. 3, 281–3.

174 See the second sections of his *Fundamenta botanica* (1735) and *Philosophia botanica* (1751), which still provide an instructive overview of the taxonomical developments of the previous one and a half centuries.

175 Accordi 1981.

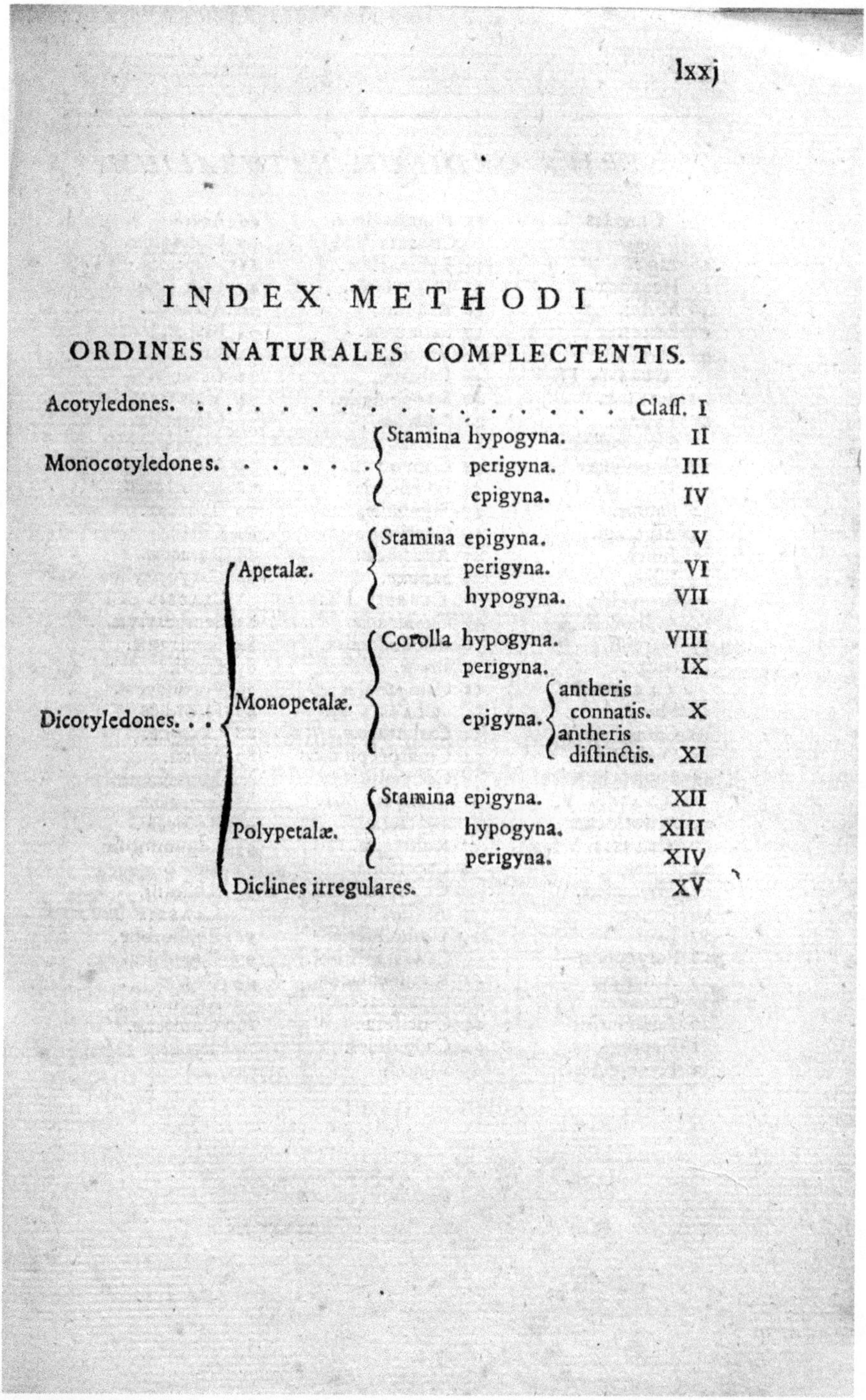
lxxj

INDEX METHODI

ORDINES NATURALES COMPLECTENTIS.

Acotyledones.				Claff. I
Monocotyledones.		Stamina hypogyna.		II
		perigyna.		III
		epigyna.		IV
Dicotyledones.	Apetalæ.	Stamina epigyna.		V
		perigyna.		VI
		hypogyna.		VII
	Monopetalæ.	Corolla hypogyna.		VIII
		perigyna.		IX
		epigyna.	antheris connatis.	X
			antheris diſtinctis.	XI
	Polypetalæ.	Stamina epigyna.		XII
		hypogyna.		XIII
		perigyna.		XIV
	Diclines irregulares.			XV

Figure 5 The macrostructure of Jussieu's *Genera plantarum* in the form of a Ramist table (p. lxxi).

zoological categories, which results in classes like 'lithophyta' ('stone plants') or 'ichthyodontes cuspidati' ('pointed fish teeth').[176]

Taxonomical aporias apart, two overarching structural principles became common over time: axiological arrangement and multi-tiered hierarchy.[177] When the material permitted it, items were often arranged in descending order of complexity and 'perfection'. For instance, books 8–16 of Nieremberg's *Historia naturae, maxime peregrinae* start with man and move down via animals and plants to minerals. In later editions of Linnaeus's *Systema naturae*, the animal part begins with man and other mammals and proceeds to birds, amphibia, fish, insects, and 'worms'. Cesalpino follows a similar order in *De plantis*: after the introductory first book, he divides his account into two main parts devoted to trees ('De arboribus', book 2) and to subshrubs and herbs ('De suffruticibus et herbis', books 4–15) respectively, with book 3 serving as a transition. In other botanical works, an ascending order was preferred, as in Matthias de L'Obel's *Plantarum seu stirpium historia* (1576) or Caspar Bauhin's *Πίναξ theatri botanici* (1623):[178] both put grass first and end with trees. Linnaeus follows a comparable logic in the plant part of the *Systema naturae* and other works: classifying plants by the number of their reproductive organs, he starts with those with one stamen and pistil only, the 'monandria' and 'monogynia'.

Early 'historiae' often featured only one hierarchical level. In Fuchs's *De historia stirpium*, for instance, 343 plants are discussed in as many consecutively numbered entries. If a second level was added, it usually corresponded to broad folk-biological categories, as in the aforementioned case of Cesalpino's *De plantis*. In the course of the seventeenth and eighteenth centuries, by contrast, the individual entries tended to be integrated into a deep hierarchy highlighted by means of layout and typography, at whose lowest or second-lowest level they were situated. In Linnaeus's *Systema naturae*, this hierarchy comprises six levels: kingdom ('regnum'), class ('classis'), order ('ordo'), genus, species (the entry level), and variety ('varietas'). There was no lack of concurring systems of broadly similar make. Brisson's *Regnum animale* (1756), for instance, features a five-tier taxonomy ('classis', 'ordo', 'sectio', 'genus', 'species'), which is explained and justified at length in the work's preface.

In marked contrast to the ideal-type described earlier, the structure and content of the entries themselves could remain quite heterogeneous far into the seventeenth century. Within one and the same work, entries might describe different kinds of items, comprise different numbers of items, and provide different types of information. In a way reminiscent of Dioscorides, the chapters of the first six books of Clusius's *Exoticorum libri decem* are devoted to plants, kinds of wood,

[176] See the unpaginated table of contents after the preface.
[177] On the second aspect, see Slaughter 1982, 55.
[178] Slaughter 1982, 57–8; Atran 1990, 135–8; Reeds 1991, 125–7.

fruits, barks, roots, animals, beaks, horns, and much more. While the first seven chapters of Thomas Moffett's *Insectorum theatrum* (1634)[179] all discuss bees, several kinds of small animals are lumped together in individual chapters later. Such diversity often corresponds to the lack of any internal subdivision of the individual entries. Especially in zoology, they often present themselves as long stretches of continuous text. This still holds true, for instance, for those in John Jonston's five-volume *Historia naturalis animalium* (1650–3),[180] where only marginalia provide some guidance for the reader.

Other authors made efforts to describe one item of the same class per entry and to follow a uniform protocol early on, although not all of them were equally consistent in doing so. Rondelet, for example, explains in part of his aforementioned arrangement chapter (p. 114) that he will start with the (Greek, Latin, and vernacular) names of his fish, move on to describe their appearance, anatomy, and behaviour, and conclude with medical and culinary uses and methods of catching and preparing them. In practice, however, this formula is varied freely, as Rondelet himself concedes. Others were more ambitious, striving for almost complete uniformity and fully exploiting the possibilities offered by layout and typography to this end. In particular, they introduced headings to visualize the structure of their chapters. A consistent and differentiated scheme is applied by Fuchs in his *De historia stirpium*: an entry title of the type 'De iride' is followed in uniform sequence by the headings 'nomina' ('names'), 'forma' ('appearance'), 'locus' ('place'), 'tempus' ('time'), 'temperamentum' (degree of hotness/coldness, dryness/wetness), and 'vires' ('medical powers', subdivided by authorities). Sometimes, varieties are inserted after 'nomina' under an additional heading 'genera', and at the end an 'appendix' may contain further remarks.[181]

As more and more 'historiae' came to be clearly structured over time, subgroups of works emerged whose entries tended to treat the same content in the same order. As a late example of such a subgroup, one can cite the large systematic works of the seventeenth and eighteenth centuries, to be discussed at the end of this subsection, as an earlier one, the extremely short entries of local floras: they only comprise a plant's names in different authors (always in Latin, sometimes also in the local vernacular) followed by its habitat, as in the following entry dealing with a maple species on the first page of Jungermann and Hofmann's *Catalogus plantarum circa Altorfium*:

[179] Bäumer 1991–6, vol. 2, 382–7; Pinon 1995, 114–15. [180] Bäumer 1991–6, vol. 2, 132–49.

[181] For sixteenth-century botanical 'historiae', the standardization of entries is documented and analysed by Habermann 2001, 274–93 (276 on Fuchs); see also Ogilvie 2006, 182–3. The introduction of headings may have been inspired by their conspicuous presence in the pharmacological section of book 2 of Avicenna's influential *Canon* (Ventura 2017, 136–8). The early modern notion of discipline-specific 'loci communes' seems to have played a role as well (Friedrich 1995, 58–65).

> Acer tenuifolia seu minor Cordi et Lob(elii). Opulus rumpotinus Colum(ellae). Gesneri Maßholter. In silvis.
>
> Cordus's and de L'Obel's 'thin-leaved or lesser maple'. Columella's 'field maple holding the vines'. Gessner's 'Maßholter'. In the woods.

In many works, the unity and uniformity of the individual entries was underlined by the addition of one illustration per entry.[182] Illustrations figured already in late antique Dioscorides codices and medieval herbals and bestiaries. Their role increased in the age of print, when they became easier to reproduce and less liable to deterioration. From the seventeenth century, even details invisible to the naked eye could be depicted with the help of the microscope.[183] Woodcuts and copper engravings of plants, animals, and fossils, especially if hand-coloured,[184] clearly constituted a major attraction for early modern readers (as they still do for historians of science today). This is shown not only by the abovementioned 'icones' titles but also by the fact that abridged versions of longer works often focused on illustrations. This is true, for instance, of the 1545 pocket edition of Fuchs's *De historia stirpium* and of those of the first two volumes of Gessner's *Historia animalium* (1553, 2nd edn 1560).[185] As late as 1795, a similar heavily illustrated epitome was produced of Albrecht von Haller's *Historia stirpium indigenarum Helvetiae* (1768).

On the other hand, the helpfulness of illustrations had already been contested by Pliny the Elder (*Naturalis historia* 25.8), who pointed to the difficulty of finding the right colours, the likely corruption of pictures in the transmission, and the problem that plants look different in different seasons. Many authors of early modern 'historiae' shared Pliny's scepticism. Ray accused bad illustrations of fabricating non-existent species and creating great confusion.[186] In addition to such arguments, good woodcuts and copper engravings were costly. As the number of species described in 'historiae' grew, full sets of illustrations became unrealistic for financial reasons and lack of space. Specimens of many animals, birds for instance, were hard to come by, and taxa like algae, lichens, or fungi turned out to be even more difficult to depict than ordinary plants. Contrary to the modern perception of the genre, which is largely marked by gorgeous illustrations, a big part of all 'historiae' featured none at all. From humble local floras to

[182] On images in the botanical tradition, see Reeds 1991, 28–32; Pavord 2005 (numerous impressive examples); Ogilvie 2006, 192–203; Egmond 2017 (who includes zoology as well).

[183] Pavord 2005, 325. Gessner had produced sketches for a projected *Historia plantarum* with the help of a magnifying glass as early as the mid-sixteenth century (Pavord 2005, 291).

[184] Christoph Plantin had some of his plant books coloured by specialists in his own workshop (Pavord 2005, 312, 316; Ogilvie 2006, 202–3). Van de Passe provides readers of his *Hortus floridus* (1614) with detailed instructions on how to colour every single engraving for themselves (sigs. A^v–$D2^r$).

[185] On Gessner's *Icones animalium*, see Bäumer 1991–6, vol. 2, 58–65.

[186] Kusukawa 2006, 95–6.

foundational works like Cesalpino's *De plantis*, Ray's *Historia generalis plantarum*, or Linnaeus's *Systema naturae*, they relied exclusively on precise wording.

Like the development of the arrangement of entries within 'historiae', their changing content bears witness to the disciplinary emancipation of natural history. In early works, which followed ancient models[187] as well as medieval herbals, bestiaries, and lapidaries in this respect, the content largely reflected the interests of other disciplines, primarily medicine and 'philologia', the early modern predecessor of the humanities. Botanical, zoological, and mineralogical facts were usually combined with information from these areas. Great importance was attached to the medicinal use of plants in particular, as evidenced by the information given under Fuchs's headings 'temperamentum', 'vires', and 'appendix'. Animals were above all seen from a cultural-historical angle. The standard headings in Gessner's *Historia animalium* are the following:[188]

> A. Names in various languages, B. Habitat and appearance, C. Behaviour and diseases, D. Character, preferences, and aversions, E. Utility, hunting, and husbandry, F. Relevance for human nutrition, G. Medical relevance, H. Role in (Greek and Latin) literature, art, and other cultural fields.

Even more extreme examples of an anthropocentric approach can be found in Ulisse Aldrovandi's animal books. In *De quadrupedibus digitatis* (1637), the long opening entry on the lion (pp. 2–63) comprises headings like 'Iniuriarum memoria et ultio' ('Memory of and revenge for unjust treatment', p. 12), 'Physiognomonica' (p. 24), 'Visiones et somnia' ('Visions and dreams', pp. 31–2), 'Mystica et allegorica' (pp. 32–4), 'Numismata' ('Coins', pp. 42–3), 'Proverbia' (pp. 51–3), and 'Usus in suppliciis' ('Use in capital punishment', p. 60).[189]

Gradually, however, 'historiae' began to focus on aspects intrinsic to the items described. Cultural-historical issues became marginal, and while medical ones remained pertinent for a longer time, they were ultimately removed as well. Precise description and classification became the main concerns, as is sometimes explicitly stated in paratexts. In the preface to the *Ornithologiae libri tres* (1676) by Francis Willughby and John Ray, one reads what sounds like a direct response to Aldrovandi (sig. a2^r):

> Cum vero species omnes accurate describere, distinguere et ad suas singulas classes seu genera reducere ad omnem obscuritatem et confusionem tollendam scopus sit in quem praecipue collineavimus, omissis plerumque homonymis et

[187] On the anthropocentric and utilitarian aspects of Theophrastus's botanical writings, see Wöhrle 2005.

[188] Bäumer 1991–6, vol. 2, 51–2; Friedrich 1995, 59–60; Pinon 2005, 248–51.

[189] On comparable examples (eagle and chicken) from Aldrovandi's *Ornithologia*, see Bäumer 1991–6, vol. 2, 81–2 and Slaughter 1982, 50.

> synonymis, moralibus, mysticis, apologis, praesagiis aliisque, quae ad theologiam, ethicam, grammaticam humanioresque litteras pertinent, ea dumtaxat proponere et persequi decrevimus, quae historiae naturalis propria sunt.
>
> But since the goal to which we aspired in the first place was to describe precisely, distinguish, and attribute to their respective classes and genera each and every species in order to remove all uncertainty and disorder, we for the most part left aside homonyms and synonyms, things related to morality and mystical sense, fables, oracles and other issues pertaining to theology, ethics, grammar, and the humanities and decided to present and go through those matters only which are peculiar to natural history.

Over time, not only were aspects foreign to natural history eliminated, but even the natural historical information itself was presented more and more selectively. To some degree, this development reflected the increasing reliance on dead specimens, which curtailed the information available about the appearance of the living being, its habitat, development, and behaviour. More importantly, as the material to be discussed grew and taxonomical concerns came to the fore, the focus of an author's attention shifted from the individual entry to the system constructed in his book. In treating a taxon, authors, therefore, omitted those traits common to all members of the superordinate taxon and focused on those distinguishing the taxon under discussion from its neighbours.[190] The way of dealing with the natural historical tradition changed as well, as authors faced a multitude of earlier descriptions of individual items and ceased to view their predecessors as authorities. While entries could be structured as veritable doxographies in an early naturalist like Brunfels (1530–6), extensive quotation and discussion were later replaced by mere referencing: occurrences of the plant or animal in question in previous literature were listed, in abbreviated form, but often with page numbers,[191] in the catalogue of synonyms opening the entry. In Bauhin's *Πίναξ theatri botanici*, the price for the most exhaustive and orderly catalogue of plants published up to this point is the complete atrophy of the single entry. Groups of related plants are presented in a few sentences, but at the lowest level only a plant's names in different authors are left. A century later, Linnaeus developed a format combining references to earlier literature with an absolute minimum of description. In his *Species plantarum* (1753), for instance, a typical entry is composed of the generic name plus the specific epithet (the latter given as a marginal note), an extremely short description, synonyms from earlier works, an indication of the habitat, and sometimes a symbol designating the lifespan (annual, biennial, etc.). A similar but even more compressed format is used in later editions of the

[190] Ogilvie 2006, 48, 210–15, 269; *BENLW* 934.

[191] Book, chapter, and page numbers are already given in Rembert Dodoens's *De frugum historia* (1552) (Bäumer 1991–6, 213).

Systema naturae. Because of its practicality and Linnaeus's authority, this style became an important model for subsequent generations of botanists.

The language of description

Finally, not only the structure and content of entries changed but also their linguistic form, which was intimately linked to these features. I would like to conclude with a few remarks on this development, focusing on the area in which it produced its most noticeable results, botany.[192] In this discipline, the Latin language evolved in a quite extraordinary way, resulting in an idiom which has been described by the leading expert in the field, albeit somewhat hyperbolically, as 'a modern Romance language of special technical application'.[193]

Botanical Latin originated as a syntactically undistinguished variety of Latin with a selective but largely non-technical vocabulary. This began to change in early modern times when the number of plants known increased and their brief, precise, and unequivocal description became a more and more challenging task. An early indication of this development is the terminological guides prefixed to 'historiae' from the sixteenth century, as in the aforementioned case of Fuchs's *De historia stirpium*. In the course of the next two centuries, the lexical specialization of botanical Latin increased through semantic restriction and redefinition of pre-existing terms, the coining of new ones, and the adoption of Greek vocabulary, as an ever-growing number of taxa had to be distinguished from one another. Concomitantly, syntax eroded, partly perhaps because plants were degraded from living creatures, whose activities could be described best in full sentences, to mere objects of study[194] but primarily for the sake of brevity.

As in many other aspects of natural history, Linnaeus played a crucial role in this process as well. On the one hand, he radicalized, systematized, standardized, and codified earlier developments in his theoretical works, particularly in the *Fundamenta botanica* (1735) and their later elaboration, the *Philosophia botanica* (1751). On the other hand, his own massive syntheses such as the *Systema naturae* (which also comprised a veritable dictionary of descriptive nomenclature in the plant volume of its later editions) and the *Species plantarum* provided models of how to apply his precepts in practice. Linnaeus's major concrete contributions to botanical Latin are the following. First (and this point equally applies to zoology), he made the binomial nomenclature for plants and animals, which had been applied occasionally and unsystematically from the sixteenth century, mandatory and gave it a precise classificatory meaning: the first term alone

[192] On the development of botanical Latin and the emergence of its modern form, see Stearn 1992, esp. 14–48; *BENLW* 933–5.

[193] Stearn 1992, 6.

[194] Cf. Ogilvie 2006, 212–15.

indicates the genus; together with the second, the specific epithet, it denotes the species.[195] In doing so, he suspended Latin grammar when it suited him: a name like *Corvus corax* for the common raven, which simply combines the Latin and the Greek word for 'raven', looks like Latin but is not. Secondly, Linnaeus introduced the diagnosis as a shorter supplement (in the case of new taxa) or an alternative to a normal description. A diagnosis exclusively serves to distinguish a taxon from its neighbours in the system and differs from a full description syntactically: while the latter is in the nominative, a diagnosis consists of a string of 'ablativi qualitatis' depending on the plant name. (At least, this is the ideal-typical situation. In practice, mixed forms exist, as in the second example to be discussed presently.[196]) Thirdly, Linnaeus's *oeuvre* codified a telegraphic, almost exclusively nominal style[197] and the consistent use of standardized layout as a substitute for syntax.

After Linnaeus, botanical Latin continued developing in the direction indicated by him. New binomials were incessantly coined, the vocabulary of plant description was enlarged further, and the meaning of individual terms stabilized. In his authoritative *Botanical Latin*, William Stearn takes his historical sketch to the mid-nineteenth century, but, in fact, the language of botany kept evolving even after that date. An important incentive was the extension of description and classification to previously neglected classes such as algae, lichens, or fungi, as the unknown structures they revealed under the microscope once more required new terms. Extremely technical and largely incomprehensible to normal Latinists, botanical Latin outlived the genre of 'historia' in which it had developed and continued to be used in the specialized journals where most descriptions of new species have been and are published in the more recent past and present. In 1906, the *Règles internationales pour la nomenclature botanique* declared a Latin description mandatory for the publication of a new-found taxon to be valid, and this rule was reinforced several times in the twentieth century.[198] The definitive codification of botanical Latin took place in Stearn's homonymous book, which was first published in 1966 and whose fourth and definitive edition appeared in 1992. It was only in 2011 that the obligation to describe new taxa in Latin was abolished and English permitted as an alternative. The option to use Latin for first descriptions

[195] Linnaeus first applied the binomial nomenclature to plants in his *Species plantarum* (1753), to animals in the tenth edition of the *Systema naturae* (1758–9). See Bäumer 1991–6, vol. 3, 265–9; on binomial names in botany before Linnaeus, see Pavord 2005, 189, 235–6, 325, 373.

[196] The diagnosis is defined in article 38.2 of the current version (2018) of the *International Code of Nomenclature for Algae, Fungi, and Plants* (https://www.iapt-taxon.org/nomen/main.php, accessed 16 February 2023) as 'a statement of that which in the opinion of its author distinguishes the taxon from other taxa'. For examples of diagnosis and description, which also show the flexibility of both notions, see Stearn 1992, 143–53 and 154–93.

[197] The major exception is the phrase 'habitat in' ('lives in') at the end of entries. However, the transition of 'habitat' from verb to noun in the major European vernaculars suggests that it soon ceased to be understood as a real verb.

[198] Stearn 1992, 8.

still exists (and is also taken in practice), as does the obligation to give new species and other taxa a Latin name.[199]

The best impression of the development of botanical Latin is perhaps conveyed by a comparison of two examples from its early and late stages. In 1576, Clusius published his *Rariorum stirpium per Hispanias observatarum historia*, which contains descriptions of about 170 conspicuous plants from the Iberian peninsula, accompanied by informative woodcuts. The following is the first part of Clusius's characterization of a variety of Iris (perhaps *Iris xiphium* L. var. *lusitanica*), which he discusses in his second book (pp. 282–4).[200]

> De Iride lusitanica seu biflora. Caput XI
>
> Huius iridis folia breviora sunt et angustiora vulgari iride, alioqui haud absimilia. In summo caule, qui pedalis est, flos e vaginulis purpurascentibus prominet, senis foliis constans ut vulgaris iridis, sed minor aliquantulum, colore omnino violaceo seu purpureo saturatiore, qui in deorsum reflexis tribus foliis intensius nitet; villosa illa fimbria, quae in vulgari lutea est, huic omnino candida est. Unicum autem plerumque in singulis caulibus florem profert, interdum, sed raro, alterum, rarissime tertium, singulari quemque vaginula inclusum, iucundissimo odore et lilium convallium aemulante. Istius flos etiam mihi conspectus est, qui quaterna haberet folia deorsum vergentia, totidem ligulas bifidas fimbriam et stamina contegentes et quaterna etiam folia sursum reflexa et invicem amplectentia, ut interdum in floribus Natura ludere solet. Semen angulosum et rugosum in siliqua triangulari satis crassa, sed raro, perficit. Radix illi subiacet nodosa ac veluti geniculata, solida, pollicari crassitudine, incurva, intus candida, ingrati saporis et paulo post fervidissimi gustus, multis fibris summa terra haerens et subinde in plures propagines e lateribus nascentes distributa.
>
> Chapter 11: The Portuguese or twice-flowering iris
>
> The leaves of this iris are shorter and narrower than those of the common iris but not unlike them otherwise. On top of the stem, which is one foot long, the flower emerges from small sheaths with a purplish tinge. It comprises six petals, like that of the common iris, but is somewhat smaller, of a completely violet or strongly saturated purple colour, which glows more intensely in the three petals that curve backwards. That shaggy seam which is deep yellow in the common iris is pure white in the present one. It mostly brings forth one flower only on its single stems—sometimes, but rarely, a second, very rarely a third, each being enclosed in a single small sheath—which is of most pleasant fragrance and challenges the lily of the valley. I have also seen a flower of it which had four petals

[199] Principle V and article 39 of the *International Code of Nomenclature for Algae, Fungi, and Plants*.
[200] The part printed here is followed by a discussion of the habitat and a comparison with two other varieties. For a German translation of the entire entry, see Korenjak 2019, 196–201. Another description by Clusius is discussed by Ogilvie 2006, 213–14.

bent downwards, as many cleaved ligules covering the seam and the stamina, and also four petals bent upwards and embracing one another, as Nature is sometimes accustomed to play in flowers. The plant produces an edged seed with a rough surface in a quite thick three-sided pod, but only rarely. It rests on a knotty root angled like a knee, so to speak, which is stout, of a thumb's thickness, bent, pure white inside, of unpleasant flavour and, a little later, a most burning taste, adhering to the surface of the earth with many fibres and being gradually divided into many shoots born from its sides.

Clusius covers all major parts of his iris and provides as many characteristics as possible in order to enable the reader to recognize the flower for certain from their concurrence. He gives a 'vicarious description', a substitute for a personal observation of the plant itself.[201] In doing so, he particularly emphasizes the aspects not covered by the accompanying woodcut: shades of colour, smell, taste, touch, size, variations in the number and form of the flowers, the seed. All of this Clusius manages to convey with the standard Latin lexicon, non-technical words used either in their normal meaning or as easily understandable metaphors, as in the case of 'fimbria', which can mean the facing of a garment in ordinary Latin. The only term that could cause a botanical layman some difficulty is 'stamen' ('thread' in standard Latin), which is used in the technical sense it still has in English today. Sometimes, ordinary language is adapted to the needs of botanical description by means of diminutives and nuancing suffixes ('vaginulis purpurascentibus'). In order to help the reader visualize the plant's shape and its dimensions (another aspect not evident from the woodcut), Clusius draws on everyday experience, comparing it to parts of the human body: a stem one foot high, a root bent like a knee and as thick as a thumb. He also compares the plant with another kind of iris which he calls 'common' and which the reader presumably knows better. Syntactically, the flower is presented in full sentences and features as the subject of several colourful verbs—'prominet', 'nitet', 'profert', 'perficit'—which underline its activity and vitality. Hypotaxis, particularly in the form of relative clauses, is generously used, as are strings of significant adjectives.

Some 300 years later, in 1861, Theodor Magnus Fries published his 'Lichenes arctoi'. Among the hundreds of descriptions of Nordic lichen species contained in this 'historia', one finds the following of a member of the genus called *Baeomyces* by Fries (pp. 176–7):

1. *B. roseus* Pers.: thallo verrucoso, albido, hypothallo glaucescente; apotheciis carneo-roseis, stipitibus cylindricis, albis.

[201] Ogilvie 2006, 183–4, 213–14.

Ad terram sterilem nudam in silvaticis et inferalpinis Lapponiae meridionalis parcius ut etiam Finmarkiae orientalis et Lapponiae Kemensis.

Asci elongati, angustissimi, cylindrici, creberrimi, paraphysibus laxe cohaerentibus, filiformibus cincti; sporae fusiformes, subincurvae, octonae, hyalinae, simplices l. dy–tetra-blastae (cfr. Koerb. Syst. p. 275), 0,18–20 mm longae et 0,0015–20 mm latae.

Technically speaking, this account is part diagnosis, part description: in the first paragraph, a string of 'ablativi qualitatis' ('thallo verrucoso', etc.) detail the traits by which the species can be told apart from other members of *Baeomyces* with the naked eye. In the second and third, the habitat and reproductive apparatus, visible under the microscope only, are described by means of nominatives ('Asci elongati', etc.). However, this difference is of secondary importance compared with the uniformity of the account in other respects. First, it is all in telegraphic style, lacking full sentences. In fact, not a single finite verb is to be found in the entire passage. Secondly, apart from the indication of the habitat, the text is dominated by specialized technical terminology. Substantives, beginning with the generic name, are mostly Latinized Greek: 'thallus' (< θαλλός, 'branch'), 'apothecium' (< ἀποθήκιον, 'pantry'), 'ascus' (< ἀσκός, 'tube'), and so on. Adjectives are mainly formed from Latin stems, but many, compounds in particular, are unknown in classical Latin. While some, such as 'filiformis' ('thread-shaped'), correspond to ordinary types of word-formation, words like 'carneo-roseus' ('incarnadine-pink'), which link two colour adjectives to indicate an intermediate colour, are nearly exclusively found in botanical Latin. 'Dy–tetra-blastus' ('with two to four blasts') cannot even be pronounced in classical Latin, which has no word to express the dash that can be rendered as 'to' in English. Abbreviations whose resolution is anything but obvious and numerical information on microscopic lengths ('0,18–20 mm longae et 0,0015–20 mm latae') complete the image of a fully grown technical language, highly efficient and indispensable to the expert but cryptic to the uninitiated.

A translation would not render Fries's account much more understandable to the non-lichenologist. Instead, I add a modern English description of the same species[202] (which meanwhile has been renamed *Dibaeis baeomyces*), because a comparison between the two texts is telling in another respect.

Thallus gray, white, or roseate tinged, crustose, smooth or powdery, covered with spherical or flattened verruculose warts with a narrowed base, in fertile plants the warts 0.1–0.3 mm broad, in sterile ones to 1 mm; **podetia** 2–6 mm tall, 1 mm thick, smooth or fissured, whitish. **Apothecia** pink, 1–4 mm broad,

[202] Thompson 1984. I have been unable to see the book, but the description can be read at https://lichenportal.org/ (accessed 16 February 2023).

> spherical, emarginate or with a very narrow margin; **cortex** of the thallus 50 μ thick, of vertically oriented hyphae with thin cell walls; algal layer with the hyphae also more or less vertically oriented; **medulla** very loose, in parts forming a paraplectenchyma; **cortex** of the warts composed of interwoven hyphae parallel to the surface; interior of the **podetium** of strands of thin-walled hyphae 2.5–3 μ in diameter, upper part fistulose, the hyphae under the hymenium thicker, to 5 μ diameter, vertically oriented and not sharply separated from the subhymenium; **subhymenium** ca. 50 μ thick, yellowish; hymenium 10μ, I+ pale blue (color soon disappearing); paraphyses 2–2.55 μ, simple, little thickened above; **asci** cylindrical, 85 × 6 μ, thin walled; **spores** fusiform, mainly 1-celled, occasionally indistinctly 2-celled, 12–26 × 2.5–3 μ.
>
> **Reactions:** thallus and podetia K+ yellow, PD-I- yellow.
>
> **Contents:** baeomycic acid, squamatic acid, and unidentified substance (Dey 1977).
>
> This species grows on soils with much clay content, in the Arctic occurring on frost boils. It is circumpolar temperate in distribution. In North America it is common east of the Mississippi River in the eastern states south to Alabama and Georgia. In Canada it is in the eastern provinces to Newfoundland and is rare across the Northwest Territories to Alaska.

Despite the differences between both descriptions—the latter is overall more comprehensive and adds some chemical information at the end—the close affinity between the technical languages used stands out at first sight. Like the Latin version, the English one lacks full sentences and verbs (except for the description of the habitat), includes cryptic Greek nouns and peculiarly formed adjectives, and features microscopic measurements and abbreviations. It is written in the same 'Botanish'. All that has changed, so to speak, is the carrier language.

Table works

The genre and its early development

In what follows, a table is understood as an arrangement of data based on the correlation of two or more lists. Its purpose is to present a large amount of data of different kinds and their relationship in a space-saving, easily searchable manner. Data may be represented by textual or non-textual elements, typically but not necessarily numbers. Textual elements are usually given in the form of keywords, as the relation between words or word groups is expressed by layout instead of syntax ('noon | rain' instead of 'At noon, there was rain'), but this may be done inconsistently, allowing for some text in full sentences. The notion of a table covers simple two-column formats, but typical tables are more complex arrangements covering full pages and featuring a grid.

In table works, tables are the main concern and such running text as there exists is subservient to them. This criterion distinguishes such works from neighbouring forms in which it is the other way round: works with a few tables within the text to clarify specific issues;[203] works with indexes; and atlas-like works with keys to their illustrations. Dictionaries are technically covered by the above definition but clearly represent a genre of their own. 'Ramist tables' are not tables in the current sense, since their logical structure is based not on coordination, but on subsumption, and their main function is not storage of data but teaching.

Historically, tables constituted the earliest non-pictorial method of displaying large amounts of information by the partial substitution of natural language with graphic means (in this case, layout), and they remained the only one for a long time. In particular, they were the only possibility for presenting large sets of numbers until the invention of the line graph in the late eighteenth century.[204] Other forms like histograms or pie charts postdate the era of Latin science altogether.

It is known today that tables of astronomical observations and ephemerides were compiled on a grand scale by Babylonian astronomers of the first millennium BC.[205] From an early modern point of view, by contrast, the first and most important ancient author of tables was Ptolemy. His *Almagest* contains tables illustrating various aspects of mathematical astronomy from the 'chords' corresponding to a given angle[206] (1.11) to the star catalogue of books 7 and 8. Books 2 to 8 of the *Geography* mainly consist of tabulated geographic positions. The *Handy Tables*, in which Ptolemy tabulated the movements of the seven planets in order to help astronomers with their computations, were the oldest pure table work. Although they are no longer extant in their original form, they exerted a crucial influence on the subsequent history of the format.[207] On their model, a fully fledged, long-lived tradition of astronomical table works developed. In the Middle Ages, this tradition comprised the Islamic genre called *zīj*, of which over 200 representatives are currently known, as well as the *Toledan* and *Alfonsine Tables* derived from it and compiled in eleventh- and thirteenth-century Spain respectively.[208] Another premodern form of astronomical table works were almanacs.[209] As early as the fourth century, Eusebius transferred the table format

[203] Such tables are found even where one would not expect them. Francesco Maurolico's cosmographical dialogues (1543) are preceded by six pages of tables presenting astronomical data needed to understand the ensuing discussion.

[204] Gross et al. 2002, 13 and 22.

[205] See Neugebauer 1955 and some of the texts in Hunger 2001.

[206] Chords were Ptolemy's functional equivalent to our trigonometric functions.

[207] Edition of Ptolemy's accompanying explanations: Heiberg 1907, 159–85. For a short description and a bibliography, see *PAL*, A.7.

[208] On the tradition of the *zījes*, see Kennedy 1989; on the *Toledan* and *Alfonsine Tables*, see *CHS* 2, 456–84, esp. 459–60, 468–70 (further literature: 479 n. 35). The tradition of astronomical tables is summarized in some later representatives of the genre, for instance in the preface to Kepler's *Tabulae Rudolphinae* (1627, pp. 1–5) and in the dedication of Philips van Lansbergen's *Tabulae motuum coelestium perpetuae* (1632, sigs. *4^r–[*6^r]).

[209] *CHS* 2, 461, 469.

to chronology, correlating the timelines of different peoples in the second book of his *Chronicon*, which soon reached the Latin West in Jerome's translation.

Early modern table works

In the manuscript age, the 'tabulae' (as they were usually titled in Latin) of Ptolemy and his successors were beset by one perennial problem: the corruption of the all-important figures in the process of transcription. Print did not solve this difficulty once and for all[210] but greatly alleviated it. Now, it sufficed to transcribe a given set of data just once—which could be done with correspondingly greater care—in order to have it reproduced and distributed hundreds of times in identical form. The time and energy that had been spent on repairing a constantly deteriorating tradition so far could be channelled into the elaboration of new, more precise tables.[211] The data stored in such works were in great demand, not only from astronomers but also from astrologers, calendar makers, chronographers, and geographers. For this reason, 'tabulae' even found a place in formal teaching in arts faculties. As becomes clear from the introductory paratexts to Johannes Schöner's *Tabulae astronomicae resolutae* (1536),[212] students learned to perform calculations on their basis as a practical counterpart to planetary theory. On the third page of a congratulatory letter to the author, Melanchthon states:

> Quod igitur hoc tempore edidisti et expoliisti tua diligentia atque industria *Tabulas resolutas*, egregiam operam navasti, quia in scholis praecipue opus est his tabulis, cum theoricas, ut vocant, motuum caelestium ad usum transferant et monstrent facilem et compendiariam rationem computandi motus caelestes.
>
> So you now have done something exceedingly helpful in publishing and polishing the *Resolved Tables* by your diligence and assiduity, because these tables are much needed, particularly in the classroom, because they put the so-called theorical reflections about the celestial movements into practice and indicate an easy shortcut towards their calculation.

Predictably, therefore, new sets of printed astronomical tables proliferated from the late fifteenth century: Schöner's *Tabulae astronomicae resolutae*, Erasmus

[210] As pointed out by contemporaries like Rasmus Bartholin (*Specimen recognitionis nuper editarum Observationum astronomicarum Tychonis Brahe*, [1668]), the value of Albert von Curtz's edition of Brahe's observational data, the *Historia coelestis* (1666), was impaired by thousands of transcription errors and misprints (Korenjak 2020, 416–17).

[211] Eisenstein 1979, 465, 467–8.

[212] They were based on fourteenth-century versions of the *Alfonsine Tables* edited for more convenient use (*CHS* 2, 469). The term 'resolutae' ('resolved') designates the result of this process, as explained in the full title ('quas vulgo, quia omni difficultate et obscuritate carent, resolutas vocant').

Reinhold's *Prutenicae tabulae* (1551), Kepler's *Tabulae Rudolphinae* (1627),[213] and Philippe de La Hire's *Tabulae astronomicae Ludovici magni iussu et munificentia exaratae* (1702)[214] are a few influential examples among many dozens. Table works tailored to specific uses such as Johannes Regiomontanus's *Tabulae directionum profectionumque* (1490, for casting horoscopes)[215] or Georg von Peuerbach's *Tabulae eclipsium* (1514, prediction of eclipses) also appeared.

Astronomy remained the most important source and field of application of tables throughout the early modern era. Alongside the tradition harking back to the *Handy Tables* discussed earlier, other strands developed as well. 'Observationes' or 'historiae caelestes'[216] tabulated, not the general movements of celestial objects, but their positions as observed at specific times and places. In doing so, they presented fresh raw data from which more precise 'tabulae' of the movements could be calculated. As early as 1544, Schöner published data collected by Regiomontanus and Bernhard Walther during several decades of observation at Nuremberg (*Scripta Regiomontani*, fols. 21^r–34^r, 36^r–60^v).[217] From the seventeenth century, such works became more common due to the increasing precision of observational instruments and the establishment of observatories. Two prominent examples are Curtz's long-awaited edition of Brahe's data (1666) and John Flamsteed's *Historia coelestis Britannica* (1725). Later, observatories themselves issued serialized observations. This was done, for example, by the observatory in Munich, whose *Observationes astronomicae in specula regia Monachiensi institutae* started in the 1820s and remained in Latin until the 1840s.[218]

While 'tabulae' fed on 'observationes', almanacs or 'ephemerides' ('day-by-day tables ') were in turn calculated on the basis of 'tabulae', whose data they used to predict the positions and movements of the planets for a specific year or series of years.[219] Also in their case, print made correct reproduction and dissemination easier. If almanacs, nonetheless, differed from one another and were often incorrect, these problems were inherited from the underlying tables and aggravated by calculation errors, as decried in Benedetto Altavilla's *Animadversiones in*

[213] *KGW* 10 (text); Reichert 2014 (text and German translation).

[214] See Souchay 2013, 96–7.

[215] Zinner 1968, 147–52. 'Directiones' and 'profectiones' are types of astrologically significant movements (Girolamo Vitali, *Lexicon mathematicum, astronomicum, geometricum*, 1668, pp. 142 and 388).

[216] 'Historia' in the second title type highlights an approach based on collecting and describing.

[217] Cf. Pomata 2011b, 49–50.

[218] In other cases, this type of publication was omitted and new observations were directly channelled into new tables, as in La Hire's aforementioned work, the full title of which mentions that its 'observationes habitae sunt ab ipso auctore'.

[219] In the preface (sig. $**^r$) to his *Tabulae motuum coelestium perpetuae* (1632), Philips van Lansbergen recommends an almanac currently being produced on the basis of them: 'Postremo lectorem monitum volo insignem logistam doctorem Davidem Goubardum Gallum occupatum esse in supputandis duorum proximorum annorum Ephemeridibus ex Tabulis nostris astronomicis easque primo quoque tempore in publicum esse daturum.'

ephemeridas (1580).[220] However, none of this detracted from the popularity of the genre. A foundational work was the *Ephemerides* of Regiomontanus (1474), the earliest astronomical table work ever printed. Originally covering the years 1475 to 1506, it was reprinted in progressively smaller editions, the earlier years being left out, until the 1490s and provided a generic model for other authors. It was continued by astronomers such as Johannes Stöffler and Jakob Pflaum, who between them issued the *Almanach nova plurimis annis venturis inservientia* (1499; numerous further editions until at least 1522), and Johann Engel (two works titled *Almanach novum atque correctum* for the years 1510 and 1512).[221] Scores of Latin almanacs in book format were produced at least until the early nineteenth century (Pietro Caturegli, *Ephemerides motuum caelestium*, 1828). As with 'observationes', independent publications were supplemented (and ultimately replaced) by periodicals issued by observatories. A long-standing example is provided by the *Ephemerides astronomicae ad meridianum Vindobonensem* (1756–1806).[222]

In various ways, the science of the stars also became responsible for the majority of table works from other fields, the most important of which was astrology. Until the seventeenth century, 'tabulae' and 'ephemerides' to a large extent served astrological ends. There is explicit evidence for this in many works. The title page of Regiomontanus's *Tabulae directionum profectionumque* advertises them as 'very useful for horoscopes' ('in nativitatibus multum utiles'). Kepler's instructions for the use of his *Tabulae Rudolphinae* conclude with a 'little gift sent to the producers of horoscopes' ('sportula genethliacis missa', pp. 121–5), which details their application to this end.[223] The section starts with the assertion that 'most people are eager for this table work for astrological reasons' ('plerique opus hoc tabularum expetunt propter astrologiam'). Astrological beliefs in turn formed the basis of prognostic almanacs, which concerned themselves with the weather during the year ahead but also with health matters and a variety of other topics of public interest.[224] An early, simple type of such prognostics was bloodletting calendars, mostly vernacular, but for about one-third also Latin broadsheets, which were printed in great numbers from the 1460s in Germany in particular.[225] After some calendric key figures like the year's position in the leap-year cycle or its golden number, such texts indicate in their main part who should be bled on what days and from what veins. The table layout is often weakly developed, but in structure and purpose bloodletting calendars are tables nonetheless. Later on,

[220] On this work and the defence of almanacs against it by Giovanni Battista Benedetti, see Omodeo 2014, 142–9.

[221] On Engel's life and works, see Knobloch 1983; on his almanacs, see Dobrzycki and Kremer 1996.

[222] Bolton 1897, 195, no. 1598 (with slightly different dates).

[223] Cf. *KGW* 10, *75–6.

[224] On this extremely diverse text type, which included Latin and vernacular, learned and popular, markedly astrological and astrology-free items alike, see Casali 2003 and, for a summary, see 2012; *EEMHO* s.v. 'Almanack'.

[225] One hundred items from the incunabula era have been collected by Heitz and Haebler 1905.

more complex booklets appeared which gave a broader preview of star-induced events in the sublunary world. In the 1560s, Cornelius Gemma apparently enjoyed some success with a series of *Ephemerides meteorographicae* for the region of Brabant. If the extant 1563 issue is representative, he used an uncommon horizontal format to coordinate the following information in six columns on a double page: (1) the days of a given month with their saints; (2) the weather; (3) human health and diseases; (4) the astronomical longitude of the moon; (5) explanations of the mechanisms of stellar influence on weather and health; (6) planetary positions and aspects as well as the rising and setting of important stars. While most columns were in keyword style, the third and fifth featured running text. The format was continued by others into the 1570s and also transferred to England, where Richard Forster published a work of the same title and in the same format for the year 1575.

The close relationship between the investigation of the stars and of the sublunary atmosphere also showed in the fact that, from the seventeenth century, astronomical observatories served for meteorological observations too. Quite logically, the title of the astronomical 'ephemerides' and their day-by-day format were transferred to the nascent science of modern meteorology. In the early eighteenth century, tabulated meteorological observations began to appear in journals. Plans forged since the seventeenth century to organize supranational networks for weather observation bore fruit in the late eighteenth.[226] Between 1783 and 1795, the Societas meteorologica Palatina printed twelve massive volumes of *Ephemerides* covering the years 1781–92. Gathering observations collected in uniform fashion at European and extra-European observatories, each of the several hundred tables that make up these tomes covers a given place and month and details no fewer than sixteen meteorological parameters such as air pressure, temperature, humidity, wind, or precipitation for each day at 7 a.m., 2 p.m., and 9 p.m. The sheer mass—about 200,000 pieces of data per volume—and homogeneity of the information collected as well as the turn from predicting the future to recording the past starkly contrast with the tables of astrological meteorology from earlier centuries.[227]

Mathematical tables proliferated in response to the need in astronomy and other disciplines like geodesy and surveying for the efficient execution of complex calculations in the absence of calculators (many of which were planned and built from the seventeenth century but remained impractical until the nineteenth). An important field in this respect was the trigonometric functions, first tabulated by Georg Joachim Rheticus in his *Canon doctrinae triangulorum* (1551), which provided a more convenient alternative to Ptolemy's tables of sines.[228] Rheticus's tables were expanded to include every single minute of a degree (instead of every tenth) in Philips van Lansbergen's *Triangulorum geometria* (1591). Nonetheless,

[226] Frisinger 1977, 101–12. [227] Cassidy 1985; Feldman 1990, 168–73.
[228] Schöbi-Fink 2010, 34; Hischer 2016, 209.

the trigonometric calculations needed in the compilation of astronomical tables remained extremely cumbersome. Reinhold, for instance, had to invest 'untold toil' ('immensum laborem') and a full six years into his *Prutenicae tabulae* (preface, sig. $\beta 3^r$). Logarithms made life much easier, because they made it possible to replace multiplications and divisions with less effortful additions and subtractions in the pertinent equations of spherical trigonometry. John Napier, who first explained the notion and presented tables of logarithms in his *Mirifici logarithmorum canonis descriptio* (1614), did so with the expressed aim of facilitating astronomical calculations.[229] The importance of the invention can be measured by the fact that Jacob Bartsch even devised *Tabulae manuales logarithmicae* (1631) customized to one particular table work, the *Tabulae Rudolphinae* of his father-in-law Kepler. Bartsch's logarithms proved helpful enough for a corrected and expanded version to appear as late as 1700.[230]

In a somewhat wider sense, chronological and geographical tables belong to the astronomical tradition too, since all three disciplines depended on one another's data. For this reason, astronomical works like the *Tabulae Rudolphinae* (part one, tables X–XII) sometimes contained chronological and geographical material. The upsurge of chronology in the sixteenth century produced volumes interspersed with calendric tables like the classicist Joseph Scaliger's *De emendatione temporum* (1583) and *Thesaurus temporum* (1606),[231] as well as pure table works. The latter were often of a less learned and more practical kind, like the *Tabulae chronologicae* (1597) by the Calvinist theologian Pantaleon Candidus. Presenting a brief and a more extensive overview of world history separated by a comparison of the chronologies of different Protestant authors, his tables remind the reader of Eusebius' *Chronicon* in their organization and layout.[232]

Full-blown geographical table works in the style of Ptolemy's *Geography* seem to have been uncommon. However, some works of descriptive geography contain table sections. At the end of Martino Martini's *Novus Atlas Sinensis* (1655), the longitudes and latitudes of all fortified places marked on his maps are tabulated on twenty pages. As a particularly extensive collection of geographical tables, Thomas Schoepf's *Inclytae Bernatum urbis delineatio chorographica* (1577) deserves mention. This text of several hundred pages is a companion piece to a printed map of the canton of Bern, but because it is much more detailed than its cartographical counterpart, the Bernese government kept it in manuscript as classified intelligence for internal use only. The individual entries contain more information than those in Ptolemy or Martini. The name and type (village, river, etc.) of each geographical item is given in a marginal note, followed by a description in full

229 Napier's notion of a logarithm does not exactly correspond to ours, but each is easily reduced to the other. See Merzbach and Boyer 2011, 286–90.

230 See Johann Caspar Eisenschmidt's preface (pp. 7–9) to the 1700 edition.

231 Grafton 1983–93, vol. 2.

232 For a number of comparable works, see Nothaft 2012, 285–8.

sentences which includes its geographical longitude and latitude, distance from other objects, and often a brief characterization.[233] The *Delineatio chorographica* looks exceptional, but the possibility that it represents a class of similar works which remained unpublished for similar reasons cannot be excluded.

Table works completely unconnected to astronomy seem to be a rarity. Linnaeus's *Systema Naturae* comes to mind, but only its first edition (1735) could perhaps be called a table work, and only with much goodwill, since the layout looks more tabular than the actual structure is: coordinated lists (of different kinds of names, of groups of animals and their characteristics) appear in the sections dedicated to stones and animals, but not in the plants section, and subdivision is more important than coordination throughout, as is only natural in a classificatory work. From the second edition onwards, even the tabular layout has gone. The work now looks like a 'historia' made up of very concise entries.

To conclude this section with a few remarks on the relationship between Latin and the vernaculars, it has already been demonstrated that the tradition remained multilingual over four centuries due to the varying level of sophistication of the works belonging to it and the different classes of readers they catered to. While German bloodletting calendars had already appeared in the fifteenth century, Latin 'tabulae' and 'ephemerides' were printed as late as the nineteenth. On top of that, some intersection and interaction between Latin and the vernaculars developed in the seventeenth and eighteenth centuries. An interesting and comparatively well-studied example of a bilingual publication is Maria Cunitz's *Urania propitia* (1650). Conceived as a less technical, more accessible (because logarithm-free) counterpart to Kepler's *Tabulae Rudolphinae*, this magnum opus by one of the few female astronomers of its time is preceded by a bilingual introduction in Latin and German.[234] Moreover, Latin tables were sometimes translated into the vernaculars. La Hire originally published his *Tabulae astronomicae* (1702) in Latin, but he also produced a French version which appeared posthumously in 1735. Already ten years earlier (1725), a German translation had come out in Nuremberg.

Tables and text

To reflect about the various languages of table works may appear otiose at first sight. Tables, one could argue, precisely nullify or minimize dependence on natural language in scientific communication, and their formalized way of presentation transcends the boundaries between speech communities. However, such reasoning would be misguided. Like all kinds of formal language, tables must be embedded

[233] German translation by Theresa Rothfuß: https://neolatin.lbg.ac.at/repository/ (accessed 16 February 2023). See also Korenjak 2013a and the thematic issue 60 (2020) of the journal *Cartographica Helvetica*.

[234] See Guentherodt 1987 and 1991; on the technical level of the work, see Swerdlow 2012.

in natural language in order to be understood. The early modern specimens under discussion are no exception. Their authors had to explain them, and they did so by premising (or, more rarely, appending)[235] instruction manuals to them. The composition of such explanations was no trivial task but a demanding exercise in scientific writing. Introductions to table works made use of a broad array of dialectical, rhetorical, and typographical techniques from definition, subdivision, and recipe style to direct address to the reader, multi-tiered layout, and diagrams.[236] Some of them grew to monographic proportions, but even with such effort, authors were not always successful in explaining themselves. Kepler, one of the most gifted scientific writers of his time, needed over 120 pages to explain the use of his *Tabulae Rudolphinae*. However, Jacob Bartsch found his instructions so unsatisfactory that he supplemented them with an appendix of his own.[237]

Introductions to table works may be said to have two main thrusts: they can be oriented towards the author and his work and/or cater to the reader's needs. Texts focusing on the former aspect may sketch the tradition leading up to the work in question, narrate its origin and formation, and summarize its structure and content. The introduction to Flamsteed's star catalogue at the beginning of the third volume of his *Historia caelestis Britannica* (1725, pp. 1–164), for instance, enumerates and characterizes all earlier major star catalogues. Flamsteed then goes on to tell the story of his own observations, detail his methods, and describe his instruments.

More important and common, however, are instructions for the reader: overviews of the tables and their content, the imparting of necessary previous knowledge, and above all exercises and calculation examples. Much of this presumably reflects common teaching practice harking back to the manuscript era and adapted to suit the needs of larger reading audiences in the age of print. Already Regiomontanus's *Ephemerides* opens with a brief explanation of its structure and the meaning of the individual columns, abbreviations, and so on. The introduction in two short books prefixed to Napier's *Mirifici logarithmorum canonis descriptio* is entirely devoted to the explanation of the novel notion of the logarithm itself, its rules of calculation, and its application to plane and spherical triangles. The inclusion of examples of calculations in printed table works can be traced back to Peuerbach. The method for using his *Tabulae directionum profectionumque* and the *Tabula primi mobilis* of his pupil Regiomontanus (1514) is demonstrated by thirty-one and sixty-three 'proble(u)mata' respectively. In later times, this device becomes ubiquitous.

What a sophisticated craft it was to work with astronomical tables and to what lengths an author had to go in order to enable his readers to do so are best shown

[235] For an explanatory dialogue concluding Rheticus's *Canon doctrinae triangulorum*, see Chapter 7.
[236] Such techniques were also applied in textbooks and will be discussed in Chapter 8.
[237] The appendix found its way into a few copies of the work only. See *KGW* 10, 257–77 (text), 77*–80* (explanations).

by a concrete example. A work that lends itself to this end is Peuerbach's comparatively simple *Tabulae eclipsium*. After an explanation of their use in general terms in seventeen 'propositiones' (sigs. bb^r–$bb5^v$), examples of calculations of a solar and a lunar eclipse are added (sigs. $bb5^v$–[$bb8^v$]). The second begins as follows (sig. [$bb7^v$]):

> Volumus invenire oppositionem veram luminarium et eclipsim lunae, quae fiet anno Domini 1460 currente in mense Decembris. Tempus numerationis Christi transactum usque ad initium illius mensis usualis est anni 1459 et November usualis. Cum quo quidem tempore dum intravero tabulas coniunctionum et oppositionum mediarum primo in annis collectis 1440, ibi invenio de tempore dies 7 horas 8 minuta 38 secunda 27, de loco solis signa 9 gradus 11 minuta 37 secunda 14, de argumento lunae signa 2 gradus 23 minuta 26 secunda 44, et de argumento latitudinis signa 4 gradus 24 minuta 58 secunda 59. Haec extra scribo. Item in annis expansis in directo 19 reperio de tempore dies 28 horas 20 minuta 12 secunda 6, de loco solis signa 11 gradus 0 minuta 58 secunda 21, de argumento lunae signa 9 gradus 11 minuta 8 secunda 16, de argumento latitudinis signa 11 gradus 6 minuta 53 secunda 42. Haec similiter extra scribo sub prius extra notatis, unumquodque sub suo simili. Deinde quia annus currens est bisextilis, intro tabulam mensium bisextilium cum Novembre. Ibi invenio de tempore dies 10 horas 3[238] minuta 55 secunda 26, de loco solis signa 10 gradus 20 minuta 10 secunda 26, de argumento lunae signa 9 gradus 13 minuta 59 secunda 6, de argumento latitudinis lunae signa 11 gradus 7 minuta 22 secunda 31. Haec quoque sub prioribus extra ponam ordinibus. Hos tres ordines aggregabo provenietque in tempore dies 46 horae 8 minuta 45 secunda 59, in loco solis signa 7 gradus 2 minuta 46 secundum 1, in argumento lunae signa 9 gradus 18 minuta 34 secunda 6, in argumento latitudinis autem signa 3 gradus 9 minuta 15 secunda 12.
>
> We want to find the true opposition of the sun and the moon and the eclipse of the moon which will take place in the December of this current year 1460.[239] The time of Christian reckoning up to the beginning of this civil month is 1459 [years] and the [months until] civil November. When I first enter the tables of the mean conjunctions and oppositions under 1440 years taken together, I find there with regard to the time 7 days 8 hours 38 minutes 27 seconds, with regard to the position of the sun 9 signs 11 degrees 37 minutes 14 seconds, with regard

[238] The print has 'signa 10 gradus 3', an obvious mistake.

[239] In what follows, 'years taken together' ('anni collecti') and 'years spread out' ('anni expansi') refer to two tables, the first of which gives the values of the items indicated for every fortieth year (1400, 1440 etc.), the second, the year-by-year differences within one period of forty years. Together, they enable the user to calculate the concrete value for any given year. The 'signs' ('signa') are the astrological signs, each of which equals 30° on the ecliptic. The 'argument of the moon' ('argumentum lunae') is the angle between its apogee and current position, the 'argument of its latitude' ('argumentum latitudinis'), the angle between its ascending node and current position (Vitali, *Lexicon mathematicum, astronomicum, geometricum*, 1668, p. 63).

> to the argument of the moon 2 signs 23 degrees 26 minutes 44 seconds, and with regard to the argument of its latitude 4 signs 24 degrees 58 minutes 59 seconds. These I write out. Further, I find under the years spread out in row 19 with regard to the time 28 days 20 hours 12 minutes 6 seconds, with regard to the position of the sun 11 signs 0 degrees 58 minutes 21 seconds, with regard to the argument of the moon 9 signs 11 degrees 8 minutes 16 seconds, with regard to the argument of its latitude 11 signs 6 degrees 53 minutes 42 seconds. These figures I write out in the same way under those noted before, each under those of the same kind. Moreover, because the current year is a leap year, I enter the table of leap years at November. There I find with regard to the time 10 days 3 hours 55 minutes 26 seconds, with regard to the position of the sun 10 signs 20 degrees 10 minutes 26 seconds, with regard to the argument of the moon 9 signs 13 degrees 59 minutes 6 seconds, with regard to the argument of the latitude of the moon 11 signs 7 degrees 22 minutes 31 seconds. These [figures] too I place outside under the earlier sets [of figures]. These three sets I will add and the result will be, for the time, 46 days 8 hours 45 minutes 59 seconds, for the place of the sun, 7 signs 2 degrees 46 minutes 1 second, for the argument of the moon, 9 signs 18 degrees 34 minutes 6 seconds, and for the argument of its latitude, 3 signs 9 degrees 15 minutes 12 seconds.

The passage contains a number of aspects worthy of mention: the simple, repetitive syntax and lexicon, the disregard of classical Latinity in favour of technical terminology and classroom slang ('intro tabulas', 'in directo', 'extra scribo'), the first person plural (at the beginning) and singular (for the rest), in which one can hear the Viennese professor perform the whole operation in front of his class. Even more salient and telling, however, are two other features: first, the sheer length and complexity of the text and the procedure described. In the present extract alone, three sets of numbers comprising twelve values and forty-eight individual figures have to be looked up in various tables, excerpted correctly, and added up—and this is only the beginning: 14 printed lines of a total of 120. Secondly, Peuerbach uses an extreme form of recipe style. He demonstrates the whole procedure in meticulous detail, including even the smallest and most banal steps, yet without giving one single word of explanation. Indeed, going back and forth between text and tables, the reader is able to calculate the eclipse correctly without understanding what he is actually doing and why. And this, one supposes, is exactly the point: to make astronomical calculations possible even for those without a firm grasp of the cosmology and geometry on which the tables are based. True, Peuerbach is an extreme case. Some of his colleagues, Kepler, for instance, try seriously to impart some true understanding as well. Nonetheless, the instructions remain bound to make a mechanical use of the tables possible. After all, if the circle of those capable of performing calculations with their help had not been much wider than the small elite capable of producing them, these books would have found no buyers.

7
Arguing

Dialogues

History of the dialogue

Since Plato, the prose dialogue, the literary 'exchange between two or more voices',[1] has been a recognized genre of Western literature. Within the ancient tradition, one may broadly distinguish four threads.[2] The Platonic dialogue is characterized by its dramatic nature (that is, the lack of a narrator outside the dialogue), philosophical content, and literary ambition, which shows in aspects such as scenery, characterization of speakers, and style. The first prominent representative of the Ciceronian dialogue seems to have been Aristotle, but for us, the form is mainly represented, as its name indicates, by Cicero's philosophical writings. While taking over from Plato most of the aforementioned characteristics, Aristotle and Cicero embedded their dialogues in a narrative frame presented by an authorial voice. Moreover, Cicero liked to feature himself as the main speaker. In this way, the author became a stronger presence in the dialogue and his position was made more unequivocally clear than was the case in Plato's dialogues. Beside these varieties, which are the most important for my purpose, there was the convivial dialogue set at a dinner party, in which the discussion mostly turned on antiquarian, philological, and generally learned questions, represented by authors like Plutarch and Athenaeus. The Lucianic dialogue, named after its main representative Lucian of Samosata, was characterized by its satirical thrust and sometimes fantastical setting.[3]

The Middle Ages produced a rich dialogic literature both in Latin and in the vernaculars.[4] Much of the material dealt with theological issues, and much was didactic, both in the sense that one of the speakers lectured the other(s) and that the texts were conceived for didactic use. Dialogues oriented towards classical models became a rarity, because most of these were little read or completely unknown. Among the well-known texts, Boethius's *De consolatione philosophiae*,

[1] This minimal definition is taken from Cox 1992, 2.
[2] Hirzel 1895; Marsh 1980, 4–8 (on their impact on the Renaissance dialogue).
[3] In Lucantonio Porzio's *Erasistratus sive de sanguinis missione* (1682), the four physicians Erasistratus, Galen, Jan Baptist van Helmont (†1644), and Thomas Willis (†1675) discuss the pros and cons of bloodletting in an imitation of Lucian's *Dialogues of the Dead*.
[4] *LMA* 3 (1986), cols. 946–65.

Latin Scientific Literature, 1450–1850. Martin Korenjak, Oxford University Press.
DOI: 10.1093/oso/9780198866053.003.0008

a Ciceronian dialogue whose two participants fulfil the role of teacher (Lady Philosophy) and pupil (Boethius) respectively, stands out for the breadth and depth of its influence. Meanwhile, a lifeworld dialogic form developed in the institutions of higher education: the academic 'disputatio'. Whether and how this practice of formalized discussion inspired the early modern revival of the literary dialogue is unclear, but some later authors of dialogues at least saw their chosen genre as a better alternative to a disputation: unbound by institutional constraints, less formalized, less contentious, and in better Latin.[5]

Whatever its relationship to the 'disputatio', the classicizing dialogue saw an early modern revival which was quantitatively substantial as well as self-conscious and programmatic: Latin and vernacular dialogues imitating ancient models were written in considerable numbers throughout the Renaissance.[6] In stark contrast to medieval habits, the generic marker 'dialogus' figures in more than half of the respective titles. Dialogues about the natural world, which had ancient predecessors in Plato's *Timaeus*, some works by Plutarch such as *De facie in orbe lunae* (translated by Kepler in 1634), and some by Galen and Caelius Aurelianus,[7] formed part of this development. On account of their learned content, most of them were written in Latin. The vogue for them started in the late fifteenth century and peaked in the sixteenth. Few examples can be found after the middle of the seventeenth, a fact in accordance with the demise of the Neo-Latin dialogue in general at that time. The reasons for this decline are unclear, but that it was a genre-specific development rather than prompted by the increasing complexity of science is suggested by the fact that other highly literary forms such as didactic poetry still flourished in the eighteenth century. While the vernacular dialogue saw a new Golden Age in the Enlightenment, its Latin counterpart did not participate in this revival.

The dialogue as a vehicle of scientific argument

The presentation of serious scientific thought in dialogic form appears strange today, but this may say more about the modern conceptualization of science than

[5] Traninger 2012, 256–70, esp. 262. Jean Fernel's *De abditis rerum causis* (1548) starts from the mention of a disputation involving some famous physicians ('quandam celebrium medicorum disceptationem') on the question of whether there is a divine element in some diseases. The discussion ended in turmoil and yielded no decision ('nec profecto tumultuoso illo congressu decidi res omnis potuit'), because the participants lacked the time to explain their points clearly and some were prejudiced (p. 6). The implication is that Fernel's dialogue will succeed where the disputation failed.

[6] On Latin dialogues, see *CNLS* 2, 229–38; Cox 2017. On fifteenth-century dialogues from Italy, see Marsh 1980; on sixteenth-century French and Italian dialogues, see Bénouis 1976 and Cox 1992 respectively; on Italian dialogues up to 1650 about meteorological, astronomical, and cosmological topics, see Paulus 2005, 235–80. Case studies on Latin as well as vernacular works are collected in Jones-Davies 1984.

[7] Hirzel 1895, vol. 2, 164–7, 171–88, 364.

about a factual incompatibility of science and dialogue.[8] While it is recognized that science is a collective enterprise and free exchange and discussion of ideas is vital to its progress, it has become customary to depersonalize scientific thought and to suppress the voice of the individual researcher in the written record. So far as discussion is still present in writing, it comes in the impersonal, neutralized form of scientific intertextuality, that is, of citation.

In the Renaissance, a personal voice was tolerated to a much higher degree in a text about nature. Moreover, many fields within the study of nature were still understood as forming part of philosophy, which was considered the most important subject of the dialogue because of the literary tradition. Under these circumstances, the form proved attractive in a number of respects. To begin with, a dialogue does not just convey results but demonstrates the process by which they are reached. To adapt the mission statement of the Collège de France, it shows 'le savoir en train de se faire', or at least in the process of being taught and grasped. For many readers, such a form, which allowed them to apprehend and learn together with the speakers, was more attractive than a plain presentation of facts and connections. This was all the more so if the process was presented in attractive literary garb, including graphic depiction of the setting, lifelike characterization of the speakers, carefully designed, sometimes surprising course of the conversation, and fluent and educated language (for which last point Cicero provided an obvious model). That this language ideally conveyed the vividness of orality made it still easier to immerse oneself in the text and to identify with the speakers. The fiction of orality also legitimized the simplification of overly technical details and allowed for entertaining digressions. All of this rendered the dialogue a genre attractive for experts and laymen alike (in fact, it could have found its place among the genres serving to publicize science as well as here). Moreover, as in antiquity, the format proved suitable for presenting the search for truth as a collaborative enterprise and for celebrating the friendship uniting those involved in it. Introducing oneself, one's colleagues, and one's friends as participants opened up ample possibilities for praising oneself and others, as could be learned from Cicero in particular. Not coincidentally, the authors of many early modern dialogues figure as their main speakers in transparent disguise or even under their real name. Finally, in an age of censorship and heated professional polemics, the dialogue gave authors a precious opportunity for addressing delicate issues without committing themselves. Although it did not help him in the end, Galilei made a conscious choice in writing the *Dialogo sopra i due massimi sistemi del mondo* (1632) in the form he did.

The genre's major Renaissance theorists, Carlo Sigonio, Sperone Speroni, Lodovico Castelvetro, and Torquato Tasso, do not have much to say about

[8] The only contribution known to me which is exclusively dedicated to the relationship between science and the dialogue is Herlitzius 1968.

science, although the last recognizes some scientific disciplines as adequate subjects among others: 'One can write dialogues on arithmetic, geometry, music, astronomy... For in all arts and sciences, one can pose questions and therefore develop conversations.'[9] By contrast, the reflections of the authors themselves in their dedications, prefaces, and openings are significant, as they explicitly address a number of the aforementioned points. Dedicating his *Botanologicon* (1534)[10] to the senate of Bremen, Euricius Cordus explains that he takes advantage of the openness of the form by declaring his opinion on most issues but leaving some others undecided to prompt the reader to reflect on them for himself (p. 5): 'Indicavi graves in quibusdam difficultates dubiumque, ut altius cogitaret, lectorem reliqui.' Jean Fernel introduces his dialogue *De abditis rerum causis* (1548) as a better alternative to an academic disputation, as already mentioned (see note 5 above). Galilei justifies his chosen form in the preface to the *Dialogo*: 'Ho poi pensato tornare molto a proposito lo spiegare questi concetti in forma di Dialogo, che per non esser ristretto alla rigorosa osservanza delle leggi Matematiche, porge campo ancora a digressioni tal'hora non meno curiose del principale argomento.' Lucantonio Porzio admits to having composed his *Erasistratus* (1682) on the contentious issue of bloodletting as a dialogue between two proponents and two opponents of the practice in order not to appear presumptuous and to avoid the hatred of his colleagues (sig. a3^{r}): 'Sed ut omnem effugiam invidiam, eam dicendi rationem sum secutus, quae minime arrogans est, dialogi scilicet...'[11]

Varieties

Like their ancient predecessors, the early modern dialogues under consideration here prove to be highly diverse in several respects. They cover most areas of thinking about nature from medicine and botany[12] to alchemy, astronomy, and mathematics, and they deal with their subjects on different levels of technicality and sophistication. Not all dialogues contribute to red-hot debates, but a topical element inheres in many. Dramatic (Platonic) and narrative (Ciceronian) dialogues alike occur, although early modern scientists prefer the former. The degree of linguistic and conversational realism, character drawing, depiction of scenery, insinuation of background action, and literary finish in general varies greatly depending on the author's ability and ambition. The latter may also be reflected in a dialogue's length. Most works run to a couple of dozen octavo pages, but

[9] *Dell'arte del dialogo* (quoted in Buck 1968, 175). On the sixteenth-century theory of the dialogue in general, see Snyder 1989.

[10] This highly original dialogue has been translated into German with an extensive introduction by Dilg 1969.

[11] The point is repeated in an address to the reader by Gregorio Rossi (unpaginated, fourth page).

[12] Some examples from these disciplines are enumerated by Dilg 1969, 53–5.

examples comprising only a handful or several hundred pages can also be found. The number of participants is not fixed either. Two are most common, but they may become quite numerous, as in the case of Hippolytus Guarinonius's *Chylosophia*, which will be discussed later. Speakers may be real persons, including the author himself, fictional persons, dummies completely reduced to their roles or even allegories (as already in Boethius). The participants' relationship within the dialogue also takes different forms: depending on their expertise or lack thereof, they may be on an equal footing or assume the roles of teacher and pupil respectively. In the latter case, the dialogue can lose its conversational character and disintegrate into a series of short questions followed by lengthy explanations. One may speak of a didactic dialogue, in which the reader is supposed to identify with the pupil, as opposed to a discussion dialogue, where he may form an opinion for himself (a distinction recognized by seventeenth-century theorists).[13] If a real discussion takes place, the participants may broadly agree and cooperate in their argumentation or face each other as intellectual opponents. Finally, dialogues may lead to a result which satisfies all participants or be open-ended in the sense that the matter remains unsettled and further discussion is said to be in order.[14]

Given the number of these variables, the freedom of authors to combine any or all of them at will, and the fluent transitions between opposites, no comprehensive classification can be established. It seems preferable to acknowledge that it was precisely the many degrees of freedom and the resulting flexibility that constituted a main strength of the dialogue and formed part of the attraction it held for authors and readers alike. In what follows, the range of variation set out abstractly so far will be illustrated by a quick look at a number of examples from different disciplines in chronological order.

Pietro Bembo's *De Aetna* (1496) is among the first humanist dialogues concerned with the natural world.[15] Dedicated to Pietro's friend Angelo Gabriele, with whom the author has just returned from a study visit to Messina, the short work (some fifty quarto pages) renders a conversation held at the villa of the author's father Bernardo in Santa Maria di Non near Padua. In this idyllic setting, Pietro informs his father about Mount Etna, which he claims to have climbed during his sojourn in Sicily. The dialogue bristles with learned references to classical authors but also describes the peculiar features of the mountain and Pietro's emotional and intellectual reactions in graphic detail, conveying a strong impression of authenticity. Apart from its descriptive and stylistic qualities, *De Aetna* is

[13] Sforza Pallavicino, *Trattato dello stile e del dialogo* (3rd edn 1662), ch. 35, pp. 343–7. See also Bénouis 1976, 194 (who calls the didactic dialogue 'dialogue d'encadrement'); Cox 1992, 2–4 (who sees a tendency towards didacticism in sixteenth-century Italy); Paulus 2005, 237.

[14] For some further classification parameters, see Cox 2017, 290–4.

[15] Modern editions: Alfieri et al. 1981; Williams 2017 (with monographic introduction). See also Luggin 2023, 127–32.

remarkable for its reversal of the expected roles of the participants. In a dialogue between father and son, one would assume the former to act as teacher, the latter, as pupil, but here it is mostly the other way round. Apart from a short passage (pp. 336–40 Williams) in which Bernardo presents a largely traditional theory of volcanism, it is primarily Pietro who instructs his father, and his personal experiences leave a stronger impression on the reader than Bernardo's conventional wisdom.[16] The youthful author, whose very first publication this is, confidently announces that he has important things to say.

Together with a programmatically didactic purpose, the peculiar role division between its two speakers also constitutes a salient feature of the *Dialogus qui inscribitur botanomethodus* (1540). Composed by the obscure 'grammaticus' (teacher of Latin and Greek) Carolus Figulus and dedicated to Hermann V. von Wied, the archbishop of Cologne, this short dramatic dialogue takes place between the author and the young physician Kornelius Sittard in the latter's house and garden in Cologne. In accordance with its title, it provides an account, not of the actual content of pharmacological botany, but of its basic categories, in order to guide the reader towards the methodical study of the discipline. This objective is advertised already on the title page by an epigram by the printer Johannes Schönstenius, which begins: 'Vendimus ecce tibi methodon, qua, lector, in hortos, / campos et silvas doctior ire queas' ('Look, reader, I am selling you a method, by which you will be better instructed when going out to gardens, fields, and woods'). In the dialogue itself, the didactic intent is underlined, at the expense of the dramatic illusion, by copious marginalia and even by tables providing overviews of the names of the colours and tastes of herbs in different languages (sigs. D^{v}–Dii^{v}, Eiv^{v}). Early modern pharmacology was a philological as well as an empirical discipline, and the roles of Figulus and Sittard as teacher and pupil change depending on the aspect under discussion: when the conversation turns on material questions, Sittard takes the lead and answers Figulus's queries. As soon as logical and terminological issues come to the fore, the roles are reversed.

It clearly is no coincidence that the *Cosmographia in tres dialogos distincta* (1543)[17] by the Messinese Benedictine and mathematician Francesco Maurolico was dedicated to Pietro Bembo. Nonetheless, the work differs sharply from *De Aetna* in most respects despite the generic and geographical points of contact. Rather than an account of a wondrous phenomenon for curious readers, it constitutes a reasonably systematic overview of elementary cosmology and astronomy, a propaedeutics to Ptolemy's *Almagest*.[18] Running to over 200 quarto pages, the

[16] The contrast between father and son is discussed in detail in Williams 2017, 221–68.

[17] Paulus 2005, 241–52; for a contextualization within Maurolico's astronomical thought and *oeuvre* as a whole, see Cioffarelli 2014.

[18] Maurolico's own judgement about his work in a letter to the Spanish viceroy Juan De Vega (quoted in Cioffarelli 2014, 556) downplays the degree of systematization to be found in the dialogue: 'Nam dialogi de forma, numero situque elementorum et caelorum, quos appellare libuit *Cosmographiam*

Cosmographia is structured as a dramatic dialogue in three parts or a series of three dialogues taking place over the course of some weeks in the summer of 1535 at different places near and in Messina. The interlocutors are the smart, curious, and sceptical lawyer Antimachus, who embodies the ideal non-expert reader, and the professional astronomer Nicomedes, an authorial figure. Both characters as well as the settings and circumstances are lovingly detailed, and the discussion comes across as lively and realistic (except for the third part, where it is replaced, for reasons that are unclear, by a long monologue by Nicomedes). To aid the reader's understanding, the work is preceded by several tables providing basic astronomical data and figures. Moreover, Maurolico, as unconcerned about dramatic illusion as Figulus, equips his text not only with marginalia but also with many dozen simple but high-quality diagrams, which prove helpful in following Nicomedes's explanations.

Shortly before his early death, the Utrecht physician and humanist Gilbert de Longueil began an ornithological dialogue which remained unfinished but was nonetheless published one year after his death, in 1544, by the English naturalist William Turner.[19] One of the first works ever to focus exclusively on birds, it makes for a zoological counterpart to Figulus's *Botanomethodus* in its double emphasis on 'res' and 'verba'. Problems of nomenclature and the identification of species appearing in classical authors are as important as the discussion of real, living birds. These are not neglected, however: quite the contrary: the appearance and behaviour of some sixty species are carefully observed and trenchantly described. The dramatic dialogue involves the author himself, who poses as an expert in the field, and his younger friend Pamphilus, while two fowlers remain 'personae mutae'. Most of the conversation takes place during an excursion, which involves a visit to a farm and a walk through the woods that provide an opportunity to observe tame and wild birds respectively. All of this is conveyed in a neat, easy, and graphic style. Many of the work's characteristics and virtues are summoned up by its full title: *Dialogus de avibus et earum nominibus Graecis, Latinis et Germanicis, non minus festivus quam eruditus et omnibus studiosis ad intelligendos poetas maxime utilis* (*A Dialogue on Birds and their Greek, Latin, and German Names, as Pleasant as it is Erudite and Extremely Useful to All Scholars for Understanding the Poets*). Once again, we are dealing with a work catering no less to humanist than to naturalist interests.

A quite peculiar specimen can be found appended to Georg Joachim Rheticus's *Canon doctrinae triangulorum* (1551), the first printed tables of all six trigonometric functions. In most works of this kind, the content and use of the tables are explained in an introduction. Here, this service is provided at the end by a short

et quos Petro Bembo dedicavi, fuere potius exercitium quoddam et repastinatio quam ordinatum ac propositionibus distinctum opus.'

[19] Faulty German translation and commentary: Vedder et al. 2016.

dialogue, presumably authored by Rheticus himself, between two generic figures called Hospes ('Stranger' or 'Guest') and Philomathes ('Lover of Mathematics').[20] In a charming metalepsis, Hospes seems to hold the *Canon* in his hand when he opens the conversation with the observation: 'Inscriptio huius libelli pollicetur magna, sed quomodo muti tantum numerorum characteres triangulorum in planitie et in globo doctrinam aliquem docuerint?' ('The title of this booklet promises great things, but how can dumb figures teach someone the doctrine of planar and spherical triangles?') In the animated discussion that follows between two mathematical experts, the use of Rheticus's tables is explained and a variety of related topics are touched upon. While most dialogues target experts and educated readers in general alike, this, like the work it complements, is meant only for those well versed in mathematics . The Latin meets humanist standards in terms of grammar and is syntactically fluent, but the terminology is forbidding for the non-initiated.

In the later sixteenth century, the medical use of chymically prepared drugs as promoted by Paracelsus was a hotly debated question. With *De iure et praestantia chymicorum medicamentorum* (1584), the English iatrochemist Thomas Moffett joined this battle.[21] The speakers of his dramatic dialogue are his alter ego Chymista and the young doctor Philerastus, whose name, 'Friend of [the fervid anti-Paracelsian Thomas] Erastus', brands him as a member of the opposing party. Their conversation is noteworthy for three aspects. First, their relationship is not static but evolves, because Philerastus's intentions change. Initially (pp. 13–14), he only feigns interest in iatrochemistry in order to involve Chymista in a discussion and to refute his opinions, but at the end (pp. 59–60), he is won over by Chymista's kindness and expertise. Secondly, the dialogue features an elaborate background action including a guided tour through Chymista's house and laboratory and a kind of initiation ritual: by taking a bath and changing his clothes, Philerastus symbolically renounces his old prejudiced self (pp. 22–3). At long last, having confessed to his former bad intentions and being allowed to stay with Chymista as his disciple, he is renamed 'Philalethes' ('Lover of truth', p. 60). Thirdly, the book contains an original generic mix, as the dialogue proper (pp. 13–60) is continued at a distance in four letters (pp. 61–103) in which Chymista, who had to go on a journey, further counsels and lectures Philalethes on various aspects of iatrochemistry.[22]

Around 1590, Galilei composed a strongly anti-Aristotelian work on kinematics titled *De motu* in which he showed, among other things, that every object has a certain weight and none moves upwards of its own accord. Although he never

[20] Schöbi-Fink 2010, 34. [21] Houliston 1989, esp. 237–40.

[22] A fifth letter addresses a certain Endymion Luddipolensis ('Sleeper from Luddipolis'), who perhaps represents the London College of Physicians, which is satirized for its conservatism (Houliston 1989, 237–8).

had it printed, he obviously cared about it, as he wrote no fewer than three versions of it. The first and second of these are plain treatises, but the last is styled as a dramatic dialogue between the two friends Dominicus and Alexander.[23] Alexander acts as Galilei's spokesman, arguing for his views and demolishing Aristotle's. The less knowledgeable Dominicus is an eager and self-confident disciple. He freely voices his doubts and questions, restrains Alexander if he proceeds too fast, and has no qualms about interrupting his friend in mid-sentence, if necessary (p. 369). The result is an unusually lively and entertaining dialogue whose lucid Latin and relative freedom from technical jargon make it an easy read. *De motu* foreshadows Galilei's masterful handling of the genre in the *Dialogo sopra i due massimi sistemi del mondo*, which for its part was translated into Latin for the sake of the international 'respublica litteraria' by Matthias Bernegger in 1635. Out of reach of the Catholic Church in Strasbourg, Bernegger calmly replaced Galilei's non-committal title by *Systema cosmicum* ('The *World System*') and made it clear in his paratexts that this single true system was the Copernican one. Like the Greek models of the genre by Plato and his successors, Galilei's *Dialogo* attained most of its initial fame and influence in Latin garb.

Like *De motu*, the *Dialogismus de natura universa maris* (1613) by the Augustinian friar Ambrosio Florido features two participants who differ greatly in their level of theoretical understanding: one, Philosophus, comes across as an academic thinker, the other, Philonauticus, as an unlearned sailor. However, the latter shows nothing of Dominicus's initiative but acts only as a straw man, which makes the text an exemplar of a didactic dialogue. Philosophus's markedly scholastic language and his frequent appeals to authority add to the classroom atmosphere pervading the conversation, which, moreover, is apparently set in the university city of Padua. Other aspects, however, notably the work's elaborate composition and its topicality, militate against the impression of a school lesson. The dialogue is artfully structured in twelve chapters, the first and last three of which deal with the sea and water as an element in general, while the six core chapters are dedicated to the question of the cause of the tides. This problem was highly topical at the time, especially so at Venice, to whose domain Padua belonged.[24] That Florido had both feet on the ground is also indicated by his eighth chapter (pp. 23–6), which is devoted to designations used by Venetian sailors for different kinds of ebb and flow.

As late as the mid-seventeenth century, one of the most ambitious experiments with the genre was made by Hippolytus Guarinonius.[25] This Tyrolean physician believed he had discovered the truth about the passage of chyle through the

[23] Galilei 1890, 367–408.
[24] A good overview of the discussion about the cause of the tides in sixteenth- and seventeenth-century Italy is given by Omodeo 2017.
[25] *TL* 571–6.

human body and refuted the prevailing Galenic theory according to which it travelled from the stomach to the intestine and from there to the liver, where blood was produced from it. He planned to publish his insights in a multivolume work titled *Chylosophia academica*, comprising a first part disproving the prevailing opinion, followed by a second part presenting his own theory. In effect, only volume one of the 'pars destruens' appeared in 1648 with a dedication to Pope Innocent X, but already this small portion of the originally conceived whole runs to six books and over 450 folio pages.[26] The dialogue contained in it is as complex as it is extensive. As specified in an 'index personarum', it features six main speakers—three doctors, two philosophers, and one theologian, led by Chylosophus ('Pulp-Wise') alias Guarinonius—an audience composed of Tyrolean students who spur the discussion on by questions and applause, a number of common people, and an allegory of the ancient theory of 'chylus' (dubbed Apochylia). The dialogue is set in Chylosophus's house near Innsbruck, where a temporary academy is installed to discuss the matter; at one point, the party undertakes an excursion to visit a mill, whose functionality is said to bear an analogy to digestion (pp. 132–9). In formal terms, the work exhibits great variety, as lectures by Chylosophus, discussions, readings from classical authors, poems, and even devices such as a Ramist table (fully integrated into the action, p. 389) ceaselessly alternate. Also in terms of content, the main subject often gives way to extended excursuses touching on an encyclopedic range of topics from disciplines such as pharmacology, optics, cosmology, or eschatology.

Prospero Alpino, *De balsamo dialogus*

The previous subsection should have given a first impression of the versatility and inventiveness of the authors who chose the dialogue as an outlet for their scientific ideas. With this as a background, I would like to conclude with a more in-depth discussion of one particular example in order to highlight one specific aspect of the genre which presumably contributed much to its success in early modern science: its usefulness in creating authority.

When in 1590, the Venetian doctor and botanist Prospero Alpino returned to his home town after several years spent in Egypt as doctor to the Venetian consul in Cairo (*c.*1580–4) and a stay in Genoa,[27] one of the first things he did was to publish a dialogue of some fifty octavo pages titled *De balsamo dialogus* (1591). At the time, the resin, fruit, and wood of the balm tree (Commiphora Gileadensis) were important commodities, and it was highly valued for its medical properties

[26] The abandonment of the project may be due to the appearance of Jean Pecquet's *Experimenta nova anatomica* (1651), whose results could either have pre-empted or refuted Guarinonius's own theory (which is carefully concealed in the published part).

[27] On Alpino's biography, see Ongaro 1961–3.

in principle, but many experts believed that the tree had become extinct and all products imported to Italy from the East under its name were fake. The matter was topical, and Alpino's biography meant that he would be listened to and could make himself known with a well-crafted contribution.

Introduced by a dedication to three Venetian supervisors of the University of Padua in which the topicality of the subject is explained and by a note to the reader and framed by two simple but lifelike and attractive woodcuts of a balm twig, Alpino's dramatic dialogue is set in the public garden el-Mattharia near Cairo. The three discussants are the Egyptian doctors Abdella and Abdachim and Alpino himself. The conversation is structured in six chapters. Chapter 1 describes the meeting of the interlocutors and their decision to converse about balm and the tree from which it comes. In chapter 2, it is agreed that, contrary to a frequently voiced opinion, the tree is not extinct, and its oriental habitats are identified. Chapter 3 is devoted to its fruit ('carpobalsamum') and sap ('opobalsamum'). Next, the distinctive characteristics of balm sap are described and the information to be found in ancient sources is reviewed (chapter 4). In chapter 5, it is discussed how true balm can be distinguished from fakes. In chapter 6, the same is done for the fruit and the wood ('xilobalsamum'). In an appendix (chapters 7–8), one finds, in Latin translation, a chapter (2.39) from Pierre Belon's *Observations de plusieurs singularitez & choses memorables* (1553) that accords with the views voiced in the dialogue.

Alpino is remarkably dexterous and inventive in the use of his chosen genre to invest himself with authority. He needs to, because his dialogue is no didactic digest of accepted knowledge. One could call it a monographic dialogue: focusing on one clearly defined subject, the work claims to offer fresh insights which challenge the prevailing opinion. As the example of Guarinonius has shown, Alpino is not alone in using a dialogue to this end. However, Alpino's self-presentation is much subtler than that of Guarinonius, who bluntly enters the stage as a renowned authority among admiring friends and disciples. At the same time, it is bolder, as will become evident shortly.

Alpino chooses the perfect setting to highlight his superior knowledge to that of his readers: since Homer's *Odyssey* (4.227–32), Egypt was renowned as a land of potent drugs, where every man was an excellent doctor. Moreover, everyone agreed that the balm tree came from Egypt or thereabouts. Most readers would never have been there, but Alpino had, as he makes clear in the first words of his preface. While the choice of the scene was to be expected, the constellation of participants is surprising. Alpino inverts the age-old convention by which the author of a dialogue introduces himself, or an authorial figure, as a kind of teacher. Instead, he assigns himself the role of the pupil.[28] While he clearly is

[28] A possible model for this reversal is Boethius's *Consolatio*, where the author is lectured by Lady Philosophy.

better informed than his readers, the real experts from whom he draws his knowledge are the Egyptian doctors.

What makes this such a bold move is the fact that neither of them is a Christian: Abdachim is a Jew, Abdella, the chief balm expert among the three, a Muslim.[29] In line with this, the reader learns in the course of the dialogue that the tree is actually not native to Egypt but to the Arabian peninsula, and that most imports reach Egypt directly from Mecca, a point emphasized over and over again (fols. 2^{r}, 3^{r}, 4^{v}, 6^{r}, 7^{v}, 8^{r}, 18^{v}–19^{r}, 21^{v}, 22^{v}). Despite their different religious affiliations, the three men discuss the topic in an atmosphere of mutual respect and friendship. Religion is a non-issue among them after its only mention in the very first sentence, where Abdella calls Abdachim 'an extremely decent man, although a Jew, and a good friend of mine' ('vir honestissimis moribus, quamvis Hebraeus, praeditus mihique multum familiaris'). The implication that religious differences should not interfere with the common search for botanical and medical truth is further bolstered by the striking similarity between the personnel of *De balsamo* and the foundation legend of the School of Salerno, which was said to result from an encounter of four doctors: one Catholic Westerner, one Orthodox Greek, one Jew, and one Muslim called Abdela.[30] The irrelevance of religious matters and the importance of access to good information are further underlined by a reversal of the traditional relationship between Muslim Easterners and the ancients. While the Arabs are usually depicted as mere transmitters, and often adulterators, of genuine Greek lore, it is exactly the other way round in the present case. Dioscorides and company are generally treated politely, but when it comes to Eastern plants, it is made clear that they have never seen them with their own eyes: they only mediate foreign knowledge and sometimes adulterate it due to their lack of personal observation. The authority firmly rests with the Easterners (fols. 3^{v}, 6^{v}–7^{r}, 18^{v}).

Given the epistemological vantage point at which he has placed himself, it is striking, and comes across as surprisingly honest, that Alpino pointedly does not claim to have seen the balm tree with his own eyes. As mentioned, the tree is not native to Egypt, and the specimens planted in el-Mattharia have withered 'through the negligence of the gardeners', ('custodum negligentia' , fol. 1^{v}). The only thing Alpino manages to see is one single twig Abdella orders a servant to fetch from his home (fols. 9^{v}, 11^{v}). For the rest, he has to trust reports from Mecca. In doing so, he places himself on the same level as, and provides a role model for, his readers, who have to take his information on trust in turn. Reasonable trust is

[29] Abdachim is called 'Hebraeus', Abdella 'Aegyptus' at the start of the dialogue. That Abdella is a Muslim is implied by the Muslim (Ottoman) rule in Egypt, the contrast to Abdachim's designation, Abdella's intimate knowledge of Mecca, and the homonymy with the Muslim in the foundation legend of the School of Salerno, to be mentioned presently.

[30] The oldest source of this story, the so-called *Chronica Elini*, is quoted in Henschel and De Renzi 1852–9, vol. 1, 106–9 n. 3. See also Kristeller 1945, 143.

the glue that keeps the community of doctors and naturalists together, Alpino seems to imply. In another respect as well, the author is careful to keep close contact with his readers despite the geographical, religious, and cultural distance between Egypt and Italy. In fact, they, particularly the sceptics who believe the balm tree extinct and the balm products available in the West fake, are present as 'remote participants' in the dialogue. Between Abdella, Abdachim, and Alpino, there is no dissent or friction, because Alpino defers to Abdella's judgement, while Abdachim mostly limits himself to moderating the conversation. Instead, the dynamics of the dialogue originate in Alpino's role as a messenger between East and West. Time and again (fols. 3^{r-v}, 3^{v}–5^{v}, 7^{r}, 10^{v}, 19^{v}–20^{r}), he reports the wrong opinions concerning the balm which are current in Europe to Abdella, who refutes them one by one.

Finally, not the least source of Alpino's authority is his ability to convey exotic knowledge in a classical language and with a literary finesse which his educated readers must have experienced as familiar and appealing. His Latin is fluent and urbane, and his descriptions have a graphic and haptic quality that conveys the impression of seeing, smelling, tasting, touching the precious stuff for oneself. Balm adulterated with resin oil, for instance, can be distinguished as follows from pure balm according to Abdella (fol. 16^{r}):

> digitis contrectatum nullo lentore vel viscositate adhaereat, quando verus balsami succus lentore ac viscositate quapiam non carere a nobis alias dictum sit a multisque antiquis scriptoribus confirmatum, oleo siquidem resinarum omnium tenacitatem abigi omnibus manifestum est. In aqua praeterea instillatum supra eius superficiem orbiculorum formis expandi nemo nostrum ignorat, quod sincerum opobalsamum non facere posterius docebitur. Demum in panno oleo infecto macula relinquitur, quae et difficile aufertur et cottidie etiam magis maiorem in circulum expanditur, quod non fit in sincero opobalsamo.

> rubbed between the fingers, it does not stick to them with sticky or gluey texture, while true balm sap does not lack a certain sticky or gluey texture, as I said elsewhere and as many ancient authors confirm, because resin oil takes away all stickiness, as is evident to everybody. Moreover, all people here know that when trickled into water it distends itself on the surface, forming little circles, something not done by pure balm sap, as I will explain later. Finally, in a piece of cloth on which the oil has been dribbled, a stain remains which is difficult to wash out and distends itself into a circle growing from day to day, which does not happen with pure balm sap.

An example of the dialogue's literary sophistication may be taken from its first lines. The work does not open as a dialogue but as a soliloquy by Abdella, who sees Abdachim approaching from a distance. This technique seems not to be taken from the ancient dialogue tradition, where it is uncommon. Rather, it has been adapted from classical drama, which often opens with one character alone on stage and revealing his plans to the audience in a monologue. Alpino elegantly uses it to partly solve a problem which arises in every dramatic dialogue, namely the quick and efficient provision of the circumstances in the absence of an authorial voice. In the present case, Abdella needs fewer than thirty words to inform the readership about the public garden where the dialogue is set ('hoc viridarium el-Mattharia'), the identity of one interlocutor ('Abdachim medicus'), and their close relationship ('mihi... multum familiaris').[31]

Picking up an exotic, medically interesting, and economically important subject and discussing it in sophisticated form, Alpino's dialogue became quite a success. Being Alpino's first publication, it paved the way for a whole publication programme focusing on things Egyptian and comprising *De medicina Aegyptiorum* (also a dialogue, 1591), *De plantis Aegypti* (1592), and *De praesagienda vita et morte aegrotantium* (1601). That the dialogue was indeed intended as an appetizer is made clear in Alpino's note to the reader, where he explains that he has written several books (*nonnullos libros*) on Egyptian botany and medicine and wanted to publish all of them at the same time, but was convinced by friends to start with *De balsamo*. At the end of this note, he promises that *De medicina Aegyptiorum* will appear soon, 'if only I note that this booklet of mine has somehow pleased those interested in medicine' ('si modo hunc meum libellum medicinae studiosis aliquo pacto arrisisse intellexerim').

De balsamo itself was appended a number of times to *De medicina Aegyptiorum* and *De plantis Aegypti*. In one such edition (1718), it is itself followed by another work on tropical medicine, Jacobus Bontius's *De medicina Indorum* (1st edn 1642). In 1750, Blasio Ugolino inserted *De balsamo* into volume 11 of his *Thesaurus antiquitatum sacrarum*. Moreover, several further self-contained editions (for instance, 1639) are on record, in which the dialogue is presented, in spite of its brevity, as a canonical work of reference by the addition of an index. While most editions appeared in Venice and Padua, several from the first half of the eighteenth century came out in Leiden: by that time, *De balsamo* had become an international medical classic. Already much earlier, its audience had been widened to non-Latinate readers by two translations, one into Italian (*Il balsamo*, together with *Le piante dell'Egitto*, 1592), the other into French (*L'histoire du baulme*, 1619). Finally, Alpino's text apparently encouraged several more balm monographs over the seventeenth century and in part determined their content. Nicolas Guibert's booklet *De balsamo* (1603), for one, in parts reads like a digest of Alpino's account and initially alerts the reader to its source (pp. 3–4).

[31] Another example of the same technique from a Neo-Latin scientific dialogue is provided by the opening (pp. 13–14) of Moffett's aforementioned *De iure et praestantia chymicorum medicamentorum*, where Philerastus reveals his evil plans in a monologue.

Dissertations

Disputation and dissertation

To begin with some terminological clarification: in many early modern book titles, 'dissertatio' means 'essay' or 'speech'. The texts are occasionally confused with those discussed in this subsection—works serving as the basis of, or at least printed in connection with, an academic 'disputatio'.[32] Besides 'dissertatio', there were alternative titles for such works, in particular 'disputatio' itself (which could also designate any other argumentative text), while the academic ritual of the 'disputatio' could be called a 'dissertatio' in turn. In short, both 'disputatio' and 'dissertatio' were used for the oral act and the written product related to it as well as for other kinds of texts. In what follows, I consistently use 'disputation' for the academic event, 'dissertation' for the corresponding literary genre.[33]

The dissertation has no ancient roots, because the disputation was a medieval and early modern practice.[34] On certain occasions, teachers and students at institutions of higher education—'studia' of religious orders, cathedral schools, gymnasia, and other higher schools, but above all universities—participated in formal discussions, defending certain positions as 'defendentes' or 'respondentes', attacking them as 'opponentes', or overseeing the whole procedure as 'praesides'. (Under certain conditions, disputations without a 'praeses' could also take place, as will become clear later.) The participants had to fulfil their respective tasks using syllogistic reasoning and speaking correct Latin. Different occasions called for different types of disputations. The most fundamental difference was between those held in class as part of ordinary teaching and those held in public, to which larger, academic and occasionally non-academic audiences were admitted. Among the latter, some were organized, like those in class, 'for the sake of exercise' ('exercitii causa'); one usually had to participate in them several times in the course of one's studies. Another form was the disputation 'pro gradu', which was mandatory for receiving a university degree.[35] In addition, there were disputations 'pro loco', by which applicants for teaching positions demonstrated their eligibility, 'disputationes valedictoriae', when students left an institution, disputations forming

[32] Even in an academic context, a 'dissertatio' was not necessarily a work of this kind. Johann Karl Anton Gerstner's *Dissertatio de salubritate aquae* (1756) originated as an oration adding some sparkle to a doctoral graduation ceremony.

[33] Dissertations in this sense should not be confused with academic 'programmata' (*HWRh* 7, cols. 154–9, s.v. 'Programm', esp. col. 156; Lesigang-Bruckmüller 2015, 361–6). Unlike dissertations, 'programmata' served to announce not only disputations but many other events as well, did not provide the subjects to be discussed in the disputations, and seem to have been restricted geographically to Germany and Sweden and chronologically to the period *c*.1580–1750.

[34] The following paragraph distils information from Horn 1893; Lawn 1993; *HWRh* 2, cols. 866–80, s.v. 'Disputation'; Grendler 2001, 152–7; Gindhart and Kundert 2010; Weijers 2013; Gindhart et al. 2016, 7–25; Füssel 2016; Chang 2021, 117–20.

[35] An overview of the academic system of exams and grades is provided by Schwinges 2007.

part of a professor's duties, and other varieties, including many regional and local ones. Disputations fulfilled various functions and served the interests of those involved in different ways: they trained students in learned argumentation and the oral use of academic Latin, constituted a common examination format in the widespread absence of written exams, displayed an institution's achievements, gave teachers a forum for presenting their skills and promoting their opinions, and were expected to help finding out the truth, or at least to establish a consensus, about difficult questions. Last but not least, disputations provided opportunities for publishing.

Medieval disputations themselves were rarely recorded in writing, but they were announced on handwritten sheets.[36] When print began to spread, these announcements came to be printed, together with the positions ('theses', 'assertiones', or the like) to be disputed, on single broadsheets. The theses could be a series of isolated statements without a common thread but showed a tendency to coalesce into a coherent text early on: they could grow into short chapters by the addition of sources and explanations and be arranged so as to form a unified argument. Along with this, broadsheets were supplemented and partly replaced by short books,[37] which appeared in steadily increasing numbers in the course of the sixteenth, seventeenth, and eighteenth centuries.[38] As these booklets were published as much for the individual reader as for the participants in the disputation, they had to be self-contained and their perusal worthwhile outside the context of the disputation. For this reason, the oral argumentation, which had found no place in the broadsheets, was increasingly integrated into the printed page,[39] which testifyies to a tension created by the genre's two different purposes and audiences which manifests itself in other ways as well. Originally, most booklets were authored or co-authored by the 'praeses'. From the eighteenth century, by contrast, more and more dissertations, in particular those 'pro gradu', were written and defended by the candidate. This gradually became the rule, starting with leading universities like Leiden, where the 'praeses' disappeared from the title page around 1700, and Göttingen, where Albrecht von Haller (professor 1736–53) granted good students access to the research infrastructure they needed to write their own dissertations.[40] From *c.*1800, the self-written and self-defended dissertation became the mandatory precondition of a doctoral degree which it still is

[36] Important work on the dissertation, its relationship to the disputation, and its history includes Allweis 1979, 17–21; *HWRh* 2, cols. 880–4, s.v. 'Dissertation'; Marti 2001; Chang 2004; Marti 2007; Chang 2010; Marti et al. 2017, 10–27; Friedenthal et al. 2021, 1–33; Chang 2021; Lindberg 2022 (regarding Sweden).

[37] Weijers 2013, 201, 231; Chang 2021, 125. For Dillingen, this process can be traced through the bibliography in Leinsle 2006, 569–616, where broadsheets become a minority from the last decades of the sixteenth century. However, the fact that broadsheets were more liable to loss (Schülling 1976, pp. iii–iv; Brockliss 2021, 194) and are less often digitized nowadays should be taken into account.

[38] Marti et al. 2017 present a selection of fifty German items from the period 1500–1800.

[39] Chang 2004, 154.

[40] Harskamp 1997; Chang 2010, esp. 351–4.

today in many disciplines. The disputation, which had long since been criticized as a sterile academic show, was restricted to the doctoral examination, while its other varieties, and the dissertations depending on them, died out.[41]

Alongside its evolution over time, the disputation-dissertation complex was markedly diverse internally at any given moment. In particular, the relationship between the oral act and the printed text varied widely and remains unclear in many cases. Only a minority of all disputations made their appearance in print, as the most frequent ones, those in class, were only rarely printed.[42] Of those disputations that were printed as dissertations, some broadsheets presented straightforwardly the questions or theses to be disputed (see the next subsection for an example), but other dissertations often had little to do with the disputation. Especially towards the end of the early modern era, 'praesides' often felt free to publish any work they wanted on such an occasion, which resulted in dissertations belonging to widely varying genres and quite ill-suited to being disputed. A Linnaean dissertation titled *Flora Alpina* (1756) and defended by Nils Nilsson Amann is, in fact, a 'historia plantarum', the second volume of Benedikt Stattler's *Mineralogiae principia physica* (1765), published in connection with a disputation 'ex universa physica', a textbook. Even in less glaring cases, a list of theses or questions with tenuous or no links to the preceding work was often appended under titles such as 'corollaria' or 'parerga' in order to supplement or replace the main text as the basis of the disputation.[43] In other cases, dissertations which were thematically closer to the corresponding disputations constituted the latter's results instead of their bases, as they were published not, as usual, before the academic ritual, but after. Their title pages do not announce that the candidate will defend ('defendet', 'propugnabit', or the like) the work on a certain day but report that he already has done so ('exponebat', 'disputavit').[44] Sometimes, this may have happened simply because the authors or the publisher were behind schedule, but sometimes it was done on purpose, in which case the text could be heavily reworked for publication. As announced on their title page, parts of Thomas Ittig's *Lucubrationes academicae de montium incendiis* (1671) had been 'presented several times before to be disputed' ('antehac ad disputandum diversis vicibus propositae'), but

[41] There were exceptions: in Helsinki, Nathanael Gerhard af Schultén published mathematical theses to be disputed by students with fellowships ('in exercitationem stipendiariorum') until the 1840s.

[42] Grendler 2001, 155–6, 342; Weijers 2013, 200.

[43] Lindberg 2021. For an explicit statement on the function of appended theses, see the letter of dedication introducing Anton Manincor, Anton Waid, and Peter Prantner's *De terraemotibus anni 1670* (1671): 'Suscepimus autem terraemotum defendendum…Adiunximus conclusiones quasdam de anima pro ampliori disputationis materia.'

[44] See, for instance, Christoph Scheiner and Johann Georg Schönperger, *Exegeses fundamentorum gnomonicorum* (1615); Johann Friedrich Blumenbach, *De generis humani varietate nativa* (1775). For another similar case and one in which publication after the disputation was promised but never happened, see Marti 2007, 271.

subsequently, they were 'reworked by the author, greatly expanded, and published as a whole' ('ab auctore recognitae, magnam partem auctae et coniunctim editae').

No hard and fast rules can be given regarding the authorship of dissertations either, despite considerable efforts made by modern scholars, particularly librarians, to determine it.[45] Before the early modern genre metamorphosed into the modern doctoral dissertation, the 'praeses' and the defendant (or defendants, as there could be several of them) were mostly named together on the title page, but in their capacity as participants in the disputation, not with regard to their contribution to the printed text. Some kind of cooperation between both parties seems to have been customary. The more experienced 'praeses' will often have acted as the main author, while the respondent had to pay for the printing. If the latter assumed a major role in the composition of the text, this could be indicated by the addition of 'auctor' after his name. However, such indications should not be trusted blindly, as some 'praesides' did the work themselves but gave credit for authorship to the respondents, a practice satirized by some contemporaries but justified by others.[46] There always remains room for further surprises. The *Κρυσταλλολογία* (1698), sometimes hailed as the birth certificate of crystallography in Switzerland, is commonly ascribed to the defendant Johann Heinrich Hottinger, because he figures as 'author' on the title page and despite the fact that he was only 17 years old at the time. However, in his self-bibliography titled 'Acarnania' (1718), Johann Jakob Scheuchzer declares that neither Johann Heinrich nor his uncle Salomon Hottinger, the 'praeses', but he himself was the main author, albeit assisted by the defendant ('quamvis haud diffitear aliqua etiam defendentem meo labori addidisse', p. 119). In some places, especially in the Netherlands, ready-made dissertations could be bought from 'dissertation factories'.[47]

Amidst all this uncertainty, one thing stayed constant throughout the early modern period: as an academic genre par excellence, 'dissertationes' (and the 'disputationes') remained almost exclusively in Latin.[48] Early modern vernacular specimens were exceptional.[49] The vernaculars were deemed admissable, if at all, only in some paratexts—particularly dedications and congratulatory poems[50]—and in quotations from thematically pertinent literature. In many countries, this

[45] Horn 1893, 51–72; Allweis 1979, 21–3; Marti 2001, 11–13; Marti 2007, esp. 260–3; Chang 2021, 130–4; Lindberg 2022, 91–104 (regarding Sweden) . If nothing specific can be said about a dissertation in this respect, I indicate its 'praeses' as well as its respondent in what follows.

[46] Füssel 2016, 42; Siegmund Jacob Apin, *Unvorgreiffliche Gedancken, wie man Dissertationes academicas sammlen soll* (1719), pp. 41–2.

[47] Chang 2021, 132.

[48] Marti 1998.

[49] Fewer than fifty are on record from the entire German-speaking world according to Marti 2018, 271. For some additional German and Swedish examples, see Weijers 2013, 230 (250 for a disputation partly held in English), and Hörstedt 2018, 21 (eleven gymnasial dissertations in Swedish from the 1770s). At Swedish universities, the vernacular was used to some degree from the 1750s in strongly practice-oriented fields such as economics (Lindberg 2022, 113–28).

[50] Marti 1998, 57–8.

state of things did not change any time soon even after the dissertation's transformation around 1800. In the German-speaking world, for instance, a general switch from Latin to German took place only in the second half of the nineteenth century. At the Friedrich-Wilhelms-Universität zu Berlin, to give but one example, German dissertations became gradually admissible alongside Latin ones between 1867 and 1883.[51]

Broadsheets

In what follows, I will focus on the developments up to *c.*1800. Although much of interest happened after that date and many worthwhile Latin specimens remain to be rediscovered,[52] in most places, the dissertation of the nineteenth century was by and large the piece of hopefully original research needed for the doctoral degree that is familiar to us today.[53] The two classes into which early modern dissertations fall with regard to their publication format, broadsheets and booklets, show many similarities. However, I will discuss them separately, starting with some remarks on, and an example of, the former before dealing more extensively with the latter, for two reasons: first, less can be said in many respects—for instance, numbers, diffusion, subjects, subforms—about broadsheets than printed books, because their loss rates are higher and their digitization rates lower and because booklets have profited far more from the recent surge of interest in disputations and dissertations as a whole. Secondly, as will become clear from the example, the simple makeup of many broadsheets makes it comparatively easy to recognize the original relation between the oral act and the printed product, thereby facilitating an understanding of how the dissertation evolved beyond this starting point in the course of its later history.

Printed broadsheets appeared alongside handwritten ones a few decades after the invention of print and persisted to varying degrees in different regions in the course of the following centuries.[54] In France, they were still standard in the seventeenth century.[55] In the Catholic world as a whole, they could become pompous media of representation from around 1600. Sometimes measuring more than one square metre, such placards usually relegated the announcement of the disputation, the theses to be disputed, and other textual elements to the bottom of the page. The greater part was taken up by a sumptuous copper

[51] Marti 2018, 279–81.

[52] For two examples by famous authors, see Ignaz Philipp Semmelweis, *De vita plantarum* (1844), and Hermann von Helmholtz's dissertation on the nervous system of invertebrates (1842), recently edited by Heideklang et al. 2021.

[53] In Scandinavia, by contrast, many dissertations were still authored by the 'praeses' in the first decades of the nineteenth century; see Krook 2022, esp. 30–4, on those defended under the presidency of Carl Adoph Agardh from the 1800s to 1830s, in particular on his *Dissertatio de metamorphosi algarum* (1820).

[54] Leinsle 2006, 39–41. [55] Brockliss 2021, 197.

engraving elevating the event and praising noble patrons. These clearly were prestige objects meant to enhance the fame of the participants and the institution.[56]

However, if only for cost concerns, most broadsheets remained simple and functional. A good example of this ordinary format is a *Dubium astrologicum* disputed at the University of Bologna on 30 March 1509 by a certain Jakob Ebelin in order to earn a paid student lectureship (Figure 6).[57] Its text consists of two parts, an announcement of the disputation itself and a presentation of its subject. The representative elements are all confined to the first part. In a few lines of running text, the university, the rector of the arts and medical faculties, and the presiding professor are named with their full titles, conveying an idea of the firm embedding of the event in the institutional structures of the university, its greatness, and the dignity of its representatives. The second part, whose sober, content-oriented character is shown typographically by its subdivision into single laconic statements, focuses on the factual question at hand: do celestial bodies shine with their own light?

> Having first received permission from the grandiose, noble lord Girolamo Tigrino de Bagnacavallo, the exceedingly honorable rector of the artists and physicians of the nurturing University of Bologna, Jakob Ebelin from Germany, master of arts, the least among the students of medicine, will try to uphold the mentioned doubtful statement below with the pertinent conclusions and corollaries under the presidency of the most distinguished doctor and master of arts and medicine, Galeotto de Bicadelli, on Friday, the second-last day of March.
>
> Astronomical Doubt
>
> Are the heavenly bodies luminous by themselves or do they receive their light from another body?
>
> First conclusion: Although the sun is situated in the middle of the planets,[58] it does not provide them with all their brightness.
>
> Second conclusion: If a celestial body is in opposition to another celestial body, this does not hinder its light.[59]
>
> Third conclusion: We are of the firm conviction that there is greater luminosity in the sun than in the other planets.
>
> Fourth conclusion: No given part of the sky is as dense or rarified as the rest.
>
> Corollary: No given part of the sky is luminous and translucent to the same degree as the rest.

[56] Appuhn-Radtke 2021; Chang 2021, 122–5.

[57] Grendler 2001, 154–5.

[58] In the context of the Ptolemaic system, this means that the sun is the fourth of the seven planets from the earth.

[59] I suppose that this somewhat cryptic statement is intended to back up the first conclusion and should be spelled out more or less as follows: 'The moon is provided with all its brightness by the sun. It is, therefore, eclipsed when it stands in opposition to the sun (because it enters the earth's shadow). But all the other celestial bodies suffer no eclipse when they stand in opposition to the sun. So they must have some light of their own.' Of course, the argument only works if one wrongly supposes the earth's shadow to be cylindrical instead of conical.

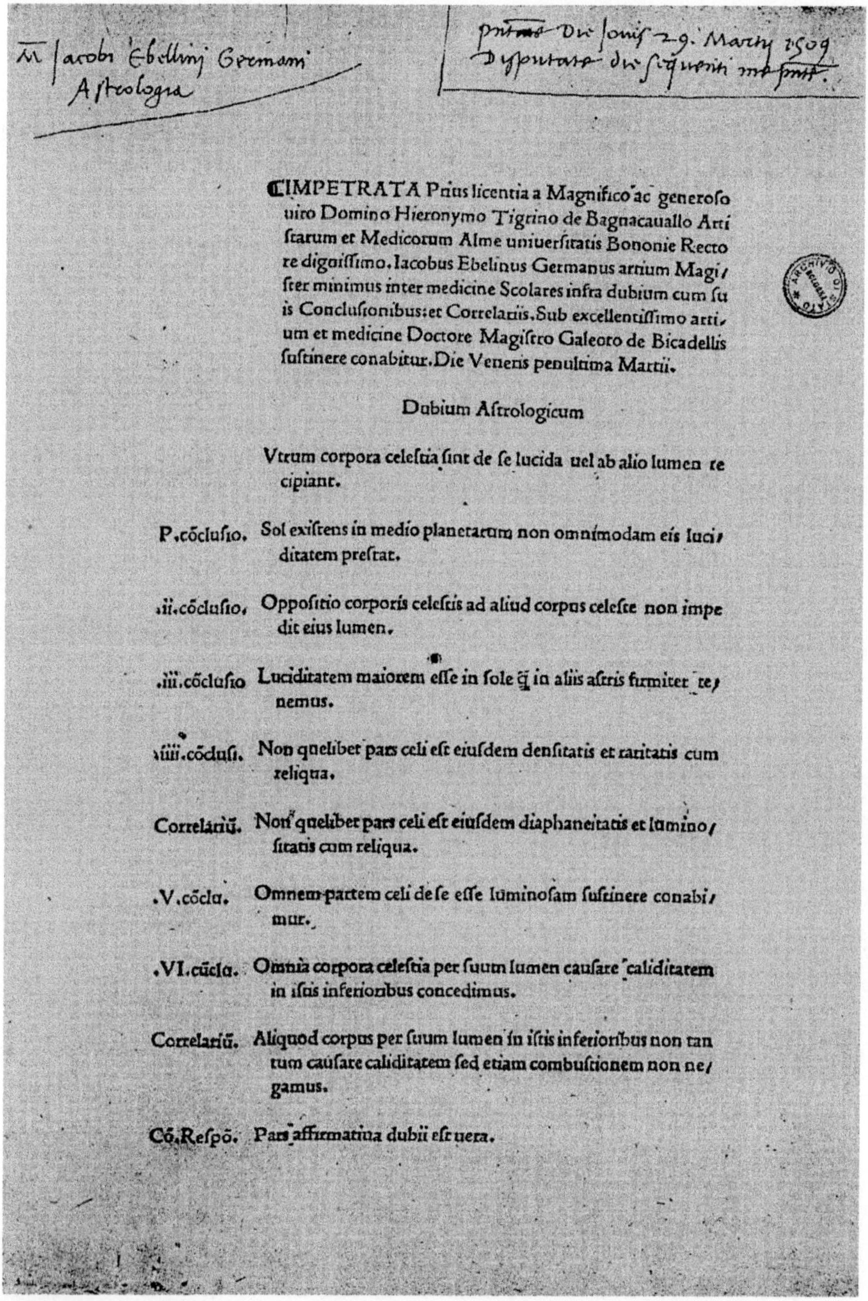

IMPETRATA Prius licentia a Magnifico ac generoso uiro Domino Hieronymo Tigrino de Bagnacauallo Artistarum et Medicorum Alme uniuersitatis Bononie Rectore dignissimo. Iacobus Ebelinus Germanus artium Magister minimus inter medicine Scolares infra dubium cum suis Conclusionibus: et Correlariis. Sub excellentissimo artium et medicine Doctore Magistro Galeoto de Bicadellis sustinere conabitur. Die Veneris penultima Martii.

Dubium Astrologicum

Vtrum corpora celestia sint de se lucida uel ab alio lumen recipiant.

P. cõclusio. Sol existens in medio planetarum non omnimodam eis luciditatem prestat.

.ii. cõclusio. Oppositio corporis celestis ad aliud corpus celeste non impedit eius lumen.

.iii. cõclusio Luciditatem maiorem esse in sole q̃ in aliis astris firmiter tenemus.

.iiii. cõclusi. Non quelibet pars celi est eiusdem densitatis et raritatis cum reliqua.

Correlariũ. Non quelibet pars celi est eiusdem diaphaneitatis et luminositatis cum reliqua.

.V. cõclu. Omnem partem celi de se esse luminosam sustinere conabimur.

.VI. cũclu. Omnia corpora celestia per suum lumen causare caliditatem in istis inferioribus concedimus.

Correlariũ. Aliquod corpus per suum lumen in istis inferioribus non tantum causare caliditatem sed etiam combustionem non negamus.

Cõ. Respõ. Pars affirmatiua dubii est uera.

Figure 6 A broadsheet announcing a disputation at Bologna: the *Dubium astrologicum*.

> Fifth conclusion: We will try to uphold that each part of the sky is luminous by itself.
>
> Sixth conclusion: We admit that all celestial bodies cause heat in the things here on earth by their light.
>
> Corollary: We do not deny that a body may cause not only heat, but even burning in the things here on earth by its light.
>
> Final response: The affirmative alternative regarding the doubt is true.

This document conveys a quite clear idea of how the disputation itself proceeded. The initial question called for a 'yes' according to the convictions of pre-Galilean astronomy, and Ebelin tried to uphold this position. On his way to that end, he had to defend a series of six assertions ('conclusiones'), two of them accompanied by supplements ('correlaria'), which collectively made the final response almost inevitable. He did so with different degrees of emphasis, presumably depending on how vital an assertion was for the position defended by him. The last 'correlarium', for instance, was an extra which had no consequences at all in this respect, so he could remain fairly indifferent regarding its truth. His opponent(s) would throw stones in his way all along by confronting him with arguments against each 'conclusio' and 'correlarium'.

The *Dubium astrologicum* is a textbook example. In other broadsheets, the relation between the oral act and the printed text can be much less obvious, since the assertions to be defended often multiply and are less focused and coherent. As already intimated, the link between disputation and dissertation becomes looser and more ambiguous still in disputation screeds in book format, to which the rest of this section will be devoted.

Booklets

In the course of the sixteenth century, short books became the dominant publication format of the dissertation and the one which exerted a major influence beyond the occasion itself. In terms of numbers, dissertations were perhaps the most popular scholarly books in the early modern period. Although we are far from having a complete overview, some figures relating to different times and places indicate that several hundred thousand of them must have been printed. For instance, some 22,000 items were published at Swedish universities between 1602 and 1852.[60] A small German university like Giessen still produced about

[60] Friedenthal et al. 2021, 13. The dissertations of the Royal Academy of Turku are available at https://www.doria.fi/handle/10024/50699 (accessed 16 February 2023). The digitization of the Uppsala dissertations is under way.

1,100 in the course of the eighteenth century.[61] In Zurich, dissertations made up around 25 per cent and 10 per cent of all books published in the course of the seventeenth and the eighteenth century respectively, while nearly 80 per cent of all seventeenth-century books in Basel are said to have been dissertations.[62] Hanspeter Marti's select bibliography of philosophical dissertations from Germany 1660–1750 comprises 9,818 items.[63]

A substantial proportion of this material, namely many dissertations from arts faculties and all medical ones, was concerned with the natural world. For instance, Basel produced some 1,600 medical dissertations between 1579 and 1829, 1,709 items published in Leiden from 1593 to 1746 are found in a collection of the British Library, and no fewer than 11,000 are extant in Paris from the period 1610–1778.[64] At the Academy of Turku 284 dissertations with natural philosophical, mathematical, and medical content were disputed from 1640 to 1713.[65] Of Giessen dissertations, 244, nearly a quarter, deal with medicine. About 20 per cent of 789 Swedish gymnasium dissertations (1620–1820) discussed in a recent PhD dissertation are devoted to the natural world, as are most of the 84 dissertations on record for the Tyrol between 1669, when the University of Innsbruck was founded, and 1773.[66] From a small subfield of medicine which was not even recognized as such at the time, namely psychiatry, 1,100 items printed before 1750 have been recorded.[67] Finally, Johann Carl Heffter's *Museum disputatorium physico-medicum* (1756–64), in which the author catalogues his own collection of dissertations, contains about 18,500 titles, although it is geographically restricted to Germany and chronologically to the years from the mid-seventeenth to the mid-eighteenth centuries.

The prominence of the German-speaking world in the above enumeration is no coincidence, since this region certainly produced most dissertations during the early modern period. However, other areas contributed substantial numbers as well. In the first place, this is true of the Netherlands with its prestigious universities and of the Swedish Empire, including not only Sweden itself and Finland, where even numerous school disputations were printed, but also the Baltic countries.[68] Some East European territories, for instance Hungary and Romania, were prominent in this field as well.[69] Western and Southern Europe were less productive, partly because universities declined there and were overshadowed

[61] Schülling 1976, esp. pp. xii, xiv.

[62] Leu 2016, 253. However, the results of a search in the *Verzeichnis der im deutschen Sprachraum erschienenen Drucke des 17. Jahrhunderts* (http://www.vd17.de/, accessed 16 February 2023) suggest that the figure given for Basel should be related to Latin books only.

[63] Marti 1982. For a few more figures, see Chang 2021, 115.

[64] Husner 1942; Harskamp 1997; Brockliss 2021, 193.

[65] Kallinen 1995.

[66] Hörstedt 2018; *TL* 833–46, esp. 833–4.

[67] Diethelm 1971, 161–206.

[68] See in general Marti 2001, 4; 2007, 251; for dissertations from Tartu, see Friedenthal and Piirimäe 2015.

[69] Friedenthal et al. 2021, 15.

as scientific institutions by the rise of academies. In France, at least medical dissertations exist in considerable numbers. Less can be said about Italy and Spain, where cataloguing is less advanced. A conspicuous hole in the dissertation landscape was England, where printed testimony of disputations was only provided in the form of broadsheets, speeches, and so-called Act and Tripos verses.[70] The situation was different again in Scotland: from eighteenth- and early-nineteenth-century Edinburgh at least, printed dissertations are on record.[71]

As this overview suggests, dissertations in book format enjoyed greater vogue in the Protestant than in the Catholic areas of Europe. However, disputations formed a vital part of academic life at Catholic institutions as well and often found their way into print. In this context, special mention should be made of the Society of Jesus, the most important order in Catholic education by far. The Jesuits greatly cared about the oral presentation of learning in general and, in particular, about the disputation, which they regulated in their *Ratio studiorum*.[72] Accordingly, their gymnasia and universities turned out dissertations in considerable numbers. For example, 312 dissertations, most of which deal with natural philosophy, were produced in the arts faculty of Dillingen from 1555 to 1648.[73]

Among the various kinds of disputations which could result in a printed dissertation, public ones were more important than those held in class, as already mentioned. Most printed dissertation are attributable to disputations 'pro gradu' and 'exercitii causa'. The former seem to have been invariably printed at many places. Of the latter, not all were accorded this honour, but the total of such disputations was higher. Disputations 'pro loco', valedictory disputations, festive disputations, and other varieties played a minor role.[74]

The popularity of the dissertation was due to its multifunctionality, a trait it shared with the disputation from which it originated. It not only provided the basis for an important technique of academic teaching and examination and for a representational ritual (in fact, this was not always the case, since some dissertations had little to do with the respective disputations) but also served a wide variety of other interests. Professors could publish dissertations for free, because the respondent(s) had to carry the costs. As publishing became more important for one's profile within the 'respublica litteraria' and the emerging scientific community,

[70] On the four last-named countries in general, see Weijers 2013, 225–8. For France in particular, see Brockliss 2021, esp. 193–6 (and Germain 1886 for examples from Montpellier). For England, see Chang 2021, 120–1, 149; Alho 2021, esp. 151–3, 155, 157–9; Barton 2021, esp. 165.

[71] Davenport 1990, 178; Ford 2000, 242; Chang 2010, 351. Medical dissertations from Edinburgh are easy to find on the internet, as many of their titles start with the words 'Dissertatio (physico-) medica inauguralis'.

[72] Weijers 2013, 209–15.

[73] See Leinsle 2006, who also offers information on disputations and dissertations in the Catholic world in general (pp. 26–85). On the dominance of natural philosophy, see pp. 157 and table 2 on p. 565.

[74] However, 'pro loco' dissertations were regularly published by candidates for chairs at Montpellier (Germain 1886, 66–182).

many professors took full advantage of this opportunity, making dissertations (alongside equally costless journal articles) the backbone of their publication record. For instance, most of the about fifty publications by the Zurich naturalist Johannes Gessner (1709–90) were dissertations.[75] For most defendants, these booklets were probably just a necessaary evil on their way to an academic degree. However, a willing student could also put a dissertation to good use as a demonstration of his progress to his family and his (prospective) patrons or as an elegant way of informing these groups about the successful conclusion of his studies. If he was ambitious and his 'praeses' generous, he could even appear as 'auctor' on the title page and claim the dissertation as his first academic publication. Friends and colleagues who contributed congratulatory pieces thereby strengthened and flaunted their friendship with the defendant. From the point of view of an academic institution, interesting dissertations formed part of its performance record and could serve as an advertisement. For the reading public, such works, depending on their type, constituted reservoirs of received knowledge, provided information about current matters and trends, or presented new facts, insights, and theories. Their novelty value in particular was much appreciated. It could be said that normal books repeated old things but dissertations contained fresh material and that professors preferred to write short dissertations about their new ideas instead of long-winded tomes.[76]

In fact, dissertations were much esteemed by contemporaries, and their impact was considerable. They were exchanged as gifts by scholars, and their content was discussed in learned correspondence, as well as in printed books.[77] Although the first editions of dissertations were mostly not sold commercially but only distributed to those involved in the disputation by the respondent, some entered the book trade later, a fact which testifies to their lasting interest.[78] Johann Gabriel Drechsler and Polycarp Michael Rechtenbach's reflections on the language of animals (*De sermone brutorum*, 1673) still found readers in 1706, as is shown by a second edition from that year. Later editions could be brought together in large compilations by the original authors or by others.[79] Georg Friedrich Sigwart, a professor of medicine in Tübingen, published thirty French dissertations of the year 1742 which he had brought back from a trip to Paris (*Quaestiones medicae Parisinae*, 1759).[80] Perhaps the best-known example of this kind is Linnaeus's *Amoenitates academicae* (1749–90), comprising 186 items in ten volumes. Occasionally, thematically related series of dissertations were revised to form a

[75] Leu 2016, 233. See also Horn 1893, 52 on a pertinent testimony by Albrecht von Haller.
[76] Allweis 1979, 23–4; Gindhart et al. 2016, 8–9; Marti 2018, 286.
[77] Chang 2021, 137, 142–3.
[78] Horn 1893, 51; Chang 2021, 133–5 (specifying that the original print run was between 100 and 300 copies), 138–42.
[79] Kronick 1976, 219–32; Friedenthal et al. 2021, 24; Chang 2021, 136–7.
[80] Brockliss 2021, 195.

unified work, for instance an encyclopedic volume or a textbook,[81] or they were written with that goal in mind from the outset. Translations made dissertations accessible to a wider, non-Latinate audience. Two translations (in five editions) of parts of the *Amoenitates Academicae* appeared between 1759 and 1781.[82] Vernacular versions were published in learned journals too, where dissertations were also excerpted and reviewed for those without access to the originals.[83] Scholars unwilling to rely on such second-hand information, by contrast, systematically exchanged and collected items and occasionally published catalogues of their treasures, as did Heffter in his *Museum disputatorium*. The physician and botanist Ernst Gottfried Baldinger, who had gathered *c.*13,000 medical dissertations, at least announced those concerned with semiotics in his *Neues Magazin für Ärzte* in 1791.[84] Collecting dissertations was popular enough to warrant a specific guide,[85] and the market was so interesting that some booksellers specialized in dissertations and issued special sales catalogues.[86]

Content and approaches

In terms of content, the huge numbers of dissertations to appear in print had two general consequences. On the one hand, such subjects as formed part of the curriculum were discussed over and over again. How completely the fields of natural philosophy and medicine were covered can by illustrated by a decidedly minor topic such as snow. The phenomenon is mentioned, together with rain, dew, rime, and hail, in Aristotle's *Meteorology* (1.11) and the pseudo-Aristotelian *De mundo* (ch. 4), texts lectured upon in the arts faculty, but the remarks make up no more than half a page in all. Nonetheless, more than two dozen dissertations considered Aristotle's opinions on the phenomenon and elaborated on them.[87] Disputations and dissertations were instrumental in the circulation and argumentative validation of standard academic knowledge.

On the other hand, authors of dissertations went far beyond the curriculum. Occasionally, they responded to perceived needs of curricular reform in doing so. A group of fourteen surgical dissertations printed in Basel around 1600 coincides with extracurricular teaching in that field and with students' requests to be awarded a degree in surgery in addition to the customary doctorate in medicine.[88]

[81] See note 92 below and, for two medical examples, see S. Schlegelmilch 2021, 288.

[82] Chang 2021, 140. [83] Marti 1998, 61–2; Chang 2021, 143–9. [84] Kronick 1976, 220.

[85] Siegmund Jacob Apin, *Unvorgreiffliche Gedancken, wie man Dissertationes academicas sammlen soll* (1719).

[86] Schülling 1976, p. iv; Friedenthal et al. 2021, 19.

[87] A dozen items are registered in the thematic indexes to the two volumes of Heffter's *Museum disputatorium* s.v. 'Nix'. An analytical overview of seventeenth-century snow dissertations is provided by Kirchler 2023.

[88] U. Schlegelmilch 2021.

Often, however, teachers and students did not care for the curricular status of a topic at all but freely roamed all disciplines concerned with the natural world and related fields in search of untrodden paths. Petrus Ekerman and Andreas Piscator investigated Cicero's interests and expertise as a mathematician (*Dissertatio academica de Cicerone mathematico*, 1759),[89] and similarly esoteric subjects can be found in great numbers in Heffter's *Museum disputatorium*. On the pages of the thematic index to his first volume, one encounters dissertations on the question of whether a man can be used as a barometer ('Aerometrum an homo'), on the wild gourd ('Colocynthis', two items), nymphomania ('Furor uterinus, Nymphomania', nine items), hand washing ('Manuluvium'), and a frog originating in a human ('Rana, in homine'). One can perhaps surmise that in tackling subjects such as these, many authors woke up to the manifold implications of seemingly low-level questions and to the interest hidden in prima facie banal topics. Together with physico-theological thinking, which detected the traces of God's planning everywhere in nature, dissertations presumably played a role in the development of an attitude towards nature which recognized each and every aspect of it as worthy of investigation.

Most non-curricular subjects were topical in one sense or another. Some authors were content to draw their inspiration from the calendar, as did Johann Georg Gross and Johann Jakob von Brunn in their *Theoremata philosophica de stella magorum* (1611), which formed the basis of a disputation held on 7 January, the day after Epiphany. That the subject was of particular interest 'at this very time of the year' ('hoc ipso anni tempore') is highlighted in the second theorem. Many dissertations, by contrast, answered to extraordinary events or developments. Christian Wolff and Georg Remus tried to come to terms with the Great Frost of 1709 in their *Consideratio physico-mathematica hiemis proxime praeterlapsae* (1709). When blood-red ice was noted in Königsberg in late 1633, Daniel Beckher and Thomas Lindemann considered the phenomenon in a *Disquisitio physica de glacie cruenta* (1634). A major earthquake hitting the Inn valley in the Tyrol gave rise to two dissertations at the newly founded University of Innsbruck in the same and the following year, Johannes Christoph Sturm and Jacob Hieronymus Lochner's *De terrae-motibus* (1670) and Anton Manincor, Anton Waid, and Peter Prantner's *De terraemotibus anni 1670* (1671). The last two subjects were geographically related to the university where they were disputed as well. Occasionally, an author also discussed a topic from his homeland, like Conrad Schindler from Glarus, who in 1755 presented at Basel a *Specimen inaugurale medicum de caseo rasili viridi* on the Schabziger, a local cheese speciality.[90]

[89] Modern edition with introduction, Swedish translation, and commentary: Bohlin 2015.
[90] See Husner 1942, 153–4 for this and some further examples.

Beyond topical events, dissertations regularly picked up novel, fashionable ideas and tenets,[91] contributing to their dissemination and evolution in doing so. Authors were alert to this aspect and sometimes recommended their work by highlighting it. In his Göttingen doctoral dissertation *De generis humani varietate nativa* (1775), which discussed whether all humans belong to one and the same or to several different species, Johann Friedrich Blumenbach drew attention to the fact that the question was being much ventilated after Linnaeus admitted 'homo' to the tenth edition of his *Systema naturae* (1758–9) and described six subspecies: he called it a thing 'which the greatest men of this century deemed worthy of their attention and scrutiny' ('quam summi huius saeculi viri attentione et disquisitione sua dignati fuerunt', p. 1) and a 'question... often discussed in our times' ('quaestio... nostris temporibus saepe agitata', pp. 39–40). A long list of comparable subjects could be drawn up, but a few more examples will suffice. From 1659 to 1663, Johannes Antonides van der Linden, professor of medicine at the University of Leiden, had his students defend a series of twenty-seven dissertations on William Harvey's discovery of the circulation of blood (1628). Uniformly titled *Hippocrates de circuitu sanguinis, exercitatio* (*I.*, *II.*, etc.) and equipped with a continuous paragraph count, these booklets added up to a monograph propagating the idea that Harvey's discovery was actually a rediscovery of an insight to be found already in the Hippocratic treatise *On Dreams*.[92] As he explains in § 2 of the first dissertation, van der Linden hoped thereby to overcome the 'odium novitatis' which still clung to Harvey's theory,[93] a thought in line with the tremendous authority Hippocrates commanded in the seventeenth century. A whole complex of hotly discussed ideas was provided by Cartesian natural philosophy. Many of Descartes's controversial tenets were frequently ventilated, for instance his view of animals as soulless automats, which gave rise to works like Johann Eberhard Schweling and Johann Heinrich Crocius's *Disquisitio philosophica de anima brutorum* (1676).[94] From the late seventeenth century, the early history of the earth became an intensely debated matter.[95] When Buffon uncoupled this history from the account in Genesis, two dissertations disputed under the

[91] Chang 2021, 128–9.

[92] Harskamp 1997, 17 and nos. 256–9, 276–84, 301, 317–22, 340–2 of his catalogue (four instalments seem to be lost). In general, serialization of disputations and dissertations was not rare. In 1646, Martin Schoock had four dissertations on phenomena involving water and cold (*Disputatio physica de nive, de glacie, de hieme, de aqua*) disputed and published at Groningen. In 1695–6, another tetralogy focusing on snow alone (*De nive*) was disputed under the presidency of Johann Andreas Planer and printed at Wittenberg. As late as 1817–26, Carl Adolph Agardh had sixteen students defend his *Aphorismi botanici* in Lund (on the phenomenon in Sweden around 1800 in general, see Lindberg 2022, 209–14).

[93] On opposition to Harvey in the seventeenth century, see English 1990, 89.

[94] On Schmeling and Crocius's work, see Sdzuj 2016; see there, pp. 179–80, and Friedenthal and Piirimäe 2015, 73, for more dissertations on the same topic. On the spread of Cartesianism through dissertations more generally by reference to the example of Duisburg, see Trevisani 2011.

[95] Poole 2010. Some pertinent dissertations will be named shortly.

presidency of Johannes Gessner introduced his ideas to Zurich in 1752 and 1756.[96] After Christiaan Huygens's *Κοσμοθεωρός* (1698) kindled new interest in the question of extraterrestrial life, this was reflected in dissertations such as Daniel van Goens's *Disputatio philosophica, an luna habitabilis sit* (1738) and Gaspare Savoy and Joseph Alphons Widmann's *De planeticolis* (1770).[97] At the end of the eighteenth century, five dissertations initiated the debate in the German-speaking world about the new medical system invented by the Scottish physician John Brown and its pivotal notion of excitability.[98]

Sometimes, the discussion of novel ideas took the form of direct, mostly critical engagement with one specific work in what could be termed 'review dissertations'. A book subject to such treatment was Thomas Burnet's *Telluris theoria sacra* (1681, 2nd, definitive edition 1689), a history of the earth which tried to harmonize the geological evidence with the biblical narrative of the Creation and the Flood.[99] In the decades after its publication, the *Theoria* was discussed and criticized in German dissertations such as Christian Wagner's *Animadversiones in T. Burnetii Telluris theoriam sacram* (1683) and Jakob Wilhelm Feuerlein and Christoph Moritz Lochner's *Montes divinitatis testes* (1729). Vernacular books in the tradition of Burnet were not immune from similar attacks: in their *Dissertatio philosophica examinans placita quaedam scripti recentissimi* (1700), Zacharias Grape and Johannes Schnabel took issue with Detlev Clüver's *Natürliche Wissenschafft von Erschaffung und Bereitung der Erd-Kugel* (1700).

As subjects differed, so did approaches to them, ranging from doxographical and encyclopedic to strictly argumentative. Although dissertations provided the basis for disputations, they did not invariably share the latter's character of exercises in academic argument. Many texts focus on the coverage of authoritative opinions rather than on their critical evaluation. As a representative example of this strand, Scheuchzer and Hottinger's *Κρυσταλλολογία* may be mentioned. The occasional identification of Hottinger as the first Swiss crystallogist on the basis of it is questionable, not only because he is not the main author but also because the *Κρυσταλλολογία* is no work of scientific research. Of its eight chapters, the first two, dealing with the names of rock crystal and the class of substances it belongs to, and the last three, discussing the mineral's medical, artisanal, and magical uses and the places where it can be found, are frankly doxographical. Even in the core of the work (chapters 3–5), where seventeen phenomena related to rock crystal are enumerated and explained, this is done on the basis of previous literature. Scheuchzer and Hottinger's own contribution is mostly confined to

[96] Leu 2016, 240.

[97] On the latter, see Callegari 1907 (who ascribes it to the respondent only, which seems improbable); *TL* 837. For a sourcebook of the discussion in general, see Crowe 2008.

[98] Tering 2021.

[99] Nicolson 1959, 184–270; Poole 2010, 55–63.

juxtaposing the divergent opinions of their authorities and passing a 'modest judgement' ('modesta ἐπίκρισις', p. 10) on them.

At the other end of the spectrum, one finds works formulating one or more specific questions or hypotheses and developing a sustained argument to answer or prove them. In such dissertations, the tradition provides no guidance but serves as a quarry of facts or as an opponent to be defeated. Such is the case of Blumenbach's *De generis humani varietate nativa*, which is entirely devoted to the question of 'whether humans from every era and every people are or have been of one and the same or of completely different kinds' ('sintne fuerintve omnis aevi omnisque gentis homines unius eiusdemque diversaeve plane speciei', p. 39). The author tries to convince his readers that the first of these alternatives is true. Such varieties of 'homo' as one may define, he argues, only constitute a device to bring some order into the confusing diversity of humankind, a memory aid ('memoriae…subsidium', p. 42). While Blumenbach's argument lacks any polemical thrust, some dissertations consciously and provocatively turn against prevailing opinions. At a time when astrology had largely lost its credibility among intellectuals and had been ousted from most universities, Franz Anton Mesmer, the later inventor of animal magnetism, tried to found a new science of stellar influence on the modern basis of gravitation in his *Dissertatio physico-medica de planetarum influxu* (1766).[100]

In accordance with the subject chosen and the argumentative effort invested, the intellectual level of dissertations varied widely. Generally speaking, it rose over time, as contemporary science was integrated into curricula, doctoral dissertations came to play a more prominent role, and the notion that a candidate should be able to contribute to scientific progress caught on. However, the majority of texts remained fairly elementary even in the eighteenth century. The place, the institution, the occasion, and the individuals involved usually outweigh the importance of the publication date in this respect.

Many dissertations do not go beyond recapitulating parts of the syllabus. Despite its specious Greek title, the *Διάσκεψις φυσικὴ de nive* (1627) defended by Johann Philipp Reusser under the presidency of Nicolaus Agerius at the University of Strasbourg is of this kind. Taking its cue from Aristotle's aforementioned remarks on snow, the text discusses the subject according to an equally Aristotelian scheme, as will soon become clear, garnishing its remarks with a handful of biblical, classical, scholastic, and early modern citations. Struggling to fill five pages of an eight-page booklet, the authors do not shy away from devoting entire sections (sections 2, 12, 17) to the statements that snow exists, is white ('Nix itaque est alba'), and mostly falls in winter. Particularly in later times, more demanding curricular dissertations can be found too, as exemplified by

[100] Stuckrad 2007, 275–6. For unclear reasons, Stuckrad cites Mesmer's dissertation as *De influxu astrorum in corpore humano* (p. 285).

Johann Kies's *Dissertatio physica de iride* (1772). In its opening paragraph, the author, a professor of physics and mathematics at Tübingen, explains that of all subjects of his physics course, students struggle hardest with the explanation of the rainbow, which is why he presents a calculus-free account of the phenomenon. The result is still a solid piece of mathematical optics, whose running text gives way to formulae in some passages (for example, §§ 16–19). Already a decade earlier (1762), a French dissertation on Newtonian physics, which had recently found its way into the curriculum, even confronted listeners and readers with the bête noire of calculus itself.[101]

Conversely, venturing beyond the syllabus was no guarantee of scientific excellence. In terms of knowledge gain, Ekerman and Piscator's dissertation on Cicero the mathematician is a failure: despite their massive use of exaggerations and paralogisms, the authors are incapable of concealing the basic fact that hardly any traces of mathematical thinking can be found in the vast *oeuvre* of the Roman orator. However, one also finds texts which make valuable contributions to contemporary discussions and sometimes present genuinely novel insights. Among the first to purposefully use dissertations to present serious scientific argument was the Jesuit Christoph Scheiner. His *Disquisitiones mathematicae* (1614) probably provides the most thorough and thought-out overview and discussion of the major world systems of its day, making use of physical and mathematical arguments, experiments, and the new observations made possible by the telescope a few years before. That Scheiner favoured the Tychonian system and that his arguments were polemically opposed in Galilei's *Dialogo* (1632) do not detract from the work's quality.[102] Another prominent example is Johann Bernoulli's 'sine praeside' dissertation for his medical doctorate at Basel, *De motu musculorum* (1694). Building on the iatromechanical work of Giovanni Alfonso Borelli (*De motu animalium*, 1680–1) but going far beyond it, Bernoulli raises the understanding of muscular contraction and the forces at work in this process to a new level. In particular, he shows that a contracting muscle fibre tends towards assuming a spherical shape in principle, even if this shape cannot be fully realized for practical reasons (§§ 10–11), and that the forces within a contracting muscle are far greater than the ones it exerts on its surroundings (§§ 12–14). Bernoulli breaks new ground in terms of methodology as well: his insights are gained through a combination of elaboration on previous anatomical and physiological research, physical considerations, and very advanced mathematics in the form of Leibnizian calculus, first published only a decade before (§ 10).

101 Brockliss 2021, 197.

102 On the *Disquisitiones*—often erroneously ascribed to their respondent Johann Georg Locher or to Locher and Scheiner in tandem—see Graney 2017 (translation with introduction and notes); Bianchi 1999, 199–214 (Galilei on the work).

In their train of thought and style of reasoning, dissertations are not categorially different from other books, but they tend to foreground these aspects to a higher degree. Explicit schematization of the line of thinking is most common in elementary specimens which directly emanate from curricular teaching. Their methods of systematically covering a given field are largely those current in academic didactics. This is true, for instance, of the habit of starting with the name of the phenomenon under discussion—including its etymology, homonyms, and synonyms—followed by its definition, as in Michael Marggraff and Johannes Bleidner's *Disputatio physica de meteoro nivis* (1667, §§ 1–4). A scheme originating from Aristotelian thinking (*Posterior Analytics* 2.1) is provided by the four questions 'an sit' ('whether something exists'), 'quid sit' ('what it is'), 'qualis sit' ('how it is'), and 'cur sit' ('why it exists'). Agerius and Reusser's *Διάσκεψις φυσικὴ de nive* proceeds according to these, as explained in its first paragraph. Under the last point, the four Aristotelian causes are subsumed, which provide a popular schematism in their own right as well. They structure the greater part of Daniel Petermann and Gottfried Schmied's *Disputatio de glacie* (1670), for instance, providing the seventeen paragraphs of its second 'pensum'.

On a more abstract level, broadly speaking, dialectical structures involving the presentation of opposing positions and the mediation between them are common. Samuel Rodigast and Johann Nikolaus Trombsdorff's *Disquisitio physica de calore nivis* (1678) falls into three chapters: the first provides information about snow and warmth in general, the second expounds the position that snow contains some warmth, and the last subjects it to a critical examination and partial refutation. The individual chapters of Grape and Schnabel's *Dissertatio philosophica examinans placita quaedam scripti recentissimi* start by sketching a position adopted by Clüver, which is followed by its rebuttal and sometimes by the discussion of possible objections to this rebuttal.

On the level of the individual argument, the hallmark of the disputation was the syllogism: in principle, the pros and cons of the positions under discussion had to be pondered according to the precepts of formal Aristotelian logic. However, this norm was applied less strictly in the early modern period than it had been in the Middle Ages, which is why dissertations rarely, if ever, rely on it consistently and many dispense with it altogether.[103] Nonetheless, syllogistic reasoning is used on occasion, again particularly in specimens with close links to classroom practice. In the second chapter of Rodigast and Trombsdorff's *Disquisitio*, the arguments for the warmth of snow are presented in this form. In § 11 of the second chapter, a modus ponens syllogism, set apart from the rest of the text by somewhat larger type, runs as follows:

[103] Chang 2021, 120; Boer 2021, 504.

> (IV.) argumentum nititur experimento vel pueris non ignoto: Si manus nivibus immersae vehementer incalescunt ac ardore quodam molesto rubent, sequitur a nive calorem prodire. Verum prius, e(rgo) et posterius.
>
> The fourth argument is based on an experience known even to boys. If hands stuck into snow become immoderately warm and red along with a certain unpleasant burning, it follows that warmth emerges from snow. The former is true, thus also the latter.

This is then elucidated further: 'That one follows from the other is obvious' ('Consequentia patet'), because becoming warm and reddening are results of warmth. 'That the antecedent applies, we are taught by boys when they have a snowball fight' ('Prius pueri nos edocent niveis se mutuo exagitantes globulis'), as they often go on for hours without complaining about the cold.

Supplementation of reasoning in words by illustrations, in contrast, is typical of more advanced dissertations, especially of those dealing with mathematical subjects, broadly speaking. Scheiner, for example, relied heavily on visual aids. His *Exegeses fundamentorum gnomonicorum* (1615) features numerous complex and well-drawn diagrams, and the same is true of his *Disquisitiones mathematicae*, which also contains an early depiction, more detailed than Galilei's in the *Sidereus nuntius*, of the moon's surface as seen through a telescope (p. 58). It is difficult to imagine how illustrations such as these could have been integrated into a disputation. They once more exemplify the tension between the two uses of a dissertation—as a template for a disputation and a reading text for a wider audience—and underline the preponderance of the second aspect in sophisticated specimens like Scheiner's.

Formal features

A typical early modern dissertation was at first glance distinguished from other types of publications by contemporary readers and is still easily recognizable today. It is a slender booklet, usually in quarto format, comprising no more than a few dozen pages. Among its paratexts, the aforementioned 'corollaria' are a peculiarity of the genre. Those shared with other genres—title page, dedication, address to the reader, congratulatory poetry[104]—are usually styled in a characteristic manner. Most distinctive is the title page, which presents itself as a hybrid of a normal book title and an announcement of an event.[105] The title of the work to be discussed comes first, followed by a relative clause: under the rector of the institution and the 'praeses' (both in the ablative), the work will be defended on a

[104] Weijers 2013, 228–9. [105] Horn 1893, 71, 82; Chang 2021, 127.

certain date (sometimes left out if it has not yet been fixed at the time of printing, to be filled in manually later) and at a certain place by the respondent(s). In the case of Remus and Wolff's *Consideratio physico-mathematica hiemis proxime praeterlapsae*, this reads as follows (Figure 7):

PHYSICAL AND MATHEMATICAL PONDERING
OF THE RECENTLY
PASSED
WINTER
which
UNDER THE RECTORATE OF THE MOST EMINENT
AND SERENE PRINCE AND LORD
LORD PHILIP WILLIAM
PRINCE OF PRUSSIA
MARGRAVE OF BRANDEN-
BURG AND SO ON
UNDER THE PRESIDENCY OF
CHRISTIAN WOLFF
public ordinary professor of mathematics
on 13 June 1709
in the morning hours
IN THE MAJOR LECTURE HALL
will present to the public judgement
of the learned:
GEORG REMUS
from Gdańsk

The dedication to notables or benefactors on the following page(s) usually takes the form of an 'ellogium'. The preface may not only introduce the topic but also explain the original social context of the work, thus bridging the gap between the disputation and its printed counterpart. Dedication and preface are generally written by the respondent or in his name. He is also the addressee of congratulations in poetry and occasionally prose right before or after the dissertation itself, which may be quite extensive; for instance, no fewer than seven epigrams conclude Drechsler and Rechtenbach's *De sermone brutorum*. All of this made the *book* the respondent's as a quid pro quo for the costs he had to carry, even if the *work* belonged in many cases to the 'praeses'.

The main text of a dissertation mostly exhibits an easily recognizable structure as well, which is more fine-grained than in a normal book. This feature seems to derive from the original broadsheets, where the single statements to be disputed were presented as short, typographically distinct units. In a booklet as well, a clear subdivision into short units will have made it easier for participants in the

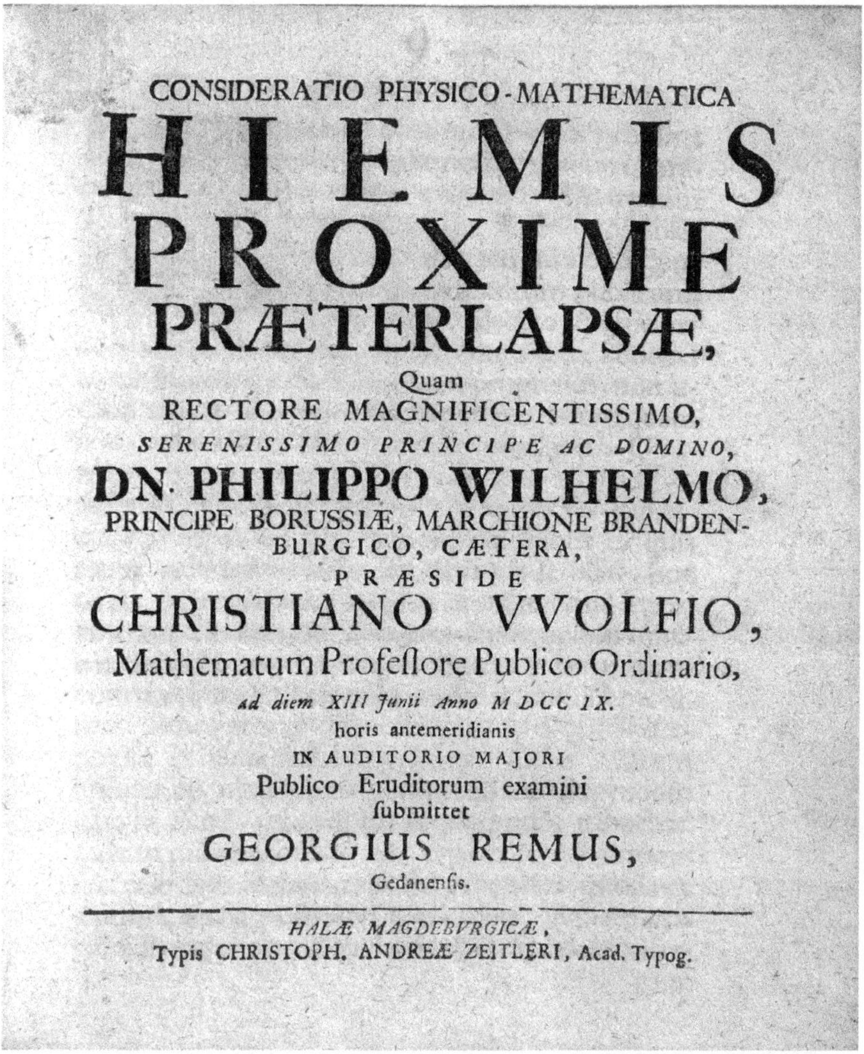

CONSIDERATIO PHYSICO-MATHEMATICA
HIEMIS
PROXIME
PRÆTERLAPSÆ,
Quam
RECTORE MAGNIFICENTISSIMO,
SERENISSIMO PRINCIPE AC DOMINO,
DN. PHILIPPO WILHELMO,
PRINCIPE BORUSSIÆ, MARCHIONE BRANDEN-
BURGICO, CÆTERA,
PRÆSIDE
CHRISTIANO VVOLFIO,
Mathematum Profeſſore Publico Ordinario,
ad diem XIII Junii Anno M DCC IX.
horis antemeridianis
IN AUDITORIO MAJORI
Publico Eruditorum examini
ſubmittet
GEORGIUS REMUS,
Gedanenſis.

HALÆ MAGDEBURGICÆ,
Typis CHRISTOPH. ANDREÆ ZEITLERI, Acad. Typog.

Figure 7 Title page of the *Considerata physico-mathematica hiemis proxime praeterlapsae*.

disputation to find points to be raised and to refer to passages quickly and precisely. It probably also helped ordinary readers to orient themselves, compensating for the general absence of tables of contents and indexes. The simplest way of structuring a dissertation was a running paragraph count. Often, however, the text was divided by subtitles into chapters, sections, 'quaestiones' (Christoph Leopold and Christoph Vian, *Philosophia historica de montibus*, 1713),[106] 'theses'

[106] Edition with German translation and commentary: Steixner 2009.

(Rudolf Jakob Camerarius and Johann David Mauchart, *De herba mimosa seu sentiente*, 1688),[107] aphorisms (Johannes Ralla and Friedrich Berger, *Chionognosia*, 1658),[108] or thematically specified units. Paragraphs and other subdivisions could be combined, as in Scheiner's *Disquisitiones mathematicae*, which features a paragraph count together with super- and subtitles indicating specific topics ('De partibus universi in specie. Terra et aqua', 'On the parts of the cosmos in particular: earth and water', p. 23) and text types ('Argumenti nucleus', 'The argument's kernel', p. 28; 'Quaestiones primae', 'First questions', p. 31). Some dissertations exhibit a multi-tiered structure. Despite its shortness (twenty pages) and the banality of it content, Petermann and Schmied's *Disputatio de glacie* falls into a 'promulsis' ('starter') and a main part structured on three hierarchical levels, namely 'sectiones', 'pensa', and paragraphs. In such cases, one gets the feeling that the structural apparatus is meant less to clarify than to impress readers by creating an air of scientificity. The same purpose seems to underlie the borrowing of words like 'definitio', 'postulatum', 'theorema', or 'propositio' from the language of Euclidean geometry, that paradigm of scientific method par excellence, and their transfer to other fields, as in Iacobus Salernus's *Positiones physio-astronomicae* (1682), a summary of the essentials of Ptolemaic cosmology, whose structural skeleton looks as follows:

Definitiones primae
Postulatum
Definitiones 2
    4 Scholia
Theorema 1. Propositio 1
    Scholion
Theorema 2. Propositio 2
    Lemmata 1–3
Theorema 3. Propositio 3
    Corollaria 1–3
Theorema 4. Propositio 4
    Corollaria 1–8
Theorema 5. Propositio 5
    Corollaria 1–2
    Scholion
Theorema 6. Propositio 6
    Corollaria 1–2
    Scholion

[107] On the work and its subject, see Giglioni 2018, esp. 422–3.
[108] See also note 92 above. Being single pithy sentences, aphorisms, like 'quaestiones' and 'theses', provided particularly convenient starting points for discussion.

Lemma
Scholion
Problema 1. Propositio 7
Problema 2. Propositio 8
Problema 3. Propositio 9
Scholion
Theorema 7. Propositio 10
Scholion
Problema 4. Propositio 11

Most of these terms are not used in the sense they have in the textual architecture of the *Elements*, and one is hard pressed to guess their precise meaning here. For instance, the fact that a full circle is divided into 360 degrees falling into 60 minutes comprising 60 seconds each, which is adduced as the only 'postulatum' in the work (p. 3), is not on a par with Euclid's 5 postulates; nor is it evident how a proposition can take the form of a problem.[109]

A similar desire to impress seems to form part of the motivation for the number of citations and quotations which clutter many dissertations, be it as marginalia, as footnotes, or, most commonly, in the main text.[110] In some cases, for instance in Scheuchzer and Hottinger's *Κρυσταλλολογία*, quotations take up about half of the entire text. While the works cited and quoted are for the most part pertinent and references become increasingly precise over time, the feeling that authors are nonetheless flaunting their learning and highlighting their membership of the academic community is reinforced by two features: the titles tend to be abbreviated so heavily that they are hard to identify,[111] and quotations from vernacular works (common due to the late flowering of the genre) as well as from Greek and Hebrew sources are mostly given in the original and, contrary to the custom in normal books, without translation. Both features seem to emphasize how easily the authors move around in a world of learning from which their uninitiated contemporaries are excluded. At all events, dissertations pre-empted and presumably did something to establish the particular form of intertextuality which constitutes one of the most conspicuous features of modern academic writing.

In terms of style, disputation manuals recommended the use of the 'genus humile', because it would not arouse the emotions of the participants and thus keep the atmosphere sober and factual, and because it was suitable for

[109] However, the 'propositiones' in Newton's *De quadratura curvarum* (1704) too can be realized as 'problemata'. There seems to be a logic behind this which I fail to grasp.
[110] Leinsle 2006, 44, 46 (for Dillingen); Boer 2021, 492–7.
[111] Admittedly, abbreviations also helped to keep dissertations short and printing costs bearable.

expounding the matters under discussion clearly and distinctly.[112] However, this simple advice proved difficult to follow in written dissertations,[113] the authors of which had to meet partly contradictory expectations which concerned all technical writing but made themselves felt with particular urgency in this genre. The solemn occasion called for the impeccable and elevated Latin that was the calling card of an early modern intellectual. At the same time, terminological precision, often implying unclassical vocabulary, was demanded by the subject, mastery of which established one's academic credentials. Finally, publication for a wider readership made it advisable to present one's thoughts in syntactically and terminologically less demanding Latin and to adopt an easy-going, attractive style of writing. To find a viable compromise between these conflicting requirements must have been taxing. Paratextual declarations show that authors were aware of the problem. In the preface to their *Philosophia historica de montibus*, Leopold and Vian comment on the need to cater to the taste of the reading public in spite of their technical subject:

> Stylo uti placuit temperantiore, ne horrorem conciperet delicatus temporum nostrorum genius, cui ipsa etiam displicet veritas, quae ornatu caret.
>
> We decided to use a well-balanced style, so that the fastidious spirit of our age, which spurns even truth itself, if it is unadorned, may feel no loathing.

A common solution was to introduce one's text with a couple of Ciceronian periods which soon gave way to a matter-of-fact style. More ambitious authors, however—often, it seems, Jesuits and students writing their own doctoral dissertations—tried to keep the uneasy balance between stylistic polish, technical terminology, and readability throughout their entire work. James Hutton's medical *Dissertatio inauguralis de sanguine et circulatione microcosmi* on the blood and its functions in the human body, defended on 12 September 1749 at Leiden, is a case in point.[114] The candidate clearly wants to demonstrate his good humanistic education as much as his medical prowess. The secretion from the blood of the 'succus vitalis', a mysterious substance thought responsible for the transmission of the life functions from the brain through the whole body, is described as follows (p. 20):

[112] Johann Friedemann Schneider, *Tractatus logicus singularis* (1718), pp. 266–7; cf. Marti 1998, 51–2.

[113] Some remarks on the style of Swedish dissertations and a few examples can be found in Benner and Tengström 1977, 86–106. On Helmholtz's dissertation, see Heideklang et al. 2021, 26–31.

[114] Similar stylistic ambitions can be found, for example, in Bernoulli's *De motu musculorum*, Franz Gaun and Johann E. Feigenpuz's *Iter per salinas Tyrolenses* (1707), and Johann Bartholomäus Adam Beringer and Georg Ludwig Hueber's *Lithographia Wirceburgensis* (1726).

§ LI. Praecipua summaeque necessitatis secretio, eodem vero tempore maxime obscura, controversa et minime demonstrabilis, ratione solummodo et analogia comprobata, in cerebro perficitur.

§ LII. Sanguis per directa<m> carotida missus impetuositatem suam inflectionibus ad cranii basin angulosis coercitam multum habet. Per substantiam deinceps cerebri et cerebelli corticalem, quae vasorum exilissimorum congeries videretur, distributus, ibi propter tenuissimam divisionem non ruber, cinericio vero apparet colore, et hic separari videtur succus iste vitalis omni functioni, actioni sensuique corporis tam necessarius, de cuius vero natura prorsus ignoramus et praeter fluiditatem subtilitatemque aliam eius aequam forte ideam in mente formare nequimus. Hunc per albidam partis medullaris substantiam fistulosam vehi imaginamur, ibique forsitan ulterius perfici, circumgiratas istas fibras tubulosas varie unitas, separatas, mixtas sibique invicem alternantes circulaturum et per continuationes earum prolongatas in omne corporis punctum, sensationis, motus muscularis nutritionisque forte usui, <se> distributurum.

§ LI. The most important discharge, which is of the utmost necessity, at the same time, however, to the highest degree unknown, disputed, and least verifiable, proved only by reason and analogy, takes place in the brain.

§ LII. Having been sent through the straight carotid, the blood has its fierceness much tamed by the contorted windings at the base of the skull. Subsequently spread through the matter of the cortex of the brain and the cerebellum, which could seem to be an accumulation of extremely thin vessels, it does not appear red but of ashen colour because of its extremely fine distribution there, and it is here that this liquid of life seems to be segregated which is so necessary to every task, act, and feeling of the body, while we are nonetheless completely unknowing regarding its nature and cannot, apart from liquidness and fineness, form any other notion of it, which would perhaps be appropriate, in our mind. We envisage that it flows through the white, porous mass of the bulbar part and possibly is brought to higher perfection there, before it is going to move around those convoluted tubular filaments, which in different ways come together, are divided, amalgamate, and take over from one another, and through their extended continuations to disseminate itself to every spot of the body, perhaps to enable feeling, movement of muscles, and nourishment.

Hutton clearly is at pains to do justice to the mystery and importance of the phenomenon in the stylistic level of his description. He writes long, complex periods and does his best to achieve the word order perceived as typically Latin, which separates syntactically matching words. Marked rhetorical figures such as tricola and tetracola ('maxime obscura, controversa et minime demonstrabilis'; 'varie unitas, separatas, mixtas sibique invicem alternantes') or inconcinnity, the wilful change of construction where parallelism would be expected ('praecipua summaeque necessitatis'), abound. This is combined with a number of technical terms like 'carotida', 'basin', or 'cerebellum', the first two of which are of Greek origin, as underlined by their case endings. Other terms are not strictly speaking technical but very rare. Some of these may even have been coined by Hutton himself in his attempt to provide a graphic description of the unobservable processes in the brain and nervous system: 'impetuositas' and 'fluiditas' (examples of the nominal style typical of scientific Latin), 'cinericius', 'circumgiratus', 'tubulosus', and so on. The result is an impressive text but also one that makes for a hard read, as factual problems are compounded by linguistic difficulties. One has to read a phrase like 'praeter fluiditatem subtilitatemque aliam eius aequam forte ideam' two or three times to understand that 'aliam' (probably) belongs, not to 'subtilitatem', but to 'ideam'. Even the author himself may occasionally have got lost in his own syntactic labyrinth: of the two places where I have intervened to arrive at a syntactically correct text, the first looks like a typical printer's error, but one can be less sure about the second.

Monographs

Notion, name, and history

The *Oxford English Dictionary* defines a monograph as 'a detailed written study of a single specialized topic (distinguished from general studies in which the topic is dealt with as part of a wider subject)'. This is a workable definition, as far as early modern science is concerned. In addition to its defining features, a monograph from this era and thematic field typically exhibits a number of further properties (although it may do so to varying degrees). The most important of these are brevity, novelty in terms of content, and argumentative penetration of the subject.

In their own time, the works dealt with in this section were not called 'monographiae' or the like. Such an expression was unknown in antiquity, when only *μονογράφος* existed (in the sense of 'notary'), and also during the Middle Ages and the first centuries of the early modern era. It apparently originated in eighteenth-century botany, perhaps under the influence of Linnaeus's seminal *Fundamenta botanica* (1735), where a class of botanical authors, those writing books about one single taxon, is defined as follows: 'Monographi…vegetabile unicum opere singulari prosecuti sunt' (aphorism 13).[115] From there, 'monographia'

[115] Linnaeus was such a 'monographus' himself (see the text before note 137 below). The first author who could lay claim to this title was Robert Morison (*Plantarum umbelliferarum distributio nova*, 1672; Pavord 2005, 387).

in the sense 'botanical work describing one single taxon' was apparently derived[116] and used in titles such as that of Nikolaus Joseph von Jacquin's *Oxalis: Monographia iconibus illustrata* (1794). However, this and similar works are actually not monographs but 'historiae plantarum' in terms of their structure and content. Subsequently, 'monographia' and its vernacular equivalents spread to thematically focused works in general. By contrast, common premodern expressions for 'monograph' were *πραγματεία* and *ὑπόμνημα* in Greek, 'exercitatio', 'disquisitio', 'dissertatio', 'tractatus', and derivations from the stem 'comment-' in Latin. Most often, however, the titles of such works simply consisted of a word indicating the subject, preceded by *περί* or 'de' ('about').

Ancient monographs[117] are not transmitted in great numbers. However, this presumably has to do with the bias of transmission, which favoured genres carrying large amounts of commonly accepted knowledge over small works on single, often contested issues. The work catalogues given in Diogenes Laertius' *Lives of Eminent Philosophers* for authors such as Aristotle (5.1.22–7), Theophrastus (5.2.42–50), or Epicurus (10.27–8) as well as for many lesser lights convey a good impression of how much there was around in this field. A sizeable number of the titles enumerated by Diogenes relate to the natural world. The Byzantine encyclopedia *Suda* provides additional information about such works. Among surviving texts, several of Aristotle's shorter zoological treatises such as *De motu animalium* or *De incessu animalium* qualify as monographs, as do some works by Theophrastus, for instance *On Sense Perception*. Archimedes's concise treatises on clearly defined mathematical and physical questions constitute prime examples. Some Galenic writings such as *De uteri dissectione*, *De victu attenuante*, or *De ptisana* are monographs too.

In the Middle Ages, pertinent items can be found too. Arab and Western authors wrote specialized treatises on a variety of topics, from finding out the geographical longitude and latitude of a place (al-Bīrūnī, *Determination of Coordinates of Cities*) and the phenomenon of speed (Bradwardine, *Tractatus de proportionibus velocitatum in motibus*) to colours, the rainbow (Grosseteste, *De colore*, *De iride*), and mathematical ratios of ratios (Oresme, *De proportionibus proportionum*).[118] One also has numerous monographs on astronomical instruments like the astrolabe, a subtradition that continued into early modern times.[119] As a rule, however, specialized questions were preferably discussed in the frame

[116] This meaning comes before the more general definition quoted initially in *OED*, which orders the meanings of a given expression chronologically. 'Monographia'/'Monograph' in the pharmaceutical sense 'description of a single drug', for instance within a pharmacopoeia, is not considered by *OED* and seems to be of more recent origin.

[117] See Asper 2007, 21–2 and his index, s.v. 'Monographie'.

[118] On these works, see *CHS* 2, 82; 414, 416, and 428; 508; 529. Critical editions with introduction and commentary on Grosseteste's work on colours and Oresme's on ratios of ratios have been produced by Dinkova-Bruun et al. 2013 and Grant 1966 respectively.

[119] Kusukawa 2000b, 139–41; *CHS* 2, 459, 464, 466–8.

of more comprehensive genres such as 'summae' and commentaries. This by and large still obtained in the early sixteenth century. When the Roman professor of medicine Bartolomeo degli Emanuelli thought he had something new to say on the definition of fever, he published his insights in the form of a supercommentary on an older commentary on Avicenna's *Canon* (*Adnotationes super dicta Gentilis super definitione de febre data ab Avicenna*, 1524).

Scientific monographs in early modern times

It thus was a paradigm shift when monographs became a major genre of scientific writing in the mid-sixteenth century and intellectuals began to routinely publish new results in this format.[120] This development stands out not only in hindsight. Already contemporaries showed an awareness of the rise of the monograph and understood it as a distinctively modern genre. In the letter of dedication to his medical bibliography, the *Biblia iatrica* (1609), Johann Georg Schenck stated that the high level of contemporary medicine was evident, inter alia, from the fact that 'to some single diseases, thick, laborious volumes have been dedicated' ('morbis nonnullis solis ingentia et operosa volumina dedicata', sig.):(3^{v}).[121] Fifty years later, Philipp Jacob Sachs von Lewenheimb was able to enumerate, as precedents to his own planned work on the vine (*Ampelographia*, 1661), some seventy natural historical monographs devoted to single substances such as antimony, plants such as wormwood, and animals such as the elk.[122] That the genre was not only seen as indicative of scientific progress but also supposed to contribute to it transpires from a number of letters addressed to Henry Oldenburg in the 1660s and early 1670s: their senders criticize monographs by members of the recently founded Academia Naturae Curiosorum for not living up to these expectations.[123]

In fact, the rise of the scientific monograph appears to be associated with two closely interrelated developments, one of which was a changing view of nature and its study. As people gradually realized how much of nature was still uncharted territory, they came to accept seemingly minor, lowly aspects of it as rewarding subjects of study (something evident also in the curious topics discussed in dissertations, mentioned in the last section). The notion of scientific progress gained ground in tandem with an understanding of science as a collaborative endeavour to which many naturalists contributed bits and pieces. The new dignity accruing

[120] Nummedal and Findlen 2000, 187–8; Grendler 2001, 352.

[121] The emphasis on the size of these books, which contradicts the quite restricted scope of most monographs, seems to be rhetorically motivated and intended to contrast with the small compass of the subjects.

[122] Letter to Johann Michael Fehr, 1 May 1658; Kefalas 2015, 70–4 provides an annotated German translation.

[123] Maclean 2019, 79 n. 10.

to these bits and pieces from their role in the large enterprise made them appear worthy of book-length treatments.

The second important development related to the book market. Printers and scholars interested in the natural world had created a market robust enough to sustain the publication of highly specialized works independently of any institutional context. In contrast to translations, commentaries, textbooks, or dissertations, which relied on demand created by the educational system, monographs depended on a stratum of private buyers willing to invest their money in fresh knowledge. That such a stratum existed is indicated by the fact that many monographs were financed not by the author or a patron, but by a publisher who found it worthwhile to take the risk upon himself. This is documented by 'sumptibus' or 'impensis' remarks on title pages: William Harvey's *De motu cordis et sanguinis* (1628) appeared 'sumptibus Guilielmi Fitzeri', Jakob Bernoulli's *Ars coniectandi* (1713) 'impensis Thurnisiorum fratrum', Moritz Anton Cappeller's *Pilati montis historia* (1767) 'typis et sumptibus Johannis Rodolphi Im-Hof et filii'.[124]

In the sixteenth, seventeenth, and eighteenth centuries, monographs appeared in almost all disciplines concerned with the natural world. In mathematics, one finds everything from foundational works like François Viète's *In artem analyticem isagoge* (1591), a milestone on the way to symbolic algebra,[125] to what could be termed 'occasional mathematics'. An example of the latter is provided by Albert von Curtz's *Problema Austriacum* (1653), which describes a method of constructing a regular fourteen-sided figure on the occasion of the election of Ferdinand IV as the fourteenth ruler of the Holy Roman Empire from the house of Habsburg.[126]

In the disciplines which later coalesced into modern physics, the boost given to optics by the invention of the telescope was reflected in a number of works. Kepler's *Dioptrice* (1611) immediately reacted to the new instrument.[127] A few years later, Christoph Scheiner used the theory of refraction to explain one particular astronomical phenomenon, the apparent elliptic deformation of the sun at sunrise and sunset (*Sol ellipticus*, 1615).[128] As a later example from another field, Euler's *Dissertatio de magnete* (1748) may be named, an attempt at explaining the persistent conundrum of magnetism with the help of Cartesian vortex theory, which the author successfully submitted as a prize essay to the Académie Royale des Sciences.[129]

In astronomy, dominated by the presentation of comprehensive world systems in various formats, reports of observations, and table works, one still finds works

[124] In addition, the very lack of a remark that a book was financed by someone else may often imply that the publisher invested his own money (Maclean 2012, 78).

[125] German translation: Reich and Gericke 1973. See also Cantor 1880–1908, vol. 2, 577–9; *CDSB* 14, 18–25; Oaks 2018, esp. 268–76.

[126] Brüser 2019 (edition) and 2021 (discussion).

[127] *KGW* 4, 327–414 (text), 462–78 (postface), 514–21 (notes).

[128] German translation with introduction and notes: Daxecker and Šubarić 1998.

[129] *LEOO*, ser. III, vol. 10. On Euler's ideas on magnetism, see Speiser 1983, 222–4.

promoting clear-cut theses on specific aspects. Girolamo Fracastoro's *Homocentrica* (1538) tries to reinstate the old Eudoxan system of concentric spheres in a more sophisticated version, doing away with the eccentrics and epicycles of the Ptolemaic system.[130] Another example of an astronomical monograph is provided by Joseph Scaliger's posthumous attack on the common interpretation of the precession of the equinoxes as due to a movement of the sphere of the fixed stars (*De aequinoctiorum anticipatione diatriba*, 1613).[131]

Geographical monographs may be devoted to the solution of long-standing puzzles like the location of the source of the Nile (Isaac Vossius, *De Nili et aliorum fluminum origine*, 1666).[132] Others are chorographical treatises, focusing on peculiar regions or phenomena (Cappeller, *Pilati montis historia*).

Being concerned with the hydrological cycle, Vossius's work also contains a meteorological component. Meteorology offered much material for monographic treatment in other respects as well. The aetiology of the tides was long a burning question which gave rise to numerous theories, such as that proposed by Pandolfo Sfondrati (*Causa aestus maris*, 1590), in which the Strait of Magellan is assigned a crucial role. Much more original and fundamental is the question at the heart of Nicolas Steno's *De solido intra solidum naturaliter contento* (1669), a work famous today for its first formulation of the principles of stratigraphy: how does it come about that a solid body of one kind can be found embedded in a solid of another kind?[133] In addition, one finds works occasioned by disasters such as earthquakes, volcano eruptions, and extreme weather events. The Ferrarese earthquakes of the years 1570–4, for instance, were particularly fruitful in this respect.[134] One century later, the 1669 eruption of Mount Etna prompted Giovanni Alfonso Borelli's *Historia et meteorologia incendii Aetnaei* (1670).

Borelli explains the behaviour of Mount Etna by the presence of various combustible substances within the mountain, which makes his text an exercise in speculative chymistry as well. In fact, the second half of the seventeenth century, which saw the emergence of a non-secretive, proto-scientific chymistry, produced the first chymical monographs as well. Georg Wolfgang Wedel's *Specimen experimenti chimici novi* (1671) argues for the presence of 'sal volatile'—a term covering different components of hartshorn salt and chemically similar substances—in plants.[135] Georg Ernst Stahl wrote the first book engaging exclusively with processes of fermentation, claiming to found an 'art of fermentation', as the neologism in his title implies (*Zymotechnia fundamentalis*, 1697).[136]

[130] *CDSB* 5, 104–8, at 106; Omodeo 2014, 79–82.
[131] Grafton 1983–93, vol. 2, 459–88.
[132] Davids 2012, 193–4, 197–8, 200–1, 203, 205.
[133] Latin edition: Maar 1910, vol. 2, 181–226 (in what follows, I will cite *De solido* from this edition); English translation with notes: Kardel and Maquet 2013, 621–60.
[134] Martin 2011, 60–79.
[135] Maclean 2019, 84, 87.
[136] A more famous neologism coined in this work is 'phlogiston' (proem, sig.)(3^{v}).

Substances and living beings from the three realms of nature were treated in numerous natural historical works, as demonstrated by Sachs von Lewenheimb's aforementioned letter. Precious and semiprecious stones such as nephrite, discussed in Outgert Cluyt's *Calsvee* (1627), provided popular subjects, as did substances extracted from animals and plants like Wedel's 'sal volatile' or opium (Daniel Winckler, *De opio tractatus*, 1635). Two specimens of botanical monographs—in the sense initially distinguished from 'monographiae'—are Johann Michael Fehr's *Hiera picra* (1668), a treatise on wormwood, and Linnaeus's *Musa Cliffortiana* (1736) on the banana plant (which Linnaeus, in contrast to modern botany, thought to be one single species).[137] Animals were discussed as well, from mammals such as the hare (Wolfgang Waldung, *Lagographia*, 1619) to fish such as the herring (Paul Neucrantz, *De harengo*, 1645) and spiders such as Lycosa tarantula (Giorgio Baglivi, *De anatome, morsu et effectibus tarantulae*, 1696).[138] Baglivi's work exemplifies a feature typical of most natural historical monographs particularly well: they devote much space to medical aspects—in this case, the tarantula's bite, its effects, and cures for it.

Medical and anatomical monographs in the strict sense of the word were very common too. In the wake of the innumerable late medieval writings on the plague, a medical microtradition was constituted by 'disease monographs' about new or particularly dangerous ailments. Syphilis treatises, which began to appear as early as 1495, form a coherent body.[139] Moreover, many physiological and anatomical discoveries were presented in monograph format. This applies to Gaspare Aselli's *De lactibus* (1627), for instance, which describes the lymphatic vessels in the mesentery,[140] and, of course, to Harvey's *De motu cordis et sanguinis*.[141]

Finally, one finds works on questions which cannot be assigned to any established early modern discipline. Kepler's *De nive sexangula* (1611), 'now hailed as the founding text of crystallography', not only discusses the hexagonal symmetry of snowflakes but also similar phenomena in honeycombs, flowers, and pomegranate grains, as well as the most space-saving way to stack cannon balls, mentioning geometers, botanists, and chymists as experts for these subjects.[142] The speculation about extraterrestrial life laid out by Christiaan Huygens in his *Κοσμοθεωρός* (1698) is another case in point.[143]

[137] Modern reprint with English translation and introduction: Freer and Müller-Wille 2007.

[138] Several more zoological monographs (intermingled with works of encyclopedic character) are discussed in Bäumer 1991–6, vol. 2, 194–220.

[139] Reprints of selected items: Sudhoff and Singer 1925. On early syphilis literature in general, see Sudhoff 1912.

[140] Gerabek et al. 2005, 110; *CDSB* 315–16, at 316.

[141] The most recent English translation is by Franklin and Wear 1993. See also *CDSB* 6, 150–62, at 152–8, and the literature cited in that entry.

[142] *KGW* 4, 261–80. The quotation is from Wootton 2015, 212.

[143] Edition with French translation: Société Hollandaise des Sciences 1888–1950, vol. 21, 653–821. See also Crowe 2008, 86–108 (selected passages from the English translation from London, 1698, plus paraphrases of the rest); Aït-Touati 2011, 95–129.

In terms of outward appearance, monographs are usually modest volumes. They rarely comprise more than one book or go beyond 200 pages, and their format is quarto at maximum, in many cases octavo or even smaller. Title pages of monographs are mostly unadorned, except perhaps for a printer's mark or publisher's signature. Paratexts are relatively few and short. Some works admittedly are dedicated to high-ranking personalities[144] and their letters are crafted with corresponding care, but otherwise, most paratexts are functional rather than representative: they aim at familiarizing the reader in advance with the work's content, composition, and argumentative structure. An item notable for its absence is the index.[145] In combination with the shortness of most monographs, the features just mentioned imply that, in contrast to many other genres, such books were usually meant to be read cover to cover. If there are any illustrations at all, they tend to be strictly functional too, although this does not imply carelessness or poor execution.[146]

Being a text type without an authoritative literary tradition, monographs occasionally borrowed the appearance, so to speak, of more clearly defined short genres. Many were styled as letters, and 'dissertatio epistolica' and the like became common titles (see Chapter 6, note 56). Even more were published as dissertations by professors, an option which had the great advantage that the printing was financed by the respondent, and later in journals. Another way of reducing publication costs was to append a monograph to a longer work, a practice applied to all kinds of short pieces in early modern times. A monographic appendix could be thematically connected to the main text, as in the case of three treatises on birth, the uterus, and conception adjoined to Harvey's *Exercitationes de generatione animalium* (1651), but this was by no means obligatory. Giovanni Battista Baffi had no qualms about adding a treatise of nine pages on the question of how early a child can be born to be viable ('Pythagorica consideratio, quo mense vivere possit foetus') to a compendium on comets (*De cometis libri tres*, 1580). There was also the option of combining several short monographs into a standard-length book, as in Hieronymus Fabricius's posthumous *Tractatus quattuor* (1624) or Nicolas Steno's *Observationes anatomicae* (1662). In such cases, a certain thematic coherence was usually sought. In the two books just mentioned, all the constituents deal with anatomical questions. Sometimes, closely related monographs even

[144] Fracastoro and Sfondrati dedicate their *Homocentrica* and *De aestu maris* to popes (Paul III and Gregory XIII respectively). Vossius addresses *De Nili et aliorum fluminum origine* to Louis XIV, Steno, *De solido intra solidum naturaliter contento* to Ferdinando II of Tuscany.

[145] There are exceptions such as Aselli's *De lactibus*, where the presence of an index is consistent with the extent and kind of reader guidance provided in general, to be discussed later.

[146] The four illustrations in Aselli's *De lactibus* constitute one of a handful early modern examples printed in (four) different colours in order to bring out the lymphatic vessels in the mesentery, a rather amorphous mass at first sight. (However, they are not the earliest illustrations of this kind, as stated by Premuda 1993, 228–34. As early as 1485, Erhard Ratdolt printed diagrams in Sacrobosco's *Sphaera* in three colours: Kusukawa 2000a, 99; Schmitz 2018, colour plate 6.)

made up a thematically unified volume on a topic such as syphilis or childhood diseases.[147] Occasionally, authors purposefully serialized very short monographs. From 1686 to 1720, Georg Wolfgang Wedel published no fewer than thirteen *Exercitationum philologo-medicarum decades* containing a total of 130 pieces of a few pages each on antiquarian and medical subjects.[148] Finally, the same work could appear both as part of a larger publication and independently. Viète's *In artem analyticem isagoge* came out as an introduction to his *Algebra nova* but also as a short volume whose title page specified that it had been 'printed separately' ('seorsim excussa') from this work. While Aselli's *De lactibus* was published on its own both in the *editio princeps* (Milan, 1627) and in an edition appearing in Basel the following year, the printer of the latter suggested that it be bound together with the *Theatrum anatomicum* (1605) by Caspar Bauhin, who was professor at that city's university.[149]

The commercial success and scientific impact of monographs varied widely. Many were published only once and left no traces—sometimes for good reasons, as in the case of Sfondrati's *Causa aestus maris*, whose attempt to explain the tides was based on a misguided idea (the water trying to follow the sun westwards is stopped at the Strait of Magellan and flows back) and presented in a barely understandable style; the book was made short shrift of by Francesco Patrizi (*Nova de universis philosophia*, 1591, fols. 139^{v}–140^{r}), and no one seems to have bothered to engage with it after that. Other monographs, by contrast, became quite successful. There were famous milestones such as Harvey's book on the circulation of blood, but numerous less well-known works found eager readers in their own day as well. Wedel's *Specimen experimenti chimici novi,* for instance, was republished three times as a stand-alone treatise until 1682 and, moreover, appeared as a journal article in volumes 4–5 (1676) of the *Miscellanea curiosa*.

Another sign of success was vernacular versions, which appeared in fields with a substantial readership without any knowledge of Latin, such as geography, chymistry, or anatomy. Vossius's book on the source of the Nile (1666) came out as *Dissertation touchant l'origine du Nile et autres fleuves* the following year. Stahl's *Zymotechnia fundamentalis* was translated into German in 1734, nearly four decades after its original publication, 'because of its indescribable usefulness' ('wegen ihres unbeschreiblichen Nutzens'), as the translation's title page has it. Translations in the opposite direction were produced as well. The first printed discussion of tracheotomy, the *Question chirurgicale* (1620) by the French surgeon Nicolas

[147] Maclean 2002, 57. [148] Maclean 2019, 87.

[149] The printer's unpaginated letter to the reader is found between the dedication and the index. Something similar seems to have happened in the case of Pierre Fermat's *De linearum curvarum cum lineis rectis comparatione* (Toulouse, 1660). This treatise is occasionally called an appendix to Antoine de La Loubère's *Veterum geometria promota*, which was published at the same place in the same year (*CDSB* 4, 566–76, at 572–3; Mahoney 1994, 267), so it apparently was sometimes bound together with that work. However, in the digitally available copies, it has a separate title page and appears as an independent work.

Habicot, was made available to learned German physicians in a shortened and adapted—and plagiarized—Latin version by Friedrich von Monau (1644; republished in 1649 and 1711).[150]

Generally speaking, the interpenetration of Latin and the vernaculars was as intense in this area as in many others, and whether a work was written in this or that language often depended on circumstances and chance. In accordance with the preferences of his patron Ferdinando II of Tuscany, Nicolas Steno began to compose his *De solido intra solidum naturaliter contento* in Italian (p. 186). Nevertheless, the work ultimately appeared in Latin, apparently because the Danish author had to leave Tuscany at short notice (p. 184), which did not give him time to finish it in a language he was just learning at the time. The first vernacular version (1671), presumably from the pen of Henry Oldenburg,[151] came out in English, as *De solido* provoked some interest in the orbit of the Royal Society.

In the course of the eighteenth century, the ratio of Latin to vernacular monographs shifted towards the latter. More importantly, the Latin and the vernacular monograph alike encountered competition from the journal article as a publication format for cutting-edge science and eventually was superseded by it. However, this did not happen overnight. When Carl Gustav Jacob Jacobi decided to give a first comprehensive presentation of the elliptic functions whose understanding had been greatly advanced by him, he still did so in a monograph (*Fundamenta nova theoriae functionum ellipticarum*, 1829).

Some parameters of variation: focus, novelty, intertextuality

Like most genres discussed in this book, monographs turn out to be more diverse than expected on closer inspection. In this sub section, three interrelated aspects will be discussed with regard to which authors could and did make different choices. The first two of these—thematic focus and novelty—are features initially invoked as constitutive or typical of the genre; the last is scientific intertextuality.

The narrow focus associated with an ideal monograph was not easy to maintain, as several factors militated against it. Most ancient, medieval, and early modern literary theorists and critics found digressions perfectly acceptable and actually desirable in a work of literature. Depending on the genre, some even tolerated multiple instances. If unity was sought and found, this happened on the level of form rather than content.[152] In ancient times, the limited capacity of the papyrus

[150] Lembke 2017.

[151] The title page says 'English'd by H. O.'. As it transpires from his preface to the reader, the translator was personally acquainted with Robert Boyle.

[152] Heath 1989, esp. 137–49 on medieval and early modern developments.

scroll, which contained a text quantity equivalent to 500–1,000 lines of poetry on average, may have helped authors of monographs to stay focused, but early modern authors were less constrained in this respect. A strict thematic focus also sat uncomfortably with the reading habits of a time when books were regarded as quarries of information and every digression as added value, while a close-knit argument was found to be less important.

For all these reasons, the well-focused monograph is a rare bird. Many works belonging to the genre in terms of their overall subject strain its borders by excursuses. Rarely does a work remain as concentrated on its main subject as Pierre Fermat's *De linearum curvarum cum lineis rectis comparatione* (1660), which discusses the determination of the length of curves in a seemless series of seven 'propositiones' and their proofs.[153] The austere style of reasoning adopted by Fermat more or less precludes any digression. However, even Fermat cannot resist the temptation to add an appendix (pp. 238–54) which does not form part of his overall argument but deals with one particular problem raised at the very end of the main text. Most authors, by contrast, felt free to digress throughout their account. In his *De lactibus*, Aselli clearly aims to provide a complete and ordered description of the lymphatic vessels in the mesentery, but this does not keep him from constantly getting off the point to discuss side issues and confute colleagues. He or his posthumous editors even draw attention to such excursuses in the summaries, introducing the individual chapters, where they may be announced by the keyword 'obiter' ('incidentally', 'in passing').[154] Others went further still in this respect, advertising digressions on their title pages. That at the start of Gottfried Voigt's *Contra nivis albedinem realem dissertatio physica* (1669), a work basically committed to the thesis that the whiteness of snow is not real but only apparent,[155] promises:

> Explicatur etiam tota nivis natura et origo colorum tum in genere, tum albedinis in specie, multis hinc inde additis problematibus iucundis atque curiosis, verbi gratia de colore Adami, coeli, aquae, incolarum frigidioris soli, item de vermibus nivis et sexangulari eius figura aliisque.
>
> Moreover, the entire nature of snow and the origin of colours in general as well as of whiteness in particular are explained and many pleasurable and peculiar questions are added from here and there, for instance about the colour of Adam, the sky, the water, and inhabitants of cold countries, along with snowworms, the hexagonal shape of snow, and other things.

[153] Modern edition: Tannery et al. 1891–1922, vol. 1, 211–54 (from which I cite in what follows); mathematical analysis: Mahoney 1994, 267–81.

[154] See the summaries of chs 2, 9, 10, 17, 22; also, in the main text, pp. 14, 23, 50.

[155] The question was much discussed. The various positions taken are summarized on pp. 34–7.

Whether in their main argument or in digressions, monographs were expected to impart new insights, as already noted. However, the amount and kind of novel information actually contained in such books varies widely, as does its presentation. Fresh content in monographs is basically of two kinds: new evidence and new ways of constructing old evidence, that is, new arguments. Works characterized by the former type—miniature pandects, so to speak—describe a small field in greater detail than has been done so far. Quite logically, they can above all be found in disciplines processing large quantities of material, such as natural history. Linnaeus's systematic enumeration of everything found out by him about the banana plant is a typical example of this brand. On the other hand, many works provide new arguments in order to answer more or less long-standing questions. Fracastoro's *Homocentrica* and Nonio Marcello Saia's *Tractatus, in quo terram esse aqua maiorem demonstratur* (1585) both respond to questions debated since antiquity, which relate to the correct way of translating the celestial phenomena into a model of the universe and the relative extent of land and sea respectively. Kepler's *De nive sexangula*, by contrast, tries to solve a problem brought up by the author himself: why do snowflakes exhibit a hexagonal symmetry? In most monographs, however, fresh evidence and fresh arguments go hand in hand, the former automatically entailing the latter. How many different topics for discussion can be provided by the discovery of one anatomical structure is exemplified by Aselli's work on the lymphatic vessels in the mesentery. Scheiner's observation of the elliptic appearance of the sun close to the horizon gives rise to a thorough argument subsumed under three questions in as many chapters: 'Ubi hoc phenomenon sit' ('Where this phenomenon is situated'), 'Quomodo sol in atmosphaera vaporumque regione sphaeroides oblongum induat' ('How the sun takes on a circlelike but oblong shape in the atmosphere and the region of the fumes'), and 'Quid secum vehat hic aspectus' ('What are the implications of this view').

The presentation of novelty took place in two complementary speech acts: advertisement and apology. Regarding the first of these, novelty clearly was a selling point and was, therefore, flaunted in titles and introductory paratexts. In fact, 'novus' became a common catchword in titles in the sixteenth and seventeenth centuries—from Tycho Brahe's *De nova et nullius aevi memoria prius visa stella* (1573) and Thomas Bartholin's *De unicornu observationes novae* (1645) to Wedel's *Specimen experimenti chimici novi* (1671) and Athanasius Kircher's *Phonurgia nova* (1673).[156] Innovation was also a standard theme in introductory paratexts such as those placed at the start of Santorio Santorio's *De statica medicina* (1614), a monograph in the form of a series of aphorisms in which the author published his discovery that the 'perspiratio insensibilis', the loss of water through the skin

[156] Thorndike 1951.

and through respiration, can be weighed and quantified.[157] Santorio emphasizes the novelty of his insight with impressive self-confidence, calling it 'new and untouched by anybody before me' ('nova quidem et ante me a nemine tacta') and even 'divine' ('divinam') in his letter of dedication (sig. a4^{v}) and elaborating on these statements on the first two pages of the preface to the reader:

> Novum atque inauditum est in medicina posse quempiam ad exactam perspirationis insensibilis ponderationem pervenire. Nec quisquam philosophorum nec medicorum umquam hanc medicae facultatis particulam aggredi ausus est. Ego vero primus periculum feci et, nisi me fallat genius, artem ratione et triginta annorum experientia ad perfectionem deduxi.
>
> It is new and unheard of in medicine that someone should arrive at a precise weighing of the imperceptible perspiration. No one among philosophers and physicians has ever dared to tackle this specific part of the medical craft. I, however, was the first to give it a try, and if my guardian spirit does not deceive me, I led this art to perfection through my reasoning and the experience of thirty years.

At the same time, novelty could cause serious communication problems between an author and his readers. It might impair the latter's understanding, provoke angry reactions by running contrary to preconceived opinions, and attract the quibblers who also haunt Santorio's dedication and preface under a multitude of names—'imperiti et malevoli homines' ('inexperienced, mischievous people'), 'Zoili', 'scioli' ('wiseacres'), 'andabatae', 'Aristarchi', and so on.[158] Most authors were more cautious than Santorio in dealing with such issues: they tried to dispose their readers favourably by displaying a modest, self-critical attitude. A common strategy was to introduce one's work as a preliminary and necessarily imperfect attempt. For this reason, a monograph was sometimes called a 'prodromus', a 'precursor' of a fuller, more accomplished work. The long title of the work commonly known as Kepler's *Mysterium cosmographicum* (1596) starts with this term, and Steno likewise calls his treatise on solids included within solids a 'dissertationis prodromus'. In his opening address to Ferdinando II of Tuscany, he explains that he has to postpone the book he really wants to write because of the unforeseen necessity of visiting his homeland Denmark and now provides a short sketch as a temporary substitute (pp. 184–5). At the very end of his account, he once more emphasizes its provisional character. The final sentence reads (p. 224):

[157] Renbourn 1960, esp. 138–40; *CDSB* 12, 101–4, at 103.

[158] The Cynic philosopher Zoilus of Amphipolis and the astronomer and mathematician Aristarchus were known as overcritical, particularly against Homer. The 'andabata' was a type of gladiator who fought blindfolded.

> Atque haec succincta, ne dicam tumultuaria, relatio est rerum praecipuarum, quae in ipsa dissertatione tum distinctius, tum fusius exponere decreveram addita locorum descriptione, ubi singula observavi.
>
> And this is my concise, not to say overhasty account of the most important subjects that I had decided to expound more precisely as well as more extensively in the monograph itself, adding also descriptions of the places where I observed the individual phenomena.

In the main body of a work, such caution may take the form of withholding judgement or emphasizing the tentative nature of one's conclusions. Good examples of this strategy can be found in Wedel's *Specimen experimenti chimici novi* (to which the author also applies the term 'prodromus', albeit as an adjective).[159] On the question why not all timbers produce the same kind and amount of soot when burnt, for instance, Wedel refuses to commit himself (p. 29):

> Quod vero non ex plantis combustis omnibus similis fuligo accrescat fornaci, forsan subtiliori adnumerandum est volatilitati, quae ab acidis lignorum spiritibus magis ligatur, licet nec illud ex toto concedamus.
>
> The fact that similar soot [as that produced by oak] does not accumulate in the oven from the burning of all kinds of plants may be ascribed to a more subtle volatility that is bound to a greater extent by the acidic spirits of the timbers—although I do not fully subscribe to that [explanation] either.

When he generalizes his earlier (pp. 23–5) reported experience that certain kinds of plants, such as scurvy grass, contain 'sal volatile', he hastens to draw the reader's attention to the dangers of such inductive reasoning (p. 33):

> Qua inducti, licet difficile sit a particulari ad universale argumentari, audemus nihilominus asserere eodem modo posse, si non ex omnibus, multis tamen aliis plantis idem perfici et ad oculum demonstrari.
>
> Induced by this [experience]—even though it is difficult to reason from the particular to the general—I nonetheless dare to state that it is possible to produce the same [salt], if not from all, but from many other plants and to make it plainly visible.

As has already transpired from the example of Santorio, the presentation of novelty has much to do with the fact that an argument never develops in a vacuum but is inevitably embedded in a research discourse: predecessors, opponents, and

[159] 'Quae omnia latius deduceremus, nisi nos prodromum scribere specimen recordaremur' (p. 32).

supporters always play an important role. Every author has to position himself vis-à-vis the tradition and current state of his discipline. Two important choices are to be made in this connection: first, one may either convey the impression of standing aloof from one's colleagues and depending entirely on oneself or openly acknowledge one's interaction with their research. In the latter case, one secondly may take a polemical stance, regarding fellow researchers as adversaries to be overcome[160] or present them as collaborators in a common quest for understanding. The degree, explicitness, and manner of scientific intertextuality apparent in an author's work varies greatly.

Among the most reticent works in this regard is Fermat's discussion of the length of curves. In an introductory note (pp. 211–12), the author baldly states that the problem of ascertaining this length has remained unsolved so far,[161] briefly discounting recent insights by Christopher Wren without naming him. After that, no other author or work is mentioned or referred to throughout the entire account. One could hardly guess from Fermat's text that it actually contributes to an ongoing, partly heated debate.[162] As a somewhat less extreme example, Johann Fabricius's *De maculis in sole observatis* (1611) may be adduced.[163] Discussing the significance of his discovery of sunspots, Fabricius mentions Giordano Bruno, Kepler, Pythagoras, Copernicus, Joseph Scaliger, and Girolamo Cardano (sigs. D2^{v}–D3^{v}), but only in passing and without providing any precise references.

On the other hand, one has monographs such as Aselli's *De lactibus*, whose author develops his entire argument in intense, outspoken dialogue with his colleagues. He constantly mentions pertinent ancient, early modern, and contemporary works in the body of his text and gives exact references, including book, chapter, paragraph, and folio numbers, in the reader-friendly form of marginalia keyed to the text by italic letters. In doing so, he shows himself combative, searching for controversial points rather than avoiding them. As soon as he touches upon a debated point, he discusses it extensively, including much polemic. In chapter 16 (pp. 35–6), for instance, he hits on the question of whether veins belong to the simple or compound parts of the human body ('substantia ista simplex sit et uniusmodi tota an composita et diversis e partibus concreta'). This problem, although by no means vital to Aselli's presentation of the lymphatic vessels, is ventilated at length, including the interpretation of pertinent passages in Galen, the philological argument that not every part of a vein is itself called a vein

160 A presumably influential model in this respect was Galen in works such as those mentioned earlier.

161 Fermat overlooked Hendrik van Heuraet's 'Epistola de transmutatione curvarum linearum in rectas', published only one year before his own work, in 1659 (Mahoney 1994, 268).

162 Anonymous 'geometri' are mentioned in passing on pp. 16 and 25. On the de facto polemical nature of Fermat's work, see Mahoney 1994, 274.

163 On the author and his work, see Berthold 1894.

(whereas a piece of a simple part like the flesh, for instance, is called 'flesh'), and a harsh attack on Gabriele Falloppio and his opposed views:

> Nam quod post omnes respondet Fallopius *r*,[164] esse quamlibet partem venae venam, si venam sumas pro substantia venosa, ex qua formatur organum, non si pro organo ex ea formato, id vero vanissimum commentum est neque huius fieri debet, *τοῦ τῶν δακτύλων ἀποκροτήματος*.... Sed nimirum faciendum, quod chirurgo, Fallopio fuit, ut in desperato ulcere tegeret linteo, quae sanare non posset.
>
> For the answer given by Falloppio, who comes at the end of this debate *r*, that every single part of a vein is a vein, if one takes 'vein' in the meaning of the venous matter from which the organ is formed, not of the organ formed from that matter, is really a most empty fabrication and must be thought to be worth even less than the proverbial 'snapping of the fingers'.... But, of course, Falloppio had to do what a surgeon does in the case of a hopeless ulcer: he covered with a piece of linen what he could not heal.

An even more polemical stance is taken by Scaliger in his *De aequinoctiorum anticipatione diatriba*. In his attempt to assert his misguided theory of the precession of the equinoxes, the author lashes out, not so much against particular opponents, but against the astronomical community as a whole: the astronomers are 'inconsiderate people devoid of judgement' ('imprudentes et iudicii expertes', p. 2), 'who think everything unknown to themselves is wrong' ('quibus falsa videntur omnia, quaecumque ignorant ipsi', p. 80); they hold 'a nonsensical opinion... worthy of old hags' ('anilis et nugatoria... opinio', p. 31) and believe in what is 'the height of absurdity' ('absurdissimum', p. 33); their views are those of one who 'is delusional' ('delirat', p. 64).

Such extremes, however, are not the rule. In fact, most authors seem to be wary of polemics. The very danger of getting entangled in professional quarrels prompts them to create a peaceful atmosphere and to formulate even criticism in friendly terms. Wedel in his work on 'sal volatile' comes closer to representing the mainstream attitude than Falloppio and Scaliger. Reporting the opinion that the salt contained in certain herbs is extremely ethereal and evaporates in a moment, he voices his disbelief with great caution (p. 24): 'de quo equidem pace eorum, quibus receptissima placet sententia, dubitare hic allubescit' ('in this respect, I personally feel here inclined to doubt this position—no offence to those who like the most widespread opinion).

164 In the corresponding marginal reference, fol. 124 of Falloppio's *Liber* [correctly: *Lectiones*] *de partibus similaribus* (1575) is given as the source. In fact, however, the reference apparently is to p. 128 of the first volume (1606) of Falloppio's *Opera genuina omnia*, where this treatise fills pp. 121–84. Its original edition has no page or folio numbers at all and does not run to 124 folios.

Striving for clarity: structure, style, and argument

While monographs can thus be quite diverse regarding some aspects of their form and content, one common denominator is provided by their striving for clarity. Since a monograph required its reader to follow an often complex and perhaps even counter-intuitive argument over a considerable distance, clarity was of the utmost importance in order to make him well disposed, explain matters to him, and convince him of the author's positions. The features to be discussed in this final subsection reflect this fact.

In order to steer the reader safely through the complexities of their arguments, most authors of monographs try to provide their text with a plain and lucid structure which is, moreover, made as explicit as possible and emphasized typographically by the printer. The reader is usually enlightened in this respect early on, either by an introductory running-text summary or, more often, by a table of contents. In the aforementioned note at the beginning of *De linearum curvarum cum lineis rectis comparatione*, Fermat summarizes the *status quaestionis*, his own argument, and the progress achieved thereby with exemplary clarity (pp. 211–12). No 'purely' ('pure') geometrical curve has been equalled to a straight line so far, he states. 'Purely' means that the curve in question must not be based on a presupposed equality of another curve with a straight line. Therefore, the recent discovery of an English mathematician that the length of a cycloid is four times the diameter of its generating circle is no counterexample: it is based on the equality of the circle's circumference with the cycloid's base line. Geometers even have declared the rectification of 'purely' geometrical curves impossible *tout court*. This is wrong, as he, Fermat, will prove and exemplify in his treatise. Tables of contents are very common from early on. Already in Fracastoro's *Homocentrica*, for instance, a particularly neat and lucid example can be found. Occasionally, as in Fehr's *Hiera picra*, a Ramist chart is used to show not only the sequence of the individual chapters but also their textual relation and coherence.

Some authors go to considerable lengths to orient their readers, combining several types of guidance. In his *De lactibus*, Aselli starts from contextualizing and summarizing the entire work in a preface immediately preceding the first chapter, and while there is no general table of contents, each of the work's thirty-five chapters opens with a detailed enumeration of its content in keyword style. The two dense quarto pages of chapter 16, for example, are outlined as follows (p. 35):

> Perpenditur horum vasorum substantia. Sunt e genere venarum. Sunt partes compositae et dissimilares, non similares et simplices. De Galeni et Aristotelis hac in re sententia cum cura inquisitum. Fuchsii, Silvii, Fallopii explosa. Partes componentes venam tres, membranae, fibrae, caro. Aristoteles iterum explicatus, notatus Capivaccius, rursus Fallopius et cum eo Vesalius.

> The substance of these vessels [the lymphatic vessels in the mesentery] is carefully considered. They belong to the class of veins. They are compound, heterogeneous parts, not homogeneous, simple ones. Diligent exploration regarding Galen's and Aristotle's opinion about this issue. Fuchs's, Dubois's, Falloppio's discarded. The parts making up a vein are three: membranes, fibres, flesh. Aristotle explained again. Criticism of Capivaccio, again of Falloppio and, with him, of Vesalius.

Fine-grained summaries such as this, one guesses, could even save readers the trouble of reading the whole book in its entirety. With their help, one could skip less interesting chapters without losing track of the train of thought and focus on those one deemed most worthwhile. That Aselli's work quite unusually also features an index fits in with the assumption that it was conceived not just for reading, but also for selective consultation.

The structure of the main text itself was basically left to the author's preferences and differed greatly from one work to the next. The quite formalized structure of Saia's *Tractatus, in quo terram esse aqua maiorem demonstratur*, for instance, is inspired by the scholastic 'quaestio disputata'.[165] Having made his own position clear in the title, the dedication, and the opening lines of the main text, Saia first adduces five arguments suited to proving the greater extension of water (fols. 1^r–4^r), followed by his own arguments to the contrary a little later (fols. 5^v–14^r). Finally, the five arguments of his opponents are revisited and refuted one by one (fols. 14^r–20^r).

More humanistically minded authors, for whom a scholastic model held little attraction, developed other solutions. Fracastoro's *Homocentrica* falls into three sections: in the first (fols. 3^r–13^r), the principles of his enhanced Eudoxan system of concentric spheres are explained, the second (fols. 13^r–34^v) discusses the sphere of the fixed stars and the planets above the sun, the third and last (fols. 35^r–66^r), the planets below the sun. Each section is subdivided into chapters, most of which are thematically analogous for most planets. In this way, Fracastoro's theory is presented thoroughly and systematically. The ten chapters of Wedel's *Specimen experimenti chimici novi* are strung together more loosely, starting with the medical powers of plants, proceeding to the occurrence of 'sal volatile' in them, ending with the subspecies of this class of substances, and devoting themselves to various aspects such as extraction methods, medical uses, and microstructure in between. However, even Wedel can become quite systematic at crucial points. In chapter 4 (pp. 22–34), where the very existence of his salt in plants is at stake, no fewer than eight pertinent arguments are enumerated one after the other and highlighted by marginalia.

[165] Lawn 1993.

A quite popular approach was dichotomous, a preliminary part being followed by the main argument. In Scheiner's *Sol ellipticus* (1615), a neat dichotomy between the presentation of observational evidence, on the one hand (pp. 1–13), and its interpretation, on the other (pp. 13–30), is established. The overall structure of Aselli's *De lactibus* (1627) likewise comprises two parts: the first (pp. 1–18) explains the state of the art, while the longer second part (pp. 19–79) presents the author's new findings.

Like dissertations, monographs dealing with mathematical subjects often apply structural categories drawn from Euclid and Greek mathematics in general. They are organized as series of relatively short segments whose titles indicate their type of content and their position in the work's textual hierarchy. The structural backbone of Newton's 'De quadratura curvarum' (1704)[166] is provided by a series of ten 'propositiones', which are realized either as 'problemata' or as 'theoremata'. A 'propositio' may contain subsections titled 'solutio', 'explicatio', 'demonstratio', and 'casus'. Moreover, the 'propositiones' are supplemented by 'corollaria', 'scholia', and 'tabulae'. Some of these categories, such as 'casus' or 'tabula', expand the inherited framework according to Newton's needs. Whatever the actual segment titles, the reader easily gains an overview of the work as a whole and always knows what kind of content awaits him in a given passage. Moreover, such a structure fulfils a rhetorical function: since classical mathematics counts as the paragon of scientific certainty, the reader tends to assume that a work following its model is methodically flawless and logically watertight.

A clear structure certainly helped getting a monograph's complex content across, but it was not enough: style played an important role too. Making the right choices in this respect was a demanding task. Part of the problem was terminological: a modicum of technical terms was usually de rigueur, and neologisms were unavoidable when novel notions had to be expressed. Most authors were conscious of this situation and strove to neutralize the harmful effects of jargon by providing some assistance for the reader. Some prefixed or appended small lexica to their works, such as the one-page 'explanation of certain words' ('verborum quorundam interpretatio') inserted before the table of contents of William Gilbert's *De magnete* (1600), where one finds, for instance, his famous 'terrella', defined as a 'magnes globosus' ('spherical lodestone'). A more common practice was to explain technical terms, especially self-coined ones, at their first occurrence. Early on in his *Homocentrica* (1.13, fol. 11^{v}), Fracastoro gives the following stipulative definition ('let us henceforth use X in the meaning Y') of two neologisms of sense that will be of great importance within his system of nested celestial orbs influencing each other's movements:

166 Introductions to the work's mathematical content are provided by Edwards 1979, 226–30 and *CDSB* 10, 50–2. On its role in the Newton–Leibniz controversy, see Hall 1980, 135–41.

> Horum autem duorum orbium maiorem illum, qui circumducit, vocemus circumducentem, minorem vero, qui circumducitur ac proprio motu successive in opposita omnia[167] movet, vocemus circitorem. Uti enim his nominibus necessarium saepe nobis erit.
>
> But of these two [types of] orbs [assigned to each planet], let us call the bigger one, which leads [the other one] around, 'the one leading around' ('circumducens'); the smaller, by contrast, which is led around and through its own movement moves everything in the opposite direction in turn, let us call it 'the one going around' ('circitor'). For we will often need to use these designations.

For the rest, most authors of monographs tried to make up for novelty and difficulty by a lucid style. Moreover, being aware that their works would usually be read cover to cover, they attempted to make the reading experience a pleasant one. Most texts are syntactically classical, and features impeding the flow for the reader—for example overlong, complex periods or clusters of abbreviated references—are as a rule avoided. Of course, the excuse of stylistic carelessness, a topos in technical texts since antiquity, can be found in early modern examples as well. However, when Wedel calls his own writing 'too sloppy and careless' ('negligentior et vacuus') on the second-last page of the preface to his *Specimen experimenti chimici novi*, this is rather meant to draw attention to the fact that his style is actually well groomed and readable. The train of thought is easy to follow, and Wedel has an agreeable manner of encapsulating single notions in medium-sized, well-structured periods, as at the beginning of chapter 4 (p. 22), where the transition from the existence of 'sal volatile' in animals to its occurrence in plants is made as follows:

> Cum concessum sit abundare in regno animali potissimum salium volatilium proventum, metallorum vero et mineralium illis affinium plerumque sal magis fixum sit (quo de specialiter alibi tractabimus), unde difficilis tincturarum paratio, videndum de vegetabili regno, num et illinc salia volatilia in medicum usum peti possint.
>
> Since it has been accepted that in the animal kingdom one has an abundant source of volatile salts but that the salt of metals and their cognate minerals is mostly more stationary (a subject I will discuss separately elsewhere), for which reason the preparation of tinctures is difficult, one must now check with respect to the plant kingdom whether volatile salts for medical purposes can be obtained from there as well.

[167] The print has 'in omnia opposita'. As shown by 'in opposita omnia' (fol. 11ʳ, last line), the printer has inadvertently reversed 'opposita' and 'omnia' in the present passage.

In this way, the reader never feels overchallenged but can nonetheless be sure to have progressed one step further when he has reached the end of a sentence.

Mathematical works are often as a special case in terms of style as in their structure, since Greek geometry serves them as a model in both regards. Although, or precisely because, early modern mathematicians went far beyond the Greeks in terms of content, they took all the more care to communicate their results in a language familiar to their readers. Fermat's *De linearum curvarum cum lineis rectis comparatione* provides a textbook example of this phenomenon: making effective use of groundbreaking infinitesimal methods associated with the calculus of the day, the work closely follows the Greek geometricians, Archimedes and Apollonius in particular, in its manner of exposition.[168] Fermat's proof that any curve can be circumscribed by two figures composed of straight lines, one of them longer, the other shorter than the curve itself, starts as follows (p. 214):

> Exponatur curva aliqua ex praedictis in 2. figura. Secetur basis AF in quotlibet portiones aequales AB, BC, CD, DE, EF, FG et a punctis B, C, D, E, F erigantur perpendiculares BQ, CV, DZ, ER, FM, quae occurrant curvae in punctis P, T, Y, N, O. Ducantur item tangentes AQ, PV, TZ, YR, NM, OI. Ex prima propositione patet tangentem AQ portione curvae AP esse maiorem, item tangentem PV portione curvae PT esse maiorem et sic de reliquis, tandemque etiam ultimam OI portione curvae OH esse maiorem. Ergo figura constans ex omnibus istis tangentium AQ, PV, TZ, YR, NM, OI portionibus curva ipsa maior erit.
>
> Let there be shown any of the aforementioned curves in figure 2. Let the base AF be cut into any number of equal parts AB, BC, CD, DE, EF, FG, and let the perpendiculars BQ, CV, DZ, ER, FM be erected, which shall meet the curve at the points P, T, Y, N, O. Likewise, let the tangents AQ, PV, TZ, YR, NM, OI be drawn. From the first proposition, it is clear that the tangent AQ is longer than the part AP of the curve, that likewise the tangent PV is longer than the part PT of the curve, and so on for the remaining parts, and that finally also the last one, OI, is longer than the part OH of the curve. Thus, the shape consisting of all these parts of the tangents AQ, PV, TZ, YR, NM, OI will be longer than the curve itself.

Just as in Fermat's Greek models, the individual steps of the construction are first prescribed in the jussive subjunctive passive. After that, the figure constructed is described, and it is explained in the indicative what can be learned from it. The technical vocabulary is restricted to a few dozen expressions, and word repetition is uninhibited. The whole account constantly refers to a diagram (the '2. figura' named at the beginning, see Figure 8), whose points and lines, indicated by

[168] Mahoney 1994, 268 (who seems to wonder at Fermat's intellectual perversity).

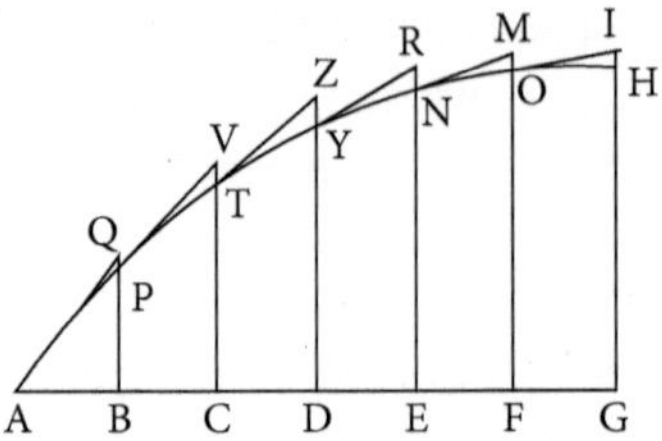

Figure 8 Fermat's '2. figura': a set of straight lines (AQ, PV, TZ, YR, NM, OI), the sum of whose lengths is greater than the length of the curve APTYNOH (Tannery et al. 1891–1922, vol. 1, 216).

capital letters and combinations of such, take the place of nouns in the function of subjects, direct and indirect objects, genitive supplements, propositional supplements, and so on.

Finally, I would like to modify and broaden the stylistic perspective taken so far by turning to 'argumentative style', as it has been called,[169] that is, by considering style in its, strictly speaking, argumentative function. Argumentative style allows for a wide variety of individual choices and solutions which are difficult to classify, and once more, the choices made by individual authors of scientific monographs differ greatly. To provide at least some impression of the range of possibilities, I will briefly discuss three examples. For the first, we can return for a moment to the above passage from Fermat, pondering how he uses the style of Greek geometry to further his argument. The function of Fermat's impersonal, compartmentalized account consists in providing the reader with a trustworthy Ariadne's thread through the labyrinth of the corresponding geometric construction and to leave him no choice but to follow that thread. He may lose track of the thrust of the argument as a whole in the process but will nonetheless be convinced of its correctness, simply because the logic of each step appears impeccable and there are no alternatives in sight.

The train of thought in Fracastoro's *Homocentrica* likewise progresses step by step, but this feature is combined with a fastidious recital of each and every single phenomenon in need of explanation. The discussion of the causes of the varying speed and the positions of Venus in chapter 3.13 (fols. 47^r–48^r) is exemplary in this respect:

> Tardiorem vero in una medietate, velociorem in alia ferri a primo circitore dari dicimus…
>
> Secunda vero inaequalitas directionis et retrogradationis a secundo circitore provenit…

[169] Eemeren 2019. His attempt to give the notion clear contours remains at a quite abstract level.

Cur autem non semper anno uno orbem expleat Venus, sed modo citius, modo tardius, causa est in circitore secundo…

Ad solem vero numquam opponi Venerem, sed semper aut coniunctam ferri aut praecedere aut sequi intervallo non maiore quam graduum 47 minutarum 32, causa praecipue pendet e circitore secundo…

Quoniam vero et illud interdum accidit, ut elongatio a sole maior etiam fiat quam gradus 45 minutae 30, quippe cum saepe graduum 47 minutarum 32 spectetur, quae omnium maxima est, hoc quidem addi a primo circitore debemus existimare.

But that [Venus] moves more slowly in one half [of its orbit], more swiftly in the other, this property, I say, is given to it by the first 'circitor'…

But the second inequality in its forward and backward movement arises from the second 'circitor'…

But the cause why Venus does not always complete its orbit in one year but sometimes more speedily, sometimes more slowly, lies with the second 'circitor'…

But as for the fact that Venus never stands in opposition to the sun but always moves along with it or precedes or follows it at a distance of not more than 47 degrees and 32 minutes, the cause in the first place depends on the second 'circitor'…

But since also the following sometimes happens, that the longitudinal difference from the sun becomes even more than 45 degrees and 30 minutes, as it is often seen to be 47 degrees and 32 minutes, which is the greatest of all [possible differences], this we must reckon to be added by the first 'circitor'.

The method used to convince the reader of the correctness of Fracastoro's homocentric system is thus simple: he enumerates all the peculiarities of the course of Venus and demonstrates that and how the two 'circitores' assigned to the planet by him are capable of providing an explanation in each case. Other than in Fermat's treatise, this hardly amounts to formal proof. However, the cumulative effect of all these explanations is still strong, if only because the reader cannot be expected to have an equally efficient counter-model at hand that he could oppose to Fracastoro's.

Both Fermat's and Fracastoro's argumentation is quite schematic: it bases itself on a preconceived formula that is repeated as many times as necessary. By way of conclusion, I would like to take a look at a more individual and inventive example, from the crucial eighth chapter (pp. 41–2) of Harvey's *Exercitatio anatomica de motu cordis et sanguinis*, in which, after much preparation, the author reveals his novel key idea.[170]

[170] The text of the original edition is faulty in a number of places, and its punctuation and subdivision into paragraphs are misleading. I follow the text printed in Korenjak 2019, 246–50, creating one extra paragraph after 'candore'.

Huc usque de transfusione sanguinis e venis in arterias et de viis, per quas pertranseat, et quomodo ex pulsu cordis transmittatur, dispensetur. De quibus forsan sunt aliqui, qui antea aut Galeni auctoritate aut Columbi aliorumve rationibus adductis assentiri se dicant mihi. Nunc vero de copia et proventu istius pertranseuntis sanguinis quae restant—licet valde digna consideratu—cum dixero, adeo nova sunt et inaudita, ut non solum ex invidia quorundam metuam malum mihi, sed verear, ne habeam inimicos omnes homines: tantum consuetudo aut semel imbibita doctrina altisque defixa radicibus, quasi altera natura, apud omnes valet et antiquitatis venerandae suspicio cogit. Utcumque, iam iacta est alea. Spes mea in amore veritatis et doctorum animorum candore.

Sane cum copia quanta fuerit, tum ex vivorum experimenti causa dissectione et arteriarum apertione, disquisitione multimoda, tum ex ventriculorum cordis et vasorum ingredientium et egredientium symmetria et magnitudine—cum natura nihil faciens frustra tantam magnitudinem proportionaliter his vasibus frustra non tribuerit—, tum ex concinno et diligenti valvularum et fibrarum artificio reliquaque cordis fabrica, tum ex aliis multis saepius mecum considerassem et animo diutius evolvissem, quanta scilicet esset copia transmissi sanguinis, quam brevi tempore ea transmissio fieret, nec suppeditare ingesti alimenti succum potuisse animadverterim, quin venas inanitas, omnino exhaustas, et arterias ex altera parte nimia sanguinis intrusione disruptas haberemus, nisi sanguis aliqua ex arteriis denuo in venas remearet et ad cordis dextrum ventriculum regrederetur, coepi egomet mecum cogitare, an motionem quandam quasi in circulo haberet.

Quam postea veram esse repperi et sanguinem e corde per arterias in habitum corporis et omnes partes protrudi et impelli a sinistri cordis ventriculi pulsu, quemadmodum in pulmones per venam arteriosam a dextri, et rursus per venas in venam cavam et usque ad auriculam dextram remeari, quemadmodum ex pulmonibus per arteriam dictam venosam ad sinistrum ventriculum, ut ante dictum est. Quem motum circularem eo pacto nominare liceat, quo Aristoteles aërem et pluviam circularem superiorum motum aemulari. Terra enim madida a sole calefacta evaporat, vapores sursum elati condensantur, condensati in pluvias rursum descendunt, terram madefaciunt, et hoc pacto fiunt hic generationes et similiter tempestatum et meteororum ortus a solis circulari motu, accessu et recessu.

So far I have spoken about the transmission of the blood from the veins into the arteries, about the ways by which it passes, and how it is pumped over and distributed by the heartbeat. Regarding these topics there may be some who, after invoking Galen's or [Realdo] Colombo's authority, will say that they agree with me. But now, when I will speak about what remains to be said regarding the quantity and source of this passing blood, this is very much worth pondering indeed but so new and undreamt of that I not only fear ill for me from the envy of certain people but am afraid that I will have all human beings as enemies: to such an extent habit or a tenet once absorbed and deeply rooted, another nature, so to speak, holds sway over all people, and such constraint is exercised by the admiration for time-honoured antiquity. Anyway, the die is cast already. I place my hope in the love of truth and the veracity of learned minds.

Indeed, I had often deliberated by myself how great the quantity [of the blood] actually was, basing myself on experimental vivisection, the opening of the arteries, and manifold investigations, on the symmetry and size of the cardiac ventricles and the vessels leading in and out—for nature does nothing in vain and has not provided the vessels with this considerable size, in which respect they correspond to one another, in vain—, on the matching up, meticulous craftmanship of the valves and muscle fibres and the structure of the heart in general, and on many other things. I had pondered for quite some time how great the quantity of the blood passing through had to be as a consequence, how little time this passing took, and had realized that the liquid generated by the food taken in could not be enough to prevent us from having the veins emptied, utterly depleted, and the arteries, on the other hand, burst by the irruption of too much blood, if the blood did not somehow return from the arteries to the veins and find its way back into the right cardiac ventricle: at this point, I began to consider privately by myself whether the blood was not possessed by a certain movement in a circle, so to speak.

This movement I found to be true afterwards: the blood is pumped and pushed from the heart through the arteries into all parts of the fabric of the body by means of the beating of the left cardiac ventricle, in the same way as into the lungs through the artery-like vein by the beating of the right one, and it flows back through the veins into the hollow vein and to the right auricle in the same way as it returns from the lungs through the artery called 'vein-like' to the left ventricle, as said before. This movement it should be legitimate to call 'circular' in the way in which Aristotle says that air and rain imitate the circular movement of the things in the heavens: the moist earth, warmed by the sun, produces fumes, the fumes are carried upwards and become denser, and having turned into rain they fall down again and moisten the earth. In this way, life is engendered here and weather conditions and meteorological phenomena arise from the sun's circular motion, its drawing nearer and moving away.

The opening paragraph of this passage serves to prepare the reader for what Harvey has in store for him—something very important ('valde digna consideratu') but also shockingly novel ('nova atque inaudita'). Harvey presents himself as a hero ready to take on the whole world, a Caesar-like figure about to cross a Rubicon ('iacta est alea'). Implicitly, the situation is presented as a test for the reader as well: will he stand with the envious and the mass or with the 'docti' in whom Harvey places his hope?

In the second paragraph, the prevailing opinion is subjected to a vivid *reductio ad absurdum*, including some drastic imagery (empty veins, bursting arteries). The reader temporarily gains access to Harvey's thought processes, his psychology of discovery, and is invited to share his reasoning. This reasoning is protracted, thorough, and profound ('cum... saepius mecum considerassem et animo diutius evolvissem'), and it is presented in a sentence of correspondingly staggering length and complexity, the structure of which I have been unable to reproduce exactly in the translation. The reader has to work hard to keep track and has to be patient: the revelation keeps being deferred, and suspense keeps rising. Only at the very end of the sentence is truth dawning on Harvey and the reader.

All of a sudden, everything resolves swiftly. What has been an indirect question just now ('an motionem quandam quasi in circulo haberet') becomes a fact, and the details of the circulation of blood are described clearly and precisely. Concomitantly, the shock of novelty is mitigated by the observation that the newly discovered truth only transfers the already known and accepted model of the pulmonary circuit to a greater scale. The reader is definitely soothed by the analogy of a much older and more venerable circular notion, the Aristotelian idea of the hydrological cycle, and by a hint at its origin in the revolutions of the heavens: everything important in the world, one feels, moves in circles, and it would be strange should the blood behave otherwise.

In the end, Harvey's early modern readership was probably won over to his position (if at all) as much by the captivating drama into which the discovery of the circulation of blood had been transformed as by scientific argument in the modern sense of the term. Clearly, Harvey was not only an exceptional anatomist but a highly gifted writer as well.

8
Summarizing knowledge

Pandects

Pandects, compendia, textbooks

The genres whose main purpose it is to store commonly accepted knowledge form a more homogeneous group than those fulfilling other functions, and the boundaries between them are more fluid. Some introductory remarks on their mutual relationship, their similarities, and their differences—on which I will elaborate later in this chapter—are, therefore, in order.

I roughly distinguish three classes of overview works, to be discussed in this order: pandects, that is, big works of an encyclopedic character; compendia, small, portable handbooks; and textbooks for institutionalized teaching. The last genre is distinguished from the other two ones by its social context. Phenomenologically, by contrast, textbooks can be very similar to compendia. Moreover, their actual use in class can be hard to verify or disprove. I am encouraged to make a categorial difference between pandects and compendia by the large number of works unequivocally attributable to the one or the other category and by the authority of Conrad Gessner, who thought long and hard about the problems of knowledge storage and retrieval which such books are designed to solve. In the preface to the third volume of his *Historia animalium* (1555, sig. [a5^{r-v}]), he defines both formats, sketches their opposing characteristics, and explains how they supplement each other. I will repeatedly have occasion to quote Gessner's reflections in what follows.[1] For the moment, it will suffice to summarize my tripartite classification by way of a schematic comparison of some salient features (Table 3).

The limits of this scheme are evident and the difficulties in assigning a given work to one of the three categories can be noteworthy. A glance at Francesco Barozzi's *Cosmographia in quattuor libros distributa* (1585), to be discussed as a compendium in the section on these, may serve as an initial warning. The *Cosmographia* is advertised as a compendium in its full title, which highlights its 'brevitas' and contrasts it with Ptolemy's massive *Almagest*. Yet 'brevitas' means a total of over 500 quarto pages in this context, and the book's content is detailed in

[1] The interest of this passage (to which I will henceforth refer by Gessner's name only) was first highlighted by Bäumer 1991–6, vol. 2, 42–3 n. 2.

Latin Scientific Literature, 1450–1850. Martin Korenjak, Oxford University Press. © Martin Korenjak 2023.
DOI: 10.1093/oso/9780198866053.003.0009

Table 3 Characteristics of pandects, compendia, and textbooks

	Pandect	Compendium	Textbook
Format	folio or quarto	quarto or smaller	quarto or smaller
Indexes?	yes	sometimes	rarely
Perusal or consultation?	mainly consultation	perusal and consultation	mainly perusal
Classroom use?	no	no	yes

an index of some 1,500 entries. At the same time, Barozzi has the ambition of replacing Sacrobosco's *Sphaera*, the standard textbook in the field, in which he notes no fewer than eighty-four errors (preface to the reader, sigs. [b7^{r}–e8^{v}]). This implies that he would like to see his own work used as an academic textbook, although there are no indications that this ever happened.

Pandects: name, notion, history

The works discussed in this section aim at providing a comprehensive overview of a certain field in running text. Despite their encyclopedic aspirations, they are not encyclopedias in the current or, necessarily, in the early modern sense of that word. The meaning then, of 'encyclopedia',[2] differed from that associated with its modern equivalent: it did not denote a lemmatized and alphabetized work of reference, a sense that became common only in the eighteenth century,[3] but a totality of knowledge conceived as a well-structured whole. (An offshoot of this was the application of the word to a form of academic teaching providing an organically structured overview for beginners.[4]) This unity was usually not imagined in the material form of a book, and 'encyclopedia' was accordingly rare as a book title. To distinguish the works which I have in mind from modern books called encyclopedias, on the one hand, and early modern forms of conceptualizing knowledge and organizing teaching, on the other, I borrow Gessner's term for them: 'pandectae'.[5]

Pandects in the sense just defined have a long history.[6] In antiquity, Varro's lost *Disciplinarum libri IX*, Celsus' partly preserved *Artes*, and Martianus Capella's

[2] On the history of word and notion, see Dierse 1977. [3] Freedman 1994, 219, 233.

[4] Dierse 1977, 41–6.

[5] Gessner had published his own, conceptually very different *Pandectae* a few years before as the second part of his great bibliographical project (see Chapter 5, 'Bibliographies'). His redefinition of the term may be inspired by the alternative title of the *Digest* (Blum 1969, cols. 1027–8 n. 21).

[6] For an overview, see Collison 1964, 1–137 (who speaks of 'encyclopaedias' but uses this designation quite loosely).

Nuptiae Mercurii et Philologiae were prominent examples, but the towering work, and at the same time the one focusing most strongly on nature, was Pliny's *Naturalis historia*. Among works dedicated to single scientific disciplines, Aristotle's *Historia animalium* and its counterpart on plants by Theophrastus, Strabo's *Geography*, and Ptolemy's *Almagest* and *Geography* come to mind. Regarding the Middle Ages, one thinks of the scholastic genre of the 'summa' and Vincent of Beauvais's *Speculum maius*. As a part of that genre, the *Speculum naturale* constitutes the major work dedicated to the natural world but by no means the only one: medieval pandects dealing with this field are numerous.[7] Many of the works just named were printed in the age of incunabula, translated into the vernaculars, and proved influential in early modern times.[8]

At the same time, new syntheses began to be compiled, as a feeling of information overload occasionally voiced since antiquity was compounded by the expanding world of the Age of Discovery, the onset of sustained scientific progress, and the faster and wider dissemination of knowledge old and new through print.[9] During the sixteenth and seventeenth centuries, pandects were a flourishing genre. There obviously was a market for such books despite their high price,[10] even though it is not clear in many cases who actually bought and used them. Gessner envisages a readership mainly composed of academics specializing in one particular field of knowledge ('quorum usus erit praecipue doctoribus et professoribus uni alicui scientiae particulariter addictis'), and such groups must indeed have constituted an important stratum of buyers as far as medicine and the disciplines taught in arts faculties were concerned.[11] In fields with no academic background, by contrast, it is hard to imagine that pandects would have been marketable on the basis of the few experts around. Less sharply defined groups of more or less educated buyers who purchased such books out of personal interest must have played a role as well. This is true, for instance, of the earth sciences, which produced a number of distinguished examples. In Giovanni Battista Riccioli's *Geographia et hydrographia reformata* (1661), a generally learned and mathematically demanding account of the terraqueous globe,[12] a long table of geographical locations (ch. 9.4, pp. 402–25) is introduced with the remark that place names are given in the vernaculars for the benefit of shipowners

[7] Meier 2002 contains a number of pertinent case studies.

[8] For the *editiones principes* of the ancient works, see *DNP* suppl. 2. Two medieval examples: Bartholomaeus Anglicus's *De proprietatibus rerum* was in print 1472–1609 (Freedman 1994, 242), Vincent's *Speculum naturale*, 1476–1624 (http://www.vincentiusbelvacensis.eu/bibl/ed2.html, accessed 16 February 2023).

[9] On premodern information overload and ways of dealing with it, see Blair 2010. A brief overview of early modern 'encyclopedias' (including my pandects) is provided by *EEMHO*, s.v. 'Encyclopedia'. For more detailed treatments, see Eybl et al. 1995 (where hardly any Latin examples are mentioned, however) and the pertinent contributions to Meier 2002.

[10] On the prices of Vesalius's *De humani corporis fabrica*, see Margócsy et al. 2018, 13–18.

[11] For Vesalius's *Fabrica*, see the ownership data provided by Margócsy et al. 2018, 6–7, 30–55.

[12] Korenjak 2018a, 76–7.

('in gratiam nauclerorum'). If this is not just naive, it suggests that a pandect could have a heterogeneous readership, different parts being of interest to different groups, including non-Latinate ones. On a larger scale, the languages of publication point in the same direction: Latin pandects had company from works in all major vernaculars (although Latin remained the leading language in the field until the end of the seventeenth century, if the statistics made for a subgroup of such works, 462 texts titled 'theatres' of knowledge, is in any way representative).[13] In addition, translations in both directions were made, and bilingual or multilingual lexicons are found appended to some works. For some buyers, however, the language presumably was immaterial or at best of interest for its prestige value: presenting themselves as stately, costly tomes, pandects must have proved particularly attractive to 'vanity purchasers'.[14]

From a publisher's point of view, a pandect first and foremost meant a huge investment. In some fields, this made publishers wary of such endeavours, especially in the case of highly technical books. For an author like Kepler, getting his magna opera, such as the *Epitome astronomiae Copernicanae* (1618–21), into print was a lifelong struggle. He even had to spend his own money to do so on occasion, as publishers anticipated that his works would not sell well.[15] Some pandects, for example Bernardo Cesi's *Mineralogia* (1636)[16] or Mario Bettini's *Aerarium philosophiae mathematicae* (1648), were never reprinted. On the other hand, a considerable number of works proved commercially successful, going through many editions and reprints, as well as being translated into different vernaculars, over several decades or even a century. Giovanni Pontano's astrological pandect *De rebus coelestibus*, for instance, was published at least ten times, alone or with other works, between 1512 and 1566. Pietro Bongo's detailed survey of numerology, the *Numerorum mysteria*, appeared in seven variously titled, continually expanding versions from 1583 until it reached its definitive form in 1599; afterwards, it saw four reprints in the early seventeenth century, and a condensed version came out as late as 1689.[17] Bernhardus Varenius's *Geographia generalis* (1650) went through some thirty complete, partial, and abridged editions in five languages (including Russian) until the end of the eighteenth century; three of these were based on a revision made by Newton.[18]

[13] Friedrich 2015, 304–5. In total, 184 of his 462 titles are in Latin (30 of 59 in the sixteenth, 134 of 289 in the seventeenth, 17 of 103 in the eighteenth century; apparently, 11 multivolume works, 3 of them in Latin, which cannot be clearly assigned to a specific century, have been left out by Friedrich for that reason). The next most important languages are German (144 titles all in all), French (43), Spanish (41), and Italian (30).

[14] The term is taken from Maclean 2016, 315. See there also pp. 307–8, 328 for this category of buyers in the eighteenth century.

[15] See Nummedal and Findlen 2000, 169–71, on the *Mysterium cosmographicum* and the *Tabulae Rudolphinae*; see Pantin 2006, 222–6 on the *Epitome*, on whose lack of popularity, see Gingerich 1990, 192.

[16] Thorndike 1929–58, vol. 7, 254–7; Mottana 2017.

[17] Reprint of the 1599 edition with introduction: Ernst 1983. See there, pp. 8–11, on the publication history.

[18] Lange 1961, 9–15. On the work in general, see Nicolson 1959, 176–9; *CDSB* 13, 583–4.

From the turn of the eighteenth century, the vernaculars started to take over in encyclopedic works. At the same time, for whatever reasons, pandects were superseded by different forms. 'Historiae litterariae', such as Daniel Georg Morhof's *Polyhistor* (1688–1708, 4th and last edition 1747), which were thematically universal but provided a first orientation and a bibliographical overview rather than exhaustive factual information, were one expression of this trend.[19] Even more important were the lemmatized and alphabetized encyclopedias written by a team of authors, the most famous example of which is the *Encyclopédie* (1751–80).[20] In both these genres, the character of the encyclopedic literature of the previous centuries had thoroughly changed.

Early modern pandects concerned with the natural world

Even apart from the fluid boundaries between pandects, compendia, and textbooks, there is much overlap between pandects and other genres, at least as far as the natural world is concerned. Commentaries, 'historiae', dissertations, and monographs frequently harbour encyclopedic aspirations. Pertinent examples from the first two genres can be found in the relevant sections. Regarding dissertations and monographs, one may think of works such as Thomas Ittig's *Lucubrationes academicae de montium incendiis* (1671), in which the reader finds everything about volcanoes on 350 octavo pages, and Thomas Bartholin's *De unicornu observationes novae* (1645), an 'encyclopedia of the unicorn' and other one-horned animals comprising over 300 octavo pages and featuring two indexes.[21] Moreover, there are books of encyclopedic dimensions which lack the unity and systematic approach of a pandect. Gaspar Schott's *Physica curiosa* (1662) and *Technica curiosa* (1664), for instance, are open-ended, anthological collections of natural and technological marvels which make no attempt at closure or systematization. The same is true, a fortiori, of compilations of several works by one or more authors: tomes like Marin Mersenne's *Cogitata physico-mathematica* (1644), 'opera omnia' editions like Giorgio Baglivi's *Opera omnia medico-practica* (1704), or the massive chymical collections from Johannes Petreius's *De alchemia* (1541) with its ten components to Jean Jacques Manget's *Bibliotheca chemica curiosa* (1702), which unites over 140 individual works.[22]

Marginal items such as these apart, pandects, strictly speaking, can be classified according to their reach. On the basis of this criterion, four classes may be

[19] Grunert and Syndikus 2015.

[20] See Collison 1964, 82–228, and the last six contributions in Eybl et al. 1995. How and why encyclopedic works turned from topical order and running text to alphabetically ordered lemmata awaits further investigation (Bogner and Vögel 1995, 305).

[21] Roling 2014, 181–6 (the quotation—'Enzyklopädie des Einhorns'—on p. 182).

[22] Chymistry was prone to such collections because it resisted authoritative codification in one single work due to its secretiveness and terminological confusion.

distinguished: thematically unrestricted pandects devoting significant space to nature and the sciences; pandects covering only the natural world in its entirety; pandects focusing on one specific discipline; and pandects dealing with only part of a discipline. My focus will be on the last two categories, which I will discuss together after some short remarks on the first and second.

In the first category, Georg Reisch's huge catechism, the *Margarita philosophica* (1503), and Johann Heinrich Alsted's *Cursus philosophici encyclopaedia* (1620) deserve pride of place. Of their twelve and twenty-seven books, books 4–10 and 8–14 respectively are dedicated to the natural world.[23] Theodor Zwinger's *Theatrum vitae humanae* (1565), the most comprehensive pandect of its time,[24] at least devotes some space to medicine.

The conceptual framework of most pandects giving an overview of the entire natural world is either provided by Aristotelian natural philosophy or results from its more or less emphatic revision. Accordingly, their structure often follows the customary order imposed on Aristotle's books in early modern times. The discussion of wondrous natural phenomena in John Jonston's *Thaumatographia naturalis* (1632), for instance, is based on this order: it starts with the sky and the cosmos as a whole (section 1), proceeds to meteorology (sections 2–4) and biology (sections 5–9), and concludes with anthropology (section 10). Among authors trying to replace Aristotle's natural philosophy with their own, Bernardino Telesio chooses a similar 'descensus' structure for his *De rerum natura iuxta propria principia* (definitive edition 1586).[25] Other anti-Aristotelian pandects adhere to this pattern less closely or not at all. In Francesco Patrizi's proudly titled *New Philosophy of Everything* (*Nova de universis philosophia*, 1591), Platonic and Aristotelian themes are arranged in four large sections termed 'panaugia', 'panarchia', 'pampsychia', and 'pancosmia' (approximately 'complete discussion of splendour, dominion, the soul, the universe'), even though the 'descensus' structure resurfaces in the final section.[26] William Gilbert's posthumous *De mundo nostro sublunari* (1651) ostentatiously focuses on the sublunary world, although some general cosmology is included as well.[27]

[23] On Reisch's work, see Collison 1964, 85–6 and Dierse 1977, 11–13; on Alsted's, see Dierse 1977, 18–20 and Hotson 2000, 29–39, 70–2, 152–77 (plus his index at pp. 253–4). Freedman 1994, 238–9, briefly discusses both. Of the 35 books of the *Encyclopaedia*'s expanded 1630 edition in 7 volumes, books 13–20, 27–30, and 33 deal with natural sciences, mathematics, and technology. A posthumous elaboration of Alsted's work in 4 volumes, the *Scientiarum omnium encyclopaedia*, appeared in 1649.

[24] Zedelmaier 2008 and 2015, 22–33.

[25] On the ordering of Aristotle's natural philosophy, see Chapter 1, 'Disciplines'; on the 'descensus' as a more general structural principle, see Hübner 2002. Telesio's work has been edited by Bondi 2009 (with Italian translation) and Giglioni 2013 (facsimile with introduction and additional index); see also Kristeller 1964, 91–109; Mocchi et al. 2012.

[26] Brickman 1941; Kristeller 1964, 110–26; further literature in *CDSB* 10, 416–17.

[27] Kelly 1965.

The single disciplines and subdisciplines did not all lend themselves equally well to an encyclopedic treatment. Two pertinent criteria, clear contours and a quantitatively manageable body of knowledge, were easiest fulfilled by some minor disciplines like numerology, physiognomy, or metoposcopy. Consequently, such fields were exhaustively treated in works such as Pietro Bongo's *Numerorum mysteria*, Giambattista Della Porta's *De humana physiognomonia* (1586), and Girolamo Cardano's posthumously published *Metoposcopia* (1658).

Among the major disciplines, those discussed exhaustively and authoritatively in one single work already in antiquity continued to attract similar presentations. The foremost example is astronomy, where Ptolemy's *Almagest* was followed by a series of works in a comparable format such as Kepler's *Epitome astronomiae Copernicanae* (1618–21), Giuseppe Biancani's *Sphaera mundi* (1620), and Giovanni Battista Riccioli's *Almagestum novum* (1651–65).[28] In astrology too, ancient and medieval authorities (Ptolemy again, Abu Ma'shar) found early modern successors such as Tommaso Campanella (*Astrologicorum libri VII*, 1629). Other authors, for instance Pedro Ciruelo (*Apotelesmata astrologiae Christianae*, 1521), Johannes Schöner (*De iudiciis nativitatum*, 1545), and Francesco Giuntini (*Speculum astrologiae*, 1573), focused on the subdiscipline(s) of (mundane and) natal astrology.[29] Geographical pandects once more followed in the footsteps of Ptolemy, whose *Geography* was no less in need of replacement than his *Almagest*. Varenius produced the most popular pandect of physical geography of the seventeenth and eighteenth centuries (*Geographia generalis*). Riccioli's *Geographia et hydrographia reformata* made a point of discussing the seas thoroughly as well as the land masses. Some works zoomed in on specific portions of the earth's surface. Johannes de Laet's *Novus orbis* (1633), which the author himself had translated from his Dutch original (1625) and expanded, provides an exhaustive account of the Americas.

In those disciplines less suited to being summarized in pandect format, problems differed from one to the other. In medicine, Avicenna's *Canon* was not replaced by another pandect of similar comprehensiveness and authority after it went out of fashion from around 1650.[30] The immense amount of material accumulated over the preceding two millennia, in combination with a number of novel and controversial developments from Paracelsianism to iatromechanics, apparently intimidated potential successors. Such pandects as appeared focused

[28] Modern edition of the *Epitome*: *KGW* 7. On the *Almagestum novum*, see Graney 2015; on its protracted gestation and publication in particular, see Remmert 2011, 76.

[29] On these and further examples of astrological pandects and their position with respect to the Ptolemaic and Arab traditions, see Hasse 2016, 254–72. The title page of Ciruelo's *Apotelesmata* explicitly states that the work is 'super duabus tantum iudiciorum partibus, hoc est, de mutationibus temporum et de genituris hominum, reiectis omnino interrogationibus et vanis electionibus falsorum astrologorum.'

[30] Siraisi 1987, 6–7. See also her chronological lists of *Canon* editions and commentaries, pp. 359–76.

on particular aspects. Andrea Cesalpino's *Praxis universae artis medicae* (1606), for instance, is only concerned with the practical side of medicine, and even for this, it presents itself as a compendium rather than an exhaustive treatment, as its 700 pages lack an index and an introductory poem underlines the work's conciseness. Johannes Dolaeus's *Encyclopaedia medica dogmatica* (1691) is only a nosology focusing on internal diseases, even though it fills some 1,300 quarto pages and covers the thinking of authorities ancient and modern from Hippocrates to Descartes.[31] One of the first works aspiring again to a truly comprehensive coverage of the discipline (albeit from a particular, namely iatromechanical point of view) is Friedrich Hoffmann's *Medicina rationalis systematica* (1718–40). The lengthy gestation of this pandect and its bulk—four tomes, the last in six parts, comprising *c.*5,000 quarto pages—make the reluctance of other authors understandable.

Things below the earth, originally the subject of meteorology and destined to become that of geology, formed an unwieldy complex which was approached from different angles. Among Georg Agricola's major works, *De natura fossilium* (1546) could be called a mineralogical pandect,[32] while *De re metallica* (1556) discusses the subterranean from the perspective of the mining industry.[33] That *De natura fossilium* is combined in print with five other texts by Agricola, whose genres range from the philosophical treatise to the dialogue and the dictionary, seems to betray an awareness that the field cannot be fully apprehended within the limits of one single work and an intention to make good for this by the creation of an encyclopedic collection similar to those mentioned before. The best seventeenth-century example of a mineralogical pandect is probably Cesi's *Mineralogia*. In his *Mundus subterraneus* (1665), Athanasius Kircher tries to come to grips with the earth's interior in the context of his 'baroque universal science'.[34]

The diverse subjects covered by the different meanings of 'mathematics' (geometry and arithmetic; the quadrivium; pure mathematics; 'mathematica mixta') were equally difficult to squeeze between two book covers. Oronce Finé's *Protomathesis* (1532), which consists of four parts dealing with practical arithmetic, geometry, cosmology, and astronomy, and some pertinent instruments (sundials, astrolabes, quadrants), is among the closest approximations to a mathematical pandect.[35] Most works are more selective. Although Bettini's *Aerarium philosophiae mathematicae* comprises three volumes and some 2,000 pages, it only concerns itself with practical applications of the geometrical books

[31] Maclean 2021a, 235.

[32] English and German translations: Bandy and Bandy 1955; Fraustadt and Prescher 1958.

[33] English and German translations: Hoover and Hoover 1912; Schiffner 1928.

[34] For an overview of the contents of the *Mundus subterraneus*, see Thorndike 1929–58, vol. 7, 567–78 and Glassie 2012, 184–95; on its conceptual foundations, see Leinkauf 1993, 221–31 (from whose title I borrow the phrase 'barocke Universalwissenschaft').

[35] See Pantin 2010 (who focuses on the work's diagrams).

(1–6, 10–13) of Euclid's *Elements*. Later, Euler produced comprehensive and at the same time highly innovative accounts of analysis (*Introductio in analysin infinitorum*, 1748) as well as of differential and integral calculus (*Institutiones calculi differentialis*, 1755; *Institutiones calculi integralis*, 1768–70).[36] Gauss's *Disquisitiones arithmeticae* (1801) 'transformed number theory from a scattering of islands into an established continent in mathematics'.[37]

Finally, there are some disciplines for which hardly any Latin pandects are on record. Botany and zoology had their own comprehensive format, the 'historia', as already mentioned. Chymistry appears to have been intrinsically incapable of producing such works (see note 22 above), and modern chemistry emerged too late, as did modern physics.

Transmitting and updating knowledge

As already mentioned, the vogue of pandects was partly due to the rapid accumulation of new knowledge in an age of scientific progress. This fact conditions an important difference between early modern pandects of the natural world, on the one hand, and their ancient and medieval predecessors, as well as the all-encompassing works by Reisch, Zwinger, and Alsted, on the other. The authors or compilers of the latter by and large perceived the knowledge assembled and ordered by them as static. If later editions of their works became more voluminous, this was simply because they had continued reading, filing, and compiling information in the meantime. In the works under consideration here, by contrast, an understanding of knowledge as changeable and evolving made itself felt. Early modern authors of pandects tried to be up to date or even at the forefront of progress, contributing to it themselves. Gessner obliges the composer of such a work to go beyond the results of his predecessors:

> Postremo non aliena tantum recitet, sed summa cura omnique studio de suis etiam observationibus, quidquid desiderari potest, quidquid utile aut iucundum cognitu adinvenerit, scriptis ac inventis veterum addat.
>
> Finally, may he not only parrot foreign knowledge but may he also add, with the greatest care and all possible diligence, to the writings and discoveries of the ancients from his own observations whatever can be perceived as lacking, whatever useful or pleasurable he has found out in addition.

[36] *LEOO*, ser. I, vols. 8, 10, and 11–13. See Merzbach and Boyer 2011, 409–15 and on the first two works, Reich 2005 and Demidov 2005 respectively (with bibliography, including editions and translations). Both Reich and Demidov understand all three works together as forming 'an analytical trilogy' (Reich 2005, 182).

[37] Neumann 2005 (with bibliography; quotation at p. 304); Merzbach and Boyer 2011, 466–70.

Thus, the accelerating evolution of knowledge about nature turned pandects from stable repositories of age-old learning into provisional and rapidly ageing records of progress and the current state of the art. Even if early modern works of this kind contain no new findings by their authors themselves, they may critically examine recent research, incorporate results deemed valid, and report contradictory data, conflicting claims, and unresolved problems.

Admittedly, they do so to widely varying degrees, depending on the discipline discussed, the date of composition, and the person of the author. Programmatic claims and practice often diverge in this respect. Pontano presents his *De rebus coelestibus* (1512) as a simple restoration of ancient astrological knowledge purified from the Arab influences that have crept in in the meantime. Despite this proclaimed aim, he actually also tries to answer the fundamental accusations recently made against the discipline in Giovanni Pico della Mirandola's *Disputationes adversus astrologiam divinatricem* (1496).[38] If Bettini's *Aerarium philosophiae mathematicae* can make any claim to novelty at all, this only applies to its applications of Euclidean results to practical problems, advertised as the 'spices and uses of the geometry of the *Elements*' ('condimenta et usus geometriae Elementorum') in an opening address to the reader (sig. d^r). However, if Bettini is not really innovative, he clearly wants to appear so. His own contributions are characterized as 'new' ('novae') twice on the title page, and the same adjective recurs several times in a short anthology of interesting results preceding the letter of dedication.

In other titles, 'novus' and similar terms figure more prominently and with better justification.[39] Two major examples are provided by the two components of Riccioli's astronomical-geographical diptych, the *Almagestum novum* and the *Geographia et hydrographia reformata*.[40] The former work's claim to novelty is founded less on new insights gained by the author himself than on a painstaking summary and evaluation of recent research. At the heart of this impressive survey of the state of the art in astronomy lies an impartial discussion of the relative merits of the Copernican and Tychonian world systems (while the Ptolemaic system has already been discounted). Riccioli avails himself of the best data obtainable and lists no fewer than 126 arguments for and against both systems.[41]

[38] On Renaissance attacks on Arab astronomy in general, see Hasse 2016, 248–92; on Pontano's reaction against Pico, see Akopyan 2021, 193–5.

[39] On 'novus' in early modern book titles in general, see Thorndike 1951.

[40] Through the publication of these two works, which form counterparts to Ptolemy's *Almagest* and *Geography*, Riccioli presents himself as a modern Ptolemy, although he never says so explicitly and lets the reader come to this conclusion by himself.

[41] See Graney 2015, who shows that Riccioli's decision in favour of Tycho Brahe rests on solid evidence and good judgement. Presupposing the Copernican system, the apparent lack of parallax of the fixed stars combined with their apparent diameter when observed through the telescope (due to technical defects of the telescopes then available, a fact unknown to contemporaries) led to the conclusion that they must be several hundred times the size of the sun, which seemed highly implausible.

No wonder the book superseded its immediate predecessor, Biancani's *Sphaera mundi*,[42] which had been perfectly up to date thirty years earlier: it was republished three times up to 1653 but not thereafter.

In his pandect of geography and hydrography, Riccioli once more shows himself keenly aware of the progress recently made in these disciplines but is somewhat more self-confident in presenting his own contributions. The title of his preface to the reader announces a discussion 'of the origin, progress, and need of reform' ('de ortu, progressu et reformationis necessitate') of geography. More than half of the preface is dedicated to 'the indication of the most important mistakes of ancient or half-ancient geography' ('geographiae veteris aut semiveteris errores praecipui indicati', subtitle of §§ VI–XIII), the latter category comprising sixteenth-century and even near-contemporary authors such as Robert Dudley (1574–1649). Great names have recently occupied themselves with geography and thalassography, Riccioli admits (§ IV), 'but no one has promoted the mathematical part of geography so far that there would not be left much for me to correct or add in practically all parts of this discipline' ('nemo tamen ita partem mathematicam geographiae excoluit, ut non supersint mihi multa vel emendanda vel addenda in omnibus fere partibus huius facultatis'). He goes on to cite the size of the earth, the height of the mountains and the atmosphere, the question of the range of vision from a given point, the gradient of rivers, and much more as examples. The work itself starts programmatically with the phrase 'the outdated beliefs of the ancients' ('obsoletas priscorum opiniones', p. 1) and concludes in the last of its twelve books (pp. 622–34) with the presentation of an instrument invented by Riccioli dubbed 'crux geographica' and designed to solve various geographical problems.

While the 'crux geographica' apparently was not a success and Riccioli's positive contributions to the knowledge of his times relate to details rather than the big picture, some pandects are thoroughly novel in terms of content and conception as a whole. In fact, many major works associated with the Scientific Revolution or counting as landmarks in early modern science more generally fall into this category. I have already had occasion to mention a number of such magna opera by Agricola, Patrizi, Kepler, Euler, and Gauss.[43] For these authors, a genre associated with systematics, thoroughness, and authority could serve to make novel insights more palatable by presenting them embedded in generally accepted knowledge and as part of a long-standing tradition. Kepler's presentation of his third law of planetary motion in the *Epitome astronomiae Copernicanae* is a case in point. The modern reader will be surprised to note how little room is

[42] On Biancani and his work, see *Dizionario Biografico degli Italiani*, vol. 10, 33–5; Paulus 2005, 286, 289–90, 291–2.

[43] One could contend that Copernicus's *De revolutionibus* is an example of an innovative pandect as well. Newton's *Principia* is less orderly and more idiosyncratic.

taken by one of the most important and famous insights of early modern science and how hard it is even to find in this massive and complex overview, known only to specialists today, of the then state of the astronomical art.[44]

Attracting and orienting the reader

In terms of form, the most obvious constant of pandects is quite simply their length, especially for the standards of the time: generally speaking, early modern books tend to be shorter than modern ones. Although pandects devoted to individual disciplines or subdisciplines are less monumental than thematically more comprehensive works such as Pliny's *Naturalis Historia* or Zwinger's *Theatrum vitae humanae*, Kepler's tripartite *Epitome astronomiae Copernicanae* arrives at the equivalent of around 250,000 Latin words of running text, Bettini's *Aerarium philosophiae mathematicae*, of around 500,000, Riccioli's *Geographia et hydrographia reformata*, of around 900,000, and his *Almagestum novum*, of around 1,900,000 words.[45] On the one hand, this was, of course, exactly what these works were about: many words made exhaustive information possible. On the other hand, such dimensions held their challenges. Readers had to be convinced that a volume this size was worth their money, time, and energy, and they needed help with orienting themselves in it.

The first job was done, to an even higher degree than in other publications, by a splendid array of paratexts, particularly by the title page, the frontispiece, and prefatory material. As is evident from the examples adduced so far, pandects often bore genre-specific and colourful titles in order to flaunt their aspirations. Two common metaphors were the book as a treasury ('aerarium', 'thesaurus')[46] and as a theatre or arena ('theatrum', 'amphitheatrum'). The former styled it as a repository of riches which the reader could take out whenever he needed them; the latter suggested the possibility of absorbing its contents as easily and pleasurably as by looking at a performance. 'Theatra' in particular remained popular from the later sixteenth to the early eighteenth century. Works thus titled were

[44] *Epitome* 4.2.1, p. 501 (*KGW* 7, 291). The first and second law are similarly hidden away in chs 58–9 and 40 respectively of the *Astronomia nova* (1609) (*KGW* 3, 364–76, esp. 366: 'Ergo ellipsis est planetae iter'; 263–70), which is, however, no pandect. Rather, Kepler's fictionalized first-person narrative of his research must be called a text *sui generis*—a highly personal response by an extraordinarily gifted writer to very particular circumstances (Voelkel 2001).

[45] These are rough estimates based on the number of words per page and the number of pages. 'The equivalent' means that space taken by figures and symbols as well as larger chunks of non-text (tables, illustrations) is converted into the words that would fit in it. By way of comparison, the present book contains the equivalent of *c.*250,000 English words, which equals *c.*146,000 Latin words, as the average lengths of English and Latin words relate in a ratio of 7:12 (Fucks 1968, 80–4).

[46] A famous example of a 'thesaurus' besides the work of Zwinger is Joseph Scaliger's *Thesaurus temporum* (1606). The full title of Cesi's *Mineralogia* continues with the phrase 'sive naturalis philosophiae thesauri'. The metaphor is particularly apt in a book dealing, inter alia, with precious stones.

published, inter alia, in chorography, geography and cartography, medicine, biology, chymistry, and natural philosophy.[47] 'Mundus' (as in Kircher's *Mundus subterraneus*) was another metaphor announcing totality and completeness, as well as implying some kind of organic unity, a notion lacking from the aforementioned titles.[48] Simply naming the discipline treated in encyclopedic manner was less flashy but carried an air of proud modesty. If the discipline could be designated by a neologism, as in Cesi's *Mineralogia*, which anticipated mineralogy's current name, this was an elegant way of suggesting that it would be discussed in its length and breadth for the first time. In the opening sentence of his work, Cesi admits to have chosen *Mineralogia* (or *De mineralibus*, the work's alternative title) for advertising reasons, 'to please the crowd' ('in gratiam vulgi', p. 1).[49] The name of the discipline could be complemented by adjectives like 'universus', 'universalis' (Kircher, *Musurgia universalis*, 1650), or, as already mentioned, 'novus'.

Frontispieces and engraved title pages are significantly more frequent in pandects than in other publications (as are illustrations in general, see at the end of this section). In a volume that was very expensive anyway, the additional costs did not carry that much weight, and one imagines that vanity purchasers in particular must have felt attracted by these beautiful and often enigmatic compositions. A fascinating item forms the frontispiece of Riccioli's *Almagestum novum* (Figure 9): Astraea, the personification of theoretical astronomy, weighs the Copernican and Tychonian world systems in her balance, finding the latter to be heavier—that is, truer—while the Ptolemaic system has already been discarded and lies on the ground. She is accompanied by the all-seeing Argus, who represents observational astronomy. Ptolemy, comfortably reclining on the ground beside his system, shows himself a good loser, as he comments: 'Erigor dum corrigor' ('I am raised while I am corrected'). The scene is firmly embedded in a theological context by God's hand in the sky and a number of biblical quotations. Riccioli explains part of its meaning in the letters of dedication to the first and second volumes of his work (sigs. A^{r}, ¶A^{r}).[50]

[47] Friedrich 2015.

[48] Another interesting image is that of the book that is tantamount to a library ('instar bibliothecae'), which Alsted has in the full title of his *Cursus philosophici encyclopaedia* and which is also found in the prefaces to the third volume of Conrad Gessner's *Historia animalium* (sig. [a5^{r}]) and to Riccioli's *Almagestum novum* (pp. xviii–xix). In it, the ideas of order, exhaustiveness, and convenience are all present. However, 'bibliotheca' as a main title usually denotes not a pandect, but a bibliography.

[49] The identity of the 'vulgus' is unclear, since the *Mineralogia*, published in Latin, quite obviously was a book for the educated. The analysis of the opening sentence by Mottana 2017, 438–9, is somewhat confused. As examples of self-coined titles which did not become lexicalized, Philipp Jacob Sachs von Lewenheimb's *Gammarologia* ('Science of Lobsters [κάμμαροι]', 1665) and Johann Paul Wurfbain's *Salamandrologia* (1683) may be adduced.

[50] Krifka 2000, 433–6; Remmert 2011, 76–88; Böhme 2016, 88–90. Remmert also briefly discusses the engraved title page of Johann Bayer's *Uranometria* (pp. 127–30; see also Hamel 2010, 51–3) and the frontispiece of Bettini's *Aerarium philosophiae mathematicae* (pp. 226–7).

Figure 9 Frontispiece of Riccioli's *Almagestum novum*.

Regarding prefatory material, one may choose the paratexts introducing Bettini's *Aerarium philosophiae mathematicae* as a telling example, because they make the need of promotion glaringly evident by their strategies and the extravagance of their claims. While the letter of dedication to the bishop of Trento, Carlo Emanuele Madruzzo, is fairly standard, the two texts preceding and following it are not. The former, which has already been mentioned in passing, is a two-page list intended to whet the reader's appetite with a preview of some of the work's more exquisite contents ('aliqua praecipua'). For the latter, the preface to the reader, the Jesuit Bettini has enlisted the services of his fellow religious, Giovanni Antonio Roffeno. Roffeno not only praises Bettini and his work much more explicitly than would have been found acceptable from Bettini himself, but also makes quite extraordinary predictions about its readership and impact. When he recommends the *Aerarium* to students ('tirones', sig. d^{r}), whom it will prepare for the reading of the more difficult mathematical classics, this already makes considerable demands on their money, time, and energy. His subsequent suggestions that the work will reach a global readership of Catholics, heretics, and pagans alike and help to proselytize the Chinese, who are easily impressed by the elegance and certitude of Western mathematics (sigs. d^{v}–[d3^{v}]), are obviously wishful thinking.[51]

The orientation needed by a reader depends on the way he or she uses a book. It is commonly, and presumably correctly, supposed that pandects were consulted on specific points rather than read from cover to cover. Gessner frankly states that they 'are not written to be leafed through in continuous perusal' ('non scripti sunt ut continua lectione evolverentur').[52] However, perusal and consultation should not be understood as mutually exclusive alternatives. In principle, a pandect allowed for selective consultation as well as for the perusal of shorter passages, longer stretches, or the entire work, and readers could choose one or more of these options. As shown by their annotations in the *c.*700 extant copies of Andreas Vesalius's *De humani corporis fabrica* (1543, 2nd edn 1555), most readers apparently consulted it on specific questions only, but the extent and depth of their engagement varied widely, and a few did indeed read and annotate the book cover to cover.[53] It is, therefore, logical that pandects offer tools to facilitate perusal and consultation alike.

[51] Roffeno's hopes regarding China are prompted by the translation of some of Christopher Clavius's mathematical works into Mandarin by Matteo Ricci at the beginning of the seventeenth century (Grendler 2022, 373).

[52] The same seems to be implied a few lines earlier by the phrase 'pandectae commentariorum potius loco sint': commentaries are typically consulted on specific points. Already in vol. 1, sig. $\beta 1^{v}$ of his own *Historia animalium,* Gessner says that this work is written, not for 'continua lectionis series', but for 'inquirendum per intervallum'. He also promises an index to render it usable as a reference work. The point is repeated in his *De libris a se editis epistola* (1566), fol. 24^{v}. See also Friedrich 1995, 34.

[53] Margócsy et al. 2018, 56–120, esp. 57 on the most diligent readers.

Consultation was made easier by tables of contents and indexes. The former device was applied already by Pliny, the first book of whose *Naturalis historia* diligently lists the work's contents, and was later recommended by Gessner.[54] In 1521, Ciruelo's *Apotelesmata astrologiae Christianae* was equipped with a rudimentary running-text table of contents in which only books and chapters are enumerated, as the print lacks any foliation or pagination. This 'tabula' was enhanced, however, by the owner of the copy in the Staats- und Stadtbibliothek Augsburg: having diligently foliated the whole work, he noted the folios on which the individual books and chapters begin in the margins. Later tables became more elaborate and sophisticated. The 'singulorum capitum et propositionum index' prefixed to Varenius's *Geographia generalis* comprises thirty-two pages and has one hierarchical level more than the text itself (its subchapters are not indicated there), which seems to underline that the book was primarily conceived as a work of reference. One pandect may even feature two tables of contents. In Oronce Finé's *Protomathesis*, a one-page 'index universalis' is followed by a detailed 'index particularis' which is introduced by accurate user instructions. Sometimes, as is the case with monographs, tables of contents take the form of Ramist tables, which provide less of a fine-grained overview but in compensation are able to show the conceptual logic of an entire work and the relationship between its different parts at a glance. An impressive example opens the 1630 edition of Alsted's *Encyclopaedia*, where the contents of the whole work and of each book are outlined in thirty-eight Ramist tables (pp. 1–26); these are supplemented by another table of contents in the form of a catechism (pp. 27–46).

Because of the partial functional analogy between tables of contents and indexes (see Chapter 4, 'The printed scientific book'), pandects equipped with extensive specimens of the former, such as those just mentioned, may lack an index. Others, by contrast, have both (this is true of Riccioli's *Almagestum novum* and his *Geographia et hydrographia reformata*, for instance) or an index only. Generally speaking, indexes are a hallmark of the genre. They are often announced on title pages, and their length, thoroughness, number, and variety can be impressive. The *c.*3,000 entries of the indexes to the *Almagestum novum*, arranged across 23-column pages, contain not just words and phrases, but also assertions and questions such as 'The right ascension of several fixed stars diminishes with growing longitude, and of which ones, and why' ('Ascensio recta stellarum aliquot fixarum decrescit crescente longitudine, et quarum et quare'; index to vol. 1, part 1). The 'index rerum et verborum' of Cesi's *Mineralogia* runs to 70 pages and comprises some 5,500 entries. Bettini's *Aerarium philosophiae mathematicae* contains a total of 36 sections titled 'index', which fill some 120 pages between them, in its first and third volumes. These, however, also exemplify the terminological and

[54] 'Congesta deinde in locos et ordines quaeque suos referat [*sc.* auctor], quorum tabulam seu brevem rationem ipse sibi conscripserit ab initio operis praefigendam.'

conceptual confusion which sometimes reigned in this field, as alphabetized indexes in the modern sense of the word are intermingled with highly specific tables of contents listing, for example, 'paradoxa' or 'corollaria' in the order in which they appear in Bettini's text.

The linear reading of a pandect or parts of it was usually facilitated by an elaborate inner structure. Gessner advises the author to take this aspect very seriously, which may indicate that his definition of pandects as reference works is not as exclusive as it sounds.[55] On the highest structural level, the occasional division into several volumes is due to practical considerations of handiness; in cases such as Hoffmann's *Medicina rationalis systematica* (1718–40), it also reflects a work's protracted gestation. The very common book division[56] usually aims at creating thematically unified entities, something occasionally underpinned by thematic titles for individual books. Riccioli names each book of his *Geographia et hydrographia reformata* in this way, freely creating neologisms like 'stadiasmicus' ('of measurements', book 2), 'altimetricus' ('concerned with measuring altitudes', book 6), or 'geomecographicus' ('describing geographical longitudes', book 8) in doing so. Kircher does much the same in his *Mundus subterraneus*. That book division could also serve other than practical ends is indicated by the commonness of works comprising twelve books: twelve, a number traditionally connoting completeness and closure, was associated with a literary magnum opus since Virgil's *Aeneid*. It certainly is chosen on purpose in works like Reisch's *Margarita philosophica*, Agricola's *De re metallica*, and Riccioli's and Kircher's aforementioned pandects. Below the book level, one often finds a deep structural hierarchy. The number of hierarchical levels varies greatly, as does the rigour with which they are applied. It sometimes happens that the structural grid is handled more flexibly on lower levels. The 'libri' of the *Mundus subterraneus* invariably fall into 'sectiones', the latter, into 'capita' (even if a 'sectio' comprises just one 'caput'). Within the single 'capita', by contrast, one encounters all kinds of further subtitles, often of Euclidean inspiration, which are no longer ordered hierarchically. The first 'caput' of the first section of book 1 (pp. 2–20), for instance, contains 'definitiones', 'corollaria', 'canones' (an obscure category), 'consectaria', 'propositiones', and 'paradoxa'.[57]

The structuring of the text by subtitles is often supplemented by marginalia. These may not only help to localize entries looked up in an index—sometimes,

[55] 'Antequam vero rem aggrederetur, diu secum meditari et cum viris doctis deliberare oporteret [*sc.* auctorem], cum de opere universo, tum eius ordine, methodo ac locis locorumque partibus et particulis, ita ut tota divisio non temere, sed ex philosophiae et dialecticae praeceptis institueretur.'

[56] Sometimes, the uppermost category of subdivision is not the book but something else. James Dalrymple of Stair's *Physiologia nova experimentalis* (1686) falls into twenty-one 'explorationes'.

[57] In pandects as in many other genres, Euclidizing terminology is quite common. Another example of this phenomenon is provided by the 'problemata', 'propositiones', 'corollaria', 'axiomata', and 'definitiones' in the more mathematical parts, such as chs 6.22–5 (pp. 223–37), of Riccioli's *Geographia et hydrographia reformata*.

for instance in Cesi's *Mineralogia*, compiled on their basis—on a large folio page, but they can also provide additional assistance in keeping track of the argument for a reader who peruses the work. Further help with continuous reading may be provided by cross references, of which some authors make extensive use. In the *Geographia et hydrographia reformata*, Riccioli anticipates information in this way ('ut paulo infra docebo', 7.23.6, pp. 309–10) and provides reminders of things discussed earlier. In doing so, he sometimes goes to considerable lengths, as in the opening sentence of book 10 (p. 426), where the discussion of the sea is introduced as follows:

> Recolenda sunt hoc loco, quae libro 1. diximus capitibus 12. et 13. de maris et Oceani divisionibus et capite 14. de utriusque sinibus et capite 15. de illorum fretis et Bosphoris Euripisque et capite 16. de isthmis eorumque perfossione, ubi etiam discussimus an mare unum sit alio altius.
>
> Here one must call to mind what I said in book 1, chapters 12 and 13, about the various parts of the sea and the Ocean, in chapter 14, on the bays of both, in chapter 15, on their straits, Bosphori and Euripi,[58] and in chapter 16, on isthmuses and their cutting, where I also discussed whether [the water level of] one sea is higher than [that of] another.

Style

Style, Gessner states, should be an author's least concern. No efforts have even to be made to standardize the different styles of one's sources:

> Nominentur igitur auctores, nec magna interim de stilo cura sit, quamquam is inaequabilis futurus est. Satis enim laboris circa cetera insumitur.
>
> Thus, the names of one's authorities should be given, but at the same time, no great care should be devoted to style, although it will be uneven, for one has enough trouble with other aspects.

Some authors programmatically adopt a similar stance regarding their own work, resorting to the old topos that technical subjects cannot or need not be discussed in ornate language. The dedication letter of Georg Agricola's *De re metallica*, for example, excuses the work's lack of elegance and polish as due to its thorny subject, the art of mining (sig. $\alpha 3^v$): 'verum quo magis ars metallica abhorret ab omni sermonis elegantia, eo minus hi mei libri sunt politi.'

[58] Two 'Bosphori', namely today's Bosporus and the Kerch Strait, are discussed at 1.15.9–10 (p. 19). 'Euripi' are defined at 1.15.7 (p. 18) as straits with strong currents.

In practice, the stylistic default option in pandects can indeed be said to be the 'genus tenue', the lowest of the three levels of style in the system of classical rhetoric. However, Gessner's and Agricola's remarks should not lead one to confuse this unassuming style with carelessness. Its aim, clarity and precision, is not easy to attain in complex subject matters. An example from Varenius's *Geographia generalis* may illustrate this point. Although Varenius relies heavily on geometrical reasoning, diagrams are few and far between in his work. By way of compensation, his geometrical considerations are formulated so clearly that a reader can easily draw the diagrams for himself by following his instructions. In Varenius's explanation of how to calculate the height of a mountain from the distance at which it is first visible, his integration of Euclidean language and syntactically simple but fluent Latin is visible at a glance (1.3.8.10, pp. 87–8).

> Propositio IV
>
> Dato intervallo, a cuius termino primum conspicitur fastigium montis, inde invenire altitudinem montis per geographiam.
>
> Sumamus montem altissimum in insula Teneriffa, una ex Canariis, qui dicitur El Pico, et sit terrae peripheria, et quidem meridianus montis illius, APCDF. Centrum R, mons ipse sit AB. Ducatur ex B recta tangens ad peripheriam BF. Erit F ultimum seu primum punctum, ex quo montis vertex B conspici poterit. Ducatur RF. Testantur autem quidam nautae, quod verticem montis illius conspiciant primo, cum adhuc quattuor gradibus ab eo absunt in meridiano (namque in hoc situ, sub eodem meridiano, facile est illis hanc in gradibus distantiam invenire). Itaque arcus AF erit quattuor graduum. Ponamus itaque veram esse hanc relationem nautarum et radium primum visivum BF venire directe a vertice B, et exquiramus, quanta sit futura montis altitudo, si ita res sese haberet. Angulus BFR est rectus, et quia FA est 4 graduum, ideo etiam angulus BRF est 4 graduum, et nota est RF semidiameter telluris et in triangulo BRF tria data.
>
> Proposition IV
>
> The distance from whose end point the top of a mountain is first visible being given, to find the height of the mountain by means of [mathematical] geography.
>
> Let us take the highest mountain, called El Pico, on the island of Tenerife, one of the Canary Islands, and let the perimeter of the earth, that is to say, the meridian of this mountain, be [the circle] APCDF. Let its centre be R, the mountain itself, AB. From B, let a straight line BF, a tangent to the perimeter, be drawn. F will be the last or first point from which the mountaintop B will be visible. Let RF be drawn. Now, certain sailors testify that they first catch sight of the top of that mountain when they are still four degrees away from it on its meridian (for in this position, on the same meridian, it is easy for them to find this distance in degrees). So the arc AF will be four degrees. So let us suppose that this account

> of the sailors is true and that the first visual line BF comes straight from the top B, and let us investigate how great the height of the mountain will be if that would be the case. The angle BFR is right, and because FA is 4 degrees, therefore the angle BRF is 4 degrees too, and the radius of the earth RF is known, and [so] three [characteristics] are given in the triangle BRF.

In the same vein, Varenius goes on to explain how the remaining characteristics of the triangle BRF can be calculated from those given and the height of Pico del Teide ascertained by subtracting the earth's radius from RB (pp. 88–9).

That the 'genus tenue' is not the same as careless writing can also be seen from the efforts made by many authors to avoid verbal repetition. Riccioli's Latin is generally simple and unobtrusive, but as a rhetorically well-trained Jesuit, he is sensitive to this aspect. In an enumeration of the first seven vessels to have circumnavigated the globe, he varies the phrase 'the nth ship to circuit the world' in a way reminiscent of Erasmus's immensely popular manual *De duplici copia rerum ac verborum* (1512) which demonstrates with a host of examples how the same notion can be expressed in ever new ways (*Geographia et hydrographia reformata* 3.24, p. 104):

> prima navis quae orbem totum quoad longitudinem obivit
> secunda navis orbis circumnavigatrix
> tertia navis Oceani circumlustrati domitrix
> quarta navis Oceani totius triumphatrix
> quinta, quae orbem circumvecta est, . . . navis
> sexta navis Oceani triumphatrix propter ambitum eius superatum
> septima navis . . . , quae orbis circuitum absolvit

This tour de force of lexical and syntactical variation with its many playfully coined female neologisms ending on '-trix' already strains the limits of the 'genus tenue'. In fact, the stylistic span covered in pandects is much wider than this register. Authors soared higher than one would expect from Gessner's remarks, but they also descended to regions below any rhetorical conception of style whatsoever. Generally speaking, the rule that introductory paratexts are written in a higher style than the technical work itself can be extended to a work's innards: the beginnings of books, sections, and chapters, as well as generalizing, introductory, and transitional passages as a whole, are usually composed in more elevated language than the informative and explanatory bulk of the text. For an example from the high end of the stylistic spectrum, one may turn to Kircher, who is always prone to break into religious pathos as soon as he leaves detailed argument behind and turns to more general reflections (*Mundus subterraneus* 3.3.2, p. 158):

> Neminem tam insensatae mentis esse puto, qui divini numinis in naturae administratione providentiam penitius considerans non in summam rapiatur admirationem. Condiderat mundi architectus et opifex mundum perpetuis quibusdam mutationis legibus constantissimum, et astralia quidem mundi corpora uti perpetua quadam et indefessa vertigine, ita et mundum elementarem perpetua elementorum pericyclosi cum incessabili quodam conflictu qualitatumque contrariarum antitechnia conservare sibi complacuit. Hinc nulla in natura rerum substantia est, quae motu proprio vel alieno non instruatur, quo privata universa rerum natura in nihilum abire necesse foret.
>
> No one, I think, is so bereft of good sense that he should not be carried away into the highest admiration when reflecting a bit more deeply on the providence of God's working in the organization of nature. The architect and artisan of the world had created it as something most stable through certain eternal laws of change and had his pleasure in maintaining not only the world's starry bodies by a certain eternal and tireless gyration but also the part of the world made of elements through an eternal circular movement accompanied by interminable strife and rivalry of opposed properties. Therefore, there is no thing in nature that would lack innate or foreign movement, without which nature as a whole would have to plunge into nothingness.

Not only are the notions voiced in this passage—the spectacle of the universe, characterized by corresponding circular motions in the heavens and on earth, the insight into God's mind at the time of the Creation, the hypothetical reversal of the 'creatio ex nihilo'—lofty here, but the language is as well. In addition to writing in periodic style, Kircher also separates grammatically corresponding words ('mundum... constantissimum') and places the finite verb at the end of the sentence, both of which traits are perceived as characteristic of high Latin. The frequent use of hendiadys ('perpetua et indefessa', 'architectus et opifex') gives the impression of being due to the impossibility of one word doing justice to the wondrous phenomena described. The paradox of persistence through change and strife is also deftly expressed in the Latin ('mutationis legibus constantissimum', 'conflictu [et] antitechnia conservare'). The lexicon is characterized by grand circumlocutions like 'astralia mundi corpora' for 'stellae' and by Greek expressions, which are not indispensable technical terms here but deliberately chosen or even coined to add some extra lustre ('pericyclosi', 'antitechnia').[59] The resulting overall impression is one of abundance, amazement, and exaltation.

Conversely, a number of factors and developments could make it difficult to satisfy even the most basic requirements of rhetorical composition. First, there

[59] The second term is not attested in classical Greek.

was the well-known problem of technical terminology. The difficulties were, of course, not restricted to pandects, but they were particularly pressing in this genre, because the vocabulary of entire disciplines had to be paraded and explained. Readers could not be expected to be familiar with this lexicon, especially if some of it was coined by the author himself, as was the case, for instance, in *De re metallica*, where Agricola had to resort to circumlocutions and neologisms in order to produce a modern Latin mining terminology that matched that already existing in German (sig. $\alpha 3^{v}$). Moreover, if a reader consulted a work selectively, he was likely to miss the introduction and explanation of new terms. For this reason, Gessner recommends the inclusion of multilingual word lists or lexicons: 'Nuda tamen rerum, de quibus agitur, praesertim si physicum argumentum fuerit, in diversis linguis poni vocabula laudarim.' ('But I would praise it if the naked expressions for the things discussed would be adduced in different languages, particularly if the subject comes from the realm of nature.') To *De re metallica*, two Latin-German dictionaries of technical vocabulary are appended, the first ordered according to the appearance of the terms within the work, the second alphabetically. Book 11 ('Onomasticus', pp. 542–622) of Riccoli's *Geographia et hydrographia reformata* is likewise composed of two dictionaries, a Latin-vernacular and a vernacular-Latin one. Comprising about 9,000 and 6,000 place names respectively, they constitute an early precursor of Graesse's *Orbis Latinus*.[60]

Two other factors called traditional notions of style into question on a more fundamental level: citation of sources and mathematization. When he asked authors to name their sources ('nominentur igitur auctores'), Gessner probably did not foresee something like the following example from Riccioli's *Almagestum novum* (9.1.7.2, p. 239):

> Sic et Michael Scotus in quaestionibus super sphaeram pag. 105. ait solidum dici tribus modis, uno quod durum est sicut terra, secundo quod continuum est, ut sunt reliqua elementa et corpora supercaelestia, tertio quod habet trinam dimensionem. Origanus in praefat. Ephemeridum, Mariana in Genesim cap. 1. versu 12. Celada ibidem, Renatus de Cartes parte 3. Philosophiae num. 24. Raphael Aversa tomo 2. q. 34. Bullialdus lib. 1. Astronomiae Philolaicae cap. 2. et 4. Tellez disput. 40. Physic. sect. 1. num. 16. et 21. et disput. 44. sect. 3. num. 5. Aristarchus Parisiis editus pag. 23. Io. Hevelius cap. 7. Selenographiae et apud eum Snellius, Stettinus, Gassendus, Crugerus, Eichstadius, Linemannus, item Ioannes Pena in praefatione Opticorum. Io. Baptista Cysatus De cometa anni 1618. affirmans cometas in limpidissima aetheris expa<n>sione suos cursus pergere. Eandemque opinionem defenderunt primum nostri professores in thesibus Parmae editis anno 1613. thesi 742., 743., 748. et Bellarminus eiusdem fuisse opinionis testatur Federicus Caesius in epistola ad Io. Fabrum apud Scheinerum in Rosa ursina lib. 4. parte 2. cap. 27.

[60] Graesse et al. 1972.

This passage, which I can leave untranslated, not only shows how much effort authors of pandects invested into being up to date and that citation of pertinent literature had become the method of choice to document this effort. It also makes clear that, in doing so, even a stylistically competent and sensitive author like the Jesuit Riccioli may silently renounce any stylistic ambition whatsoever in the traditional sense of the word, burying his text under an avalanche of author and work names, abbreviations, dates, and numbers of books, sections, chapters, and pages. In passages such as the present one, a gulf opens up between technical writing and normal prose that goes beyond anything to be found in ancient and medieval texts.

The results are comparable when text in natural language becomes more and more interspersed with mathematical formulae. This phenomenon rose to prominence over time through three interrelated developments: the mathematization of formerly non-mathematical disciplines; the arithmetization and algebraization of mathematics itself at the expense of geometry; and the development of mathematical formulaic language.[61] To be sure, mathematical texts, which are demanding in terms of content and shot through with references to points and lines in geometrical diagrams (as in the example of Varenius), had been hard reading since antiquity. At the beginning of the seventeenth century, Kepler depicts the situation of mathematical authors and readers in gloomy terms (*Astronomia nova*, 1609, sig. (***)2^{r}). In doing so, however, he still speaks about problems attached to a highly technical variety of natural language: it is 'the characteristic sublety of propositions, instructions, proofs, and conclusions' ('genuina subtilitas propositionum, instructionum, demonstrationum, conclusionum') that makes the perusal of mathematical works 'most tedious' ('morosissima'). One hundred and fifty years later, in a work like Euler's *Institutiones calculi differentialis*, the situation has thoroughly changed. Now, more than half of the space is taken up by equations and other formulaic expressions, while the function of natural language is often reduced to leading from one cluster of these to the next.

Illustrations

As prestigious publications, pandects can a priori be expected to be well equipped with illustrations in general, in addition to the frontispieces discussed earlier. Their visual appeal presumably was a major selling point in many cases, an assumption backed up by pertinent advertisements on title pages. The eighteen books of Laet's *Novus orbis* are announced as being 'illustrated by new geographical maps and different images of animals, plants, and fruits' ('novis tabulis geographicis et variis animantium, plantarum fructuumque iconibus illustrati'). However, in this, as in other respects, the range of possibilities was wide.

[61] On the last phenomenon, see Cajori 1928–9.

Numerous pandects were illustrated sparingly, if at all. This is understandable in a field like natural philosophy, where a high degree of abstraction militated against the inclusion of images. It is less expected in cases where one feels that copious illustrations clearly would have furthered the reader's understanding. However, images could play a subordinate role in an anatomical pandect such as André Du Laurens's *Historia anatomica* (1600),[62] while other works of the same kind renounced them completely as late as the eighteenth century.[63] Cesi's *Mineralogia* likewise is entirely devoid of illustrations. In the preface to his *Institutiones calculi differentialis* (p. xx), Euler even proudly draws attention to the fact that he has carried the arithmetization of his subject so far as to render any diagrams superfluous: 'Hic autem omnia intra analyseos purae limites continentur, ut ne ulla quidem figura opus fuerit ad omnia huius calculi praecepta explicanda.'[64]

On the other hand, many pandects were exceptionally well illustrated. Figure illustrations range from impressive technical drawings in Agricola's *De re metallica* to depictions of exercizing Greeks and Romans in the second, expanded edition of Girolamo Mercuriale's medico-antiquarian *De arte gymnastica* (1573) and the imaginative engravings of the bowels of the earth and erupting volcanoes in Kircher's *Mundus subterraneus*. An example distinguished by its numerous diagrams is provided by Oronce Finé's *Protomathesis*. The work's 280 diagrams, taken over from different sources but adapted and stylistically unified, continued to be revised and elaborated throughout its long publication history. Underlining the scientific aspirations of the *Protomathesis* through their austere, abstract style, they inaugurated a tradition of high-quality illustrations of mathematical prints in Paris.[65] Of course, one work could also contain different kinds of non-text. Kepler's *Epitome* has everything from crude sketches to refined geometrical diagrams.[66] Riccioli's *Geographia et hydrographia reformata* features several drawings of geodetic instruments, numerous geometrical diagrams, and many short tables, which provide a convenient way of presenting numerical data.

In geography, astronomy, and anatomy, there even developed a subform of the pandect that was and is defined by the role played by illustrations: the atlas.[67] In

[62] All illustrations are inserted between books 5 and 6. In a note to the reader (p. 213), Du Laurens explains that he, unlike some contemporaries, does not find images totally useless ('non omnino inutilem') in anatomy, that his illustrations are derivative, that all errors have to be blamed on the illustrators ('pictori et sculptori'), and that he places the images in the middle of the work to please the reader ('figuras omnes simul et in medio fere operis excudendas iussi, ut legentium oculos oblectarem'). The reasoning behind the last remark is unclear to me.

[63] Kusukawa 2006, 92. [64] Demidov 2005, 197.

[65] Pantin 2010. [66] Pantin 2006, 226–36.

[67] The genre's name originated with Gerardus Mercator's *Atlas sive Cosmographicae meditationes* (1595) and remained primarily attached to geographical works. This is illogical, because the mythological Atlas was renowned as an early astronomer, and came about unintentionally. Mercator had planned a predominantly astronomical and cosmological work, but due to the vagaries of life and publication, the book ended up as a collection of terrestrial maps. I hope to explain this more fully in a future article. For the moment, see Remmert 2007; Krogt 2009. The transfer of the title *Atlas* to astronomical and anatomical works is secondary. An early example of a work titled *Atlas coelestis* is

an atlas, the relationship between text and image markedly differs from what is customary. As Marcel Peter René van den Broecke remarks on the first geographical atlas, Abraham Ortelius's *Theatrum orbis terrarum* (1570): 'Cartographic publications prior to Ortelius contain cartographic material as illustrations of the text. Ortelius' *Theatrum* was the first publication in which the texts were intended as comments on the maps.'[68] This has to be taken with a pinch of salt, however: the text of the *Theatrum*—one page per map—is more than a simple commentary, as it contains much information missing from the maps. The two components mutually complement each other. In general, texts still played a more important role in early modern atlases than in modern ones. With Johann Bayer's *Uranometria* (1603), the format was transferred to astronomy.[69] In anatomy, the genre had already made an independent start before its emergence in geography. However, even more than in that discipline, the first anatomical examples present intermediate forms between a normal pandect and an atlas. Although Vesalius's *De humani corporis fabrica* is first and foremost famous for Jan Steven van Calcar's woodcuts and is routinely called an atlas in the literature, it also contains extensive self-contained descriptions of the parts of the human body alongside keys to the images. Its readers interacted more closely with and gained more information from the book's text than its illustrations.[70] Adriaan van de Spiegel's homonymous successor work (1627) has over 300 pages of running text, to which the illustrations by Daniel Bucretius with their keys are appended. Only in more modern examples such as Robert Froriep and Friedrich August Andorff's *Atlas anatomicus* (1851) have the textual elements been more or less reduced to keys.

Compendia

Notion, names, and history

Compendia provide short summaries of whole disciplines or parts of them. Like encyclopedic works (from which they sometimes derive as abridgements), they

the one by John Flamsteed (1729); what perhaps is the first *Atlas anatomicus* appeared in 1851 (see main text).

[68] Broecke 2009, 16.

[69] Facsimile edition with companion volume: Berberich 2010 and Hamel 2010 (who discusses four precursors to Bayer's atlas from the second half of the sixteenth century on pp. 31–8).

[70] English translations: Richardson and Carman 1998–2009; Garrison and Hast 2014 (translation of both the 1543 and the 1555 editions, including a historical introduction by Vivian Nutton with a good description of the book and its impact in vol. 1, pp. lxxv–ciii). A census of all extant copies of the first two editions is provided by Margócsy et al. 2018; for the greater importance of the text and readers' difficulties with the images, see there at pp. 4, 59, 62, 64, 78–80, 95. On the woodcuts in particular, see Kusukawa 2006, esp. 84–91.

react to a widespread feeling that there is 'too much to know'[71] and aim at documenting the state of the art. However, they try to come to terms with the problem of information overload, not by comprehensiveness, but on the contrary by selectiveness. In contrast to textbooks, which they often resemble in form and content, compendia are not meant for formal teaching but for the world outside the classroom, where they can be used in two ways: first, for self-teaching (including learning by heart in some cases);[72] secondly, for quick consultation, for instance to identify a flower in a field or the disease of a patient.

Many compendia bear generic titles. Besides 'compendium' itself, perhaps the most common, one finds ἐγχειρίδιον, εἰσαγωγή, and ἐπιτομή plus their transliterations and, in the first two cases, the loan translations 'manuale' and 'introductio', as well as 'summula', 'elementa', and many similar expressions.[73] In addition, there are designations restricted to certain disciplines or special forms, some of which will be mentioned in what follows.

Texts of this kind were well known in antiquity and the Middle Ages, when in addition to the pressure of information overload, writing materials were expensive and book production laborious. In antiquity, such texts existed in all kinds of disciplines from music theory (Cleonides, *Introduction to Harmonics*) to historiography (Florus, *Epitoma de Tito Livio*).[74] Among medieval factual texts, various kinds of summaries and excerpt collections based on Aristotelian works, some of them translations from Arabic, loom large.[75] Comparable short versions were produced in other fields such as law and theology as well.[76]

The early modern period: an age of compendia

For a variety of reasons, compendia benefited particularly from the introduction of print. Already during the manuscript age scholars had produced summaries

[71] The phrase is borrowed from the title of Blair 2010. On compendia as abbreviations of larger works, see Paulus 2005, 281–3, 300, 313–21.

[72] On compendia as a help for private study, see Paulus 2005, 280. On learning by heart, see pp. 6–7 of Georg Henisch's letter of dedication to an anonymous *Enchiridion medicinae* (1573): 'Quamvis in enchiridio magnitudo spectari non debet, cum quo minus est, eo animos studiosorum ad ediscendum evolvendumque magis alliciat et provocet.' In general, the idea of knowing scientific works by heart did not strike contemporaries as absurd. In an unpaginated letter dedicating the *Epitoma in Almagestum Ptolemaei* (1496) to Bessarion, Regiomontanus claims that Peuerbach had memorized more or less the whole *Almagest* ('quem ille quasi ad litteram memoriae tenebat', second page). See also note 102 below on aphorisms.

[73] Paulus 2005, 301.

[74] Several examples are analysed in Fuhrmann 1960 as textbooks (see the beginning of the next section for the problems attached to that label). Galen characterizes some of his own works as compendia, calling them ἐπιτομή or σύνοψις (Von Staden 1998, 73, 93).

[75] Grabmann 1939, 54–104. The material is collected, together with the other Aristotelian commentary literature of the Middle Ages, in Lohr 2010–13 (for the different types, see vol. 1, p. VII, nos. 4–6).

[76] Grabmann 1939, 54.

such as those mentioned above as personal working tools. Now, it became comparatively easy to get such digests published. In doing so, one served the scholarly community and enhanced one's own prestige without investing too much extra work in so doing. Samuel Schönborn's *Manuale medicinae practicae Galenico-chymicae* (1637), Thomas Burnet's *Hippocrates contractus* (1685), which started life as a personal 'easily portable Hippocrates' ('Hippocrates facile portatilis') for sickbed visits, and Johann Samuel Carl's three *Ichnographiae* of clinical practice, chymistry, and anatomy (1722) are cases in point.[77] At the same time, readers' demand for compendia was on the rise. The knowledge explosion brought about by print intimidated many readers, sparking a desire to learn the really important things quickly and without too much effort. In satisfying this desire, compendia boosted autodidacticism,[78] thereby creating a need for even more compendia.

The readers for whom such works were written came from different social strata. They comprised merchants and craftsmen, interested amateurs,[79] independent scholars, mathematical practitioners, doctors and pharmacists, professors, and students. Content and level of technicality varied accordingly. So did the language. Latin pharmacopoeias, for instance, had had vernacular counterparts for medical laymen since medieval times.[80] In sixteenth- and seventeenth-century Italy, comparable traditions of compendia of astronomy, cosmology, meteorology, and the other parts of Aristotelian natural philosophy evolved in Latin and in the vernacular.[81] In some other fields, the parallelism between Latin and vernacular items was even more striking, as they were similar in everything except the language. 'Algorithmi', guides to elementary arithmetic, closely conformed to German 'Rechenbücher'. Short dietary guides like Conrad Gessner's *Sanitatis tuendae praecepta* ([1560]) were exactly matched by vernacular items of the same text type as well. Robert Grospré's *Regimen sanitatis* (1538) also appeared in an Italian version appended to a cookbook, Domenico Romoli's *La singolare dottrina* (1560).[82] In general, it was not rare for the same compendium to be published in Latin as well as the vernacular. Petrus Apianus's successful and influential introduction to cosmography, the *Cosmographicus liber* ([1524]), which went through thirty-five Latin editions, was also translated into Dutch, French, and Spanish.[83]

[77] See the prefaces of Schönborn's, Burnet's, and Carl's works at sigs. A5^{r-v}, the second to third unnumbered page, and pp. 6–7, 11–12 respectively. Another 'bedside compendium' is Alfonsus Morescottus's *Compendium totius medicinae* (1584), whose letter of dedication to the author's son declares at p. 4: 'Hoc igitur opusculum tibi semper adsit velut promptuarium, ex quo valeas cuique morbo suum fabricare medicamentum atque receptum.'

[78] Crupi 2019, 22. For a pertinent statement by an early modern reader, see the main text following note 91 below.

[79] See Gaida 2016 on 'mathematical amateurs' as the target readership of Petrus Apianus's *Cosmographicus liber* ([1524]).

[80] Schmitz et al. 1998–2005, vol. 1, 368, vol. 2, 119–23.

[81] Paulus 2005, 280–357.

[82] Fols. 359^{v}–376^{r}. On Grospré's booklet, see Elaut 1961; on early modern dietetics and dietary guides in general, see *EEMHO*, s.v. 'Dietetics'.

[83] Van Ortroy 1902, 25–66; Gaida 2016, 290.

The increasing importance of compendia was mirrored in their recognition and revaluation as a distinct genre. When describing pandects at the beginning of the third volume of his *Historia naturalis* (1555), Conrad Gessner also discussed the compendium, which he called 'epitome' and described as a small-scale, less learned, more fact-oriented, stylistically neater version of the former, relating to it as plain text relates to commentary, catering to a readership of learners and destined to be read cover to cover[84] (a characterization too narrow to cover the genre's internal variety, as will become clear shortly). Books were promoted as belonging to this genre in various ways. They were advertised as 'compendia', 'enchiridia', 'epitomae', and 'manualia' on their title pages. Promises that they were short, handy, readily understandable, and convenient for autodidactic learners and on-the-spot consultation abounded in titles, dedications, prefaces, and poems of praise. The full title of Francesco Barozzi's *Cosmographia in quattuor libros distributa* (1585) assures the reader that he will be introduced 'in perfect order and wonderfully easily and briefly to the *Almagest* and astronomy at large' ('summo ordine miraque facilitate ac brevitate ad Magnam Ptolemaei Mathematicam constructionem ad universamque astrologiam').[85] A friend of Giovanni Battista Baffi advertises the latter's *De cometis libri tres* (1580) as follows (sig. (ii^r)):

> Te non ergo decet libros vagari
> per multos veterum, ast iter tenendum
> est longe brevius, laborque nullus
> scandendo superas domos Olympi
> reddet fessum animum via patenti.
>
> So you should not wander through many ancient books. Stay on this much shorter track: no toil will tire your mind, while you will ascend to the abodes of the sky above on this open road.

The promotional character of such claims made in titles and paratexts is underlined by their occasionally dubious or outright contrafactual nature. Pietro Andrea Mattioli's *Compendium de plantis omnibus* (1571) belies its name by filling over 900 quarto pages. In his opening address to the reader on the title page of Andrés de Laguna's *Epitome Galeni operum* (1551), the printer Michael Isengrin

[84] Sigs. [a5^{r-v}]: 'Epitomen vero [appello] compendium illius maioris operis, in quo omnia certius breviusque proponantur et rerum explicationis maior quam verborum habeatur cura.... Stili vero elegantia et aequalitas ad epitomen magis pertinet, quae quidem scribetur ut serie continua legatur... Stilus esto purus, elegans, et aequabilis, elocutio brevis et arguta.... Ab hac epitome lector ad pandectas tamquam commentarios se recipiet, ubi voluerit. Haec discentibus aptior erit, illa docentibus.'

[85] See also the ninth page of his unpaginated preface to the reader, where the *Almagest* is contrasted with 'compendiariae institutiones' in general.

claims that Galen's works have been reduced to a 'small body' ('exiguum quoddam corpus')—meaning a tome of 640 folio pages in small and dense two-column print. Despite the inevitable difficulty of Giuseppe Biancani's *Sphaera mundi* (1620), a new overview of cosmology and astronomy accounting for the developments of the preceding decades, the reader is assured on the title page that he will find the subject 'presented according to an easy method' ('facili methodo tradita'). On the first page of his preface, Biancani promises to explain everything with 'most simple...proofs' ('demonstrationibus...facillimis') so that things will become clear to everyone, even if he knows almost no geometry ('quibusvis paene etiam ageometretis').[86]

In guardians of higher learning, the popularity of compendia occasionally provoked strong reactions. Professors seem to have been afraid of losing listeners, whose numbers were important for their reputation,[87] if students preferred such shortcuts, and spoke of these books in terms reminiscent of what students are told about Wikipedia today. In an address to first-year medical students delivered around 1570 in Padua, Girolamo Mercuriale warns his audience to distrust claims that medicine can be learned quickly from compendia and to avoid them like the plague ('compendia haec et epitomas, quibus nonnulli sese brevi artem docturos pollicentur, ut perniciem summam fugatis'). Only experienced doctors may consult them now and then to help their memory.[88] Later on, one finds testimonies to a more nuanced awareness that compendia were a double-edged sword. In 1722, Carl opens the preface to his *Ichnographiae* with a dispassionate statement: 'Compendiorum usum et abusum esse et ratio et experientia dictitat.' ('Reason and experience alike keep telling us that compendia can be used to good and bad ends.') As he goes on to explain (pp. 3–4), such books provide focus and efficiency through the pruning of superfluous detail but may become obscure through excessive brevity and harbour the seduction of laziness.

Disciplines

Compendia were published in most disciplines dealing with the natural world, often in parallel to developments in the textbook sector. Following medieval predecessors, summaries of Aristotle's natural philosophy began to appear in print in the sixteenth century. As a successful example, Ermolao Barbaro's *Compendium*

[86] Other authors were more cautious, claiming comparative shortness and simplicity only. The dedication of Peuerbach and Regiomontanus's refurbished *Almagest* (1496) explains that Bessarion asked for a 'shorter and clearer' ('breviorem lucidioremque', second page) version of the work. David Gregory composed his summary of Newtonian astronomy (1702) with the intention of rendering the new celestial mechanics 'a bit easier' ('paulo facilius', first page of preface).

[87] Grendler 2001, 150–1.

[88] *De modo studendi*, ed. Durling 1990, lines 174–92 (quotation: lines 175–6).

scientiae naturalis ex Aristotele can be cited. Composed in the late fifteenth century, published in 1545, and re-edited at least seven times north of the Alps until 1597, it provides an overview of Aristotle's natural philosophy from fundamentals and cosmology to meteorology and the soul in some 150 octavo pages. In a 1548 edition by Conrad Gessner, the text was supplemented by another compendium by Hieronymus Wildenberger, whose by and large similar structure is more explicitly based on Aristotle's single works, epitomizing the *Physics*, *De caelo*, *De generatione et corruptione*, *Meteorology*, and *De anima* in turn. A kernel of this setup could be retained until the late seventeenth century, but at this time, the structure was handled flexibly and accommodated much new material. Johann von Muralt's *Scientiae naturalis seu physicae compendium* (1694) starts in the traditional way with chapters on physics in general, cosmology/astronomy, and meteorology (pp. 5–47), but the subsequent, far longer part (pp. 47–208) replaces psychology with an enlarged version of the three realms of nature, descending from anthropology to zoology, botany, and mineralogy.[89] The reader learns of microscopes, the circulation of the blood, and vortices (pp. 15, 16, 31), and the whole account is strongly imbued with physico-theology. Subfields of the Aristotelian encyclopedia were presented in compendium format as well, as in Baffi's aforementioned summary of the traditional knowledge about comets.

Outside the Aristotelian framework, the science of the stars had become the subject of separate handbooks already in the incunabula era, when Peuerbach and Regiomontanus produced their updated abridgement of Ptolemy's *Almagest* (1496). Later on, astronomical compendia quickly responded to challenges posed by rapid scientific progress. Ten years after a telescope was first pointed at the sky, Biancani's *Sphaera mundi* took full account of the discoveries made with its help and established 'telescopium' as the standard term for the new instrument,[90] besides discussing the opinions of Copernicus, Brahe, Galilei, Kepler, and others. As early as 1702, David Gregory introduced a broader stratum of readers to the transformation of astronomy through Newton's celestial mechanics. As announced in its title and at the beginning of its preface, his *Astronomiae physicae et geometricae elementa* largely dispenses with calculations ('astronomia arithmetica sive calculatoria') in favour of a purely geometrical approach but nonetheless provides a complete overview of the solar system according to Newton. Some compendia of astronomy's twin discipline astrology appeared as well. Girolamo Cardano's collection of astrological aphorisms (1547; see later in this section) is a prominent example.[91] In the sixteenth century, Apianus's *Cosmographicus liber* inaugurated a trend for combining astronomy with cosmology and geography (both mathematical

[89] This replacement could be inspired by the structural equivalence between Aristotle's hierarchy of psychological faculties and the four classes of beings named in the main text.

[90] McColley 1938.

[91] For a handbook of astrometeorology, see Antoine Mizauld's *Ephemerides aëris perpetuae* (1554).

and descriptive) in cosmographical compendia. Further exponents of this strand include Franciscus Bordinus's *Quaesitorum et responsionum mathematicae disciplinae chilias* (1573), which adds some geometry and time reckoning, and Barozzi's *Cosmographia in quattuor libros distributa*, intended as a replacement for Sacrobosco and a new standard account of the field for the general reader. Compendia restricted to geography alone, such as Joachim Vadianus's *Epitome trium terrae partium* (1534), are a rarer phenomenon.

As most higher mathematics was learned and used in the context of astronomy, and as Euclid's prominence stifled an independent geometrical tradition, purely mathematical handbooks were basically concerned with elementary arithmetic. The 'algorithmi' to be discussed in the textbook section were perhaps also appropriated by self-learners, although their conciseness must have created difficulties. By contrast, the comparative detail of Gemma Frisius's *Arithmeticae practicae methodus facilis* (1540), which went through over sixty editions, proved very helpful for autodidacts. Jacques Peletier, who produced a revised edition in 1545, states that he had previously used the book 'instead of a teacher' ('pro praeceptore', fol. 59^r) himself.

Medicine clearly was the most prolific discipline in this genre, as in many others. A search in *USTC*, where results are classified by disciplines (the pertinent categories being 'Medical Texts', 'Science', 'Astrology and Cosmography', and 'Mathematics'), provides a rough impression of its prominence. Going by typical titles, one finds 98 editions of Latin medical 'compendia' against 210 in the three other categories combined. For 'epitome/epitoma', the ratio is 79:98, for 'enchiridion' and its Latin equivalent 'manuale', 65:20, for all three types of titles together, 242:328. The medical tradition was long-lived, starting with incunabula prints like Saladinus de Asculo's *Aromatariorum compendium* (1488) and crossing the Atlantic three centuries later in the form of the *Vade-mecum medicum* (1798) by the Virginian William Tazewell, a copy of which elicited a letter of thanks from Thomas Jefferson in 1803.[92] However, as in the case of medical pandects, few works aspired to an overview of the discipline at large. Most focused on practically important subfields, particularly pharmacology. The two examples just adduced are typical in this respect: the former addresses pharmacists ('aromatarii') in its title, while a nosology (pp. 1–70) is followed by a more extensive pharmacopoeia (pp. 71–200) in the latter. Many works presented as medical compendia in their titles, such as the anonymous *Enchiridion medicinae* (1573),[93] were pharmacopoeias pure and simple. Another area where doctors appreciated some guidance was iatromathematics. As exemplified by Walther Hermann Ryff's *Iatromathematicae enchiridion* (1542), a publication pieced together from three

[92] Abstracted at https://founders.archives.gov/documents/Jefferson/01-39-02-0331 (accessed 16 February 2023).
[93] Šimon 2018.

earlier compendia of similar construction, there was a demand for concise information on how to adapt medical intervention to planetary influence.[94] The dietary guides mentioned in the preceding subsection primarily catered to laymen. Short, elementary, and mostly commonsensical, such booklets provided the deceptive but reassuring impression that staying healthy was easy. Five editions of Grospré's *Regimen sanitatis* appeared within two years in Paris in addition to the later Italian version.

The appearance of botanical compendia in the sixteenth century was due to the discipline's integration into the study of medicine and the concomitant emergence of botanical field trips, which created a demand for short 'historiae plantarum' one could take along to identify the plants encountered. As early as 1534, this use is found illustrated in Euricius Cordus's dialogue *Botanologicon*: when going out into the author's garden, the interlocutors take a 'lesser Dioscorides' ('minorem Dioscoridem', p. 27) with them.[95] Some decades later, Mattioli defines the *Compendium de plantis omnibus*, the abbreviated and generically changed version of his famous Dioscorides commentary (see Chapter 5, 'Commentaries'), as a handbook for botanical fieldwork. He declares that he published the work for the benefit of neophytes 'so that under its guidance they could safely go to all places where plants grow in order to come to know and study them' ('ut eo duce quaecumque loca, ubi stirpes proveniant, tuto ad eas noscendas indagandasque adire possent', first page of dedication).

Compendia in other areas of natural history followed suit later, as exemplified by John Jonston's *Notitia regni mineralis*, a booklet of some 100 duodecimo pages which appeared in tandem with a more extensive *Notitia regni vegetabilis* in 1661. So did chymical works of this kind, which presupposed the emergence of a non-secretive chymistry. Carl's *Ichnographia chymiae fundamentalis* (1722), a summary of the theories of Johann Joachim Becher and Georg Ernst Stahl, is an early example.

Formal pluralism

As mentioned initially, many compendia formally resemble textbooks. It can indeed be impossible to assign a given work confidently to either category in the absence of explicit information about its purpose and use. In view of the fuller treatment of the common standard format shared by both genres in the next section, I will not dwell on it extensively here. Restricting myself to a few short

[94] An overview of iatromathematical literature in general is provided by Sudhoff 1902; see there, pp. 52–3 on Ryff; and see pp. 25–8, 29–31, 45–6 on the publications plagiarized by him.

[95] The identity of this work is unclear. Presumably on account of its illustrations, Brunfels's massive *Herbarum vivae eicones* (1530–6) is taken along as well.

remarks, I will quickly move on to the plethora of peculiar forms which also exist among early modern compendia.

The prototypical compendium is a short, handy, and portable book in octavo or smaller format. Standard paratexts such as dedication, address to the reader, table of contents, or index may be present, but they are less numerous and lavish than in big encyclopedic works. The account itself is tightly structured. In many cases, one notices a compartmentalized, multilayered subdivision which is meant to mirror the thematic structure of the field under discussion itself. This setup is highlighted by typographical means. The reasoning tends to proceed by definition, subdivision, and distinction. In terms of speech acts, informing and prescribing predominate at the expense of argumentation. The language may contain a certain number of formulaic elements, while—despite Gessner's recommendations to the contrary—stylistic aspirations in the proper sense of the word are lacking. Technical terminology is present, but it is introduced in a reader-friendly way. This standard type is well represented from the sixteenth to the eighteenth centuries. Among works already mentioned, Ryff's *Iatromathematicae enchiridion* (1542), Barbaro's *Compendium scientiae naturalis ex Aristotele* (1545), Muralt's *Scientiae naturalis seu physicae compendium* (1694), or Carl's *Ichnographia chymiae fundamentalis* (1722) conform closely to the pattern.

Of the numerous special forms, some are typical of certain disciplines, while others are thematically unrestricted. Among the former, the structure of abbreviated 'historiae', such as Jonston's two *Notitiae*, is based on the full-sized format. In mathematics, 'algorithmi' also served as textbooks, in which capacity they will be discussed in the next section.

A little more should be said about pharmacopoeia, a major medical form which reaches back to Egypt and the Ancient Near East, shows a continuous tradition in antiquity and the Middle Ages, and received new impetus towards the end of the latter period through the emergence of pharmacy as a profession in its own right.[96] From the sixteenth century, some pharmacopoeias were sanctioned by city governments;[97] from the eighteenth, they could become legally binding in principalities and empires, a development accompanied by attempts at selection and reduction to counter the genre's innate tendency to swell to unwieldy dimensions. The Latin tradition of such official specimens, whose elaboration was usually entrusted to commissions of physicians and pharmacists, extended to the second edition of the *Pharmacopoea Germanica* (1882), the predecessor of the present *Deutsches Arzneibuch*.

[96] See Schmitz et al. 1998–2005, vol. 1, 364–85, vol. 2, 195–210, 404–8, and the indexes s.vv. 'Arzneibuch', 'dispensatorium', 'pharmacopoea'; on the Eastern tradition up to late Byzantine times, see Lehmhaus and Martelli 2017.

[97] For London and Edinburgh, see English 1990, 86.

The lists of medications offered by pharmacopoeias may be ordered according to different criteria and also differ in their individual presentation. Regarding the former, recurrent principles of order are alphabetization, the distinction between simples and composite drugs, indication, and dosage form. In practice, these structuring devices are applied selectively and can be combined with others in various ways. In Schönborn's *Manuale medicinae*, for instance, body parts 'a capite ad calcem' provide the uppermost structural level, followed by the simple/composite dichotomy, the differentiation between warming and cooling, external and internal medications, and finally by different parts of plants for simples and different dosage forms for composite drugs. The *Pharmacopoea Austriaco-provincialis* (1794), the official pharmacopoeia of the Habsburg monarchy, is organized in a simpler way, enumerating first simples, then composite drugs (pp. 5–44, 45–146) in alphabetical order.

The recipes for the individual composite drugs become increasingly standardized, syntactically reduced, and concise (including abbreviations) over time. In its fullest form, a recipe includes the name(s) of a drug, indication, ingredients (with their respective quantities), preparation, and application in this or another order, but in practice, one or more of these sections are usually left out.[98] Schönborn's *Manuale medicinae* and the *Pharmacopoea Austriaco-provincialis* make for an instructive contrast in this respect as well. The former mostly provides lists of names and nothing more, leaving anything else to the doctor to whom it is addressed; he will have procured the ready-made medications from his pharmacist and applied them at his own discretion. In the latter, each recipe features the same tripartite structure, which includes the name as well as precise information on ingredients and preparation. In this case, the lack of indication and application conforms with the fact that the addressees are exclusively pharmacists.

Perhaps the most significant development of the recipe in early modern times relates to its typographical presentation. This aspect can also be illustrated well by a comparison between Schönborn and the *Pharmacopoea*: such recipes as are given in fuller form at all in the *Manuale* appear in running text. They are typographically indistinguishable from the rest of the text, and their inner structure is difficult to grasp. In the *Pharmacopoea*, a uniform recipe layout has been introduced. By virtue of capitalization, blank lines, line breaks, and indentation, each recipe is immediately discernible as a separate unit and its different sections can be distinguished at a glance (Figure 10).

[98] On the history of the recipe as a text type (which overlaps with the history of the pharmocopoeia, of course), no comprehensive account seems to exist. An overview of sorts may be pieced together from the index of Schmitz et al. 1998–2005, s.v. 'Rezept'. For the ancient Near East and Greece, see Goltz 1974; for summaries of subsequent developments, see *LMA* 7, 778–9 and Telle 2003, 252 and 268–9 nn. 4–6 (further literature); the remarks on Galen's pertinent reflections in Von Staden 1998, 73 and 93 are revealing as well.

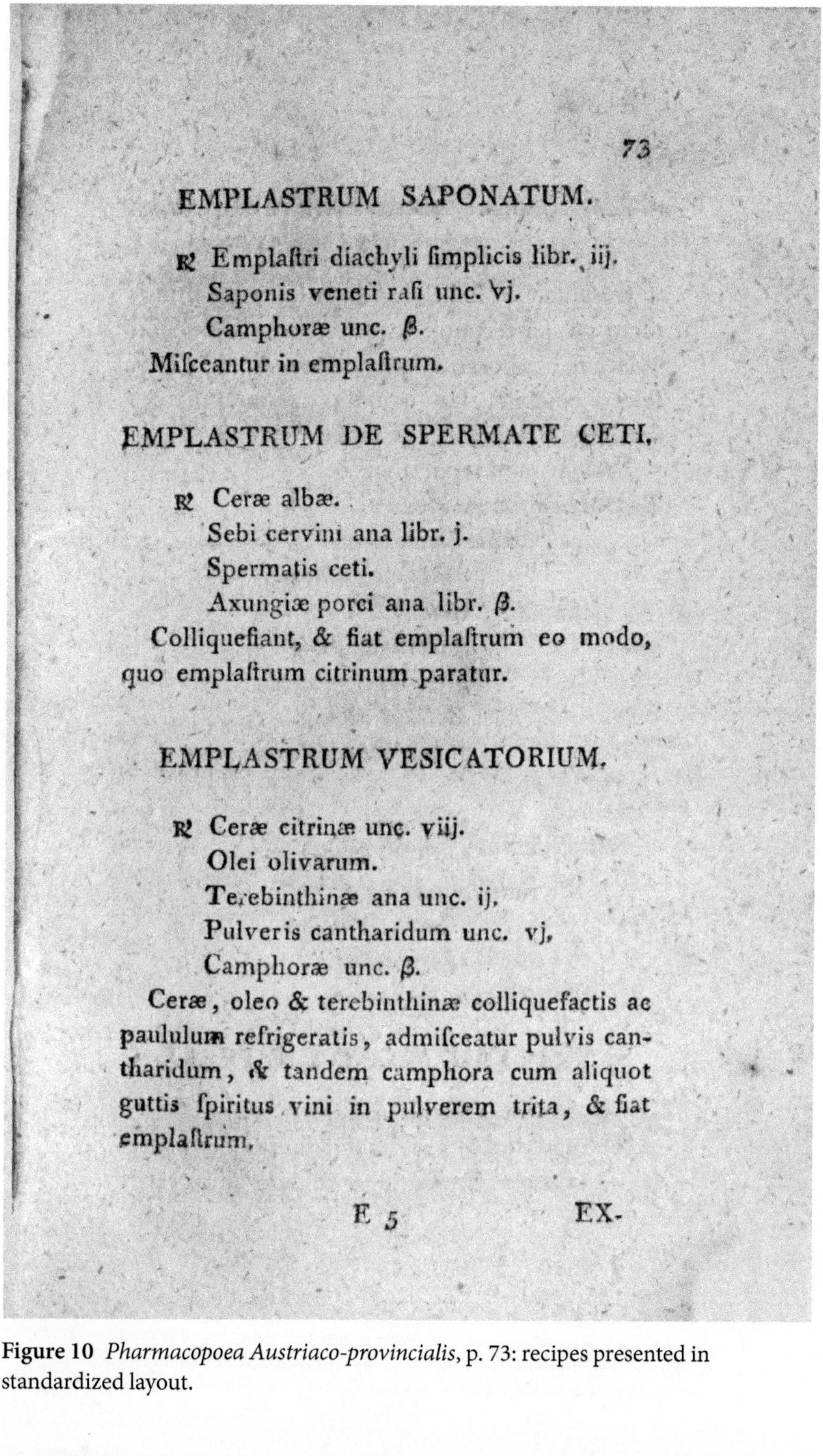

73

EMPLASTRUM SAPONATUM.

℞ Emplaſtri diachyli ſimplicis libr. iij.
Saponis veneti raſi unc. vj.
Camphoræ unc. β.
Miſceantur in emplaſtrum.

EMPLASTRUM DE SPERMATE CETI.

℞ Ceræ albæ.
Sebi cervini ana libr. j.
Spermatis ceti.
Axungiæ porci ana libr. β.
Colliqueſiant, & fiat emplaſtrum eo modo, quo emplaſtrum citrinum paratur.

EMPLASTRUM VESICATORIUM.

℞ Ceræ citrinæ unc. viij.
Olei olivarum.
Terebinthinæ ana unc. ij.
Pulveris cantharidum unc. vj.
Camphoræ unc. β.
Ceræ, oleo & terebinthinæ colliquefactis ac paululum refrigeratis, admiſceatur pulvis cantharidum, & tandem camphora cum aliquot guttis ſpiritus vini in pulverem trita, & fiat emplaſtrum.

E 5 EX-

Figure 10 *Pharmacopoea Austriaco-provincialis*, p. 73: recipes presented in standardized layout.

Beyond discipline-specific subgenres, authors were even more inventive, producing a confusing variety of forms. Ramist tables and catechisms, popular as textbook formats, were occasionally used for compendia too. Among these, Bordinus's *Quaesitorum et responsionum mathematicae disciplinae chilias*, for instance, presents one thousand questions and answers arranged in three sections, focusing on cosmology, geography, and spherical astronomy respectively. More talented or aspiring authors tried to develop such mechanical exchanges into true dialogues. In the *Rei medicae totius compendiosa traditio* (1558) by Georg Pictorius, Podalirius, an experienced physician, reveals the fundamentals of medicine and pharmacology to a medical novice named 'Tyro' (that is, 'tiro', 'beginner') in a series of six conversations.

Perhaps unsurprisingly in view of its attractiveness and mnemotechnical advantages, poetry could be put into the service of compendious information as well. In Josse de Harchies's materia medica manual *Enchiridion medicum* (1573), some 360 plants, animals, and minerals are in each case presented in an epigram followed by some additional information in prose.

Finally, a quite substantial group of works is culled in part or *in toto* from classical authors. Some of these compilations are quite modest, like Gessner's *Sanitatis tuendae praecepta*, which despite its brevity (twenty-three octavo pages) unites three different components: extracts from ancient authorities (Pliny, Cicero, Celsus), comments on them by Gessner, and two chapters composed by Gessner himself. On the other hand, one finds a more aspiring type which may be dubbed 'short collected works': attempts to extract the gist of the massive unwieldy *oeuvres* of first-rank authorities and to present it in a handy format. An early work of this kind was produced by Moses Maimonides (twelfth century), who assembled some 1,500 extracts of a few sentences each from Galen in Judaeo-Arabic translation, together with some additions of his own. This *Fusul Musa* ('Moses' Aphorisms'), as the result was called,[99] was in turn translated into Latin and printed as a compendium of Galenic medicine in 1489 (*Aphorismi secundum doctrinam Galeni*). In the mid-sixteenth century, Laguna produced his admirable *Epitome Galeni operum*. This folio volume, which reduces the bulk of the Corpus Galenicum to about one-third but whose scope still stretches the boundaries of the term 'compendium', is superior to the *Aphorismi* not only because of the medical and philological expertise of its compiler but even more so because of its overall design: instead of a series of excerpts, Laguna presents each and every Galenic work in a heavily abridged but nonetheless fully coherent and easily readable version. It is understandable that the *Epitome* was reissued at least four times up to 1643. Less ambitious, but even more successful was a late counterpart to the aforementioned works for Hippocrates, Burnet's *Hippocrates contractus*. Burnet

[99] An English translation with introduction and notes is provided by Rosner 1989.

returns to the anthology style and thus manages to reduce the entire Corpus Hippocraticum (in the form of the 'opera omnia' edition of Frankfurt 1624, whose eight-part structure and translation he silently adopts) to a little over 200 octavo pages. More than half a dozen later editions, published over a period of eighty years in England, the Netherlands, Germany, and Italy, are on record. As late as 1791, Johann Matthias Alexander Ecker translated the booklet into German as *Geist Hippokrats* for the benefit of non-Latinate army surgeons.

Aphorisms

The titles of the *Fusul Musa* and its Latin translation reflect the contemporary popularity of aphorism collections.[100] I conclude this section with a look at this remarkable subform of the compendium. To the modern reader, who tends to associate aphorisms with philosophy or belles-lettres, the popularity of scientific aphorism collections may come as a surprise, but for premodern authors and readers interested in the natural world, the form held several attractions.[101] It responded to a pervasive predilection for 'sententiae', pithy, pointed phrases that were easy to memorize and impressive to quote.[102] Collections of such phrases circulating under the name of wise men, from the 'logia' of Jesus to the *Disticha Catonis*, had been popular since antiquity. The terseness and incisiveness of aphorisms proved useful for condensing knowledge about a certain field. The form seemed to avoid the dangers of system constraint and promised direct access to the phenomena themselves. Authors found the generic *ipse dixit* style of aphorisms convenient, as it enabled them to state their opinions bluntly without providing any arguments and to set themselves up as unquestionable authorities in doing so. (In fact, aphorisms significantly contributed to the fame of several great names.[103]) The generic conventions of such collections also saved them the trouble of structuring their work, if they did not want to: disorder was not obligatory, but it could be appreciated as pleasant 'varietas' by readers.

Of course, it also contributed to the popularity of aphorisms in science that the genre had such an illustrious founder: the seven sections of the Hippocratic *Aphorisms* were the first collection of 'sententiae' to go by that title and the only

[100] The history of the premodern aphorism remains to be written. For some preliminary information, see *HWRh* 1, cols. 773–90, s.v. 'Aphorismus'.

[101] Some of the advantages enumerated in what follows (no system constraints, direct access to the phenomena, no time wasted on structuring one's work) are addressed by Giorgio Baglivi, *De praxi medica* (1696), pp. 85–6, in a praise of the aphoristic form which is somewhat redundant and difficult to understand in places. Cf. Maclean 2021a, 257–8 with n. 139.

[102] On 'sententiae', see *HWRh* 8, cols. 847–67, s.v. 'Sentenz'. Aphorisms were indeed learned by heart (Troupeau 1997, 319–20, 324).

[103] To the examples discussed in what follows, one can add authors such as Francis Bacon ('Novum Organum', 1620) and Herman Boerhaave (*Aphorismi de cognoscendis et curandis morbis*, 1709).

important one for a long time. The work was extremely popular in antiquity and in the Arab world and became a staple text in medieval and early modern medical faculties.[104] In 1679, the medical bibliographer Martin Lipenius could list 138 printed commentaries on the collection.[105] Among the first works composed in imitation of it were some 130 medical aphorisms by Yuhanna ibn Masawaih (ninth century) and the 100 astrological sayings of the pseudo-Ptolemaic *Centiloquium*, a work of obscure but probably also Arab origin.[106] Both texts were Latinized in the Middle Ages and became highly successful.[107] In the course of the Hippocratic revival, new collections of aphorisms came to be written by Western physicians and other naturalists from the sixteenth century.[108]

An early example of this trend comes from a famous author who took full opportunity of the genre's aforementioned advantages. Cardano's collection of astrological 'sententiae', the 'Aphorismorum astronomicorum segmenta septem' (1547),[109] flaunts its Hippocratic character in its very first aphorism, which is closely modelled on Hippocrates' famous opening statement, as well as by its seven-part structure.[110] In a concluding note, Cardano puts himself on the same level as his model even more explicitly by styling himself as a founding father of astrology: in his opinion, his work constitutes a full vindication of the discipline against its critics and transforms it into an 'eternal art' ('ars sempiterna', fol. 307^{r}). Hippocrates' laconic style and haughty tone are imitated throughout the work, as is the occasional obscurity for which the father of medicine was famous. At times, even the reaction of uncomprehending readers is anticipated: 'Qui autem haec, quae hic tradita sunt, non intelligent, ut difficillima<m> destituent.' ('But those who will not understand what is presented here will abandon [astrology] as exceedingly difficult', aphorism 3.107). Most interestingly, Cardano seems to use the genre-typical hotchpotch structure to play hide and seek with readers and censors (the Roman Inquisition had been founded a few years earlier, in 1542). One must read quite carefully through the entire work, for instance, to ascertain his predilection for the Copernican system. Besides some inconclusive remarks,[111]

[104] For medieval Latin translations and commentaries, see Kibre 1985, 29–90.

[105] *Bibliotheca realis medica*, pp. 36–40; Maclean 2021a, 222–3. See also Maclean 2023 on the importance of the collection in the seventeenth century in general.

[106] Lemay 1978; Rinaldi 1999.

[107] On their reception in the West, see Troupeau 1997 and Rinaldi 1999 respectively. An edition of Yuhanna's aphorisms, including two medieval Latin translations and an early modern French one, is provided by Jacquart and Troupeau 1980.

[108] Nutton 1989; Maclean 2002, 114.

[109] Briefly discussed by Grafton 1999, 250–1, and Pompeo Faracovi 2012a, 130.

[110] Another example of an aphorism collection in seven sections is Santorio Santorio's *De statica medicina* (1614). Also in this case, the structure is announced already on the title page ('aphorismorum sectionibus septem comprehensa'). On the second page of the unpaginated preface to the reader, the author expressly states that he was moved to use the aphoristic form by the example of Hippocrates ('in imitationem magni nostri dictatoris, cuius vestigiis insistere gloriosum semper duxi').

[111] Grafton 1999, 250–1 nn. 23–4, notes aphorisms 4.65 (Copernicus is probably right in making the moon the only celestial body circling the earth) and 5.32 (enigmatic: 'Copernici autem nondum perspecta est recte sententia; vix enim, quae vellet, dicere visus est'; 'visus' may be a typographical error for 'ausus').

an incontrovertible statement to this effect is hidden in aphorism 2.12: 'Sol omnibus planetis praestat immobilitate, magnitudine, lucis primigeniae natura.' ('The sun is superior to all planets by its immobility, its size, and the nature of its light, which originates from it.')

Another, late example is atypical of the genre in some respects but all the more shows its flexibility and potential in the right hands. Linnaeus's *Fundamenta botanica* (1735),[112] which he published at the age of 28, in the same year as the first edition of his *Systema naturae*, is perhaps the most ambitious aphorism collection ever. Like Cardano, Linnaeus aims at nothing less than putting an entire discipline on solid foundations once and for all, but his attempt is much more serious and successful than Cardano's. On only 35 duodecimo pages, the *Fundamenta* feature 365 consecutively numbered aphorisms in 12 chapters, each of which is dedicated to a clearly defined topic: botanical literature in general (I. *Bibliotheca*), earlier classificatory attempts (II. *Systemata*), general morphology (III. *Planta*), morphology of flowers and fruits (IV. *Fructificatio*), reproduction (V. *Sexus*), classification (VI. *Characteres*), generic names (VII. *Nomina*), species epithets (VIII. *Differentiae*), varieties and their names (IX. *Varietates*), synonyms (X. *Synonyma*), plant descriptions (XI. *Adumbrationes*), and medical properties (XII. *Vires*). As a coda, Linnaeus draws twelve general 'conclusiones', one from each chapter.

Compared with Cardano's and many other collections, the *Fundamenta* are special in two respects: they are no hotchpotch but on the contrary are extremely systematic and coherent,[113] and they exploit to the fullest the potential of the form for creating an authoritative speaker persona. Order is established not only by the arrangement of the single aphorisms in strictly thematic chapters, within which one follows the other in logical succession, but also by a thorough system of cross references. Most items redirect the reader to one or more earlier ones on which they build, suggesting a quasi-Euclidean deductive coherence:

> 132. Initio rerum ex omni specie viventium (3) unicum sexus par creatum fuisse suadet ratio.
>
> 132. Reason leads one to believe that one single male and female couple of every species of living beings (3) was created at the beginning of the world.

Turning to aphorism 3, the reader can be confident of finding the fundamental notion of a living being determined once and for all there. The resulting sense of closure is enhanced by numerical symbolism: 12 chapters and 365 aphorisms obviously suggest the completeness and unity associated with the circle of the year.[114]

[112] The title page has 1736, but see Stafleu et al. 1976–, vol. 3, 77–8 (no. 4710), where bibliographical data for all later editions are given as well.

[113] Three precursors in this respect were the collections by Santorio (see note 110 above), Bacon, and Boerhaave (see note 103 above).

[114] This does not make the work a calendar, as claimed by Chang 2011. On the literary symbolism of the number twelve by itself, see the main text following note 56 above.

At least as striking is the aplomb with which Linnaeus—clearly a rising star, but still a very young man at the time—acts as a lawgiver for his older fellow botanists. The dedication of the work to twelve colleagues from seven different countries, many of them Linnaeus's own age or only slightly older, is a first step towards making readers accept the role he arrogates, as it implies that he speaks for an entire generation of successful young botanists. The assertion made in this dedication of having worked seven years on the *Fundamenta* and examined 8,000 flowering plants ('opusculum hoc aliquot foliorum, verum septem annorum, 365 aphorismis constans, at examine florum 8000') invalidates the possible charge of lacking experience. On this basis, Linnaeus permits himself an extremely authoritative style in the work itself. Wasting no space on explanations, he defines, subdivides, states, and decrees, and he does so as concisely as possible. These are the words of a man who is sure of himself and his knowledge and has no time to waste:

> 151. Fundamentum botanices (4) duplex est: dispositio et denominatio.
> Botany (4) rests on a double foundation: classification and terminology.
>
> 156. Filum Ariadneum botanices est systema (155), sine quo chaos.
> Ariadne's thread in botany is the system (155), without which there is a mess.
>
> 297. Nomen specificum nec comparativum nec superlativum sit.
> A species name may not be comparative or superlative.

The cumulative effect of such phrases is overpowering, and readers were duly impressed. The *Fundamenta* went through half a dozen editions in their original form until 1753. In 1751, Linnaeus, now a celebrity, saw fit to republish the work as one publishes a classic, namely with a commentary, renaming it *Philosophia botanica* (see Chapter 5, note 152). In this, as in the original version, it remained in print until the nineteenth century and has been so again since the mid-twentieth.[115]

Textbooks

What is a textbook?

> Hi sunt magistri, qui nos instruunt sine virgis et ferula, sine verbis et cholera, sine pannis et pecunia. Si accedis, non dormiunt; si inquirens interrogas, non abscondunt; non remurmurant, si oberres; cachinnos nesciunt, si ignores.

[115] See Stafleu et al. 1976–, vol. 3, 90–1 (no. 4760); Freer 2003.

> These are the teachers who instruct us without rods and ruler, without insult and anger, without asking for clothes and money. If you approach them, they are not asleep; if you have a question to ask them, they do not hide their knowledge; they do not reprimand you, if you make a mistake; they do not think of laughing at you, if you fail to know something.

This enthusiastic praise of books as teachers in Richard de Bury's *Philobiblon* (1344)[116] is more than just a nice personification. It has a real-life background, and it implies that problems arise when one speaks of textbooks in premodern and early modern times.[117] With teaching being much less institutionalized, autodidacticism much commoner, and the boundary between them more fluid than today, virtually any book might be used as a learning tool by a curious reader. On the other hand, even books written for and used in the classroom could be bought and read just for fun by any given individual. In the age of print, boundaries at first became even more blurred, as books became more generally available.

This uncertainty has resulted in a widely diverging usage of the term 'textbook' in modern scholarship. In most cases, the word is applied to all kinds of books that appear somewhat learner-friendly, in particular compendia.[118] This contrasts with strict definitions such as the following: 'In the most general way a textbook can be said to be a book specifically designed for classroom use.'[119] None of these two extremes is viable for my purpose. The former tends to equate textbooks with books *tout court*. The latter makes it difficult to identify a book as a textbook in many cases, as will become evident shortly. Under these circumstances, a middle course will be steered. In what follows, 'textbook' will designate a self-contained work (as opposed to a commentary) designed to be used *in connection with* formal teaching. This definition sets textbooks apart from other kinds of books. At the same time, it does justice to our lack of information about the exact intended use of many early modern books.

Since this last point, in fact, severely limits our knowledge and understanding of textbooks about the natural world, some further remarks on it are in order at the outset. Circumstantial evidence, such as lecture announcements and study regulations, is seldom available, and even if it is, it is usually of limited help.[120]

[116] Ed. Thomas 1889, 13–14.

[117] On early modern scholarly textbooks in general, see Campi et al. 2008. Within that volume, Lines 2008, De Angelis 2008, Leu 2008, Siraisi 2008, and Keller and Steinke 2008 specifically deal with scientific textbooks (the study of which was advocated already by Sarton 1948). On philosophical textbooks, see Schmitt 1988.

[118] In Sarton 1948, for instance, 'textbook' seems to mean something like 'simplified, popularizing account of a certain field'. Many contributions to Campi et al. 2008 take a similar approach. For the resulting difficulties, see Grafton 2008.

[119] Schmitt 1988, 792.

[120] On lecture announcements in Germany, see Bruning 2011. Digitized announcements reaching back to before 1750 are available from the Universities of Braunschweig, Giessen, Helmstedt, Königsberg, and Leipzig. They occasionally mention the textbooks used, as do the announcements

Book titles cannot be trusted. While a given *Compendium*, *Cursus*, *Epitome*, *Isagoge*, *Introductio*, *Institutio*, or *Systema* may indeed be a textbook, one cannot be sure, as other overviews have the same names. The overall design of a book is normally open to divergent interpretations. The same is true of readers' annotations.[121] In the comparatively rare cases in which an author chooses to speak about the purpose of his textbook in the preface, the remarks are our most valuable source. Rembert Dodoens, for instance, explains that his *Physi<o>logices*[122] *medicinae partis tabulae* (1581) is to prepare students for the perusal of Galen's and other authors' physiological works (pp. 336–7). Ignaz Weinhart states that his *Quaestiones et responsa ex arithmetica vulgari, algebra et geometria* (1765) will spare his students at the arts faculty of the University of Innsbruck the trouble of taking notes during his classes (fol. 2^{r}). An exceptionally informative text is the preface to Petrus van Musschenbroek's *Epitome elementorum physico-mathematicorum* (1726), an early textbook of Newtonian physics, which explains in rare detail how the work will be used in the author's own teaching at the University of Utrecht. The students—who, it is silently assumed, will buy the book—will be relieved from taking notes, which also frees up time that is better invested in teaching itself (sig. [$*7^{r}$]). In order not to overwhelm the youngsters who only have some elementary geometry, no higher mathematics is required. Proofs, experiments, and illustrations are all left out, because the author will provide them in class (sigs. $*5^{r-v}$, [$*6^{v}$]); he accepts that the work thereby becomes difficult to understand for ordinary readers (sig. [$*6^{v}$]). The names of Musschenbroek's source authors are given in the text, as they may guide the students towards further reading (sigs. [$*6^{v}$–$*7^{r}$]).

In most cases, one is much less lucky and has to speculate about many questions for oneself. Was a given book linked to a specific didactic setting or not? If so, how exactly did it relate to the actual teaching? Was it a template for a lecture course, the published version of one,[123] or both? How exactly was a textbook used

found, among many other documents of intramural academic communication, in the *Scripta publice proposita* (1546–72) from Wittenberg (on this corpus, see now Domtera-Schleichardt 2021). I do not know whether similar announcements are extant from other German universities or outside Germany. A disappointing example of a study regulation is the Jesuit *Ratio studiorum* (1599), whose rules for the 'professor mathematicae' barely fill one page. The only books and contents named are 'Euclid's *Elements*' and 'something about geography or the sphere or of things pupils usually listen to with pleasure' (Lukács 1986, 402).

[121] However, they can be helpful on occasion. In a copy of the Zurich 1548 edition of Johannes Honterus's *Rudimenta cosmographica* (1542), a pupil noted the date when Conrad Gessner began a lecture course based on them: 'D. Conradus Gesnerus praeceptor meus pluri<m>um observandus auspicatus est haec Cosmographica rudimenta 25 die Februarii anno 1563' (quoted in Leu 2008, 240).

[122] That the title's 'physilogices' is a typographical error is shown by the fact that Dodoens invariably uses 'physiologice' within the work.

[123] Lectures could be published by the professor or his listeners. The former holds, for instance, for Johannes Widmann's six short *Algorithmi* dealing with various parts of mathematics ([*c.*1490]), Melanchthon's *Initia doctrinae physicae dictata in Academia Vuitebergense* (1549), Pierre Gassendi's *Institutio astronomica* (1647), Georg Horn's *Orbis politicus* (1668), or Herman Boerhaave's *Praelectiones publicae de morbis oculorum* (1746), the latter for Girolamo Mercuriale's *De morbis*

in class? Did the teacher read it out loud ('praelegere') and add explanations in the same way in which he presented a classical text?[124] Did he lecture on the entire textbook or (again, as in the case of many classics) only on part of it? Did all or some pupils possess the book? Did they have it in front of them in class? Did they use it to prepare or to rehearse lessons and to prepare themselves for exams? Did they read it through or consult it selectively? (Some textbooks have indexes; most do not.) Was a textbook bought and used also or predominantly by interested amateurs? Was a work intended as a textbook used as such at all?[125] Many such questions are bound to remain open. Generally speaking, there was probably a broad range of variation between different books, while one and the same book could be used in widely different ways as well.

Historical overview

Classical antiquity knew a class of books which are called textbooks in modern literature. They cover such diverse fields as rhetoric, grammar, law, agriculture, music theory, architecture, surveying, and medicine. Manfred Fuhrmann[126] has provided an overview of this genre and demonstrated that the works are characterized by the prominence of certain logical operations and corresponding formal features, which will be discussed later.

In terms of social context and function, however, the texts in question do not form such a well-defined group as in terms of their argumentative structure and its linguistic realization. The question of actual use in formal teaching is even harder to answer for most of them than for their early modern successors. Only some of the above disciplines were formally taught on a large scale, and even in those cases, we do not know whether any of the books analysed by Fuhrmann

cutaneis (1572), Isaac Barrow's mathematical *Lectiones* (1683), or Boerhaave's *Praelectiones academicae* (1739–44; see note 151 below). For an unauthorized edition, see note 152 below . An example of a course which only survived in manuscript, a dissection class held in January 1540 in Padua, where Matthaeus Curtius lectured while Andreas Vesalius acted as 'prosector', was published by Eriksson 1959.

[124] On the first page of his *Rudimentorum cosmographiae libri duo* (1530, sig. Aii^r), Honterus explains: 'Quae vero demonstrationibus ac viva voce magis indigere videbantur, fido praeceptori committimus, nos enim praeter compendiariam introductionem nihil supervacaneum noviciis tironibus obtrudendum censuimus.' This can either mean that he thinks of the book as a text to be used in class and commented upon by the teacher or as a text used by pupils as a preparation for class. See also note 125 below.

[125] Martin Borrhaus, who let an astronomical and geographical compendium originally written for students at Wittenberg be rearranged and edited for the schools by Achilles Pirmin Gasser, was unsure about the success of the operation: 'Sed hunc defectum [i.e. lack of geometrical diagrams] praefectoris industria facile sarciet praestabitque, si modo hoc scriptum quisquam schola in qua vivat dignabitur' (*Μαθημάτων στοιχεῖα*, 1539, p. 4). Classroom use ('publicum scandere suggestum') and private perusal by students ('a iunioribus cum fructu legi') are presented as the two possible uses of a medical textbook on the third page of Pieter Pauw's preface to Johan van Heurne's *Institutiones medicinae* (1592).

[126] Fuhrmann 1960.

were actually used in class. On the other hand, should concise, systematic, and comprehensive overviews not conforming to the formal pattern identified by Fuhrmann be, *ipso facto*, denied textbook status? Cleomedes' *On the Circular Motion of the Celestial Bodies*, a Stoic introduction to astronomy and cosmology, for instance, is more discursive than Fuhrmann's examples but would be elementary enough for beginners, and it does not seem improbable a priori that its subject should have been taught in a Stoic school of the Roman Empire. There are, however, few comparable candidates dealing with subjects from the natural sciences.

In the (especially later) Middle Ages, higher teaching was for a large part based on authoritative works of an originally non-didactic character, which were explained by commentary. Concomitantly, however, a number of genuinely didactic works were produced in the mathematical disciplines in particular. These comprised various instructions for calculating with Hindu-Arabic numerals, Sacrobosco's astronomical and cosmological *Sphaera*, the *Theorica planetarum* ascribed to Gerard of Cremona, a textbook of planetary astronomy,[127] Jean de Meur's *Musica* for musical theory, and John Pecham's *Perspectiva communis* for optics.[128] When such works became famous and authoritative, they were equipped with commentaries as well.

From the Late Middle Ages to the first centuries of the early modern period, a number of developments contributed to the birth of a large-scale textbook tradition in the disciplines concerned with the natural world. Apart from the changed conditions brought about by print, two aspects seem decisive: teaching in general became more of an issue, and scientific progress an undeniable reality. As for the former, humanism attributed an outstanding importance to education, thought indispensable for moulding a person into a truly human being. Humanism's pedagogical impetus was soon picked up by the Church and the early modern state and made to serve the needs of these institutions. Accordingly, questions of what and how to teach became the object of intense reflection and debate, a trend epitomized and amplified by pedagogical thinkers such as Erasmus, Petrus Ramus, or Amos Comenius. Educational institutions proliferated and diversified, and teaching became more organized, systematic, and standardized, also in the sciences. Regulations got stricter and more detailed, including those not only for subjects and timetables, but also for the books to be used. A document like the Jesuit *Ratio studiorum* would have been inconceivable a few centuries earlier.[129]

The accelerated accumulation of new and obsolescence of old knowledge made themselves felt from the sixteenth century. For a time, novel truths could be

[127] Gerard's authorship is disputed but cannot be excluded (Aiton 1987, 6 n. 4).

[128] For a more complete overview of the books that were mandatory in Central European arts faculties towards the end of the Middle Ages, see Lorenz 1985.

[129] For an overview of these developments, see Korenjak 2016, 117–40.

ignored or introduced by means of new commentaries on dated classics when teaching the young, but at some point this did not seem sensible any more. Around 1700 it was commonly accepted that scientific progress created a need to update teaching materials. Lorenz Heister, professor of medicine at Altdorf, justified his *Compendium anatomicum* (1717), a new textbook of anatomy, as follows at the beginning of his preface:

> Etenim quoniam scientia anatomica ita est comparata, ut industria atque improbo multorum clarissimorum virorum labore in dies incrementa sumat atque etiam a pauco tempore largissima sumpserit, hinc harum rerum gnaris facile erit manifestum in hucusque usitatis compendiis et syntagmatibus anatomicis quam plurima iam deficere, quae ab aliquo tempore detecta atque inventa sunt quaeque studiosae iuventuti et imprimis tironibus medicinae scitu sunt necessaria.
>
> For since the science of anatomy is of such a nature that it is being augmented daily—even very much so recently—by the diligence and drudgery of many illustrious men, it will easily be evident to those skilled in these things that in the digests and compilations in use so far, a great number of facts are missing which were discovered and found out some time ago and which are necessary for students to know, in particular beginners in medicine.

Under these circumstances, it is sometimes said, the commentary, the main didactic genre in the Middle Ages, died out and was replaced by the textbook.[130] This is broadly true but in need of qualification. First of all, the didactic commentary died hard. In fact, it flourished until the middle of the seventeenth century in medicine and astronomy, when texts like Avicenna's *Canon* and Sacrobosco's *Sphaera* attracted an incessant stream of explanations. On the other hand, the textbook as such was no early modern invention, as just explained. However, in the Middle Ages, little difference was made between books written specifically for teaching purposes and other authoritative works, as both kinds were used together with commentaries on them. Now, textbooks became more self-reliant. Although many of them remained dependent on authoritative works in a certain respect, being abridgements, paraphrases, or expansions of these,[131] and although oral explanations could be added to them in class, they were usually not equipped with written commentaries any more. Authors tried to compose them so that they were not in need of such complements. If a textbook turned out to be too

[130] To a small degree, lecture announcements document the introduction of certain textbooks in certain places. For example, the *Arithmeticae practicae methodus facilis* (1545) of Gemma Frisius and Jacques Peletier is first mentioned as a textbook at Helmstedt in the academic year 1595/96 (*Catalogus praelectionum…, quae… in inclyta Academia Iulia, quae est Halmstadii, publice proponuntur et tractantur*, 1595, sig. A4^{v}).

[131] Cf. Paulus 2005, 281('Reduktions-, parallele, Amplifikationsformen').

difficult or became outdated, this was not mended by a commentary either. Rather, a new simplified or updated edition was produced, or the deficient textbook simply was replaced by a new one.

In the course of these developments, the textbook gained sharper contours and grew into a distinct genre, differentiating itself more clearly from other kinds of writing. It developed a clearer profile in terms of structure, language, and form in general. Features taken over from ancient and medieval predecessors were applied in a more systematic and sophisticated way, and new ones were added with the help of print, as will be explained later. Moreover, while the difference between university textbooks, schoolbooks, and books for amateurs was still not as clear-cut as it is today, textbooks nonetheless began to be geared more specifically to certain settings and audiences and became more outspoken in this respect. When the Leiden professor Franco Burgersdijck published a version of Sacrobosco's *Sphaera* that was completely rewritten in classical Latin, he titled it *Sphaera in usum scholarum* (1626) to indicate that his work targeted not the arts faculties any more, but the Latin schools. The development culminated in the composition of textbooks for their own courses by many professors, sometimes in order to comply with specific regulations issued by the authorities. The Innsbruck professor Franz Seraphin von Zallinger zum Thurn composed his *Praelectiones ex physica theoretica et experimentali* (1805) in reaction to a writ arriving from Vienna 'when the academic year was already about to begin' ('imminente iam anno scholastico') which stipulated that physics should be taught in Latin and separately from mathematics. No textbook fitting such specific demands was available, so Zallinger had to write one for himself (preface to the reader, pp. iii–iv).[132]

That a generic consciousness developed as well is indicated by the fact that textbooks were increasingly criticized. In the preface to a Sacrobosco edition published in 1531 in Wittenberg, Melanchthon notes that schoolbooks are especially liable to criticism.[133] Often, it was authors of later textbooks who found fault with their predecessors, deeming their works ill-structured, outdated, or otherwise unfit. An example from Michael Maestlin's *Epitome astronomiae* (1582) will be discussed later.[134] Alternatively, it was sometimes explained why an otherwise good book could not work as a textbook. In his *Introductio ad veram astronomiam*

[132] The strict regulation of academic teaching in Austria often yielded similar results in the later eighteenth and early nineteenth centuries. In the 1760s, Benedikt Stattler was appointed to teach a class on 'the natural foundations of mineralogy and metallurgy' at the arts faculty at Innsbruck. 'In the great dearth of some fitting treatment of this subject' he saw it as his duty to compose *Mineralogiae principia physica* (1765) (preface, sig. *2^{r-v}). Leopold Biwald's *Physica generalis* (1767) (see Lippitsch et al. 2008) was written for a specific course which students in the arts faculty in Graz had to attend in their second year.

[133] Sig. aii^{v}: 'Videmus enim paucissima scripta vetustatem ferre, praesertim in scholis, ubi morosissime iudicari solet.'

[134] See also, for instance, the third page of Pauw's preface to van Heurne's *Institutiones medicinae*: earlier textbooks are too long-winded, too obscure, or written in a vulgar style that puts students off.

seu lectiones astronomicae (1718, pp. xiii–xiv), John Keill spoke with the highest respect of the *Astronomiae physicae et geometricae elementa* (1702) by his own teacher David Gregory, the first compendium of astronomy to take Newton's findings into account. However, its combination of traditional astronomy and celestial mechanics would, in Keill's eyes overwhelm, students.

As to their language, textbooks for subjects of higher education remained predominantly in Latin until the eighteenth century, in some fields such as medicine even into the nineteenth. In other areas such as elementary arithmetic, Latin and vernacular traditions developed in parallel, as mentioned in the previous section. Moreover, from around 1700, Latin textbooks were occasionally translated into various vernaculars. Johann Christoph Sturm's schoolbook *Mathesis iuvenilis* (1699–1701) came out in German and English (1705; 1708–9), and Heister's academic *Compendium anatomicum* knew numerous later editions not only in Latin but in German, English, and French as well.

Textbooks across the disciplines

The development of the early modern textbook started with the disciplines of the trivium, and the genre became particularly productive in these areas. Scores of grammars and rhetorical textbooks in particular were written from the second half of the fifteenth century.[135] Analogous works dealing with aspects of the natural world did not lag behind much in terms of chronology, although they were produced in far lower numbers. The composition of textbooks for specific sciences has been said to have been inspired by the model of Aristotelian textbooks, which began to replace the masters' own texts in arts faculties from the sixteenth century and whose *φυσικά* sections grew disproportionately over time.[136] Such textbooks continued to be written for a long time and absorbed much modern science over the centuries.[137] That they influenced non-Aristotelian textbooks for specific disciplines is plausible a priori but not universally valid. Because institutionalized teaching was not introduced at the same time and to the same degree in all fields, some disciplines developed no robust textbook tradition during the early modern era at all. Between those that did, the differences were considerable, and each disciplinary microtradition was informed by its own previous history. Generally speaking, the disciplines that had already been academic in the Late Middle Ages

[135] For an overview of the sixteenth century and case studies, see Leonhardt 1997. For grammar in particular, see Ax 2001; for rhetoric, see Green and Murphy 2006; Mack 2011.

[136] Schmitt 1983, 50–2 and 1988, esp. 802–3. An influential early author of such textbooks in Protestant Germany was Melanchthon (e.g. *De anima*, 1540, and *Initia doctrinae physicae*, 1549; see Helm 1997 and Bauer 1997 respectively).

[137] A good late example is Biwald's *Physica generalis*. Biwald mostly steers clear of mathematics but nonetheless teaches much Newtonian physics, chemistry, and astronomy.

were at the forefront: the quadrivium, connected to the arts faculty but sometimes also taught in Latin schools, and medicine, which had a faculty of its own.

Within the quadrivium, Euclid's *Elements* remained the standard textbook for geometry.[138] Ramus's attempt to replace them with his own abridgement (1569), which he reduced to practically useful aspects and organized in a better way, as he believed, enthused some of his contemporaries for mathematics and found followers in other fields,[139] but he did not oust the original from the classroom. Musical theory apparently was little taught and learned. By contrast, arithmetic and in particular the art of calculating with Hindu-Arabic numerals, introduced to Europe during the preceding centuries, attracted great interest. It was a subject in arts faculties and in Latin schools, and textbooks were written for both. Such works, which started to appear in the thirteenth century, were often titled 'algorit(h)mus' or '-ismus', a malapropism of the name of the Persian mathematician Al-Khwarizmi who played an important role in transmitting the new numerals to the West.[140] Restricting themselves to the basic arithmetic operations, these books strongly contrasted with Boethius's philosophically minded *De institutione arithmetica*, which had dominated the teaching of the subject in the medieval quadrivium. Those for 'artistae', for instance six short treatises by Johannes Widmann ([c.1490]) or Johannes Huswirt's *Enchiridion novus algorismi* ([1501]), belong to the earliest textbooks of a scientific discipline ever printed. Schoolbooks such as Lucas Lossius's *Arithmetices erotemata puerilia* ([1557]), whose intended use is explained on the final pages of the preface, followed somewhat later. In content and structure, there is no great difference between the two. They are also similar to the 'Rechenbücher' composed in German lands for vernacular schools and perhaps autodidacts, except that these, which targeted representatives of practical professions like merchants, contain more examples.[141] A closer look at this subgenre will be provided later. From the 1530s, the elementary tradition was supplemented by more demanding and wider-ranging works.[142] An egregious, much later example is Sturm's *Mathesis iuvenilis*. Written for the schools of Württemberg, this work comprises some 1,800 octavo pages in two volumes and besides arithmetic and algebra covers a wide range of fields from applied mathematics. Even more astonishing is its success, which manifested itself in two further editions and the aforementioned translations. One would like to know whether and how it was actually used in the schools. If it was, Württemberg's

[138] Steck 1981.

[139] Verdonk 1968. Ramus's impression that Euclid was confused resulted from the way in which the *Elements* was presented in the classroom, namely as a collection of theorems. The proofs, which would have connected the theorems and justified their order, were thought unimportant (and even deemed an addition by Euclid's ancient editor Theon) and disregarded.

[140] Merzbach and Boyer 2011, 206, 227–9. Numerous examples of 'algorithmi' can be found with the help of the index of Smith 1908 (who also gives a succinct description of the genre at pp. 5–7), s.vv. 'Abacus', 'Algorism', 'Counters'.

[141] Gärtner 2000.

[142] Kusukawa 2000b, 129–30.

pupils must have been better mathematicians than Tyrolean students nearly one century later. In Innsbruck, Ignaz Weinhart's *Quaestiones et responsa* (1765) was the prescribed textbook in the arts faculty until 1779. This mathematical catechism discusses arithmetic, algebra, and geometry on a fairly elementary level. Its introductory part, which starts with the notion of number and reviews the basic operations, still resembles the works by Widmann and Huswirt.

Astronomy presents a special case in two respects.[143] First, it had already developed a mature textbook tradition in the later Middle Ages. This tradition was bipartite, comprising spherical astronomy, on the one hand, and theorical (planetary) astronomy, on the other. The former was represented most prominently by Sacrobosco's *Sphaera*, in which the doctrine of the sphere, set out in book 3, was supplemented by two introductory books and some remarks on the planets in book 4. The standard textbook for planetary astronomy, however, was the *Theorica planetarum* attributed to Gerard of Cremona. At the very end of the Middle Ages, this book was replaced by Georg Peuerbach's *Theoricae novae planetarum*, which was printed in 1472 by Regiomontanus, became a new classic, was reprinted over fifty times, and attracted much commentary up to the mid-seventeenth century.[144] Secondly, astronomical knowledge not only expanded massively in early modern times, but was also fundamentally changed, as happened in no other discipline. Alongside a mass of commentaries on the medieval classics, a stream of new textbooks, therefore, appeared from the sixteenth century. At first, all of them kept to the medieval bipartite structure, covering one (Heinrich Brucaeus, *De motu primo*, 1573) or both of the fields (Maestlin, *Epitome astronomiae*). Also, all sixteenth-century textbooks remained geocentric. The introduction of heliocentrism took time. The matter was delicate (Copernicus's idea was initially embraced by few, and *De revolutionibus* entered the *Index* in 1616, to be removed only two centuries later), and teachers hesitated to confuse students with discussions involving several world systems, given the difficulty of understanding one.[145] Kepler, it is true, openly embraced Copernicanism, but his *Epitome astronomiae Copernicanae* (1618–21) is a very advanced and idiosyncratic textbook, if it can be called this at all. It failed to exert much influence.[146] The first successful textbook to discuss the Copernican and Tychonian systems openly was Pierre Gassendi's *Institutio astronomica* (1647), which enlarged the traditional two-part structure with a third book dedicated to that very question.

[143] For a concise overview of astronomical textbooks from Sacrobosco to the twentieth century, see Gingerich 1990.

[144] English translation with introduction and commentary: Aiton 1987. Edition with French translation and monographic treatment: Malpangotto 2020. From a letter by Melanchthon to Johannes Schöner prefixed to the latter's *Tabulae resolutae astronomicae* (1536), it becomes clear that the *Theoricae* was used in tandem with table works in the classroom (see Chapter 8, note 212).

[145] Melanchthon (*Initia doctrinae physicae*, 2nd edn 1550, fols. 39^v–40^v) suggests that discussion of Copernicus should be reserved for more mature students for this reason. See Methuen 1996, 237.

[146] Edition and postface: *KGW* 7. Lack of influence: Gingerich 1990, 192.

It had a dozen editions and reprints until 1727.[147] By that time, Copernicanism was endorsed as a matter of course within the framework of Newtonianism, which posed a new challenge through the introduction of celestial mechanics and an unprecedented degree of mathematization. Textbook authors reacted to this situation in different ways. Gregory, for one, integrated celestial mechanics into his *Astronomiae physicae et geometricae elementa*, while trying to maintain the traditional basic structure of an astronomical textbook. Book 2 discusses spherical astronomy, books 3–5, the solar system. The whole is preceded by an introductory book demonstrating the truth of Newtonian cosmology and followed by another, which paints a vivid picture of the universe by sketching its appearance as seen from the other planets, the sun, and the fixed stars. Keill's *Introductio ad veram astronomiam*, on the other hand, is traditional in content but innovative in form. Newtonian physics is left out as too complicated for young students, but the arrangement of the thirty lectures on which the work is based radically breaks with the bipartite structure. At this point, the astronomical textbook tradition had left its medieval predecessors behind for good.

A discipline often taught in tandem with astronomy was geography, in particular mathematical geography, closely linked to the former through the structural analogy between the celestial and terrestrial spheres and the use of astronomical data to determine positions on earth. Astronomical textbooks often supplemented their information about the sky with some geographical sections. Already Sacrobosco did so in the final parts of his first book. He was later followed by Maestlin, for instance. Concomitantly, a tradition of schoolbooks evolved in which a mixture of cosmology, astronomy, and geography was discussed on a more elementary level. Apianus's *Cosmographicus liber* (1524) provides a 100-page summary of all three disciplines with a focus on geography.[148] Most of the astronomical information precedes the geographical, an order which becomes quite strict in other examples such as Johannes Honterus's *Rudimentorum cosmographiae libri duo* (1530) and their poetic version, the *Rudimenta cosmographica* (1542),[149] or Martin Borrhaus's *Μαθηματων στοιχεῖα* (1539). Descriptive geography, on the other hand, could be presented together with history and politics. In the very successful *Orbis politicus* (1668) by Georg Horn, a geographical overview is appended to an outline of the political conditions across the globe and an anecdotal world history.

In medicine, the classical authors—above all, Galen and Avicenna—held their ground for a long time: they were taught by commentary until far into the seventeenth century. The first original textbooks containing some medical information stem from the aforementioned Aristotelian tradition. Melanchthon's *Commentarius*

[147] Pantin 2012.

[148] Lindgren 1998, 248, states that Apianus's work became a model for textbooks of geography until the twentieth century, an assertion I have been unable to verify.

[149] Modern edition: Offner et al. 2017.

de anima (1540)—inspired by Aristotle's *De anima*, of course—pays great attention to the somatic foundations of psychology and in its second edition (*Liber de anima*, 1553) draws heavily on Vesalius's recent findings. In some Protestant universities, the work was to be read 'by a professor of medicine or natural philosophy' ('a medico vel a physico').[150] Freshly written textbooks exclusively dedicated to medicine appeared from the later sixteenth century. Often, they provided a first introduction to the discipline as a whole and were connected to pertinent lecture courses, as is the case with the *Institutiones medicinae* (1592) by the Leiden professor Johan van Heurne. A prominent later example from the same university is Herman Boerhaave's *Institutiones medicae* (1708), which provided a digest of and served as a companion to a one-year introduction.[151] Similar works with similar titles, written by professors of medicine, structured according to Galen's five parts of medicine or some adaptation of these, based on introductory lectures, and often addressed to students in their prefaces, appeared in considerable numbers. Their readers could use them as a preparation for their lectures, as aides-mémoires, and perhaps also to follow the lectures themselves better with a short printed version in front of them.[152] Soon, some *Institutiones* grew to intimidating proportions. Daniel Sennert's *Institutiones medicinae* (1611) already ran to 1,200 pages. Towards the end of the Latin tradition, such works had definitely crossed the line between textbook and encyclopedia, as in the case of the ten volumes of Pietro Aloisio Valentini's *Institutiones medicinae practicae* (1823–47).

In addition to *Institutiones*, thematically restricted formats presented help with specific aspects. Guides to dissections helped confused students to make sense of what they saw on the table: Johann Vesling's *Syntagma anatomicum* (1641) was structured according to the progress of the procedure so that students 'could attend public dissections with greater gain for their studies' ('ut publicis dissectionibus maiore cum studiorum emolumento interessent', preface, sig. § 3^{r-v}). There were also terminological introductions and, particularly popular, bibliographical guides for beginners (see Chapter 5, 'Dictionaries', and note 257 there respectively). A textbook tradition in chymistry originated when the discipline began to be taught as an auxiliary science to medicine. Its starting point was Jean Béguin's often reprinted and translated *Tyrocinium chymicum* (1610; 2nd, much expanded edition 1612).[153]

150 Schmitt 1988, 797–8; Eckart 1998, 192–9.

151 The degree of abbreviation may be estimated by the *Praelectiones academicae* (1739–44), Albrecht von Haller's edition of the entire course from Boerhaave's manuscripts (admittedly with some additions) in no fewer than six volumes.

152 After his introductory lectures to chemistry had suffered a pirate edition by unscrupulous students, Boerhaave felt particularly embittered by later generations bringing the booklet with them to the very same lecture course in order to follow his presentation more easily: 'Ita ferre ipse debui, quod auditores odioso sane spectaculo librum in manibus oculisque gererent coram mihique ostenderent ita docenti, dum dicentis verba expendi ad textum turpis libelli praesens quotidie cernerem' (*Elementa chemiae*, 1732, sig. †ii^{v}).

153 Patterson 1937; Thorndike 1929–58, vol. 8, 104–69; *CHS* 3, 508–10.

In other disciplines, by contrast, Latin textbooks are hard to find or seem to be entirely absent. Botany, zoology, mineralogy, and the earth sciences were introduced into higher education too late or remained subordinate to other disciplines for too long for a robust textbook tradition to develop. The first three, moreover, belonged to natural history, whose enumerative and descriptive approach and weak systematization did not lend themselves easily to formal teaching. Not before the eighteenth century does one encounter works like Benedikt Stattler's two volumes of *Mineralogiae principia physica* (1765), written for his students at Innsbruck, or the *Fundamenta et termini botanici* (1789), a guide to botanical description and taxonomy, by the Leiden professor Franz Joseph Märter.

Formal features

The definition given at the beginning of this section contains the phrase '*designed* to be used in connection with formal teaching'. This implies that a textbook fulfils its function by virtue of its formal features, and to a certain degree, that is indeed the case. As Fuhrmann has shown for antiquity, the basic aim, the systematic and intelligible presentation of a body of knowledge, is realized through the systematic use of a set of logical operations and specific ways to clad these operations in words. Although, as already mentioned, he has nothing to say about the function served by these features in the ancient texts, they are, nonetheless, of interest in the present context, as many of them recur in early modern textbooks.

The three most important logical operations found by Fuhrmann are definition, subdivision, and distinction. They usually do not occur in isolation but work closely together: a field of knowledge (for instance, astronomy) is defined and divided into subfields (spherical, theorical astronomy), which are defined in turn. Next, a category playing a role in such a subfield (say, great circles) is defined and subdivided. The members of the category are again defined and at the same time distinguished from one another by the addition of a specific difference: the ecliptic is the great circle described by the sun in the course of a year, the celestial equator the great circle whose plane is perpendicular to the line through the celestial poles, and so on. The definitions may or may not be followed by further characterization. When definition of a subfield or category has been exhausted, the account moves on to the next. On the linguistic and stylistic levels, these procedures are typically expressed in stereotypical formulas, including definitions beginning with the *definiendum* and the copula ('A great circle is…'), catalogues of species resulting from subdivisions, and meta-terminology highlighting logical entities and operations ('The *category* of the great circles may be *subdivided* as follows…'). Transitional formulas lead over from one topic to the next.

Recapitulations and announcements keep the reader oriented. Uniform structuring of coordinate sections serves the same end.

Of course, this basic structure can be realized in purer or more diluted forms. Some works more or less restrict themselves to a bare skeleton of definitions, subdivisions, and distinctions. Wilhelm Adolf Scribonius's *Isagoge sphaerica methodica* (1580), which presents the whole of spherical astronomy on twenty-four octavo pages, approaches the ideal of the bare-bones textbook quite closely. In other examples, such as Vesling's *Syntagma anatomicum*, the logical substructure is harder to see, as it is presented in fairly discursive garb.

Since the general features just mentioned will be concretized by the example of Maestlin's *Epitome astronomiae* at the end of this section, I will not dwell on them any longer here. Instead, I supplement them with a number of less fundamental traits which nonetheless, are regularly encountered in the works under discussion as well. Most, if not all of them are also known from earlier eras and other genres, but they are deployed with particular frequency and efficiency in early modern textbooks. The features to be discussed in what follows are the typographical highlighting of structure, visual aids, the catechism format, and a number of discipline-specific formats.

Regarding the first point, subtitles, almost absent in the first prints, quickly became an important structuring device. Before the end of the fifteenth century, Widmann clearly separated the individual sections of his *Algorithmi* in this way. Soon, different kinds of subtitles were used to establish multi-tiered subdivisions. The resulting sections could be numbered. Running numbers for paragraphs might be used as well. In Zallinger's *Praelectiones ex physica theoretica*, not only do chapters and subchapters have roman numerals, but also all the work's paragraphs are numbered consecutively from 1 to 892, which permits precise cross references. As for paragraphs, these may become short and pithy. In cases such as Borrhaus's *Μαθημάτων στοιχεῖα*, a paragraph typically comprises one sentence only (Figure 11). Each sentence is thus marked typographically as something important, worthy of the pupil's full attention and of commentary by the teacher.[154] Finally, textual hierarchies are sometimes made evident typographically not only through subtitles in various type sizes but also through differentiation between the absolutely indispensable skeleton of definitions and subdivisions, on the one hand, and further explanations, on the other. In Scribonius's *Isagoge sphaerica methodica*, the small part of the text which belongs to the latter category is distinguished from the former, which is set in roman type, by italics.

As an ideal-typical example of the typographical sophistication used to make textbooks user-friendly and easily accessible in the later part of the early modern period, one may look at Leopold Biwald's *Physica generalis* (1767). The book is

[154] With regard to Greek law inscriptions, this otherwise seldom noticed technique of 'Hervorhebung des Einzelsatzes' is discussed by Asper 2007, 176–7.

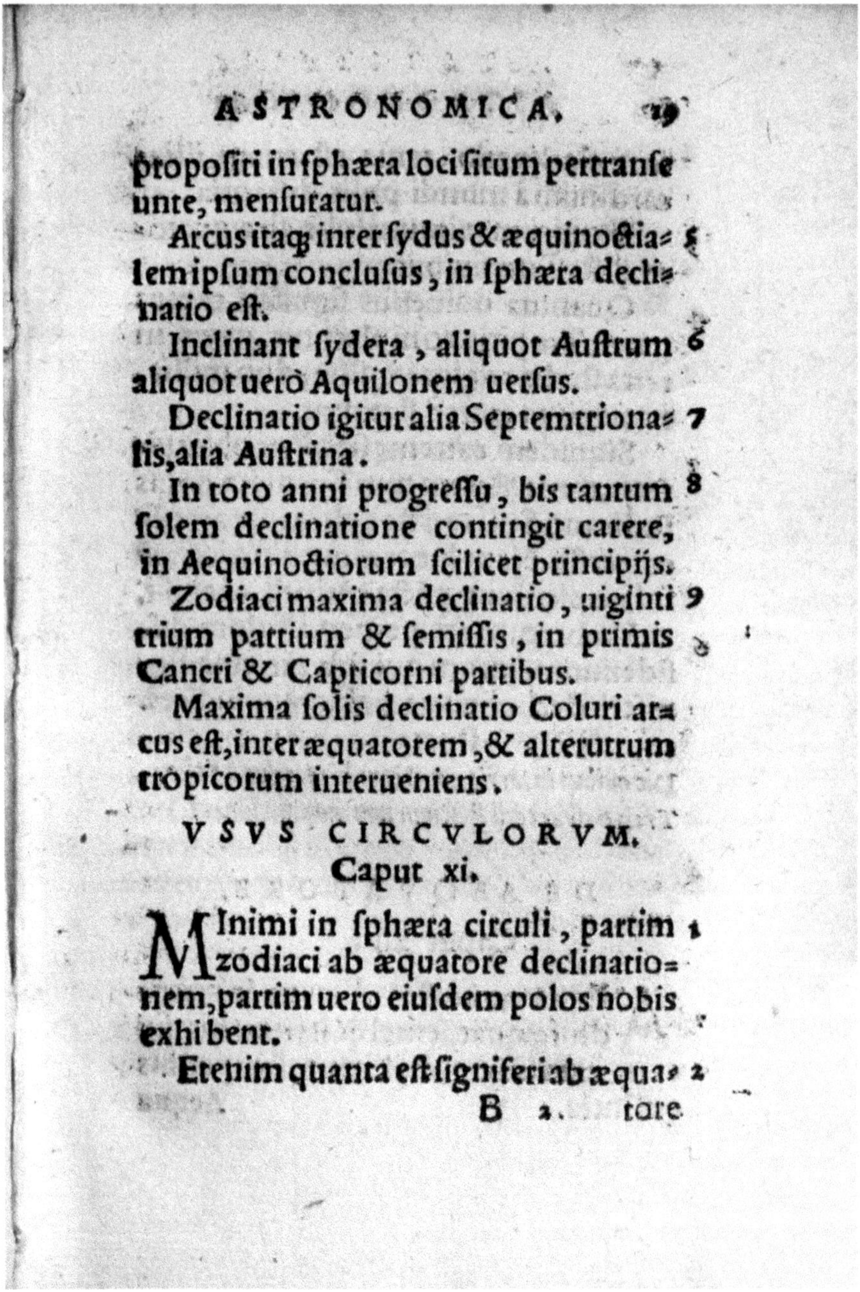

ASTRONOMICA. 19

propoſiti in ſphæra loci ſitum pertranſe
unte, menſuratur.
Arcus itaq; inter ſydus & æquinoctia- 5
lem ipſum concluſus, in ſphæra decli-
natio eſt.
Inclinant ſydera, aliquot Auſtrum 6
aliquot uerò Aquilonem uerſus.
Declinatio igitur alia Septemtriona- 7
lis, alia Auſtrina.
In toto anni progreſſu, bis tantum 8
ſolem declinatione contingit carere,
in Aequinoctiorum ſcilicet principijs.
Zodiaci maxima declinatio, uiginti 9
trium partium & ſemiſſis, in primis
Cancri & Capricorni partibus.
Maxima ſolis declinatio Coluri ar-
cus eſt, inter æquatorem, & alterutrum
tropicorum interueniens.

VSVS CIRCVLORVM.
Caput xi.

MInimi in ſphæra circuli, partim 1
zodiaci ab æquatore declinatio-
nem, partim uero eiuſdem polos nobis
exhibent.
Etenim quanta eſt ſigniferi ab æqua- 2
B 2. tore

Figure 11 A typical page from Borrhaus's *Μαθημάτων στοιχεῖα*: one sentence per paragraph.

divided into three 'partes', the 'partes' into 'sectiones', the 'sectiones' into 'articuli'. The titles of each are set in capitals of different font size. The 'articuli' fall into paragraphs numbered in Arabic numerals (whereas at the higher levels, the unities are numbered in roman numerals). In many paragraphs, the main text is followed by 'corollaria' or 'scholia', additional remarks or explanations set in a smaller font. New terminology is always highlighted by italics at the point where it is introduced and defined. There are marginalia. The illustrations, which are not integrated into the text but appended at the end as folding maps, are referenced in the margins as well (Figure 12).

Visual aids such as the ones just mentioned became more widespread and sophisticated in early modern times in general, but their help was particularly appreciated in textbooks. Between one-half and two-thirds of them contain some non-text. That illustrations were highly valued and attracted buyers is particularly evident from the frequent advertisement of them on title pages. There are many examples of this phenomenon, the earliest and most interesting of which may be provided by Peuerbach's *Theoricae novae planetarum*. Its first printing, carried out by Johannes Regiomontanus in his own printing shop in 1472, not only illustrates the geometrically demanding theory of planetary motion by twenty-nine diagrams taking up some 40 per cent of the overall space, but Regiomontanus also seems to advertise this aspect in his title by a reinterpretation of the strange term 'theorica'. In the *Theorica planetarum* ascribed to Gerard of Cremona, the word is used I the singular and may mean something like 'theory'. In the title of the new work, it is plural, while every single diagram within the work is titled 'theorica solis', 'theorica lunae', etc. Apparently, Regiomontanus, and perhaps Peuerbach before him, understood the word as 'diagram' ('something to look at', according to its Greek root 'theōr-') and called the book *New Diagrams of the Planets*.[155] Later editions are always copiously illustrated as well.[156]

One kind of illustration merits special attention in the present context, both as an early modern innovation and because it ties in particularly well with the striving for rigorous subdivision so characteristic of textbooks: the so-called Ramist tables, a method of presenting an overview of a discipline or field and an analysis of its inner structure via a hierarchical system of brackets. The system, in fact, predated Petrus Ramus, and how much he contributed to its

[155] To understand the title as *New Theories of the Planets*, as in Aiton 1987, does not make sense, since just one theory is proposed. In Malpangotto's translation, *Les nouvelles théoriques des planètes*, 'théoriques' seems little more than a transliteration, since no such substantive exists in French.

[156] Many of these are reproduced in Malpangotto 2020. Illustrations were also added and advertised to make later editions of an originally unillustrated work more attractive. The first edition of Borrhaus's *Μαθημάτων στοιχεῖα* (1539) is devoid of illustrations, but the second (1551) is 'figuris . . . opportunissimis illustratum', as the title page has it.

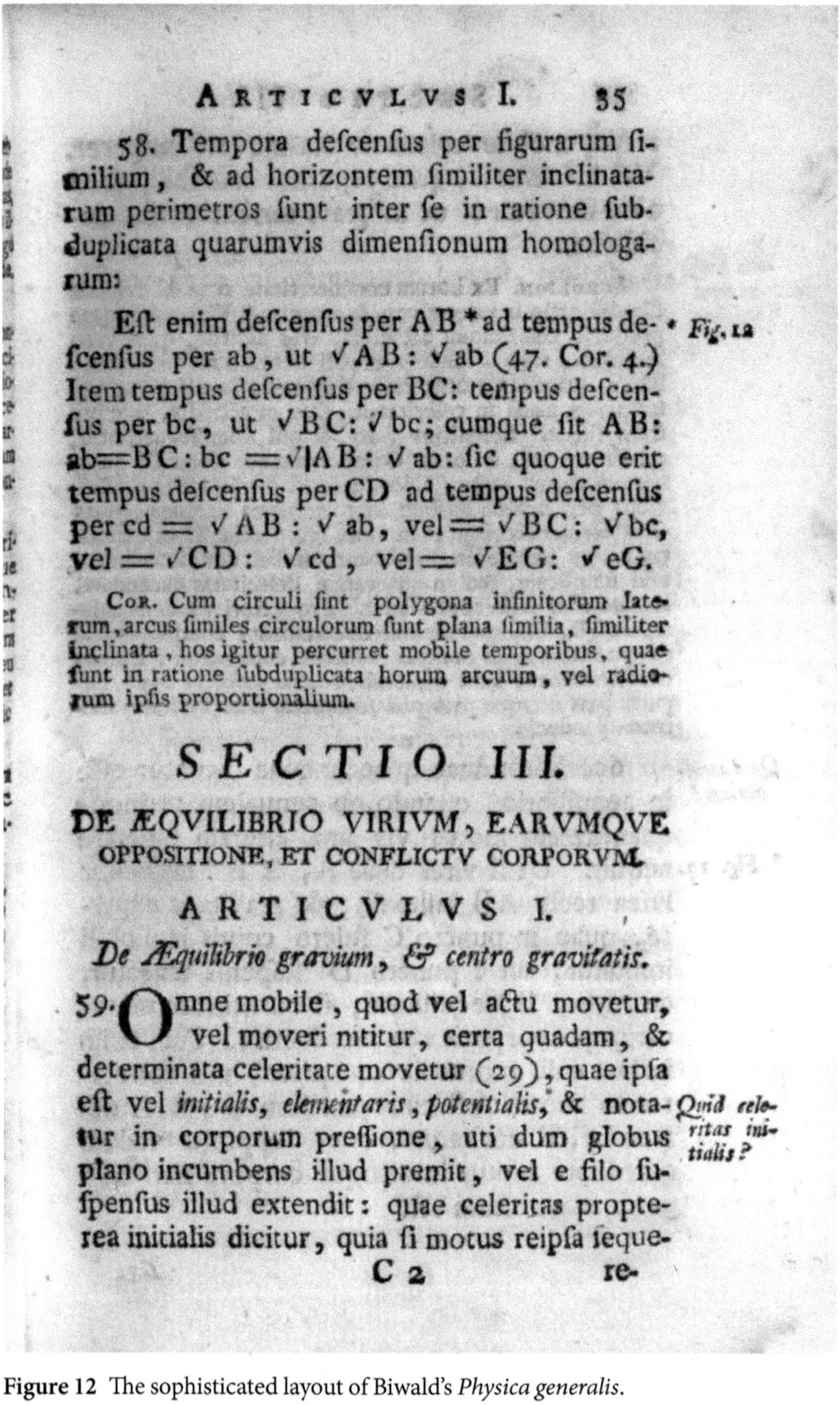

ARTICVLVS I. 35

58. Tempora deſcenſus per figurarum ſimilium, & ad horizontem ſimiliter inclinatarum perimetros ſunt inter ſe in ratione ſubduplicata quarumvis dimenſionum homologarum:

Eſt enim deſcenſus per AB *ad tempus deſcenſus per ab, ut √AB : √ab (47. Cor. 4.) Item tempus deſcenſus per BC: tempus deſcenſus per bc, ut √BC: √bc; cumque ſit AB: ab=BC : bc =√AB : √ab: ſic quoque erit tempus deſcenſus per CD ad tempus deſcenſus per cd = √AB : √ab, vel= √BC: √bc, vel= √CD: √cd, vel= √EG: √eG. * Fig. 12

Cor. Cum circuli ſint polygona infinitorum laterum, arcus ſimiles circulorum ſunt plana ſimilia, ſimiliter inclinata, hos igitur percurret mobile temporibus, quae ſunt in ratione ſubduplicata horum arcuum, vel radiorum ipſis proportionalium.

SECTIO III.

DE ÆQVILIBRIO VIRIVM, EARVMQVE OPPOSITIONE, ET CONFLICTV CORPORVM.

ARTICVLVS I.

De Æquilibrio gravium, & centro gravitatis.

59. Omne mobile, quod vel actu movetur, vel moveri nititur, certa quadam, & determinata celeritate movetur (29), quae ipſa eſt vel *initialis, elementaris, potentialis,* & notatur in corporum preſſione, uti dum globus plano incumbens illud premit, vel e filo ſuſpenſus illud extendit: quae celeritas propterea initialis dicitur, quia ſi motus reipſa ſeque- *Quid celeritas initialis?*

C 2 re-

Figure 12 The sophisticated layout of Biwald's *Physica generalis*.

spread seems doubtful.[157] In any case, it fitted his predilection for subdivision as a didactic procedure. In works dealing with the natural world, especially in natural philosophical and medical publications, Ramist tables became common from the 1530s and remained popular until the early seventeenth century. In most books, they were applied inconsistently. Tables of contents were often styled in this way, which could clarify not just the sequence but also the inner coherence of the various parts of a book (see Figure 5). Francesco Barozzi, for example, anticipates the structure of his *Cosmographia in quattuor libros distributa* (1585) in a lucid one-page Ramist table (sig. b^{v}). Some authors, by contrast, harboured greater ambitions: they published works made up wholly of Ramist tables. Theodor Zwinger reduced Galen's *Ars medica* to this form (*In artem medicinalem Galeni tabulae et commentarii*, 1561). Ernst Friedrich Fabricius even tried to contrast the whole of Galenic and Paracelsian medicine on a dozen folio pages (*Medicinae utriusque Galenicae et Hermeticae anatome philosophica*, 1626). One is inclined to doubt the serious didactic purpose, or at least the efficiency, of such works, and rather to regard them as virtuoso exercises in formalized subdivision.[158] Similar reservations do not apply to Dodoens's *Physi<o>logices medicinae partis tabulae*, a work analysing physiology, the doctrine of the nature of the human body and the first part of medicine according to Galen, where the use of Ramist tables is extensive, but not exclusive. Rather, longer and shorter tables, which take up about half of the space, are embedded in an equal amount of running text and simple lists. In his dedication to a former pupil, Dodoens reflects on the didactic usefulness of this form of presentation: It prepares students for the perusal of Galen's works and may also remind them of what they have already learned (pp. 336–7).

One format adopted with remarkable frequency in scientific textbooks was of religious origin: the catechism. The term 'catechismus' in the meaning 'religious instruction' goes back to antiquity, and there was a medieval tradition of written catechisms. However, the catechism as a genre—instruction in the didactic format of questions posed by a teacher and answered by a pupil—rose to unprecedented popularity in the age of confessionalization. Confessional catechisms were

[157] On Ramus and his influence on early modern didactics in general, see Ong 1958; Feingold et al. 2001; Hotson 2007; Reid and Wilson 2011; for examples and discussions of Ramist tables, Ong 1958, 199–202 and his index s.v. 'Dichotomized tables', Höltgen 1965 (medicine), Schmitt 1983, 56–9 (Aristotelian works), Tebeaux 1997, 50–73 (English technical literature), Maclean 2001a (class logical thinking in law and medicine), and Hallett 2011 (spatial configuration of thought in general).

[158] On his title page, Zwinger claims that his work is of interest not only to doctors and philosophers but also to logicians, who can learn from it how to lay bare the conceptual structure of a discursive account ('artificiosam ordinis definitivi dialysin'). In any case, not all works composed of Ramist tables were textbooks. Felix Platter claims his *De corporis humani structura et usu* (1583) is based on personal research and contains many new insights, sharply separating the work from his teaching as professor of medicine in Basel (letter of dedication, sig. $\alpha 2^{r-v}$). Georg Wolfgang Wedel's *Tabulae synopticae de compositione medicamentorum extemporanea* (1677) and *Tabulae chimicae XV* (1692) read like compendia, not textbooks, of iatrochemistry.

printed hundreds of times, used as schoolbooks for learning to read, write, and master the Latin language, and left their stamp on the mentality of the era.[159]

Given the ubiquity of questions and answers in teaching in general, it is understandable that the genre also became popular in fields other than religion, science being one of them. The sample at my disposal suggests that about one-fifth of all early modern scientific textbooks use the catechism format. Examples from various centuries and disciplines include a pharmaceutical catechism intended for the examination of aspiring pharmacists by doctors and forming the first part of Saladinus de Asculo's *Aromatariorum compendium* (1488), Melanchthon's Aristotelian textbook *Initia doctrinae physicae* (1549),[160] Lucas Lossius's *Arithmetices erotemata puerilia* ([1557]), Kepler's *Epitome astronomiae Copernicanae* (1618–21), Sturm's *Mathesis iuvenilis* (1699–1701), and Weinhart's *Quaestiones et responsa ex arithmetica vulgari, algebra et geometria* (1765). If this list and the collection on which it is based are in any way representative, catechisms were particularly popular in mathematics and schoolbooks for younger children. In medicine, taught in one of the higher academic faculties, they seem to have found little favour. Authors were clearly aware of the form's didactic advantages, which they highlighted in their titles, as in the cases of Lossius, Weinhart, and Kepler (whose *Epitome* is called 'usitata forma quaestionum et responsionum conscripta'), and, more elaborately, in their prefaces. As Kepler explained, a catechism was easier to understand for beginners, and it helped advanced readers to cope with the mass of dry definitions occurring in a textbook and prevented them from experiencing mental exhaustion while going through the more difficult parts.[161] Weinhart, who wrote for young university students, repeated the first point and added that the question and answer format was particularly useful because it favoured the less gifted and was thoroughly familiar to all from their school and university experience ('materiam aliter quam sic propositam hebetiores non tam facile ita digerunt, ut quae sibi inde quaestiones formari possint semper praevideant...hic modus...est familiarissimus in circulis, disputationibus, tentaminibus et examinibus mathematicis', sigs. $*2^{v}$–$*3^{r}$).

So much for theory. In practice, the catechetic form was often deployed in a dry and schematic way and amounted to little more than a convention. Saladinus, for example, has long strings of questions of the type 'Quid est rodomel?' ('What is rhodomel?', sig. $a4^{v}$), followed by the definition to be given by the pharmacist. For Lossius as for many others, the questions are just a convenient structuring device and typographically relate to their answers like a paragraph title to its paragraph. The answer to the question 'Numeri in divisione quomodo appellantur?' ('How are the numbers in a division called?', fol. 15^{v}) could as well be titled

159 *EEMHO*, s.v. 'Catechism'.

160 On the catechism format in philosophical textbooks in general, see Schmitt 1988, 798.

161 *KGW* 7, 8. In what follows, page numbers refer to this edition.

'De nominibus numerorum in divisione' ('On the designations for the numbers in a division'). In cases like this, even the roles of the interlocutors fade away. While in religious catechisms, the teacher is supposed to ask and the pupil to answer, in Lossius there is no way of telling, and it does not matter, which part belongs to whom.

Sometimes, however, the two voices are on the contrary endowed with a marked identity and engage in a lively exchange which tends towards a genuine dialogue. Kepler's *Epitome* fully realizes this possibility.[162] In this work, which by its very length and high level admittedly situates itself on the fringes of textbook literature, the traditional roles of questioner and respondent are completely reversed.[163] The latter clearly impersonates Kepler himself,[164] whereas the former comes across as an intelligent and interested disciple. For instance, he asks for further explanation regarding the difference between the movements of the fixed stars and the planets: 'Velim clarius explices discrimen hoc motus fixarum a motu planetarum, ut id incurrit in oculos' (p. 30). He demands examples: 'Ostende tamen rem exemplo' (p. 40). He doubts that heavy things fall downwards of their own accord, given that lower places are less dignified than those more elevated, and clads his disbelief in the ironic form of a rhetorical question: 'At fugient gravia ab extremis ut a loco nobiliori transibuntque ad intima ut ad locum ignobiliorem Naturae ingenio?' (p. 77). Finally, he even formulates hypotheses by himself, as when he speculates that the daily rotation of the earth may be exclusively caused by some force emanating from the sun: 'Fortasse vis omnis turbinationis huius effectrix in unico sole est, nulla in aliquo principio motus separatim terrae insito?' (p. 317). By introducing such an independent-minded pupil, Kepler forces himself to discuss his topic from every possible standpoint, to refute erroneous opinions, resolve problems, point out open questions, and give the fullest possible account of his subject.

Alongside formats applicable to all or most fields like the catechism, individual disciplines also developed ways of presentation which specifically responded to their needs and whose application was restricted to them accordingly. Some of these have already been mentioned in passing: many astronomical textbooks featured a bipartite structure, spherical astronomy being discussed first, followed by theorical. In an analogous way, textbooks on cosmography fell into an astronomical and a geographical part. These two traditions overlapped to some degree, giving rise to mixed forms. *Institutiones medicae*, which usually originated as lectures, provided a first overview of medicine whose structure often followed the Galenic fivefold division of the discipline or part of it. Bibliographical guides

[162] As noted in *KGW* 7, 549.

[163] Kepler may have taken his cue from Erasmus's catechism *Explanatio symboli* (1533), where the same reversal takes place and the dialogue is equally lively.

[164] This is proven by occasional self-citations such as 'Vide…librum II. Harmonicorum meorum' (p. 273).

introduced students to medicine by giving an overview of the essential literature, pre-empting the approach of the later, more wide-ranging 'historiae litterariae'.

A particularly striking example of a subject-bound style of exposition is provided by the 'algorithmi' developed for the teaching of elementary arithmetic based on Hindu-Arabic numerals. These books have been mentioned before, but their makeup merits some additional remarks. In nearly all cases, it is characterized by conciseness, strict order, simple, repetitive language, and a recipe style.

As to the first point, Huswirt manages to cover the whole field on forty quarto pages, although he inserts many examples of calculations. In other cases, for instance in Widmann's booklets, examples are omitted completely. The 'algorithmi' were obviously meant to be affordable for many. The teacher no doubt had much to add in class to make the account comprehensible.

All 'algorithmi' show a strong familiy resemblance in terms of content and order of presentation. The centrepiece is formed by the basic arithmetic operations, whose canon may admittedly vary to some extent: in addition to the four operations whose names are still used today, it occasionally also includes things like counting or doubling. This core may be preceded by an introduction to numbers in general and to Hindu-Arabic numerals and followed by more complex operations, for instance dealing with fractions, powers, and roots, or the rule of three ('regula detri'). Keeping to this order is essential because the more complex operations presuppose mastery of the simpler ones. Formulas such as 'as explained above' ('ut supra declaratum est') are very common.

The language is extremely standardized and repetitive in terms of both lexicon and syntax. The lexical repetitiveness is due to the comparatively small number of elements and operations in elementary arithmetic in combination with the limited use or utter absence of mathematical notation. In Huswirt, the latter is restricted to the numerals themselves and their spatial arrangement on the page in sample additions, subtractions, and so on, which partly corresponds to modern usage. He does not use, for example, the operators '+', '– ', '·', and ':' or the equals sign. Widmann (who otherwise played a significant role in the introduction of '+' and '–') has nothing at all, not even the numerals.[165]

The syntactic monotony is due to the most conspicuous feature of this text group, their pure recipe style. The 'algorithmi' describe in meticulous detail *how* a certain operation is to be performed, distinguishing the different cases that may occur by conditional clauses followed by jussive main clauses exactly enumerating the steps to be taken in each case. But not a word of explanation is given *why* the operation must be executed in this way to reach a correct result. Whether explanations were added by the teacher or whether pupils were simply content to learn the recipe by heart (as is still often the case today) is hard to tell, but the

[165] On the development of the symbols for the elementary operations and Widmann's contribution in this context, see Cajori 1928–9, vol. 1, 229–309, esp. 231–4.

latter may have been more common. This, for instance, is all Widmann has to say on the addition of two fractions in the chapter on these in his *Algorithmus minutiarum vulgarium* (sig. [Aiv^{r-v}]):

> In additione, si minutiae fuerint simplices et eiusdem denominationis, numerator unius numeratori alterius arte integrorum addatur et aggregatum eiusdem denominatoris numerator constituatur. Si vero dissimilium fuerint denominationum, ipsis ad eandem denominationem ductis, ut prius, simul addantur. Si vero minutiae minutiarum simul fuerint addendae vel minutiae minutiarum minutiis simplicibus vel quibuslibet aliis, semper prius reductis, ut supra declaratum est, fiat operatio ut in simplicibus et tandem ut in integris. Et tantum de additione.
>
> Regarding addition, if fractions are simple and of the same denominator, the numerator of one fraction shall be added to the numerator of the other by the technique of [adding] integers and the result shall be made the numerator of the same denominator. But if they have different denominators, they shall be brought to the same denominator, as before, and be added together. But if fractions of fractions are to be added together, or fractions of fractions to simple fractions, or to other [numbers] whatsoever, they shall always be reduced [to simple fractions], as explained above, and then the operation shall be executed as with simple fractions and finally as with integers. So much about addition.

Little has changed in the style of presentation nearly 300 years later in Weinhart's *Quaestiones et responsa*, whose remarks about the addition of fractions start as follows (p. 10):

> Si hae habeant denominatores aequales, soli numeratores inter se adduntur subscripto iis semper denominatore communi. Si vero diversos, prius reducendae sunt ad eundem denominatorem.
>
> If they have the same denominators, only the numerators are added to each other and their common denominator is always written below. But if [they have] different denominators, they must first be brought to the same denominator.[166]

A textbook example

I would like to conclude with a closer look at one particular textbook which is in many respects typical of its genre and illustrates a number of recurring features of

[166] The treatment of the subject in Huswirt's *Enchiridion novus algorismi* ('tractatus tertius, de additione, capitulum primum') and Gemma Frisius and Peletier's *Arithmeticae practicae methodus facilis* (1545, fol. 21^{v}) is similar. Lossius does not discuss fractions.

such works as well as some difficulties one encounters in dealing with them. The work in question is Michael Maestlin's *Epitome astronomiae* (1582), the most advanced and at the same time didactically most accomplished astronomical textbook of its day.[167] Its author is known best as teacher and mentor of Kepler; yet he was also a highly capable mathematician and astronomer in his own right.[168] Born in 1550 in the small Swabian town of Göppingen, he studied at the University of Tübingen, attending mathematical and astronomical lectures by Philipp Apian and graduating as 'magister artium' in 1571. In 1580, he became professor of mathematics in Heidelberg, four years later, professor in Tübingen, where he remained for the rest of his long life. He died in 1631. The *Epitome astronomiae* was published in 1582 while Maestlin was still at Heidelberg; however, he already harboured hopes of returning as professor to the place of his studies, as will become clear shortly. In what follows, three noteworthy aspects of the *Epitome* will be discussed, namely the information given in its introductory paratexts, its content and structure, and its success.

The paratexts comprise a dedication to Louis III, Duke of Württemberg, whose state university was Tübingen, an address to the reader, and two laudatory elegies on Maestlin and his work. While the elegies are phrased in general terms and quite uninformative, the author is remarkably outspoken about his purpose and approach in the dedication and the letter to the reader. In the former (sigs. *2^{r}–**3^{v}), he informs the Duke that in Tübingen, many students learn astronomy from lecture manuscripts which are hopelessly corrupted through many steps of uncomprehending transcription.[169] Worse still, such manuscripts are conflated by incompetent private teachers who use the resulting conflations as a basis for their instruction. Students taught in this way of course teach a strange kind of astronomy in the Latin schools of Württemberg in turn (sigs. [*8^{r}]–**^{r}). Maestlin originally wrote the *Epitome* only for the pupils and students of the duchy in order to improve this situation. However, the general desire to make such a book available led him to publish it 'for all beginners and students of astronomy and for the entire republic of letters' (sigs. [**^{v}]–**2^{r}):

> Eam etsi initio singulariter in usum eorum, quos Celsitudo vestra vel in coenobiis vel in augustissimo Tubingensi suo collegio fovet…conscribere coeperam, quia tamen illiusmodi astronomiae elementa passim ubique desiderantur, in lucem

[167] Some context and a detailed summary are provided by Rex 2002.

[168] *Allgemeine Deutsche Biographie*, vol. 20, 575–80; *CDSB* 9, 167–70; Betsch and Hamel 2002.

[169] This was a persistent problem. Paulus Aicardius, the editor of Girolamo Mercuriale's extracurricular lectures *De morbis cutaneis* (1572), explains in his preface that he has taken it upon himself to preserve Mercuriale's reputation, which was endangered by confused and erroneous student transcripts. Horn published his *Orbis politicus* as a better alternative to such transcripts (preface to the reader, first two pages). Boerhaave's lectures on chemistry were published by students from their notes without his consent, against which 'turpis libellus' he composed his authorized version (*Elementa chemiae*, preface to the reader, sigs. **2^{r}–**3^{r}).

publice datam omnibus astronomiae tironibus et studiosis totique rei publicae litterariae hanc meam epitomen fruendam propono.

Thus, despite his obvious desire to highlight the usefulness of his textbook for the formal teaching of astronomy in Württemberg, Maestlin also shows himself aware that by the act of publication it becomes a book for an indeterminate audience of autodidacts as well.

Quite logically, then, his subsequent letter (sigs. **4^{r}–[**8^{v}]) is addressed to students of astronomy and the gentle reader in general ('astronomiae studiosis candidoque lectori'). In it, the author discusses widespread deficiencies in the arrangement of astronomical textbooks as well as his own decisions in this respect. Pertinent works are legion, but he has found none that would satisfy the needs of beginners. Some authors lack any sense of order ('methodus').[170] Some focus on irrelevances while leaving out the essential. Others think they can achieve order by restricting themselves to bare subdivisions without giving any further explanations ('praeter ullam rationem nudas et concisas divisiones existimant methodicum quid esse'). Finally, there is much factual error (sig. **4^{r-v}). The best order is still that followed by Sacrobosco, which explains his long-standing success despite mistakes ascribable to his time and made worse by uncomprehending commentators: the cosmological fundamentals of astronomy in book 1, the theory of the celestial circles in book 2, spherical and theorical astronomy proper in the last two books. Maestlin adopts Sacrobosco's structure, improving it in some respects and supplementing his deficient last book with the help of Peuerbach's *Theoricae novae planetarum* (sigs. **4^{v}–[**7^{v}]). In doing so, he puts himself in the shoes of beginners, 'who hear everything for the first time' ('quibus omnia sunt inaudita'), and resists the temptation to be overly concise. He also provides a kind of auto-commentary with additional information in a smaller font ('loco scholiorum nonnulla scriptura magis gracili scripta... apponi curavi')[171] as well as illustrations, because seeing for oneself is as essential to understanding as verbal explanation in astronomy ('nam haec res non est contenta nudis verbis sine oculari delineatione doceri'). Finally, Maestlin emphasizes that he has done much independent research for his book. For this reason, and not because of negligence, his dates and figures often differ from those found in the earlier tradition (sigs. **7^{v}–8^{v}).

The structure and content of the *Epitome* itself broadly conform to the author's announcements. The core text is indeed divided into four books after the model of Sacrobosco's *Sphaera*. They deal with the subject of astronomy and its

[170] The early modern use of 'methodus' as a didactic key term originated in medicine; see Maclean 2002, 200–3.

[171] These remarks (pp. 1–2, 12–13, 26–8, 29–30, 36–7, etc.) are set not only in smaller type but also in roman font, whereas the rest is in italics.

fundamental notions (book 1, pp. 20–88), the representation of the celestial circles on the armillary sphere ('sphaera materialis', depicted on the title page; book 2, pp. 89–182), 'astronomia sphaerica' (book 3, pp. 183–292), and 'astronomia theorica' (book 4, pp. 293–486). However, the very full treatment of planetary astronomy, which starkly contrasts with Sacrobosco's meagre account, implies that the overall structure of Maestlin's book is actually a novel one in traditional disguise: for the first time, a full treatment of spherical and theorical astronomy is provided in one textbook.[172] In all of this, Ptolemy's geocentric system is adopted, although Maestlin was personally convinced of the superiority of heliocentrism and orally taught his pupils this.[173] Not mentioned in the letter to the reader is a preliminary section (pp. 1–19) dedicated to the geometrical foundations of astronomy along with the relevant terminology and reminiscent of dictionary-like introductions to mathematics such as Konrad Dasypodius's *Λεξικὸν seu dictionarium mathematicum* (1573). Moreover, Maestlin follows the early modern tendency to combine astronomy and geography in the same textbook by including four geographical appendices to the individual books. The first of these (pp. 81–8) deals with the dimensions of the earth as a whole; the second describes the zones (pp. 156–61); the third discusses the 'climata', sections of roughly the same latitude, as well as 'perioeci', 'antoeci' and 'antipodes', people living on the same or opposite circle of latitude or meridian (pp. 278–92); the fourth, which concludes the work (pp. 487–95), teaches how to calculate distances on the earth's surface.

In terms of presentation, the text is a catechism, but one looks in vain for such lively dialogue as characterizes Kepler's *Epitome astronomiae Copernicanae*. In fact, questions and answers mainly serve as the most important building blocks of an overarching structure characterized by rigorous subdivision, which starts at the book level but contains many lower-level steps. The four books fall into three, three, four, and three 'partes' respectively. A 'pars' may in turn be subdivided into 'propositiones' (book 1, part 3, pp. 42–80) or into units defined by topics such as the various celestial circles (book 2, part 2, pp. 98–155), the single planets (book 4, part 1, pp. 294–391), or their different 'passiones' (for example, their apparent movements or positions in relation to one another; book 4, part 2, pp. 392–475). Within these smaller parts, thematic sections may be indicated by Roman numerals, before one reaches the individual questions and answers. Even at this point, one still has not reached the bottommost level in all cases, as answers may be partitioned into numbered paragraphs and supplemented by explanatory notes. Maestlin and his printer-publisher Jakob Müller take care to distinguish the various levels of this complex hierarchy by means of font size, centred type setting, indentation, and italic and roman type. Moreover, Maestlin sometimes explicitly explains the structure of his book and discloses its rationale. Book 4,

[172] Pantin 2012, 74. [173] Methuen 1996.

for instance, begins with the question 'How many parts does this fourth book consist of?' ('Quot sunt libri huius quarti partes?'). After the answer that the book falls into three parts, the first of which starts with a discussion of the sun, it is asks: 'Quare initium fit a theoria solis?', and four reasons are provided: the comparative simplicity of the theory of solar motion, its helpfulness in dealing with the other planets, the sun's fundamental role in time measurement, and the authority of Ptolemy (pp. 293–5).

The sophisticated subdivision suggested by textual structure and layout is only partly matched on the conceptual level. At the beginning of the work and most individual sections, the field is indeed carefully subdivided, as indicated by the types of questions asked. Often, these are 'qui / quae / quod', 'quot', or 'quotuplex' questions: into what and how many species does a given genus fall? After a subdivision, the resulting parts are defined, which is signalled by 'quid est' or 'quid vocas' questions. Definitions of two or more coordinate species are often marked by parallelism coupled with antithesis in terms of the specific difference. Conversely, technical terms for already known entities are introduced by questions beginning with 'quomodo nominatur' or the like. A typical example of the whole procedure is provided by the introduction and subdivision of the notion of 'stella' (p. 35).

> Quid est stella?
>
> Est densior pars sui orbis lucida et globosa existens, orbi suo in certo loco affixa, a quo etiam statis et certis legibus circulariter circumfertur.
>
> Quotuplices sunt stellae?
>
> Duplices, fixae vel inerrantes et planetae seu erraticae stellae.
>
> Quae sunt fixae stellae?
>
> Stellae octavi orbis, hoc est omnes praeter 7 planetas.
>
> What is a star?
>
> It is a denser part of its orb which is radiating and round, attached to its orb in a certain place and carried around by it according to fixed and certain laws.
>
> How many kinds of stars are there?
>
> Two kinds: fixed or non-wandering and planets or wandering stars.
>
> Which ones are fixed stars?
>
> The stars of the eighth sphere, that is, all except the 7 planets.

However, the more advanced and technical the account becomes, the more the interplay of subdivision, definition, and establishment of terminology is interrupted by other kinds of information, and new types of questions appear accordingly. 'Quomodo' or 'cuiusmodi' questions ask how something behaves or works. 'Cur' and 'quare' questions demand reasons. Sometimes, questions ask for the development of certain inferences ('porismata' questions, for example pp. 342, 359) or quite indefinitely for more precise information ('explica hoc' questions, pp. 324, 349–50). Yes-no questions occur as well (p. 397). Some questions qualify as such only by virtue of a concluding question mark; otherwise, they are better termed requests, or even objections (p. 69). The sections containing additional information in a smaller font, often used to explain the numerous diagrams, become more frequent and extensive in the course of the account as well. These are in treatise-like style uninterrupted by any questions.

One may conclude that Maestlin's fine-grained textual structure serves a double function: in the introductory, more general sections, it indeed mirrors the fabric of the field covered, reflecting basic subdivisions and laying bare the relationship between crucial notions. When the account becomes more sophisticated, by contrast, the structure covers up, rather, the absence of clear-cut conceptual hierarchies and dichotomies and reassures the reader by breaking up a complex argument into manageable parts.

Readers were obviously satisfied with Maestlin's presentation, and the *Epitome* proved highly successful for nearly half a century. Seven editions appeared up to 1624 in total[174] (all but the very first in Tübingen, testifying to the work's close connection to Maestlin's lifelong teaching there). Twice, in 1588 and 1610, Maestlin took the opportunity of revising his text. In the 1610 edition, an instruction on how to use spherical trigonometry to compute astronomical tables was advertised on the title page and added to the third book (pp. 200–6). Didactically more important, however, were some seemingly minimal additions in the edition of 1588, which was otherwise left untouched except for the correction of a few misprints: Maestlin, who had become aware that much of his content was too difficult for beginners, decided to mark such questions as he deemed indispensable for them with an asterisk. Moreover, he often separated the elementary part of an answer from the advanced portion by square brackets. The *Epitome* thus became two books in one, so to speak, as it could be used differently by readers of different expertise. The author rightly thought this change important enough to explain it in a second, one-page address 'Ad lectorem' which he inserted immediately before the main text. This note is of interest also because in it Maestlin showed himself convinced that teachers would know better than himself which passages they should cover with which pupils: 'Quamquam autem quilibet, qui discipulis suis haec nostra praecepta ediscenda tradit, pro sua prudentia noverit inter quaestiones primis incipientibus et paulo adultioribus proponendas discernere.' In doing so, he assumed it was self-evident that the *Epitome* was indeed being used at various levels of formal teaching, which confirmed its status as a textbook in the strictest sense of the word.

174 *DSB* 9, 167–70, at 169; Methuen 1996, 234.

One last aspect of the book's success is its model function for Kepler's *Epitome astronomiae Copernicanae.* Although this monumental work is at least as much a summary of its author's lifetime achievement as a textbook strictly speaking, Kepler clearly had Maestlin's work in mind when conceiving his own. He took up its title, to which only the important word 'Copernicanae' was added, basically stuck to the division into fundamentals (book 1, supplemented by the celestial mechanics explained in book 4), the armillary sphere (book 2), spherical (book 3), and theorical astronomy (books 5–7), and adopted (and enlivened) the catechetism format. Maestlin's book, which consciously followed in the footsteps of Sacrobosco's *Sphaera,* thus became part of a succession of astronomical didactic works starting in the High Middle Ages and leading straight into the Scientific Revolution.

9
Publicizing science

Didactic poems

Historical background

Because of its mnemotechnical advantages, poetry is an age-old didactic medium, predating in this function prose, and even writing itself, in many cultures. In the Western tradition, the notion that teaching was a main function of poetry remained basically unquestioned until the Romantic movement. Didactic poetry in the strict sense of the word may be taken to mean poetry that is essentially non-narrative (although it may contain narrative inserts) and professes a clearly recognizable didactic intent. With respect to Latin works, 'didactic poetry' is sometimes used synonymously with 'didactic epic', that is, long didactic poems written in hexameters. Indeed, epic is the most prominent form of didactic poetry, and my account will focus on it, although not to the exclusion of other forms.

In the Greco-Roman tradition, didactic poetry was hardly an object of reflection in literary theory, but people were clearly aware of it as a genre and it was widely practised.[1] In fact, it was among the first genres committed to writing. The earliest didactic poems extant, Hesiod's *Theogony* and *Works and Days*, seem to be roughly contemporary with the Homeric epics. Later, Greek didactic poetry flourished particularly during Hellenism, when Aratus wrote his famous astronomical poem *Phainomena* and Nicander his pharmacological works *Theriaka* and *Alexipharmaka*. In the first centuries AD, Dionysius Periegetes' description of the known world is noteworthy. As these examples show, the natural world was an important subject of didactic poetry early on. This did not change when the genre was taken over by the Romans. In fact, two of the three most important and influential Roman representatives deal with natural science in the stricter or wider sense of the word: Lucretius' *De rerum natura*, an overview of Epicurean physics, and Manilius's *Astronomica*. (The third and most famous of all, Virgil's *Georgics*, is likewise concerned with the natural world but does not adopt a scientific perspective.) Among poems written in metres other than the hexameter, one may mention Ovid's *Ars amatoria* in elegiac couplets and, from the fourth

[1] Overviews are provided by Effe 1977 and Volk 2002.

Latin Scientific Literature, 1450–1850. Martin Korenjak, Oxford University Press. © Martin Korenjak 2023.
DOI: 10.1093/oso/9780198866053.003.0010

century AD, Avienus' *Ora maritima*, a description of the coastline of the known world in iambic trimeters.

In many ancient didactic poems, poetic ambition prevails over didactic intent. However, this is not the case in a number of examples exhibiting a simpler, often serial structure and lacking much of the poetic ornament common in the more ambitious poems. A good example of a didactic poem primarily intended for practical use is the *Liber medicinalis* (perhaps fourth century AD) by a certain Quintus Serenus, which lists a series of medications in sixty-four chapters ordered (very roughly) 'a capite ad calcem'. This pragmatic strand came to the fore in the medieval tradition, when numerous didactic poems were written primarily for use in formal teaching.[2] The phytotherapeutic *Macer Floridus* by Odo Magdunensis (eleventh century) is an influential example from the realm of medicine. Alexander de Villa Dei's *Doctrinale* (*c.*1200) became the most important advanced Latin grammar throughout the Late Middle Ages and beyond.

Scientific didactic poetry in the early modern era

Early modern times witnessed a boom in didactic poetry, not only in Latin but in the vernaculars as well.[3] Latin didactic poets went far beyond their ancient predecessors with regard to their themes: every conceivable topic from coffee to logic and from the art of conversation to the Northern Lights could be treated in Latin verse. A considerable part of this ample production was concerned with the sciences. Of scientific didactic epic alone, over one hundred specimens are known at present.[4] The first early modern poem of this kind seems to be Basinio Basini's *Astronomica* (*c.*1455, first printed in 1794), but the genre peaked as late as the didactically minded Enlightenment. Some 60 per cent of all didactic epics on aspects of the natural world that are extant were written in the eighteenth century, and examples are also known from the nineteenth and even the twentieth. The doctor Giuseppe Favaro published a *Mors laniata*, on dissection, as late as 1936. Among the authors, the Jesuits were most prominent. Around 40 per cent of all pertinent texts were written by members of the Society of Jesus.

It is not always easy to distinguish didactic from panegyrical poetry. The account of scientific subjects is often combined with praise of their importance

[2] On Latin didactic poetry in the Middle Ages, see Haye 1997.

[3] Kühlmann 2016 provides an overview for sixteenth-century Germany which encompasses Latin and vernacular forms alike. On vernacular didactic poetry in early modern France and Germany, see Schmidt 1970 and Siegrist 1974 respectively.

[4] On Latin didactic epic in the early modern period in general, see Roellenbleck 1975; Ludwig 1982; *CNLS* 2, 38–45; Haskell and Hardie 1999; Haskell 2003; *BENLW* 437–48; Moul 2017b, 180–99. On scientific didactic epic, see Haskell 2003, 178–244 and Korenjak 2019 (including a list of poems, their reissues, translations, and modern editions, as well as additional secondary literature). Much of what follows is taken from the last-mentioned publication.

and of the figures who played a role in their exploration. When scientific content is presented by telling the story of particular research, the line between the didactic and the panegyrical mode becomes particularly blurred. This is the case, for instance, in Robert Grove's account of William Harvey's discovery of the circulation of the blood (*Carmen de sanguinis circuitu*, 1685) and in György Alajos Szerdahélyi's *Historia Uraniae Musae* (1787), which tells the discovery of Uranus by William Herschel.

Didactic poetry covered practically all early modern scientific disciplines. To begin with the apparently most unpoetic, mathematics and numerology, these provided the subjects for Giordano Bruno's *De minimo* and *De monade* (both 1591).[5] Not only Newtonian physics as a whole, as in Benedict Stay's *Philosophia recentior* (1755–92), but also physical subdisciplines from optics (Carlo Noceti, *Iris*, 1730 or 1731)[6] and acoustics (Bernardo Zamagna, *Echo*, 1764)[7] to hydrostatics (Loup Thomas, 'Barometrum', 1749), electrostatics (Giuseppe Maria Mazzolari, *Electrica*, 1767), and magnetism (Thomas Bernard Fellon, *Magnes*, 1696) received poetic treatment. Many astronomical and cosmological poems presented the Ptolemaic universe in the wake of Manilius' *Astronomica*. Giovanni Pontano's 'Urania' (1505) was perhaps the most famous specimen, while Bruno's insistence on an infinite and eternal universe (*De immenso*, 1591[8]) was an important exception. The sublunary world was treated, for example, in meteorological (Pontano, 'Meteora', 1505), hydrological (Girolamo Lagomarsini, 'De origine fontium', 1749), and seismological (François Antoine le Febvre, *Terrae-motus*, 2nd edn 1704) works. Botanical, zoological, and mineralogical poems reflected the boom in natural history from the sixteenth century. János Molnár's *Physiologicon* (1781), a triptych comprising a 'Zoologicon', a 'Phytologicon', and an 'Oryctologicon', is a particularly systematic and exhaustive example.[9] Medicine was popular as well. As an influential early example, Girolamo Fracastoro's *Syphilis* (1530) can be cited.[10] Besides certain diseases, many other areas, for instance paediatrics (Giulio Alessandrini, *Paedotrophia*, 1547), dietetics (the first part of Franz Adolph Habersack's *Medicina Austro-Viennensis versice exposita*, 1731), or psychosomatics (Malcolm Flemyng, *Neuropathia*, 1740) were discussed as well.[11] Chymistry was explained in didactic verse from Giovanni Aurelio Augurello's *Chrysopoeia* (1515) onwards. Technical applications elucidated in poetic form include scientific instruments such as the barometer (Joseph Addison, *Barometri descriptio*, 1698) as well as technologies such as gunpowder (François Tarillon,

[5] Italian translation with notes: Monti 1980, 71–292 and 293–414.

[6] On the early publication history of the poem, which is somewhat obscure, see Tautschnig 2017, 12 n. 55.

[7] Schindler 2023.

[8] Modern edition: Fiorentino and Tocco 1879–91, vol. 1, parts 1–2; Italian translation with notes: Monti 1980, 415–809; book-by-book interpretation: Granada and Tessicini 2020.

[9] Roling 2023, 236–50.

[10] Modern edition: Eatough 1984.

[11] Seidenschnur 1847 is a treasure trove of medical didactic poetry.

Pulvis pyrius, 1699) or the steam machine (Giuseppe Giacoletti, *De lebetis materie et forma*, 1863).

In exploring such a diversity of topics, scientific didactic poetry was also quite varied in terms of form. Besides the main variety, the didactic epic, there were several other options. The second most popular metre after the hexameter was elegiac couplets. The works in this form are typically short and thematically focused. Sometimes, they are inserted into more comprehensive publications, as in the case of Joachim Camerarius's translation of the 'Galene', a Greek poem a universal antidote in eighty-seven elegiac couplets by Andromachus the Elder, which forms part of the translator's collection titled *De theriacis et mithridateis commentariolus* ([1533]). By contrast, Hans Frandsen's *De oculorum fabricatione* (1556), an elegy on the anatomy of the eye, was published as a self-contained small volume. Metrically mixed forms exist as well. Of Habersack's extensive *Medicina Austro-Viennensis*, the first part, devoted to dietetics, is in elegiacs, the second, therapeutic part, in hexameters.

An English speciality is represented by the so-called Act and Tripos verses produced at Oxford and Cambridge respectively from the sixteenth to the late nineteenth century.[12] These short poetic accounts of various learned topics in elegiac couplets, hexameters, lyric, and other metres were composed and distributed or read out by students on the occasion of academic disputations on the particular subjects, to which they formed light-hearted companion pieces. Originally, they were printed in the form of single sheets. Some have survived as such, in manuscript, or in printed collections. The latter include the *Carmina comitialia Cantabrigensia* (1721) from Cambridge, the two volumes of the Oxonian *Carmina quadragesimalia* (1723–48), and the *Musae Anglicanae* (1741), in which Act and Tripos verses from both universities can be found among other poems. A substantial portion of such compositions, namely those from the medical and some of those from the arts faculty, deals with natural phenomena. The authors strive to present their subjects in an elegant and witty way but also to demonstrate their factual understanding.

Information about nature could also be conveyed in collections of epigrams.[13] The form was often used to poeticize prose templates already made up of such units. Especially popular were translations of the Hippocratic *Aphorisms* in various metres. Students of medicine were grateful for presentations of this important work in a form that was appealing and easy to memorize. No less tha fewer n eighteen renderings into (Greek and) Latin epigrams are known.[14] In 1558, François Boussuet rendered Guillaume Rondelet's influential encyclopedia of water animals, the *Aquatilium historia* (1554–5), into a series of epigrams, each of

[12] See Hall 2009 (including a catalogue of poems from Cambridge). Two case studies are provided by Barton 2021 and 2023.

[13] On Neo-Latin epigrams in general, see Beer et al. 2009.

[14] Seidenschnur 1847, 782–5.

which is accompanied by its woodcut from the source text. Together with a title naming the animal in question, the resulting format is reminiscent of an emblem book, a genre invented only a few decades earlier and hugely popular at the time[15] (Figure 13). Real emblem books could draw inspiration from the natural world as well,[16] and not all of them restricted themselves to its moral allegorization. In his *Emblemata de purpura* (1671), Johann Daniel Major combines a moral and a scientific perspective on shellfish and snails. Forms featuring a mix of poetry and prose can also be found. Josse de Harchies's *Enchiridion medicum* (1573) is an epigrammatic pharmaceutical handbook. Each of the *c.*360 simples contained in it is treated in an epigram followed by supplementary remarks in prose.

Even if one leaves such special forms aside and focuses on didactic epic, one notices considerable internal variation, most obviously in terms of length. The span reaches from poems of a few dozen lines such as Addison's elegant *Barometri descriptio* to monumental works like Camillus Eucherius de Quintiis's balneological poem *Inarime* (1726) in six books, the nine books of Melchior de Polignac's *Anti-Lucretius* (1747), or Stay's *Philosophia recentior*, presumably the longest and most difficult Latin didactic poem ever written, which provides a detailed overview of Newtonian physics and cosmology in ten books and *c.*24,000 lines. However, comparably few poems exceed two or three books. A common type is what may be called the poetic monograph, covering one clearly circumscribed topic such as the rainbow or the tea plant in a single book of some 500 to 1,000 lines.[17] As was the case with the aforementioned poems in elegiacs, very short pieces were often published as companion pieces to more substantial works. The *c.*300 hexameters of Dionysius's 'De oculis carmen', for example, were appended to the author's poetic rendering of Hippocrates' *Aphorisms*.

Regarding literary ambition and polish, the poetic strand once more prevailed over the pragmatic. Examples of the latter are not altogether lacking. The *Rudimenta cosmographica* by the Transylvanian polymath Johannes Honterus, for instance, is basically a cosmographical and geographical inventory of the world, which presents its contents with a minimum of poetical ornament. Much is in the form of dry catalogues, as in the case of the most important rivers of northern Anatolia: 'Parthenius posthac et Haly vicinior Iris / Sangariusque petunt Euxini litora Ponti' ('After that, the Bartın, the Yeşilırmak, quite close to the Kızılırmak, and the Sakarya head towards the shores of the Black Sea', book 3, sig. d3^{r}). This style did not prevent the *Rudimenta* from becoming hugely successful. First published in its definitive form in 1542,[18] it saw some one hundred editions and reprints until the end of the seventeenth century. Nonetheless, most authors

[15] Enenkel and Visser 2003; Daly 2008. [16] Enenkel and Smith 2017.

[17] In antiquity, that was the standard length of a 'book', that is, a papyrus scroll, of poetry. Many early modern poetry books kept within these limits out of convention.

[18] Facsimile edition with translations into German, Hungarian, and Romanian: Offner et al. 2017.

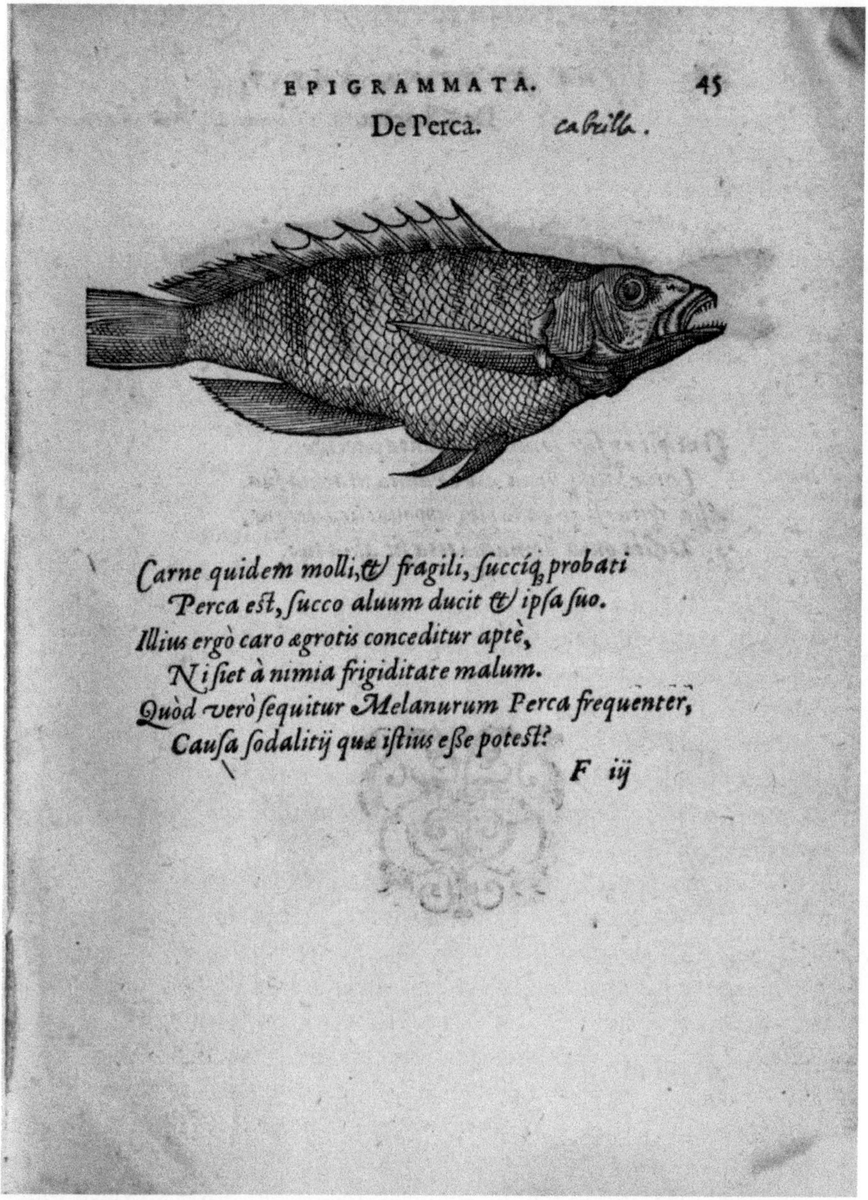

EPIGRAMMATA. 45

De Perca. cabrilla.

Carne quidem molli, & fragili, ſuccíq; probati
Perca eſt, ſucco aluum ducit & ipſa ſuo.
Illius ergò caro ægrotis conceditur aptè,
Ni ſiet à nimia frigiditate malum.
Quòd verò ſequitur Melanurum Perca frequenter,
Cauſa ſodalitij quæ iſtius eſſe poteſt?

F iij

Figure 13 Presentation of the 'perca', a kind of bass, in the style of an emblem in Boussuet's *De natura aquatilium carmen*.

aspired to the linguistic and compositional standards of the ancient classics, orienting themselves in particular towards Lucretius's *De rerum natura*, Virgil's *Georgics*, or a combination of both. The advantages of the pragmatic strand, simplicity and clarity, could also be achieved by means of layout and paratexts, as will be explained later.

Scientific didactic poetry, epic in particular, was much in demand. More than half of all epics went through two or more editions. A few, such as Fracastoro's *Syphilis* or the *Rudimenta*, became long-term bestsellers. Many more, such as Raphael Thorius's poem on smoking, the *Hymnus tabaci* (1626), or Claude Quillet's paediatric *Callipaedia* (1655), were reissued several times in the decades after their initial publication. A telling case is Stay's *Philosophia recentior*. When this magnum opus, equipped with a rich commentary by Roger Joseph Boscovich, was finally completed with the appearance of the last of its three volumes in 1792, a one-volume edition without commentary was simultaneously produced. As explained in its preface (p. iv), this was done 'for the benefit of some people who find it less annoying to run into some difficulty of understanding now and then than to interrupt the reading of the verses so often to consult the commentary, and of those so well-exercised in scientific reasoning that they do not need... any explanations' ('ut nonnullis gereretur mos, et iis scilicet, qui minus moleste ferant in aliquam interdum obscuritatem incurrere quam tam crebro versuum lectionem inspiciendis commentariis interrumpere, et iis, qui in philosophicis rationibus multum exercitati nullis indigeant... explicationibus'). As late as the end of the eighteenth century, a publisher thus found it reasonable to produce two distinct editions of a very long and apparently hopelessly difficult poem for different classes of readers.

In addition to reissues in the original Latin, translations were a common phenomenon. As a rule, they were combined with the original text in a bilingual edition, which suggests that they were meant not only for the completely non-Latinate but also for those who only needed a little help with the Latin. Demetrius de LaCroix's *Connubia florum* (1728) appeared with a facing French translation from the outset. More often, a translation followed the first edition at some chronological distance. Editions including versions in Italian blank verse were produced of Francesco Eulalio Savastano's *Botanica* (1712) in 1749 and of Noceti's *Iris* and *Aurora borealis* (1747) in 1755. Scévole de Sainte-Marthe's *Paedotrophia* (1584) had to wait for over a century before it was provided with a French translation in 1698.

In the eighteenth century, when interest in scientific didactic poetry reached its apogee, several anthologies appeared. Those of Act and Tripos verses have already been mentioned. The most imposing collection of didactic poems was compiled by François Oudin in 1749. Comprising three volumes and some 1,200 pages, his *Poemata didascalica* contains more than thirty poems, no fewer than fourteen of which are dedicated to science and its technical applications such as gunpowder or glassmaking. An expanded edition, which features six additional scientific poems, was published as late as 1813. Finally, the enthusiasm of the Enlightenment for didactic poems also left traces in the periodical literature of the time. New publications in the field were greeted by reviews in journals such as the *Journal de Trévoux*, the *Acta eruditorum*, or the *Gentleman's Magazine*.[19]

Didacticism I: layout and paratexts

The task of rendering science understandable and palatable to the average educated reader was achieved in two ways: by means linked to the medium of the printed book and by literary techniques inherent in the poems themselves. The first point comprises the skilful application of layout and typography as well as the use of paratexts. From an early date, poems were divided not only into books but also into chapters or paragraphs. This could be done by subtitles, as already in Conrad Celtis's 'Germania generalis' (1502), or simply by indentation, as in many editions of Honterus's *Rudimenta cosmographica*. In the course of the sixteenth century, marginalia became another common means of disclosing a poem's fine structure.

Explanatory paratexts already accompanied medieval representatives of the genre.[20] In the early modern era, they became more and more common and extensive. In the eighteenth century, a plain-text edition might even call for a specific justification, as shown by the example of Stay's *Philosophia recentior*. For some readers, the translations constituted an efficient way to enhance understanding. Other types of paratext were less laborious. A preceding summary of the main content in verse or prose, a so-called 'argumentum', could help readers keep track of the train of thought. For example, 'argumenta' in hexameters introduce the individual books of George Buchanan's *Sphaera* (1586). Stay's *Philosophia recentior* is preceded by a prose summary of a dozen pages in small and dense print, which is perhaps also intended to enable the reader to select passages of particular interest to him. Notes were often added, either at the bottom of the page or as endnotes. They could grow into full-blown commentaries, as in the case of a lengthy manuscript by the Scotsman Adam King (1616) explaining Buchanan's work, which was presumably used in the teaching of astronomy at the University of Edinburgh,[21] and of Roger Joseph Boskovich's annotations to the *Philosophia recentior* as well as to Noceti's *Iris* and *Aurora borealis* (1747), which equal or exceed their respective source texts in length. Boskovich even

[19] Reviews of the first volume of Stay's *Philosophia recentior* and of Boskovich's *De solis ac lunae defectibus* appeared in the *Journal de Trévoux* (1756) and in the *Nova acta eruditorum* (1761) respectively. Boskovich's poem was discussed again in his obituary in the *Gentleman's Magazine* (1787). On some reviews of Zamagna's *Echo*, see Schindler 2023, 193 n. 24.

[20] Haye 1997, 348–59.

[21] McOmish 2017, 65–72.

commented upon his own *De solis ac lunae defectibus* (1760), 'so that also those unacquainted with astronomy and mathematics in general may understand the poem itself more easily' ('ut ab iis etiam, qui in astronomia et universa mathesi sunt rudes, facilius intelligi possent ipsa carmina', preface to the reader, p. ii). Much freer forms of auto-commentary are found in Nicolas Biese's *De universitate libri tres* (1556) as well as Bruno's *De minimo*, *De monade*, and *De immenso*.[22] In these poems, the poetic sections simple alternate with explanatory prose passages, which not only provide explanations but also supplement and continue the reflections presented in the poetry. Finally, paratexts may also include non-text, namely diagrams. Antoine Mizauld's cosmographical overview titled *De mundi sphaera* (1552) is an early case in point: facts and phenomena from the structure of the Ptolemaic universe to a lunar eclipse are illustrated by clearly drawn diagrams with captions (Figure 14).

Despite its lack of notes or commentaries, *De mundi sphaera* may also be cited as presenting a textbook example of a sparing but efficient paratextual apparatus, the interplay of whose single components renders the work's perusal pleasant and easy. The work is introduced by detailed summaries of each of its three books, both in a short paragraph of continuous text and in a more extended series of keywords. The poetic treatment itself is divided into chapters by inserted headings, which mostly respond to the keywords in the introductory summaries and often grow into a short paraphrase of the subsequent section. It is further subdivided by marginalia. Both headings and marginalia are set apart by roman type from the body text, which is in italics.

Didacticism II: literary techniques

The literary didactic techniques applied may roughly be subdivided according to the two classical functions of poetry, 'docere et delectare'. For both of these, the poetic tradition held a rich arsenal of methods in store. To take the second half of the formula first: from antiquity, didactic poems exhibited decorative text blocks such as the proem, the address to the poet's patron, the invocation of one or more Muses, and mythological digressions which could develop into self-contained narratives, so-called 'epyllia'. They used a didactic variety of the epic language characterized by artful circumlocutions, ornamental epithets, metaphors, and similes. The aesthetic effect of this language could be enhanced by archaisms borrowed from Homer, Hesiod, or (in the case of Lucretius) Old Latin. All of these elements were taken up in early modern didactic poetry. Some were adapted to new circumstances—God, Christ, or the Virgin might stand in for the Muses,

[22] Granada and Tessicini 2020, 13–23.

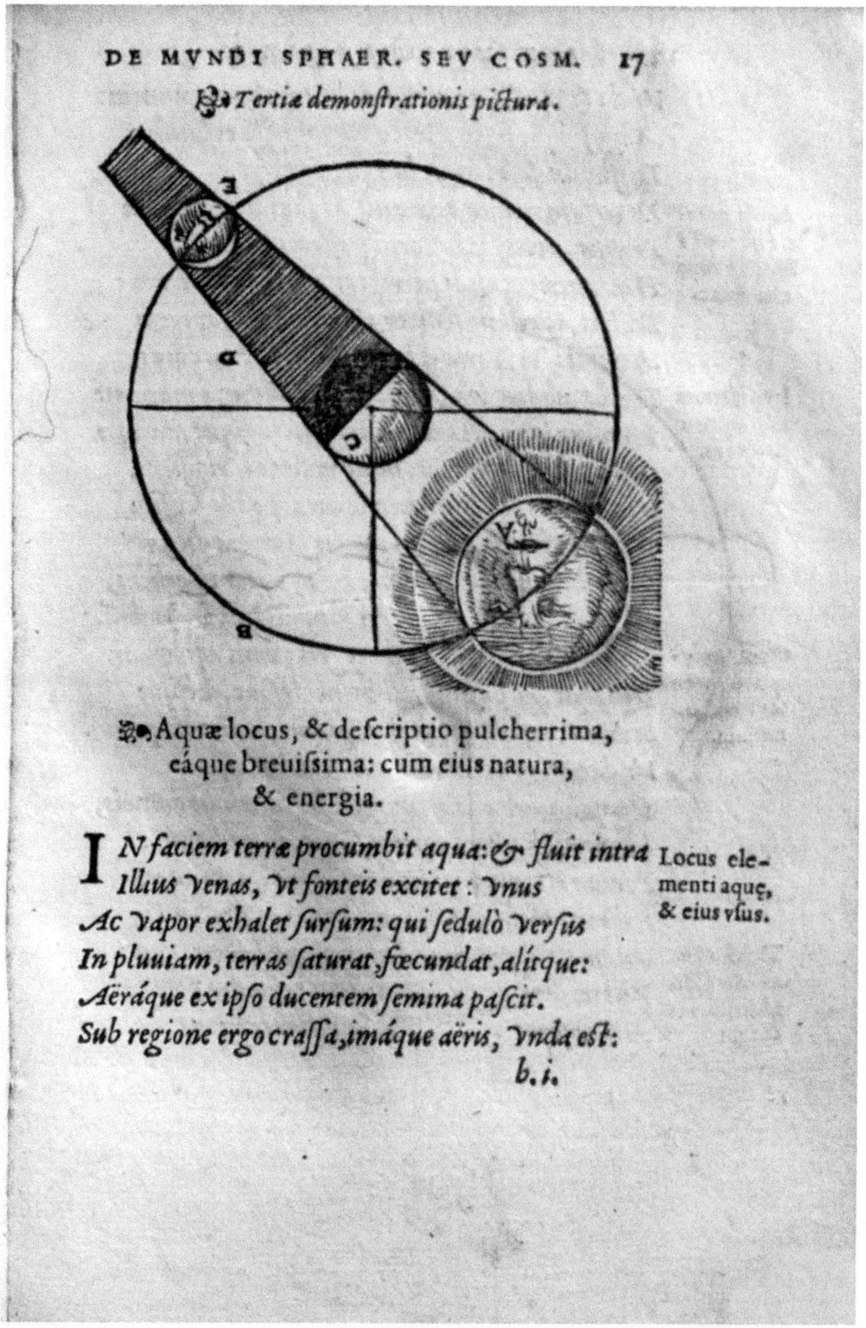

DE MVNDI SPHAER. SEV COSM. 17

Tertiæ demonſtrationis pictura.

Aquæ locus, & deſcriptio pulcherrima, eáque breuiſsima: cum eius natura, & energia.

IN faciem terræ procumbit aqua: & fluit intra
Illius venas, vt fonteis excitet: vnus
Ac vapor exhalet ſurſum: qui ſedulò verſus
In pluuiam, terras ſaturat, fœcundat, alitque:
Aëráque ex ipſo ducentem ſemina paſcit.
Sub regione ergo craſſa, imáque aëris, vnda eſt:

Locus elementi aquę, & eius vſus.

b. i.

Figure 14 Diagram illustrating the occurrence of a lunar eclipse from Mizauld's *De mundi sphaera*.

heathen myths be replaced by ones of transparent allegorical meaning invented by the author—others adopted unchanged. To see such devices as pure ornament which could be left out without any damage would be misguided. Rather, they are recreational in the literal sense of the word: in giving the reader some rest and pleasure, they keep him fresh and willing to follow the more technical explanations. As Boscovich declares in the preface to his *De solis ac lunae defectibus*: 'epeisodia immiscui sane multa ad demulcendum animum severiorum contemplatione defatigatum' ('I have mixed in a fair number of short stories in order to soothe the mind tired from reflecting on more demanding topics', p. ii).

How, to turn from 'delectare' to 'docere', were these demanding topics themselves conveyed? In order to have a common frame of reference for the examples, my discussion will be limited to one single poem, Noceti's *Iris* (1730).[23] This short poem comprising one single book and some 500 lines, which aims at providing the reader with an understanding of the rainbow in terms of Newtonian optics,[24] is in many ways a typical representative of early modern didactic epic, particularly of its eighteenth-century Jesuit variety. The work starts with a refutation of older explanations of the phenomenon (vv. 1–69). Next, some optical fundamentals (reflexion, refraction, the composite nature of sunlight, different angles of refraction for different colours) are set out (vv. 70–172). The path of a beam of light within a single raindrop is analysed, and it is demonstrated how the series of refractions and reflections results in a rainbow on the macroscopic level (vv. 173–444). As a corollary, the secondary rainbow is explained (vv. 445–516). All of this is garnished with numerous recreational elements, from an opening invocation of Phoebus Apollo (in his capacity as the sun god and the god of poetry alike, vv. 6–12) to an epyllion about an unhappy love story between the goddess Iris and a certain Xanthus (vv. 389–444). In what follows, however, I focus on five strictly speaking didactic techniques: emphatic structuring, careful use of technical terminology, graphic description, explanatory metaphor, and close contact with the reader.

Iris is not only clearly structured, as the summary shows: it actively foregrounds its structure, so that the reader may never lose sight of the logical unfolding of the argument. Each step in the argument is signposted throughout. After the optical fundamentals, for example, the transition to the discussion of the path of the beam in a raindrop is announced as follows (vv. 173–5):

> Hoc autem posito fundamine percipe porro,
> cur quae plexa simul radii simulacra recondunt,
> ordine ab aëriis videas effulgere guttis.

[23] *Iris* was revised and republished several times in the eighteenth century. I follow the text of the edition by Tautschnig 2017, which is based on that in Oudin's *Poemata didascalica*. For complementary examples from other poems, see Korenjak 2019, 149–61.

[24] Newton's *Opticks* was published in 1704 and became widely accessible on the Continent through its Latin translation as *Optice* in 1706.

> But now that we have laid this basis, learn further why we see the colours, interwoven and hidden in the sun's rays, gleam one after the other from the drops in the air.[25]

Building upon similar complaints made by Lucretius (*De rerum natura* 1.136–9, 1.830–2, 3.258–60), Noceti justifies the need to use terms unknown in classical Latin because of the language's lexical poverty early on in the poem: 'Antiquis numquam tentata Latinis / nomine multa carent' ('Many things never attempted by the ancient Latin authors lack a name', vv. 74–5). When he has to coin a new term, he makes sure that his readers will understand it by clearly defining it beforehand. What, for instance, should one call the direction in which the greatest part of the light of a certain colour is inflected, making that colour appear most saturated? Noceti decides on the expression 'strong rays' ('radii valentes') and explains its meaning as follows (vv. 235–7, his emphasis):

> Atque uno tramite plures,
> quod sequitur, coeunt radii; qui robore iuncto
> plus possunt, radiosque ideo dixere *valentes*.
>
> And in one particular direction, several rays come together, as follows [from the above]. Their strength united, they have greater power, which is why they are called *strong* rays.

Because scientific truths tend to be abstract in and of themselves, non-specialists are thankful for anything they are able to imagine in a concrete way. Noceti helps his readers by describing unfamiliar phenomena and objects as precisely and graphically as possible, resorting to the techniques of ancient rhetoric.[26] The prism, for instance, a crucial instrument in a poem on optics, is presented and its use is explained in a carefully crafted ekphrasis (vv. 151–60):

> Est ternam in faciem, quarum levissima quaeque,
> porrectum mediam vitrum non amplius ulnam,
> nomine quod Graii dixerunt prisma vocantes,
> aptum opus et tales fabricatum munus in usus:
> hoc ubi ad apricae solem opposuere fenestrae,
> atque ubi multiplici lux est discreta refractu,
> tectum introgressae diverso tramite partes
> haec violas, haec secum undas, herbam illa virentem,
> ista crocum, tulit illa rosas, propriisque refulget
> pariete in adverso variata coloribus iris.

[25] For further elaborate transitions, see vv. 195–7, 254–5, 444–5, for simpler transitional formulae, see, e.g. vv. 13, 36, 55, 93, 198.

[26] On the techniques productive of 'evidentia', see *HlRh*, §§ 810–19.

> There is a piece of glass, which extends itself on three sides—each of which is perfectly smooth—for no more than half a cubit. The Greeks called and named it a 'prisma': a convenient object, an instrument designed to be used as follows: when, the windows being sunlit, the prism is opposed to the sun and the light is divided by manifold refraction, its parts enter the room on different paths. This one brings with itself violets, this one water, another one green grass, yet others crocus and roses—and on the opposite wall, a rainbow gleams in its characteristic different colours.

Inasmuch as they express the unfamiliar in familiar terms, metaphors are another means of conveying eidetic understanding (or quasi-understanding) of thoughts that are hard to grasp in the abstract. Noceti happily exploits this possibility too. How different colours coexist within white light was probably no less difficult to imagine in the eighteenth century than it is today. To help the reader out, Noceti devises a metaphor from weaving (vv. 97–9): the sunbeams are said to consist of seven 'threads' ('stamina'), and light appears as a fabric 'woven' ('texta') from these. The thread metaphor is used as a kind of leitmotif throughout the poem.[27] On one occasion, Noceti expands it into two similes, which bring out some of its implications. First, God is said to have 'combed' the fabric of light, when he was creating the world ('Ipse pater prima nascentis origine mundi / versicoloratam percurrens pectine telam', vv. 135–6), so as to ensure that light should be resolved in an orderly fashion into its various colours when refracted. A little later, light is compared specifically to a silk robe woven from threads of different colours: looking at the robe as a whole, these colours are invisible, but when the fabric is dissolved, they reappear (vv. 144–8).

Among the most striking traits of Noceti's text is the close contact he maintains with his reader by way of second person singular verbs, imperatives, personal pronouns, and combinations of these. On average, Noceti addresses the reader every five lines, safely escorting him through his entire account.[28] This does not imply that the reader is assigned a passive role. He is, on the contrary, emphatically urged to participate actively in his instruction. Throughout the poem, he is exhorted to pay attention, consult his own experience, refrain from trusting authorities blindly, or to conclude an unfinished train of thought by himself (vv. 170–2):

> Atque ea, quae in viridi contingere cernimus herba,
> quamlibet ad lucis partem discretaque fila
> transfer et ad cunctos prudens extende colores.

[27] See v. 182 ('stamina'); vv. 121, 133, 171, 191, 241, 270, 469 ('fila').

[28] By way of comparison, Lucretius, the most reader-oriented of all ancient didactic poets, turns to his dedicatee Memmius or to the general reader once every twenty to thirty lines.

> And what we see happen in the case of green grass, carry it over to any part of the light and to its different threads; look ahead and generalize it for all colours.

When a particularly difficult point is reached, the poet even prescribes in detail, step by step, the mental operations to be performed. In order to grasp the origin of the rainbow colours on a microscopic level, the reader has to 'paint in his mind' ('pinge animo', v. 177) that half of a raindrop which faces the sun, to 'cut it in the middle and separate it into two parts' ('medioque resectam / in partes distingue duas', vv. 177–8), to leave the other colours out of consideration ('sepone', v. 181), and to consider ('perpende', v. 182) red only. To apprehend the angle of 40–42° between the rainbow and the antisolar point, he must perform an even more complex sequence of steps in imaginative reasoning (vv. 266–85). Finally, at the end of a long, complex argument, Noceti encourages the reader, as a trainer would spur on an exhausted athlete, to make one last effort: 'Tum vero optatam victor iamiam arripe metam' ('At that point, you have won—now seize the mark you have hoped for!', v. 231).

Panegyrical poems

Historical overview

To give praise to those who deserve it has been attested as a key function of poetry for millennia. In the Western tradition, its power to convey fame (κλέος) is highlighted already in the *Iliad*, and a thriving tradition of praise poetry developed in antiquity. The texts did not constitute a genre but a poetic mode making use of various generic possibilities, from the *Victory Odes* of Pindar to the small occasional poetry of Statius' *Silvae* and Claudian's propaganda epics for his patron Stilicho. Natural philosophy or science as a subject of poetic praise, by contrast, was slow to appear. Lucretius' praise of his philosophical leader Epicurus in the opening passages of *De rerum natura* 1, 3, and 5 is the most famous ancient specimen, although it is integrated into a major poem whose main thrust goes elsewhere. Another example is an epigram ascribed to Ptolemy and transmitted in the *Anthologia Palatina* (9.577) in which the author, picking up a notion already expressed by Ovid (*Fasti* 1.297–308), extols the spiritually uplifting effects of astronomy. But these are isolated cases. No real tradition of such poems developed in antiquity or the Middle Ages.

The body of texts presented in this section is thus an early modern novelty. Why this should be so is easy to explain. The three main factors are the recovery of the ancient tradition of occasional poetry of praise, the spread of print, and the growing reputation of research into the natural world. The first development provided authors with the techniques and models enabling them to produce praise of any given subject, which resulted in an unprecedented output of panegyrical poetry in general.[29] The second not only gave this poetry a wider audience and greater impact, it also led to a growing book market and an advertising industry which was for the greater part integrated into the publications themselves in the form of paratexts. The third guaranteed that scientific works profited as much from this trend as other types of learned books. The apogee of panegyrical poetry on science was reached in the sixteenth and seventeenth centuries, when print had become robust, paratextual book advertising had developed clear-cut forms, and science was on the rise but had not yet completely diverged from belles-lettres, a process under way by the eighteenth century. In this century, self-contained poems were still written on recognized scientists (of whom a canon had begun to form, headed by Newton), but they became rare as a means of book advertising, as scientific publications in general took on a more sober look. Over time, the chorus of Latin poetry was joined by an increasing number of vernacular voices. Often, one and the same person or event was celebrated in different languages. Panegyrical poems were translated in both directions, and vernacular and Latin poems rubbed shoulders in collections of academic poetry.

Setting and publication context

In contrast to modern poetry, most poetry of earlier periods was occasional.[30] Panegyrical poetry in general and on science in particular was no exception. A number of standard occasions more or less imperatively called for exaltation in verse. The most common of these, as will soon become clear, was the publication of a new work. Another was a scientific breakthrough of any kind. New insights, inventions, and discoveries were regularly extolled in triumphalist poetry. As late as 1781, when Wilhelm Herschel discovered Uranus, this feat prompted György Alajos Szerdahélyi to immortalize it in his semi-didactic, semi-panegyrical *Historia Uraniae Musae* (1787). In the life of scientists, steps in their academic career such as a successful disputation or the conferral of a degree were acclaimed by friends and colleagues. In doing so, they celebrated the candidate, tightened or created bonds of friendship, enhanced the occasion, and commemorated it for posterity. When Hieronymus Leopold Bacchetoni received his medical doctorate at the University of Innsbruck in 1726, a collection of Latin and Italian poems titled *Applausus poetici* was published to immortalize the event.[31] Finally, the death of a scientist called for public commemoration as well, at least in the case of renowned figures. Theodor Reysmann lamented the mathematician, astronomer,

[29] *CNLS* 2, 100–4; *BENLW* 1142–6, 1148–50.
[30] Leonhardt 2020, esp. 111, 115–16 (with older literature at 115 n. 16).
[31] *TL* 639.

astrologer, and manufacturer of scientific instruments Johannes Stöffler in an elegy titled *De obitu Iohannis Stoefler* (1531).[32] Constantijn Huygens commented on the death of Descartes in a number of Latin (and French) epigrams and longer poems.[33] Claude Quillet appended an elegiac 'Lugubre encomium' to the 1656 edition of his *Callipaedia* (pp. 91–4) in reaction to Pierre Gassendi's demise the year before. Nicolas Fatio de Duillier's *Neutonus ecloga* (1728)[34] was published to celebrate the first anniversary of Newton's death.

Literary motivations that were independent of a concrete occasion came to the fore only rarely. One case in point is connected to Pope's famous epitaph for Newton, 'Nature and Nature's laws lay hid in night. / God said: "Let Newton be!" and all was light.' In a 1741 issue of the *Gentleman's Magazine* (vol. 11, p. 548) a Greek version of it was accompanied by a call for Latin translations. No fewer than twenty-two submissions (most of them anonymous) were sent in and printed (vol. 11, pp. 601, 663).[35] One of the more convincing examples, by a certain A.B., reads:

Naturae et lex Naturae latuere tenebris.
Sit Newtonus, ait Numen, et omne nitet.

Regarding the context of publication, there seems at first sight to be considerable variety. A few poems were published autonomously, some as booklets, such as Reysmann's elegy on Stöffler, some as broadsheets, as in the case of Francesco Vigilio Grammatica's *Cometa poeticus* (1726) celebrating a disputation of his fellow student Hieronymus Leopold Bacchetoni (the broadsheet was glued into some copies of his dissertation).[36] Others were collected to form a volume dedicated to a certain occasion. When the French mathematician and cartographer Oronce Finé died in 1555, twelve friends of his published a short volume titled *Funebre symbolum* and comprising twenty-six funerary poems in Latin, Greek, and French (as well as a bibliography of the deceased). Anthologies proved a good outlet for texts that were difficult to publish otherwise. Joseph Addison's praise of Thomas Burnet's spectacular scientific explanation of the biblical prehistory of the earth, the *Telluris theoria sacra* (1681, definitive edition 1689), first came out in an anthology of British Latin poetry (*Examen poeticum duplex*, 1698, pp. 49–54) before it found its way into the author's collected works (vol. 1, pp. 333–4) in 1811. One could also append a shorter panegyrical piece to a major work of one's own, as did Quillet with his elegy on Gassendi. The 1571 edition of Antoine Mizauld's *Cosmologia*, a didactic poem on the universe, concludes with

[32] Modern edition with German translation: Kottke 2013.
[33] Modern edition: Worp 1892–9, vol. 4, 232–8.
[34] See Figala and Petzold 1987 with appended text and translation.
[35] Fara and Money 2004, 559–60.
[36] *TL* 639–40.

an elegant praise of the astronomer and astronomy in general ('Encomium docti astronomi', pp. 65–72). Publication in a journal was an option from the later seventeenth century, as the translations of Pope's epigram demonstrate. Lucretius' praise of Epicurus in *De rerum natura*, besides providing an influential model for lauding men of science in general, in particular prompted many poets to incorporate analogous panegyrical passages into scientific didactic epics. Roger Joseph Boscovich decided to open the last book of his astronomical epic with an invocation of Newton as an inspirational deity (*De solis ac lunae defectibus*, 1760, 5.1–98).[37] Finally, there was the phenomenon of secondary publication in a new context, as when a portrait with an accompanying epigram was included in a collection of such pieces (see note 53 below) or a poem about a scientist found its way into his biography. For instance, three of the early modern biographies of Copernicus contain Latin poems on him.[38] The long *Tychonis Brahei vita* (1654) by Pierre Gassendi quotes, besides numerous poems by Brahe himself, no fewer than fifteen items celebrating the great astronomer and his work.[39]

In terms of sheer numbers, however, all options enumerated so far dwindle in comparison with the scores of poems composed to introduce scientific books. These poems were written by friends and colleagues, mostly on the request of the author of the book, sometimes also on their own initiative.[40] Their composition was a service the members of the republic of letters rendered each other. Occasionally, a learned printer put pen to paper as well, as did Gérard Morrhy when publishing Oronce Finé's mathematical textbook *Protomathesis* (1532). A substantial proportion of all self-contained scientific publications is equipped with such poems, which often form sizeable groups. Pietro Bongos's *Numerorum mysteria* (1599), an encyclopedia of numerology, features no fewer than eighteen items. Intriguingly, the inclusion of Latin poems was not restricted just to Latin books. The paratextual apparatus to Johann Georg Triegler's German version of Sacrobosco's *Sphaera* (1614) features seven elegies and epigrams extolling the very act of translation and its beneficial results. Did some of Triegler's readers actually have decent Latin and only found it more convenient to read astronomy in the vernacular? Or did the prestige conveyed by Latin poems depend on their not being understood at all?

Be that as it may, it is clear that such texts were the early modern forerunners of the blurbs and extracts from friendly reviews which decorate the covers of present-day scientific and scholarly books: they served to promote the work in question. The prospective customer who encountered a book in a bookshop was informed by these poems of its virtues in a more outspoken way than the author

37 On a comparable passage in Benedict Stay's *Philosophia recentior* (1755–92), see Haskell 2003, 213–14.

38 Kühne and Kirschner 2004, nos. 14, 15, 19.

39 See pp. 123–4, 142–4, 187, 210–11, 230–2, 252.

40 See Tautschnig 2022 on nine poems in praise of Johannes Hevelius's *Selenographia* (1647).

himself could do, for reasons of modesty, in his letter of dedication or preface to the reader. At least as importantly, the very fact that colleagues were willing to contribute a recommendation in writing documented the esteem in which the author and his work were held by the learned community.

Subjects of praise

In terms of scientific disciplines, astronomy with its lofty image, already extolled in antiquity, seems to have attracted most praise, followed by medicine, the discipline with the strongest institutional and professional background, and the mathematical disciplines. However, hardly any field remained unsung: there are poems on subjects from natural philosophy, physics, botany, zoology, geography, chymistry, and even decidedly minor fields like numerology. Conversely, the range of subjects praised is limited in that most praise focuses on a comparatively small number of aspects of the scientific enterprise. Although these aspects usually blur into each other, they are enumerated separately in what follows for the sake of clarity.

Some poems primarily extol new scientific results. In doing so, they do not limit themselves to those inviting a celebration in verse from a modern point of view, a fact which testifies to the new esteem in which scientific progress per se was held, independently of the dignity of its object. In the only poem introducing Nicolas Steno's *Observationes anatomicae* (1662), Matthias Jacobaeus admires Steno's contribution to the discovery of the lacrimal and salivary glands and ducts, and has no qualms about speaking of the bodily fluids, tears, saliva, and mucus. Of course, however, more rewarding subjects, for instance spectacular applications of the new science, were celebrated as well. One year after the first flights of the Montgolfier brothers' hot air balloon, Giuseppe Antonio Taruffi immortalized the new invention in his elegy *Montgolferii machina volans* (1784). At this late date, his Latin voice was one of few in a chorus of vernacular ones.[41]

Other texts praise not a particular scientific achievement, but a scientific discipline in general. Mizauld's 'Encomium docti astronomi' does so with regard to astronomy by describing, in the wake of Ovid and Ptolemy, the happiness of the true astronomer whose sublime mind dwells high up in the sky with the stars and has nothing but contempt for the mundane concerns of average mortals. Moreover, as Mizauld humorously adds towards the end (pp. 70–2), the sky also offers the astronomer all kinds of less serious pastimes, from hunting constellations like the Bears or Lepus to flirting with Andromeda and Cassiopeia and drinking from Crater. Cornelius Grapheus introduces Rembert Dodoens's *Cosmographica*

[41] Wiegand 1984, 37–9; Bitzel 1997, 226–7. On the vernacular tradition, see Jacobius 1909.

in astronomiam et geographiam isagoge (1548) with two complementary poems, the first of which, titled 'Commendatio divinae Mathematices', details the merits of the mathematical disciplines at large, while the second one focuses on the book at hand and its author.

Generous patrons may earn their share of praise as well. An appealing, novel, and well-researched example, in which science is turned from a mere object to a medium of praise, is the *Μελισσογραφία* (1625) offered to Pope Urban VIII by the members of the Accademia dei Lincei. On a broadsheet, the anatomy of three bees as revealed under the microscope is depicted in a brilliant engraving. The allusion to the coat of arms of the Pope's family, the Barberini, is made explicit in the accompanying epigram: revealing the secrets of that noble insect, the bee, implies ennobling the papal coat of arms as well.[42]

Most poems, however, focus on a particular book or scientist. In both cases, a set of panegyrical topoi established itself on which poets could draw but which they could also supplement by inventions tailored to the case at hand. Poems in praise of scientific books are usually included in the book's paratextual apparatus, although some independent items such as Addison's aforementioned praise of Burnet's work exist, as do a few handwritten poems entered in famous scientific books by their (occasionally equally famous) readers.[43] The merits most commonly highlighted are comprehensiveness and clarity (the latter often despite the book's profound and difficult content). In Guillaume Rondelet's *Aquatilium historia* (1554), for instance, Petrus Costus asserts that Rondelet has brought *all* water animals onto dry land. With regard to a collection of chymical treatises compiled and arranged by Andreas Libavius, the *Alchemia Libavii* (1597), Martin Praetorius exclusively focuses on the assertion that the book has done away with the conceptual and terminological confusion of the discipline (illustrated by a long list of monstrous terms) and rendered it understandable. However, one cannot say that the majority of all poems in praise of books address these topics. The range of virtues extolled is in fact much wider; often several aspects are combined within one poem. Johannes Sapidus, for example, finds many different features to commend in Brunfels's *Herbarum vivae eicones* (1530–6): no such convenient herbal has ever been produced before. The book is like a garden full of plants brought together from different places. Their medical properties are described so well that even personal experience could add nothing to it. The style is concise, but not to

[42] Freedberg 2002, 160–3; Finocchiaro 2004; Guardo 2004, 128–9; Brevaglieri 2015, 146–9. Science could be turned into a medium of praise in prose works as well, as shown by Albert von Curtz's *Problema Austriacum* (see Chapter 7, note 126).

[43] Melanchthon left a Latin elegy on the human body as an admirable work of God in a copy of Vesalius's *Fabrica*, from which it found its way into several more copies (Schullian 1945, 437–9; Margócsy et al. 2018, 73, 112). A Greek elegy on Copernicus's *De revolutionibus* by Joachim Camerarius, extant in two copies of the work, was translated into Latin by Kepler and prefixed to his personal copy (Gingerich 2002, 355–61).

the detriment of comprehensibility. The illustrations are deceivingly lifelike. The book is printed with great diligence and accuracy. Every doctor should buy it.

Praise of scientists can be found both in paratextual book promotion and in self-contained odes. Considerable numbers of such poems are dedicated to renowned figures such as Stöffler, Paracelsus,[44] Gessner, Brahe, Galileo, Harvey, Gassendi, Huygens, Burnet, Newton, or Herschel. Newton in particular is easily the most-praised early modern scientist. Numerous Latin and vernacular (mainly English) poems, sometimes quite extended and of high quality, were written in his honour.[45] The major exception to the focus on celebrities is academic congratulatory poetry, which is usually dedicated to lesser figures such as the aforementioned Bacchetoni.

In many respects, the presentation of scientists seamlessly continues humanist strategies of praise for intellectuals in general. In doing so, it focuses mainly on social virtues. The man of science is depicted as one link in a long chain of tradition and as an exemplary member of the republic of letters. He embodies, practises, and furthers universally recognized virtues and values such as learning ('doctrina'), hard work ('labor'), carefulness ('diligentia'), friendship ('amicitia'), and usefulness ('utilitas'). A well-liked trope is the comparison with a famous predecessor, who is equalled or outshone by the actual object of praise. This technique has the additional advantage of creating a direct link to a venerable past and presenting science as a timeless concern of humankind. The numerologist Pietro Bongo, for instance, is declared to be superior to Pythagoras in many of the introductory poems to his *Numerorum mysteria*. A related trope compares the scientist in question to a celebrity from another field. An elegy by Cyprian Kinner introducing Hevelius's *Selenographia* (1647) compares Hevelius's charting of the moon to Columbus's discovery of America: each of them has shown his contemporaries a new world ('novos orbes manifestavistis uterque', v. 13).

On the other hand, an anti-integrative strategy could be adopted. Instead of incorporating the person lauded into the commonwealth of science, the poet could pointedly set him apart from and above his peers. Inspiration for this approach was provided by Lucretius who had openly called Epicurus a god ('deus', *De rerum natura* 5.9). This could hardly be repeated verbatim in a Christian context, but a great scientist could at least be presented as a 'hērōs' in the ancient sense of the word, a larger-than-life personality. 'O hero, radiating by virtue of your strain and high descent' ('O fulgens genere et celsis natalibus heros') is how

[44] A collection of poetry in praise (and criticism) of Paracelsus, including thirty-three Latin poems, has been arranged by Telle and Limbeck 2008.

[45] This body of works includes, for instance, Halley's 'In Newtoni opus hocce mathematico-physicum' prefixed to the *Principia* (see Albury 1978), Pope's epitaph with its Greek and Latin translations, Fatio de Duillier's *Ecloga*, and Alfonso Niccolai's 'Carmen elegiacum' (1756). On the Latin tradition, see Fara and Money 2004; on the vernacular one, see Nicolson 1946 and Wagner 1974 (who also includes painting and architecture).

Kepler addresses Brahe in a poem preceding his *Astronomia nova* (1609). Such a man can be said to be of divine origin ('claro divorum sanguine cretus'[46]) and appear as a counterpart to great figures from classical mythology: Galileo is called the 'Oedipus of nature' in a fictitious epitaph at the start of a late edition (1699) of the *Systema cosmicum*, the Latin version of his *Dialogo*. The scientific hero's intellect appears to be superhuman ('longe exsuperare videris, / humanum, Newtone, genus'; Stay, *Philosophia recentior* 3.1968–9). Through his achievements, he approximates divinity and earns everlasting and universal glory: Newton is worshipped like an ancient god with temples, altars, and sacrifices by the 'studiosa iuventus' according to Boskovich's *De solis ac lunae defectibus* (1760, 5.89–94). In this way, panegyrical poetry contributed to a vision of science and scientific progress as the work of a small group of quasi-superhuman beings that is still popular today.

Formal features

Nowadays, panegyrical poetry has a bad reputation for ethical as well as for aesthetic reasons: it is thought to be insincere and is devalued as uninspired and unoriginal. Both parts of this judgement are unjust and result from an anachronistic application of post-Romantic criteria to premodern poetry. The first fails to recognize the initially mentioned view of poetry as a vehicle of praise. The second turns a blind eye to the fact that poetry was not seen as an expression of a poet's extraordinary personality but rather as a craft to be learned and applied according to the needs of the situation. In particular, it ignores the fact that panegyrical poets were usually pressed for time and thereby forced to recur to structural templates and recycle text modules from other people's production as well as from their own.[47] To compare the results with the poetry of famous authors who took years to polish their works creates expectations which are bound to be disappointed.

Even so, there is no denying that the quality of panegyrical poetry, about science as well as about other subjects, varies greatly. Many poems are indeed of little worth—one or two commonplaces cast into a few elegiac couplets and provisionally adapted to the occasion at hand. That they were commissioned at short notice and written in a hurry often results in inconsistencies and obscurities.[48] That being so, it should all the more be highlighted that a good

[46] This is how Harvey is imagined describing Boyle in Robert Grove's *Carmen de sanguinis circuitu* (1685), v. 396.

[47] On this aspect, see the profound study of Schirg 2016, esp. 155–276.

[48] In Costus's poem on Rondelet's *Aquatilium historia*, the author claims that sea monsters are leaving the sea 'offensos...tuis, victor Neptune, quadrigis' (v. 3). Why should they take offence at Neptune's chariot, and why is the sea god, whose realm is pillaged by Rondelet according to Costus,

number of the texts in question are actually well crafted, witty, and inventive. It was exactly the challenge of quickly creating an attractive poem for a recurring social occasion on the basis of a limited set of conventionalized ideas that incited good poets to come up with interesting solutions.

As a result, panegyrical poetry makes full use of its generic freedom and exhibits great formal variety. In terms of length, anything is possible from two lines to several hundred verses, although short poems dominate. With regard to metres and genres, elegiac couplets are the default option in this as in most other kinds of Latin occasional poetry. They are the most common metre for epigrams but are used in full-blown elegies as well. (Both genres merge into each other without a clear division.) There are even a few panegyric-didactic hybrids in elegiacs, perhaps inspired by the use of the metre in lighter didactic in the wake of Ovid's *Ars amatoria*, for instance Taruffi's *Montgolferii machina volans* and Szerdahélyi's celebration of Herschel's discovery of Uranus. Similar poems could also be written in hexameters, as Robert Grove did in his account of Harvey's discovery of the circulation of the blood (*Carmen de sanguinis circuitu*, 1685). Generally speaking, more ambitious compositions tended to be composed in this metre, which ranked above elegiacs in the literary hierarchy and carried heroic overtones as the metre of epic. However, as if to mitigate such high-flown pretensions, hexameter poems may carry titles such as *Idyllion* (Johann Baptist Hebenstreit, 'In astro-poecilopyrgium Keplerianum idyllion', 1627) or *Ecloga* (Fatio de Duillier, *Neutonus ecloga*), which are associated with the bucolic tradition and in any case denote less ambitious compositions.

Hendecasyllables were also quite popular. Through Catullus, they were associated with short, witty poems, and through his first poem in particular, with the introduction and advertising of a book. Martin Praetorius logically resorted to them when recommending the *Alchemia Libavii* to the reader ('Phaleuci ad Libavium de chymicis'). Somewhat less common was another kind of short stichic verses, namely iambic dimeters, used, for instance, by Morrhy to extol Finé's mathematical textbook *Protomathesis*. With his *Carmina*, Horace had bequeathed to posterity models of lyric poetry in a dozen different metres. Among the most popular were Alcaic stanzas. Addison extolled Burnet's *Telluris theoria sacra* in such strophes, which were perceived as manly and emotional. More eccentric was Mizauld's decision to praise the profession of the astronomer in 270 lesser Asclepiads in his 'Encomium'. At the border with poetry, one finds so-called 'ellogia', texts which transfer the style and layout of stone inscriptions to paper. Like real inscriptions, they come across visually as poetry, or at least as non-prose, through

called a 'vanquisher'? In v. 7, fish use the same porticoes ('porticibus') as men: what porticoes? That the same riverbed ('alveus') should contain all sea and land (v. 8), is also a strange thought. In vv. 13 and 15, the British are 'swift' ('celeres'), the Arabs 'doubting' ('dubius'), epithets whose point remains obscure.

their centred lines of uneven length, although they do not follow any metrical rules.[49] The 'Epitaphium' introducing the 1699 edition of Galileo's *Systema cosmicum* is in this style.

Last but not least, panegyrical passages could be woven into non-panegyrical poems of all metres and genres, not only in didactic epics, as already mentioned. Even a poem such as the Alcaic ode 'Adulatio perniciosa' (1620) by Maffeo Barberini, the later Pope Urban VIII, which discusses the dangers of flattery and seems to be totally alien to science at first sight, in fact praises Galileo's telescopic observations of the moons of Jupiter and of sunspots and draws a moral lesson from them.

Apart from metres and genres, the literary forms and techniques adopted are too many and various to allow for any general statements. A few remarks on some conspicuous features that occur more frequently must suffice. While modern poetry tends to take the form of an interior monologue, most premodern poems have an addressee, and panegyrical ones are no exception. The addressee's presence may be quite conspicuous (he is often already indicated in the title), which gives the poem a lively and immediate character. Paratextual poems may be addressed to a book's prospective buyer and reader. Sapidus's elegy on Brunfels's *Herbarum vivae eicones* addresses the ideal reader in the first line as 'you who want to know the innate forces of herbs' ('qui cupis ingenitas herbarum noscere vires') and in the last couplet urges him to purchase the work. Even more frequently, the poet speaks to a (fellow) scientist and congratulates him on his achievements, so that the reader appears to be overhearing a conversation between two experts. Such is the case, for instance, in Praetorius's hendecasyllables on Libavius's *Alchemia*. As in Sapidus's elegy, the addressee is invoked at the beginning and end, being placed most prominently and at the same time holding the poem together through ring composition.

In terms of structure and train of thought in general, shorter poems, epigrams in particular, often catch the reader's attention by paradox, surprise, and point. In the introductory part of Kepler's *Epitome astronomiae Copernicanae* (1618), a certain Saxirupius Virginianus elevates Kepler over Ptolemy—but not on scientific grounds: Ptolemy, it is asserted, only moves the sky, which is easy because of its lightness, but Kepler manages to move the heavy earth. The final couplet drives the point home by contrasting children playing with feathers, straw, and bubbles with a grown man moving the sea and the mountains: 'A pueris plumae, stipulae bullaeque rotantur: / At maria et montes qui rotat, ille vir est.'

A certain playfulness is in evidence in other respects as well. In particular, poets are prone to wordplay. As already noted, Praetorius takes pleasure in confronting the reader with a list of unintelligible names for obscure alchemistic

[49] Tilg 2019.

notions. Who on earth could know 'what Noble Lily is, what Gold of Life, what Famous Stone, what Egg of the Wise, what Green Lion and Rebis, what Virgin's Milk and Red Lizard's Blood, or what Realgar and Xenechdon' ('quid sit nobile lilium, quid aurum / vitae, quid celebris lapis, quid ovum / sophorum, leo quid virens rebisque, / quid lac virginis et ruber lacertae / sanguis quidve realgar et xenechdon', vv. 9–14)? In the paratext to Carolus Clusius's *Rariorum stirpium per Pannoniam et Austriam observatarum historia* (1583), Justus Lipsius plays with the author's last name, which suggests a connection to 'cludere', 'to lock up': 'Omnia naturae dum, Clusi, arcana recludis, / Clusius haud ultra sis, sed Aperta, mihi.' Clusius has got the wrong name, Lipsius suggests. He does not lock anything up, but on the contrary opens up ('*re*cludis') all the secrets of nature. Therefore, he should no longer be called 'Clusius', but rather 'Aperta' ('Opener'). As the latter is a sobriquet of Apollo, the god of medicine (to which discipline Clusius's botanical research contributes through its pharmacological implications), Clusius is even playfully deified by his new name. Some poems are preceded by anagrammatic mottos. This is true, for instance, of seven epigrams by a certain Samuel Pollucius which introduce, together with a number of other laudatory poems, the *Urania propitia* (1650) of the Silesian astronomer Maria Cunitz, a simplified, bilingual (Latin and German) version of Kepler's *Tabulae Rudolphinae* (1627). The mottoes introducing each epigram, for instance 'I, VINCIT AMARA' ('GO, SHE DEFEATS THE BITTERNESS') are anagrams of the Latinized name of the author. They make only limited sense in themselves but are subsequently incorporated into the poem in quite a meaningful manner. A final epigram features another popular device, an acrostic: the first letters of each of its hexameters and pentameters spell the Latinized name of the author, 'MARIA CUNITIA', again.

In contrast with such frivolity, more ambitious compositions strive to do justice to the importance of their subjects and to overwhelm the reader by an elevated, emotional style, resorting to the 'genus grande' of ancient rhetoric. This strategy is exemplified, for instance, by Addison's Alcaic ode on Burnet's *Telluris theoria sacra* and Edmund Halley's hymnic praise of Newton's *Philosophiae naturalis principia mathematica* (whose publication he brought about with great personal commitment).[50] The latter poem, which comprises forty-six hexameters, presents a selection of the *Principia*'s content, accentuates the epoch-making importance of a work granting humankind mental access to the sky, and exhorts the reader to join in the praise of Newton, who has come as close to the gods as is possible for a human being. The monumentality of Newton's achievement is underlined already by the metre. Halley's hexameters sound grave and weighty due to a large number of spondaic feet (in which two short syllables are replaced

[50] Howard 2022, 162–7.

by a long one). Elaborate circumlocutions abound. Nouns are supplied with exquisite epithets; the human mind, for instance, is 'caeligena' ('sky-created', v. 28). The content of the *Principia* and the previous cultural achievements surpassed by Newton are presented in the form of extended poetic catalogues (vv. 10–25, 29–38). The impression of plenty and abundance conveyed by this device is enhanced by ubiquitous anaphora. Personifications serve to enliven the account: the Sun sits on his throne ('Sol solio residens', v. 7), Mathematics chases away the cloud that blocks human understanding ('nubem pellente Mathesi', v. 23). While a single reader is addressed at the beginning ('En tibi . . .', 'Here you have . . .', v. 1), the audience is later enlarged to encompass humankind as a whole: 'Surgite, mortales' ('Rise up, you mortals', v. 27). Although Halley steers clear of the more extravagant linguistic mannerisms of Lucretius (who embraced an archaizing idiom quite distinct from the Latin of authors such as Virgil and Ovid a few decades later), a Lucretian flavour is provided by typical expressions borrowed from *De rerum natura*, such as 'primordia rerum' ('the first constituents of things', v. 2). This is fitting, since Lucretius's praise of Epicurus constitutes an obvious point of reference for the entire poem; Halley's comparison of Newton with earlier culture heroes as well as his near-divinization of Newton are directly inspired by *De rerum natura* 5.1–54.

While most panegyrical poetry is plain text, some poems work in tandem with images. The most popular combination of text and image is the portrait accompanied by an inscription and an epigram, a form by no means restricted to men of science or even to intellectuals in general, but applied to all kinds of important personalities. The inscription usually mentions the name and age (or date of birth) of the man or (rarely) woman portrayed and often runs around the portrait, which takes the form of a medallion. The epigram is placed beneath.[51] This overall structure somewhat resembles, and might be inspired by, that of the emblem book, a genre emerging in the early sixteenth century and remaining highly popular over the next two hundred years. The poem almost invariably mentions the portrait. In the case of intellectuals, a standard topos highlights the contrast between the outward appearance of the person portrayed, which can be depicted by the artist, and the mental greatness, which cannot. Besides, many other possibilities for referring to the image are exploited. Johannes Cornelius Uterverius, for instance, saves the reader from the erroneous belief that Ulisse Aldrovandi's portrait at the start of the his *De avibus* (1599; Figure 15) represents Aristotle, the greatest zoological authority of ancient times, but hastens to add

[51] Many such portraits of physicians can be found in Berghaus et al. 1979. See nos. 3–4, 6–7, 17, 19, 21, 23–4, 27, 30–1, 36, 39, 42, 44–5, 48–50, 52, 56–7, 62, 65, 68–9, 71, 78 of the catalogue. For portraits-cum-epigrams of Copernicus, see Kühne and Kirschner 2004, portraits P3, image no. 4, and P17, image no. 11; for those of Paracelsus, see Telle and Limbeck 2008, 18, 20–1, 23–9, 41–5, 49–50, 63–4, 68–9, 79, 83–4, 86–8.

Figure 15 Portrait of Aldrovandi with epigram: 'This is not your image, Aristotle, but the one of Ulisse. / Your faces are different but your genius is equal.' The text circling the portrait says 'Ulisse Aldrovandi from Bologna in the 74th year of his life'. What the text at the bottom—a quotation from Virgil, *Eclogues* 3.54: 'Remember this most deeply; it is no mean thing'—refers to is unclear.

that Aldrovandi nonetheless shares Aristotle's genius ('ingenium').[52] In his subscription to the image of Descartes in the Latin version of the *Géometrie* (1659), Constantijn Huygens the Younger points out that a man who will be famous forever should be visually present for all time as well ('omnia ut aspicerent, quem saecula nulla tacebunt'). As in these cases, portraits-cum-epigrams were usually prefixed to works by renowned authors. However, they were also distributed as broadsheets, collected, and glued into albums. Their popularity even led to a distinct genre of portrait books such as—to adduce one thematically pertinent example among nearly two hundred—Johannes Sambucus's *Icones veterum aliquot ac recentium medicorum philosophorumque* (1574).[53] In addition to portrait epigrams, scattered examples of other kinds of text–image combinations, such as the Lincei's *Μελισσογραφία*, occur as well.

Finally, there are some texts set *as* images, in other words, pattern poems. I conclude the present section with an example of this phenomenon, namely Grammatica's broadsheet for Bacchetoni, in order to demonstrate how fancifully and extravagantly some authors could deal with hackneyed occasions and topics. The broadsheet was produced on the occasion of Bacchetoni's defence, at the University of Innsbruck, of an anatomical textbook published earlier in 1726 by Professor Franz Karl Anton Egloff von Staadthoff and features a pattern poem titled *Cometa poeticus, qui Morti ferale illuxit* (*A Comet in Verse Which Deadly Shone on Death*; Figure 16). The poem is arranged in the form of a comet with ten thin projections, each of which is formed by one elegiac couplet, and a tail of one more. Read clockwise, starting from the uppermost projectionand ending with the tail, the story it tells can be summarized as follows: dire Death ('Mors', feminine in Latin) riots all over the earth, sparing nobody. Will this calamity never come to an end? Hippocrates (who seems to be imagined as a kind of pagan patron saint of medicine) watches the disaster from the sky, becomes angry at Death and decides to send a strong opponent against her. He assigns the task to Bacchetoni, who duly enters the fight (that is, he undergoes his disputation). Victory is his, because Death remembers that Bacchetoni has already beaten her twice before (in two earlier disputations, as explained in a footnote) and gives up. May she now lament, while the world rejoices! May Bacchetoni take up the crown due to him from Apollo!

An additional message is expressed by the first and last letter of each verse plus one in the middle, set in upper case and arranged in three concentric circles within the comet. Read clockwise from the outside to the inside, they proclaim: 'BACCHETONI RUTILUM MEDICINAE SIDUS VIVE ET SANA'

[52] See Findlen 1994, 309–11 (who seems to ascribe the authorship of the anonymous epigram to Aldrovandi himself).

[53] On Renaissance portrait books in general, see Pelc 2002. Of the 173 books in his catalogue, none focuses exclusively on men of science, but around 20 include some among other intellectuals.

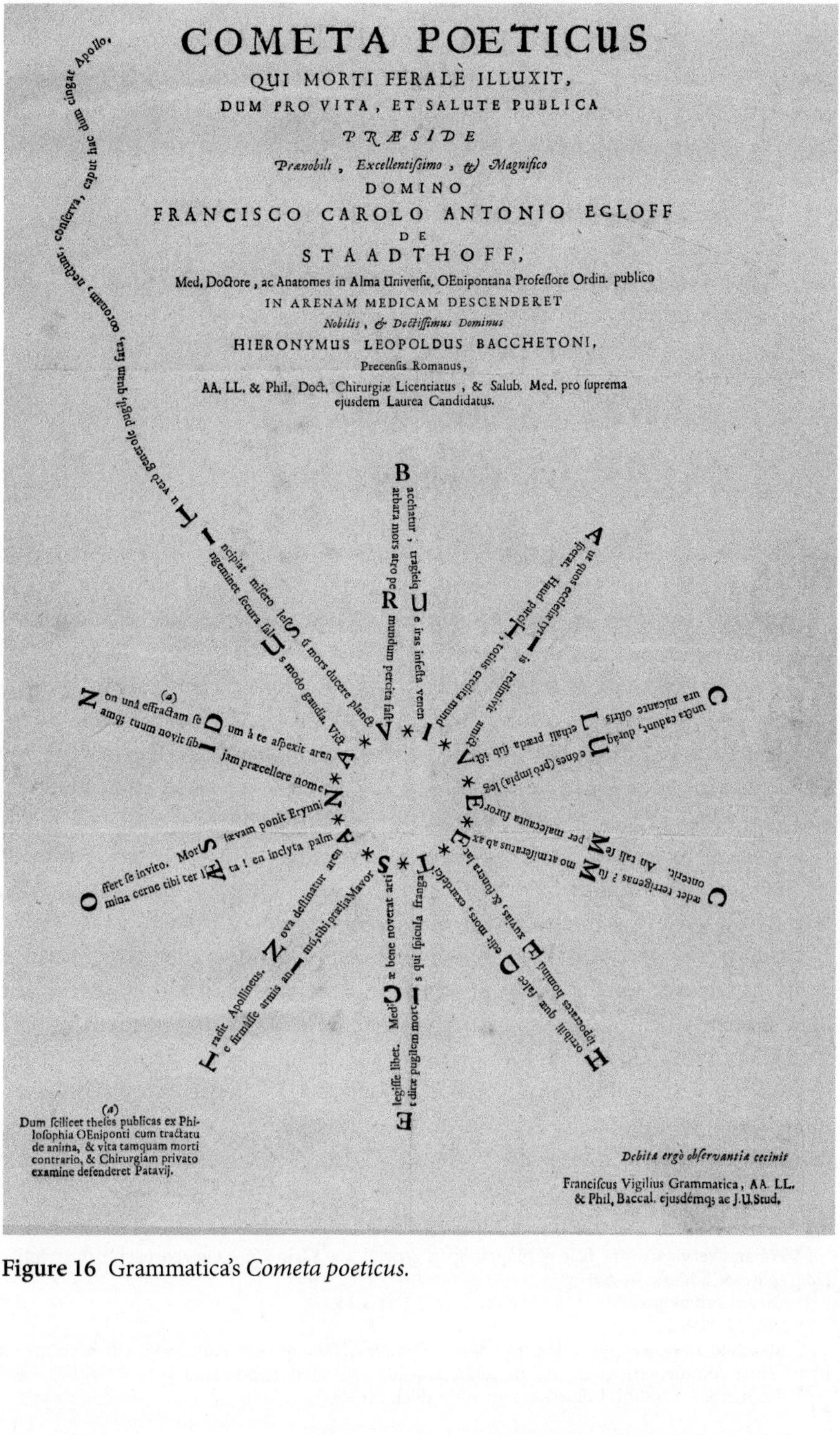

Figure 16 Grammatica's *Cometa poeticus.*

('BACCHETONI, RED-GLOWING STAR OF MEDICINE, LIVE AND HEAL!'). Together with the title, this insert reveals the basic idea of the poem: the text's graphic form corresponds to the fact that the defendant himself is styled as a comet. Comets are generally regarded as bad omens, and Bacchetoni is such an omen too—but not for humankind, as is usually the case, but for Death. He thus turns out to be a true medical superstar. His person even seems to be associated with Christ himself, who has vanquished Death and whose birth was greeted by the Star of Bethlehem, usually visualized as a comet. Of course, no reader will have taken such hyperbolic praise seriously, but most will have found the idea witty and engaging.

Biographies

Ancient and medieval biography

Among the most accessible aspects of science are the personalities active in the field. Biographies of scientists are, therefore, popular today, and they were already in the early modern era, when some men of science first became public figures through the new medium of print.[54]

The literary genre of biography (*βίος*, 'vita') dates back to antiquity.[55] Drawing on anecdotes, panegyrical (especially funeral) oratory, historiography, and philosophical interest in human personality, it always remained a flexible form. Many extant ancient lives, for instance those by Cornelius Nepos, Suetonius, Plutarch, and Diogenes Laertius, were arranged in series. This provided a model for medieval collections such as the *Liber Pontificalis* as well as for early modern ones, starting with Petrarch and Boccaccio.[56] For didactic reasons, short biographies introduced ancient editions of and commentaries on classical authors. In medieval times, they were often the first item of a set of preliminary remarks known as 'accessus ad auctores'.[57] In terms of structure, Suetonius's *Twelve Caesars* became most influential. It features a chronological narrative beginning with an emperor's family and birth and ending with his death and funeral, which frames a thematic section concerned with his appearance and character.[58] Besides political leaders, intellectuals were favourite subjects of 'vitae'. In addition to the lives prefixed to editions, there are extant works by Nepos, Suetonius, and Diogenes specifically

[54] For an overview of the field of biography in general, see Klein 2009. On approaches to scientific biography, see Söderqvist 2007.

[55] On ancient biography, see, for instance, *DNP* 2, cols. 692–8; Ehlers 1998.

[56] *CNLS* 2, 204–5.

[57] Mansfeld 1994 (see his index, s.v. 'bios'), Korenjak 2003, 63–4; Quain 1945. On 'accessus' in Renaissance commentaries, see Maclean 2002, 210–12.

[58] On Suetonius and his influence, see Conte 1994, 546–50.

devoted to historians, grammarians and poets, and philosophers respectively. Christianity produced the most famous lives of all, the Gospels. Christian hagiography became the most popular brand of medieval biography.[59]

While biographies of intellectuals in general were common, only a few meagre examples of biographies of students of nature are available from premodern times in the Western world. A life of Hippocrates by Soranus, biographies of philosophers interested in natural phenomena in Diogenes, and biographies of personalities like Archimedes, Ptolemy, and Galen in pseudo-Walter Burley's late medieval *De vita et moribus philosophorum*[60]—that is about as good as it gets. No biography is extant, for instance, of a figure as important as Euclid. By contrast, the voluminous biographical dictionaries compiled in the Arab world contain numerous lives of physicians, astronomers, and experts in other scientific disciplines.[61]

Scientific biography in early modern times: an overview

From the later fifteenth century, material from Arabic works found its way into the West. Interest in scientific biography began to grow, a development which formed part of the proliferation of biographical forms in general brought about by a renewed interest in history and in personal character and behaviour as well as by the invention of print.[62] Biographies of scientists formed a subsection, not always clearly distinguished from the rest, of biographies of intellectuals, a type recognized as such in early modern times.[63] In accordance with their subject, 'vitae' of learned men and scientists remained predominantly in Latin until the eighteenth century, while biography in general flourished in the vernaculars as well.

As lives of representatives of the natural sciences began to appear in greater numbers, they also became more diverse, differing in numerous respects. The most significant of these may briefly be mentioned.

In terms of occasion, the single most important event that could prompt the composition of a biography was the subject's demise. Numerous biographies are connected to a death. This includes self-contained books like Josias Simmler's *Vita Conradi Gesneri* (1566) as well as, later on, obituaries in journals such as the

[59] On medieval hagiography, see Nahmer 1994; on its early modern successor, see Ditchfield 1995, 117–34. On medieval biography in general, see Berschin 1986–2004.

[60] Edited by Knust 1886. For a characterization, see Blum 1983, cols. 164–8.

[61] Hasse 2016, 28–68, esp. 47–8.

[62] For a concise overview of early modern biography up to the eighteenth century, see Berschin 2009, Albrecht 2009, Schnicke 2009; on Renaissance and baroque biography, see Berschin 1993; on Neo-Latin lives, see *CNLS* 2, 200–17.

[63] See Albrecht 2009, 231; Werner 2009, 267. Johann Georg Wiggers includes a chapter 'Über die Biographie der Gelehrten' in his *Über die Biographie* (1777), pp. 116–56.

Miscellanea curiosa (1670–), the *Nova litteraria eruditorum* (1718–23), or the *Commentarii de rebus in scientia naturali et medicina gestis* (1752–1806). Still, the greater part of biographical production did not respond directly to the biographee's death. For many lives, no specific occasion is discernible at all.

Biographies could be of oral or written origin. Some of them were initially funeral orations, belonging to a genre constituting the Latin, secular counterpart of the vernacular funerary sermon which became immensely popular in the Reformed part of Europe from the mid-sixteenth century. Others originated as academic orations. Most, however, were originally composed in writing.

The chronological relation between biographer and biographee could also be widely different. In most cases, the lifetime of the latter dates either back to antiquity or the Middle Ages or belongs to the recent past.[64] Lives of figures from a few generations before, such as Friedrich Otto Mencke's *De vita, moribus, scriptis meritisque Fracastorii* (1731) are less common.

Regarding length and aspirations, anything was possible as well. The spectrum ranges from polished, comprehensive accounts in several books such as Pierre Gassendi's *Tychonis Brahei vita* (1654) to a few lines recording some essential data.

Biographies might occupy any point on a scale leading from 'hagiography' on one end to critical historiography on the other. Many early modern scientific biographies approached their sources in a fairly uncritical manner, heaped praise on the biographee and extolled the importance of his work as well as of his discipline. As historical awareness grew in the course of the early modern period, biographies appeared that were more critical of their sources and subjects alike. In the sciences, such texts collectively helped to prepare the emergence of disciplinary history in the seventeenth century.[65]

As a biography traditionally focuses on the achievements of the person whose life is described, which in the case of an intellectual are usually documented in his writings, biographies could be combined with bibliographies in different ways and to varying degrees. The practice goes back to antiquity: Porphyry's biography of his teacher is titled *On the Life of Plotinus and the Arrangement of his Books*. As an early modern example, Simmler's *Vita Conradi Gesneri* may be cited, which dwells at length on Gessner's *Historia animalium* and has a self-bibliography by Gessner appended to it. Bibliographical parts were also included in serialized biographies of physicians, some of which will be mentioned shortly.[66]

[64] On this dichotomy in medicine, see Siraisi 2007, 119–27.

[65] With regard to medicine, see Siraisi 2007, 106–33, esp. 122–7 (two critical biographies of two famous physicians, Jean Fernel and Guillaume Rondelet, from sixteenth-century France), 132–3 (preparation of disciplinary history). By the eighteenth century, a panegyrical approach could be felt to be inappropriate in a serious scholarly context. In their opening statement, the editors of the *Commentarii de rebus in scientia naturali et medicina gestis* ask authors of obituaries to keep them sober and focus on the deceased's professional work (vol. 1, 1752, pp. xiv–xv).

[66] Symons 1990, 241–2. On the link between biography and bibliography in general, see also Chapter 5, 'Bibliographies', and Maclean 2009, 407, 425.

Finally, while biographer and biographee were usually not identical, one also finds autobiographies in considerable numbers. Such works comprise book-length specimens such as Girolamo Cardano's *De propria vita* (1643)[67] alongside short and dry CVs like those appended to nineteenth-century dissertations.

As these parameters could be combined in all possible ways, no clear-cut classification results from them. The most workable criterion for grouping biographies is another, namely their publication context. If one leaves aside some minor groups such as the aforementioned obituaries published in journals, the three main types emerging from such a classification are, in ascending order of length and ambition, biographies assembled into collections, biographies introducing works by their biographees, and biographies published as self-contained books. These groups differ markedly not only in length but also in structure and style.

Collections

As already mentioned, serialization had been a popular way of organizing biographies since ancient times. In science, this model was particularly attractive, because too little was known about many of its representatives for a monographic treatment. Biographical collections predated other forms at the beginning of early modernity. In most cases, representatives of a specific discipline were collected. In Italy, doctors formed distinct subsections within collections of lives of famous men in general as early as the fourteenth century.[68] The first printed collections of lives of doctors and mathematicians, that is, scholars active within the quadrivium, appeared shortly after 1500. Examples of this brand of biographical writing are provided by Georg Tannstetter Collimitius's 'Viri mathematici' (1514),[69] Otto Brunfels's *Catalogus illustrium medicorum* (1530), Giuseppe Biancani's 'Clarorum mathematicorum chronologia' (1615), and Melchior Adam's *Vitae Germanorum medicorum* (1620). The last work once more forms just one part of a larger project, in this case a five-volume collection dedicated to German intellectuals in general and structured according to the four academic faculties.[70] Lives of scientists can also be found scattered in thematically unspecific collections.[71]

While Collimitius's geographical focus is even narrower than Adam's—he enumerates mathematicians from his home university Vienna only—Brunfels and Biancani know no geographical or chronological restrictions at all, listing doctors

[67] Translated and introduced by Stoner and Grafton 2002. [68] Siraisi 2001, 157–83.

[69] Edition with German translation: Graf-Stuhlhofer 1996, 156–71.

[70] Intellectuals active in the realm of the quadrivium (for instance, Copernicus) are found in the tome dedicated to 'philosophers' (*Vitae Germanorum philosophorum*, 1615). The second edition (1603) of Johannes Sambucus's portrait collection *Icones veterum aliquot ac recentium medicorum philosophorumque* may be mentioned as a further collection containing medical biographies, since it is introduced by a series of biobibliographical sketches of its subjects.

[71] On lives of doctors in such collections, see Siraisi 2007, 109–13.

and mathematicians from all times and places. Moreover, they are also generously inclusive in terms of disciplinary affiliation, listing as many celebrities as possible for the greater glory of their disciplines. Brunfels has no qualms in devoting a lengthy discussion to Aristotle as a physician (pp. 42–50). Biancani includes Sappho among his mathematicians on account of her contributions to music (p. 41). In terms of structure, Brunfels follows neither a chronological nor a thematic order consistently but tends to copy out en bloc what he finds in his sources. Four writers mentioned together in Pliny's *Natural History* 14.120, for instance, appear under the common heading 'On those who published on vines and wine types' ('De his, qui vites celebraverunt et vina', p. 35).[72] Collimitius and Biancani, on the other hand, proceed chronologically.

Regardless of their structure, the proper subject in biographies of this type is the discipline rather than its individual representatives. The individual entries are more numerous (*c*.330 in Brunfels, 250 in Biancani) and much shorter than in authors like Suetonius or Boccaccio. At most, they run to a couple of pages. More often, they are compressed to a few lines. This is what Biancani has to say about the mathematician and astronomer of late antiquity Theon of Alexandria (p. 53):

> THEON Alexandrinus, cuius sunt Graeca Commentaria in Magnam Syntaxim Ptolemaei, praetera De arithmetica, De ortu Caniculae, De Nili ascensu, Commentaria in parvum astrolabium. Ex Suida. Fuit Pappi synchronus.
>
> THEON of Alexandria, by whom there is a Greek Commentary on Ptolemy's Almagest, further On Arithmetic, On the Rise of the Sirius, On the Flooding of the Nile, Commentary on the Small Astrolabe (Suda). He was a contemporary of Pappus.

Not only the reference but also the overall design of this article—minimal biographical information plus bibliography, presented in dry keyword style—reminds one of a lexicon. In fact, not only is the information on Theon taken from the *Suda*, but the entry is simply an abridged translation from this Byzantine lexicon (*Θ* 205). While Biancani's chronological order makes his work appear like an embryonic history of mathematics, an alphabetical one would have made it into a biographical dictionary of the kind that began to appear in the eighteenth century. Indeed, collections of short biographies can be viewed as forerunners of both types of publication.[73]

[72] Moreover, they are called 'medicorum ordinis et clarissimi scriptores', whereas Pliny says no such thing.

[73] On the link between collections of biographies and disciplinary and literary history, see *CNLS* 2, 197–8; Berschin 2009, 227 (on Jerome's *De viris illustribus*). On such collections and lexica, see Richter and Hamacher 2009, 138–9; Werner 2009, 267.

Authors' biographies

The contrast between Biancani's treatment of Theon and the two other variants of scientific biography is strong. Both authors' biographies and autonomous biographies are rarer but longer and richer in substance than serialized lives. At least in some cases, they are also demarcated more strongly from biographies of other intellectuals in that they make a serious attempt to do justice to the biographee's scientific achievements, even going into technical details if necessary. When Gassendi reports on the implications of Brahe's discovery of the 1572 supernova, he gives many exact figures for its angular distance from other stars and discusses its lack of observable parallax in detail (*Tychonis Brahei vita*, pp. 26–7). Most such biographies loosely follow Suetonius in terms of structure, combining a chronological narration with thematic insertions on topics such as the biographee's appearance, character, scientific achievements, or literary works. Both authors' biographies and autonomous biographies have panegyrical intentions: they promote books and praise scientific heroes. In both, the dignity they ascribe to their subjects is reflected in a polished, periodic, and classicizing style.

In the wake of ancient and medieval models, printed editions of important authors frequently featured a biography among their introductory paratexts. This happened particularly often in the case of major works such as Avicenna's *Canon*, whose editions were introduced by three different lives in the sixteenth century,[74] 'opera omnia' editions like that of Pierre Gassendi (to be mentioned shortly),[75] and substantial groups of minor works such as Newton's *Opuscula mathematica, philosophica et philologica* (1744), which begins with a 'De vita Isaaci Newtoni commentariolus'. Regarding the authors whose lives were described, lack of information proved a problem for ancient and medieval ones. In many cases, it prevented the publication of a proper 'vita'. Just a few scraps of biographical information are given, for instance, in George of Trebizond's proem to his commentary on the *Centiloquium* ascribed to Ptolemy (1544, sigs. [A6^{v}–A7^{r}]). In the rare case in which an ancient biography existed, it was, of course, used. Latin versions of Soranus's *Life of Hippocrates* introduced early modern 'opera omnia' editions of the father of medicine.[76] If no satisfying account but a sufficient amount of scattered information existed, an editor could try his hand at a biography himself. Gessner's 'Prolegomena' to a variorum translation of Galen's collected works (1562) features a biography composed on the basis of statements by Galen himself (who was prone to talk about himself). Combined with a bibliography of Galen's works and secondary literature, this life runs to sixty folio pages and is

[74] Siraisi 1987, 162–3. [75] See in general *CNLS* 2, 207.

[76] This is true, for instance, of the editions by Favio Calvo (first complete Latin edition, 1525) and Janus Cornarius (1546). On early modern lives of and biographical interest in Hippocrates in general, see Siraisi 2007, 79–89.

real added value, providing preliminary orientation while also promoting the edition itself by highlighting the excellence of the author edited. Others were less fortunate with their sources than Gessner. This applies to an anonymous biographer of Sacrobosco, whose biography was placed at the end (not at the beginning, as customary) of a number of Parisian editions of the *Sphaera* from 1545 on and was later incorporated in revised form in the letter of dedication of an edition by Élie Vinet (1551, reprinted until 1620). What the biographer was able to find out about the most influential European astronomer of the Middle Ages barely fills one and a half octavo pages.

More information was available for contemporary authors, in particular if the biographer was personally acquainted with the biographee, as was often the case. Accordingly, such 'vitae' are as a rule more expansive. Often, collected works of modern authors were published shortly after their death, in which case the introductory biography constituted a kind of obituary in addition to its informative function. After the demise of Pierre Gassendi (1592–1655), for instance, his friend Samuel Sorbière wrote a 'Praefatio, in qua de vita et moribus Petri Gassendi disseritur' to introduce the deceased's *Opera omnia* (1658). A somewhat earlier case is Gerardus Mercator's monumental *Atlas*, published the year after his death (1595), in whose prefatory section the cartographer's life and works are lovingly detailed by his friend Walter Ghymm.[77] By contrast, the three tomes of Newton's *Opuscula mathematica* appeared seventeen years after his demise and their editor Giovanni Salvemini had not known Newton in person. For his introductory 'De vita Isaaci Newtoni commentariolus', he drew instead on the literature about the scientific hero that had accumulated in the meantime, diligently identifying his sources in twenty-five footnotes. Girolamo Cardano died in 1576, but Gabriel Naudé's extended 'De Cardano iudicium', which first appeared in 1643 as a companion piece to his edition of Cardano's *De propria vita*, found its place at the head of the belated *Opera omnia* only in 1663.

Biographical monographs

When men of science attained higher public visibility from the sixteenth century, this was reflected in book-length biographies devoted to some of them.[78] Such books were a novelty, and they were not written frequently or lightly. A special occasion was usually needed. Often, as already mentioned, it was the death of the biographee that prompted a friend or acquaintance to take up the pen. The way to a self-contained publication could be indirect, many texts being originally conceived as

[77] Text and German translation with introduction and notes: Gesk 1962.

[78] The trend seems hardly to have extended to ancient and medieval figures. Gassendi's *De vita et moribus Epicuri* (1647) is a rare exception, if one is prepared to count Epicurus as a man of science.

something different. Some, for instance Johann von Jessen's biography of Brahe and the *Vita Danielis Bernoulli* written by the biographee's homonymous nephew, originated as funeral speeches. In addition, Bernoulli's biography took the form of a journal article as well as of a book. Read out to the Societas physico-medica Basiliensis on the first anniversary of Bernoulli's death, 17 March 1783, it was published as a book the same year, and republished in the first volume of the *Nova acta Helvetica* after the book had become a rarity. The *Elogium* on the Tyrolean peasant cartographer Peter Anich, composed by his Jesuit teacher and patron Ignaz Weinhart and the Viennese astronomer Maximilian Hell, went in the opposite direction—from journal contribution to booklet. First printed in Vienna as an appendix to the *Ephemerides astronomicae* for the year 1767 by Hell, who based himself on two manuscripts by Weinhart, the text was reissued separately two years later in the Tyrolean capital Innsbruck, where interest in Anich could be expected to be greatest. A biography could even originate from an academic disputation, as in the case of a dissertation about the life of Kepler published in 1770 by Johann Heinrich Rümelin, a candidate for the 'magister philosophiae' in Tübingen. The author clearly took care to present this biography, whose full bibliography and lengthy footnotes take up more than half of its twenty-two pages between them, as a work of solid scholarship.

There remain a few biographies which were conceived as self-contained, full-length books from the outset. Unsurprisingly, all the texts concern personalities who count among the early modern heroes of science and have remained famous until the present day. In reaction to Conrad Gessner's death, his younger friend Josias Simmler issued his *Vita Conradi Gesneri*. Gassendi's *Tychonis Brahei vita*, by contrast, was composed over fifty years after the death of its hero. The text comprises no fewer than six books and 250 pages.[79] Appended to it, one finds a shorter, but still substantial life of Copernicus—who was also the subject of eight more Latin and an even larger number of vernacular biographies from the sixteenth to the eighteenth centuries[80]—and a double biography of Peuerbach and Regiomontanus. Another pertinent example is Mencke's *De vita, moribus, scriptis meritisque Fracastorii* (1731). In both Gassendi and Mencke, a panegyrical intention is combined with scholarly thoroughness, as both apply the increasingly high standards of historiography to their biographical projects. Gassendi inserts numerous verbatim source quotations into the body of his text. Of Mencke's *c.*200 pages, only some twenty are taken up by the actual narrative. In each chapter, this part is followed by a series of notes providing evidence and further reflections which are about ten times as long.

However, men like Gassendi and Mencke were not only scholars but also capable writers who wanted to impress as well as to inform. The artful and

[79] On the long tradition of Brahe biographies starting from Gassendi's, see Hansen 2007.
[80] Collected in Kühne and Kirschner 2004. The Latin biographies are nos. 3, 5, 10–12, 14–15, 19, 21.

spirited way in which early modern biographers could go about their task is illustrated by a crucial passage from Gassendi's life of Brahe: the narration of Brahe's discovery of the 1572 supernova (p. 21):

> Anno itaque insinuato MDLXXII ac die Novembris xi. vesperi, paulo ante cenam, redibat Tycho lento gradu ex seposito illo suo, ut diximus, laboratorio (ad quod inspiciendi ordinandique gratia bis terve dietim divertebatur), cum invitante serenitate, qua spes fiebat observationum peracta cena continuandarum, coepit oculos circumquaque versus caelum convertere. Ecce autem insolens iubar in asterismo Cassiopeiae circa verticem tum exstante ita oculos radianti fulgore magnitudineque perstrinxit, ut ratus aliunde nihil simile in eo asterismo fuisse prius visum plane obstupuerit et vix sibi ipsi attendenti attonitoque <h>abendam fidem censuerit, quousque famulos sciscitatus et rusticos forte curribus transvectos percontatus est, num quid simile cernerent, ac illi uno ore et cernere se ingentem stellam professi sunt et talem fuisse numquam a se illic conspectam sunt contestati.
>
> So in the aforementioned year 1572, on the evening of 11 November, a little before dinner, Tycho was returning with slow step from his laboratory, which was quite remote, as I said, and which he used to visit two or three times a day to inspect it and tidy it up. Because the clear sky, which gave hope that observations could be resumed after dinner, invited him to do so, he began to turn his eyes towards it and to gaze all around it. But look! In the constellation of Cassiopeia, which was then close to the zenith, a strange glow so much impressed his eyes with its shining blaze, that he, convinced never to have seen anything similar in that constellation otherwise, was simply shocked and thought he should hardly trust himself, tense and baffled as he was. In the end, he asked his servants and questioned some peasants accidentally driving by on their chariots, whether they saw something similar to what he did. With one voice, they asserted that they saw an enormous star and had never beheld anything of this kind in that place before.

Gassendi's literary ambitions are unmistakable. Brahe's momentous discovery is told in distinctly lofty style. The whole narrative is packed into two long, syntactically complex periods (which had to be cut up in the translation). Gassendi avoids word repetition and demonstrates the rich vocabulary at his command by the use of synonyms such as 'iubar' and 'fulgore' or 'sciscitatus' and 'percontatus'. Rhetorical figures of sense and sound abound: the shine of the nova is magnified in hyperboles ('glow', 'shining blaze'). Brahe's attentive gazing around is underlined by massive alliteration ('*c*ena *c*ontinuandarum', '*c*oepit o*c*ulos *c*ir*c*umquaque versus *c*aelum *c*onvertere'), his amazement by assonance ('attendenti attonitoque'). For all its self-conscious flourish, the passage does not appear stiff or pompous. The scene comes alive through telling details such as the exact date and time, the position of Cassiopeia and the nova, and Brahe's habits. Brahe in particular is finely characterized: he walks slowly, presumably because he is in a reflective mood, and wastes no thought on dinner but already looks forward to his stargazing after it. His emotional reaction to the apparition is graphically described as well. Ironically, he finally has to seek help from the most unlearned (but presumably also the most unprejudiced): a group of peasants on their way home. These do not figure in Brahe's own account of the incident (*De nova stella*, 1573, sig. A^{r}), where it is only said that the star is seen by 'others' ('aliis') as well, and one suspects that Gassendi introduces them not just to confirm Brahe's observation: in combination with the miraculous star, they evoke the shepherds of the Christmas story. A new age, Gassendi seems to say, begins with Tycho's discovery as well.

Orations

Speeches oral and written

While the importance of print, and writing in general, continued to grow in the course of the early modern period, the role of oral eloquence was still considerable, although strongly focused on epideictic to the detriment of juridical and political oratory. Ambassadors addressed rulers, clerics churchgoers, and friends of those deceased their fellow mourners in often carefully premeditated, artful speeches.[81] Under the influence of ancient rhetorical theory and of the model of Cicero, classicizing oratory replaced late medieval forms of eloquence. A substantial part of public speaking was done in the language of the Romans, especially on prestigious occasions and in front of learned audiences.[82] Some of the thematic fields disclosed in this context were innovations going beyond the traditional subjects of ancient oratory. This is also true of scholarship in general and the natural sciences in particular.

Not everything that could technically be classified as a speech or was titled 'oratio', 'sermo', 'declamatio', or the like was actually an oration in the ordinary sense of the word. Many examples were, conceptually or factually, written texts. The papers read at the sessions of academies and later published, presumably in heavily edited form, in their *Acta* have already been discussed (see Chapter 6, note 125). Moreover, just as many letters were never sent (see Chapter 6, 'Letters'), so not all alleged 'orationes' were actually delivered. In principle, every prose text

[81] Funeral speeches and sermons in particular were exceedingly popular genres (Albrecht 2009, 230–1; Werner 2009, 265–6): 250,000 items are known from Germany alone between 1550 and 1750 (Manger 1993, 279).

[82] *CNLS* 2, 165–79; *OHNL* 119–34; Poel 2017.

could be disguised as an oration, the freshness and immediacy associated with a viva voce speech being employed to enhance its message. Leonhart Fuchs intervened in an argument between Pierre Brissot and Jérémy de Dryvère about the right way of bloodletting with a book styled a 'defence speech' for the former against the latter's allegations (*Apologia*, 1534). Konrad Dasypodius's *Oratio de disciplinis mathematicis* (1579), addressed to King Frederick II of Denmark, was used instead of a conventional letter of dedication to introduce a mathematical lexicon. An even more patent fiction was the stylization of Pierre Gassendi's 'Philosophiae Epicuri syntagma' (1658) as a speech by Epicurus himself. The stratagem allowed Gassendi to provide a full account of Epicurus' dangerous philosophy while keeping his distance.

To turn to orations *sensu stricto*, their manuscripts and prints hold a special place in early modern scientific literature as the most important group of texts originally meant for oral delivery and aural reception. However, not even in these cases should the conceptual difference from normal reading texts be overstated. On the one hand, literature in general was rhetorically informed and often still read aloud. On the other hand, oratory was based on written theory and ancient models transmitted in writing. Most speeches were presumably read out from a manuscript as today, not pronounced extemporaneously or from memory as in antiquity. This is presupposed by an anecdote about the busy Melanchthon who wrote many academic speeches to be delivered by others and allegedly finished them while they were already being pronounced, handing each finished sheet to the speaker from behind a curtain.[83] Most importantly, speeches were received as written texts once published. In this way, they reached not only a limited group mostly composed of men of science but also a broader educated audience. To adjust an oration to this new public, its original text would in most cases have been altered for publication, an aspect sometimes mentioned in the print version itself. Of his speech 'De generatione ex primaevo semine' (1694), originally delivered at a ceremony for the awarding of 'magister' degrees, Rudolf Jakob Camerarius writes that he published it 'having left out ... what seems not to belong to my present purpose' ('omissis ... quae huc non spectare videntur', pp. 48–9).

Although the transmitted texts form a sizeable body,[84] they represent just a small fraction of all the speeches that were actually given. Of this fraction, a good proportion is presumably preserved only in manuscript and therefore largely unknown at present. In order to track down these texts, one would have to scan the archives of universities, academies, learned societies, and local museums.

[83] Eckart 1998, 185.

[84] For medicine, Siraisi 2004, 205–11, lists twenty-seven (groups of) speeches in alphabetical order as well as according to time, place, and occasion. A dozen mathematical orations are listed in Imhausen and Remmert 2006, 148–50.

A case in point is the *Sermo dictus in creatione baccalaureorum* pronounced in 1779 by Ignaz Weinhart, professor of physics and mathematics, at the University of Innsbruck and preserved in the Tiroler Landesmuseum Ferdinandeum.[85] With its many corrections, this text does not suggest itself as a fair copy for publication, which in any case never happened, but rather as a simple lecture manuscript.

For a speech to be printed after delivery, a special reason such as the extraordinary importance of the speaker, occasion, or topic was usually required. Notably, an oration by a famous scientist had a much better chance of finding a publisher than one given by a rank-and-file speaker. It is no coincidence that among the authors of printed orations, one finds many great names such as Tycho Brahe, Herman Boerhaave, or Carl Linnaeus. A special case is Melanchthon, who was not, of course, an important scientist in his own right but had a decisive influence on the establishment of the sciences in the university curricula of Protestant Germany.[86] The four hefty tomes of *Selectae declamationes* ascribed to him and propagating his vision of the Protestant university, which appeared in various editions in Strasbourg from 1541 and were later republished in volumes 11 and 12 of the *Corpus Reformatorum*, contain close to thirty orations on scientific, especially medical topics.[87] However, some of these pieces were presumably not only delivered (as the phrase 'partim aliis recitandas exhibuit' in the full title of the *Selectae declamationes* admits) but also written by members of Melanchthon's entourage. This is suggested by the quite technical content of some orations as well as by their publication history.[88] The 'Oratio de Avicennae vita', for instance, was first published under the name of the German physician Jakob Milich (1550) before it was incorporated into the *Selectae declamationes*. That Melanchthon was suggested as the sole author of a corpus which probably came only in part from his pen underlines how helpful a famous name on the title page must have been for the publisher. Johann von Jessen's funeral oration on Tycho Brahe illustrates that the printed version of a speech could also sell well on account of its subject and occasion. Pronounced at the funeral on 4 November 1601 in front of Brahe's body lying in repose at the Church of Our Lady before Týn in Prague, it was printed immediately afterwards in that city and republished the same year in Hamburg. Later, it was reprinted in 1654 by Gassendi as a companion piece to his *Tychonis Brahei vita*, for which he also used it as a source. Building on Brahe's pre-existing fame, the speech contributed to making him one of the first celebrities of early modern science.[89]

[85] Dipauliana 978/10, pp. 15–23. See *TL* 947–8.

[86] Frank and Rhein 1998.

[87] Of the twenty-seven pieces translated in Kusukawa and Salazar 1999, over one-third deal with the natural sciences. All twenty-two medical orations from the *Corpus Reformatorum* plus three more orations omitted from it are printed in German translation and in the Latin original respectively by Hofheinz 2001, 101–305. Lists of the medical orations with bibliography are given by Eckart 1998, 184–5 and Siraisi 2004, 202 n. 42. For an amply contextualized case study, see Rütten and Rütten 1998.

[88] For a cautious overview of the question of authorship, see Hofheinz 2001, 36–7.

[89] On the tradition of Brahe biographies, see Hansen 2007.

Institutions and occasions

The majority of orations concerned with the natural sciences were given at universities, where ceremonial occasions were manifold and speeches were routinely delivered at many of them from medieval times.[90] How common this practice was is demonstrated by collections of standard orations.[91] However, the specific event cannot be determined for every speech. It is unclear, for instance, on what occasion a 'most splendid gathering' ('concio ornatissima') and 'excellent fathers' ('patres optimi') assembled to listen to Federico Grisogono's 'Oratio in Academia Patavina habita' (1507)[92] or when Milich praised the Arab scientist and philosopher Avicenna as a model physician in a talk dedicated to the his biography ('Oratio de Avicennae vita', 1550).

When the occasion is clear, it is sometimes directly linked to the speaker. In many places, it was customary to deliver an inaugural lecture upon taking up to a professorship. When Luca Gaurico was appointed professor of mathematics at the Studio di Ferrara in 1507, he gave a speech in praise of astronomy ('Oratio de laudibus astrologiae'), which he submitted to print twenty-four years later. Pietro Catena performed his *Oratio pro idea methodi* in favour of the notion of mathematical method when taking up the chair of mathematics at the same university in 1548 (published in 1563). From Herman Boerhaave, one has two speeches on chemistry, one given when he took up a professorship in the discipline in 1718, the other when he retired as a professor of botany and chemistry in 1729.[93] The 'Oratio praefatoria' given by Isaac Barrow, the first holder of the Lucasian chair at Cambridge in 1664, sounds like an inaugural speech as well, although it later was prefixed to a published lecture course (*Lectiones*, 1683).

Indeed, courses were also often opened by a thematically related speech. An early testimony to this practice comes from Johannes Regiomontanus. When teaching a course on Alfraganus's summary of Ptolemaic astronomy in 1464 in Padua, he introduced it with the 'Oratio habita Patavii in praelectione Alfragani' (printed 1537).[94] Tycho Brahe's *De disciplinis mathematicis oratio* (1574, printed 1610) inaugurated a mathematics course at the Royal Academy of Copenhagen.

Among the most common academic ceremonies was the awarding of degrees. The solemn character of the occasion was often highlighted by a formal speech. An oration by Camerarius on oak apples ('De quercuum gallis', 1694) and

[90] In general, see Schirrmeister 2018; for medical orations, see Siraisi 2004, 193–4 and 2007, 113–14. Buchwald and Herrle 1921 provide a collection of speeches from fifteenth-century Leipzig at the awarding of academic degrees, from bachelor of arts to doctor of philosophy. A number of orations from sixteenth- and seventeenth-century Montpellier can be found in Germain 1886, 20–3, 24–5, 26–9, 182–90 (bestowal of degrees), and 69–70 (thanks for election to a chair).

[91] Bertalot 1975. [92] For the quotations, see Girardi-Karšulin et al. 2007, 8 and 22.

[93] Dutch translation: Cohen and Renkema 1918, 80–100 and 101–15.

[94] Facsimile edition: Schmeidler 1972. See also Rose 1975, 95–8; Byrne 2006; Hasse 2016, 60–1, 331–3.

Weinhart's *Sermo dictus in creatione baccalaureorum* graced the awarding of the baccalaureate to a number of students from the arts faculties in Tübingen and Innsbruck respectively. The former author also gave his oration on procreation (1694) in the context of the awarding of 'magister' degrees, as mentioned above. A speech on Hippocrates by Melanchthon (or one of his medical friends at Wittenberg) was given at a conferral of several doctorates.[95] Linnaeus delivered his *Oratio de telluris habitabilis incremento* when Johannes Westmann graduated as a doctor of medicine at Uppsala in 1743.

Celebrations concerning a university as a whole were also enhanced by speeches. When the University of Turku celebrated its fiftieth anniversary in 1690, Professor Daniel Achrelius praised the university and scholarship in general in his *Scientiarum magnes*. Boerhaave's *Sermo academicus de comparando certo* (1715) was delivered when the speaker vacated the rectorate of the University of Leiden which he had held during the previous academic year.[96]

In addition to the universities, there were the scientific academies and societies which established themselves from the seventeenth century. They held regular meetings but also congregated on special occasions. Both types of meeting could give rise to formal speeches. In 1764, the young Horace-Bénédict de Saussure gave an (untitled) oration on glaciers to the Académie de Genève which was not printed in his lifetime.[97] At the first anniversary of Daniel Bernoulli's (1700–82) death, the members of the Societas physico-medica Basiliensis met to listen to a *Vita Danielis Bernoulli* recited by the homonymous nephew of the deceased (published 1783/1787).

This last example leads us to funeral speeches in which men of science were eulogized in front of large, at least partly non-academic audiences.[98] Von Jessen's aforementioned oration for Brahe is one such case. Further examples include Georg Hizler's *Oratio de vita et morte Leonharti Fuchsii* (1566) and Aelius Everardus Vorstius's 'Oratio funebris in obitum Caroli Clusii' (1609, published 1611).

Even when the occasion of an oration is known in general terms, the information on the specific place, time, audience, style of delivery, and reception of a given speech is usually minimal. Sometimes, however, paratexts to the printed version prove helpful. With regard to his oration on oak apples, Camerarius explains in the long letter to which it is appended to it that he tried to make the subject easy for the young 'baccalaurei' who made up a good part of his audience ('captui auditorum accomodavi', p. 19). In other cases, the speaker refers to some circumstances in his exordium. Particularly telling in this respect is the opening

[95] On the oration, see Rütten and Rütten 1998. They seem to think of one candidate only, namely Jakob Milich (see pp. 29, 48), but several new 'doctores' are addressed at the end of the speech.
[96] Cohen and Renkema 1918, 55–79 (Dutch translation); Lindeboom 1968, 100–2.
[97] Edited and translated by Carozzi and Newman 1995, 41–54.
[98] Siraisi 2004, 197, 200; 2007, 118–19.

(fol. 5^{r-v}) of the inaugural oration *De podagrae laudibus* given in November 1552 in Padua by the physician Johannes Carnarius and printed the following year. There, it is specified that not only students of medicine but also students from the arts and law faculties ('bonarum litterarum aut etiam iuris prudentiae studiosi') used to attend medical inaugural speeches. If a speaker avoided the customary praise of medicine and turned straightaway to the explanation of some medical classic, part of the audience would not understand anything. They would show their exasperation by interrupting the speaker and hissing at him ('Quo fit, ut saepe in medio orationis cursu ab imperitis earum rerum hominibus... interpellerentur exsibilenturque').[99] Occasionally, additional data may be supplied by a picture. Figure 17 shows Herman Boerhaave performing his leave-taking speech as rector in what looks like the main auditorium of the University of Leiden. Standing on a raised lectern and gesturing rhetorically, he delivers his oration to an audience of several hundred, composed of professors, students, and perhaps also non-academic citizens. Of course, this frontispiece intends to visualize Boerhaave's fame and the power of his eloquence, but without some basis in fact, it would have appeared ridiculous and unhelpful as an advertisement of the book.[100]

Content and structures

The topics discussed in scientific oratory show great diversity. In terms of disciplines, those taught in universities from the Middle Ages, medicine, astronomy/astrology, and mathematics, take the lion's share.[101] Others like botany, chymistry, geography, or modern physics, which became part of the curriculum later and more sporadically, are addressed less often. Approaches to these subjects cover the entire spectrum from the panegyrical to the factual. Praise of the disciplines, their long history, usefulness, loftiness, and capacity to grasp God's creation are the most pervasive ingredient. Mathematics is also extolled as a propaedeutics to philosophy.[102] In many cases, praise of science is complemented by praise of individual scientists but also of other helpful persons such as patrons or donors. Occasionally, a speaker programmatically takes sides with regard to a topical question. There are also orators who see the delivery of a speech as an occasion to discuss a specific scientific question in which they are personally interested.

[99] Cf. Siraisi 2004, 201.

[100] A similar illustration adorns the first page of Luca Gaurico's 'Oratio de laudibus astrologiae' in the first tome of his *Opera omnia* (1575). There, the assembly is smaller (which may be due to the smaller format of the image) and Gaurico is not standing but sitting at his lectern.

[101] On medical orations, see Siraisi 2004 and 2007, 113–19. Case study of a mathematical oration: Imhausen and Remmert 2006.

[102] Imhausen and Remmert 2006, 75.

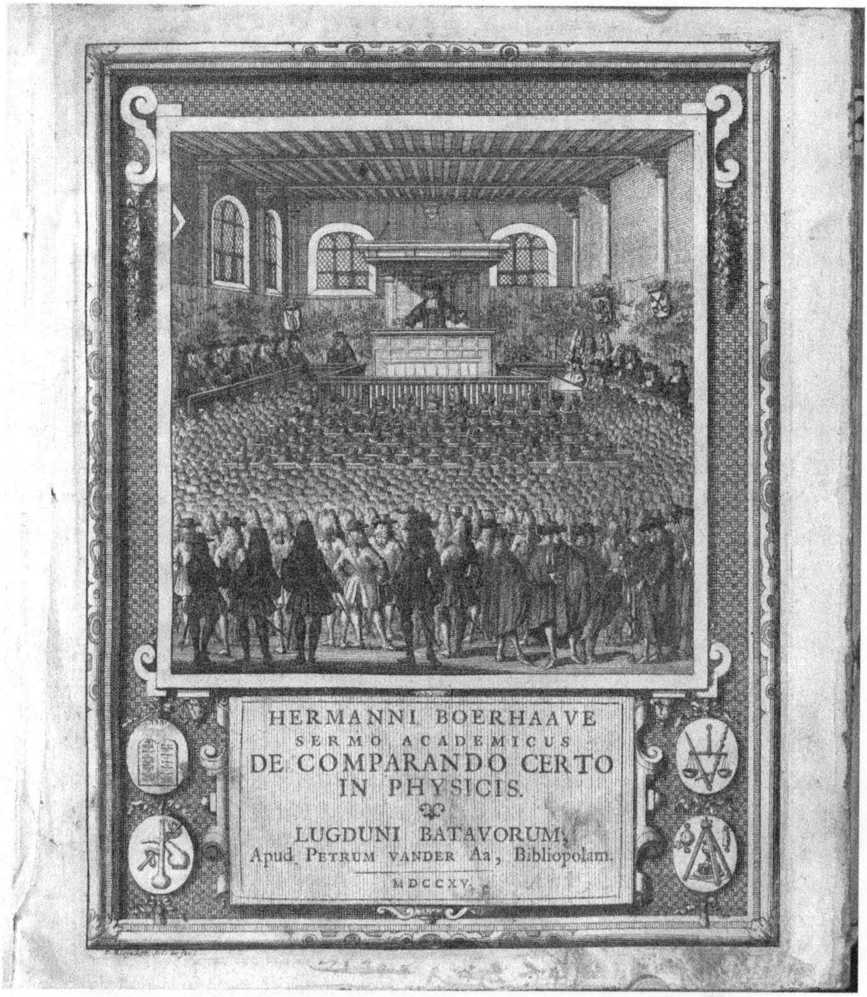

Figure 17 Herman Boerhaave delivering his speech on the lack of certainty in the natural sciences (frontispiece of *Sermo academicus de comparando certo in physicis*, 1715).

The following examples may give an impression of how these aspects can be combined within speeches.[103]

Regiomontanus's Paduan oration offers a historical sketch of the mathematical disciplines supplemented with lavish praise of them, especially of astronomy. Melanchthon and Georg Joachim Rheticus's 'Oratio de astronomia et geographia' (1544) praises astronomy on account of the knowledge of God which it imparts. In this respect, it is typical of many 'Melanchthon orations' which extol the study

[103] For additional medical examples, see Siraisi 2004, 196 n. 16.

of nature as a way to a better understanding of the Creation and the Creator Himself. Another example among these orations, Milich's 'Oratio de Avicennae vita', sets up Avicenna as an exemplary man of science, whose model should be followed by the audience. Catena's *Oratio pro idea methodi* (1563) takes sides in the so-called 'controversia de certitudine mathematicarum' ('dispute over the certainty of the mathematical disciplines')[104] by stating that mathematics is a real science in the Aristotelian sense of the word, since it is capable of causal explanation; at the same time, the speaker also extols the discipline in general. Brahe's *De disciplinis mathematicis oratio* likewise indulges in the praise of mathematics, which is coupled with a vindication of astronomy. Barrow's 'Oratio praefatoria' first celebrates the merits of Henry Lucas, the founder of Barrow's chair, and justifies the speaker's exchange of the chair of Greek for the Lucasian chair; after that, it concludes with an exhortation to the students to apply themselves to the study of mathematics and with a hymnal laudation of this discipline. Achrelius extols the University of Turku and all sciences in his *Scientiarum magnes*, devoting part of his speech to the natural sciences and using the metaphor of magnetism to describe the sway held by scholarship over the human mind. Boerhaave's *Sermo academicus de comparando certo* refrains from panegyrics in favour of programme. In an anti-Cartesian vein, Boerhaave insists on the impossibility of any real insight into the principles of nature but concludes with a short coda to the effect that scientific progress is, nonetheless, possible. Saussure's oration on glaciers is an important starting point for scientific glaciology, explaining for the first time how glacier ice is formed from snow over the years and how glaciers flow downhill, but it also celebrates the beauty of the phenomenon. Weinhart's *Sermo dictus in creatione baccalaureorum* inspires an audience of young students by extolling the pleasures of experimental physics.

Depending on the occasion and the speaker, content such as this could be treated quite succinctly or more elaborately. About one half of the extant texts comprise between ten and twenty octavo or duodecimo pages, while the longest speeches run to a multiple of this figure. Not coincidentally, the most extensive example in my corpus, Achrelius's *Scientiarum magnes*, with its seventy octavo pages, celebrates a highlight in the history of the speaker's university. If one can take the printed versions as a guide, the speaking time may have fallen between ten minutes for the shortest and well over an hour for the longest items.

In terms of the structure of the orations, there was great freedom, as there were no ancient prescriptions for or models of scientific oratory to be followed. Erasmus's *Encomium artis medicae* (1518), one of many speeches in praise of medicine,[105] adopts the model of the classical court speech (exordium, narration,

[104] On this debate, see Schöttler 2012 and cf. *CHS* 3, 154–7, with earlier literature.

[105] English translation with introduction and notes: McGregor 1989, 31–50, 431–7. On the subgenre of 'laus medicinae' and its prehistory in a debate about the pre-eminence of medicine or law, see Rütten and Rütten 1998, 30, with earlier literature.

proof, refutation of counterarguments, peroration),[106] but it seems to be exceptional in doing so. Orations on important scholars may partly follow the rules of the personal encomium or the funeral speech laid down by ancient authors such as Menander Rhetor or orient themselves towards biographical models like Suetonius. Von Jessen's funeral oration on Brahe combines these templates in proceeding from initial despair to consolation at the end while describing and praising Brahe's life, achievements, and personality in its main part. For the rest, only an exordium and a peroration are apparently seen as obligatory.[107] In accordance with ancient theory, in the former the speaker may address his listeners in flattering terms, refer to the occasion of the speech, profess his incapacity and timidity, ask for the audience's benevolence, and summarize his subsequent remarks.[108] Many of these topics are packed into the opening sentence of Regiomontanus's Paduan speech:[109]

> Nisi primitias lectionum facturo mihi mansuetudo vestra aspiraret orationisque tremebundae filum dirigeret, praestantissimi viri, siluisse prorsus mallem quam pulpitum hoc philosophicum audentius conscendisse, praesertim cum novitate coepti facinoris, tanto tamque celebri clarissimorum hominum conventu, diuturna demum a scholasticis exercitiis abstinentia facile deterreri potuerim.
>
> If your kindness were not supporting me, as I am embarking on the start of my lectures, and guiding the trembling thread of my speech, most eminent men, I would really rather prefer to have remained silent than to have overboldly climbed this pedestal of the philosophers. In particular, I could easily be deterred by the novelty of what I have set out to do, by this very numerous and renowned assembly of most famous men, and finally by my long-standing abstention from school exercises [such as this very speech].

Alternatively, the exordium may serve to embed the subject in a larger context. Rheticus opens his praise of astronomy and geography with a lament on the decline of Germany, which could only be remedied by the arts and sciences.

The peroration often leads back to the situation at hand, as in Barrow's 'Oratio praefatoria', where this move is given a nicely self-ironic twist (p. xxxvii): the tedium caused by his long speech, Barrow says, is a good preparation for what the students will have to endure during his lessons. But since mathematicians are in general bad orators (an idea of unclear origin), his rhetorical incapacity is at least to his credit as a mathematician.

[106] *HlRh*, § 262.

[107] Even these parts may be skipped on occasion, as in Linnaeus's *Oratio de telluris habitabilis incremento*, which begins by plunging *in medias res* and only has a brief exhortation to the audience instead of a formal 'peroratio' (§§ 1 and 100).

[108] Cf. *HlRh*, §§ 266–79.

[109] Already quoted as an example of Regiomontanus's humanist style by Byrne 2006, 45.

As far as the main part is concerned, most speakers freely adapt the structure of their texts to the requirements of the subject and occasion, including sometimes idiosyncratic choices. Barrow's aforementioned speech falls into three little related parts, as already seen. Camerarius presents his 'De generatione ex primaevo semine' as a commentary on some passages from Aristotle's (and pseudo-Aristotle's) biological writings. Linnaeus's *Oratio de telluris habitabilis incremento* is an extended auto-commentary on a single aphorism from his *Fundamenta botanica* (1735), and he overlays it with a decidedly written structure, dividing it into one hundred paragraphs.

Style and the audience

Like early modern oratory in general, scientific eloquence stylistically conformed to the classicizing taste of humanism. Humanistically informed speeches are evident as early as the first decades of the fifteenth century from the medical faculty at Padua.[110] A mid-fifteenth century oration on astrology by Gregory Tifernas, the original date and place of delivery of which are unknown, combines medieval characteristics such as an opposition of pros and cons and syllogistic argumentation reminiscent of scholastic disputations with a generally classicizing style, examples from and allegorical interpretations of the Bible and ancient mythology, and quotations from Roman poets.[111] By the sixteenth century, the new style had become standard everywhere. Therefore, most speeches displayed a fair measure of Ciceronian periodization, while technical vocabulary was used with caution. In the case of more general and non-technical topics, the freedom for rhetorical stylization was greatest. However, Camerarius, for one, was able to define even an oak apple in a carefully phrased Ciceronian period ('De quercuum gallis', p. 90):

> Gallae sunt morbosae excrescentiae vi depositi ovi a turbata plantarum compage et vitiato humorum motu excitatae, quibus inclusa ova et animalcula velut in utero foventur et augentur, donec manifestatis firmatisque propriis partibus quantum satis exoriantur novam exoptantia auram.
>
> Oak apples are morbid outgrowths caused by the force of an egg laid [at the specified place], by which the plant tissue is disturbed and the movement of humours impaired [—outgrowths,] in which the eggs and small animals included therein are protected and grow, until, their proper parts sufficiently formed and strengthened, they emerge, longing for the fresh air.

[110] Bertalot 1975, 227. On the early transformation of medieval into humanist oratory in Italian medical faculties in general, see Siraisi 2007, 114.

[111] Edition: Müllner 1970, 173–82.

To listeners and readers alike, this stylistic choice imparted a message that was pivotal to the publicizing function of orations: there was no categorial difference between natural science and other, more traditional kinds of learning. Science was nothing radically new and potentially threatening. The nascent scientific community simply constituted another province of the 'respublica litteraria'.

Beyond this basic classicism, two rhetorical strategies strike the eye as recurring with particular frequency: first, the emphatic attempt to establish close contact with the audience[112] in order to create solidarity between them and the speaker and to boost a common group identity; secondly, the creation of pathos in order to involve the audience emotionally in science. We have already seen an example of the first strategy, which is particularly in evidence at the beginning of a speech, from Regiomontanus's Padua oration. For the second, one may turn to the following paragraph from Saussure's oration on glaciers, in which he describes the collapse of a glacier serac as a true doomsday scenario (Carozzi and Newman, 1995, 48):

> Ipse ego pluries vidi, nec etiam interritus memini, immensa glacierum fragmenta sacras hasce aedes mole superantia e culmine rupium uno temporis puncto ruentia cum fragore fulmineo in alia fragmina disici, mox torrentis effusi more per praecipitia loca volutantia rupes obvias secum abripere tandemque caliginoso oboriente fumo ex ipsis confractis saxis glacieque confestim evanescere, ut in densa illa caligine et terribili strepitu fracti et illabentis orbis ruinas pavidus forsan reformides.
>
> Several times, I have seen with my own eyes—I am still scared at the memory—giant ice boulders, larger than this church [the Cathédrale Saint-Pierre at Geneva, where the speech was given], fall in a moment's time from the summit of a cliff. They burst into fragments with a crash as from a lightning blast, then plunged over steep terrain like a mountain torrent, carrying along rock fragments in their way, and finally suddenly vanished from sight in dark smoke rising from these broken rocks and the ice. You may dread the collapse of the broken, crumbling world in such thick darkness and terrible uproar!

At first hearing, this passage sounds above all frightening, and Saussure's listeners may even have looked up for a moment to make sure that the church itself was not crumbling on top of them. At the same time, however, the elucidation of such staggering powers and phenomena is presented as an exciting challenge, while the allusion to the Last Judgement calls to mind the theological implications of natural science—an aspect to which the audience will have been receptive in an age of physico-theology.

[112] See *HlRh*, §§ 758–78, where the techniques are classified as 'Figuren der Publikumszugewandtheit'.

By way of conclusion, I would like to present an excerpt which combines an emphatic address to the audience with great pathos and, moreover, exemplifies how the same words could affect listeners and readers in different ways. It stems from a peroration, the part of the speech in which both strategies just mentioned were traditionally combined for a strong finish. Grisogono concludes his *Oratio in Academia Patavina habita* in praise of the quadrivium, especially astronomy, with a conspicuous and highly emotional figure of speech, namely a 'prosopopoeia': the mathematical disciplines are personified and introduced as speaking.[113] In an address which is itself filled with strong figures such as exclamations, rhetorical questions, and litotes, they berate humankind for neglecting them and exhort them to study them (pp. 22–4 Girardi-Karšulin, Perić, and Čepulić).

> Quibus si loquendi facultas daretur, nonne mortales omnes lacrimosa voce sic compellare<n>t: 'O crudeles homines, vanis deliciis dediti a propria perfectione vestra revocantibus! In sola enim cognitione intellectiva Dei imagines estis, scire tamen et intelligere non quaeritis non solum secreta Dei et naturae, verum etiam nec manifesta in natura. Quid enim manifestius est quam solem atque lunam singula luce horoscopalem partem modo occupare, modo in Oceano se abscondere? Et tamen ortus et occasus est vobis ignotus?...Quare, o vos omnes, elevate animas vestras. Quaerite perfectionem vestram. Sequimini divinas mathematicas. Quae solae eminenter continent facultatem, sanitatem, animam et totam perfectionem humanam humanitus acquisitam.'
>
> Would they be given the ability to speak, would they not appeal to all mortals with tearful voice as follows: 'Oh cruel men, given to empty pleasures which divert you from your own perfection! For only in your intellectual understanding are you God's images; yet you do not want to know and understand not only God's and nature's secrets but even the things which are obvious in nature. For what is more manifest than that on every single day, the sun and the moon now occupy their rising place, now conceal themselves in the sea? And still their rise and setting are unknown to you?...Therefore, oh all of you, raise your souls! Search for your perfection! Follow the divine disciplines of mathematics! For they alone contain within themselves the practical dexterousness, the health, the soul, as well as the complete perfection of humankind, as far as it is attainable for humans.'

The introduction of superhuman figures, who are, moreover, weeping and desperate about the present state of affairs, not only creates an emotional scene but also allows Grisogono to formulate his plea for mathematics in stronger terms

[113] On the affective character of the 'peroratio' and on the 'prosopopoeia', see *HlRh*, §§ 436–9 and 826–9 respectively.

than he could do if he were speaking in his own person. Grisogono's original audience, consisting mainly of Paduan professors and students, presumably felt encouraged to continue in their efforts rather than reproached. The mathematical laymen among the readers, by contrast, will have felt strongly urged to pay more attention to mathematics, or at least to venerate and help those who did. Passages like these demonstrate that academic orations could be more than boring set exercises. They could deploy great emotional power and may well have been instrumental in enthusing susceptible minds for the study of nature in one way or another.

Conclusion

Such results as this study may have yielded do not lend themselves well to summarization. Some general characteristics of its subject have already been sketched in Chapter 3 above: in stark contrast to its modern counterpart, marked by a clear demarcation from other kinds of writing, a restricted range of genres, and a uniform, impersonal, self-effacing style, early modern scientific literature in Latin seamlessly merged into the literature of other fields, featured a rich array of genres, openly displayed its communicative function, and used many different styles. Instead of elaborating on these findings, I will, rather, briefly look ahead in two directions. Distilling and systematizing some remarks made at various places in my account, I will first note some tendencies in early modern texts about nature which paved the way for the scientific writing of the present. After that, I will point out some desiderata and possibilities for future research.

Towards modern scientific literature

To a considerable extent, Latin scientific writing retained its literary character until the end of the early modern era. Even within this framework, however, some developments took place which prepared the rise of modern scientific prose.

On the level of the work as a whole, a tendency towards structural standardization is noteworthy. Standardized structures existed in scientific writing since antiquity, as exemplified by the presentation of theorems and proofs in Euclid's *Elements* or of plants, animals, and stones in Dioscorides' *Materia medica*. In early modern times, models such as these were freely adopted and adapted, the former in a broad variety of works with a mathematical slant, the latter particularly in herbals and 'historiae'. In addition, other genres featuring a standardized structure came to prominence, for instance pharmacopoeias, dictionaries, bibliographies, and table works. All these genres are discrete texts serializing small units which exhibit the same structure, are formally unconnected, and do as a rule not build upon one another, at least not explicitly.[1] In these respects, such forms differ from the modern journal article, in which the four parts of the IMRaD scheme—introduction, methods, results, and discussion—are structurally

[1] The propositions of Euclid's *Elements* do so implicitly (Asper 2007, 114–16). This aspect is sometimes made explicit by references in Euclid's early modern successors.

Latin Scientific Literature, 1450–1850. Martin Korenjak, Oxford University Press. © Martin Korenjak 2023.
DOI: 10.1093/oso/9780198866053.003.0011

and functionally distinct and based upon one another. However, the early modern forms anticipate the article on a more basic level, that is, in their very use of a ready-made structure which takes the burden of organizing the text off the author's shoulders. In all such texts, structure functions as a vessel into which any reasonably suitable content can be poured.

Regarding style, authors' frequent denials of having any stylistic ambitions simply repeated an ancient commonplace. Independently of such talk, however, one indeed notices some trends heralding the quiet demise of artistic prose. At least one of these can be linked to the preceding point: structural standardization often came in tandem with the use of standardized formulas, something at odds with the lexical and constructional variation that was an indispensable ingredient of good style according to humanistic ideas (as shown by the enormous success of Erasmus's *De duplici copia rerum ac verborum*, 1512). A further factor limiting variation was the development of technical terminology, which implied the establishment of one-to-one relations between things and words and the elimination of synonyms. Additionally, a combination of expressive economy, time-saving, and financial constraints could lead to the cluttering of texts with long strings of abbreviated citations, other abbreviations, figures, mathematical formulas, and astronomical, chymical, and pharmaceutical symbols. This practice can even be observed in prestigious folio volumes by authors who were otherwise capable stylists. In adopting it, they quietly recognized that their chosen kind of writing was governed by considerations categorially different from those of classical rhetoric and poetics.

Both trends sketched so far were accompanied and, in the second case, amplified by a partial replacement of syntax—whose artful complexity, modelled on Cicero and other classical authors, was seen as another hallmark of high-quality writing—by visual features. Standardized structures tended to be expressed by standardized use of layout, typography, and illustrations. One may again think of mathematical works and 'historiae', where single argumentative or descriptive steps were set apart by line breaks and subtitles and each proof or species was illustrated by a diagram or image. Geographical and anatomical atlases, in which the non-textual element became even more dominant, constitute further examples of this tendency. In all such cases, syntax tended to be eroded: steps of construction were described elliptically, plants in keywords, single place names were put on maps, and anatomical structures were named in long keys to illustrations. The function of syntax, the grammatical expression of relations between different entities, was taken over by the spatial presentation of words and non-text on the page. The same is true of devices which were not part of standardized structures but were occasionally inserted into works dominated by running text, for instance Ramist tables.

Sophisticated syntax also came under pressure from another direction, namely from the aforementioned growth of technical terminology. Technical terms were massively present in non-syntactic modules. In running text parts, they not only violated the rules of artistic prose but also impacted sentence construction. Being predominantly nouns and adjectives, they tended to come with colourless verbs which basically functioned as placeholders, initiating a development which ultimately led to the nominal style and syntactic reduction of modern scientific texts. Together with this, the massive and sometimes—particularly in natural history—systematic creation of new words generated new disciplinary languages which were marked by precise and unambiguous referentiality and had something of the artificiality of planned languages. Texts composed in such a register were transparent only to an in-group, opaque to everyone outside, and invested scientific writing with the hermetic aura that surrounds it today.

A last respect in which early modern scientific writing foreshadowed its modern counterpart regards intertextuality. A variety of this ubiquitous phenomenon developed that was characterized by explicit, content-related, precise, elliptic, and typographically marked references. Most of these features were not completely new in themselves, and all were shared with learned writing in other fields. Explicitness and content-relatedness were characteristics of scholarly as opposed to literary intertextuality (which usually left the recognition of an allusion to the reader and in which form was at least as important as content) in general. Precision going beyond the level of author, work, and book was already achieved in medieval theology and law but became more common in the age of print along with the routine numbering of chapters, sections, folios, pages, page sections, and columns. Numbering the parts of a book and thereby rendering them easily traceable also made it acceptable just to indicate a particular passage and to dispense with verbatim quotation in the interest of expressive and spatial economy. Typographical separation of references from the main text as marginalia and footnotes resulted in greater clarity and expressed a textual hierarchy that fitted the ethos of the new science: the author's own argument took centre stage, references were only there to support it. The reliable and intuitive referencing system resulting from the interplay of all these factors helped to resolve the tension between scientific empiricism and the fact that scientists actually learned most of what they knew from books: written information that could be reliably traced back to its empirical origin could count as tantamount to personal observation.

Taken together, the aforementioned characteristics simplified communication about the natural world and made it more efficient. However, they only did so for those authors and readers who were familiar with the rules of the game. In practice, this implied that the identity of the reader and his relationship to the author began to change. Whereas readers of scientific works originally were a heterogeneous group, including people of different levels of expertise, from leading figures in the field to interested laymen, the readership became more homogeneous and expert in the measure in which texts became more technical.

Ultimately, the two groups of authors and readers merged into one group of scientific author-readers. This group, closed in on itself, was what we call the scientific community today.

What next?

The present book is not the last word on a field but in the best case opens it up to further discussion. Some ways in which further research could confirm, refute, elaborate on, or build on its conclusions may be worth sketching out.

A broader basis of easily accessible texts extending beyond recognized magna opera and the works of famous authors is needed. Texts of interest for their literary form, content, or both should be published on paper or online. Editions should include, besides the original text,[2] an introduction, a translation, and a commentary or notes.

All genres discussed in the individual chapters of this book merit in-depth treatment in article or monograph format. The same applies a fortiori to those which have no sections of their own but are discussed only incidentally or not at all.[3] This is true, for instance, of astrological and medical broadsheet calendars[4] and almanacs in book format,[5] geniture collections,[6] medical 'consilia',[7] eye-witness narratives,[8] aphorisms, 'variae lectiones',[9] catalogues of collections (from cabinets of curiosities to botanical gardens), pharmacopoeias, 'quaestiones', polemical literature, science satires, and the numberless and largely neglected derivative genres such as paraphrases, summaries, and anthologies.[10]

Although more work has been done on early modern vernacular than on Latin scientific literature, no true overview of the former exists either.[11] As vernacular writing about nature developed under markedly different conditions from country to country, such an overview should probably be elaborated piecemeal, at least as a first stage. It should help to answer a number of questions which have to remain open at the moment: how far did vernacular literature follow in the footsteps of Latin, and how far did it develop its own forms? In what way were these

[2] Digital help with transcription is now provided by the 'NOSCEMUS General Model 6' of the software Transkribus (https://readcoop.eu/model/print-latin-texts-15-th-19th-htr/, accessed 30 June 2023).

[3] On the importance of 'minor' genres in general, see Frasca-Spada and Jardine 2000, 4–5. How fruitful the analysis of such genres can be is shown in exemplary fashion by the discussion of the early modern 'observatio' in Pomata 2010 and 2011a.

[4] Heitz and Haebler 1905.

[5] *EEMHO*, s.v. 'Almanack'; for Italy, see Casali 2003, esp. 35–60.

[6] Grafton 2000.

[7] On their late medieval history, see Agrimi and Crisciani 1994.

[8] Siraisi 2007, 69–72.

[9] Blair 2006 (on the genre in general); Siraisi 2007, 43, 67 (on its use in medicine).

[10] Maclean 2012, 82–91; for Renaissance Aristotelianism in particular, see Schmitt 1983, esp. 44–5, 53–4, 59–62, and Lohr 1988, esp. pp. xiv–xv.

[11] Olschki 1919–27 focuses on Italy, up to the seventeenth century, and on big names.

forms conditioned by the fact that vernacular texts targeted other groups of readers than Latin ones? How closely were the Latin and the vernacular sides of the market linked through translations and adaptations? How did vernacular literature come to outgrow and marginalize its Latin predecessor? And why? Ideally, an overall picture of the interplay of Latin and the vernaculars in early modern science would ultimately emerge.

Just as the early modern study of nature was linked to other fields of learning much more closely than its modern counterpart, so the respective literature constituted only one segment of learned writing in general. It overlapped with other segments of this field in terms of content and shared many formal features with them. Overviews of the literary forms in which subjects like theology, law, philosophy, history, antiquarianism, and philology were presented to readers could highlight differences, establish common ground, and pave the way for a history of early modern learned literature in general.

In recent decades, the history of science has moved away from the analysis of great books and the ideas expounded in them in themselves and has striven to incorporate the material, sociological, economic, political, and religious aspects of its subject instead. The present book, in which texts once more take centre stage, while wider cultural contexts are sketched only in part and lightly, may look somewhat anachronistic in this respect. However, it is not my intention to turn back the clock. I simply suggest that a higher proportion of texts (all of them, not just the famous ones) should again be added to the mix of important other aspects. Texts, through which we know most about the propositional content of past science, and scientific culture in the widest sense, without which texts and their content cannot be fully understood, should be studied in tandem in order to arrive at as comprehensive a picture as possible of early modern science.

Bibliography

Abbreviations

BENLW: Philip Ford et al. (eds), *Brill's Encyclopaedia of the Neo-Latin World*, 2 vols (Leiden and Boston, MA, 2014).

CDSB: Charles C. Gillispie et al. (eds), *Complete Dictionary of Scientific Biography*, 27 vols (Detroit, MI, 2008).

CHS: *The Cambridge History of Science*, 8 vols (Cambridge, 2003–20).

CNLS: Joseph IJsewijn and Dirk Sacré, *Companion to Neo-Latin Studies*, 2 vols (Leuven, 1990–8).

CTC: *Catalogus Translationum et Commentariorum* (Washington DC, 1960–).

DNP: *Der Neue Pauly: Enzyklopädie der Antike* (Stuttgart and Weimar, 1996–).

EEMHO: *Encyclopedia of Early Modern History Online* (https://referenceworks.brillonline.com/browse/encyclopedia-of-early-modern-history-online/alpha/c, accessed 16 February 2023).

EMLO: Early Modern Letters Online (http://emlo.bodleian.ox.ac.uk/home, accessed 16 February 2023).

HlRh: Heinrich Lausberg, *Handbuch der literarischen Rhetorik: Eine Grundlegung der Literaturwissenschaft*, 2 vols, 3rd edn (Stuttgart, 1990).

HWRh: Gerd Ueding (ed.), *Historisches Wörterbuch der Rhetorik*, 11 vols (Tübingen, 1992–2011).

KGW: Johannes Kepler, *Gesammelte Werke*, 22 vols (Munich, 1937–2017).

LEOO: *Leonhardi Euleri Opera Omnia*, 80 vols (Leipzig, 1911–2022).

LMA: *Lexikon des Mittelalters*, 10 vols (Munich and Zurich, 1980–99).

OHNL: Sarah Knight and Stefan Tilg (eds), *The Oxford Handbook of Neo-Latin* (Oxford, 2015).

PAL: Ptolemaeus Arabus et Latinus (https://ptolemaeus.badw.de/start, accessed 16 February 2023).

TL: Martin Korenjak, Florian Schaffenrath, Lav Šubarić, and Karlheinz Töchterle (eds), *Tyrolis Latina: Geschichte der lateinischen Literatur in Tirol*, 2 vols (Innsbruck, 2012).

USTC: Universal Short Title Catalogue (https://www.ustc.ac.uk/, accessed 16 February 2023).

Sources

This part of the bibliography comprises only those early modern works and editions which I have seen myself and which are individually named and discussed in the text. Ancient and medieval works are quoted from the leading modern editions, except where specific early modern editions are mentioned in the text. Early modern works not printed in their time or not seen by me in their early modern editions are likewise cited after their modern editions (which are listed under 'secondary literature'). Journals are not listed here (for some important ones, see Table 2 in Chapter 6), but journal articles and reviews are. Titles are usually given in a less abbreviated form than in the text. I have tried to apply a consistent and intuitive citation style conforming more or less to that used for the secondary literature, but the variety of early modern title forms and space constraints often put a spoke in my wheels.

A.B., ['Epitaphium in Newtonum'], *Gentleman's Magazine*, 11 (1741), p. 601.

Achrelius, Daniel, *Scientiarum magnes recitatus publice anno 1690, die 25 Novembris, cum regis optimi natalem et pari ardore primum suum celebraret iubilaeum Academia Aboensis* ([Turku], 1690).

Adam, Melchior, *Vitae Germanorum philosophorum, qui saeculo superiori et quod excurrit... floruerunt* (Heidelberg, 1615).

Adam, Melchior, *Vitae Germanorum medicorum, qui saeculo superiori et quod excurrit claruerunt* (Heidelberg, 1620).

Addison, Joseph, 'Ad Burnettum Sacrae telluris theoriae auctorem', in *Examen poeticum duplex sive Musarum Anglicanarum delectus alter* (London, 1698), pp. 49–54; *The Works of the Right Honourable Joseph Addison*, vol. 1 (London, 1811), pp. 333–4.

Addison, Joseph, 'Barometri descriptio', in *Examen poeticum duplex sive Musarum Anglicanarum delectus alter* (London, 1698), pp. 75–8.

Agardh, Carl Adolph, *Aphorismi botanici*, 16 instalments (Lund, 1817–26).

Agardh, Carl Adolph, and Joachim Åkerman, *Dissertatio de metamorphosi algarum* (Lund, 1820).

Agerius, Nicolaus, and Johann Philipp Reusser, *Διάσκεψις φυσικὴ de nive* (Strasbourg, 1627).

Agricola, Georg, *Bermannus sive de re metallica* (Basel, 1530).

Agricola, Georg, *De ortu et causis subterraneorum libri V, De natura eorum, quae effluunt ex terra, libri IIII, De natura fossilium libri X, De veteribus et novis metallis libri II, Bermannus sive de re metallica dialogus, Interpretatio Germanica vocum rei metallicae* (Basel, 1546).

Agricola, Georg, *De re metallica libri XII* (Basel, 1556).

Aldrovandi, Ulisse, *Ornithologiae, hoc est de avibus historiae, libri XII*, 3 vols (Bologna, 1599–1603).

Aldrovandi, Ulisse, *De quadrupedibus digitatis viviparis libri tres et de quadrupedibus digitatis oviparis libri duo* (Bologna, 1637).

Alessandrini, Giulio, *Paedotrophia sive de puerorum educatione* (Venice, 1547).

Ali ibn Ridwan (comm.), *Liber quadripartiti Ptholemei... cum commento Haly Heben Rodan* (Venice, 1493).

Alpino, Prospero, *De balsamo dialogus* (Venice, 1591); (Padua, 1639); together with *De medicina Aegyptiorum* and Bontius, *Medicina Indorum* (Leiden, 1718); in Blasio Ugolino (ed.), *Thesaurus antiquitatum sacrarum*, vol. 11 (Venice, 1750), cols. dclvii–dccxxvi.

Alpino, Prospero, *De medicina Aegyptiorum libri quattuor* (Venice, 1591).

Alpino, Prospero, *De plantis Aegypti liber* (Venice, 1592).

Alpino, Prospero, *Le piante dell'Egitto: Il balsamo* (Venice, 1592).

Alpino, Prospero, *De praesagienda vita et morte aegrotantium libri septem* (Frankfurt am Main, 1601).

Alpino, Prospero, 'L'histoire du baulme', in Antoine Colin (tr., ed.), *Histoire des drogues, espiceries, et de certains medicamens simples*, 2nd edn (Lyons, 1619), book 4, separately paginated.

Alsted, Johann Heinrich, *Cursus philosophici encyclopaedia libris XXVII complectens universae philosophiae methodum* (Herborn, 1620); 2nd edn in 7 vols (Herborn, 1630); posthumous edn: *Scientiarum omnium encyclopaedia*, 4 vols (Lyons, 1649).

Altavilla, Benedetto, *Animadversiones in ephemeridas... ad rerum caelestium studiosos* (Turin, 1580).

Annio da Viterbo, *Commentaria... super opera diversorum auctorum de antiquitatibus loquentium confecta* (Rome, 1498).

'[Announcement of the *Philosophical Transactions*]', *Le journal des sçavans* (1665), p. 156.

Apianus, Petrus, *Cosmographicus liber* ([Landshut, 1524]).

Apin, Siegmund Jacob, *Unvorgreiffliche Gedancken, wie man so wohl Alte als Neue Dissertationes Academicas mit Nutzen sammlen... soll* (Nuremberg and Altdorf, 1719).

Applausus poetici, quos Musae Apollinea comitatae sunt lyra, dum... medicinae laurea... condecoraretur... Bacchetoni (Innsbruck, 1726).

Archimedes, *Ἀρχιμήδους τοῦ Συρακοσίου τὰ μέχρι νῦν σωζόμενα ἅπαντα/Archimedis Syracusani... opera, quae quidem extant, omnia... Graece et Latine* (Basel, 1544).

Aselli, Gaspare, *De lactibus sive lacteis venis, quarto vasorum mesaraicorum genere... dissertatio* (Milan, 1627); 2nd edn (Basel, 1628).

Augurello, Giovanni Aurelio, *Chrysopoeiae libri III et Geronticon liber primus* (Venice, 1515).

Bacon, Francis, 'Pars secunda operis, quae dicitur Novum organum sive indicia vera de interpretatione Naturae', in *Instauratio magna* (London, 1620), pp. 35–360.

Bacon, Francis, *Sylva Sylvarum or A Naturall History in Ten Centuries* (London, 1626).

Bacon, Francis, 'New Atlantis', in *Sylva Sylvarum* (London, 1626), separately paginated.

Bacon, Francis, 'Nova Atlantis, fragmentorum alterum', in *Operum moralium et civilium tomus* (London, 1638), pp. 351–86.

Baffi, Giovanni Battista, *De cometis libri tres* (Perugia, 1580).

Baffi, Giovanni Battista, 'Pythagorica consideratio, quo mense vivere possit foetus', in *De cometis libri tres* (Perugia, 1580), unpaginated.

Baglivi, Giorgio, *De praxi medica ad priscam observandi rationem revocanda libri duo* (Rome, 1696).

Baglivi, Giorgio, 'Dissertatio I., De anatome, morsu et effectibus tarantulae', in *De praxi medica ad priscam observandi rationem revocanda* (Rome, 1696), separately paginated.

Baglivi, Giorgio, *Opera omnia medico-practica et anatomica* (Lyons, 1704).

Baier, Johann Jakob, *Ὀρυκτογραφία Norica sive rerum fossilium et ad minerale regnum pertinentium... succincta descriptio* (Nuremberg, 1708).

Báñez, Domingo, *Commentaria et quaestiones in duos libros De generatione et corruptione* (Salamanca, 1585).

Barbaro, Ermolao (tr.), *Dioscoridis De materiali medicina* (Venice, 1516).

Barbaro, Ermolao, *Compendium scientiae naturalis ex Aristotele* (Venice, 1545).

Barberini, Maffeo, 'Adulatio perniciosa', in *Poemata* (Paris, 1620), pp. 46–9.

Barclay, John, *Icon animorum* (Frankfurt, 1625).

Barozzi, Francesco (tr.), *Procli... in primum Euclidis Elementorum librum commentariorum... libri IIII* (Padua, 1560).

Barozzi, Francesco, *Cosmographia in quattuor libros distributa summo ordine miraque facilitate ac brevitate ad magnam Ptolemaei mathematicam constructionem ad universamque astrologiam instituens* (Venice, 1585).

Barrow, Isaac, *Mathematicae professoris Lucasiani lectiones habitae in scholis publicis Academiae Cantabrigensis anno Domini MDCLXIV* (London, 1683).

Barrow, Isaac, 'Oratio praefatoria', in *Mathematicae professoris Lucasiani lectiones* (London,1683), pp. i–xxxvii.

Bartels, Matthias, *Biblionomia historico-politico-geographica* (Venice, 1682).

Bartholin, Rasmus, *Specimen recognitionis nuper editarum observationum astronomicarum nobilissimi viri Tychonis Brahe* (Copenhagen, [1668]).

Bartholin, Rasmus, *Experimenta crystalli Islandici disdiaclastici, quibus mira et insolita refractio detegitur* (Copenhagen, 1669).
Bartholin, Thomas, *De unicornu observationes novae* (Padua, 1645).
Bartholomaeus de Montagna, [*Consilia*] ([Padua, 1476]).
Basini, Basinio, 'Astronomicon libri duo', in *Opera praestantiora nunc primum edita* (Rimini, 1794), pp. 295–342.
Bauhin, Caspar, *Theatrum anatomicum* (Frankfurt am Main, 1605).
Bauhin, Caspar, *Πίναξ theatri botanici... sive index in Theophrasti, Dioscoridis, Plinii et botanicorum qui a saeculo scripserunt opera* (Basel, 1623).
Bauhin, Johann, *De plantis a divis sanctisve nomen habentibus* (Basel, 1591).
Bayer, Johann, *Uranometria omnium asterismorum continens schemata nova methodo delineata, aereis laminis expressa* (Augsburg, 1603).
Bayer, Michael, *Epistola admonitoria... ad Franciscum Cimam medicum Venetum* (Padua, 1662).
Beckher, Daniel, and Thomas Lindemann, *Disquisitio physica de glacie cruenta anno 1633 mense Decembri Regiomonti observata* (Königsberg, 1634).
[Béguin, Jean], *Tyrocinium chymicum* (Paris, 1610); 2nd edn, under the author's name (Paris, 1612).
Belon, Pierre, *Histoire naturelle des estranges poissons marins* (Paris, 1551).
Belon, Pierre, *De aquatilibus libri duo cum eiconibus* (Paris, 1553).
Belon, Pierre, *Les observations de plusieurs singularitez et choses memorables* (Paris, 1553).
Belon, Pierre, *La nature et diversité des poissons avec leurs pourtraicts* (Paris, 1555).
Bembo, Pietro, *De Aetna ad Angelum Gabrielem liber* (Venice, 1496).
Berengario da Carpi, Jacopo, *Commentaria cum amplissimis additionibus super Anatomia Mundini* (Bologna, 1521).
Beringer, Johann Bartholomäus Adam, and Georg Ludwig Hueber, *Lithographiae Wirceburgensis ducentis lapidum figuratorum, a potiori insectiformium, prodigiosis imaginibus exornatae specimen primum* (Würzburg, 1726).
Bernegger, Matthias (tr.), *Systema cosmicum authore Galilaeo Galilaei* (Strasbourg, 1635).
Bernoulli, Daniel, the Younger, *Vita immortalis viri Danielis Bernoulli* (Basel, 1783); republished in *Nova Acta Helvetica*, 1 (1787), pp. 1–32.
Bernoulli, Jacob, *Ars coniectandi, opus posthumum* (Basel, 1713).
Bernoulli, Johann, *Dissertatio inauguralis physico-anatomica de motu musculorum* (Basel, 1694).
Beroaldo, Filippo, 'Annotationes Galeni', in *Opusculum de terrae motu et pestilentia* (Bologna, 1505), sigs. Eiiii^{v}–Gii^{v}.
Bettini, Mario, *Aerarium philosophiae mathematicae*, 3 vols (Bologna, 1648).
Beughem, Cornelis à, *Bibliographia mathematica et artificiosa novissima, perpetuo continuanda* (Amsterdam, 1688).
Beyer, Hartmann, *Quaestiones novae in libellum de sphaera Iohannis de Sacro Busto* (Frankfurt am Main, 1549).
Biancani, Giuseppe, 'Clarorum mathematicorum chronologia', in *Aristotelis loca mathematica ex universis ipsius operibus collecta et explicata* (Bologna, 1615).
Biancani, Giuseppe, *Sphaera mundi seu cosmographia demonstrativa ac facili methodo tradita* (Bologna, 1620).
Bicadelli, Galeotto de, and Jakob Ebelin, *Dubium astrologicum* (Bologna, 1509).
Biese, Nicolas, *De universitate libri tres, quibus universa de natura philosophia continetur* (Antwerp, 1556).
Bilfinger, Georg Bernhard, 'De tracheis plantarum ex melone observatio', *Commentarii Academiae scientiarum imperialis Petropolitanae*, 4 (1735), pp. 182–7.

Biwald, Leopold, *Physica generalis, quam auditorum philosophiae usibus accomodavit Leopoldus Biwald* (Graz, 1767).

Blankaart, Steven, *Lexicon medicum Graeco-Latinum* (Amsterdam, 1679).

Blankaart, Steven, *The Physical Dictionary* (London, 1693).

Blumenbach, Johann Friedrich, *De generis humani varietate nativa* (Göttingen, 1775).

Blumenbach, Johann Friedrich, *Introductio in historiam medicinae litterariam* (Göttingen, 1786).

Blumenbach, Johann Friedrich, 'Decas [prima–sexta] collectionis suae craniorum diversarum gentium illustrata', *Commentationes Societatis regiae scientiarum Gottingensis*, 10 (1791), pp. 3–27; 11 (1793), pp. 59–71; 12 (1796), pp. 38–51; 14 (1800), pp. 35–48; 16 (1808), pp. 199–216; 18 (1820), pp. 159–74.

Bock, Hieronymus, *New Kreütter Bůch von underscheydt, würckung und namen der kreüter so in Teütschen landen wachsen* (Strasbourg, 1539).

Bock, Hieronymus, *De stirpium... usitatis nomenclaturis propriisque differentiis neque non temperaturis ac facultatibus commentariorum libri tres* (Strasbourg, 1552).

Boehmer, Georg Rudolf, *Handbuch der Naturgeschichte/Bibliotheca scriptorum historiae naturalis*, 9 vols (Leipzig, 1785–9).

Boerhaave, Herman, *Institutiones medicae in usus annuae exercitationis domesticos digestae* (Leiden, 1708).

Boerhaave, Herman, *Aphorismi de cognoscendis et curandis morbis* (Leiden, 1709).

Boerhaave, Herman, *Sermo academicus de comparando certo in physicis* (Leiden, 1715).

Boerhaave, Herman, *Methodus discendi medicinam* (Amsterdam, 1726).

Boerhaave, Herman, *Elementa chemiae, quae anniversario labore docuit in publicis privatisque scholis*, 2 vols (Leiden, 1732).

Boerhaave, Herman, *Praelectiones academicae in proprias Institutiones rei medicae. Edidit et notas addidit Albertus Haller*, 6 vols (Göttingen, 1739–44).

Boerhaave, Herman, *Praelectiones publicae de morbis oculorum ex codice manu scripto editae* (Göttingen, 1746).

Boissier de la Croix de Sauvages, François, *Nosologia methodica sistens morborum classes, genera et species iuxta Sydenhami mentem et botanicorum ordinem* (Amsterdam, 1763).

Bolyai, János, 'Appendix scientiam spatii absolute veram exhibens', in Farkas Bolyai, *Tentamen iuventutem studiosam in elementa matheseos purae... introducendi*, vol. 1 (Târgu Mureş, 1832), separately paginated.

Bonet, Théophile (tr.), *Iacobi Rohaulti Tractatus physicus Gallice emissus et recens Latinitate donatus* (Geneva, 1674).

Bongo, Pietro, *De mystica quaternarii numeri significatione* (Bergamo, 1583); final edition: *Numerorum mysteria... Postrema hac editione ab auctore ipso copioso indice et ingenti appendice auctum* (Bergamo, 1599); abridgement: *Praecipuae numerorum notae et earum valor* (Parma, 1689).

Bontius, Jacobus, *De medicina Indorum libri quattuor* (Leiden, 1642); see also Alpino, *De balsamo dialogus*.

Bordinus, Franciscus, *Quaesitorum et responsionum mathematicae disciplinae ad totius universi cognitionem spectantium chilias* (Bologna, 1573).

Borel, Pierre, *Bibliotheca chimica seu Catalogus librorum philosophicorum Hermeticorum* (Paris, 1654).

Borelli, Giovanni Alfonso, *Historia et meteorologia incendii Aetnaei anni 1669* (Reggio di Calabria, 1670).

Borelli, Giovanni Alfonso, *De motu animalium... opus posthumum*, 2 vols (Rome, 1680–1).

Börner, Friedrich, *Bibliothecae librorum rariorum physico-medicorum historico-criticae specimen primum/secundum*, 2 vols (Helmstedt, 1751–2).

Borrhaus, Martin, *Μαθημάτων στοιχεῖα, hoc est elementale cosmographicum* (Strasbourg, 1539).
Boscovich, Roger Joseph, 'De proximo Veneris sub sole transitu. Ad Londinensem Regiam Societatem', *Philosophical Transactions*, 51 (1759), pp. 865–88.
Boscovich, Roger Joseph, *De solis ac lunae defectibus libri V* (London, 1760).
Bouelles, Charles de, *Géométrie practique*, 2nd edn (Paris, 1551).
Boussuet, François, *De natura aquatilium carmen in universam Gulielmi Rondeletii... historiam* (Lyons, 1558).
Boyle, Robert, 'A Proëmial Essay...with Some Considerations Touching Experimental Essays in General', in *Certain Physiological Essays and Other Tracts Written at Distant Times, and on Several Occasions* (London, 1661), pp. 1–40.
Boyle, Robert, 'Commentatio prooemialis', in *Tentamina quaedam physiologica diversis temporibus et occasionibus conscripta* (London, 1661), pp. 1–34.
Boym, Michał Piotr, *Flora Sinensis* (Vienna, 1656).
Boym, Michał Piotr, *Clavis medica ad Chinarum doctrinam de pulsibus* (Nuremberg, 1686).
Brahe, Tycho, *De nova et nullius aevi memoria prius visa stella* (Copenhagen, 1573).
Brahe, Tycho, *De mundi aetherei recentioribus phaenomenis liber secundus* (Uraniborg, 1588).
Brahe, Tycho, *Epistolarum astronomicarum libri, quorum primus hic* (Uraniborg, 1596).
Brahe, Tycho, *Astronomiae instauratae progymnasmata, quorum haec prima pars* (Uraniborg and Prague, 1602).
Brahe, Tycho, *De disciplinis mathematicis oratio publice recitata in Academia Hafniensi anno 1574* (Copenhagen, 1610).
Brasavola, Antonio Musa, *Index refertissimus in omnes Galeni libros, qui ex Iunctarum sexta editione extant* (Venice, 1586).
Bresadola, Giacomo, et al., *Iconographia mycologica*, 28 vols (Milan, 1917–60).
Breyn, Johann Philipp, *Epistola de melonibus petrefactis montis Carmel vulgo creditis* (Leipzig, 1722).
Brisson, Mathurin-Jacques, *Regnum animale in classes IX distributum* (Paris, 1756).
Brisson, Mathurin-Jacques, *Ornithologia/Ornithologie*, 6 vols (Paris, 1760).
Brucaeus, Heinrich, *De motu primo libri tres* (Rostock, 1573).
Brudzewski, Albert, *Commentum in Theoricas planetarum Georgii Purbachii* (Milan, 1495).
Brunfels, Otto, *Catalogus illustrium medicorum sive de primis medicinae scriptoribus* (Strasbourg, 1530).
Brunfels, Otto, *Herbarum vivae eicones ad naturae imitationem summa cum diligentia et artificio effigiatae*, 3 vols (Strasbourg, 1530–6).
Brunfels, Otto, *Contrafayt Kreüterbůch* (Strasbourg, 1532–7).
Brunfels, Otto, *Ὀνομαστικὸν medicinae continens omnia nomina* (Strasbourg, 1534).
Bruno, Giordano, *De monade, numero et figura liber. Consequens quinque De minimo, magno et mensura. Item De innumerabilibus, immenso et infigurabili, seu de universo et mundis libri octo* (Frankfurt am Main, 1591).
Buchanan, George, *Sphaera...quinque libris descripta* (Herborn, 1586).
Burgersdijck, Franco, *Sphaera Iohannis de Sacro-Bosco...in usum scholarum...recensita* (Leiden, 1626).
Burnet, Thomas, *Telluris theoria sacra originem et mutationes generales orbis nostri, quas aut iam subiit aut olim subiturus est, complectens* (London, 1681); 2nd edn (1689).
Burnet, Thomas (physician), *Hippocrates contractus, in quo magni Hippocratis...opera omnia in brevem epitomen...redacta habentur* (Edinburgh, 1685).
Cabeo, Niccolò, *In quattuor libros Meteorologicorum Aristotelis commentaria et quaestiones quattuor tomis comprehensa*, 4 vols (Rome, 1646).

Caius, John, *De canibus Britannicis liber unus. De rariorum animalium et stirpium historia liber unus. De libris propriis liber unus* (London, 1570).
Calepino, Ambrogio, *Dictionarium* (Reggio Emilia, 1502).
Callard de La Ducquerie, Jean Baptiste, *Lexicon medicum etymologicum, in quo ad tria etymologiarum milia* (Caen and Paris, 1692).
Calzolari, Francesco, 'Iter Baldi, civitatis Veronae montis', in Pietro Andrea Mattioli, *Compendium de plantis omnibus* (Venice, 1571), unpaginated.
Camerarius, Joachim (tr.), *Alberti Dureri... de symmetria partium in rectis formis humanorum corporum libri in Latinum conversi* and *Alberti Dureri de varietate figurarum et flexuris partium ac gestibus imaginum libri duo*, 2 vols (Nuremberg, 1532–4).
Camerarius, Joachim (tr.), 'Galene Andromachi', in *De theriacis et mithridateis commentariolus* (Nuremberg, [1533]), sigs. [c8^{r}]–d3^{r}.
Camerarius, Joachim, *Commentarii utriusque linguae, in quibus est... diligens exquisitio nominum, quibus partes corporis humani appellari solent* (Basel, 1551).
Camerarius, Rudolf Jakob, *Ad Thessalum, dominum Michaelem Bernardum Valentini... de sexu plantarum epistola* (Tübingen, 1694).
Camerarius, Rudolf Jakob, 'De generatione ex primaevo semine', in *De sexu plantarum epistola* (Tübingen, 1694), pp. 98–110.
Camerarius, Rudolf Jakob, 'De quercuum gallis', in *De sexu plantarum epistola* (Tübingen, 1694), pp. 85–97.
Camerarius, Rudolf Jakob, and Johann David Mauchart, *Disquisitio botanica de herba mimosa seu sentiente* (Tübingen, 1688).
Campanella, Tommaso, *Astrologicorum libri VI* [read *VII*], *in quibus astrologia omni superstitione Arabum et Iudaeorum eliminata physiologice tractatur* (Lyons, 1629).
Candidus, Pantaleon, *Tabulae chronologicae continentes seriem annorum mundi* (Strasbourg, 1597).
Cappeller, Moritz Anton, *Pilati montis historia in pago Lucernensi Helvetiae siti* (Basel, 1767).
Cardano, Girolamo, 'Aphorismorum astronomicorum segmenta VII', in *Libelli quinque* (Nuremberg, 1547), fols. 207^{v}–309^{v}.
Cardano, Girolamo, *In Claudii Ptolemaei Pelusiensis IIII de astrorum iudiciis... libros commentaria* (Basel, 1554).
Cardano, Girolamo, 'De libris propriis', in *Somniorum Synesiorum omnis generis insomnia explicantes libri IIII* (Basel, 1562), separately paginated.
Cardano, Girolamo, *De propria vita liber. Ex bibliotheca Gabrielis Naudaei* (Paris, 1643).
Cardano, Girolamo, *Metoposcopia libris tredecim et octingentis faciei humanae eiconibus complexa* (Paris, 1658).
Carl, Johann Samuel, *Ichnographia praxeos clinicae duplicis methodi pro usu in tyrocinio practico memoriae et iudicii concinnata. Accedit Ichnographia anatomiae et chymiae nec non formularum* (Büdingen, 1722) (the single *Ichnographiae* were also printed separately).
Carmina comitialia Cantabrigensia (London, 1721).
Carmina quadragesimalia ab Aedis Christi Oxoniensis alumnis composita et... publice recitata, 2 vols (Oxford, 1723–48).
Carnarius, Johannes, *De podagrae laudibus oratio habita in celeberrimo gymnasio Patavino... in initio lectionum, III. Idus Novembris 1552* (Padua, 1553).
Castelli, Bartolomeo, *Lexicon medicum Graecolatinum... ex Hippocrate et Galeno desumptum* (Messina, 1598).
Castro, Pedro de, *Bibliotheca medici eruditi* (Padua, 1654).
Catalogus praelectionum..., quae... in inclyta Academia Iulia, quae est Halmstadii, publice proponuntur et tractantur (Helmstedt, 1595).

Catena, Pietro, *Oratio pro idea methodi. Ad nobilem Federicum Sforciam* (Padua, 1563).

Caturegli, Pietro, *Ephemerides motuum caelestium ex anno 1829 in annum 1832* (Bologna, 1828).

Celsius, Anders, 'Observationes meteorologicae anno 1731 Upsaliae habitae', *Miscellanea Berolinensia ad incrementum scientiarum*, 5 (1737), pp. 132–56.

Celsius, Anders, 'Observationes meteorologicae habitae Upsaliae anno 1740', *Acta Societatis regiae scientiarum Upsaliensis ad annum MDCCXL* (Uppsala, 1744), pp. 38–47.

Celtis, Conrad, 'Germania generalis', in *Quattuor libri amorum secundum quattuor latera Germaniae* (Nuremberg, 1502), fols. lxxiii^{v}–[lxxvii^{v}].

Cesalpino, Andrea, *De plantis libri XVI* (Florence, 1583).

Cesalpino, Andrea, *Praxis universae artis medicae* (Treviso, 1606).

Cesi, Bernardo, *Mineralogia sive naturalis philosophiae thesauri* (Lyons, 1636).

Champier, Symphorien, 'De medicinae claris scriptoribus', in *Libelli duo* ([Lyons, 1506]), fols. iii^{r}–xl^{r}.

Champier, Symphorien, *Vocabulorum medicinalium epitoma* (Lyons, 1508).

[Chrysoloras, Manuel, and Jacopo d'Angelo] (trs.), *Cosmographia Ptolemaei* (Vicenza, 1475).

Ciruelo, Pedro, *Apotelesmata astrologiae Christianae... super duabus tantum iudiciorum partes* (Alcalá de Henares, 1521).

Clarke, Samuel (tr., comm.), *Jacobi Rohaulti Physica* (London, 1697); 4th edn (London, 1718).

Clarke, Samuel (tr.), *Optice sive de reflexionibus, refractionibus, inflexionibus et coloribus lucis libri tres. Auctore Isaaco Newton* (London, 1706).

Clavius, Christoph, *In Sphaeram Ioannis de Sacro Bosco commentarius* (Rome, 1570).

Clerck, Carl, *Svenska spindlar/Aranei Suecici* (Stockholm, 1758).

Clusius, Carolus, *Rariorum aliquot stirpium per Hispanias observatarum historia libris duobus expressa* (Antwerp, 1576).

Clusius, Carolus, *Rariorum aliquot stirpium per Pannoniam et Austriam observatarum historia quattuor libris expressa* (Antwerp, 1583).

Clusius, Carolus, *Exoticorum libri decem, quibus animalium, plantarum, aromatum aliorumque peregrinorum fructuum historiae describuntur* (Leiden, 1605).

Clüver, Detlev, *Geologia sive Natürliche Wissenschafft von Erschaffung und Bereitung der Erd-Kugel* (Rostock, 1700).

Cluyt, Outgert, *Calsvee sive dissertatio lapidis nephritici... naturam, proprietates et operationes exhibens* (Rostock, 1627).

Collins, John, et al., *Commercium epistolicum... de analysi promota iussu Societatis Regiae in lucem editum*, 2nd edn (London, 1722).

Columbus, Christopher, and Leander de Cosco (tr.), *Epistola... de insulis Indiae supra Gangem nuper inventis... ad magnificum dominum Gabrielem Sanches* (Rome, 1493).

Comenius, John Amos, *Orbis sensualium pictus/Die sichtbare Welt* (Nuremberg, 1658).

Commandino, Federico (tr., comm.), *Apollonii Pergaei Conicorum libri quattuor* (Bologna, 1566).

Commandino, Federico (tr., comm.), *Pappi Alexandrini Mathematicae collectiones* (Pesaro, 1588).

Copernicus, Nicolaus, *De revolutionibus orbium coelestium libri VI* (Nuremberg, 1543).

Cordus, Euricius, *Botanologicon* (Cologne, 1534).

Cornarius, Janus (ed., tr.), *Hippocratis Coi... opera quae apud nos extant omnia* (Paris, 1546).

Costus, Petrus, 'Ad Gulielmum Rondeletium medicum praestantissimum', in Guillaume Rondelet, *Libri de piscibus marinis* (Lyons, 1554–5), vol. 1, sig. α3^{r}.

Crato, Johannes, *Assertio... pro libello suo Germanico modo tertium recuso* (Frankfurt am Main, 1585).

Crüger, Peter, *Uranodromus cometicus: Ein außführlicher Tractat vom grossen Cometen deß 1618 Jahrs* (Gdańsk, 1619).

Cruser, Hermann (tr.), *Hippocratis Coi De natura humana liber cum duobus commentariis Galeni. Eiusdem De victus ratione salubri liber cum Galeni commentario* (Paris, 1534).

Cunitz, Maria, *Urania propitia sive tabulae astronomicae mire faciles* (Byczyna and Oleśnica, 1650).

Curtz, Albert von, *Problema Austriacum, super data recta linea quattuor-decangulum aequiangulum et aequilaterum constituere* (Munich, 1653).

Curtz, Albert von (ed.), *Historia coelestis ex libris, commentariis, manuscriptis observationum vicennalium Tichonis Brahe* (Augsburg, 1666).

Dalechamps, Jacques, *Historia generalis plantarum in libros XVIII per certas classes artificiose digesta*, 2 vols (Lyons, 1586).

Dalrymple of Stair, James, *Physiologia nova experimentalis* (Leiden, 1686).

Dasypodius, Konrad (ed., tr.), 'Ex scriptis Hieronis Alexandrini de geometricis definitionibus selecta quaedam', in *Εὐκλείδου Στοιχείων τὸ πρῶτον/Euclidis Elementorum liber primus* (Strasbourg, 1570), fols. 55^{v}–75^{r}.

Dasypodius, Konrad, *Λεξικὸν seu dictionarium mathematicum* (Strasbourg, 1573).

Dasypodius, Konrad, *Oratio...de disciplinis mathematicis...Hieronis Alexandrini nomenclaturae vocabulorum geometricorum translatio...Lexicon mathematicum ex diversis collectum antiquis scriptis* (Strasbourg, 1579).

Degli Emanuelli, Bartolomeo, *Adnotationes super dicta Gentilis de Fulgineo dicta ab illo super definitione de febre data ab Avicenna* (Rome, 1524).

Della Porta, Giambattista, *De humana physiognomonia libri IIII* (Vico Equense, 1586).

'De pulsibus libri quattuor e Sinico translati', in Andreas Cleyer (ed.), *Specimen medicinae Sinicae sive opuscula medica ad mentem Sinensium* (Frankfurt am Main, 1682), separately paginated.

De Rycke, Josse, et al., *Μελισσογραφία* (Rome, 1625).

Descartes, René, 'La géometrie', in *Discours de la méthode* (Leiden, 1637), pp. 296–413.

Diderot, Denis, and Jean Baptiste Le Rond d'Alembert (eds), *Encyclopédie ou dictionnaire raisonné des sciences, des arts et des métiers*, 35 vols (Paris, 1751–80).

Dionysius, Paullus (tr.), *Aphorismi Hippocratis versibus redditi* (Verona, 1599).

Dionysius, Paullus, 'De oculis carmen', in *Aphorismi Hippocratis* (Verona, 1599), pp. 89–99.

Dioscorides, [*De materia medica*] (Colle, 1478).

Dodoens, Rembert, *Cosmographica in astronomiam et geographiam isagoge* (Antwerp, 1548).

Dodoens, Rembert, *De frugum historia liber unus* (Antwerp, 1552).

Dodoens, Rembert, *Physi<o>logices medicinae partis tabulae* (Cologne, 1581).

Dolaeus, Johannes, *Encyclopaedia medica dogmatica* (Frankfurt am Main, 1691).

Dorn, Gerhard, *Dictionarium Theophrasti Paracelsi continens obscuriorum vocabulorum, quibus in suis scriptis passim utitur, definitiones* (Frankfurt am Main, 1573); 2nd edn (1583).

Draud, Georg, *Bibliotheca classica* (Frankfurt am Main, 1611); 2nd edn (1625).

Drechsler, Johann Gabriel, and Polycarp Michael Rechtenbach, *Dissertatio historico-physica de sermone brutorum* (Leipzig, 1673); 2nd edn (Erfurt, 1706).

Dubois, Jacques, *Ordo et ordinis ratio in legendis Hippocratis et Galeni libris* (Basel, [1535]).

Du Laurens, André, *Historia anatomica humani corporis* (Paris, 1600).

Dürer, Albrecht, *Vier Bücher von menschlicher Proportion* (n.p., 1528).

Ecchelensis, Abraham (tr.), and Giovanni Alfonso Borelli (comm.), *Apollonii Pergaei Conicorum libri V, VI, VII paraphraste Abalphato Asphahanensi, nunc primum editi* (Florence, 1661).

Ecker, Johann Matthias Alexander (tr.), *Geist Hippokrats. Nach dem Lateinischen des Burnets aus dem griechischen Urtexte* (Vienna, 1791).

Εἰς τὴν Τετράβιβλον τοῦ Πτολεμαίου ἐξηγητὴς ἀνώνυμος/In Claudii Ptolemaei Quadripartitum enarrator ignoti nominis (Basel, 1559).

Ekerman, Petrus, and Andreas Piscator, *Dissertatio academica de Cicerone mathematico* (Uppsala, 1759).

Elsholtz, Johann Sigismund, *De phosphoris quattuor observatio* (Berlin, 1676).

Enchiridion medicinae medicamentorum tam simplicium quam compositorum in certos titulos distinctam silvam continens (Basel, 1573).

Engel, Johannes, *Almanach novum atque correctum... super anno Domini 1510* (Vienna, [1509]).

Engel, Johannes, *Almanach novum atque correctum... super anno Domini 1512* (Vienna, [1511]).

'Epitaphium eius', in Galileo Galilei, *Systema cosmicum* (Leiden, 1699), sig. **2^{r-v}.

Erasmus of Rotterdam, *De duplici copia rerum ac verborum... De ratione studii... De puero Iesu concio scholastica... et quaedam carmina* ([Paris], 1512).

Erasmus of Rotterdam, *Encomium artis medicae* (Leuven, 1518).

Erasmus of Rotterdam, *Dilucida et pia explanatio symboli... et decalogi praeceptorum* (Antwerp, 1533).

Estienne, Henri, *Dictionarium medicum vel expositiones vocum medicinalium* ([Geneva], 1564).

Estienne, Henri, *Thesaurus Graecae linguae*, 5 vols ([Geneva], 1572).

Euclid and Campanus of Novara (tr.), *Opus elementorum... in geometriam artem, in id quoque... perspicacissimi commentationes* (Venice, 1482).

Euler, Leonhard, 'Dissertatio de magnete', in *Pièces qui ont remporté le Prix de l'Académie Royale des Sciences en 1743 et 1746* (Paris, 1748), separately paginated.

Euler, Leonhard, *Introductio in analysin infinitorum*, 2 vols (Lausanne, 1748).

Euler, Leonhard, *Institutiones calculi differentialis cum eius usu in analysi finitorum ac doctrina serierum* (St Petersburg, 1755).

Euler, Leonhard, *Institutionum calculi integralis volumen primum/secundum/tertium*, 3 vols (St Petersburg, 1768–70).

Euler, Leonhard, 'Observationes circa bina biquadrata, quorum summam in duo alia biquadrata resolvere liceat', *Novi commentarii Academiae scientiarum imperialis Petropolitanae*, 17 (1772/73), pp. 64–9.

Eustachi, Bartolomeo, 'Epistula de auditus organis', in *Opuscula anatomica* (Venice, 1564), pp. 148–64.

Faber, Basilius, *Thesaurus eruditionis scholasticae sive ratio docendi ac discendi* (Leipzig, 1571).

Fabri, Wenceslaus (ed., comm.), *Opusculum Johannis de Sacro Busto spericum cum notabili commento atque figuris textum declarantibus utilissimis* ([Leipzig, *c.*1495]).

Fabricius, Ernst Friedrich, *Medicinae utriusque, Galenicae et Hermeticae, anatome philosophica* (Hamburg, 1626).

Fabricius, Hieronymus, *Tractatus quattuor* (Frankfurt am Main, 1624).

Fabricius, Johann, *De maculis in sole observatis et apparente earum cum sole conversione narratio* (Wittenberg, 1611).

Fahrenheit, Daniel Gabriel, 'Experimenta et observationes de congelatione aquae in vacuo factae', *Philosophical Transactions*, 33/382 (1724), pp. 78–84.

Falloppio, Gabriele, *Observationes anatomicae ad Petrum Mannam* (Paris, 1562).

Falloppio, Gabriele, *Lectiones de partibus similaribus humani corporis* (Nuremberg, 1575).

Falloppio, Gabriele, *Opera genuina omnia, tam practica quam theorica* (Venice, 1606).

Fatio de Duillier, Nicolas, *Neutonus ecloga* (Oxford, 1728).
Favaro, Giuseppe, *Mors laniata* (Amsterdam, 1936).
Fehr, Johann Michael, *Hiera picra vel de absinthio analecta* (Leipzig, 1668).
Fehr, Johann Michael, 'Epistola invitatoria ad celeberrimos Europae medicos', *Miscellanea curiosa*, 1 (1670), pp. 1–8.
Fellon, Thomas Bernard, *Magnes* (Lyons, 1696).
[Fermat, Pierre, under the pseudonym] M.P.E.A.S., *De linearum curvarum cum lineis rectis comparatione* (Toulouse, 1660).
Fernel, Jean, *De abditis rerum causis libri duo* (Paris, 1548).
Feuerlein, Jakob Wilhelm, and Christoph Moritz Lochner, *Montes divinitatis testes contra Lucretium et Burnetium* (Altdorf, 1729).
Figulus, Carolus, *Dialogus, qui inscribitur botanomethodus sive herbarum methodus* (Cologne, 1540).
Finé, Oronce, *Protomathesis* (Paris, 1532).
Flamsteed, John, *Historia coelestis Britannica*, 3 vols (London, 1725).
Flamsteed, John, *Atlas coelestis* (London, 1729).
Flemyng, Malcolm, *Neuropathia sive de morbis hypochondriacis et hystericis libri tres: Poema medicum, cui praemittitur dissertatio epistolaris prosaica eiusdem argumenti* (York, 1740).
Florianus, Johannes (tr.), *Ioannis Leonis Africani De totius Africae descriptione libri IX* (Antwerp, 1556).
Florido, Ambrosio, *Dialogismus de natura universa maris ac eius genesi et de causa fluxus et refluxus eiusdem* (Padua, 1613).
Foësius, Anutius, *Oeconomia Hippocratis alphabeti serie distincta* (Frankfurt am Main, 1588).
Forster, Richard, *Ephemerides meteorographicae... ad annum domini 1575* ([London, 1574]).
Fracastoro, Girolamo, *Syphilis sive morbus Gallicus* (Verona, 1530).
Fracastoro, Girolamo, *Homocentrica* ([Venice], 1538).
Frandsen, Hans, *De oculorum fabricatione et coloribus liber elegiaco carmine conscriptus* (Wittenberg, 1556).
Fratta, Giovanni, *Della dedicatione de' libri, con la correttion dell' abuso... introdotto* (Venice, 1590).
Fries, Theodor Magnus, 'Lichenes arctoi Europae Groenlandiaeque', *Nova acta Regiae societatis scientiarum Upsaliensis*, 3rd series, 3 (1861), pp. 103–398.
Frisius, Andreas (tr.), *Francisci Redi... Experimenta circa generationem insectorum* (Amsterdam, 1671).
Frisius, Gemma, *Arithmeticae practicae methodus facilis* (Antwerp, 1540); annotated and expanded by Jacques Peletier (Paris, 1545).
Froriep, Robert, and Friedrich August Andorff, *Atlas anatomicus partium corporis humani per strata dispositarum imagines... exhibens* (Weimar, 1851).
Fuchs, Leonhart, *Errata medicorum recentiorum numero LX adiectis eorundem confutationibus* (Hagenau, 1530).
Fuchs, Leonhart, *Apologia... contra Hieremiam Thriverum Brachelium* (Hagenau, 1534).
Fuchs, Leonhart, *Paradoxorum medicinae libri III* (Basel, 1535).
Fuchs, Leonhart, *De historia stirpium commentarii insignes* (Basel, 1542).
Fuchs, Leonhart, *Primi de stirpium historia tomi vivae imagines* (Basel, 1545).
Füchsel, Georg Christian, 'Historia terrae et maris ex historia Thuringiae per montium descriptionem eruta', *Acta Academiae electoralis Moguntinae*, 2 (1761), pp. 44–254.

Funebre symbolum virorum aliquot illustrium de... Orontio Finaeo (Paris, 1555).

Galilei, Galileo, *Sidereus nuntius magna longeque admirabilia spectacula pandens* (Venice, 1610).

Galilei, Galileo, *Dialogo... sopra i due massimi sistemi del mondo* (Florence, 1632).

Gallus, Paschalis, *Bibliotheca medica sive catalogus illorum, qui... artem medicam... scriptis illustrarunt* (Basel, 1590).

Galvani, Luigi, 'De viribus electricitatis in motu musculari commentarius', *De Bononiensi scientiarum et artium instituto atque academia commentarii*, 7 (1791), pp. 363–418; book ed. *cum Joannis Aldini dissertatione et notis* (Modena, 1792).

[*Gart der Gesundheit*] (Mainz, 1485).

Gassendi, Pierre, *Mercurius in sole visus et Venus invisa Parisiis anno 1631* (Paris, 1632).

Gassendi, Pierre, *De vita et moribus Epicuri libri octo* (Lyons, 1647).

Gassendi, Pierre, *Institutio astronomica iuxta hypotheseis tam veterum quam Copernici et Tychonis* (Paris, 1647).

Gassendi, Pierre, *Tychonis Brahei, equitis Dani, astronomorum coryphaei, vita... Accessit Nicolai Copernici, Georgii Peurbachii et Ioannis Regiomontani, astronomorum celebrium, vita* (Paris, 1654).

Gassendi, Pierre, *Opera omnia in sex tomos divisa*, 6 vols (Lyons, 1658).

Gassendi, Pierre, 'Philosophiae Epicuri syntagma', in *Opera omnia* (Lyons, 1658), vol. 3, pp. 1–94.

Gaun, Franz, and Johann E. Feigenpuz, *Iter per salinas Tyrolenses publicae concertationi propositum* (Innsbruck, 1707).

Gaurico, Luca [in fact, William of Moerbeke] (tr.), 'Archimedis Syracusani tetragonismus', in *Tetragonismus, id est circuli quadratura* (Venice, 1503), fols. 15^r–29^v.

Gaurico, Luca, 'Oratio de laudibus astrologiae habita in Ferrariensi Academia', in *Spherae tractatus Ioannis de Sacro Busto* (Venice, 1531), sigs. aii^r–avi^v; republished in *Tomus I. operum omnium* (Basel, 1575), pp. 1–8.

Gauss, Carl Friedrich, *Disquisitiones arithmeticae* (Leipzig, 1801).

Gaza, Theodorus (tr.), [Aristotle's *Historia animalium, De partibus animalium, De generatione animalium*] (Venice, 1476).

Gaza, Theodorus (tr.), *Theophrasti De historia plantarum, De causis plantarum* (Treviso, 1483).

Gemma, Cornelius, *Ephemerides meteorographicae anni MDLXIII ad directionem horizontis Brabantici... institutae* (Antwerp, 1563).

George of Trebizond (tr., comm.), *In Claudii Ptolemaei centum aphorismos commentarius* (Cologne, 1544).

[Gerard of Cremona] (tr.), *Almagestum Claudii Ptolemaei... astronomorum principis* (Venice, 1515).

Gerstner, Johann Karl Anton, *Dissertatio medica inauguralis de salubritate aquae* (Innsbruck, 1756).

Gessner, Conrad, *Catalogus plantarum Latine, Graece, Germanice et Gallice* (Zurich, 1542).

Gessner, Conrad, *Bibliotheca universalis sive catalogus omnium scriptorum locupletissimus in tribus linguis, Latina, Graeca et Hebraica* (Zurich, 1545).

Gessner, Conrad (ed.), *Naturalis scientiae totius compendium ex Aristotele... Hermolao Barbaro... autore... Cui accessit Hieronymi Wildenbergii Aurimontani in universam Aristotelis physicam epitome* (Basel, 1548).

Gessner, Conrad, *Pandectarum sive partitionum universalium... libri XXI*, 2 vols (Zurich, 1548–9).

Gessner, Conrad, *Historia animalium*, 4 vols (Zurich, 1551–8); supplemented by *Historiae animalium liber V., qui est de serpentium natura* (Zurich, 1587) and Thomas Moffett, *Insectorum sive minimorum animalium theatrum* (London, 1634).

Gessner, Conrad, *Icones animalium quadrupedum viviparorum et oviparorum, quae in Historia animalium Conradi Gesneri describuntur* (Zurich, 1553); 2nd edn (Zurich, 1560).

Gessner, Conrad, 'Descriptio Montis Fracti sive Montis Pilati', in *De raris et admirandis herbis* (Zurich, 1555), pp. 43–67.

Gessner, Conrad, *Nomenclator aquatilium animantium: icones animalium aquatilium in mari et dulcibus aquis degentium plus quam DCC* (Zurich, 1560).

Gessner, Conrad (ed.), *Sanitatis tuendae praecepta* (Zurich, [1560]).

Gessner, Conrad, 'Prolegomena in Galenum in tres partes divisa', in *Claudii Galeni Pergameni omnia, quae extant, in Latinum sermonem conversa*, vol. 1 (Basel, 1562), sigs. α†4^{r}–C†4^{r}.

Gessner, Conrad, 'De libris a se editis epistola ad Guilielmum Turnerum', in Josias Simmler, *Vita Conradi Gesneri Tigurini* (Zurich, 1566), fols. 20^{v}–31^{r}.

Ghymm, Walther, 'Vita…Gerardi Mercatoris', in Gerardus Mercator, *Atlas sive cosmographicae meditationes* (Duisburg and Düsseldorf, 1595), sigs. †^{r}–†3^{r}.

Giacoletti, Giuseppe, *De lebetis materie et forma eiusque tutela in machinis vi vaporis agentium carmen didascalicum* (Amsterdam, 1863).

Gilbert, William, *De magnete magneticisque corporibus et de magno magnete tellure* (London, 1600).

Gilbert, William, *De mundo nostro sublunari philosophia nova* (Amsterdam, 1651).

Giovio, Paolo, *Elogia veris clarorum virorum imaginibus apposita* (Venice, 1546).

Giuntini, Francesco, *Speculum astrologiae, quod attinet ad iudiciariam rationem nativitatum atque annuarum revolutionum* (Lyons, 1573).

Gmelin, Georg, *Flora Sibirica sive historia plantarum Sibiriae*, 4 vols (St Petersburg, 1747–69).

Goens, Daniel van, *Disputatio philosophica inauguralis comprehendens examen, an luna habitabilis sit* (Utrecht, 1738).

Grammatica, Francesco Vigilio, *Cometa poeticus, qui morti ferale illuxit* (Innsbruck, 1726).

Grape, Zacharias, and Johannes Schnabel, *Dissertatio philosophica examinans placita quaedam scripti recentissimi sub titulo Geologia* (Rostock, 1700).

Grapheus, Cornelius, 'Commendatio divinae Mathematices', in Rembert Dodoens, *Cosmographica isagoge* (Antwerp, 1548), sig. Aii^{r}.

Gregory, David, *Astronomiae physicae et geometricae elementa* (Oxford, 1702).

Greiling, Johann Christoph, 'Einige allgemeine Grundsätze zu einer Theorie der Recensionen', *Archiv für die Physiologie*, 3 (1799), pp. 349–85.

Grischow, Augustin, 'Hyetometri descriptio una cum sexennio observationum hyetometricarum Berolini…habitarum', *Miscellanea Berolinensia ad incrementum scientiarum*, 4 (1734), pp. 349–76.

Grospré, Robert, *Regimen sanitatis* (Ghent, 1538).

Grospré, Robert, 'Un breve et notabile trattato del reggimento della sanità', in Domenico Romoli, *La singolare dottrina…dei condimenti di tutte le vivande* (Venice, 1560), pp. 359–76.

Gross, Johann Georg, and Johann Jakob von Brunn, *Theoremata philosophica de stella magorum* (Basel, 1611).

Großinger, Johann Baptist, *Universa historia physica regni Hungariae secundum tria regna naturae digesta*, 5 vols (Bratislava, 1793–7).

Grove, Robert, *Carmen de sanguinis circuitu a Gulielmo Harvaeo Anglo primum invento* (London, 1685).

Gruter, Jacob (tr.), *Sylva sylvarum sive historia naturalis in decem decurias distributa, Anglice olim conscripta a Francisco Bacone* (Leiden, 1648).

[Guarino Veronese and Gregory Tifernas (trs)], [Strabo's *Geography*] (Rome, [1469]).
Guarinonius, Hippolytus, *Chylosophiae academicae...primi...tomi pars prima* (Innsbruck, 1648).
Guericke, Otto von, *Experimenta nova (ut vocantur) Magdeburgica de vacuo spatio* (Amsterdam, 1672).
Guibert, Nicolas, *De balsamo eiusque lacrimae, quod opobalsamum dicitur, natura, viribus et facultatibus admirandis* (Strasbourg, 1603).
Habersack, Franz Adolph, *Medicina Austro-Viennensis versice exposita* (Vienna, 1731).
Habicot, Nicolas, *Question chirurgicale* (Paris, 1620).
Haller, Albrecht von, *Iter Helveticum anni MDCCXXXVIII* (Göttingen, 1740).
Haller, Albrecht von, *Historia stirpium indigenarum Helvetiae inchoata*, 3 vols (Bern, 1768).
Haller, Albrecht von, *Bibliotheca botanica*, 2 vols (Zurich, 1771–2).
Haller, Albrecht von (ed.), *Epistolarum ab eruditis viris ad Albertum Hallerum scriptarum pars I. Latinae*, 6 vols (Bern, 1773–5).
Haller, Albrecht von, *Bibliotheca chirurgica*, 2 vols (Bern and Basel, 1774–5).
Haller, Albrecht von, *Bibliotheca anatomica*, 2 vols (Zurich, 1774–7).
Haller, Albrecht von, *Bibliotheca medicinae practicae*, 4 vols (Bern and Basel, 1776–88).
Haller, Albrecht von, *Icones plantarum Helvetiae. Ex ipsius Historia stirpium denuo recusae* (Bern, 1795).
Halley, Edmond, 'In viri praestantissimi Isaaci Newtoni opus hocce mathematico-physicum', in Isaac Newton, *Principia mathematica* (London, 1687), unpaginated.
Halley, Edmond (tr., comm.), *Apollonii Pergaei Conicorum libri octo et Sereni Antissensis De sectione cylindri et coni libri duo* (Oxford, 1710).
Halley, Edmond, 'Methodus singularis, qua solis parallaxis sive distantia a terra...tuto determinari poterit', *Philosophical Transactions*, 29/348 (1716), pp. 454–64.
Harchies, Josse de, *Enchiridion medicum* (Basel, 1573).
Hartner, Willy, 'Ptolemäische Astronomie im Islam und zur Zeit des Regiomontanus', in Günther Hamann (ed.), *Regiomontanus-Studien* (Vienna, 1980), pp. 109–124.
Harvey, William, *Exercitatio anatomica de motu cordis et sanguinis in animalibus* (Frankfurt am Main, 1628).
Harvey, William, *Exercitationes de generatione animalium* (London, 1651).
Hebenstreit, Johann Baptist, 'In astro-poecilo-pyrgium Keplerianum...tabularum opus depictum exhibens...idyllion', in Johannes Kepler, *Tabulae Rudolphinae* (Ulm, 1627), fols. [5^{v}–7^{v}].
Heffter, Johann Carl, *Museum disputatorium physico-medicum tripartitum*, 2 vols (Zittau, 1756–64).
Heister, Lorenz, *Compendium anatomicum veterum recentiorumque observationes et inventa brevissime complectens* (Altdorf, 1717).
Hell, Maximilian, *Observatio transitus Veneris ante discum solis die 5ta Iunii 1761* (Vienna, [1761]).
Helmholtz, Hermann von, *De fabrica systematis nervosi evertebratorum dissertatio inauguralis anatomica* (Berlin, 1842).
Helmont, Jan Baptist van, *Ortus medicinae, id est initia physicae inaudita* (Amsterdam, 1648).
Hemmer, Johann Jakob, 'Conductorum fulmineorum vim egregiam tribus recentioribus exemplis docet Johannes Jacobus Hemmer', *Acta Academiae Theodoro-Palatinae, 6, pars physica* (1790), pp. 516–32.
Heuraet, Hendrik van, 'Epistola de transmutatione curvarum linearum in rectas', in Frans van Schooten and Florimond de Beaune, *Geometria a Renato des Cartes edita* (Amsterdam, 1659–61), vol. 1, pp. 517–20.

Heurne, Johan van, *Institutiones medicinae exceptae ex dictantis ore* (Leiden, 1592).

Hevelius, Johannes, *Selenographia sive lunae descriptio* (Gdańsk, 1647).

Hizler, Georg, *Oratio de vita et morte... Leonharti Fuchsii* (Tübingen, 1566).

Hofer, Johannes, 'Dissertatio epistolaris de possibilitate physica longaevitatis patriarcharum antediluvianorum', *Acta Helvetica*, 3 (1758), pp. 169–211.

Hoffmann, Friedrich, *Medicina rationalis systematica*, 4 vols (Halle, 1718–40).

Honterus, Johannes, *Rudimentorum cosmographiae libri duo* (Kraków, 1530).

Honterus, Johannes, *Rudimenta cosmographica*, 2nd, definitive edn (Braşov, 1542).

Horn, Georg, *Orbis politicus imperiorum, regnorum, principatuum, rerumpublicarum* (Leipzig, 1668).

Horne, Johannes van, *Suarum circa partes generationis in utroque sexu observationum prodromus* (Leiden, 1668).

Horrocks, Jeremiah, 'Excerpta ex epistolis... ad Gulielmum Crabtrium, suum in studiis astronomicis socium', in *Opera posthuma* (London, 1673), pp. 241–338.

Hottinger, Johann Heinrich, 'Montium glacialium Helveticorum descriptio', *Miscellanea curiosa*, decuriae III. annus nonus et decimus (1706), pp. 41–75.

Huet, Pierre-Daniel, *De interpretatione libri duo, quorum prior est de optimo genere interpretandi, alter de claris interpretibus* (Paris, 1661).

Humboldt, Alexander von, *Florae Fribergensis specimen plantas cryptogamas, praesertim subterraneas, exhibens* (Berlin, 1793).

Humphrey, Lawrence, *Interpretatio linguarum seu de ratione convertendi et explicandi autores tam sacros quam profanos libri tres* (Basel, 1559).

Huswirt, Johannes, *Enchiridion novus algorismi* ([Cologne, 1501]).

Hutton, James, *Dissertatio physico-medica inauguralis de sanguine et circulatione microcosmi* (Leiden, 1749).

Huygens, Christiaan, 'De ratiociniis in ludo aleae', in Frans van Schooten, *Exercitationum mathematicarum libri quinque* (Leiden, 1657), pp. 521–34.

Huygens, Christiaan, *Κοσμοθεωρὸς sive de terris caelestibus* (The Hague, 1698).

Huygens, Constantijn, the Younger, 'Primus inaccessum...', in Frans van Schooten and Florimond de Beaune, *Geometria a Renato des Cartes edita* (Amsterdam, 1659–61), at beginning of vol. 1.

al-Idrīsī, Muḥammad Ibn-Muḥammad, tr. Gabriel Sionita and Johannes Hesronita, *Geographia Nubiensis, id est accuratissima totius orbis... descriptio* (Paris, 1619).

'L'imprimeur au lecteur', *Le journal des sçavans* (1665), unpaginated.

Index librorum prohibitorum (Venice, 1564).

Ingrassia, Giovanni Filippo, *In Galeni librum de ossibus... commentaria* (Palermo, 1603).

'Ioannis de Sacrobosco vita', in *Sphaera Ioannis de Sacrobosco typis auctior quam antehac* (Paris, 1545), sig. [Giii^{v}]; revised version: 'Elias Vinetus de Ioanne de Sacro Bosco ad Ioannem Tacitum φιλίατρον', in *Sphaera Ioannis de Sacrobosco typis auctior quam antehac... cum annotationibus et scholiis... Eliae Vineti* (Paris, 1551), fol. 2^{r-v}.

'Ioannis Garlandii... Synonymorum in arte alchimistica expositio', in *Compendium Alchimiae Ioannis Garlandii Angli philosophi doctissimi* (Basel, 1560), pp. 53–71.

Ittig, Thomas, *Lucubrationes academicae de montium incendiis* (Leipzig, 1671).

Jacobaeus, Matthias, 'In novos salivae et lacrimarum ductus', in Nicolas Steno, *Observationes anatomicae* (Leiden, 1662), sig. [*6^{v}].

Jacobi, Carl Gustav Jacob, *Fundamenta nova theoriae functionum ellipticarum* (Königsberg, 1829).

Jacquin, Nikolaus Joseph von, *Oxalis. Monographia iconibus illustrata* (Vienna etc., 1794).

Javellus, Chrysostomus, *Super tres libros Aristotelis De anima quaestiones subtilissimae* (Venice, 1552).

Jessen, Johann von, *De vita et morte illustris et generosi viri domini Tychonis Brahei... oratio funebris* (Prague, 1601).

John of Genoa, *Catholicon* (Mainz, [*c.*1460]).

Jonston, John, *Thaumatographia naturalis in decem classes distincta* (Amsterdam, 1632).

Jonston, John, *Historia naturalis animalium*, 5 vols (Frankfurt am Main, 1650–3).

Jonston, John, *Notitia regni mineralis seu subterraneorum catalogus* (Leipzig, 1661).

Jonston, John, *Notitia regni vegetabilis seu plantarum a veteribus observatarum... series* (Leipzig, 1661).

Juncker, Christian, *Schediasma historicum de ephemeridibus sive diariis eruditorum* (Leipzig, 1692).

Jungermann, Ludwig, and Caspar Hofmann, *Catalogus plantarum circa Altorfium Noricum et vicinis quibusdam locis* (Altdorf, 1615).

Junius, Hadrianus, *Nomenclator omnium rerum propria nomina variis linguis explicata indicans* (Antwerp, 1567).

Jurin, James, 'Invitatio ad observationes meteorologicas communi consilio instituendas', *Philosophical Transactions*, 32/379 (1723), pp. 422–7.

Jussieu, Antoine Laurent de, *Genera plantarum secundum ordines naturales disposita* (Paris, 1789).

Keill, John, *Introductio ad veram astronomiam seu lectiones astronomicae habitae in schola astronomica Academiae Oxoniensis* (Oxford, 1718).

Kentmann, Johannes, 'Catalogus rerum fossilium', in Conrad Gessner (ed.), *De omni rerum fossilium genere* (Zurich, 1565), separately paginated.

Kepler, Johannes, *Prodromus dissertationum cosmographicarum continens mysterium cosmographicum* (Tübingen, 1596); 2nd edn (Frankfurt am Main, 1621).

Kepler, Johannes, *De stella nova in pede serpentarii* (Prague, 1606).

Kepler, Johannes, *Astronomia nova αἰτιολογητὸς seu physica coelestis* (Heidelberg, 1609).

Kepler, Johannes, 'Respondet author operis', in *Astronomia nova* (Heidelberg, 1609), sigs. [(**)6^{r}]–(***)^{r}.

Kepler, Johannes, *Dissertatio cum Nuntio sidereo nuper ad mortales misso a Galilaeo Galilaeo* (Prague, 1610).

Kepler, Johannes, *Dioptrice seu demonstratio eorum, quae visui et visibilibus propter conspicilla non ita pridem inventa accidunt* (Augsburg, 1611).

Kepler, Johannes, *Strena seu de nive sexangula* (Frankfurt am Main, 1611).

Kepler, Johannes, *Epitome astronomiae Copernicanae usitata forma quaestionum et responsionum conscripta*, 3 vols (Linz and Frankfurt am Main, 1618–21).

Kepler, Johannes, *Tabulae Rudolphinae, quibus astronomicae scientiae temporum longinquitate collapsae restauratio continetur* (Ulm, 1627).

Kepler, Johannes, *Somnium seu opus posthumum de astronomia lunari* (Żagań and Frankfurt am Main, 1634).

Kepler, Johannes (tr.), 'Plutarchi... libellus de facie, quae in orbe lunae apparet', in *Somnium* (Żagań and Frankfurt am Main, 1634), pp. 97–184.

Kepler, Johannes, and Jacob Bartsch, *Tabulae manuales logarithmicae ad calculum astronomicum, in specie Tabularum Rudolphinarum, compendiose tractandum mire utiles*, 2nd edn (Strasbourg, 1700).

Kies, Johann, *Dissertatio physica de iride* (Tübingen, 1772).

Kinner, Cyprian, 'In Selenographiam Hevelio-Gedanensem ad auctorem... epigramma', in Johannes Hevelius, *Selenographia* (Gdańsk, 1647), sig. [***4^{v}].

Kircher, Athanasius, *Musurgia universalis sive ars magna consoni et dissoni*, 2 vols (Rome, 1650).
Kircher, Athanasius, *Mundus subterraneus in XII libros digestus*, 2 vols (Amsterdam, 1665); 2nd and 3rd edns (1678).
Kircher, Athanasius, *Phonurgia nova* (Kempten, 1673).
Kircher, Athanasius, *Arca Noe in tres libros digesta* (Amsterdam, 1675).
Kircher, Athanasius, *d'Onder-Aardse Weereld*, 2 vols (Amsterdam, 1682).
Kirchmaier, Georg Caspar, *De locustis insolitis... dissertatio epistolica* (Wittenberg, 1693).
Klein, Jacob Theodor, *Stemmata avium quadraginta tabulis aeneis ornata / Geschlechtstafeln der Vögel mit vierzig Kupfern erläutert* (Leipzig, 1759).
Krause, Johann Gottlieb, 'Lectori litterarum elegantiorum amanti', *Nova litteraria... in supplementum Actorum eruditorum divulgata* (1718), sigs. A[r]–A2[v].
Kraz, Georg, 'Dissertatio pro praemio: De ratione motus medii et distantiae mediae lunae a terra ad vires, quibus in lunam premitur', *Abhandlungen der Churfürstlich-baierischen Akademie der Wissenschaften*, 4 (1767), pp. 271–307.
Kühn, Carl Gottlieb, *Bibliotheca medica continens scripta medicorum omnis aevi ordine methodico disposita* (Leipzig, 1794).
Lachmund, Friedrich, *De ave Diomedea dissertatio cum vera eius effigie* (Amsterdam, 1674).
LaCroix, Demetrius de, *Connubia florum Latino carmine demonstrata* (Paris, 1728).
Laet, Johannes de, *Nieuwe Wereldt ofte Beschrijvinghe van West-Indien* (Leiden, 1625).
Laet, Johannes de, *Novus orbis seu descriptionis Indiae occidentalis libri XVIII* (Leiden, 1633).
Laet, Johannes de, *L'histoire du nouveau monde ou description des Indes occidentales* (Leiden, 1640).
Laet, Johannes de (tr., comm.), 'Theophrasti De lapidibus liber', in *De gemmis et lapidibus libri duo, quibus praemittitur Theophrasti liber de lapidibus* (Leiden, 1647), separately foliated.
Lagomarsini, Girolamo, 'De origine fontium / Dell' origine de' fonti', in Francesco Eulalio Savastano and Giampietro Bergantini (tr.), *I quattro libri delle cose botaniche* (Venice, 1749), pp. 1–35.
Laguna, Andrés de, *Epitome Galeni Pergameni operum in quattuor partes digesta* (Basel, 1551).
La Hire, Philippe de, *Tabulae astronomicae Ludovici magni iussu et munificentia exaratae et in lucem editae* (Paris, 1702).
La Hire, Philippe de, and Johann Albrecht Klimm (tr.), *Astronomische Tabellen* (Nuremberg, 1725).
La Hire, Philippe de, *Tables astronomiques* (Paris, 1735).
La Loubère, Antoine de, *Veterum geometria promota in septem de cycloide libris* (Toulouse, 1660).
Lansbergen, Philips van, *Triangulorum geometriae libri quattuor* (Leiden, 1591).
Lansbergen, Philips van, *Tabulae motuum coelestium perpetuae* (Middelburg, 1632).
Ledesma, Miguel Jerónimo (tr.), *Prima primi Canonis Avicennae sectio* (Valencia, 1547).
Leeuwenhoek, Antonie van, 'Observationes... de natis e semine genitali animalculis', *Philosophical Transactions*, 12/142 (1679), pp. 1040–3.
Leeuwenhoek, Antonie van, *Anatomia et contemplatio nonnullorum naturae invisibilium secretorum* (Leiden, 1685).
Leeuwenhoek, Antonie van, *Anatomia seu interiora rerum... ope et beneficio exquisitissimorum microscopiorum detecta* (Leiden, 1687).
Leeuwenhoek, Antonie van, *Arcana naturae detecta* (Delft, 1695).
Leeuwenhoek, Antonie van, *Continuatio Arcanorum naturae detectorum* (Delft, 1697).
Leeuwenhoek, Antonie van, *Opera omnia*, 4 vols (Leiden, 1715–22).

Le Febvre, François Antoine, *Terrae-motus. Carmen*, 2nd edn (Paris, 1704) (1st edn probably from the same year, apparently lost).

Leibniz, Gottfried Wilhelm, 'Nova methodus pro maximis et minimis, item tangentibus... et singulare pro illis calculi genus', *Acta eruditorum* (1684), pp. 467–73.

Leo Africanus, 'Descrittione dell'Africa', in Gian Battista Ramusio (ed.), *Navigationi et viaggi*, vol. 1 (Venice, 1550), fols. 1^r–103^v.

Leoniceno, Niccolò (tr.), 'Aristotelis Quaestiones mechanicae', in *Opuscula nuper in lucem edita* (Venice, 1525), fols. xxiii^r–liiii^v.

Leopold, Christoph, and Christoph Vian, *Philosophia historica de montibus* (Innsbruck, 1713).

Le Seur, Thomas, François Jacquier, and Jean Louis Calandrini (eds, comm.), *Philosophiae naturalis principia mathematica auctore Isaaco Newtono... perpetuis commentariis illustrata*, 3 vols (Geneva, 1739–42).

Lesser, Friedrich Christian, *De sapientia, omnipotentia et providentia divina ex partibus insectorum cognoscenda epistolaris disquisitio* (Nordhausen, 1735).

Lhuyd, Edward, *Lithophylacii Britannici ichnographia* (London, 1699).

Lhuyd, Edward, 'Epistola VI. De fossilium marinorum et foliorum mineralium origine ad... Joannem Rajum', in *Lithophylacii Britannici ichnographia* (London, 1699), pp. 131–45.

Libavius, Andreas (ed.), *Alchemia Libavii* (Frankfurt am Main, 1597).

Lichtenberg, Georg Christoph, 'Observationes astronomicae per annum 1772 et 1773 ad situm Hannoverae, Osnabrugi et Stadae determinandum institutae', *Novi commentarii Societatis regiae scientiarum Gottingensis*, 7 (1777), pp. 210–32.

Linacre, Thomas (tr.), *Procli Diadochi Sphaera astronomiam discere incipientibus utilissima* (Venice, 1499).

Linden, Johannes Antonides van der, *De scriptis medicis libri duo, quibus praemittitur ad dominum Petrum Tulpium manuductio ad medicinam* (Amsterdam, 1637).

Linden, Johannes Antonides van der, *Hippocrates de circuitu sanguinis, exercitatio I., II., ..., XXVII.* (Leiden, 1659–63).

Linnaeus, Carl, *Fundamenta botanica, quae... theoriam scientiae botanicae per breves aphorismos tradunt* (Amsterdam, 1735).

Linnaeus, Carl, *Systema naturae sive regna tria naturae systematice proposita* (Leiden, 1735); 10th edn, 2 vols (Stockholm, 1758–9); 12th edn, 4 vols (Stockholm, 1766–8).

Linnaeus, Carl, *Musa Cliffortiana florens Hartkampi 1736 prope Harlemum* (Leiden, 1736).

Linnaeus, Carl, *Critica botanica, in qua nomina plantarum... examini subiciuntur* (Leiden, 1737a).

Linnaeus, Carl, *Genera plantarum eorumque characteres naturales* (Leiden, 1737b).

Linnaeus, Carl, *Hortus Cliffortianus* (Amsterdam, 1737c).

Linnaeus, Carl, *Oratio de telluris habitabilis incremento* (Leiden, 1744).

Linnaeus, Carl, *Flora Suecica exhibens plantas per regnum Sueciae crescentes systematice* (Stockholm, 1745).

Linnaeus, Carl, *Fauna Suecica sistens animalia Sueciae regni* (Stockholm, 1746).

Linnaeus, Carl, *Amoenitates academicae seu dissertationes variae physicae, medicae, botanicae antehac seorsim editae*, 10 vols (Leiden, Stockholm, and Erlangen, 1749–90).

Linnaeus, Carl, *Philosophia botanica, in qua explicantur Fundamenta botanica* (Stockholm, 1751).

Linnaeus, Carl, *Species plantarum* (Stockholm, 1753).

Linnaeus, Carl, and Nils Nilsson Amann, *Flora Alpina* (Uppsala, 1756).

Lipsius, Justus, 'Ad Clusii nomen lusus', in Carolus Clusius, *Rariorum stirpium per Pannoniam et Austriam observatarum historia* (Antwerp, 1583), sig. *3^v.

L'Obel, Matthias de, *Plantarum seu stirpium historia* (Antwerp, 1576).

L'Obel, Matthias de, and Pierre Pena, *Stirpium adversaria nova* (London, 1571).
Longueil, Gilbert de, and William Turner (ed.), *Dialogus de avibus* (Cologne, 1544).
Lonicer, Johannes (tr.), *Nicandri… Theriaca et Alexipharmaca cum scholiis* (Cologne, 1531).
Lossius, Lucas, *Arithmetices erotemata puerilia… in gratiam et usum scholarum puerilium Latinarum collecta* (Frankfurt an der Oder, [1557]).
Lubieniecki, Stanisław, *Theatrum cometicum* (Amsterdam, 1668).
Ludwig, Christoph Gottlieb, et al., 'Instituti ratio', *Commentarii de rebus in medicina et scientia naturali gestis*, 1 (1752), pp. iii–xviii.
Lusitanus, Amatus, *Curationum medicinalium centuria prima*[*–septima*] (Florence, etc., 1551–66).
Luther, Martin, *Ein sendbrieff… von Dolmetschen und fürbitt der heiligenn* ([Nuremberg], 1530).
Maestlin, Michael, *Epitome astronomiae* (Heidelberg, 1582); 2nd edn (Tübingen, 1588); 3rd edn (Tübingen, 1610).
Magini, Giovanni Antonio, *Confutatio diatribae Scaligeri de aequinoctiorum praecessione* (Rome, 1617).
Maimonides, *Aphorismi secundum doctrinam Galeni* (Bologna, 1489).
Major, Johann Daniel, *Emblemata de purpura* (Kiel, 1671).
Major, Johann Daniel, 'Observatio VI', *Miscellanea curiosa*, 3 (1672), p. 11.
Major, Johann Daniel, and Philipp Jacob Sachs von Lewenheimb, *Dissertatio epistolica de cancris et serpentibus petrefactis… Responsoria dissertatio historico-medica… de miranda lapidum natura* (Jena, 1664).
Manardi, Giovanni, *Opus… de erroribus Symonis Pistoris de Lypczk circa morbum Gallicum* ([Nuremberg, c.1500]).
Manardi, Giovanni, *Epistolae medicinales, in quibus multa recentiorum errata et antiquorum decreta reserantur* (Ferrara, 1521).
Manget, Jean Jacques (ed.), *Bibliotheca chemica curiosa seu rerum ad alchemiam pertinentium thesaurus instructissimus*, 2 vols (Geneva, 1702).
Manincor, Anton, Anton Waid, and Peter Prantner, *Disputatio philosophica de terraemotibus anni 1670* (Innsbruck, 1671).
Marggraff, Michael, and Johannes Bleidner, *Disputatio physica the meteoro nivis* (Leipzig, 1667).
Märter, Franz Joseph, *Fundamenta et termini botanici* (Brussels, 1789).
Martini, Martino, *Novus Atlas Sinensis* (Amsterdam, 1655).
Mattioli, Pietro Andrea (tr., comm.), *Di Pedacio Dioscoride… libri cinque della historia e materia medicinale* (Venice, 1544).
Mattioli, Pietro Andrea, *Commentarii in sex libros Pedacii Dioscoridis Anazarbei de medica materia* (Venice, 1554); definitive edn (1565).
Mattioli, Pietro Andrea, *Compendium de plantis omnibus una cum earum iconibus, de quibus scripsit in Commentariis suis in Dioscoridem editis* (Venice, 1571); 2nd edn titled *De plantis epitome utilissima* (Frankfurt am Main, 1586).
Maurolico, Francesco, *Cosmographia… in tres dialogos distincta* (Venice, 1543).
Mazzolari, Giuseppe Maria, *Electricorum libri VI* (Rome, 1767).
Melanchthon, Philip, *Elementorum rhetorices libri duo* (Wittenberg, 1531).
Melanchthon, Philip (ed.), *Liber Iohannis de Sacro Busto de sphaera* (Wittenberg, 1531).
Melanchthon, Philip, *Commentarius de anima* (Strasbourg, 1540); 2nd edn titled *Liber de anima* (Wittenberg, 1553).
Melanchthon, Philip, *Initia doctrinae physicae dictata in Academia Vuitebergensi* (Wittenberg, 1549); 2nd edn (Frankfurt am Main, 1550).

Melanchthon, Philip, and Georg Joachim Rheticus, 'Oratio de astronomia et geographia', in *Selectarum declamationum Philippi Melanchthonis... tomus primus* (Strasbourg, 1544), pp. 350–61.

Memoir's for a Natural History of Animals Containing the Anatomical Descriptions of Several Creatures Dissected by the Royal Academy of Sciences at Paris (London, 1688).

Memus, Johannes Baptista (tr.), *Apollonii Pergaei... opera... de Graeco in Latinum traducta* (Venice, 1537).

Mencke, Friedrich Otto, *De vita, moribus, scriptis meritisque... Fracastorii* (Leipzig, 1731).

Mercator, Gerardus, *Atlas sive cosmographicae meditationes de fabrica mundi et fabricati figura* (Duisburg and Düsseldorf, 1595).

Mercuriale, Girolamo, *De morbis cutaneis et omnibus humani corporis excrementis tractatus* (Venice, 1572).

Mercuriale, Girolamo, *De arte gymnastica libri sex* (Venice, 1573).

Mercuriale, Girolamo, *Censura de Hippocratis operibus* (Venice, 1583).

Merian, Maria Sibylla, *Metamorphosis insectorum Surinamensium* (Amsterdam, 1705).

Merian, Maria Sibylla, *Metamorphosis insectorum Surinamensium ofte verandering der surinaamsche insecten* (Amsterdam, [1705]).

Merian, Maria Sibylla, *Dissertatio de generatione et metamorphosibus insectorum Surinamensium/Dissertation sur la génération et les transformations des insectes de Surinam* (The Hague, 1726).

Merklin, Georg Abraham (ed.), *Lindenius renovatus sive Johannis Antonidae van der Linden De scriptis medicis libri duo* (Nuremberg, 1686).

Mersenne, Marin, *Cogitata physico-mathematica, in quibus tam naturae quam artis effectus admirandi certissimis demonstrationibus explicantur* (Paris, 1644).

Mesmer, Franz Anton, *Dissertatio physico-medica de planetarum influxu* (Vienna, 1766).

Milich, Jakob, 'Oratio... de Avicennae vita', in *Oratio de consideranda sympathia et antipathia* (Wittenberg, 1550), sigs. [Biii^r–Ciii^r]; also in *Selectarum declamationum Philippi Melanchthonis... tomus tertius* (Strasbourg, 1559), pp. 460–72.

Mizauld, Antoine, *De mundi sphaera seu cosmographia: libri tres* (Paris, 1552).

Mizauld, Antoine, *Ephemerides aëris perpetuae seu popularis et rustica tempestatum astrologia ubique terrarum et vera et certa* (Paris, 1554).

Mizauld, Antoine, 'Encomium docti astronomi et periti caeli interpretis', in *Cosmologia historiam caeli et mundi... colligens et dilucide proponens* (Paris, 1571), pp. 65–72.

Mizler, Lorenz Christoph, 'Epistola ad Polonos litteratos', *Acta litteraria regni Poloniae et magni ducatus Lithuaniae*, 1 (1755), pp. 9–13.

Mizler, Lorenz Christoph, '[Review of Jacques Émilien Camuset, *Traité des fièvres malignes épidemics* (Dresden, 1738)]', *Acta litteraria regni Poloniae et magni ducatus Lithuaniae*, 1 (1755), pp. 62–7.

Moffett, Thomas, *De iure et praestantia chymicorum medicamentorum dialogus apologeticus* (Frankfurt am Main, 1584).

Moffett, Thomas, *Insectorum sive minimorum animalium theatrum, olim ab Edoardo Wottono, Conrado Gesnero Thomaque Pennio inchoatum* (London, 1634).

Molnár, János, *Physiologicon complexum historiae naturalis tria regna* (Buda, 1780).

Monau, Friedrich von, *Bronchotome, nimirum gutturis artificiose aperiendi ἐγχείρησις* (Königsberg, 1644); 2nd edn (Greifswald, 1649); 3rd edn (Jena, 1711).

Mondella, Luigi, *Epistolae medicinales variarum quaestionum et locorum insuper Galeni difficilium expositionem continentes* (Basel, 1543).

Montalbani, Ovidio, *Bibliotheca botanica* (Bologna, 1657).

Montanus, Johannes Fabricius, 'Galandae montis longe altissimi... stirpium enumeratio', in Conrad Gessner (ed.), *Valerii Cordi Simesusii Annotationes in Pedacii Dioscoridis Anazarbei De medica materia libros V* (Strasbourg, 1561), fol. 236ᵛ.
Monte, Giovanni Battista da, and Walenty z Lublina (ed.), *In nonum librum Rhasis ad Mansorem regem Arabum expositio* (Venice, 1554).
Monteux, Sebastien de, *Annotatiunculae in Errata recentiorum medicorum per Leonardum Fuchsium collecta* (Lyons, 1533).
Morescottus, Alfonsus, *Compendium totius medicinae* (Frankfurt am Main, 1584).
Morhof, Daniel Georg, *Polyhistor sive de notitia auctorum et rerum commentarii*, 3 vols (Lübeck, 1688–1708); 4th and last edn, 3 vols (1747).
Morison, Robert, *Plantarum umbelliferarum distributio nova* (Oxford, 1672).
Morrhy, Gérard, 'In operis commendationem epigramma', in Oronce Finé, *Protomathesis* (Paris,1532), unpaginated.
Müller, Otto Frederik, *Vermium terrestrium et fluviatilium... succincta historia*, 2 vols (Leipzig and Copenhagen, 1773–4).
Müller von Löwenstein, Friedrich, *Lexicon medico-galeno-chymico-pharmaceuticum* (Frankfurt am Main, 1661) (two otherwise identical versions with German and Latin paratexts respectively).
Münster, Sebastian, *Cosmographia. Bschreibung aller Lender* (Basel, 1544).
Münster, Sebastian, *Cosmographiae universalis libri VI* (Basel, 1552).
Muralt, Johann von, *Scientiae naturalis seu physicae compendium* (Zurich, 1694).
Musae Anglicanae sive poemata quaedam melioris notae, 2 vols (London, 1741).
Musschenbroek, Petrus van, *Epitome elementorum physico-mathematicorum conscripta in usus academicos* (Leiden, 1726).
Napier, John, *Mirifici logarithmorum canonis descriptio eiusque usus... explicatio* (Edinburgh, 1614).
Naudé, Gabriel, *Advis pour dresser une bibliothèque* (Paris, 1627).
Naudé, Gabriel, 'De Cardano iudicium', in Girolamo Cardano, *De propria vita liber* (Paris, 1643), sigs. aiiiiʳ–*viiᵛ; also in Girolamo Cardano, *Opera omnia*, 10 vols (Lyons, 1663), sigs. iʳ–62ᵛ.
Nemnich, Philipp Andreas, *Lexicon nosologicum polyglotton* (Hamburg, 1801).
Neucrantz, Paul, *De harengo exercitatio medica* (Lübeck, 1645).
Newton, Isaac, 'A Letter... Containing His New Theory about Light and Colors', *Philosophical Transactions*, 6/80 (1672), pp. 3075–87.
Newton, Isaac, 'Letter of April 13. 1672... Being an Answer to the Fore-Going Letter of P. Pardies', *Philosophical Transactions*, 7/84 (1672), pp. 4091–3.
Newton, Isaac, *Philosophiae naturalis principia mathematica* (London, 1687); 3rd edn (London, 1726).
Newton, Isaac, *Opticks or, a Treatise of the Reflexions, Refractions, Inflexions and Colours of Light* (London, 1704).
Newton, Isaac, 'Tractatus de quadratura curvarum', in *Opticks* (London, 1704), pp. 165–211.
Newton, Isaac, *The Method of Fluxions and Infinite Series* (London, 1736).
Newton, Isaac, *Opuscula mathematica, philosophica et philologica*, 3 vols (Lausanne and Geneva, 1744).
Newton, Isaac, 'Methodus fluxionum et serierum infinitarum', in *Opuscula* (Lausanne and Geneva, 1744), vol. 1, pp. 31–200.
Niccolai, Alfonso, 'Carmen elegiacum', in *Arcadum carmina. Pars altera* (Rome, 1756), pp. 243–51.

Nieremberg, Juan Eusebio, *Historia naturae, maxime peregrinae, libris XVI distincta* (Antwerp, 1635).

Noceti, Carlo, *De iride et aurora boreali carmina...cum notis Josephi Rogerii Boscovich* (Rome, 1747).

Noceti, Carlo, and Antonio Ambrogi (tr.), *L'Iride e l'Aurora boreale descritte in verso latino...e tradotte in verso toscano* (Florence, 1755).

Non-entia chymica (Frankfurt am Main, 1645).

Nuñez, Pedro, *Opera* (Basel, 1566).

Nuñez, Pedro, 'In Theoricas planetarum Georgii Purbachii annotationes aliquot', in *Opera* (Basel, 1566), pp. 197–307.

[Obituary of Roger Joseph Boscovich], *Gentleman's Magazine*, 57 (1787), pp. 274–5.

'Observationes quaedam meteorologicae in Pennsylvania habitae', *Miscellanea Berolinensia ad incrementum scientiarum*, 5 (1737), pp. 123–31.

Oddi, Oddo degli, *In primam totam fen primi libri Canonis Avicennae...expositio* (Venice, 1575).

O[ldenburg], H[enry] (tr.), *The Prodromus to a Dissertation Concerning Solids Naturally Contained within Solids...by Nicolaus Steno* (London, 1671).

Olhoff, Johann Erich (ed.), *Excerpta ex litteris illustrium et clarissimorum virorum ad... Iohannem Hevelium...perscriptis* (Gdańsk, 1683).

Ortelius, Abraham, *Theatrum orbis terrarum* (Antwerp, 1570).

Oudin, François (ed.), *Poemata didascalica nunc primum vel edita vel collecta*, 3 vols (Paris, 1749); 2nd edn, 3 vols (1813).

Pallavicino, Sforza, *Trattato dello stile e del dialogo*, 3rd edn (Rome, 1662).

Pardies, Ignace Gaston, 'A Latin Letter...Containing Some Animadversions upon Mr. Isaac Newton...His Theory of Light', *Philosophical Transactions*, 7/84 (1672), pp. 4087–90.

Passe, Crispijn van de, the Younger, *Hortus floridus, in quo rariorum et minus vulgarium florum icones...exhibentur* (Arnhem, 1614).

Patrizi, Francesco, *Nova de universis philosophia, in qua Aristotelica methodo non per motum, sed per lucem et lumina ad primam causam ascenditur* (Ferrara, 1591).

Paulli, Simon, *Flora Danica, det er: Dansk Urtebog* (Copenhagen, 1648).

Paulli, Simon, *Orbis terraqueus in tabulis geographicis et hydrographicis descriptus* (Strasbourg, 1670).

Pauw, Pieter, *Hortus publicus Academiae Lugduno-Batavae* (Leiden, 1601).

Peano, Giuseppe, *Arithmetices principia nova methodo exposita* (Turin, 1889).

Pecquet, Jean, *Experimenta nova anatomica* (Paris, 1651).

Peletier, Jacques, *L'algèbre...departie en deus livres* (Lyons, 1554).

Peletier, Jacques, *De occulta parte numerorum, quam algebram vocant, libri duo* (Paris, 1560).

Petermann, Daniel, and Gottfried Schmied, *Disputationem de glacie sub praesidio magistri Danielis Petermanni examini publico exponit Gottfried Schmied* (Leipzig, 1670).

Petreius, Johannes (ed.), *In hoc volumine de alchemia continentur haec:...* (Nuremberg, 1541).

Peuerbach, Georg von, *Theoricae novae planetarum* ([Nuremberg, 1472]).

Peuerbach, Georg von, and Franciscus Capuanus (comm.), *Theoricae novae planetarum cum commento* (Venice, 1495).

[Peuerbach, Georg von, and] Johannes Regiomontanus, *Epitoma...in Almagestum Ptolemaei* (Venice, 1496).

Peuerbach, Georg von, and Johannes Regiomontanus, *Tabulae eclipsium magistri Georgii Peurbachii. Tabula primi mobilis Joannis de Monteregio* (Vienna, 1514).

Pharmacopoea Austriaco-provincialis emendata (Vienna, 1794).

Pharmacopoea Germanica (Berlin, 1872); 2nd edn (1882).

Pico della Mirandola, Giovanni, *Disputationes... adversus astrologiam divinatricem* (Bologna, 1496).

Pictorius, Georg, *Rei medicae totius compendiosa traditio* (Basel, 1558).

Piso, Willem, and Georg Marggraf, *Historia naturalis Brasiliae* (Leiden and Amsterdam, 1648).

Planer, Johann Andreas, et al., *De nive dissertatio prima/secunda/tertia/quarta*, 4 vols (Wittenberg, 1695–6).

Platner, Johann Zacharias, 'Relatio de machinis chirurgicis novissime inventis', *Acta eruditorum* (1729), pp. 347–9.

Platter, Felix, *De corporis humani structura et usu... libri III* (Basel, 1583).

Ploucquet, Wilhelm Gottfried, *Literatura medica digesta sive repertorium medicinae practicae, chirurgiae atque rei obstetriciae*, 4 vols (Tübingen, 1808–9).

Polignac, Melchior de, *Anti-Lucretius sive de deo et natura libri novem* (Paris, 1747).

Pollucius, Samuel, 'Anagrammata quaedam', in Maria Cunitz, *Urania propitia* (Byczyna and Oleśnica, 1650), sigs. []):(6^{v}]–):():(^{r}.

Pomponazzi, Pietro, *Dubitationes in quartum Meteorologicorum Aristotelis librum* (Venice, 1563).

Pontano, Giovanni, *Opera* (Venice, 1505).

Pontano, Giovanni, 'Meteororum liber unus', in *Opera* (Venice, 1505), sigs. [ov^{r}]–s^{v}.

Pontano, Giovanni, 'Urania sive de stellis libri quinque', in *Opera* (Venice, 1505), sigs. aii^{r}–oiiii^{v}.

Pontano, Giovanni, *Commentationes super centum sententiis Ptolemaei* (Naples, 1512).

Pontano, Giovanni, *De rebus coelestibus* (Naples, 1512).

Porzio, Lucantonio, *Erasistratus sive de sanguinis missione* (Rome, 1682).

Porzio, Simone, *De conflagratione agri Puteolani... epistola* (Florence, 1551).

Praetorius, Martin, 'Phaleuci ad Libavium de chymicis', in Andreas Libavius, *Alchemia Libavii* (Frankfurt am Main, 1597), sigs. [b3^{v}–b4^{r}].

Pritzel, Georg August, *Thesaurus literaturae botanicae omnium gentium*, 2nd edn (Leipzig, 1872).

Quillet, Claude, *Callipaedia seu de pulchrae prolis habendae ratione poema didacticon* (Paris, 1655); 2nd edn (1656).

Quillet, Claude, 'In obitu Petri Gassendi... lugubre encomium', in *Callipaedia* (Paris, 1656), pp. 91–4.

Quintiis, Camillus Eucherius de, *Inarime seu de balneis Pithecusarum libri VI* (Naples, 1726).

Ralla, Johannes, and Friedrich Berger, *Σὺν ἐμπεδοχρησοπονήτῳ Chionognosian... praeside magistro Friderico Bergern... publice proponet Johannes Ralla* (Leipzig, 1658).

Ramus, Petrus, *Arithmeticae libri duo, geometriae septem et viginti* (Basel, 1569).

Ray, John, *Historia plantarum species hactenus editas aliasque insuper multas noviter inventas et descriptas complectens*, 3 vols (London, 1686–1704).

Redi, Francesco, *Esperienze intorno alla generazione degl'insetti* (Florence, 1668).

Regiomontanus, Johannes, [*Ephemerides*] (Nuremberg, 1474).

Regiomontanus, Johannes, *Tabulae directionum profectionumque... in nativitatibus multum utiles* (Augsburg, 1490).

Regiomontanus, Johannes, 'Oratio habita Patavii in praelectione Alfragani', in *Rudimenta astronomica Alfragani* (Nuremberg, 1537), sigs. [α4^{r}–β5^{r}].

Reinhold, Erasmus (ed., comm.), *Theoricae novae planetarum Georgii Purbacchii... pluribus figuris auctae et illustratae scholiis* (Wittenberg, 1542).

Reinhold, Erasmus, *Prutenicae tabulae caelestium motuum* (Tübingen, 1551).
Reisch, Georg, *Margarita philosophica* (Freiburg im Breisgau, 1503).
Renieri, Vincenzo, *De Etruscarum antiquitatum fragmentis Scornelli prope Vulterram repertis disquisitio astronomica* (Florence, 1638).
Reuss, Jeremias David, *Repertorium commentationum a societatibus litterariis editarum*, 16 vols (Göttingen, 1801–21).
[Review of Boskovich, *De solis et lunae defectibus* (1760)], *Nova acta eruditorum* (1761), pp. 168–74.
[Review of Hebenstreit, Johann Ernst, *Anthropologia forensis* (Leipzig, 1751)], *Commentarii de rebus in scientia naturali et medicina gestis*, 1 (1752), pp. 365–78.
[Review of Lambert, Johann Heinrich, *Pyrometrie* (Berlin, 1779)], *Nova acta eruditorum* (1776) (*sic*—the *Nova acta* were far behind schedule), pp. 162–72.
[Review of Moscati, Pietro, *Discorso academico dei vantaggi della educazione filosofica nello studio della chimica* (Milan, 1784)], *Commentarii de rebus in scientia naturali et medicina gestis*, 27 (1785), pp. 491–2.
[Review of Stay, Benedict, *Philosophia recentior* (Rome, 1755–92), vol. 1], *Journal de Trévoux* (1756), pp. 81–108.
[Review of *Verhandelingen uitgegeeven door de Hollandse Maatschappy der Weetenschappen te Harlem*, 1 (1754)], *Commentarii de rebus in scientia naturali et medicina gestis*, 6 (1757), pp. 309–23.
[Review of *Verhandelingen*, 3 (1757)], *Commentarii de rebus in scientia naturali et medicina gestis*, 9 (1761), pp. 3–22.
[Review of *Verhandelingen*, 5 (1760)], *Commentarii de rebus in scientia naturali et medicina gestis*, 11 (1763), pp. 120–39.
Rheticus, Georg Joachim, *De libris revolutionum…Nicolai Copernici…narratio prima* (Gdańsk, 1540).
Rheticus, Georg Joachim, *Canon doctrinae triangulorum* (Leipzig, 1551).
Rheticus, Georg Joachim, 'Dialogus de Canone doctrinae triangulorum', in *Canon doctrinae triangulorum* (Leipzig, 1551), sigs. Ci^{v}–Civ^{r}.
Riccioli, Giovanni Battista, *Almagestum novum astronomiam veterem novamque complectens*, 2 vols in 3 parts (Bologna, 1651–65).
Riccioli, Giovanni Battista, *Geographiae et hydrographiae reformatae libri duodecim* (Bologna, 1661).
Roberval, Gilles Personne de, *Aristarchi Samii De mundi systemate, partibus et motibus eiusdem libellus* (Paris, 1644).
Rodigast, Samuel, and Johann Nikolaus Trombsdorff, *Disquisitio physica de calore nivis* (Jena, 1678).
Rohault, Jacques, *Traité de physique* (Paris, 1671).
Rondelet, Guillaume, *Libri de piscibus marinis*, 2 vols (vol. 2 titled *Aquatilium historia*) (Lyons, 1554–5).
Rösel von Rosenhof, August Johann, *Historia naturalis ranarum nostratium/Die natürliche Historie der Frösche hiesigen Landes* (Nuremberg, 1758).
Ruel, Jean (tr.), *Pedacii Dioscoridis Anazarbei De medicinali materia libri quinque* (Paris, 1516).
Ruel, Jean (tr.), *Veterinariae medicinae libri II* (Paris, 1530).
Ruel, Jean, *De natura stirpium libri tres* (Paris, 1536).
Ruf, Jacob, *Ein schön lustig Trostbüchle von den empfengknussen und geburten der menschen unnd jren vilfaltigen zuofälen und verhindernussen* (Zurich, 1554).
Ruf, Jacob, and Wolfgang Haller (tr.), *De conceptu et generatione hominis et iis, quae circa haec potissimum considerantur, libri sex* (Zurich, 1554).

Ruland, Martin, the Younger, *Lexicon Alchemiae sive dictionarium alchemisticum* (Frankfurt am Main, 1612).
Rümelin, Johann Heinrich, *Dissertatio historico-mathematica de vita, scriptis et inventis Ioannis Kepleri* (Tübingen, 1770).
Rumpf, Georg Eberhard, and Johannes Burmann (ed., tr.), *Herbarium Amboinense/Het amboinsch kruid-boek*, 6 vols (Amsterdam, 1741–50).
Ryff, Walther Hermann, *Iatromathematicae, hoc est medicationis accomodatae ad astrologicam rationem, enchiridion* (Strasbourg, 1542).
Sachs von Lewenheimb, Philipp Jacob, *Ampelographia sive vitis viniferae eiusque partium consideratio* (Leipzig, 1661).
Sachs von Lewenheimb, Philipp Jacob, *Gammarologia sive gammarorum, vulgo cancrorum, consideratio* (Frankfurt am Main and Leipzig, 1665).
Saia, Nonio Marcello, *Tractatus, in quo adversus antiquorum et praecipue Peripateticorum opinionem terram esse aqua maiorem... demonstratur* (Paris, 1585).
Sainte-Marthe, Scévole de, *Paedotrophiae libri III* (Paris, 1584).
Sainte-Marthe, Scévole de, and Abel de Sainte-Marthe (tr.), *La manière de nourrir les enfans à la mammelle* (Paris, 1698).
Saladinus de Asculo, *Aromatariorum compendium* (Bologna, 1488).
Salernus, Iacobus, *Positiones physio-astronomicae de sphaera elementari* (Naples, 1682).
Salvemini, Giovanni, 'De vita Isaaci Newtoni commentariolus', in Isaac Newton, *Opuscula* (Lausanne and Geneva, 1744), vol. 1, pp. xxi–xxxiii.
Salviani, Ippolito, *Aquatilium animalium historiae liber primus* (Rome, 1554).
Sambucus, Johannes, *Icones veterum aliquot ac recentium medicorum philosophorumque elegiolis suis editae* (Antwerp, 1574); 2nd edn ([Leiden], 1603).
Santorio, Santorio, *Ars... de statica medicina aphorismorum sectionibus septem comprehensa* (Venice, 1614).
Sapidus, Johannes, 'Qui cupis ingenitas herbarum noscere vires...', in Otto Brunfels, *Herbarum vivae eicones* (Strasbourg,1530–6), vol. 1, sig. A4^{r}.
Saumaise, Claude, *Exercitationes de homonymis hyles iatricae* (Utrecht, 1689).
Savastano, Francesco Eulalio, *Botanicorum seu institutionum rei herbariae libri IV* (Naples, 1712).
Savastano, Francesco Eulalio, and Giampietro Bergantini (tr.), *I quattro libri delle cose botaniche... colla traduzione in verso sciolto Italiana* (Venice, 1749).
Savoy, Gaspare, and Joseph Alphons Widmann, *Dissertatio philosophica de planeticolis* (Trento, 1770).
Scaliger, Joseph, *In Manilii quinque libros Astronomicon commentarius et castigationes* (Paris, 1579).
Scaliger, Joseph, *Opus novum de emendatione temporum in octo libros tributum* (Paris, 1583).
Scaliger, Joseph, *Thesaurus temporum Eusebii Pamphili* (Leiden, 1606).
Scaliger, Joseph, *De aequinoctiorum anticipatione diatriba* (Paris, 1613).
Scaliger, Joseph, 'In lexicorum compilatores', in *Poemata omnia ex museio Petri Scriverii* ([Leiden], 1615), p. 35; reprinted with German adaptation in Johann Christoph Sommerhoff, *Lexicon pharmaceutico-chymicum* (Nuremberg, 1701), second-last page of prefatory matter.
Scaliger, Julius Caesar (ed., comm.), *Ἀριστοτέλους περὶ ζῴων ἱστορίας/Aristotelis Historia de animalibus* (Toulouse, 1619).
Scheiner, Christoph, *Disquisitiones mathematicae de controversiis et novitatibus astronomicis* (Ingolstadt, 1614).
Scheiner, Christoph, *Exegeses fundamentorum gnomonicorum* (Ingolstadt, 1615).

Scheiner, Christoph, *Sol ellipticus, hoc est novum et perpetuum solis contrahi soliti phaenomenon* (Augsburg, 1615).
Schenck, Johann Georg, *Biblia iatrica sive bibliotheca medica* (Frankfurt am Main, 1609).
Scheuchzer, Johann Jakob, *Charta invitatoria quaestionibus, quae historiam Helvetiae naturalem concernunt, praefixa* and *Einladungs-Brief zu Erforschung natürlicher Wunderen, so sich im Schweitzer-Land befinden* ([Zurich, 1699]).
Scheuchzer, Johann Jakob (tr.), *Specimen geographiae physicae, quo agitur de terra et corporibus terrestribus, speciatim mineralibus... authore Johanne Woodwardo* (Zurich, 1704).
Scheuchzer, Johann Jakob, *Bibliotheca scriptorum historiae naturali omnium terrae regionum inservientium, Historiae naturalis Helvetiae prodromus* (Zurich, 1716).
Scheuchzer, Johann Jakob, *Helvetiae historia naturalis oder Natur-Historie des Schweitzerlandes*, 3 vols (Zurich, 1716–18).
Scheuchzer, Johann Jakob, 'Acarnania sive relatio eorum, quae hactenus elaboravit Acarnan, Johannes Jacobus Scheuchzer Tigurinus', *Miscellanea Lipsiensia*, 8 (1718), pp. 117–55; 2nd edn in *Οὐρεσιφοιτὴς Helveticus* (Leiden, 1723), pp. 609–35.
Scheuchzer, Johann Jakob, *Οὐρεσιφοιτὴς Helveticus sive itinera per Helvetiae Alpinas regiones facta* (Leiden, 1723).
Scheuchzer, Johann Jakob, *Physica sacra... iconibus aeneis illustrata* and *Kupfer-Bibel*, 4 vols (Augsburg and Ulm, 1731–5).
[Scheuchzer, Johann Jakob, and] Johann Heinrich Hottinger, *Κρυσταλλολογία seu dissertatio de crystallis* (Zurich, 1698).
Schickard, Wilhelm, *Pars responsi ad epistolas Petri Gassendi... de Mercurio sub sole viso* (Tübingen, 1632).
Schindler, Conrad, *Specimen inaugurale medicum de caseo rasili viridi* (Basel, 1755).
Schneider, Johann Friedemann, *Tractatus logicus singularis, in quo processus disputandi seu officia aeque ac vitia... exhibentur* (Halle, 1718).
Schönborn, Samuel, *Manuale medicinae practicae Galenico-chymicae* (Gdańsk, 1637).
Schöner, Johannes, *Tabulae astronomicae, quas vulgo, quia omni difficultate et obscuritate carent, resolutas vocant* (Nuremberg, 1536).
Schöner, Johannes (ed.), *Scripta clarissimi mathematici magistri Ioannis Regiomontani* (Nuremberg, 1544).
Schöner, Johannes, *De iudiciis nativitatum libri tres* (Nuremberg, 1545).
Schoock, Martin, et al., *Disputatio physica de nive/de glacie/de hieme/de aqua*, 4 vols (Groningen, 1646).
Schooten, Frans van (tr., comm.), and Florimond de Beaune (comm.), *Geometria a Renato des Cartes anno 1637 Gallice edita*, 2 vols, 2nd, definitive edn (Amsterdam, 1659–61); 3rd edn (1683).
Schott, Kaspar, *Physica curiosa sive mirabilia naturae et artis libris XII comprehensa* (Würzburg, 1662).
Schott, Kaspar, *Technica curiosa sive mirabilia artis libris XII comprehensa* (Nuremberg, 1664).
Schreckenfuchs, Erasmus Oswald, *Commentaria in Novas theoricas planetarum Georgii Purbachii* (Basel, 1556).
Schulze, Johann Heinrich, 'Scotophorus pro phosphoro inventus', in Johannes Christophorus Franck (ed.), *Bibliotheca novissima observationum ac recensionum* (Halle, 1721), pp. 234–40.
Schweling, Johann Eberhard, and Johann Heinrich Crocius, *Disquisitio philosophica de anima brutorum* (Bremen, 1676).
Schwenckfeld, Caspar, *Stirpium et fossilium Silesiae catalogus* (Leipzig, 1600).

Schwenckfeld, Caspar, *Theriotropheum Silesiae* (Legnica, 1603).

Scribonius, Wilhelm Adolf, *Isagoge sphaerica methodica* (Frankfurt am Main, 1580).

Scripta quaedam in Academia Witenbergensi... publice proposita (and variants of this title), 11 vols (Wittenberg, 1546–72).

Semmelweis, Ignaz Philipp, *Dissertatio inauguralis sistens tractatum de vita plantarum* (Vienna, 1844).

Sennert, Daniel, *Institutionum medicinae libri V* (Wittenberg, 1611).

Sfondrati, Pandolfo, *Causa aestus maris* (Ferrara, 1590).

Sigwart, Georg Friedrich (ed.), *Quaestiones medicae Parisinae* (Tübingen, 1759).

Simmler, Josias, *Vita clarissimi philosophi et medici excellentissimi Conradi Gesneri Tigurini* (Zurich, 1566).

Societas meteorologica Palatina, 'Monitum ad observatores Societatis meteorologicae Palatinae', *Ephemerides Societatis meteorologicae Palatinae: Historia et observationes anni 1781* (1783), pp. 8–17.

Sommerhoff, Johann Christoph, *Lexicon pharmaceutico-chymicum Latino-Germanicum et Germanico-Latinum* (Nuremberg, 1701).

Sorbière, Samuel, 'Praefatio, in qua de vita et moribus Petri Gassendi disseritur', in Pierre Gassendi, *Opera omnia* (Lyons, 1658), vol. 1, sigs. a^r–[$i5^v$].

Spener, Christian Maximilian, 'Disquisitio de crocodilo in lapide scissili expresso aliisque lithozois', *Miscellanea Berolinensia ad incrementum scientiarum*, 1 (1710), pp. 99–118.

Spener, Christian Maximilian, 'Recensio *Itineris Alpini* Ioannis Iacobi Scheuchzeri', *Miscellanea Berolinensia ad incrementum scientiarum*, 1 (1710), pp. 139–42.

Spiegel, Adriaan van de, *De humani corporis fabrica libri decem* (Venice, 1627).

Sprat, Thomas, *The History of the Royal-Society of London* (London, 1667).

Sprengel, Kurt, *Literatura medica externa recentior* (Leipzig, 1829).

Stahl, Georg Ernst, *Zymotechnia fundamentalis seu fermentationis theoria generalis* (Halle, 1697).

Stahl, Georg Ernst, *Zymotechnia fundamentalis oder allgemeine Grund-Erkänntniß der Gährungs-Kunst* (Frankfurt am Main and Leipzig, 1734).

Stattler, Benedikt, *Mineralogiae et metallurgiae chemicae principia physica*, 2 vols (Innsbruck, 1765).

Stay, Benedict, *Philosophiae recentioris versibus traditae... libri decem*, 3 vols (Rome, 1755–92); one-vol. edn (Rome, 1792).

Stay, Christoph, 'De poesi didascalica dialogus', in Benedict Stay, *Philosophia recentior* (Rome, 1755–92), vol. 3, separately paginated.

Steno, Nicolas, *Observationes anatomicae* (Leiden, 1662).

Steno, Nicolas, 'De vitelli in intestina pulli transitu epistola', in *De musculis et glandulis observationum specimen* (Amsterdam, 1664), pp. 76–90.

Steno, Nicolas, *De solido intra solidum naturaliter contento dissertationis prodromus* (Florence, 1669).

Stifel, Michael, *Arithmetica integra* (Nuremberg, 1544).

Stöffler, Johannes, and Jakob Pflaum, *Almanach nova plurimis annis venturis inservientia* (Ulm, 1499).

Stromeyer, Friedrich, 'De gas hydrogenii arseniati natura atque indole commentatio', *Commentationes Societatis regiae scientiarum Gottingensis*, 16 (1808), pp. 141–68.

Struś, Józef (tr.), *Galeni De urinis liber* (Venice, 1535).

Sturm, Johann Christoph, *Mathesis iuvenilis*, 2 vols (Nuremberg, 1699–1701).

Sturm, Johann Christoph, *Mathesis iuvenilis. Das ist: Anleitung der Jugend zur Mathesin* (Nuremberg, 1705).

Sturm, Johann Christoph, *Mathesis iuvenilis: or a Course of Mathematics for Young Students*, 3 vols (London, 1708–9).

Sturm, Johann Christoph, and Jacob Hieronymus Lochner, *De terrae-motibus eorumque praecipuis accidentibus e causis suis genuinis ac naturalibus demonstrativa methodo deductis* (Altdorf, 1670).

Sturm, Reginald, *In septem libros Aphorismorum Hippocratis paraphrasis poetica ad illorum memoriam summe utilis* (Lyons, 1582).

Swammerdam, Jan, *Miraculum naturae sive uteri muliebris fabrica, notis in domini Johan van Horne prodromum illustrata* (Leiden, 1672).

Swammerdam, Jan, Herman Boerhaave (ed.), and Hieronymus David Gaub (tr.), *Bybel der Natuure of historie der insecten/Biblia Naturae sive historia insectorum*, 2 vols (Leiden, 1737).

Syllanus de Nigris, *In Almansoris nonum collectiones* ([Padua], 1476).

Szerdahélyi, György Alajos, *Historia Uraniae Musae, quam…recens detexit Herschelius* (Vienna, 1787).

Tannstetter Collimitius, Georg, 'Viri mathematici, quos inclytum Viennense gymnasium ordine celebres habuit', in Georg von Peuerbach and Johannes Regiomontanus, *Tabulae eclipsium* (Vienna, 1514), sigs. aa3^{v}–aa5^{v}.

Tarillon, François, *Pulvis pyrius, carmen* (Parma, 1699).

Tartaglia, Niccolò (ed.), *Opera Archimedis Syracusani philosophi et mathematici ingeniosissimi* (Venice, 1543).

Taruffi, Giuseppe Antonio, *Montgolferii machina volans. Carmen elegiacum* (n.p., 1784).

Tazewell, William, *Vade-mecum medicum* (Paris, Philadelphia, PA, and Edinburgh, 1798).

Telesio, Bernardino, *De rerum natura iuxta propria principia libri IX* (Naples, 1586).

Thomas, Loup, 'Barometrum', in François Oudin, *Poemata didascalica* (Paris, 1749), vol. 3, pp. 192–209.

Thorer, Alban (trans.), *Pauli Aeginetae, medici insignis, opus divinum* (Basel, 1532).

Thorius, Raphael, *Hymnus tabaci* (London, 1626).

Tilli, Michaele Angelo, *Catalogus plantarum horti Pisani* (Florence, 1723).

Toxites, Michael, and Johann Fischart, 'Onomasticon alterum, Theophrasti Paracelsi', in *Onomastica duo* (Strasbourg, 1574), pp. 383–490.

Triegler, Johann Georg, *Sphaera. Das ist: Ein kurtzes Astronomisches Tractätlein* (Leipzig, 1614).

Trithemius, Johannes, *Liber de scriptoribus ecclesiasticis* (Basel, 1494).

Turcq, Pascasius Justus, *Alea sive de curanda ludendi in pecuniam cupiditate, libri II* (Basel, 1561); 3rd edn (Amsterdam, 1642).

Uterverius, Johannes Cornelius, 'Non tua, Aristoteles…', in Ulisse Aldrovandi, *Ornithologia* (Bologna, 1599–1603), at the beginning of vol. 1.

Vadianus, Joachim (ed., comm.), *Pomponii Melae Hispani libri de situ orbis tres* (Vienna, 1518); 2nd edn (Basel, 1522).

Vadianus, Joachim, *Epitome trium terrae partium, Asiae, Africae et Europae* (Zurich, 1534).

Valentini, Michael Bernhard (tr.), *Amphitheatrum zootomicum tabulis aeneis quam plurimis exhibens historiam animalium anatomicam* (Frankfurt am Main and Giessen, 1720).

Valentini, Pietro Aloisio, *Institutiones medicinae practicae*, 10 vols (Rome, 1823–47).

Valla, Giorgio (tr.), 'In Ptolemaei ad Syrum Apotelesmata commentarium', in *Commentationes* (Venice, 1502), sigs. Aii^{r}–[Giv^{r}].

Varenius, Bernhardus, *Geographia generalis, in qua affectiones generales telluris explicantur* (Amsterdam, 1650).

Varolio, Constanzio, and Girolamo Mercuriale, *De nervis opticis nonnullisque aliis praeter communem opinionem in humano capite observatis* (Padua, 1573).

Venusti, Antonio Maria, *Consilia medica* (Venice, 1571).

Vesalius, Andreas (tr.), *Paraphrasis in nonum librum Rhazae, medici Arabis clarissimi, ad regem Almansorem* (Basel, 1537).
Vesalius, Andreas, *De humani corporis fabrica libri septem* (Basel, 1543); 2nd edn (1555).
Vesling, Johann, *Syntagma anatomicum. Publicis dissectionibus in auditorum usum diligenter aptatum* (Padua, 1641).
Vespucci, Amerigo, *Mundus novus* (Augsburg, 1504).
Vettori, Benedetto, *Commentaria in Tractatum proportionum Alberti de Saxonia* (Bologna, 1506).
Viète, François, *In artem analyticem isagoge* (Tours, 1591); also in *Opus restitutae mathematicae analyseos seu algebra nova* (Tours, 1591), separately paginated.
Virginianus, Saxirupius, 'Comparatio Ptolemaei et Kepleri', in Johannes Kepler, *Epitome astronomiae Copernicanae* (Linz and Frankfurt am Main, 1618–21), vol. 1, sig. [*6^r].
Vitali, Girolamo, *Lexicon mathematicum, astronomicum, geometricum* (Paris, 1668).
Vives, Juan Luis, *De ratione dicendi libri tres* (Leuven, 1533).
Voigt, Gottfried, *Contra nivis albedinem realem dissertatio physica* (Güstrow, 1669).
Vorstius, Aelius Everardus, 'Oratio funebris in obitum Caroli Clusii', in Carolus Clusius, *Curae posteriores* ([Leiden], 1611), separately paginated.
Vossius, Isaac, *De Nili et aliorum fluminum origine* (The Hague, 1666).
Vossius, Isaac, *Dissertation touchant l'origine du Nile et autres fleuves* (Paris, 1667).
Wagner, Christian, *Animadversiones in Thomae Burnetii Telluris theoriam sacram* (Leipzig, 1683).
Waldung, Wolfgang, *Lagographia. Natura leporum... Liber singularis* (Amberg, 1619).
Wedel, Georg Wolfgang, *Specimen experimenti chimici novi de sale volatili plantarum* (Frankfurt am Main, 1671); 4th edn as a self-contained book (Jena, 1682); also in *Miscellanea curiosa*, 4–5 (1676), pp. 215–47.
Wedel, Georg Wolfgang, *Tabulae synopticae de compositione medicamentorum extemporanea* (Jena, 1677).
Wedel, Georg Wolfgang, 'De latere Christi aperto', in *Exercitationum medico-philologicarum decas tertia* (Jena, 1687), pp. 1–6.
Wedel, Georg Wolfgang, *Tabulae chimicae XV in synopsi universam chimiam exhibentes* (Jena, 1692).
Weidler, Johann Friedrich, *Bibliographia astronomica* (Wittenberg, 1755).
Weinhart, Ignaz, *Quaestiones et responsa ex arithmetica vulgari, algebra et geometria* (Innsbruck, 1765).
Weinhart, Ignaz, and Maximilian Hell, 'Elogium rustici Tyrolensis celeberrimi Petri Anich', in Maximilian Hell, *Ephemerides astronomicae anni 1767* (Vienna, 1766), separately paginated; also as a self-contained book (Innsbruck, 1768).
Welsch, Georg Hieronymus, *Sylloge curationum et observationum medicinalium* (Ulm, 1667).
[Widmann, Johannes], *Algorithmus minutiarum vulgarium* ([Leipzig, *c.*1490]).
Wiggers, Johann Georg, *Über die Biographie* (Jelgava, 1777).
Willughby, Francis, and John Ray, *Ornithologiae libri tres* (London, 1676).
Winckler, Daniel, *De opio tractatus* (Leipzig, 1635).
Winter, Johann (trans.), *Claudii Galeni Pergameni De anatomicis administrationibus libri novem* (Basel, 1531).
Winter, Johann (trans.), *Pauli Aeginetae Opus de re medica nunc primum integrum Latinitate donatum* (Paris, 1532).
Winther, Mathias, *Bibliotheca Danorum medica* (Copenhagen, 1832).
Wolff, Christian, and Georg Remus, *Consideratio physico-mathematica hiemis proxime praeterlapsae* (Halle, 1709).

Woodward, John, *An Essay towards a Natural History of the Earth and Terrestrial Bodies, Especially Minerals* (London, 1695).
Wotton, Edward, *De differentiis animalium libri decem* (Paris, 1552).
Wurfbain, Johann Paul, *Salamandrologia, hoc est descriptio historico-philologico-philosophico-medica salamandrae, quae vulgo in igne vivere creditur* (Nuremberg, 1683).
Xylander, Wilhelm (tr.), *Diophanti Alexandrini Rerum arithmeticarum libri sex* (Basel, 1575).
Zallinger zum Thurn, Franz Seraphin von, *Praelectiones ex physica theoretica et experimentali* (Innsbruck, 1805).
Zamagna, Bernardo, *Echo, libri duo* (Rome, 1764).
Zetzner, Lazarus (ed.), *Theatrum chemicum*, 6 vols (Oberursel and Strasbourg, 1602–61).
Zwinger, Friedrich, 'Fungus nigricans rarus...', in Daniel Bruckner, *Versuch einer Beschreibung historischer und natürlicher Merkwürdigkeiten der Landschaft Basel. III. Stück* (Basel, 1749), pp. 271–2; republished in expanded form as 'Observatio botanica de fungo peculiari autumni tempore reperto', *Acta Helvetica*, 1 (1751), pp. 50–1.
Zwinger, Theodor, *In artem medicinalem Galeni tabulae et commentarii* (Basel, 1561).
Zwinger, Theodor, *Theatrum vitae humanae* (Basel, 1565).

Secondary literature

Modern editions of ancient and medieval works are only cited if specific reasons so require.

About, Pierre-José (ed.), *La Correspondence de Blaise Pascal et de Pierre de Fermat: La Géométrie du hasard ou le début du calcul des probabilités* (Fontenay aux Roses, 1983).
Abraham, Lyndy, *A Dictionary of Alchemical Imagery* (Cambridge, 1998).
Abrams, Meyer H., *The Mirror and the Lamp: Romantic Theory and the Critical Tradition* (New York, 1953).
Accordi, Bruno, 'Tentativi di classificazione delle pietre e delle gemme nei secoli XVI e XVII', *Physis: Rivista internazionale di storia della scienza*, 23 (1981), 311–24.
Adrados, Francisco R., *Geschichte der griechischen Sprache von den Anfängen bis heute* (Tübingen, 2001).
Agrimi, Jole, and Chiara Crisciani, *Les 'consilia' médicaux* (Turnhout, 1994).
Ahrens, Gerhard, *Medizinisches und naturwissenschaftliches Latein, mit latinisiertem griechischem Wortschatz* (Leipzig, 1988).
Aiton, Eric J., 'Peurbach's *Theoricae Novae Planetarum*: A Translation with Commentary', *Osiris*, 3 (1987), 4–43.
Aït-Touati, Frédérique, *Fictions of the Cosmos: Science and Literature in the Seventeenth Century* (Chicago, 2011).
Akopyan, Ovanes, *Debating the Stars in the Italian Renaissance: Giovanni Pico della Mirandola's Disputationes adversus astrologiam divinatricem and Its Reception* (Leiden and Boston, MA, 2021).
Albrecht, Ruth, 'Frühe Neuzeit', in Christian Klein (ed.), *Handbuch Biographie: Methoden, Traditionen, Theorien* (Stuttgart, 2009), 230–2.
Albury, William R., 'Halley's Ode on the *Principia* of Newton and the Epicurean Revival in England', *Journal of the History of Ideas*, 39 (1978), 24–43.
Alfieri, Vittorio Enzo, Marcello Carapezza, and Leonardo Sciascia (eds), *Pietro Bembo, De Aetna* (Palermo, 1981).
Alho, Tommi, 'Disputations at Seventeenth-Century Oxford', in Meelis Friedenthal, Hanspeter Marti, and Robert Seidel (eds), *Early Modern Disputations and Dissertations in an Interdisciplinary and European Context* (Leiden, 2021), 145–63.

Allgemeine Deutsche Biographie (Leipzig, 1875–1912).

Allweis, Werner, 'Von der Disputation zur Dissertation', in Rudolf Jung and Paul Kaegbein (eds), *Dissertationen in Wissenschaft und Bibliotheken* (Munich, 1979), 13–27.

Alvermann, Dirk, 'Von steifen Matronen und tanzenden Amazonen: Latein und Deutsch als Gelehrtensprachen in der Greifswalder Universitätsgeschichte (17.–19. Jh.)', in Michael Prinz and Jürgen Schiewe (eds), *Vernakuläre Wissenschaftskommunikation: Beiträge zur Entstehung und Frühgeschichte der modernen deutschen Wissenschaftssprachen* (Berlin and Boston, MA, 2018), 15–46.

Ammannati, Francesco, 'Book Prices and Monetary Issues in Renaissance Europe', *Italian Journal of Library, Archives, and Information Science*, 9/2 (2018), 179–91.

Appuhn-Radtke, Sibylle, 'Formen und Funktionen des Thesenblattes: Programm, Plakat und Memorialbild', in Meelis Friedenthal, Hanspeter Marti, and Robert Seidel (eds), *Early Modern Disputations and Dissertations in an Interdisciplinary and European Context* (Leiden, 2021), 65–101.

Arrizabalaga, Jon, *The 'Articella' in the Early Press, c.1476–1534* (Cambridge and Barcelona, 1998).

Ashworth, William B., 'The Habsburg Circle', in Bruce T. Moran (ed.), *Patronage and Institutions: Science, Technology, and Medicine at the European Court, 1500–1750* (Woodbridge, 1991), 137–67.

Asper, Markus, *Griechische Wissenschaftstexte: Formen, Funktionen, Differenzierungsgeschichten* (Stuttgart, 2007).

Asper, Markus (ed.), *Writing Science: Medical and Mathematical Authorship in Ancient Greece* (Berlin, 2013).

Asúa, Miguel de, and Roger French, *A New World of Animals: Early Modern Europeans on the Creatures of Iberian America* (Aldershot, 2005).

Atran, Scott, *Cognitive Foundations of Natural History: Towards an Anthropology of Science* (Cambridge, 1990).

Avramov, Jordan, 'Men of Science as Objects of Observation in the Correspondence of Henry Oldenburg, 1641–77', in Dirk van Miert (ed.), *Communicating Observations in Early Modern Letters (1500–1675)* (London and Turin, 2013), 199–222.

Ax, Wolfram (ed.), *Von Eleganz und Barbarei: Lateinische Grammatik und Stilistik in Renaissance und Barock* (Wiesbaden, 2001).

Babeş, Ovidiu, 'Playing with the Ancients: The Cosmology of Gilles Personne de Roberval', *Perspectives on Science*, 30 (2022), 950–81.

Baier, Thomas, 'Das Bildungsprogramm des Camerarius', in Baier (ed.), *Camerarius Polyhistor: Wissensvermittlung im deutschen Humanismus* (Tübingen, 2017), 77–93.

Baldasso, Renzo, 'La stampa dell' editio princeps degli *Elementi* di Euclide (Venezia, Erhard Ratdolt, 1482)', in Lisa Pon and Craig Kallendorf (eds), *The Books of Venice/Il libro veneziano* (Venice, 2008), 61–100.

Bandy, Mark Chance, and Jean A. Bandy, *De natura fossilium (Textbook of Mineralogy) by Georgius Agricola: Translated from the First Latin Edition of 1546* (New York, 1955).

Banks, David, *The Development of Scientific Writing: Linguistic Features and Historical Context* (London, 2008).

Barker, Nicolas, 'Typography and the Meaning of Words: The Revolution in the Layout of Books in the Eighteenth Century', in Giles Barber and Bernhard Fabian (eds), *Buch und Buchhandel in Europa im achtzehnten Jahrhundert/The Book and the Book Trade in Eighteenth-Century Europe* (Hamburg, 1981).

Barton, William, 'Singing the Study of Sound: Literary Engagement with Natural Philosophy in the Act and Tripos Verses of Oxford and Cambridge', in Meelis

Friedenthal, Hanspeter Marti, and Robert Seidel (eds), *Early Modern Disputations and Dissertations in an Interdisciplinary and European Context* (Leiden, 2021), 164–87.

Barton, William, 'The Mountain at Nature's Peak: The Case of Eighteenth-Century Oxford's Lenten Verse', in Martin Korenjak and Irina Tautschnig (eds), *Die antike Literatur und die Wissenschaftliche Revolution* (Baden-Baden, 2023), 207–32.

Bauer, Barbara, 'Gott, Welt, Mensch und Sterne in Melanchthons *Initia doctrinae physicae*', in Jürgen Leonhardt (ed.), *Melanchthon und das Lehrbuch des 16. Jahrhunderts* (Rostock, 1997), 149–73.

Bäumer, Änne, *Geschichte der Biologie*, 3 vols (Frankfurt am Main etc., 1991–6).

Bäumer-Schleinkofer, Änne, *Die Geschichte der beobachtenden Embryologie: Die Hühnchenentwicklung als Studienobjekt über zwei Jahrtausende* (Frankfurt am Main, 1993).

Bäumer-Schleinkofer, Änne, 'Die Enzyklopädien als Etablierung des Forschungsgebietes Zoologie in der Renaissance', in Wolfgang Kullmann, Jochen Althoff, and Markus Asper (eds), *Gattungen wissenschaftlicher Literatur in der Antike* (Tübingen, 1998), 141–59.

Bazerman, Charles, *Shaping Written Knowledge: The Genre and Activity of the Experimental Article in Science* (Madison, WI, and London, 1988).

Beer, Susanna de, 'The Survival of Pliny in Padua: Transforming Classical Scholarship during the Botanical Renaissance', in Karl A. E. Enenkel (ed.), *Transformations of the Classics via Early Modern Commentaries* (Leiden and Boston, MA, 2014), 327–62.

Beer, Susanna de, Karl A. E. Enenkel, and David Rijser (eds), *The Neo-Latin Epigram: A Learned and Witty Genre* (Leuven, 2009).

Benner, Margareta, and Emin Tengström, *On the Interpretation of Learned Neo-Latin: An Explorative Study Based on Some Texts from Sweden (1611–1716)* (Gothenburg, 1977).

Bénouis, Mustapha K., *Le Dialogue philosophique dans la littérature française du seizième siècle* (The Hague and Paris, 1976).

Berberich, Winfried (ed.), *Uranometria von Johannes Bayer 1603, Faksimile* (Gerchsheim, 2010).

Berghaus, Peter, Hans-Dietrich von Diepenbroick-Grüter, and Axel H. Murken (eds), *Porträt 2. Der Arzt: Graphische Bildnisse aus dem Porträtarchiv Diepenbroick* (Münster, 1979).

Berrens, Dominik, 'The Meaning of Flora', *Humanistica Lovaniensia*, 68 (2019), 237–49.

Berrens, Dominik, 'Names and Things: Latin and German Mining Terminology in Georgius Agricola's *Bermannus*', *Antike und Abendland*, 65/66 (2020), 232–43.

Berrens, Dominik, *Naming New Things and Concepts in Early Modern Science: The Case of Natural History* (unpublished habilitation thesis, University of Innsbruck, 2022).

Berrens, Dominik, 'Die Erhellung des Labyrinths: Pierre Gilles' Aelian-Übersetzung', in Martin Korenjak and Irina Tautschnig (eds), *Die antike Literatur und die Wissenschaftliche Revolution* (Baden-Baden, 2023), 61–80.

Berrens, Dominik, 'Camerarius und die medizinische Terminologie des 16. Jahrhunderts', in Marion Gindhart and Joachim Hamm (eds), *Camerarius im Kontext* (forthcoming).

Berschin, Walther, *Biographie und Epochenstil im lateinischen Mittelalter*, 5 vols (Stuttgart, 1986–2004).

Berschin, Walter (ed.), *Biographie zwischen Renaissance und Barock* (Heidelberg, 1993).

Berschin, Walther, 'Mittelalter', in Christian Klein (ed.), *Handbuch Biographie: Methoden, Traditionen, Theorien* (Stuttgart, 2009), 227–9.

Bertalot, Ludwig, 'Eine Sammlung Paduaner Reden des XV. Jahrhunderts', in *Studien zum italienischen und deutschen Humanismus*, vol. 2 (Rome, 1975), 209–35.

Berthold, Gerhard, *Der Magister Johann Fabricius und die Sonnenflecken neben einem Excurse über David Fabricius* (Leipzig, 1894).

Besson, Alain (ed.), *Thornton's Medical Books, Libraries and Collectors: A Study of Bibliography and the Book Trade in Relation to the Medical Sciences*, 3rd edn (Aldershot, 1990a).

Besson, Alain, 'Private Medical Libraries', in Besson (ed.), *Thornton's Medical Books, Libraries and Collectors* (Aldershot, 1990b), 267–300.

Besterman, Theodore, *A World Biography of Biographies and of Bibliographical Catalogues, Calendars, Abstracts, Digests, Indexes, and the Like*, 4 vols, 3rd edn (Geneva, 1955–6).

Betsch, Gerhard, and Jürgen Hamel (eds), *Zwischen Copernicus und Kepler: M. Michael Maestlinus Mathematicus Goeppingensis 1550–1631: Vorträge auf dem Symposium, veranstaltet in Tübingen vom 11. bis 13. Oktober 2000 von der Fakultät für Physik der Universität Tübingen* (Frankfurt am Main, 2002).

Bettini, Maurizio, *Vertere: Un' antropologia della traduzione nella cultura antica* (Turin, 2012).

Beullens, Pieter, 'Aristotle, His Translators, and the Formation of Ichthyologic Nomenclature', in Michèle Goyens, Pieter de Leemans, and An Smets (eds), *Science Translated: Latin and Vernacular Translations of Scientific Treatises in Medieval Europe* (Leuven, 2008), 105–22.

Beullens, Pieter, and Allan Gotthelf, 'Theodore Gaza's Translation of Aristotle's *De Animalibus*: Content, Influence and Date', *Greek, Roman, and Byzantine Studies*, 47 (2007), 469–513.

Beyer, Jürgen, 'How Complete Are the German National Bibliographies for the Sixteenth and Seventeenth Centuries (VD16 and VD17)?', in Malcolm Walsby and Graeme Kemp (eds), *The Book Triumphant: Print in Transition in the Sixteenth and Seventeenth Centuries* (Leiden and Boston, MA, 2011), 57–77.

Bianchi, Luca, 'Galileo fra Aristotele, Clavio e Scheiner: La nuova edizione del *Dialogo* e il problema delle fonti', *Rivista di Storia della Filosofia*, 54 (1999), 189–227.

Bienkowski, Piotr, and Alan Millard (eds), *British Museum Dictionary of the Ancient Near East* (London, 2000).

Biermann, Kurt-Reinhard, Hans Wußing, and Olaf Neumann (eds), *Mathematisches Tagebuch 1796–1814 von Carl Friedrich Gauß*, 5th edn (Frankfurt am Main, 2005).

Bitzel, Diane (ed.), *Bernardo Zamagna, Navis aeria: Eine Metamorphose des Lehrgedichts im Zeichen des technischen Fortschritts* (Frankfurt am Main, 1997).

Black, Crofton, 'Leo Africanus' *Descrittione dell' Africa* and Its Sixteenth-Century Translations', *Journal of the Warburg and Courtauld Institutes*, 65 (2002), 262–72.

Blair, Ann, 'La Persistance du latin comme langue de science à la fin de la Renaissance', in Roger Chartier and Pietro Corsi (eds), *Sciences et langues en Europe* (Paris, 1996), 21–42.

Blair, Ann, 'Mosaic Physics and the Search for a Pious Natural Philosophy in the Late Renaissance', *Isis*, 91 (2000a), 32–58.

Blair, Ann, 'Annotating and Indexing Natural Philosophy', in Marina Frasca-Spada and Nick Jardine (eds), *Books and the Sciences in History* (Cambridge, 2000b), 69–89.

Blair, Ann, 'The Collective Commentary as a Reference Genre', in Ralph Häfner and Markus Völkel (eds), *Der Kommentar in der Frühen Neuzeit* (Tübingen, 2006), 115–31.

Blair, Ann, *Too Much to Know: Managing Scholarly Information before the Modern Age* (New Haven, CT, 2010).

Blair, Ann, and Kaspar von Greyerz (eds), *Physico-Theology: Religion and Science in Europe, 1650–1750* (Baltimore, MD, 2020).

Bloemendal, Jan (ed.), *Bilingual Europe: Latin and Vernacular Cultures, Examples of Bilingualism and Multilingualism c.1300–1800* (Leiden and Boston, MA, 2015).

Blum, Rudolf, 'Bibliographia. Eine wort- und begriffsgeschichtliche Untersuchung', *Archiv für Geschichte des Buchwesens*, 10 (1969), cols. 1009–1246.

Blum, Rudolf, 'Die Literaturverzeichnung im Altertum und Mittelalter: Versuch einer Geschichte der Biobibliographie von den Anfängen bis zum Beginn der Neuzeit', *Archiv für Geschichte des Buchwesens*, 24 (1983), cols. 1–256.

Boer, Jan-Hendryk de, 'Kommentar', in Boer, Marian Füssel, and Maximilian Schuh (eds), *Universitäre Gelehrtenkultur vom 13.–16. Jahrhundert: Ein interdisziplinäres Quellen- und Methodenhandbuch* (Stuttgart, 2018), 265–318.

Boer, Jan-Hendryk de, 'Being Entitled to Dispute: On Disputations in Duisburg in the Second Half of the 17th Century', in Meelis Friedenthal, Hanspeter Marti, and Robert Seidel (eds), *Early Modern Disputations and Dissertations in an Interdisciplinary and European Context* (Leiden, 2021), 478–509.

Bogner, Ralf Georg, and Herfried Vögel, 'Protokoll der Diskussionen', in Franz M. Eybl et al. (eds), *Enzyklopädien der Frühen Neuzeit. Beiträge zu ihrer Erforschung* (Tübingen, 1995), 297–307.

Bohlin, Erik (ed.), *Om Cicero som matematiker: En 1700-talsdissertation under presidium av Petrus Ekerman: Inledning, översättning och förklarande noter* (Uppsala, 2015).

Böhme, Hartmut, '*Nur leerer Raum und Schatten*: Phantasien und Bilder über Weltraum und Vakuum, Planeten und Meteoriten', in Philipp Weber, Tim Sparenberg, and Reto Rössler (eds), *Kosmos und Kontingenz: Eine Gegengeschichte* (Paderborn, 2016), 77–99.

Bolton, Henry C., *Catalogue of Scientific and Technical Periodicals, 1665–1895: Together with Chronological Tables and a Library Check-List*, 2nd edn (Washington DC, 1897).

Bondi, Roberto (ed.), *Bernardino Telesio: La natura secondo i suoi principi* (Milan, 2009).

Bonelli, Federico, and Lucio Russo, 'The Origin of Modern Astronomical Theories of Tides: Chrisogono, de Dominis and their Sources', *British Journal for the History of Science*, 29 (1996), 385–401.

Boran, Elizabethanne, '*A Cloud of Witnesses*: Scientific Observation in the Correspondence of James Ussher (1581–1656)', in Dirk van Miert (ed.), *Communicating Observations in Early Modern Letters (1500–1675)* (London and Turin, 2013), 179–97.

Boscani Leoni, Simona, 'La ricerca nella montagna nel Settecento sotto nuove prospettive: Il network anglo-elvetico-alpino', *Histoire des Alpes/Geschichte der Alpen*, 12 (2007), 201–13.

Boschung, Urs, Hubert Steinke, and Wolfgang Proß (eds), *Repertorium zu Albrecht von Hallers Korrespondenz 1724–1777* (Basel, 2002).

Botley, Paul, *Latin Translation in the Renaissance: The Theory and Practice of Leonardo Bruni, Giannozzi Manetti and Desiderius Erasmus* (Cambridge, 2004).

Bouras-Vallianatos, Petros, and Barbara Zipser (eds), *Brill's Companion to the Reception of Galen* (Leiden and Boston, MA, 2019).

Bracke, Wouter, and Herwig Deumens (eds), *Medical Latin from the Late Middle Ages to the Eighteenth Century: Proceedings of the European Science Foundation, Exploratory Workshop in the Humanities, Organized under the Supervision of Albert Derolez in Brussels on the 3 and 4 September 1999* (Brussels, 2000).

Brammall, Sheldon, 'Laurence Humphrey, Gabriel Harvey, and the Place of Personality in Renaissance Translation Theory', *Review of English Studies*, 69 (2018), 56–75.

Brevaglieri, Sabina, 'Die Wege eines Chamäleons und dreier Bienen: Naturgeschichtliche Praktiken und Räume der politischen Kommunikation zwischen Rom und dem Darmstädter Hof zu Beginn des Dreißigjährigen Krieges', in Arndt Brendecke (ed.), *Praktiken der Frühen Neuzeit: Akteure, Handlungen, Artefakte* (Cologne, 2015), 131–50.

Brickman, Benjamin, *An Introduction to Francesco Patrizi's Nova de universis philosophia* (New York, 1941).

Broc, Numa, *La Géographie de la Renaissance: 1420–1620* (Paris, 1986).

Brock, William. H., 'Scientific Bibliographies and Bibliographers, and the History of the History of Science', in Andrew Hunter (ed.), *Thornton and Tully's Scientific Books, Libraries and Collectors,* 4th edn (Aldershot, 2000), 298–332.

Brockliss, Laurence, 'Printed Theses in Seventeenth- and Eighteenth-Century France', in Meelis Friedenthal, Hanspeter Marti, and Robert Seidel (eds), *Early Modern Disputations and Dissertations in an Interdisciplinary and European Context* (Leiden, 2021), 191–206.

Broecke, Marcel Peter René van den, *Ortelius' Theatrum Orbis Terrarum (1570–1641): Characteristics and Development of a Sample of 'on verso' Map Texts* (Utrecht, 2009).

Broman, Thomas, 'Periodical Literature', in Marina Frasca-Spada and Nick Jardine (eds), *Books and the Sciences in History* (Cambridge, 2000), 225–38.

Browne, Janet, *The Secular Ark: Studies in the History of Biogeography* (New Haven, CT, and London, 1983).

Bruning, Jens, 'Vorlesungsverzeichnisse', in Ulrich Rasche (ed.), *Quellen zur frühneuzeitlichen Universitätsgeschichte* (Wiesbaden, 2011), 269–92.

Brüning, Volker F., *Bibliographie der Kometenliteratur* (Stuttgart, 2000).

Brüning, Volker F., *Bibliographie der alchemistischen Literatur*, 3 vols (Munich, 2004–7).

Brüser, Clemens (ed.), 'Albert Curtz: "Problema Austriacum": Text mit Einleitung, Übersetzung, Anmerkungen und didaktischen Überlegungen' (unpublished diploma thesis, University of Innsbruck, 2019).

Brüser, Clemens, 'Mathematics and Panegyric: The *Problema Austriacum* by Albert Curtz', *Lias*, 48 (2021), 35–61.

Bryant, John, *The Fluid Text: A Theory of Revision and Editing for Book and Screen* (Ann Arbor, MI, 2002).

Buchwald, Georg, and Theo Herrle (eds), *Redeakte bei Erwerbung der akademischen Grade an der Universität Leipzig im 15. Jahrhundert: Aus Handschriften der Leipziger Universitätsbibliothek herausgegeben* (Wiesbaden, 1921).

Buck, August, 'Der humanistische Beitrag zur Ausbildung des naturwissenschaftlichen Denkens', in Buck (ed.), *Die humanistische Tradition in der Romania* (Bad Homburg, 1968), 165–81.

Buck, August, 'Die *studia humanitatis* im italienischen Humanismus', in Wolfgang Reinhard (ed.), *Humanismus im Bildungswesen des 15. und 16. Jahrhunderts* (Weinheim, 1984), 11–24.

Buck, August, and Otto Herding (eds), *Der Kommentar in der Renaissance* (Boppard, 1975).

Burke, Peter, '*Heu domine, adsunt Turcae*: A Sketch for a History of Post-Medieval Latin', in Burke (ed.), *The Art of Conversation* (Cambridge, 1993), 34–65.

Burke, Peter, *A Social History of Knowledge: Based on the First Series of Vonhoff Lectures Given at the University of Groningen (Netherlands)* (Cambridge, 2000).

Burke, Peter, *Languages and Communities in Early Modern Europe* (Cambridge, 2004).

Burke, Peter, 'Translations into Latin in Early Modern Europe', in Peter Burke and R. Po-chia Hsia (eds), *Cultural Translation in Early Modern Europe* (Cambridge, 2007), 65–80.

Burke, Peter, and R. Po-chia Hsia (eds), *Cultural Translation in Early Modern Europe* (Cambridge, 2007).

Burnett, Charles, 'Revisiting the 1552–1550 and 1562 Aristotle-Averroes Edition', in Anna Akasoy and Guido Giglioni (eds), *Renaissance Averroism and Its Aftermath: Arabic Philosophy in Early Modern Europe* (Dordrecht, 2013), 55–64.

Busard, Hubert L. L. (ed.), *Campanus of Novara and Euclid's Elements* (Stuttgart, 2005).

Byrne, James S., 'A Humanist History of Mathematics? Regiomontanus's Padua Oration in Context', *Journal of the History of Ideas*, 67 (2006), 41–61.

Cajori, Florian, *A History of Mathematical Notations*, 2 vols (Chicago, 1928–9).

Callegari, Guido, 'Le idee sulla pluralità dei mondi di G. A. Widmann di Coredo (Val di Non)', *Atti dell'imperial regia Accademia Roveretana*, 13 (1907), 249–67.

Camargo, Martin, *Ars dictaminis, Ars dictandi* (Turnhout, 1991).

Campi, Emilio, Simone De Angelis, Anja-Silvia Goeing, and Anthony Grafton (eds), *Scholarly Knowledge: Textbooks in Early Modern Europe* (Geneva, 2008).

Cantor, Moritz B., *Vorlesungen über Geschichte der Mathematik*, 4 vols (Leipzig, 1880–1908).

Carozzi, Albert V., and John K. Newman, *Horace-Bénédict de Saussure, Forerunner in Glaciology* (Geneva, 1995).

Casali, Elide, *Le spie del cielo: Oroscopi, lunari e almanacchi nell'Italia moderna* (Turin, 2003).

Casali, Elide, 'Pronostici, almanacchi, libri di ventura', in Germana Ernst and Guido Giglioni (eds), *Il linguaggio dei cieli: Astri e simboli nel Rinascimento* (Rome, 2012), 271–85.

Cassidy, David C., 'Meterology in Mannheim: The Palatine Meterological Society, 1780–1795', *Sudhoffs Archiv*, 69 (1985), 8–25.

Céard, Jean, 'Theory and Practices of Commentary in the Renaissance', in Judith R. Henderson (ed.), *The Unfolding of Words* (Toronto, 2012), 3–23.

Chang, Han-Liang, 'Calendar and Aphorism: A Generic Study of Carl Linnaeus' *Fundamenta Botanica* and *Philosophia Botanica*', in Britt L. Gunnarsson (ed.), *Languages of Science in the Eighteenth Century* (Berlin and Boston, MA, 2011), 263–78.

Chang, Kevin, 'From Oral Disputation to Written Text: The Transformation of the Dissertation in Early Modern Europe', *History of Universities*, 19 (2004), 129–87.

Chang, Kevin, 'Collaborative Production and Experimental Labor: Two Models of Dissertation Authorship in the Eighteenth Century', *Studies in History and Philosophy of Biological and Biomedical Sciences*, 41 (2010), 347–55.

Chang, Kevin, 'For the Love of Truth: The Dissertation as a Genre of Scholarly Publication in Early Modern Europe', *Know*, 5 (2021), 113–66.

Cioffarelli, Giovanni, 'La *Cosmographia* di Maurolico: Le novità nell'insegnamento dell'astronomia', in Vincenzo Fera, Daniela Gionta, and Antonio Rollo (eds), *Archimede e le sue fortune: Atti del convegno di Siracusa—Messina, 24–26 giugno 2008* (Messina, 2014), 551–65.

Clagett, Marshall (ed.), *Archimedes in the Middle Ages*, 5 vols (Madison, WI, 1964–84).

Clair, Pierre, *Jacques Rohault 1618–1672: Bio-bibliographie. Avec l'édition critique des entretiens sur la philosophie* (Paris, 1978).

Clough, Cecil, 'The Cult of Antiquity: Letters and Letter Collections', in Clough (ed.), *Cultural Aspects of the Italian Renaissance: Essays in Honour of Paul Oskar Kristeller* (Manchester and New York, 1976), 33–67.

Codazzi, Angela, *Le edizioni quattrocentesche e cinquecentesche della Geografia di Tolomeo* (Milan and Venice, 1950).

Cohen, Ernst, and Margareta Renkema, *Herman Boerhaave en zijne Beteekenis voor de Chemie* (Utrecht, 1918).

Cohen, Hendrik F., 'From Philosophia Naturalis to Science, from Latin to the Vernacular', in Jan Bloemendal (ed.), *Bilingual Europe* (Leiden and Boston, MA, 2015,) 144–60.

Collison, Robert, *Encyclopaedias: Their History throughout the Ages. A Bibliographical Guide with Extensive Historical Notes on the General Encyclopaedias Issued throughout the World from 350 B.C. to the Present Day* (New York and London, 1964).

Commentaria in Aristotelem Graeca (Berlin, 1882–1909).

Constable, Giles, *Letters and Letter-Collections* (Turnhout, 1976).

Conte, Gian Biagio, *Latin Literature: A History*, tr. Joseph B. Solodow, rev. Don Fowler and Glenn W. Most (Baltimore, MD, 1994).

Cook, Harold J., and Sven Dupré (eds), *Translating Knowledge in the Early Modern Low Countries* (Münster, 2012).

Coolidge, William A., *Josias Simler et les origines de l'alpinisme jusqu'en 1600* (Grenoble, 1904).

Cooper, Alix, *Inventing the Indigenous: Local Knowledge and Natural History in Early Modern Europe* (Cambridge, 2007).

Copeland, Rita, *Rhetoric, Hermeneutics and Translation in the Middle Ages: Academic Traditions and Vernacular Texts* (Cambridge, 1991).

Cormack, Lesley B., *Creating an Empire: Geography at the English Universities, 1580–1620* (Chicago and London, 1997).

Cortesi, Mariarosa (ed.), *Tradurre dal greco in età umanistica: Metodi e strumenti: Atti del Seminario di studio, Firenze, Certosa del Galluzzo, 9 settembre 2005* (Florence, 2007).

Cox, Virginia, *The Renaissance Dialogue: Literary Dialogue in Its Social and Political Contexts, Castiglione to Galileo* (Cambridge, 1992).

Cox, Virginia, 'Dialogue', in Victoria Moul (ed.), *A Guide to Neo-Latin Literature* (Cambridge, 2017), 289–307.

Crosland, Maurice P., *Historical Studies in the Language of Chemistry* (London, 1962).

Crowe, Michael J. (ed.), *The Extraterrestrial Life Debate, Antiquity to 1915: A Source Book* (Notre Dame, IN, 2008).

Crowther, Kathleen M., and Peter Barker, 'Training the Intelligent Eye: Understanding Illustrations in Early Modern Astronomy Texts', *Isis*, 104 (2013), 429–70.

Crupi, Gianfranco, 'Volvelles of Knowledge: Origin and Development of an Instrument of Scientific Imagination (13th–17th centuries)', *JLIS.it*, 10/2 (2019), 2–27.

Cuomo, Serafina, *Pappus of Alexandria and the Mathematics of Late Antiquity* (Cambridge, 2000).

D'Alessandro, Paolo, and Pier Daniele Napolitani, *Archimede latino: Iacopo da San Cassiano e il Corpus Archimedeo alla metà del quattrocento: Con edizione della Circuli dimensio e della Quadratura parabolae* (Paris, 2012).

Daly, Lloyd W., *Contributions to a History of Alphabetization* (Brussels, 1967).

Daly, Peter M. (ed.), *Companion to Emblem Studies* (New York, 2008).

Danielson, Dennis, *The First Copernican: Georg Joachim Rheticus and the Rise of the Copernican Revolution* (New York, 2006).

Davenport, Geoffrey, 'Medical Books of the Nineteenth Century', in Alain Besson (ed.), *Thornton's Medical Books, Libraries and Collectors* (Aldershot, 1990), 160–220.

Davids, Karel, 'In the Shadow of Jesuits: Isaac Vossius and Geography', in Eric Jorink and Dirk van Miert (eds), *Isaac Vossius (1618–1689) between Science and Scholarship* (Leiden and Boston, MA, 2012), 189–206.

Daxecker, Franz, and Lav Šubarić (eds), *Christoph Scheiners Sol ellipticus* (Innsbruck, 1998).

De Angelis, Simone, 'From Text to the Body: Commentaries on *De anima*, Anatomical Practice and Authority around 1600', in Emilio Campi et al. (eds), *Scholarly Knowledge* (Geneva, 2008), 205–27.

Dear, Peter R., '*Totius in verba*: Rhetoric and Authority in the Early Royal Society', *Isis*, 76 (1985), 145–61.

Dear, Peter R., *Discipline and Experience: The Mathematical Way in the Scientific Revolution* (Chicago, 1995).

Dear, Peter R., 'Experience and Experiment', in Wilbur Applebaum (ed.), *Encyclopedia of the Scientific Revolution from Copernicus to Newton* (New York and London, 2000), 340–7.

Delisle, Candice, '*The Spices of our Art*: Medical Observation in Conrad Gessner's Letters', in Dirk van Miert (ed.), *Communicating Observations in Early Modern Letters (1500–1675)* (London and Turin, 2013), 27–42.

Demidov, Sergei S., 'Leonhard Euler, Treatise on the Differential Calculus (1755)', in Ivor Grattan-Guinness (ed.), *Landmark Writings in Western Mathematics 1640–1940* (Amsterdam, 2005), 191–8.

Deneire, Tom (ed.), *Dynamics of Neo-Latin and the Vernacular: Language and Poetics, Translation and Transfer* (Leiden, 2014).

Deneire, Tom, 'Neo-Latin Literature and the Vernacular', in Victoria Moul (ed.), *A Guide to Neo-Latin Literature* (Cambridge, 2017), 35–51.

Dierse, Ulrich, *Enzyklopädie: Zur Geschichte eines philosophischen und wissenschaftstheoretischen Begriffs* (Bonn, 1977).

Diethelm, Oskar, *Medical Dissertations of Psychiatric Interest Printed before 1750* (Basel, 1971).

Dietz, Bettina (ed.), 'Translating and Translations in the History of Science', *Annals of Science*, 73 (2016), 117–223.

Dilg, Peter (ed.), 'Das Botanologicon des Euricius Cordus: Ein Beitrag zur botanischen Literatur des Humanismus' (unpublished dissertation, University of Marburg, 1969).

Dilg, Peter, 'Die botanische Kommentarliteratur Italiens um 1500 und ihr Einfluß auf Deutschland', in August Buck and Otto Herding (eds), *Der Kommentar in der Renaissance* (Boppard, 1975), 225–52.

Dinkova-Bruun, Greti, et al. (eds), *The Dimensions of Colour: Robert Grosseteste's De colore* (Toronto, 2013).

Dirckx, John H., *The Language of Medicine: Its Evolution, Structure, and Dynamics*, 2nd edn (New York, 1983).

Dispenza, Francesco, et al., 'The Discovery of Stapes', *Acta otorhinolaryngologica Italica*, 33 (2013), 357–9.

Ditchfield, Simon, *Liturgy, Sanctity and History in Tridentine Italy: Pietro Maria Campi and the Preservation of the Particular* (Cambridge, 1995).

Dizionario biografico degli Italiani, 100 vols (Rome, 1960–2020).

Dobrzycki, Jerzy, and Richard L. Kremer, 'Peurbach and Maragha Astronomy? The Ephemerides of Johannes Angelus and their Implications', *Journal for the History of Astronomy*, 27 (1996), 187–237.

Domtera-Schleichardt, Christiane, *Die Wittenberger 'Scripta publice proposita' (1540–1569): Universitätsbekanntmachungen im Umfeld des späten Melanchthon* (Leipzig, 2021).

Doody, Aude, Sabine Föllinger, and Liba Taub (eds), 'Structures and Strategies in Ancient Greek and Roman Technical Writing', *Studies in History and Philosophy of Science*, 43 (2012), 233–420.

Dooley, Brendan (ed.), *A Companion to Astrology in the Renaissance* (Leiden and Boston, MA, 2014).

Dreyer, John L. E. (ed.), *Tychonis Brahe Dani opera omnia*, 15 vols (Copenhagen, 1913–29).

Duhr, Bernhard (ed.), *Die Studienordnung der Gesellschaft Jesu* (Freiburg im Breisgau, 1896).

Duncan, Dennis, *Index, a History of the: A Bookish Adventure* (London, 2021).

Dupré, Sven, and Fernand Hallyn (eds), 'Early Modern Cosmography: Proceedings of a Conference Organized at the Centre for History of Science, Ghent University', *Archives Internationales d'Histoire des Sciences*, 59 (2009), 419–603.

Dupré, Sven, et al., 'Focus: Translating Science over Time', *Isis*, 109 (2018), 302–45.

Duris, Pascal (ed.), *Traduire la science: Hier et aujourd'hui* (Pessac, 2008).

Durling, Richard J., 'A Chronological Census of Renaissance Editions and Translations of Galen', *Journal of the Warburg and Courtauld Institutes*, 24 (1961), 230–305.

Durling, Richard J., 'Girolamo Mercuriale's *De modo studendi*', *Osiris*, 6 (1990), 181–95.

Eatough, Geoffrey (ed.), *Fracastoro's Syphilis: Introduction, Text, Translation and Notes* (Liverpool, 1984).

Eckart, Wolfgang U., 'Philipp Melanchthon und die Medizin', in Günther Frank and Stefan Rhein (eds), *Melanchthon und die Naturwissenschaften seiner Zeit* (Sigmaringen, 1998), 183–202.

Edney, Matthew H., and Mary S. Pedley, *The History of Cartography*, vol. 4. *Cartography in the European Enlightenment*, 2 vols (Chicago and London, 2019).

Edwards, Charles H., *The Historical Development of the Calculus* (New York, 1979).

Edwards, William F., 'Niccolò Leoniceno and the Origins of Humanist Discussion of Method', in Edward P. Mahoney (ed.), *Philosophy and Humanism: Renaissance Essays in Honor of Paul Oskar Kristeller* (Leiden, 1976), 283–305.

Eemeren, Frans H. van, 'Argumentative Style: A Complex Notion', *Argumentation*, 33 (2019), 153–71.

Effe, Bernd, *Dichtung und Lehre: Untersuchungen zur Typologie des antiken Lehrgedichts* (Munich, 1977).

Egmond, Florike, *The World of Carolus Clusius* (London and New York, 2010).

Egmond, Florike, 'Observing Nature: The Correspondence Network of Carolus Clusius (1526–1609)', in Dirk van Miert (ed.), *Communicating Observations in Early Modern Letters (1500–1675)* (London and Turin, 2013), 43–72.

Egmond, Florike, *Eye for Detail: Images of Plants and Animals in Art and Science, 1500–1630* (London, 2017).

Ehlers, Widu-Wolfgang (ed.), *La Biographie antique: Huit exposés suivis de discussions: Vandœuvres-Genève 25–29 août 1997* (Geneva, 1998).

Eichholz, David E. (ed.), *Theophrastus: De lapidibus. Edited with Introduction, Translation and Commentary* (Oxford, 1965).

Eisenstein, Elizabeth L., *The Printing Press as an Agent of Change: Communications and Cultural Transformations in Early Modern Europe*, 2 vols (Cambridge, 1979).

Elaut, Leon, 'De Hygienische verhandeling van Robertus Gropetius', *Scientiarum Historia: Tijdschrift voor de Geschiedenis van de Wetenschappen en de Geneeskunde*, 3/2 (1961), 74–81.

Enenkel, Karl A. E., 'The Species and Beyond: Classification and the Place of Hybrids in Early Modern Zoology', in Enenkel and Paul J. Smith (eds), *Zoology in Early Modern Culture: Intersections of Science, Theology, Philology, and Political and Religious Education* (Leiden and Boston, MA, 2014a), 57–148.

Enenkel, Karl A. E. (ed.), *Transformations of the Classics via Early Modern Commentaries* (Leiden and Boston, MA, 2014b).

Enenkel, Karl A. E., *Die Stiftung von Autorschaft in der neulateinischen Literatur (ca. 1350–ca. 1650): Zur autorisierenden und wissensvermittelnden Funktion von Widmungen, Vorworttexten, Autorporträts und Dedikationsbildern* (Leiden, 2015).

Enenkel, Karl A. E., and Henk Nellen, *Neo-Latin Commentaries and the Management of Knowledge in the Late Middle Ages and the Early Modern Period (1400–1700)* (Leuven, 2013).

Enenkel, Karl A. E., and Paul J. Smith (eds), *Emblems and the Natural World* (Leiden and Boston, MA, 2017).

Enenkel, Karl A. E., and Arnoud S. Q. Visser (eds), *Mundus emblematicus: Studies in Neo-Latin Emblem Books* (Turnhout, 2003).

English, Christine R., 'Seventeenth-Century Medical Books', in Alain Besson (ed.), *Thornton's Medical Books, Libraries and Collectors* (Aldershot, 1990), 83–115.

Eriksson, Ruben (ed.), *Andreas Vesalius' First Public Anatomy at Bologna 1540: An Eyewitness Report by Baldasar Heseler, medicinae scolaris, together with His Notes on Matthaeus Curtius' Lectures on Anatomia Mundini* (Uppsala, 1959).

Ernst, Germana, and Guido Giglioni (eds), *Il linguaggio dei cieli: Astri e simboli nel Rinascimento* (Rome, 2012).

Ernst, Ulrich (ed.), *Petrus Bungus: Numerorum Mysteria* (Hildesheim, 1983).

Eybl, Franz M., et al. (eds), *Enzyklopädien der Frühen Neuzeit: Beiträge zu ihrer Erforschung* (Tübingen, 1995).

Fahnestock, Jeanne, *Rhetorical Figures in Science* (New York, 1999).

Fara, Patricia, and David Money, 'Isaac Newton and Augustan Anglo-Latin Poetry', *Studies in History and Philosophy of Science*, 35 (2004), 549–71.

Farrell, Joseph, *Latin Language and Latin Culture from Ancient to Modern Times* (Cambridge, 2001).

Favaro, Antonio (ed.), *Carteggio inedito di Ticone Brahe, Giovanni Keplero e di altri celebri astronomi e matematici dei secoli XVI. e XVII. con Giovanni Antonio Magini* (Bologna, 1886).

Febvre, Lucien, and Henri-Jean Martin, *The Coming of the Book: The Impact of Printing, 1450–1800* (London and New York, 1976).

Feingold, Mordechai (ed.), *Jesuit Science and the Republic of Letters* (Cambridge, MA, 2003).

Feingold, Mordechai, 'Between Teaching and Research: The Place of Science in Early Modern English Universities', in Feingold and Giulia Giannini (eds), *Institutionalization of Science in Early Modern Europe* (Leiden, 2019), 3–19.

Feingold, Mordechai, Joseph S. Freedman, and Wolfgang Rother (eds), *The Influence of Petrus Ramus: Studies in Sixteenth and Seventeenth Century Philosophy and Sciences* (Basel, 2001).

Feingold, Mordechai, and Giulia Giannini (eds), *Institutionalization of Science in Early Modern Europe* (Leiden, 2019).

Feldhay, Rivka, 'The Jesuits, Transmitters of the New Science', in Massimo Bucciantini, Michele Camerota, and Franco Giudice (eds), *Il caso Galileo: Una rilettura storica, filosofica, teologica: Convegno Internazionale di Studi, Firenze, 26–30 maggio 2009* (Florence, 2011), 47–74.

Feldman, Theodore S., 'Late Enlightenment Meteorology', in Tore Frängsmyr, John L. Heilbron, and Robin E. Rider (eds), *The Quantifying Spirit in the 18th Century* (Berkeley, CA, 1990), 143–77.

Fery-Hue, Françoise, and Fabio Zinelli (eds), *Habiller en latin: La Traduction de vernaculaire en latin entre moyen âge et Renaissance* (Paris, 2018).

Figala, Karin, and Ulrich Petzold, 'Physics and Poetry: Fatio de Duillier's *Ecloga* on Newton's *Principia*', *Archives internationales d'histoire des sciences*, 37 (1987), 316–49.

Findlen, Paula, *Possessing Nature: Museums, Collecting, and Scientific Culture in Early Modern Italy* (Berkeley, CA, 1994).

Finocchiaro, Giuseppe, 'Dall' Apiarium alla *ΜΕΛΙΣΣΟΓΡΑΦΙΑ*: Una vicenda editoriale tra propaganda scientifica e strategia culturale', *Rendiconti dell'Accademia nazionale dei Lincei. Classe di Scienze morali, storiche e filologiche*, 9/15 (2004), 767–79.

Fiorentino, Francesco, and Felice Tocco (eds), *Jordani Bruni Nolani opera Latine conscripta*, 3 vols (Naples and Florence, 1879–91).

Fleischmann, Wolfgang Bernhard, 'Zum Anti-Lucretius des Kardinals de Polignac', *Romanische Forschungen*, 77 (1965), 42–63.

Forbes, Eric G., Lesley Murdin, and Frances Willmoth (eds), *The Correspondence of John Flamsteed, First Astronomer Royal*, 3 vols (Bristol and Philadelphia, PA, 1995–2002).

Ford, Brian J., 'Eighteenth-Century Scientific Publishing', in Andrew Hunter (ed.), *Thornton and Tully's Scientific Books, Libraries and Collectors*, 4th edn (Aldershot, 2000), 216–57.

Fortuna, Stefania, 'The Latin Editions of Galen's *Opera omnia* (1490–1625) and Their Prefaces', *Early Science and Medicine*, 17 (2012), 391–412.

Fortuna, Stefania, 'Editions and Translations of Galen from 1490 to 1540', in Petros Bouras-Vallianatos and Barbara Zipser (eds), *Brill's Companion to the Reception of Galen* (Leiden and Boston, MA, 2019), 437–52.

Frank, Günther, and Stefan Rhein (eds), *Melanchthon und die Naturwissenschaften seiner Zeit* (Sigmaringen, 1998).

Franklin, Kenneth J., and Andrew Wear (eds), *William Harvey: The Circulation of the Blood and Other Writings* (London, 1993).

Fransen, Sietske, Niall Hodson, and Karl A. E. Enenkel (eds), *Translating Early Modern Science* (Leiden and Boston, MA, 2017).

Frasca-Spada, Marina, and Nick Jardine (eds), *Books and the Sciences in History* (Cambridge, 2000).

Fraustadt, Georg, and Hans Prescher (eds), *Georgius Agricola: De natura fossilium libri X: Die Mineralien* (Berlin, 1958).

Freedberg, David, *The Eye of the Lynx: Galileo, His Friends, and the Beginnings of Modern Natural History* (Chicago and London, 2002).

Freedman, Joseph S., 'Encyclopedic Philosophical Writings in Central Europe during the High and Late Renaissance (ca. 1500–ca. 1700)', *Archiv für Begriffsgeschichte*, 37 (1994), 212–56.

Freer, Stephen (ed.), *Linnaeus' Philosophia botanica* (Oxford, 2003).

Freer, Stephen, and Staffan Müller-Wille (eds), *Carl Linnaeus: Musa Cliffortiana: Clifford's Banana Plant* (Ruggell, 2007).

French, Roger K., 'Berengario da Carpi and the Use of Commentary in Anatomical Teaching', in Andrew Wear, Roger K. French, and Ian M. Lonie (eds), *The Medical Renaissance of the Sixteenth Century* (Cambridge, 1985), 42–74.

Friedenthal, Meelis, Hanspeter Marti, and Robert Seidel (eds), *Early Modern Disputations and Dissertations in an Interdisciplinary and European Context* (Leiden, 2021).

Friedenthal, Meelis, and Pärtel Piirimäe, 'Philosophical Disputations at the University of Tartu 1632–1710: Boundaries of a Discipline', *Studia Philosophica Estonica*, 8/2 (2015), 65–90.

Friedlein, Gottfried (ed.), *Procli Diadochi in primum Euclidis Elementorum librum Commentarii* (Leipzig, 1873).

Friedrich, Markus, 'Frühneuzeitliche Wissenstheater', in Frank Grunert and Anette Syndikus (eds), *Wissensspeicher der Frühen Neuzeit: Formen und Funktionen* (Berlin, 2015), 297–327.

Friedrich, Udo, *Naturgeschichte zwischen 'artes liberales' und frühneuzeitlicher Wissenschaft: Conrad Gessners 'Historia animalium' und ihre volkssprachliche Rezeption* (Tübingen, 1995).

Friedrich, Udo, 'Grenzen des Ordo im enzyklopädischen Schrifttum des 16. Jahrhunderts', in Christel Meier (ed.), *Die Enzyklopädie im Wandel vom Hochmittelalter bis zur frühen Neuzeit* (Munich, 2002), 391–408.

Frisinger, Howard H., *The History of Meteorology to 1800* (New York, 1977).

Fritz, Gerd, 'Scholarly Reviewing', in Annette Leßmöllmann, Marcelo Dascal, and Thomas Gloning (eds), *Science Communication* (Boston, MA, and Berlin, 2020), 289–310.

Fück, Johann, *Die arabischen Studien in Europa bis in den Anfang des 20. Jahrhunderts* (Leipzig, 1955).

Fucks, Wilhelm, *Nach allen Regeln der Kunst: Diagnosen über Literatur, Musik, bildende Kunst: Die Werke, ihre Autoren und Schöpfer* (Stuttgart, 1968).

Fuhrmann, Manfred, *Das systematische Lehrbuch: Ein Beitrag zur Geschichte der Wissenschaften in der Antike* (Göttingen, 1960).

Fulton, John F., *The Great Medical Bibliographers: A Study in Humanism* (Philadelphia, PA, 1951).

Fulton, John F., *A Bibliography of the Honourable Robert Boyle, Fellow of the Royal Society*, 2nd edn (Oxford, 1961).

Fumaroli, Marc, *L'Âge de l'éloquence: Rhétorique et res literaria de la Renaissance au seuil de l'époque classique* (Geneva, 1980).

Fumaroli, Marc (ed.), *Histoire de la rhétorique dans l'Europe moderne 1450–1950* (Paris, 1999).

Funder, Lærke Maria Andersen, 'Meeting Expectations and Revealing Aspirations: Ole Worm and the Emergent Genre of Museum Catalogues', in Marianne Pade and Camilla Horster (eds), *Cultural Encounter and Identity in the Neo-Latin World* (Rome, 2018), 103–19.

Füssel, Marian, 'Universität', in Marianne Sommer, Staffan Müller-Wille, and Carsten Reinhardt (eds), *Handbuch Wissenschaftsgeschichte* (Stuttgart, 2017), 193–201.

Füssel, Stephan, 'Die Praxis der Disputation: Heuristische Zugänge und theoretische Deutungsangebote', in Marion Gindhart, Hanspeter Marti, and Robert Seidel (eds), *Frühneuzeitliche Disputationen* (Cologne, 2016), 27–48.

Gaida, Margareth, 'Reading Cosmographia: Peter Apian's Book-Instrument Hybrid and the Rise of the Mathematical Amateur in the Sixteenth Century', *Early Science and Medicine*, 21 (2016), 277–302.

Galilei, Galileo, *Le opere: Edizione nazionale*, vol. 1 (Florence, 1890).

Garin, Eugenio (ed.), *Giovanni Pico della Mirandola, Disputationes adversus astrologiam divinatricem* (Florence, 1946–52).

Garrison, Daniel, and Malcolm H. Hast (eds), *Andreas Vesalius: De humani corporis fab rica libri septem: The Fabric of the Human Body: An Annotated Translation of the 1543 and 1555 Editions*, 2 vols (Basel, 2014).

Gärtner, Barbara, *Johannes Widmanns 'Behende und hubsche Rechenung': Die Textsorte 'Rechenbuch' in der Frühen Neuzeit* (Tübingen, 2000).

Gascoigne, Robert M., *A Historical Catalogue of Scientists and Scientific Books* (New York and London, 1984).

Gascoigne, Robert M., *A Historical Catalogue of Scientific Periodicals, 1665–1900: With a Survey of Their Development* (New York and London, 1985).

Gaskell, Philip, *A New Introduction to Bibliography* (New Castle, DE, 1995).

Geerlings, Wilhelm, and Christian Schulze (eds), *Der Kommentar in Antike und Mittelalter*, 2 vols (Leiden and Boston, MA, 2002–4).

Genette, Gérard, *Paratexts: Thresholds of Interpretation*, tr. Jane E. Lewin, foreword by Richard Macksey (Cambridge, 1997).

Gerabek, Werner, et al. (eds), *Enzyklopädie Medizingeschichte* (Berlin, 2005).

Germain, Alexandre, *Les Anciennes Thèses de l'école de Montpellier: Collations de grades et concours professoraux* (Montpellier, 1886).

Gibson, Roy K., and Christina S. Kraus (eds), *The Classical Commentary: Histories, Practices, Theory* (Leiden and Boston, MA, 2002).

Giesecke, Michael, *Der Buchdruck in der frühen Neuzeit* (Frankfurt am Main, 1991).

Giglioni, Guido (ed.), *Bernardino Telesio: De rerum natura iuxta propria principia libri IX (Orazio Salviano, Napoli 1586)* (Rome, 2013).

Giglioni, Guido, 'Touch Me Not: Sense and Sensibility in Early Modern Botany', *Early Science and Medicine*, 23 (2018), 420–43.

Gindhart, Marion, and Ursula Kundert (eds), *Disputatio 1200–1800: Form, Funktion und Wirkung eines Leitmediums universitärer Wissenskultur* (Berlin and New York, 2010).

Gindhart, Marion, Hanspeter Marti, and Robert Seidel (eds), *Frühneuzeitliche Disputationen: Polyvalente Produktionsapparate gelehrten Wissens* (Cologne, 2016).

Gingerich, Owen, 'Five Centuries of Astronomical Textbooks and their Role in Teaching', in Jay M. Pasachoff and John R. Percy (eds), *The Teaching of Astronomy: Proceedings of IAU Colloq. 105, Held in Williamstown, MA, 27–30 July 1988* (Cambridge, MA, 1990), 189–95.

Gingerich, Owen, *An Annotated Census of Copernicus' De revolutionibus (Nuremberg, 1543 and Basel, 1566)* (Leiden, 2002).

Girardi-Karšulin, Mihaela, Olga Perić, and Tomislav Čepulić (eds), *Federicus Chrysogonus/ Federik Grisogono: Speculum astronomicum/Astronomsko zrcalo* (Zagreb, 2007).

Glassie, John, *A Man of Misconceptions: The Life of an Eccentric in an Age of Change* (New York, 2012).

Glei, Reinhold, 'Durchs wilde Bsurdistan: Zur Terminologie und räumlichen Darstellung höherer Potenzen in Michael Stifels *Arithmetica integra* (1544)', *Neulateinisches Jahrbuch*, 19 (2017), 494–503.

Gmelig-Nijboer, Caroline Aleid, *Conrad Gesner's 'Historia animalium': An Inventory of Renaissance Zoology* (Meppel, 1977).

Goltz, Dietlinde, *Studien zur altorientalischen und griechischen Heilkunde: Therapie—Arzneibereitung—Rezeptstruktur* (Wiesbaden, 1974).

Göpferich, Susanne, *Textsorten in Naturwissenschaften und Technik: Pragmatische Typologie—Kontrastierung—Translation* (Tübingen, 1995).

Göpferich, Susanne, 'Fachtextsorten der Naturwissenschaften und der Technik: Ein Überblick', in Lothar Hoffmann, Hartwig Kalverkämper, and Herbert E. Wiegand (eds), *Fachsprachen: Ein internationales Handbuch zur Fachsprachenforschung und Terminologiewissenschaft*, 2 vols (Berlin, 1998–9), vol. 1, 545–56.

Gordin, Michael D., *Scientific Babel: How Science Was Done before and after Global English* (Chicago, 2015).

Goulet-Cazé, Marie-Odile, et al. (eds), *Le Commentaire entre tradition et innovation: Actes du colloque international de l'Institut des traditions textuelles (Paris et Villejuif, 22–25 septembre 1999)* (Paris, 2000).

Goyens, Michèle, Pieter de Leemans, and An Smets (eds), *Science Translated: Latin and Vernacular Translations of Scientific Treatises in Medieval Europe* (Leuven, 2008).

Grabmann, Martin, *Methoden und Hilfsmittel des Aristoteles-Studiums im Mittelalter* (Munich, 1939).

Graesse, Johann G. T., Friedrich Benedict, and Helmut Plechl, *Orbis Latinus: Lexikon lateinischer geographischer Namen des Mittelalters und der Neuzeit* (Braunschweig, 1972).

Graf-Stuhlhofer, Franz, *Humanismus zwischen Hof und Universität: Georg Tannstetter (Collimitius) und sein wissenschaftliches Umfeld im Wien des frühen 16. Jahrhunderts* (Vienna, 1996).

Grafton, Anthony, *Joseph Scaliger: A Study in the History of Classical Scholarship*, 2 vols (Oxford, 1983–93).

Grafton, Anthony, *Defenders of the Text: The Traditions of Scholarship in an Age of Science, 1450–1800* (Cambridge and London, 1991).

Grafton, Anthony, *The Footnote: A Curious History* (Cambridge, MA, 1997).

Grafton, Anthony, *Cardanos Kosmos: Die Welten und Werke eines Renaissance-Astrologen* (Berlin, 1999).

Grafton, Anthony, 'Geniture Collections, Origins and Uses of a Genre', in Marina Frasca-Spada and Nick Jardine (eds), *Books and the Sciences in History* (Cambridge, 2000), 49–68.

Grafton, Anthony, 'Textbooks and the Disciplines', in Emilio Campi et al. (eds), *Scholarly Knowledge* (Geneva, 2008), 11–36.

Grämiger, Gregory, 'Reconstructing Order: The Spatial Arrangement of Plants in the Hortus Botanicus of Leiden University in Its First Years', in Hubertus Fischer, Volker R. Remmert, and Joachim Wolschke-Bulmahn (eds), *Gardens, Knowledge and the Sciences in the Early Modern Period* (Basel, 2016), 235–51.

Granada, Miguel Á., and Dario Tessicini (eds), *Giordano Bruno: De immenso: Letture critiche* (Pisa and Rome, 2020).

Graney, Christopher M., *Setting Aside All Authority: Giovanni Battista Riccioli and the Science against Copernicus in the Age of Galileo* (Notre Dame, IN, 2015).

Graney, Christopher M. (ed.), *Mathematical Disquisitions: The Booklet of Theses Immortalized by Galileo* (Notre Dame, IN, 2017).

Grant, Edward (ed.), *Nicole Oresme: De proportionibus proportionum and Ad pauca respicientes: Edited with Introductions, English Translations, and Critical Notes* (Madison, WI, 1966).

Grant, Leonard W., 'European Vernacular Works in Latin Translation', *Studies in the Renaissance*, 1 (1954), 120–56.

Grattan-Guinness, Ivor (ed.), *Landmark Writings in Western Mathematics 1640–1940* (Amsterdam, 2005).

Green, Jonathon, *Chasing the Sun: Dictionary-Makers and the Dictionaries They Made* (London, 1996).

Green, Lawrence D., and James J. Murphy, *Renaissance Rhetoric Short-Title Catalogue 1460–1700* (Aldershot, 2006).

Greene, Thomas, *The Light in Troy: Imitation and Discovery in Renaissance Poetry* (New Haven, CT, 1982).

Grendler, Paul F., *The Universities of the Italian Renaissance* (Baltimore, MD, 2001).

Grendler, Paul F., 'The *Sphaera* in Jesuit Education', in Matteo Valleriani and Andrea Ottone (eds), *Publishing Sacrobosco's De sphaera in Early Modern Europe* (Cham, 2022), 369–404.

Gross, Alan G., *The Rhetoric of Science*, 2nd edn (Cambridge, MA, 1996).

Gross, Alan G., Joseph E. Harmon, and Michael Reidy, *Communicating Science: The Scientific Article from the 17th Century to the Present* (Oxford, 2002).

Grunert, Frank, and Anette Syndikus, 'Historia literaria', in Grunert and Syndikus (eds), *Wissensspeicher der Frühen Neuzeit: Formen und Funktionen* (Berlin, 2015), 243–94.

Grunert, Frank, and Friedrich Vollhardt (eds), *Historia literaria: Neuordnungen des Wissens im 17. und 18. Jahrhundert* (Berlin, 2007).

Guardo, Marco, 'L'ape e le api: Il paratesto linceo e l'omaggio ai Barberini', *Paratesto*, 1 (2004), 121–36.

Gudger, Eugene Willis, 'The Five Great Naturalists of the Sixteenth Century: Belon, Rondelet, Salviani, Gesner and Aldrovandi: A Chapter in the History of Ichthyology', *Isis*, 22 (1934), 21–40.

Guentherodt, Ingrid, '*Urania propitia* (1650) – in zweyerley Sprachen: Lateinisch- und deutschsprachiges Compendium der Mathematikerin und Astronomin Maria

Cunitz', in Sebastian Neumeister and Conrad Wiedemann (eds), *Res Publica Litteraria: Die Institutionen der Gelehrsamkeit in der frühen Neuzeit*, vol. 2 (Wiesbaden, 1987), 619–40.

Guentherodt, Ingrid, 'Maria Cunitia: *Urania propitia*: Intendiertes, erwartetes und tatsächliches Lesepublikum einer Astronomin des 17. Jahrhunderts', *Daphnis*, 20 (1991), 311–53.

Guicciardini, Niccolò, 'Editing Newton in Geneva and Rome: The Annotated Edition of the *Principia* by Calandrini, Le Seur and Jacquier', *Annals of Science*, 52 (2015), 337–80.

Gunnarsson, Britt L. (ed.), *Languages of Science in the Eighteenth Century* (Berlin and Boston, MA, 2011).

Habermann, Mechthild, *Deutsche Fachtexte der frühen Neuzeit: Naturkundlich-medizinische Wissensvermittlung im Spannungsfeld von Latein und Volkssprache* (Berlin and New York, 2001).

Häfner, Ralph, and Markus Völkel (eds), *Der Kommentar in der Frühen Neuzeit* (Tübingen, 2006).

Hall, Alfred R., *Philosophers at War: The Quarrel between Newton and Leibniz* (Cambridge, 1980).

Hall, Alfred R., and Marie B. Hall (eds), *The Correspondence of Henry Oldenburg*, 13 vols (Madison, WI, and London, 1965–86).

Hall, J. J., *Cambridge Act and Tripos Verses 1565–1894* (Cambridge, 2009).

Hallett, Raphael, 'Ramus, Printed Loci, and the Re-Invention of Knowledge', in Steven J. Reid and Emma A. Wilson (eds), *Ramus, Pedagogy and the Liberal Arts: Ramism in Britain and the Wider World* (Farnham and Burlington, VT, 2011), 89–112.

Hallyn, Ferdinand, 'Dialectique et rhétorique devant la nouvelle science du XVIIe siècle', in Marc Fumaroli (ed.), *Histoire de la rhétorique dans l'Europe moderne 1450–1950* (Paris, 1999), 601–28.

Hamel, Jürgen, *Die Himmelsvermessung des Johannes Bayer: Begleitbuch zur Uranometria von Johann Bayer* (Gerchsheim, 2010).

Hansen, Signe L., 'Received Wisdom in Biography: Tycho-Biographies from Gassendi to Christianson', in Thomas Söderqvist (ed.), *The History and Poetics of Scientific Biography* (Burlington, VT, 2007), 269–80.

Harris, Randy A. (ed.), *Landmark Essays on Rhetoric of Science: Case Studies*, 2nd edn (New York and London, 2018).

Harris, Stephen J., 'Transposing the Merton Thesis: Apostolic Spirituality and the Estabilishment of the Jesuit Scientific Tradition', *Science in Context*, 3 (1989), 29–65.

Harrison, Peter, *The Bible, Protestantism, and the Rise of Natural Science* (Cambridge, 1998).

Harskamp, Jaap, *Dissertatio medica inauguralis... Leyden Medical Dissertations in the British Library, 1593–1746* (London, 1997).

Hartmann, Florian, *Ars dictaminis: Briefsteller und verbale Kommunikation in den italienischen Stadtkommunen des 11. bis 13. Jahrhunderts* (Ostfildern, 2013).

Haskell, Yasmin, *Loyola's Bees: Ideology and Industry in Jesuit Latin Didactic Poetry* (Oxford, 2003).

Haskell, Yasmin, and Philip Hardie (eds), *Poets and Teachers: Latin Didactic Poetry and the Didactic Authority of the Latin Poet from the Renaissance to the Present* (Bari, 1999).

Hasse, Dag Nikolaus (ed.), *Georg Christoph Lichtenberg: Observationes: Die lateinischen Schriften* (Göttingen, 1997).

Hasse, Dag Nikolaus, *Success and Suppression: Arabic Sciences and Philosophy in the Renaissance* (Cambridge, MA, and London, 2016).

Hassig, Debra, *Medieval Bestiaries: Text, Image, Ideology* (Cambridge, 1995).

Hatch, Robert, 'Between Erudition and Science: The Archive and Correspondence of Ismael Boulliau', in Michael Hunter (ed.), *Archives of the Scientific Revolution: The Formation and Exchange of Ideas in Seventeenth-Century Europe* (Woodbridge, 1998).

Hausmann, Franz J., et al. (eds), *Wörterbücher: Ein internationales Handbuch zur Lexikographie/Dictionaries. An International Encyclopedia of Lexicography/Dictionnaires. Encyclopédie internationale de lexicographie*, 3 vols (Berlin and New York, 1989–91).

Haye, Thomas, *Das lateinische Lehrgedicht im Mittelalter: Analyse einer Gattung* (Leiden, 1997).

Heard, Stephen B., 'On Whimsy, Jokes, and Beauty: Can Scientific Writing be Enjoyed?', *Ideas in Ecology and Evolution*, 7 (2014), 64–72.

Heath, Malcolm, *Unity in Greek Poetics* (Oxford and New York, 1989).

Heiberg, Johan L. (ed.), *Claudii Ptolemaei opera quae exstant omnia*, II: *Opera astronomica minora* (Leipzig, 1907).

Heideklang, Julia, Hans-Joachim Pflüger, and Helmut Kettenmann (eds), *De fabrica systematis nervosi evertebratorum: Die kommentierte Dissertation von/Commented thesis by Hermann Helmholtz* (Darmstadt, 2021).

Heitz, Paul, and Konrad Haebler (eds), *Hundert Kalender-Inkunabeln* (Strasbourg, 1905).

Helander, Hans, *Latin och Grekiska i naturvetenskaperna* (n.p., 2021).

Hellyer, Marcus, *Catholic Physics: Jesuit Natural Philosophy in Early Modern Germany* (Notre Dame, IN, 2005).

Helm, Jürgen, 'Zwischen Aristotelismus, Protestantismus und zeitgenössischer Medizin: Philipp Melanchthons Lehrbuch *De anima*', in Jürgen Leonhardt (ed.), *Melanchthon und das Lehrbuch des 16. Jahrhunderts* (Rostock, 1997), 175–91.

Henderson, Judith R., *The Unfolding of Words: Commentary in the Age of Erasmus* (Toronto, 2012).

Henry, John, *The Scientific Revolution and the Origins of Modern Science* (Basingstoke, 2002).

Henschel, August W. E. T., and Salvatore De Renzi (eds), *Collectio Salernitana: ossia Documenti inediti, e trattati di medicina appartenenti alla scuola medica salernitana*, 5 vols (Naples, 1852–9).

Herberger, Maximilian, *Dogmatik: Zur Geschichte von Begriff und Methode in Medizin und Jurisprudenz* (Frankfurt am Main, 1981).

Herlitzius, Erwin, 'Der Dialog in der Geschichte der Wissenschaft', *Deutsche Zeitschrift für Philosophie*, 16 (1968), 571–80.

Hirai, Hiro, 'Medicina e astrologia: Aspetti della medicina astrale platonica', in Germana Ernst and Guido Giglioni (eds), *Il linguaggio dei cieli* (Rome, 2012), 205–21.

Hirai, Hiro, 'The New Astral Medicine', in Brendan Dooley (ed.), *A Companion to Astrology in the Renaissance* (Leiden and Boston, MA, 2014), 267–86.

Hirschi, Caspar, 'Akademie', in Marianne Sommer, Staffan Müller-Wille, and Carsten Reinhardt (eds), *Handbuch Wissenschaftsgeschichte* (Stuttgart, 2017), 211–24.

Hirzel, Rudolf, *Der Dialog: Ein literarhistorischer Versuch* (Leipzig, 1895).

Hischer, Horst, *Mathematik—Medien—Bildung: Medialitätsbewusstsein als Bildungsziel: Theorie und Beispiele* (Wiesbaden, 2016).

Hofheinz, Rolf D., *Philipp Melanchthon und die Medizin im Spiegel seiner akademischen Reden* (Herbolzheim, 2001).

Høgel, Christian, and Elisabetta Bartoli (eds), *Medieval Letters: Between Fiction and Document* (Turnhout, 2015).

Höltgen, Karl J., 'Synoptische Tabellen in der medizinischen Literatur und die Logik Agricolas und "Ramus"', *Sudhoffs Archiv*, 49 (1965), 371–90.

Hoogendoorn, Klaas, *Bibliography of the Exact Sciences in the Low Countries from ca. 1470 to the Golden Age (1700)* (Leiden and Boston, MA, 2018).

Hoover, Herbert C., and Lou H. Hoover (eds), *Georgius Agricola: De re metallica: Translated from the First Latin Edition of 1556* (London, 1912).

Horn, Ewald, *Die Disputationen und Promotionen an den deutschen Universitäten vornehmlich seit dem 16. Jahrhundert* (Leipzig, 1893).

Hörstedt, Axel, *Latin Dissertations and Disputations in the Early Modern Swedish Gymnasium: A Study of a Latin School Tradition, c.1620–c.1820* (Gothenburg, 2018).

Hort, Arthur (ed.), *Theophrastus: Enquiry into Plants and Minor Works on Odours and Weather Signs*, 2 vols (London, 1916).

Hotson, Howard, *Johann Heinrich Alsted, 1588–1638: Between Renaissance, Reformation, and Universal Reform* (Oxford, 2000).

Hotson, Howard, *Commonplace Learning: Ramism and Its German Ramifications, 1543–1630* (Oxford and New York, 2007).

Houdt, Toon van, et al. (eds), *Self-Presentation and Social Identification: The Rhetoric and Pragmatics of Letter Writing in Early Modern Times* (Leuven, 2002).

Houliston, Victor H., 'Sleepers Awake: Thomas Moffet's Challenge to the College of Physicians of London, 1584', *Medical History*, 33 (1989), 235–46.

Houzeau, Jean C., and Albert Lancaster, *Bibliographie générale de l'astronomie jusqu' en 1880*, 2 vols (London, 1882–7).

Hoven, René, and Karin Renard-Jadoul, *Lexique de la prose latine de la Renaissance/ Dictionary of Renaissance Latin from Prose Sources* (Leiden, 2006).

Howard, Nicole, *Loath to Print: The Reluctant Scientific Author, 1500–1750* (Baltimore, MD, 2022).

Hübner, Wolfgang, 'Der *descensus* als ordnendes Prinzip in der *Naturalis historia* des Plinius', in Christel Meier (ed.), *Die Enzyklopädie im Wandel vom Hochmittelalter bis zur frühen Neuzeit* (Munich, 2002), 25–41.

Hugonnard-Roche, Henri, et al. (eds), *Georgii Joachimi Rhetici Narratio prima: Édition critique, traduction française et commentaire* (Wrocław, 1982).

Hultsch, Friedrich (ed.), *Pappi Alexandrini Collectio*, 3 vols (Berlin, 1876–8).

Hunger, Hermann (ed.), *Astronomical Diaries and Related Texts from Babylonia*, vol. 5 (Vienna, 2001).

Hunter, Andrew (ed.), *Thornton and Tully's Scientific Books, Libraries and Collectors: A Study of Bibliography and the Book Trade in Relation to the History of Science*, 4th edn (Aldershot, 2000).

Hupka, Werner, 'Das enzyklopädische Wörterbuch', in Franz J. Hausmann et al. (eds), *Wörterbücher*, 3 vols (Berlin and New York, 1989–91), vol. 1, 988–99.

Hurwitz, Brian, 'Form and Representation in Clinical Case Reports', *Literature and Medicine*, 25 (2006), 216–40.

Husner, Fritz, *Verzeichnis der Basler medizinischen Universitätsschriften von 1575–1829* (Basel, 1942).

Hyrtl, Joseph, *Onomatologia anatomica: Geschichte und Kritik der anatomischen Sprache der Gegenwart; mit besonderer Berücksichtigung ihrer Barbarismen, Widersinnigkeiten, Tropen, und grammatikalischen Fehler* (Vienna, 1880).

Imhausen, Anette, and Volker R. Remmert, 'The Oration on the Dignity and the Usefulness of the Mathematical Sciences of Martinus Hortensius (1634): Text, Translation and Commentary', *History of Universities*, 21/1 (2006), 71–150.

Jackson, Heather J., *Marginalia: Readers Writing in Books* (New Haven, CT, and London, 2001).

Jacobius, Helene, *Luftschiff und Pegasus: Der Widerhall der Erfindung des Luftballons in der zeitgenössischen Literatur* (Halle, 1909).

Jacquart, Danielle, and Gérard Troupeau (eds), *Abū-Zakarīyā Yūḥannā ibn-Māsawaih: Le Livre des axiomes médicaux (Aphorismi): Édition du texte arabe et des versions latines avec traduction française et lexique* (Geneva and Paris, 1980).

Janson, Tore, *Latin Prose Prefaces: Studies in Literary Conventions* (Stockholm, 1964).

Janson, Tore, *A Natural History of Latin* (Oxford, 2004).

Jardine, Nick, 'Books, Texts, and the Making of Knowledge', in Marina Frasca-Spada and Nick Jardine (eds), *Books and the Sciences in History* (Cambridge, 2000), 393–407.

Jaumann, Herbert, 'Iatrophilologia: *Medicus philologus* und analoge Konzepte in der frühen Neuzeit', in Ralph Häfner (ed.), *Philologie und Erkenntnis: Beiträge zu Begriff und Problem frühneuzeitlicher Philologie* (Tübingen, 2001), 151–76.

Jensen, Kristian, 'Description, Division, Definition: Caesalpinus and the Study of Plants as an Independent Discipline', in Marianne Pade (ed.), *Renaissance Readings of the Corpus Aristotelicum* (Copenhagen, 2001), 185–206.

Johns, Adrian, 'The Past, Present, and Future of the Scientific Book', in Marina Frasca-Spada and Nick Jardine (eds), *Books and the Sciences in History* (Cambridge, 2000), 408–26.

Jones, Peter Murray, 'Medical Libraries and Medical Latin, 1400–1700', in Wouter Bracke and Herwig Deumens (eds), *Medical Latin from the Late Middle Ages to the Eighteenth Century* (Brussels, 2000), 115–135.

Jones, William P., *The Rhetoric of Science: A Study of Scientific Ideas and Imagery in Eighteenth-Century English Poetry* (London, 1966).

Jones-Davies, Marie-Thérèse (ed.), *Le Dialogue au temps de la Renaissance* (Paris, 1984).

Jönsson, Ann-Mari, 'Linnaeus' International Correspondence: The Spread of a Revolution', in Britt L. Gunnarsson (ed.), *Languages of Science in the Eighteenth Century* (Berlin and Boston, MA, 2011), 171–91.

Joy, Lynn S., *Gassendi the Atomist* (Cambridge, 1987).

Kallinen, Maija, *Change and Stability: Natural Philosophy and the Academy of Turku (1640–1713)* (Helsinki, 1995).

Kardel, Troels, and Paul Maquet, *Nicolaus Steno: Biography and Original Papers of a 17th Century Scientist* (Berlin and Heidelberg, 2013).

Kay, Charles D., *William Gilbert's Renaissance Philosophy of the Magnet* (unpublished dissertation, University of Pittsburgh, 1981).

Kefalas, Katharina (ed.), 'Philipp Jacob Sachs von Lewenhaimb (1627–1672) und die Academia Naturae Curiosorum: Eine Edition seiner Briefe in chronologischer Reihenfolge sowie seines Reisetagebuches' (unpublished dissertation, University of Heidelberg, 2015).

Keller, Hildegard E., and Hubert Steinke, 'Jakob Ruf's *Trostbüchlein* and *De conceptu* (Zurich 1544): A Textbook for Midwives and Physicians', in Emilio Campi et al. (eds), *Scholarly Knowledge* (Geneva, 2008), 307–32.

Kelly, Suzanne, *The De Mundo of William Gilbert* (Amsterdam, 1965).

Kennedy, Edward S., *Islamic Astronomical Tables*, 2nd edn (Philadelphia, PA, 1989).

Kibre, Pearl, *Hippocrates Latinus: Repertorium of Hippocratic Writings in the Latin Middle Ages*, rev. edn (New York, 1985).

Kirchler, Urban, ‚Schnee von gestern. Neulateinische Disputationsschriften über Schnee aus dem 17. Jahrhundert' (unpublished dissertation, University of Innsbruck, 2023).

Klebs, Arnold C., 'Incunabula scientifica et medica', *Osiris*, 4 (1938), 1–359.

Klein, Christian (ed.), *Handbuch Biographie: Methoden, Traditionen, Theorien* (Stuttgart, 2009).

Klein, Felix (ed.), 'Gauß' wissenschaftliches Tagebuch 1796–1814', *Mathematische Annalen*, 57 (1903), 1–34.

Klein, Wolf Peter, 'Formen der Fachlexikographie in vorindustrieller Zeit: eine historische Übersicht', in Lothar Hoffmann, Hartwig Kalverkämper, and Herbert E. Wiegand (eds),

Fachsprachen: Ein internationales Handbuch zur Fachsprachenforschung und Terminologiewissenschaft, 2 vols (Berlin, 1998–9), vol. 2, 1910–25.

Klein, Wolf Peter, *Die Geschichte der meteorologischen Kommunikation in Deutschland: Eine historische Fallstudie zur Entwicklung von Wissenschaftssprachen* (Hildesheim, 1999).

Kleinert, Andreas, 'Galileis Verbrechen: Kepler, Galilei und das *crimen laesae humanitatis*', *Berichte für Wissenschaftsgeschichte*, 44 (2021), 325–9.

Knight, David, *Voyaging in Strange Seas: The Great Revolution in Science* (New Haven, CT, 2014).

Knobloch, Eberhard, 'Astrologie als astronomische Ingenieurskunst des Hochmittelalters: Zum Leben und Wirken des Iatromathematikers und Astronomen Johannes Engel (vor 1472–1512)', *Sudhoffs Archiv*, 67 (1983), 129–44.

Knufmann, Helmut, 'Das deutsche Übersetzungswesen des 18. Jahrhunderts im Spiegel von Übersetzer- und Herausgebervorreden', *Börsenblatt für den deutschen Buchhandel: Frankfurter Ausgabe* 91, *Anhang: Archiv für Geschichte des Buchwesens* 61 (1967), 2676–2717.

Knust, Hermann (ed.), *Gualteri Burlaei Liber de vita et moribus philosophorum: Mit einer altspanischen Übersetzung der Eskurialbibliothek* (Tübingen, 1886).

Korenjak, Martin, '*Tityri sub persona*: Der antike Biographismus und die bukolische Tradition', *Antike und Abendland*, 49 (2003), 58–79.

Korenjak, Martin, 'Das Gebirge als Erlebnisraum: Johann Jakob Scheuchzers *ΟΥΡΕΣΙΦΟΙΤΗΣ Helveticus sive Itinera per Helvetiae Alpinas regiones*', *Daphnis*, 41 (2012), 203–31.

Korenjak, Martin, '*Inclitae Bernatum urbis delineatio chorographica*: Der Text zu Thomas Schoepfs Karte des Bernischen Staatsgebiets (1578)', *Cartographica Helvetica*, 47 (2013a), 27–36.

Korenjak, Martin, '*Pulcherrimus foecundissimusque Naturae hortus*: Berichte über botanisch motivierte Bergbesteigungen im 16. Jahrhundert', *Neulateinisches Jahrbuch*, 15 (2013b), 197–218.

Korenjak, Martin, *Geschichte der neulateinischen Literatur: Vom Humanismus bis zur Gegenwart* (Munich, 2016).

Korenjak, Martin, 'Humanist Demography: Giovanni Battista Riccioli on the World Population', *Journal of Early Modern Studies*, 7/2 (2018a), 73–104.

Korenjak, Martin, 'Regional Latinity: A Worm's-Eye View on Neo-Latin Literature', in Neven Jovanović et al. (eds), *Neo-Latin Contexts in Croatia and Tyrol: Challenges, Prospects, Case Studies* (Vienna, 2018b), 51–60.

Korenjak, Martin, 'Explaining Science in Hexameters: Latin Didactic Epic in the Age of the Scientific Revolution', *Humanistica Lovaniensia*, 68 (2019), 101–41.

Korenjak, Martin, 'Curtz, Albert', in Stefanie Arend et al. (eds), *Frühe Neuzeit in Deutschland 1620–1720: Literaturwissenschaftliches Verfasserlexikon*, vol. 2 (Berlin and Boston, MA, 2020), 413–21.

Korenjak, Martin, and Irina Tautschnig (eds), *Die antike Literatur und die Wissenschaftliche Revolution* (Baden-Baden, 2023).

Koskenniemi, Heikki, *Studien zur Idee und Phraseologie des griechischen Briefes bis 400 n.Chr.* (Helsinki, 1956).

Kottke, Dirk (ed.), *De obitu Iohannis Stoefler Iustingani mathematici Tubingensis elegia: Ein Gedicht auf den Tod des Tübinger Astronomen Johannes Stöffler (1452–1531): Edition, Übersetzung und Kommentar mit einem Verzeichnis der poetischen Werke Reysmanns* (Hildesheim, 2013).

Koyré, Alexandre, Bernard I. Cohen, and Anne Whitman (eds), *Isaac Newton's Philosophiae naturalis principia mathematica*, 3rd edn (Cambridge, MA, and Cambridge, 1972).

Krebs, David, 'Latein als Medium wissenschaftlicher Kommunikation bei Albrecht von Haller', in Martin Stuber, Stefan Hächler, and Luc Lienhard (eds), *Hallers Netz* (Basel, 2005), 351–70.

Krebs, David, 'Die lateinische Korrespondenz zwischen Albrecht von Haller und Johann Georg Gmelin, 1743–1755' (unpublished dissertation, University of Bern, 2008).

Kremer, Richard L., 'Galileo in Danzig, as Portrayed in Peter Crüger's *Schreibkalender*', in Andrea Albrecht, Giovanna Cordibella, and Volker R. Remmert (eds), *Tintenfass und Teleskop: Galileo Galilei im Schnittpunkt wissenschaftlicher, literarischer und visueller Kulturen im 17. Jahrhundert* (Berlin and Boston, MA, 2014), 103–24.

Kretzenbacher, Heinz L., '"Just Give us the Facts!" The Connection between the Narrative Taboo, the Ego Taboo and the Metaphor Taboo in Scientific Style', *Lingua e stile*, 29 (1994), 91–106.

Krifka, Sabine, 'Zur Ikonographie der Astronomie', in Hans Holländer (ed.), *Erkenntnis, Erfindung, Konstruktion: Studien zur Bildgeschichte von Naturwissenschaften und Technik vom 16. bis zum 19. Jahrhundert* (Berlin, 2000), 409–48.

Kristeller, Paul Oskar, 'The School of Salerno: Its Development and its Contribution to the History of Learning', *Bulletin of the History of Medicine*, 17 (1945), 138–94.

Kristeller, Paul Oskar, *Eight Philosophers of the Italian Renaissance* (Stanford, CA, 1964).

Kristeller, Paul Oskar, *Humanismus und Renaissance*, 2 vols (Munich, 1975).

Krogt, Peter C. J. van der, 'Gerhard Mercator and his Cosmography: How the *Atlas* Became an Atlas', *Archives Internationales d'Histoire des Sciences*, 59 (2009), 465–83.

Kronick, David A., *A History of Scientific and Technical Periodicals: The Origins and Development of the Scientific and Technological Press 1665–1790* (Metuchen, NJ, 1976).

Kronick, David A., *Scientific and Technical Periodicals of the Seventeenth and Eighteenth Centuries: A Guide* (Metuchen, NJ, 1991).

Krook, Magnus, *Nature and Metamorphosis: A Study of Carl Adolph Agardh's Latin Dissertations and Monographs on Botany and Biology* (Gothenburg, 2022).

Kühlmann, Wilhelm, *Wissen als Poesie: Ein Grundriss zu Formen und Funktionen der frühneuzeitlichen Lehrdichtung im deutschen Kulturraum des 16. und 17. Jahrhunderts* (Berlin and Boston, MA, 2016).

Kühn, Carl Gottlob (ed.), *Galeni opera omnia*, 20 vols (Leipzig, 1821–33).

Kühn, Peter, 'Typologie der Wörterbücher nach Benutzungsmöglichkeiten', in Franz J. Hausmann et al. (eds), *Wörterbücher*, 3 vols (Berlin and New York, 1989–91), vol. 1, 111–27.

Kühne, Andreas, and Stefan Kirschner (eds), *Biographia Copernicana: Die Copernicus-Biographien des 16. bis 18. Jahrhunderts: Texte und Übersetzungen: Mit einem Katalog der frühen Copernicus-Porträts von Gudula Metze* (Berlin, 2004).

Kullmann, Wolfgang, Jochen Althoff, and Markus Asper (eds), *Gattungen wissenschaftlicher Literatur in der Antike* (Tübingen, 1998).

Kunitzsch, Paul, *Arabische Sternnamen in Europa* (Wiesbaden, 1959).

Küster, Hansjörg, and Ulf Küster (eds), *Garten und Wildnis: Landschaft im 18. Jahrhundert* (Munich, 1997).

Kusukawa, Sachiko, 'Illustrating Nature', in Marina Frasca-Spada and Nick Jardine (eds), *Books and the Sciences in History* (Cambridge, 2000a), 90–113.

Kusukawa, Sachiko, 'Incunables and Sixteenth-Century Books', in Andrew Hunter (ed.), *Thornton and Tully's Scientific Books, Libraries and Collectors*, 4th edn (Aldershot, 2000b), 115–63.

Kusukawa, Sachiko, 'The Uses of Pictures in the Formation of Learned Knowledge: The Cases of Leonhard Fuchs and Andreas Vesalius', in Kusukawa and Ian Maclean (eds), *Transmitting Knowledge* (Oxford, 2006), 73–96.

Kusukawa, Sachiko, *Picturing the Book of Nature: Image, Text, and Argument in Sixteenth-Century Human Anatomy and Medical Botany* (Chicago and London, 2012).

Kusukawa, Sachiko, and Ian Maclean (eds), *Transmitting Knowledge: Words, Images, and Instruments in Early Modern Europe* (Oxford, 2006).

Kusukawa, Sachiko (ed.) and Christine F. Salazar (tr.), *Philip Melanchthon: Orations on Philosophy and Education* (Cambridge, 1999).

Laeven, Augustinus Hubertus, *The Acta eruditorum under the Editorship of Otto Mencke (1644–1707): The History of an International Learned Journal between 1682 and 1707*, tr. from the Dutch by Lynne Richards (Amsterdam, 1990).

Lange, Gottfried, 'Das Werk der Varenius: Eine kritische Gesamtbibliographie', *Erdkunde*, 15 (1961), 1–18.

Lawee, Eric, 'From Sepharad to Ashkenaz: A Case Study in the Rashi Supercommentary Tradition', *AJS Review*, 30 (2006), 393–425.

Lawn, Brian, *The Rise and Decline of the Scholastic Quaestio disputata: With Special Emphasis on Its Use in the Teaching of Medicine and Science* (Leiden, 1993).

Lawrence-Mathers, Anne, *Medieval Meteorology: Forecasting the Weather from Aristotle to the Almanac* (Cambridge, 2020).

Leeuwen, Joyce van, 'Image, Word and Translation in Niccolò Leoniceno Tomeo's *Quaestiones Mechanicae*', in Sietske Fransen, Niall Hodson, and Karl A. E. Enenkel (eds), *Translating Early Modern Science* (Leiden and Boston, MA, 2017), 182–206.

Lehmhaus, Lennart, and Matteo Martelli (eds), *Collecting Recipes: Byzantine and Jewish Pharmacology in Dialogue* (Boston, MA, and Berlin, 2017).

Leibniz, Gottfried Wilhelm, *Sämtliche Werke und Briefe* [Akademie-Ausgabe] (Berlin, 1923–).

Leijenhorst, Cees, Christoph Lüthy, and Johannes M. M. H. Thijssen (eds), *The Dynamics of Aristotelian Natural Philosophy from Antiquity to the Seventeenth Century* (Leiden, 2002).

Leinkauf, Thomas, *Mundus combinatus: Studien zur Struktur der barocken Universalwissenschaft am Beispiel Athanasius Kirchers SJ (1602–1680)* (Berlin, 1993).

Leinsle, Ulrich G., *Dilinganae Disputationes: Der Lehrinhalt der gedruckten Disputationen an der Philosophischen Fakultät der Universität Dillingen 1555–1648* (Regensburg, 2006).

Lemay, Richard, 'Origin and Success of the Kitab Thamara of Abu Ja'far ibn Yusuf ibn Ibrahim: From the Tenth to the Seventeenth Century in the World of Islam and the Latin West', in Aḥmad Yūsuf al-Ḥasan (ed.), *Proceedings of the First International Symposium for the History of Arabic Science*, vol. 2 (Aleppo, 1978), 91–107.

Lembke, Frank, 'Ein lebensrettendes Plagiat? Friedrich von Monaus Trakat über den Luftröhrenschnitt (1644)', in Arne Jönsson and Gregor Vogt-Spira (eds), *The Classical Tradition in the Baltic Region: Perceptions and Adaptations of Greece and Rome* (Hildesheim, 2017), 163–85.

Leonhardt, Jürgen (ed.), *Melanchthon und das Lehrbuch des 16. Jahrhunderts* (Rostock, 1997).

Leonhardt, Jürgen, *Latin: Story of a World Language* (Cambridge, MA, 2013).

Leonhardt, Jürgen, 'Lateinische Dichtung zwischen Kommunikation und nicht-Kommunikation: Überlegungen zur Rolle des Gelegenheitsgedichts im 18. Jahrhundert', in Valerio Sanzotta (ed.), *Una lingua morta per letterature vive: Il dibattito sul latino come lingua letteraria in età moderna e contemporanea: Atti del convegno internazionale, Roma, 10–12 dicembre 2015* (Leuven, 2020), 107–19.

Lerner, Michel-Pierre, et al. (eds), *Nicolas Copernic: De Revolutionibus orbium coelestium/ Des révolutions des orbes célestes*, 3 vols (Paris, 2015).

Lesigang-Bruckmüller, Annemarie, '*Prolusio academica* und *Programma*: Zwei akademische Textsorten im Vergleich', in Astrid Steiner-Weber and Karl A. E. Enenkel (eds), *Acta*

Conventus Neo-Latini Monasteriensis: Proceedings of the Fifteenth International Congress of Neo-Latin Studies (Münster 2012) (Leiden and Boston, MA, 2015), 359–70.

Lestringant, Frank, *Mapping the Renaissance World* (Padstow, 1994).

Leu, Urs B., 'Textbooks and Their Uses: An Insight into the Teaching of Geography in 16th Century Zurich', in Emilio Campi et al. (eds), *Scholarly Knowledge* (Geneva, 2008), 229–48.

Leu, Urs B., '*Praeterit enim Species hujus mundi*: Die paläontologischen Zürcher Dissertationen von Johannes Gessner (1709–1790)', in Marion Gindhart, Hanspeter Marti, and Robert Seidel (eds), *Frühneuzeitliche Disputationen* (Cologne, 2016), 229–53.

Levere, Trevor H., *Transforming Matter: A History of Chemistry from Alchemy to the Buckyball* (Baltimore, MD, and London, 2001).

Licht, Tino, 'Zu Entstehung und Überlieferung der *Nova Atlantis* des Francis Bacon anläßlich ihrer Neuausgabe (Mailand 1996)', in Hermann Wiegand (ed.), *Strenae nataliciae: Neulateinische Studien, Wilhelm Kühlmann zum 60. Geburtstag* (Heidelberg, 2006), 113–26.

Lind, Levi Robert, *Studies in Pre-Vesalian Anatomy: Biography, Translations, Documents* (Philadelphia, PA, 1975).

Lindberg, Bo, 'Corollaries and Disputations', in Meelis Friedenthal, Hanspeter Marti, and Robert Seidel (eds), *Early Modern Disputations and Dissertations in an Interdisciplinary and European Context* (Leiden, 2021), 649–80.

Lindberg, Bo, *Disputation, dissertation, avhandling: Historien om en genre* (Stockholm, 2022).

Lindeboom, Gerrit A., *Herman Boerhaave: The Man and his Work* (London, 1968).

Lindenfeld, David F., *The Practical Imagination: The German Sciences of State in the Nineteenth Century* (Chicago and London, 1997).

Lindgren, Uta, 'Philipp Melanchthon und die Geographie', in Günther Frank and Stefan Rhein (eds), *Melanchthon und die Naturwissenschaften seiner Zeit* (Sigmaringen, 1998), 239–52.

Lines, David A., 'Teaching Physics in Louvain and Bologna: Frans Titelmans and Ulisse Aldrovandi', in Emilio Campi et al. (eds), *Scholarly Knowledge* (Geneva, 2008), 183–203.

Lippitsch, Max, Franz Römer, and Sonja Schreiner, 'Latin Physics: Made in Styria: Literary Ambition and Scientific Development in Gottlieb Leopold Biwald's *Physica Generalis* and *Physica Particularis*', in Peter M. Schuster and Denis Weaire (eds), *Proceedings of the First European History of Physics (EHoP) Conference…, Graz/Austria, September 18–21, 2006* (Pöllauberg, 2008), 207–20.

Livesey, Steven, 'Transitions 1: Scientific Writing in the Latin Middle Ages', in Andrew Hunter (ed.), *Thornton and Tully's Scientific Books, Libraries and Collectors*, 4th edn (Aldershot, 2000), 72–98.

Loewe, Richard, 'Gas', *Zeitschrift für vergleichende Sprachforschung auf dem Gebiete der indogermanischen Sprachen*, 63 (1936), 118–22.

Lohr, Charles H., *Latin Aristotle Commentaries*, vol. 2: *Renaissance Authors* (Florence, 1988).

Lohr, Charles H., 'The Social Situation of the Study of Aristotelian Natural Philosophy in the Sixteenth and Early Seventeenth Centuries', in Cees Leijenhorst, Christoph Lüthy, and Johannes M. M. H. Thijssen (eds), *The Dynamics of Aristotelian Natural Philosophy from Antiquity to the Seventeenth Century* (Leiden, 2002), 343–8.

Lohr, Charles H., *Latin Aristotle Commentaries: Medieval Authors*, 2 vols (Florence, 2010–13).

Long, Pamela O., *Artisan/Practitioners and the Rise of the New Science, 1400–1600* (Corvallis, OR, 2011).

Lo Presti, Roberto, '"Der Seele und dem Körper gemeinsam": Das Forschungsprogramm der *Parva naturalia* und die Begründung der *scientia de animalibus* in den *praefationes* zu den italienischen Kommentaren des 16. Jahrhunderts', in Jochen Althoff (ed.),

Aristoteles, Parva naturalia: Akten der 18. Tagung der Karl und Gertrud Abel-Stiftung vom 30. September bis 2. Oktober 2015 in Mainz (Berlin and Boston, MA, 2021), 247–79.

Lorenz, Sönke, '*Libri ordinarie legendi*: Eine Skizze zum Lehrplan der mitteleuropäischen Artistenfakultät um die Wende vom 14. zum 15. Jahrhundert', in Wolfgang Hogrebe (ed.), *Argumente und Zeugnisse* (Frankfurt am Main, 1985), 204–58.

Lowood, Henry E., and Robin E. Rider, 'The Scientific Book as a Cultural and Bibliographical Object', in Andrew Hunter (ed.), *Thornton and Tully's Scientific Books, Libraries and Collectors,* 4th edn (Aldershot, 2000), 1–25.

Lowry, Martin, *The World of Aldus Manutius: Business and Scholarship in Renaissance Venice* (Ithaca, NY, 1979).

Ludwig, Walther, 'Neulateinische Lehrgedichte und Vergils Georgica', in D. H. Green et al. (eds), *From Wolfram and Petrarch to Goethe and Grass: Studies in Literature in Honour of Leonard Forster* (Baden-Baden, 1982), 151–80.

Luggin, Johanna, 'Wie beschreibt man *mirabilia*? Antike und neuzeitliche literarische Erkundungen des Ätna', in Martin Korenjak and Irina Tautschnig (eds), *Die antike Literatur und die Wissenschaftliche Revolution* (Baden-Baden, 2023), 119–41.

Lukács, Ladislaus (ed.), *Ratio atque institutio studiorum Societatis Jesu 1586, 1591, 1599* (Rome, 1986).

Lüthy, Christoph, and Alexis Smets, 'Words, Lines, Diagrams, Images: Towards a History of Scientific Imagery', *Early Science and Medicine*, 14 (2009), 398–439.

Maar, Vilhelm (ed.), *Nicolai Stenonis opera philosophica* (Copenhagen, 1910).

McCabe, Anne, *A Byzantine Encyclopaedia of Horse Medicine: The Sources, Compilation, and Transmission of the Hippiatrica* (Oxford, 2007).

McColley, Grant, 'Josephus Blancanus and the Adoption of our Word "Telescope"', *Isis*, 28 (1938), 364–5.

MacDonald, Michael J. (ed.), *The Oxford Handbook of Rhetorical Studies* (Oxford, 2017).

McElduff, Siobhán, *Roman Theories of Translation: Surpassing the Source* (London, 2013).

McGregor, Brian (ed.), 'Erasmus Desiderius: Oration in Praise of the Art of Medicine', in Elaine Fantham, Erika Rummel, and Jozef IJsewijn (eds), *Collected Works of Erasmus: Literary and Educational Writings 7* (Toronto, 1989), 31–50, 431–7.

Mack, Peter, *A History of Renaissance Rhetoric: 1380–1620* (Oxford, 2011).

Mack, Peter, *Rhetoric's Questions: Reading and Interpretation* (Cham, 2017).

McLaughlin, Martin, *Literary Imitation in the Italian Renaissance: The Theory and Practice of Literary Imitation in Italy from Dante to Bembo* (Oxford, 1996).

Maclean, Ian, *Interpretation and Meaning in the Renaissance: The Case of Law* (Cambridge, 1992).

Maclean, Ian, 'Logical Division and Visual Dichotomies: Ramus in the Context of Legal and Medical Writing', in Mordechai Feingold, Joseph S. Freedman, and Wolfgang Rother (eds), *The Influence of Petrus Ramus: Studies in Sixteenth and Seventeenth Century Philosophy and Sciences* (Basel, 2001), 228–47.

Maclean, Ian, *Logic, Signs and Nature in the Renaissance: The Case of Learned Medicine* (Cambridge, 2002).

Maclean, Ian (ed.), *Girolamo Cardano: De libris propriis: The Editions of 1544, 1550, 1557, 1562 with Supplementary Material* (Milan, 2004).

Maclean, Ian, 'The Medical Republic of Letters Before the Thirty Years War', *Intellectual History Review*, 18 (2008), 15–30.

Maclean, Ian, *Learning and the Market Place: Essays in the History of the Early Modern Book* (Leiden and Boston, MA, 2009).

Maclean, Ian, *Scholarship, Commerce, Religion: The Learned Book in the Age of the Renaissance* (Cambridge, MA, and London, 2012).

Maclean, Ian, 'The Thesauruses of Otto and Meerman as Publishing Enterprises: Legal Humanism in its Last Phase, 1725–1780', in Paul J. du Plessis and John W. Cairns (eds), *Reassessing Legal Humanism and Its Claims: Petere Fontes?* (Edinburgh, 2016), 299–346.

Maclean, Ian, 'Die Arbeiten von Hercules: Georg Wolfgang Wedel, die Leopoldina und der norddeutsche Büchermarkt um 1680', *Acta Historica Leopoldina*, 74 (2019), 75–91.

Maclean, Ian, 'The Reception of Hippocrates by Physicians at the End of the Seventeenth Century: A Comparative Study', in Dmitri Levitin and Ian Maclean (eds), *The Worlds of Knowledge and the Classical Tradition in the Early Modern Age* (Leiden and Boston, MA, 2021a), 217–74.

Maclean, Ian, *Episodes in the Life of the Early Modern Learned Book* (Leiden and Boston, MA, 2021b).

Maclean, Ian, 'Sacrobosco at the Book Fairs, 1576–1624', in Matteo Valleriani and Andrea Ottone (eds), *Publishing Sacrobosco's De sphaera in Early Modern Europe* (Cham, 2022), 187–224.

Maclean, Ian, 'Bacon, Hippocrates, and the Aphorism: Seventeenth-Century Critiques', in Martin Korenjak and Irina Tautschnig (eds), *Die antike Literatur und die Wissenschaftliche Revolution* (Baden-Baden, 2023), 99–118.

McOmish, David, 'A Community of Scholarship: Latin Literature and Scientific Discourse in Early Modern Scotland', in Steven J. Reid and David McOmish (eds), *Neo-Latin Literature and Literary Culture in Early Modern Scotland* (Leiden and Boston, MA, 2017), 40–73.

Mahoney, Michael S., *The Mathematical Career of Pierre de Fermat, 1601–1665*, 2nd edn (Princeton, NJ, 1994).

Malherbe, Abraham J., *Ancient Epistolary Theorists* (Atlanta, GA, 1988).

Maloney, Gilles, and Raymond Savoie, *Cinq cents ans de bibliographie hippocratique, 1473–1982* (Quebec, 1982).

Malpangotto, Michaela, *Theoricae novae planetarum Georgii Peurbachii dans l'histoire de l'astronomie* (Paris, 2020).

Mandelbrote, Giles, 'Scientific Books and their Owners: A Survey to *c.*1720', in Andrew Hunter (ed.), *Thornton and Tully's Scientific Books, Libraries and Collectors*, 4th edn (Aldershot, 2000), 333–66.

Manger, Klaus, 'Nekrolog als Biographie: Lohensteins Rede von 1679 auf Hoffmannswaldau', in Walter Berschin (ed.), *Biographie zwischen Renaissance und Barock* (Heidelberg, 1993), 277–309.

Manning, Patrick, and Abigail Owen (eds), *Knowledge in Translation: Global Patterns of Scientific Exchange, 1000–1800 CE* (Pittsburgh, PA, 2018).

Mansfeld, Jaap, *Prolegomena: Questions to be Settled before the Study of an Author, or a Text* (Leiden, 1994).

Marcus, Hannah, *Forbidden Knowledge: Medicine, Science, and Censorship in Early Modern Italy* (Chicago and London, 2020).

Margócsy, Dániel, Mark Somos, and Stephen N. Joffe, *The Fabrica of Andreas Vesalius: A Worldwide Descriptive Census, Ownership, and Annotations of the 1543 and 1555 Editions* (Leiden, 2018).

Marsh, David, *The Quattrocento Dialogue: Classical Tradition and Humanist Innovation* (Cambridge, MA, 1980).

Marti, Hanspeter, *Philosophische Dissertationen deutscher Universitäten 1660–1750: Eine Auswahlbibliographie* (Munich, 1982).

Marti, Hanspeter, 'Lateinsprachigkeit: Ein Gattungsmerkmal der Dissertationen und seine historische Konsistenz', *Jahrbuch für internationale Germanistik*, 30/1 (1998), 50–63.

Marti, Hanspeter, 'Dissertation und Promotion an frühneuzeitlichen Universitäten des deutschen Sprachraums: Versuch eines skizzenhaften Überblicks', in Rainer A. Müller (ed.), *Promotionen und Promotionswesen an deutschen Hochschulen der Frühmoderne* (Cologne, 2001), 1–20.

Marti, Hanspeter, 'Von der Präses- zur Respondentendissertation: Die Autorschaftsfrage am Beispiel einer frühneuzeitlichen Literaturgattung', in Rainer C. Schwinges (ed.), *Examen, Titel, Promotionen* (Basel, 2007), 251–74.

Marti, Hanspeter, 'Disputation und Dissertation in der Frühen Neuzeit und im 19. Jahrhundert: Gegenstand der Wissenschaftssprachgeschichte?', in Michael Prinz and Jürgen Schiewe (eds), *Vernakuläre Wissenschaftskommunikation: Beiträge zur Entstehung und Frühgeschichte der modernen deutschen Wissenschaftssprachen* (Berlin and Boston, MA, 2018), 271–92.

Marti, Hanspeter, Reimund B. Sdzuj, and Robert M. Seidel (eds), *Rhetorik, Poetik und Ästhetik im Bildungssystem des Alten Reiches: Wissenschaftshistorische Erschließung ausgewählter Dissertationen von Universitäten und Gymnasien 1500–1800* (Cologne, 2017).

Martin, Craig, 'Scientific Terminology and the Effects of Humanism: Renaissance Translations of *Meteorologica* IV and the Commentary Tradition', in Michèle Goyens, Pieter de Leemans, and An Smets (eds), *Science Translated: Latin and Vernacular Translations of Scientific Treatises in Medieval Europe* (Leuven, 2008), 155–80.

Martin, Craig, *Renaissance Meteorology: Pomponazzi to Descartes* (Baltimore, MD, 2011).

Martin, Henry-Jean, *The History and Power of Writing* (Chicago and London, 1994).

Meier, Christel (ed.), *Die Enzyklopädie im Wandel vom Hochmittelalter bis zur frühen Neuzeit: Akten des Kolloquiums des Projekts D im Sonderforschungsbereich 231 (29.11.-1.12.1996)* (Munich, 2002).

Merzbach, Uta C., and Carl B. Boyer, *A History of Mathematics* (Hoboken, NJ, 2011).

Mesnard, Jean (ed.), *Blaise Pascal: Œuvres complètes*, vol. 3: *Œuvres diverses (1654–1657)* (Paris, 1991).

Methuen, Charlotte, 'Maestlin's Teaching of Copernicus: The Evidence of His University Textbook and Disputations', *Isis*, 87 (1996), 230–47.

Meyer, Frederick J., Emily E. Trueblood, and John L. Heller (eds), *The Great Herbal of Leonhart Fuchs: De historia stirpium commentarii insignes*, 2 vols (Stanford, CA, 1999).

Meyer, Heinz, 'Ordo rerum und Registerhilfen in mittelalterlichen Enzyklopädiehandschriften', *Frühmittelalterliche Studien*, 25 (1991), 315–39.

Miazek, Monika (ed.), *Flora Sinensis Michała Boyma* (Gniezno, 2005).

Michel, Paul, *Physikotheologie: Ursprünge, Leistung und Niedergang einer Denkform* (Zurich, 2008).

Miert, Dirk van (ed.), *Communicating Observations in Early Modern Letters (1500–1675): Epistolography and Epistemology in the Age of the Scientific Revolution* (London and Turin, 2013a).

Miert, Dirk van, 'Philology and Empiricism: Observation and Description in the Correspondence of Joseph Scaliger (1540–1609)', in Miert (ed.), *Communicating Observations in Early Modern Letters (1500–1675)* (London and Turin, 2013b), 89–113.

Mocchi, Giuliana, Sandra Plastina, and Emilio Sergio (eds), *Bernardino Telesio: Tra filosofia naturale e scienza moderna* (Pisa and Rome, 2012).

Monfasani, John, 'The Pseudo-Aristotelian *Problemata* and Aristotle's *De Animalibus* in the Renaissance', in Anthony Grafton and Nancy G. Siraisi (eds), *Natural Particulars: Nature and the Disciplines in Renaissance Europe* (Cambridge, MA, and London, 1999), 79–121.

Monti, Carlo (ed.), *Opere latine di Giordano Bruno: Il triplice minimo e la misura. La monade, il numero e la figura. L'immenso e gli innumerevoli* (Turin, 1980).

Morrow, Glenn R. (ed.), *Proclus: A Commentary on the First Book of Euclid's Elements* (Princeton, NJ, 1970).

Morton, Alan G., *History of Botanical Science: An Account of the Development of Botany from Ancient Times to Present Day* (London and New York, 1981).

Morton, Leslie T., 'The Growth of Medical Periodical Literature', in Alain Besson (ed.), *Thornton's Medical Books, Libraries and Collectors* (Aldershot, 1990), 221–38.

Mosley, Adam, 'Astronomical Books and Courtly Communication', in Marina Frasca-Spada and Nick Jardine (eds), *Books and the Sciences in History* (Cambridge, 2000), 114–31.

Mosley, Adam, 'Tycho Brahe's *Epistolae Astronomicae*: A Reappraisal', in Toon van Houdt et al. (eds), *Self-Presentation and Social Identification: The Rhetoric and Pragmatics of Letter Writing in Early Modern Times* (Leuven, 2002), 449–68.

Mosley, Adam, 'The Cosmographer's Role in the Sixteenth Century: A Preliminary Study', *Archives Internationales d'Histoire des Sciences*, 59 (2009), 423–39.

Mosley, Adam, 'Reading the Heavens: Observation and Interpretation of Astronomical Phenomena in Learned Letters circa 1600', in Dirk van Miert (ed.), *Communicating Observations in Early Modern Letters (1500–1675)* (London and Turin, 2013), 115–34.

Moss, Ann, *Printed Commonplace-Books and the Structuring of Renaissance Thought* (Oxford, 1996).

Moss, Jean D., '[Renaissance Rhetoric:] Rhetoric and Science', in Michael J. MacDonald (ed.), *The Oxford Handbook of Rhetorical Studies* (Oxford, 2017), 423–35.

Most, Glenn (ed.), *Commentaries—Kommentare* (Göttingen, 1999).

Mottana, Annibale, 'Bernardo Cesi (Caesius) and His *Mineralogia* (1636): Naming a New Science from an Indiscriminate Piling of Mineral Accounts', *Rendiconti Lincei: Scienze Fisiche e Naturali*, 28 (2017), 435–48.

Moul, Victoria (ed.), *A Guide to Neo-Latin Literature* (Cambridge, 2017a).

Moul, Victoria, 'Didactic Poetry', in Moul (ed.), *A Guide to Neo-Latin Literature* (Cambridge, 2017b), 180–99.

Moureau, François (ed.), *Les Presses grises: La Contrefaçon du livre (XVIe–XIXe siècles)* (Paris, 1988).

Müller, Peter O., *Deutsche Lexikographie des 16. Jahrhunderts* (Tübingen, 2001).

Müller-Jahncke, Wolf-Dieter, 'Der Höhepunkt der Iatromathematik', *Berichte zur Wissenschaftsgeschichte*, 4 (1981), 41–50.

Müllner, Karl (ed.), *Reden und Briefe italienischer Humanisten: Nachdruck der Ausgabe Wien 1899: Mit einer Einleitung, analytischer Inhaltsübersicht, Bibliographie und Indices von Barbara Gerl* (Munich, 1970).

Munske, Horst H., and Alan Kirkness (eds), *Eurolatein: Das griechische und lateinische Erbe in den europäischen Sprachen* (Tübingen, 1996).

Müsch, Irmgard, *Geheiligte Naturwissenschaft: Die Kupfer-Bibel des Johann Jakob Scheuchzer* (Göttingen, 2000).

Myers, Robin, Michael Harris, and Giles Mandelbrote (eds), *Under the Hammer: Book Auctions since the Seventeenth Century* (New Castle, DE, and London, 2001).

Nahmer, Dieter von der, *Die lateinische Heiligenvita: Eine Einführung in die lateinische Hagiographie* (Darmstadt, 1994).

Naumann, Friedrich, *Georgius Agricola: Berggelehrter, Naturforscher, Humanist* (Erfurt, 2007).

Nellen, Henk, 'La Correspondance savante au XVIIe siècle', *XVIIe siècle*, 45 (1993), 87–98.

Neue Deutsche Biographie, 28 vols (Berlin, 1953–2023).

Neugebauer, Otto (ed.), *Astronomical Cuneiform Texts: Babylonian Ephemerides of the Seleucid Period for the Motion of the Sun, the Moon, and the Planets*, 3 vols (London, 1955).

Neumann, Olaf, 'Carl Friedrich Gauß, *Disquisitiones arithmeticae* (1801)', in Ivor Grattan-Guinness (ed.), *Landmark Writings in Western Mathematics 1640–1940* (Amsterdam, 2005), 303–15.

Newman, William R., and Lawrence M. Principe, 'Alchemy versus Chemistry: The Etymological Origins of a Historiographic Mistake', *Early Science and Medicine*, 3 (1998), 32–65.

Nicolson, Marjorie Hope, *Newton Demands the Muse: Newton's Opticks and the Eighteenth Century Poets* (Princeton, NJ, 1946).

Nicolson, Marjorie Hope, *Mountain Gloom and Mountain Glory: The Development of the Aesthetics of the Infinite* (Ithaca, NY, 1959).

Niedling, Justus, *Die mittelalterlichen und frühneuzeitlichen Kommentare zur Techne des Galenos* (Paderborn, 1924).

Norton, Glyn P., *The Ideology and Language of Translation in Renaissance France and Their Humanist Antecedents* (Geneva, 1984).

Nothaft, Philippe C. E., *Dating the Passion: The Life of Jesus and the Emergence of Scientific Chronology (200–1600)* (Leiden and Boston, MA, 2012).

Nummedal, Tara, and Paula Findlen, 'Words of Nature: Scientific Books in the Seventeenth Century', in Andrew Hunter (ed.), *Thornton and Tully's Scientific Books, Libraries and Collectors*, 4th edn (Aldershot, 2000), 164–215.

Nutton, Vivian, 'Hippocrates in the Renaissance', in Gerhard Baader and Rolf Winau (eds), *Die Hippokratischen Epidemien: Theorie—Praxis—Tradition* (Stuttgart, 1989), 420–39.

Nutton, Vivian, 'John of Alexandria Again: Greek Medical Philosophy in Latin Translation', *Classical Quarterly*, 41 (1991), 509–19.

Nutton, Vivian, 'The Changing Language of Medicine, 1450–1550', in Olga Weijers (ed.), *Vocabulary of Teaching and Research between Middle Ages and Renaissance: Proceedings of the Colloquium London, Warburg Institute, 11–12 March 1994* (Turnhout, 1995), 184–98.

Nybakken, Oscar E., *Greek and Latin in Scientific Terminology* (Ames, IA, 1959).

Oaks, Jeffrey A., 'François Viète's Revolution in Algebra', *Archive for History of Exact Sciences*, 72 (2018), 245–302.

Offner, Robert, et al. (eds), *Johannes Honterus, Rudimenta Cosmographica: Grundzüge der Weltbeschreibung (Corona/Kronstadt 1542), ins Deutsche, Rumänische und Ungarische übersetzte und kommentierte Faksimile-Ausgabe*, 2nd edn (Hermannstadt and Bonn, 2017).

Ogilvie, Brian W., *The Science of Describing: Natural History in Renaissance Europe* (Chicago, 2006).

Oldroyd, David R., *Thinking about the Earth: A History of Ideas in Geology* (Cambridge, MA, 1996).

Olschki, Leonardo, *Geschichte der neusprachlichen wissenschaftlichen Literatur*, 3 vols (Heidelberg, 1919–27).

O'Malley, John W., *Praise and Blame in Renaissance Rome: Rhetoric, Doctrine, and Reform in the Sacred Orators of the Papal Court, c.1450–1521* (Durham, NC, 1979).

Omodeo, Pietro D., *Copernicus in the Cultural Debates of the Renaissance. Reception, Legacy, Transformation* (Leiden and Boston, MA, 2014).

Omodeo, Pietro D., 'Modelli esplicativi delle maree nel Rinascimento: Una rassegna', *Galilaeana*, 14 (2017), 97–114.

Ong, Walter J., *Ramus, Method and the Decay of Dialogue* (Cambridge, MA, 1958).

Ongaro, Giuseppe, 'Contributi alla biografia di Prospero Alpino', *Acta Medicae Historiae Patavina*, 8–9 (1961–3), 79–168.

Osler, William, *Incunabula medica: A Study of the Earliest Printed Medical Books, 1467–1480* (Oxford, 1923).

Ostler, Nicholas, *Ad infinitum: A Biography of Latin* (New York, 2007).

Overmier, Judith, 'Scientific Book Collectors and Collections, Public and Private, 1720 to Date', in Andrew Hunter (ed.), *Thornton and Tully's Scientific Books, Libraries and Collectors*, 4th edn (Aldershot, 2000), 367–91.

Pade, Marianne (ed.), *On Renaissance Commentaries* (Hildesheim, 2005).

Pantin, Isabelle, 'Latin et langues vernaculaires dans la littérature scientifique européenne au début de l'époque moderne (1550–1635)', in Roger Chartier and Pietro Corsi (eds), *Sciences et langues en Europe* (Paris, 1996), 43–58.

Pantin, Isabelle, 'Kepler's *Epitome*: New Images for an Innovative Book', in Sachiko Kusukawa and Ian Maclean (eds), *Transmitting Knowledge* (Oxford, 2006), 217–37.

Pantin, Isabelle, 'The Astronomical Diagrams in Oronce Finé's *Protomathesis* (1532): Founding a French Tradition?', *Journal for the History of Astronomy*, 41 (2010), 287–310.

Pantin, Isabelle, 'Pierre Gassendi's *Institutio Astronomica* and the Diffusion of the Arguments in Defence of the Earth's Movement after the Condemnation of Galileo', *Galilaeana*, 9 (2012), 65–90.

Patterson, Thomas S., 'Jean Béguin and his *Tyrocinium chymicum*', *Annals of Science*, 2 (1937), 243–98.

Paulsen, Friedrich, *Geschichte des gelehrten Unterrichts auf den deutschen Schulen und Universitäten vom Ausgang des Mittelalters bis zur Gegenwart*, 2 vols, 3rd edn (Leipzig, 1919–21).

Paulus, Sybille, *Wissenschaftliche Textsorten in der italienischen Renaissance: Der Sprachwechsel aus dem Lateinischen in der astronomischen, meteorologischen und kosmologischen Literatur* (Tübingen, 2005).

Pavlinov, Igor Y., *Biological Systematics: History and Theory* (Boca Raton, FL, 2021).

Pavord, Anna, *The Naming of Names: The Search for Order in the World of Plants* (London, 2005).

Pelc, Milan, *Illustrium imagines: Das Porträtbuch der Renaissance* (Leiden, 2002).

Pera, Marcello, and William R. Shea (eds), *Persuading Science: The Art of Scientific Rhetoric* (Canton, MA, 1991).

Perfetti, Stefano, '*Cultius atque integrius*: Teodoro Gaza, traduttore umanistico del *De partibus animalium*', *Rinascimento*, 35 (1995), 253–86.

Perfetti, Stefano, '*Docebo vos dubitare*: Il commento inedito di Pietro Pomponazzi al *De partibus animalium* (Bologna 1521–24)', *Documenti e studi sulla tradizione filosofica medievale*, 10 (1999), 439–66.

Perfetti, Stefano, *Aristotle's Zoology and Its Renaissance Commentators (1521–1600)* (Leuven, 2000).

Perfetti, Stefano (ed.), *Pietro Pomponazzi: Expositio super primo et secundo De partibus animalium* (Florence, 2004).

Pesenti, Tiziana, 'How Did Early Printers Choose Medical Commentaries for the Press?', in Wouter Bracke and Herwig Deumens (eds), *Medical Latin from the Late Middle Ages to the Eighteenth Century* (Brussels, 2000), 67–92.

Pettegree, Andrew, *The Book in the Renaissance* (New Haven, CT, 2010).

Pfeiffer, Jeanne, and Jean-Pierre Vittu, 'Les Journaux savants, formes de la communication et agents de la construction des savoirs (17e–18e siècles)', *Dix-huitième siècle*, 40 (2008), 281–300.

Pfeiffer, Rudolf, *History of Classical Scholarship*, 2 vols (Oxford, 1968–76).

Piepenbring-Thomas, Carola, 'Garden Visits, Observations, Reading and Excerpts: Martin Fogel (1634) and His Techniques of Acquiring Knowledge', in Hubertus Fischer,

Volker R. Remmert, and Joachim Wolschke-Bulmahn (eds), *Gardens, Knowledge and the Sciences in the Early Modern Period* (Basel, 2016), 253–65.

Pinon, Laurent, *Livres de zoologie de la Renaissance: Une anthologie (1450–1700)* (Paris, 1995).

Pinon, Laurent, 'Conrad Gessner and the Historical Depth of Renaissance Natural History', in Gianna Pomata and Nancy Siraisi (eds), *Historia: Empiricism and Erudition in Early Modern Europe* (Cambridge, MA, 2005), 241–68.

Pitkäranta, Reijo, *Lexicon Neolatinum dissertationum Academiae Aboensis (1642–1828) / Uuslatinen sanasto Turun akatemian väitoskirjoissa (1642–1828)* (Helsinki, 2018).

Poccetti, Paolo, Diego Poli, and Carlo Santini, *Eine Geschiche der lateinischen Sprache: Ausformung, Sprachgebrauch, Kommunikation* (Tübingen and Basel, 2005).

Poel, Marc van der, 'Humanist Rhetoric in the Renaissance: Classical Mastery?', in Wim Verbaal, Maes Yanick, and Jan Papy (eds), *Latinitas perennis*, vol. 1: *The Continuity of Latin Literature* (Leiden and Boston, MA, 2007), 119–37.

Poel, Marc van der, 'Oratory and Declamation', in Victoria Moul (ed.), *A Guide to Neo-Latin Literature* (Cambridge, 2017), 272–88.

Pomata, Gianna, '*Observatio* ovvero *Historia*: Note su empirismo e storia in età moderna', *Quaderni storici*, 91 (1996), 173–98.

Pomata, Gianna, 'Sharing Cases: The *Observationes* in Early Modern Medicine', *Early Science and Medicine*, 15 (2010), 193–236.

Pomata, Gianna, 'A Word of the Empirics: The Ancient Concept of Observation and Its Recovery in Early Modern Medicine', *Annals of Science*, 68 (2011a), 1–25.

Pomata, Gianna, 'Observation Rising: Birth of an Epistemic Genre, 1500–1650', in Lorraine Daston and Elizabeth Lunbeck (eds), *Histories of Scientific Observation* (Chicago and London, 2011b), 45–80.

Pomata, Gianna, 'The Medical Case Narrative: Distant Reading of an Epistemic Genre', *Literature and Medicine*, 32/1 (2014), 1–23.

Pomata, Gianna, and Nancy Siraisi (eds), *Historia: Empiricism and Erudition in Early Modern Europe* (Cambridge, MA, 2005).

Pompeo Faracovi, Ornella, 'Girolamo Cardano e il ritorno a Tolomeo', in Germana Ernst and Guido Giglioni (eds), *Il linguaggio dei cieli* (Rome, 2012a), 125–38.

Pompeo Faracovi, Ornella, 'La polemica antiastrologica di Giovanni Pico della Mirandola', in Germana Ernst and Guido Giglioni (eds), *Il linguaggio dei cieli* (Rome, 2012b), 91–107.

Pontani, Filippomaria, and Stefan Weise (eds), *The Hellenizing Muse: A European Anthology of Poetry in Ancient Greek from the Renaissance to the Present* (Berlin and Boston, MA, 2022).

Poole, William, *The World Makers: Scientists of the Restoration and the Search for the Origins of the Earth* (Oxford, 2010).

Pörksen, Uwe, *Deutsche Naturwissenschaftssprachen: Historische und kritische Studien* (Tübingen, 1986).

Porter, Roy, *The Making of Geology: Earth Science in Britain, 1660–1815* (Cambridge, 1977).

Premuda, Loris, *Storia dell'iconografia anatomica*, 2nd edn (Origgio, 1993).

Principe, Lawrence M., *The Aspiring Adept: Robert Boyle and His Alchemical Quest* (Princeton, NJ, 1998).

Prinz, Michael, and Jürgen Schiewe (eds), *Vernakuläre Wissenschaftskommunikation: Beiträge zur Entstehung und Frühgeschichte der modernen deutschen Wissenschaftssprachen* (Berlin and Boston, MA, 2018).

Quain, Edwin A., 'The Medieval *Accessus ad auctores*', *Traditio*, 3 (1945), 215–64.

Ralph, Thomas S. (ed.), *Opuscula omnia botanica Thomae Johnsoni, Pharmaceuticae societatis Londinensis socii* (London, 1847).

Randles, William G. L., *Geography, Cartography and Nautical Science in the Renaissance: The Impact of the Great Discoveries* (Aldershot, 2000).

Rautenberg, Ursula, 'Die Entstehung und Entwicklung des Buchtitelblatts in der Inkunabelzeit in Deutschland, in den Niederlanden und in Venedig', *Archiv für Geschichte des Buchwesens*, 62 (2008), 1–105.

Rautenberg, Ursula, *Reclams Sachlexikon des Buches*, 3rd edn (Stuttgart, 2015).

Reeds, Karen M., *Botany in Medieval and Renaissance Universities* (New York and London, 1991).

Reich, Karin, 'Leonhard Euler, *Introduction to Analysis* (1748)', in Ivor Grattan-Guinness (ed.), *Landmark Writings in Western Mathematics 1640–1940* (Amsterdam, 2005), 181–90.

Reich, Karin, and Helmuth Gericke (eds), *François Viète: Einführung in die Neue Algebra* (Munich, 1973).

Reichert, Jürgen (ed.), *Ioannis Kepleri Tabulae Rudolphinae: Rudolphinische Tafeln von Johannes Kepler: Originaler lateinischer Text und deutsche Übersetzung* (Würzburg, 2014).

Reid, Steven J., and Emma A. Wilson (eds), *Ramus, Pedagogy and the Liberal Arts: Ramism in Britain and the Wider World* (Farnham and Burlington, VT, 2011).

Remmert, Volker R., '*Docet parva pictura, quod multae scripturae non dicunt*: Frontispieces, Their Functions, and Their Audiences in Seventeenth-Century Mathematical Sciences', in Sachiko Kusukawa and Ian Maclean (eds), *Transmitting Knowledge* (Oxford, 2006), 239–70.

Remmert, Volker R., 'Visual Legitimisation of Astronomy in the 16th and 17th Centuries: Atlas, Hercules and Tycho's Nose', *Studies in History and Philosophy of Science*, 38 (2007), 327–62.

Remmert, Volker R., *Picturing the Scientific Revolution: Title Engravings in Early Modern Scientific Publications* (Philadelphia, PA, 2011).

Renbourn, Edward T., 'The Natural History of Insensible Perspiration: A Forgotten Doctrine of Health and Disease', *Medical History*, 4 (1960), 135–52.

Rener, Frederick M., *Interpretatio: Language and Translation from Cicero to Tytler* (Amsterdam and Atlanta, GA, 1989).

Reuss, Jeremias David, *Repertorium commentationum a societatibus litterariis editarum* (Göttingen, 1801–21).

Rex, Friedemann, 'Keplers Lehrer Michael Mästlin und sein Lehrbuch der Astronomie (1582)', in Gerhard Betsch and Jürgen Hamel (eds), *Zwischen Copernicus und Kepler: M. Michael Maestlinus Mathematicus Goeppingensis 1550–1631* (Frankfurt am Main, 2002), 11–32.

Rhodes, Dennis E., 'Medical Incunabula', in Alain Besson (ed.), *Thornton's Medical Books, Libraries and Collectors* (Aldershot, 1990), 30–42.

Richardson, William F., and John B. Carman (eds), *Andreas Vesalius: On the Fabric of the Human Body* (San Francisco, 1998–2009).

Richter, Myriam, and Bernd Hamacher, 'Biographische Kleinformen', in Christian Klein (ed.), *Handbuch Biographie: Methoden, Traditionen, Theorien* (Stuttgart, 2009), 137–42.

Rinaldi, Michele, 'Pontano, Trapezunzio ed il *Graecus Interpres* del *Centiloquio* pseudo-tolemaico', *Atti dell' Accademia Pontaniana*, 48 (1999), 125–71.

Rinaldi, Michele, 'La lettera di dedica a Federico da Montefeltro del primo libro delle *Commentationes in centum sententiis Ptolemaei* di Giovanni Gioviano Pontano', *Cahiers de recherches médiévales et humanistes/Journal of medieval and humanistic studies*, 25 (2013), 341–55.

Roberts, Ernest S., and John Venn (eds), *The Works of John Caius, M.D., Second Founder of Gonville and Caius College and Master of the College 1559–1573* (Cambridge, 1912).

Robinson, Richard, *Definition* (Oxford, 1954).
Roellenbleck, Georg, *Das epische Lehrgedicht Italiens im 15. und 16. Jahrhundert* (Munich, 1975).
Roelli, Philipp, *Latin as the Language of Science and Learning* (Berlin and Boston, MA, 2021).
Roelli, Philipp, 'Sinn aus Unsinn? Der Umgang mit der Überlieferung spätantiker medizinischer Literatur bis zum 16. Jahrhundert', in Martin Korenjak and Irina Tautschnig (eds), *Die antike Literatur und die Wissenschaftliche Revolution* (Baden-Baden, 2023), 15–36.
Rogg, Ignaz, *Bibliotheca mathematica sive criticus librorum mathematicorum, qui inde ab rei typographicae exordio ad anni 1830mi usque finem excusi sunt, index... Sectio I libros arithmeticos et geometricos complectens/Handbuch der mathematischen Literatur vom Anfange der Buchdruckerkunst bis zum Schlusse des Jahres 1830: Erste Abtheilung, welche die arithmetischen und geometrischen Wissenschaften enthält* (Tübingen, 1830).
Roling, Bernd, *Physica sacra: Wunder, Naturwissenschaft und historischer Schriftsinn zwischen Mittelalter und Früher Neuzeit* (Leiden and Boston, MA, 2013).
Roling, Bernd, 'Der Wal als Schauobjekt: Thomas Bartholin (1616–80), die dänische Nation und das Ende der Einhörner', in Karl A. E. Enenkel and Paul J. Smith (eds), *Zoology in Early Modern Culture: Intersections of Science, Theology, Philology, and Political and Religious Education* (Leiden and Boston, MA, 2014), 172–96.
Roling, Bernd, 'Linné Demands the Muse: János Molnár, Erasmus Darwin und das *Systema naturae*', in Martin Korenjak and Irina Tautschnig (eds), *Die antike Literatur und die Wissenschaftliche Revolution* (Baden-Baden, 2023), 233–68.
Roller, Duane, *The De magnete of William Gilbert* (Amsterdam, 1959).
Rose, Paul L., *The Italian Renaissance of Mathematics: Studies on Humanists and Mathematicians from Petrarch to Galileo* (Geneva, 1975).
Rosen, Edward (ed.), *Kepler's Conversation with Galileo's Sidereal Messenger: First Complete Translation, with an Introduction and Notes* (New York, 1965).
Rosenberg, Rainer, 'Eine verworrene Geschichte: Vorüberlegungen zu einer Biographie des Literaturbegriffs', *LiLi: Zeitschrift für Literaturwissenschaft und Linguistik*, 20/77 (1990), 36–65.
Rosner, Fred (ed.), *The Medical Aphorisms of Moses Maimonides* (Haifa, 1989).
Rowland, Ingrid D., *The Scarith of Scornello: A Tale of Renaissance Forgery* (Chicago, 2004).
Rüegg, Walter (ed.), *Geschichte der Universität in Europa*, 4 vols (Munich, 1993–2010).
Rütten, Thomas, and Ulrich Rütten, 'Melanchthons Rede *de Hippocrate*', *Medizinhistorisches Journal*, 33 (1998), 19–55.
Sahr, Eva, 'Vorbild oder Feindbild: Averroes' Kommentare zu den *Zweiten Analytiken* im 16. Jahrhundert in Italien', in Martin Korenjak and Irina Tautschnig (eds), *Die antike Literatur und die Wissenschaftliche Revolution* (Baden-Baden, 2023), 81–98.
Santamaría Hernández, Teresa, '*Nicander Latinus*: La difusión Latina de Nicandro en el siglo XVI', *Humanistica Lovaniensia*, 53 (2004), 125–73.
Santamaría Hernández, Teresa (ed.), *Pedro Jaime Esteve: Nicandri Theriaca: Traducción latina en verso (1552)* (Cuenca, 2018).
Sarton, George, 'Rumphius, Plinius Indicus (1628–1702)', *Isis*, 27 (1937), 242–57.
Sarton, George, 'The Scientific Literature Transmitted through the Incunabula', *Osiris*, 5 (1938), 41–245.
Sarton, George, 'The Study of Early Scientific Textbooks', *Isis*, 38 (1948), 137–48.
Savino, Christina, 'Galenic Forgeries of the Renaissance: An Overview of Commentaries Falsely Attributed to Galen', in Petros Bouras-Vallianatos and Barbara Zipser (eds), *Brill's Companion to the Reception of Galen* (Leiden and Boston, MA, 2019), 453–71.

Schiffner, Carl (ed.), *Georg Agricola: Zwölf Bücher vom Berg- und Hüttenwesen… sowie sein Buch von den Lebewesen unter Tage* (Berlin, 1928).

Schindler, Claudia, 'Didactic Poetry as Elitist Poetry: Christopher Stay's *De poesi didascalica dialogus* in the Context of Classical and Neo-Latin Didactic Discourse', in Alexander Winkler and Florian Schaffenrath (eds), *Neo-Latin and the Vernaculars* (Leiden and Boston, MA, 2018), 232–50.

Schindler, Claudia, 'Die gefangene Nymphe: Naturerfahrung und wissenschaftlicher Fortschritt in Bernardo Zamagnas *Echo* (1764)', in Martin Korenjak and Irina Tautschnig (eds), *Die antike Literatur und die Wissenschaftliche Revolution* (Baden-Baden, 2023), 189–205.

Schirg, Bernhard, *Die Ökonomie der Dichtung: Das Lobgedicht des Pietro Lazzaroni an den Borgia-Papst Alexander VI. (1497): Einleitung, Interpretation, kritische Erstedition und Kommentar* (Hildesheim, 2016).

Schirrmeister, Albert, 'Universitätsreden', in Jan-Hendryk de Boer, Marian Füssel, and Maximilian Schuh (eds), *Universitäre Gelehrtenkultur vom 13.–16. Jahrhundert: Ein interdisziplinäres Quellen- und Methodenhandbuch* (Stuttgart, 2018), 559–77.

Schlegelmilch, Sabine, 'The Scientific Revolution in Marburg', in Meelis Friedenthal, Hanspeter Marti, and Robert Seidel (eds), *Early Modern Disputations and Dissertations in an Interdisciplinary and European Context* (Leiden, 2021), 288–311.

Schlegelmilch, Ulrich, 'Surgical Disputations in Basel at around 1600', in Meelis Friedenthal, Hanspeter Marti, and Robert Seidel (eds), *Early Modern Disputations and Dissertations in an Interdisciplinary and European Context* (Leiden, 2021), 255–87.

Schmeidler, Felix (ed.), *Joannis Regiomontani Opera collectanea: Faksimiledrucke von 9 Schriften Regiomontans und einer von ihm gedruckten Schrift seines Lehrers Purbach* (Osnabrück, 1972).

Schmidt, Albert-Marie, *La Poésie scientifique en France au seizième siècle* (Lausanne, 1970).

Schmiedel, Irina, 'Gardens on Canvas and Paper: Cataloguing Botanical Abundance in Late Medici Tuscany', in Hubertus Fischer, Volker R. Remmert, and Joachim Wolschke-Bulmahn (eds), *Gardens, Knowledge and the Sciences in the Early Modern Period* (Basel, 2016), 211–33.

Schmied-Kowarzik, Walther (ed.), *G. W. Leibniz: Deutsche Schriften*, vol. 1: *Muttersprache und völkische Gesinnung* (Leipzig, 1916).

Schmitt, Charles B., *Aristotle and the Renaissance* (Cambridge, MA, and London, 1983).

Schmitt, Charles B., 'The Rise of the Philosophical Textbook', in Schmitt et al. (eds), *The Cambridge History of Renaissance Philosophy* (Cambridge, 1988), 792–804.

Schmitz, Rudolf, Christoph Friedrich, and Wolf-Dieter Müller-Jahncke, *Geschichte der Pharmazie*, 2 vols (Eschborn, 1998–2005).

Schmitz, Rudolf, and Gundolf Keil (eds), *Humanismus und Medizin* (Weinheim, 1984).

Schmitz, Rudolf, and Fritz Krafft (eds), *Humanismus und Naturwissenschaften* (Boppard, 1980).

Schmitz, Wolfgang, *Grundriss der Inkunabelkunde: Das gedruckte Buch im Zeitalter des Medienwechsels* (Stuttgart, 2018).

Schnicke, Falko, '18. Jahrhundert', in Christian Klein (ed.), *Handbuch Biographie: Methoden, Traditionen, Theorien* (Stuttgart, 2009), 234–42.

Schöbi-Fink, Philipp, 'Rheticus: Der erste Kopernikaner', in Gerhard Wanner and Philipp Schöbi-Fink (eds), *Rheticus: Wegbereiter der Neuzeit* (Feldkirch, 2010), 7–44.

Scholar, Richard, 'Introduction', in Sachiko Kusukawa and Ian Maclean (eds), *Transmitting Knowledge* (Oxford, 2006), 1–9.

Schönberger, Rolf, et al., *Repertorium edierter Texte des Mittelalters aus dem Bereich der Philosophie und angrenzender Gebiete*, 4 vols, 2nd edn (Berlin, 2011).

Schöner, Christoph, *Mathematik und Astronomie an der Universität Ingolstadt im 15. und 16. Jahrhundert* (Berlin, 1994).

Schottenloher, Karl, *Die Widmungsvorrede im Buch des 16. Jahrhunderts* (Münster, 1953).

Schöttler, Tobias, 'From Causes to Relations: The Emergence of a Non-Aristotelian Concept of Geometrical Proof out of the *Quaestio de certitudine mathematicarum*', *Society and Politics*, 6/2 (2012), 29–47.

Schröder, Konrad (ed.), *Fremdsprachenunterricht 1500–1800* (Wiesbaden, 1992).

Schullian, Dorothy M., 'Old Volumes Shake Their Vellum Heads', *Bulletin of the Medical Library Association*, 33 (1945), 413–48.

Schülling, Hermann, *Die Dissertationen und Habilitationsschriften der Universität Gießen im 18. Jahrhundert* (Giessen, 1976).

Schwetschke, Carl G., *Codex nundinarius Germaniae literatae bisecularis: Meß-Jahrbücher des deutschen Buchhandels*, 2 vols (Halle, 1850–77).

Schwinges, Rainer C. (ed.), *Examen, Titel, Promotionen: Akademisches und staatliches Promotionswesen vom 13. bis zum 21. Jahrhundert* (Basel, 2007).

Scudder, Samuel H., *Catalogue of Scientific Serials of All Countries, Including the Transactions of Learned Societies, in the Natural, Physical and Mathematical Sciences* (Cambridge, MA, 1879).

Sdzuj, Reimund B., 'Zum Cartesianismus am Bremer Gymnasium illustre: Johann Eberhard Schwelings Dissertation *De anima brutorum* (1676)', in Marion Gindhart, Hanspeter Marti, and Robert Seidel (eds), *Frühneuzeitliche Disputationen* (Cologne, 2016), 179–98.

Seele, Astrid, *Römische Übersetzer: Nöte, Freiheiten, Absichten: Verfahren des literarischen Übersetzens in der griechisch-römischen Antike* (Darmstadt, 1995).

Seidenschnur, Otto, 'Beiträge zur Kenntniss medicinischer Gedichte', *Janus: Zeitschrift für Geschichte und Literatur der Medicin*, 2 (1847), 772–806.

Selosse, Philippe, 'The Role of Carolus Clusius (1526–1609) as Translator in the Emergence of a Taxonomic Terminology in Botany', in Harold J. Cook and Sven Dupré (eds), *Translating Knowledge in the Early Modern Low Countries* (Münster, 2012), 349–68.

Serfati, Michel, 'René Descartes, *Géometrie*, Latin Edition (1649), French Edition (1637)', in Ivor Grattan-Guinness (ed.), *Landmark Writings in Western Mathematics 1640–1940* (Amsterdam, 2005), 1–22.

Shapin, Steven, 'Pump and Circumstance: Robert Boyle's Literary Technology', *Social Studies of Science*, 14 (1984), 481–519.

Shapin, Steven, *The Scientific Revolution* (Chicago, 1996).

Shapin, Steven, and Simon Schaffer, *Leviathan and the Air-Pump: Hobbes, Boyle, and the Experimental Life* (Princeton, NJ, 1985).

Shapiro, Barbara J., *A Culture of Fact: England, 1550–1720* (Ithaca, NY, and London, 2000).

Siegrist, Christoph, *Das Lehrgedicht der Aufklärung* (Stuttgart, 1974).

Šimon, František, '*Enchiridion medicinae* by Georg Henisch (1549–1618)', *Ceská a Slovenská farmacie*, 67 (2018), 85–90.

Siraisi, Nancy G., *Avicenna in Renaissance Italy: The Canon and Medical Teaching in Italian Universities after 1500* (Princeton, NJ, 1987).

Siraisi, Nancy G., *Medicine and the Italian Universities* (Leiden, 2001).

Siraisi, Nancy G., 'Oratory and Rhetoric in Renaissance Medicine', *Journal of the History of Ideas*, 65 (2004), 191–211.

Siraisi, Nancy G., *History, Medicine, and the Tradition of Renaissance Learning* (Ann Arbor, MI, 2007).

Siraisi, Nancy G., '*Medicina practica*: Girolamo Mercuriale as Teacher and Textbook Author', in Emilio Campi et al. (eds), *Scholarly Knowledge* (Geneva, 2008), 287–305.

Slaughter, Mary M., *Universal Languages and Scientific Taxonomy in the Seventeenth Century* (Cambridge, 1982).

Smith, David E., *Rara arithmetica: A Catalogue of the Arithmetics Written before the Year MDCI with a Description of Those in the Library of George Arthur Plimpton of New York* (Boston, MA, and London, 1908).

Smith, Margaret M., *The Title-Page: Its Early Development 1460–1510* (London and New Castle, DE, 2000).

Snyder, Jon R., *Writing the Scene of Speaking: Theories of Dialogue in the Late Italian Renaissance* (Stanford, CA, 1989).

Societé Hollandaise des Sciences (ed.), *Les Œuvres complètes de Christiaan Huygens*, 22 vols (The Hague, 1888–1950).

Söderqvist, Thomas (ed.), *The History and Poetics of Scientific Biography* (Burlington, VT, 2007).

Sohm, Walter, *Die Schule Johann Sturms und die Kirche Straßburgs in ihrem gegenseitigen Verhältnis 1530–1581: Ein Beitrag zur Geschichte deutscher Renaissance* (Munich and Berlin, 1912).

Souchay, Jean, 'La Hire à l'Observatoire de Paris', in Antonio Becchi, Hélène Rousteau-Chambon, and Joel Sakarovitch (eds), *Philippe de La Hire, 1640–1718, entre architecture et sciences: Actes du colloque, Paris, 24–26 juin 2010* (Paris, 2013), 93–105.

Speiser, David, 'Eulers Schriften zur Optik, zur Elektrizität und zum Magnetismus', in Kanton Basel-Stadt (ed.), *Leonhard Euler 1707–1783: Beiträge zu Leben und Werk* (Basel, 1983), 215–28.

Stackelberg, Jürgen von, 'Das Bienengleichnis: Ein Beitrag zur Geschichte der literarischen Imitatio', *Romanische Forschungen*, 68 (1956), 271–93.

Stafleu, Frans A., et al., *Taxonomic Literature: A Selective Guide to Botanical Publications and Collections with Dates, Commentaries and Types*, 2nd edn (Utrecht, 1976–).

Stannard, Jerry, 'P. A. Mattioli: Sixteenth Century Commentator on Dioscorides', in University of Kansas (ed.), *Bibliographical Contributions I* (Lawrence, KS, 1969), 59–81.

Stearn, William T., *Botanical Latin: History, Grammar, Syntax, Terminology and Vocabulary* (London, 1966); 4th edn (1992).

Steck, Max, *Bibliographia Euclideana* (Hildesheim, 1981).

Steinmann, Holger, *Absehen—Wissen—Glauben: Physikotheologie und Rhetorik 1665–1747* (Berlin, 2008).

Steixner, Roland (ed.), *Philosophia historica de montibus: Eine Dissertationsschrift der Universität Innsbruck aus dem Jahr 1713. Text, Übersetzung, Kommentar* (Vienna, 2009).

Stöger, Johannes Nepomuk, *Scriptores provinciae Austriacae Societatis Jesu* (Vienna, 1855).

Stolberg, Michael (ed.), *Archiv der europäischen Lexikographie: Fach-Enzyklopädien Geschichte der Medizin* (microfiche edn, 1993–2006).

Stoner, Jean, and Anthony Grafton (eds), *Girolamo Cardano: The Book of My Life. De Vita Propria Liber* (New York, 2002).

Stotz, Peter, *Handbuch zur lateinischen Sprache des Mittelalters*, 5 vols (Munich, 1996–2004).

Stowers, Stanley K., *Letter Writing in Greco-Roman Antiquity* (Philadelphia, PA, 1986).

Stroh, Wilfried, *Latein ist tot, es lebe Latein! Kleine Geschichte einer großen Sprache* (Berlin, 2007).

Struik, Dirk J. (ed.), *A Source Book in Mathematics (1200–1800)* (Cambridge, MA, 1969).

Stuber, Martin, Stefan Hächler, and Luc Lienhard (eds), *Hallers Netz: Ein europäischer Gelehrtenbriefwechsel zur Zeit der Aufklärung* (Basel, 2005).

Stuckrad, Kocku von, *Geschichte der Astrologie: Von den Anfängen bis zur Gegenwart*, 2nd edn (Munich, 2007).

Sudhoff, Karl, *Bibliographia Paracelsica: Besprechung der unter Hohenheims Namen 1527–1893 erschienenen Druckschriften* (Berlin, 1894).

Sudhoff, Karl, *Iatromathematiker vornehmlich im 15. und 16. Jahrhundert* (Bratislava, 1902).

Sudhoff, Karl, *Aus der Frühgeschichte der Syphilis: Handschriften- und Inkunabelstudien, epidemiologische Untersuchung und kritische Gänge* (Leipzig, 1912).

Sudhoff, Karl, and Charles Singer (eds), *The Earliest Literature on Syphilis, Being Ten Tractates from the Years 1495–1498*, 12 vols (Florence, 1925).

Sumillera, Rocío G., 'Reading Scientific Translations in the First Half of Sixteenth-Century Europe through Hernando Colón's Library', in Sumillera, Jan Surman, and Katharina Kühn (eds), *Translation in Knowledge, Knowledge in Translation* (Amsterdam and Philadelphia, PA, 2020), 17–40.

Sumillera, Rocío G., Jan Surman, and Katharina Kühn (eds), *Translation in Knowledge, Knowledge in Translation* (Amsterdam and Philadelphia, PA, 2020).

Suter-Meyer, Katharina, *Wissensvermittler—Kritiker—Autor: Joachim Vadians Kommentare zu Pomponius Mela (Basel, 1522)* (Zurich, 2020).

Swerdlow, Noel M., '*Urania Propitia, Tabulae Rudophinae faciles redditae a Maria Cunitia:* Beneficent Urania, the Adaptation of the Rudolphine Tables by Maria Cunitz', in Jed Z. Buchwald (ed.), *A Master of Science History: Essays in Honor of Charles Coulston Gillispie* (Dordrecht, 2012), 81–121.

Symons, John, 'Medical Bibliographies and Bibliographers', in Alain Besson (ed.), *Thornton's Medical Books, Libraries and Collectors* (Aldershot, 1990), 239–66.

Tannery, Paul, Charles Henry, and Cornélis de Waard (eds), *Œuvres de Fermat*, 5 vols (Paris, 1891–1922).

Taub, Liba, *Science Writing in Greco-Roman Antiquity* (Cambridge, 2017).

Tautschnig, Irina, 'Carlo Noceti, Iris: Einleitung, Text, Übersetzung und Anmerkungen' (unpublished diploma thesis, University of Innsbruck, 2017).

Tautschnig, Irina, 'Constructing Authority in the Paratext: The Poems to Johannes Hevelius's *Selenographia*', *Perspectives on Science*, 30 (2022), 1005–41.

Taylor, John R., *Linguistic Categorization*, 3rd edn (Oxford, 2003).

Tebeaux, Elizabeth, *The Emergence of a Tradition: Technical Writing in the English Renaissance, 1475–1640* (Amityville, NY, 1997).

Telle, Joachim, 'Das Rezept als literarische Form: Zum multifunktionalen Gebrauch des Rezepts in der deutschen Literatur', *Berichte zur Wissenschaftsgeschichte*, 26 (2003), 251–74.

Telle, Joachim, and Sven T. Limbeck (eds), *Paracelsus im Gedicht: Theophrastus von Hohenheim in der Poesie des 16. bis 21. Jahrhunderts* (Hürtenwald, 2008).

Tering, Arvo, 'On the Early Reception of John Brown's Medical Theory on the Example of Doctoral Dissertations Defended in Jena 1794–1795', in Meelis Friedenthal, Hanspeter Marti, and Robert Seidel (eds), *Early Modern Disputations and Dissertations in an Interdisciplinary and European Context* (Leiden, 2021), 312–39.

Tessicini, Dario, 'Defining Cosmology in the Early Modern System of Knowledge, 1530–1621', *Perspectives on Science*, 30 (2022), 1–18.

Thijssen, Johannes M. M. H., and Henricus A. G. Braakhuis (eds), *The Commentary Tradition on Aristotle's De generatione et corruptione: Ancient, Medieval, and Early Modern* (Turnhout, 1999).

Thomas, Ernest C. (ed.), *The Philobiblon of Richard de Bury, Bishop of Durham, Treasurer and Chancellor of Edward III* (New York, 1889).

Thompson, John W., *American Arctic Lichens 1: The Macrolichens* (New York, 1984).

Thorndike, Lynn, *A History of Magic and Experimental Science*, 8 vols (London and New York, 1929–58).

Thorndike, Lynn (ed.), *The Sphere of Sacrobosco and Its Commentators* (Chicago, 1949).

Thorndike, Lynn, 'Newness and Craving for Novelty in Seventeenth-Century Science and Medicine', *Journal of the History of Ideas*, 12 (1951), 584–98.

Thorndike, Lynn, and Pearl Kibre, *A Catalogue of Incipits of Medieval Scientific Writings in Latin: Revised and Augmented Edition* (Cambridge, MA, 1963).

Thraede, Klaus, *Grundzüge griechisch-römischer Brieftopik* (Munich, 1970).

Tilg, Stefan, 'Die argute Inschrift als barocke Form des freien Verses', in Tilg and Benjamin Harter (eds), *Neulateinische Metrik: Formen und Kontexte zwischen Rezeption und Innovation* (Tübingen, 2019), 133–47.

Todd, Robert B., 'The Manuscripts of the Pseudo-Proclan *Sphaera*', *Revue d'histoire des textes*, 23 (1993), 57–71.

Tonelli, Giorgio, *A Short-Title List of Subject Dictionaries of the Sixteenth, Seventeenth and Eighteenth Centuries as Aids to the History of Ideas* (London, 1971).

Topham, Jonathan, 'A Textbook Revolution', in Marina Frasca-Spada and Nick Jardine (eds), *Books and the Sciences in History* (Cambridge, 2000), 317–37.

Tophinke, Doris, 'Zum Problem der Gattungsgrenze: Möglichkeiten einer prototypentheoretischen Lösung', in Barbara Frank, Thomas Haye, and Doris Tophinke (eds), *Gattungen mittelalterlicher Schriftlichkeit* (Tübingen, 1997), 161–82.

Toribio, Pablo, 'The Latin Translations of the *Philosophical Transactions* (1671–1681)', in Rocío G. Sumillera, Jan Surman, and Katharina Kühn (eds), *Translation in Knowledge, Knowledge in Translation* (Amsterdam and Philadelphia, PA, 2020), 123–43.

Touber, Jetze, 'Sacred Medicine in Early Modern Europe', in Dmitri Levitin and Ian Maclean (eds), *The Worlds of Knowledge and the Classical Tradition in the Early Modern Age* (Leiden and Boston, MA, 2021), 179–216.

Tournoy, Gilbert, and Terence O. Tunberg, 'On the Margins of Latinity? Neo-Latin and the Vernaculars', *Humanistica Lovaniensia*, 45 (1996), 134–75.

Touwaide, Alain, 'Printing Greek Medicine in the Renaissance: Scholars, Collections, Opportunities, and Challenges: Introduction', *Early Science and Medicine*, 17 (2012), 371–7.

Traninger, Anita, *Deklamation—Disputation—Dialog: Medien und Gattungen europäischer Wissensverhandlungen zwischen Scholastik und Humanismus* (Stuttgart, 2012).

Trappen, Stefan, 'Fiktionsvorstellungen der Frühen Neuzeit: Über den Gegensatz zwischen *fabula* und *historia* und seine Bedeutung für die Poetik: Mit einem Exkurs zur Verbreitung und Deutung von Laktanz, *Divinae institutiones* I 11, 23–25', *Simpliciana*, 20 (1998), 137–63.

Trepp, Anne-Charlott, *Von der Glückseligkeit alles zu wissen: Die Erforschung der Natur als religiöse Praxis in der Frühen Neuzeit* (Frankfurt am Main, 2009).

Trevisani, Francesco, *Descartes in Deutschland: Die Rezeption des Cartesianismus in den Hochschulen Nordwestdeutschlands* (Zurich and Berlin, 2011).

Troupeau, Gérard, 'Les *Aphorismes* de Jean Mésué, médecin du Calife Haroun al-Rashid, et leur diffusion en occident', *Histoire des sciences médicales*, 31 (1997), 317–26.

Udías, Augustin, *Jesuit Contribution to Science: A History* (Heidelberg, 2015).

Valleriani, Matteo, 'The Tracts on the *Sphere*: Knowledge Constructed over a Network', in Valleriani (ed.), *The Structures of Practical Knowledge* (Cham, 2017), 416–73.

Valleriani, Matteo (ed.), *De sphaera of Johannes de Sacrobosco in the Early Modern Period: The Authors of the Commentaries* (Cham, 2020).

Valleriani, Matteo, and Andrea Ottone (eds), *Publishing Sacrobosco's De sphaera in Early Modern Europe: Modes of Material and Scientific Exchange* (Cham, 2022).

Van Helden, Albert, 'The Importance of the Transit of Mercury of 1631', *Journal for the History of Astronomy*, 7 (1976), 1–10.

Van Ortroy, Fernand, *Bibliographie de l'œuvre de Pierre Apian* (Besançon, 1902).

Vedder, Fritz, Hans-Jürgen Hoffmann, and Axel Peuster (eds), *Der Dialog über die Vögel von G. Longolius (Köln 1544): Lateinischer Originaltext, deutsche Übersetzung, Kommentierung* (Cologne, 2016).

Ventura, Iolanda, 'Translating, Commentating, Re-Translating: Some Considerations on the Latin Translations of the Pseudo-Aristotelian *Problemata* and Their Readers', in Michèle Goyens, Pieter de Leemans, and An Smets (eds), *Science Translated* (Leuven, 2008), 123–54.

Ventura, Iolanda, 'Classification Systems and Pharmacological Theory in Medieval Collections of *Materia Medica*: A Short History from the Antiquity to the End of the 12th Century', in Tanja Pommenering and Walter Bisang (eds), *Classification from Antiquity to Modern Times. Sources, Methods, and Theories from an Interdisciplinary Perspective* (Berlin and Boston, MA, 2017), 101–66.

Verdonk, Johannes J., 'Über die Geometrie des Petrus Ramus', *Sudhoffs Archiv*, 52 (1968), 371–81.

Vickers, Brian, 'Epideictic Rhetoric in Galileo's *Dialogo*', *Annali dell' Istituto e Museo di storia della scienza di Firenze*, 8 (1983), 69–102.

Vickers, Brian, *In Defence of Rhetoric* (Oxford, 1988).

Vinne, Theodore L. de, *A Treatise on Title-Pages: With Numerous Illustrations in Facsimile and Some Observations on the Early and Recent Printing of Books* (New York, 1902).

Vittu, Jean-Pierre, 'Les Contrefaçons du *Journal des Savants* de 1665 à 1714', in François Moureau (ed.), *Les Presses grises: La Contrefaçon du livre (XVIe–XIXe siècles)* (Paris, 1988), 303–31.

Vittu, Jean-Pierre, 'Du *Journal des savants* aux *Mémoires pour l'histoire des sciences et des beaux-arts*: L'Esquisse d'un système européen des périodiques savants', *Dix-septième siècle*, 228 (2005), 527–46.

Voelkel, James R., *The Composition of Kepler's Astronomia nova* (Princeton, NJ, and Oxford, 2001).

Volk, Katharina, *The Poetics of Latin Didactic: Lucretius, Vergil, Ovid, Manilius* (Oxford, 2002).

Völlnagel, Jörg, *Alchemie: Die königliche Kunst* (Munich, 2012).

Von Staden, Heinrich, 'Gattung und Gedächtnis: Galen über Wahrheit und Lehrdichtung', in Wolfgang Kullmann, Jochen Althoff, and Markus Asper (eds), *Gattungen wissenschaftlicher Literatur in der Antike* (Tübingen, 1998), 65–94.

Von Staden, Heinrich, '"A Woman Does Not Become Ambidextrous": Galen and the Culture of Scientific Commentary', in Roy K. Gibson and Christina S. Kraus (eds), *The Classical Commentary: Histories, Practices, Theory* (Leiden and Boston, MA, 2002), 109–39.

Vossen, Carl, *Latein: Muttersprache Europas*, 3rd edn (Düsseldorf, 1979).

Waard, Cornelis de, et al. (eds), *Correspondance du P. Marin Mersenne, religieux minime*, 17 vols (Paris, 1932–88).

Waddell, Mark A., *Jesuit Science and the End of Nature's Secrets* (Farnham and Burlington, VT, 2015).

Wagner, Fritz, *Zur Apotheose Newtons: Künstlerische Utopie und naturwissenschaftliches Weltbild im 18. Jahrhundert* (Munich, 1974).

Wallisch, Robert (ed.), *Kolumbus: Der erste Brief aus der Neuen Welt: Lateinisch—deutsch, mit dem spanischen Text des Erstdrucks im Anhang* (Stuttgart, 2000).

Wallisch, Robert (ed.), *Der Mundus novus des Amerigo Vespucci: Text, Übersetzung und Kommentar* (Vienna, 2002).

Walmsley, Peter, '[Early Modern and Enlightenment Rhetoric:] Rhetoric and Science', in Michael J. MacDonald (ed.), *The Oxford Handbook of Rhetorical Studies* (Oxford, 2017), 547–57.

Walravens, Hartmut, 'Der Jesuit Michael Boym (1612–1659) und seine *Flora sinensis*', in Ingrid Kästner et al. (eds), *Erkunden, Sammeln, Notieren und Vermitteln: Wissenschaft im Gepäck von Handelsleuten, Diplomaten und Missionaren* (Aachen, 2014), 289–306.

Waquet, Françoise, 'Qu'est-ce que la République des Lettres? Essai de sémantique historique', *Bibliothèque de l'école des chartes*, 147 (1989), 473–502.

Waquet, Françoise, *Latin or the Empire of a Sign: From the Sixteenth to the Twentieth Centuries* (London and New York, 2001).

Weijers, Olga, *In Search of the Truth: A History of Disputation Techniques from Antiquity to Early Modern Times* (Turnhout, 2013).

Weise, Stefan, '*Ἑλληνίδ' αἶαν εἰσιδεῖν ἱμείρομαι*: Neualtgriechische Literatur in Deutschland (Versuch eines Überblicks)', *Antike und Abendland*, 62 (2016), 114–81.

Wels, Volker (ed.), *Philipp Melanchthon: Elementa rhetorices: Grundbegriffe der Rhetorik* (Berlin, 2001).

Wenskus, Otta, 'The Use of Greek in Selected Writings of Kepler (*Astronomia nova, Epitome astronomiae Copernicanae* and *Harmonice mundi*)', in Martin Korenjak and Irina Tautschnig (eds), *Die antike Literatur und die Wissenschaftliche Revolution* (Baden-Baden, 2023), 167–88.

Werner, Lukas, 'Deutschsprachige Biographik', in Christian Klein (ed.), *Handbuch Biographie: Methoden, Traditionen, Theorien* (Stuttgart, 2009), 265–77.

Wiegand, Hermann, '*Icarus hinc absit*...Das Technikmotiv in der neulateinischen Dichtung', in *Karl-Friedrich-Gymnasium Mannheim. Jahresbericht 1983–1984* (Mannheim, 1984), 35–45.

Williams, Gareth D., *Pietro Bembo on Etna: The Ascent of a Venetian Humanist* (Oxford, 2017).

Winkler, Alexander, and Florian Schaffenrath (eds), *Neo-Latin and the Vernaculars: Bilingual Interactions in the Early Modern Period* (Leiden and Boston, MA, 2018).

Witt, Ronald, 'Medieval *Ars Dictaminis* and the Beginnings of Humanism: A New Construction of the Problem', *Renaissance Quarterly*, 35 (1982), 1–35.

Witty, Francis J., 'The Beginnings of Indexing and Abstracting: Some Notes towards a History of Indexing and Abstracting in Antiquity and the Middle Ages', *The Indexer*, 8 (1973), 193–8.

Wöhrle, Georg, *Theophrasts Methode in seinen botanischen Schriften* (Amsterdam, 1985).

Wöhrle, Georg, 'Vom Nutzen der Pflanzen für den Menschen: Anthropozentrische Perspektiven in Theophrasts botanischen Schriften', *Antike Naturwissenschaft und ihre Rezeption*, 15 (2005), 73–89.

Woodward, David (ed.), *The History of Cartography*, vol. 3: *Cartography in the European Renaissance*, 2 vols (Chicago, 2007).

Wootton, David, *The Invention of Science: A New History of the Scientific Revolution* (New York, 2015).

Worp, Jacob A. (ed.), *De gedichten van Constantijn Huygens, naar zijn handschrift uitgegeven*, 9 vols (Groningen, 1892–9).

Worstbrock, Franz J., 'Die Antikerezeption in der mittelalterlichen und der humanistischen *Ars dictandi*', in August Buck (ed.), *Die Rezeption der Antike: Zum Problem der Kontinuität zwischen Mittelalter und Renaissance* (Hamburg, 1981), 187–207.

Yang, Shou-zhong (ed.), *The Pulse Classic: A Translation of the Mai Jing by Wang Shu-he* (Boulder, CO, 1997).

Zedelmaier, Helmut, 'Navigieren im Textuniversum: Theodor Zwingers *Theatrum vitae humanae*', *metaphorik.de*, 14 (2008), 113–35.

Zedelmaier, Helmut, *Werkstätten des Wissens zwischen Renaissance und Aufklärung* (Tübingen, 2015).

Zgusta, Ladislav, 'Copying in Lexicography: Monier-Williams' Sanskrit Dictionary and Other Cases (Dvaikośyam)', *Lexicographica*, 4 (1988), 145–64.

Zgusta, Ladislav, et al., *Manual of Lexicography* (Prague, 1971).

Zilsel, Edgar, 'The Sociological Roots of Science', *American Journal of Sociology*, 47 (1942), 544–62.

Zinner, Ernst, *Leben und Wirken des Johannes Müller von Königsberg, genannt Regiomontanus*, 2nd edn (Osnabrück, 1968).

Index of Names

Note: Tables and figures are indicated by an italic '*t*' and '*f*', respectively, following the page number.

For the benefit of digital users, indexed terms that span two pages (e.g., 52–53) may, on occasion, appear on only one of those pages.

Abbas, Haly 78–9
Achrelius, Daniel 423, 425–6
Adam, Melchior 413–14
Addison, Joseph 384–6, 397–8, 400–1, 403–6
Aelian 92
Agardh, Carl Adolph 269n.53, 278n.92
Agerius, Nicolaus 280–2
Agricola, Georg 33, 46–7, 68n.65, 137–9, 144, 322, 325–6, 331–3, 335–6, 338
Aicardius, Paulus 376n.169
Albert of Saxony 108–10, 116–21, 124
Albertus Magnus 123
Albinus 176–8
al-Bīrūnī 291–2
Alciati, Andrea 186–7
Alciati, Francesco 186–8
Aldrovandi, Ulisse 226–7, 227nn.169–170, 233, 406–8, 407*f*
Alessandrini, Giulio 384–5
Alexander de Villa Dei 383
Alexander of Aphrodisias 105–6, 123
Alhazen 78–9
al-Idrīsī, Muḥammad Ibn-Muḥammad 81n.18
Ali ibn Ridwan 110
Alpino, Prospero 260–5
Alsted, Johann Heinrich 136, 320, 323, 327n.48, 330
Altavilla, Benedetto 243–4
Amann, Nils Nilsson 267–8
Ambrose 163
Andorff, Friedrich August 338–9
Andromachus the Elder 385
Annio da Viterbo 111n.143
Antimachus 256–7
Apian, Philipp 375–6
Apianus, Petrus 341, 344–5, 364
Apin, Siegmund Jacob 268n.46, 276n.85
Apollonius 84, 86n.52, 105–6, 110, 113, 122, 309
Apuleius 70n.71
Aratus 382–3
Archimedes 77n.2, 78–9, 84, 90, 101–3, 105–6, 291, 309, 411
Aristarchus 80n.15, 110–11, 301n.158, 336
Aristotle 13–14, 24, 30–3, 35, 79n.8, 82n.24, 83–6, 91–2, 98n.91, 105–10, 108n.124, 113, 116–24, 132n.199, 133n.206, 143, 149–50, 225–7, 251, 258–9, 276, 280–1, 291, 298, 305–6, 312–13, 316–17, 320, 343–4, 364–5, 406–8, 407*f*, 413–14, 428
Aselli, Gaspare 295–7, 296nn.145–146, 299–300, 303–7
Ashmole, Elias 168*t*
Asper, Markus 49
Athenaeus 251
Aubrey, John 168*t*
Augurello, Giovanni Aurelio 384–5
Averroes 78–9, 106–7, 110, 123, 131n.198
Avicenna 2n.15, 27–8, 45, 64n.50, 67n.61, 68n.65, 78–83, 92, 106, 108–10, 113n.154, 113n.156, 115–16, 124–5, 129–30, 181–2, 231n.181, 291–2, 319, 321–2, 354, 359–60, 364–5, 415–16, 422, 425–6
Avienus 382–3

Bacchetoni, Hieronymus Leopold 396–8, 401, 408–10
Bacon, Francis 87–8, 191, 351n.103, 353n.113
Bacon, Roger 82n.24
Baffi, Giovanni Battista 296–7, 342–4
Baglivi, Giorgio 295, 319, 351n.101
Baier, Johann Jakob 221–2
Baldinger, Ernst Gottfried 275–6
Báñez, Domingo 123
Barbaro, Ermolao 83, 343–4, 347
Barberini, Maffeo (Pope Urban VIII) 400, 404
Barbette, Paul 186n.65
Barclay, John 154
Barlaeus, Caspar 168*t*
Barozzi, Francesco 81n.20, 91–2, 94, 96–9, 107, 122n.173, 125–6, 131–2, 315–16, 342, 344–5, 369–71

Barrow, Isaac 356n.123, 422, 425–8
Bartels, Matthias 149
Bartholin, Rasmus 154, 192, 242n.210
Bartholin, Thomas 157–8, 208*t*, 300–1, 319
Bartholomaeus Anglicus 317n.8
Bartholomaeus de Montagna 181–2
Bartholomew of Messina 82n.24
Bartsch, Jacob 245–8
Basini, Basinio 383
Bauhin, Caspar 137–9, 178, 226–7, 230, 234–5, 296–7
Bauhin, Johann 178
Bayer, Johann 46–7, 218n.134, 327n.50, 338–9
Bayer, Michael 184
Beaune, Florimond de 114
Beckher, Daniel 277
Beeckman, Isaac 173n.24
Béguin, Jean 365
Belon, Pierre 220–1, 224, 228–30, 261
Bembo, Bernardo 255–6
Bembo, Pietro 56, 194–5, 255–7
Benedetti, Giovanni Battista 244n.220
Berengario da Carpi, Jacopo 108–10, 129n.192
Berger, Friedrich 284–6
Beringer, Johann Bartholomäus Adam 288n.114
Bernard, Edward 168*t*
Bernegger, Matthias 86–7, 168*t*, 258–9
Bernoulli, Daniel 216n.129, 416–17, 423
Bernoulli, Jacob 114, 293
Bernoulli, Johann 204, 281, 288n.114
Beroaldo, Filippo 104n.104
Bettini, Mario 318, 322–4, 326, 327n.50, 329–31
Beughem, Cornelis à 149
Beyer, Hartmann 123n.175
Biancani, Giuseppe 321, 324–5, 342–5, 413–15
Bicadelli, Galeotto de 270
Biese, Nicolas 112–13, 116–21, 117*f*, 389–90
Bilfinger, Georg Bernhard 192
Biwald, Leopold 360n.132, 361n.137, 367–9, 370*f*
Blankaart, Steven 136–7, 140–1
Blasius, Gerard 92
Bleidner, Johannes 282
Blumenbach, Johann Friedrich 147n.241, 214–15, 267n.44, 278–80
Boccaccio, Giovanni 410–11, 414
Bock, Hieronymus 224, 227
Boehmer, Georg Rudolf 64
Boerhaave, Herman 98n.89, 150n.258, 176–7, 351n.103, 353n.113, 356n.123, 364–5, 376n.169, 421–6, 425*f*
Boethius 77–8, 79n.8, 251–2, 254–5, 261n.28, 362–3
Boissier de la Croix de Sauvages, François 218n.134
Bolyai, János 43
Bonet, Théophile 99–100
Bongo, Pietro 123n.203, 318, 321, 398, 401
Bontius, Jacobus 264
Borch, Ole 138, 144–5
Bordinus, Franciscus 344–5, 350
Borel, Pierre 147–9, 150n.255
Borelli, Giovanni Alfonso 113, 281, 294
Börner, Friedrich 150–1
Borrhaus, Martin 357n.125, 364, 367, 368*f*, 369n.156
Borromeo, Carlo 187
Boscovich, Roger Joseph 112–13, 206–7, 212–13, 388–92, 389n.19, 397–8
Bouelles, Charles de 41n.30
Boussuet, François 385–6, 387*f*
Boyle, Robert 87–8, 137, 168*t*, 191, 192n.73, 195, 196n.86, 197–8, 298n.151, 402n.46
Boym, Michał Piotr 100n.95, 222–3
Bradwardine, Thomas 291–2
Brahe, Tycho 2, 19, 59–60, 166n.6, 167–70, 168*t*, 172, 178–9, 180n.42, 182, 193–4, 196, 201, 242n.210, 243, 300–1, 324n.41, 344–5, 397–8, 401–2, 415–19, 421, 423, 425–7
Brasavola, Antonio Musa 71
Bresadola, Giacomo 218n.132
Breyn, Johann Philipp 184n.56
Brisson, Mathurin-Jacques 220–1, 230
Brissot, Pierre 419–20
Broecke, Marcel Peter René van den 338–9
Brown, John 278–9
Brucaeus, Heinrich 363–4
Bruckner, Daniel 213n.124
Brudzewski, Albert 111–12
Brunfels, Otto 86n.53, 133–6, 149–50, 220–1, 224, 234–5, 346n.95, 400–1, 404, 413–14
Bruni, Leonardo 79–80, 82n.24
Brunn, Johann Jakob von 277
Bruno, Giordano 28–9, 136, 303, 384–5, 389–90
Buchanan, George 389–90
Budé, Guillaume 125n.180
Burgersdijck, Franco 104n.104, 360
Burley, Walter 411
Burman, Johannes 224
Burnet, Thomas (physician) 92, 340–1, 350–1
Burnet, Thomas (theologian) 23, 279, 397–8, 400–1, 403–6

Cabeo, Niccolò 124–6
Caelius Aurelianus 78–9, 252
Caius, John 221n.146, 222–3, 227–8
Calandrini, Jean Louis 114

Caldani, Leopoldo Marcantonio 172n.21
Calepino, Ambrogio 133
Callard de LaDucquerie, Jean Baptiste 145
Calzolari, Francesco 87n.54, 194n.80, 200
Camerarius, Joachim 86–7, 100n.94, 138–9, 385, 400n.43
Camerarius, Rudolf Jacob 104n.104, 284–6, 420, 422–4, 428
Campanella, Tommaso 321
Campanus of Novara 82–3, 121n.170
Candidus, Pantaleon 246
Capella, Martianus 316–17
Cappeller, Moritz Anton 293–4
Capuanus, Franciscus 111–12, 115–21, 117*f*, 124–5
Cardano, Girolamo 6n.27, 104n.103, 113, 125, 153–4, 303, 321, 344–5, 352–3, 413, 416
Carl, Johann Samuel 340–1, 343, 346–7
Carnarius, Johannes 423–4
Castelli, Bartolomeo 136–7
Castelvetro, Lodovico 253–4
Castro, Pedro de 150–1
Catena, Pietro 422, 425–6
Caturegli, Pietro 243–4
Celsius, Anders 192, 215n.128, 216–17
Celsus 52, 335, 350–1
Celtis, Conrad 389
Cesalpino, Andrea 73n.87, 219n.137, 225–8, 227n.169, 230, 232–3, 321–2
Cesi, Bernardo 318, 322, 326–7, 330–2, 337–8
Champier, Symphorien 135–6, 138–9, 149–50, 182
Chrysoloras, Manuel 82–3
Cicero 34–5, 50–1, 77–8, 97, 180–1, 184, 251, 253, 281, 350–1, 419, 433
Ciruelo, Pedro 321, 330
Clarke, Samuel 87, 89n.65, 95, 99–100, 129–31
Claudian 395
Clavius, Christopher 126, 131, 329n.51
Cleomedes 340, 357–8
Clerck, Carl 221–2
Cleyer, Andreas 100–1
Clifford, George 222–3
Clusius, Carolus 166n.7, 167, 171, 173n.24, 182n.47, 222–4, 227n.170, 230–1, 237–8, 237n.200, 404–5
Clüver, Detlev 279, 282
Cluyt, Outgert 295
Collins, John 168*t*
Columbus, Christopher 183–4, 194–5, 401
Columbus, Ferdinand 82n.22
Comenius, John Amos 145, 358
Commandino, Federico 84–6, 92n.72, 110–11, 121–2, 124–5, 129n.192
Conon 102
Copernicus, Nicolaus 2, 28–9, 38, 61n.23, 169–70, 303, 325n.43, 344–5, 352n.111, 363–4, 397–8, 400n.43, 406n.51, 413n.70, 417
Cordus, Euricius 253–4, 346
Cordus, Valerius 182
Cornarius, Janus 415n.76
Costus, Petrus 400–1, 402n.48
Crato, Johannes 41n.35
Cristini, Bartolomeo 172
Crocius, Johann Heinrich 278–9
Crüger, Peter 40
Cruser, Hermann 110, 116–21
Cunitz, Maria 90, 247, 404–5
Curtius, Matthaeus 356n.123
Curtz, Albert von 154, 242n.210, 243, 293, 400n.42

Dalechamps, Jacques 227–8
d'Alembert, Jean Baptiste le Rond 44n.45
Dalrymple, James 331n.56
d'Angelo, Jacopo 82–3
Darwin, Charles 169–70
Dasypodius, Konrad 84, 107n.123, 138, 140, 143, 145, 377–8, 419–20
Dear, Peter R. 191
de Beaune, Florimond 114
degli Emanuelli, Bartolomeo 131–2, 291–2
Della Porta, Giambattista 321
Descartes, René 52–3, 87–8, 94–5, 114, 129–31, 167–9, 168*t*, 278–9, 321–2, 396–7, 406–8
De Vega, Juan 256n.18
Diogenes Laertius 291, 410–11
Dionysius, Paullus 93, 386
Dionysius Periegetes 382–3
Diophantus 83–4
Dioscorides 20, 27n.62, 65–6, 82–3, 87, 108–10, 126, 128–9, 219, 221–2, 226–7, 230–2, 262, 346, 432–3
Dodoens, Rembert 234n.191, 355–6, 369–71, 399–400
Dolaeus, Johannes 321–2
Dorn, Gerhard 135–6, 143
Draud, Georg 147–9
Drechsler, Johann Gabriel 275–6, 284
Dryvère, Jérémy de 419–20
Dubois, Jacques 73n.87, 306
Du Laurens, André 337–8
Dürer, Albrecht 86–7, 100n.94

Ebelin, Jakob 270, 272
Ecchelensis, Abraham 113
Ecker, Johann Matthias Alexander 350–1

Edward III, King 201
Egloff von Staadthoff, Franz Karl Anton 408
Eisenstein, Elisabeth 57
Ekerman, Petrus 276–7, 281
Elsholtz, Johann Sigismund 195
Engel, Johannes 243–4
Epicurus 291, 395, 397–8, 401–2, 405–6, 416n.78, 419–20
Erasmus, Desiderius 125n.180, 153, 165, 334, 358, 373n.163, 426–7, 433
Erotianus 133–4, 140
Estienne, Henri 89, 135–6, 139–40, 143
Euclid 54, 56, 72, 78–9, 82–3, 91–2, 94, 96, 105–6, 121n.170, 122, 125–6, 131–2, 143, 214, 218–19, 287, 307, 322–3, 345, 355n.120, 362–3, 411, 432–3
Euler, Leonhard 2, 166n.6, 167–9, 168*t*, 172n.22, 173n.24, 214–16, 218, 293, 322–3, 325–6, 337–8
Eusebius 241–2, 246
Eustachi, Bartolomeo 176, 186–90, 189n.67
Eutocius 110

Faber, Basilius 145n.233
Fabri, Wenceslaus 121
Fabricius, Ernst Friedrich 369–71
Fabricius, Hieronymus 296–7
Fabricius, Johann 193–4, 196, 200–1, 303
Fahrenheit, Daniel Gabriel 161, 206–7, 213–14
Falloppio, Gabriele 185, 192, 202–3, 303–4, 306
Fatio de Duillier, Nicolas 396–7, 401n.45, 403
Favaro, Giuseppe 383
Fehr, Johann Michael 183, 292n.122, 295, 305
Feigenpuz, Johann E. 288n.114
Fellon, Thomas Bernard 384–5
Ferdinand IV, King 293
Ferdinando II of Tuscany 86n.52, 184–5, 296n.144, 298, 301
Fermat, Pierre de 16n.10, 168*t*, 172, 297n.149, 299, 303, 305, 309–11, 310*f*
Fernel, Jean 252n.5, 253–4, 412n.65
Feuerlein, Jakob Wilhelm 279
Figulus, Carolus 256–7
Finé, Oronce 321–3, 330, 338, 397–8, 403–4
Fischart, Johann 142
Flamsteed, John 166n.6, 169–70, 243, 248, 338n.67
Flemyng, Malcolm 384–5
Florianus, Johannes 94n.79
Florido, Ambrosio 259
Florus 340
Foësius, Anutius 136, 140
Fogel, Martin 6n.25
Forster, Richard 244–5
Fracastoro, Girolamo 293–4, 296n.144, 300, 305–7, 310–11, 384–5, 388
Francis I, King 96
Frandsen, Hans 385
Fratta, Giovanni 69–70
Fries, Theodor Magnus 214, 216–17, 238–9
Frisius, Andreas 87, 89, 195
Frisius, Gemma 59–60, 111–12, 121, 345, 359n.130, 375n.166
Froriep, Robert 338–9
Fuchs, Leonhart 154n.268, 223–7, 230–3, 235, 419–20
Füchsel, Georg Christian 213–17
Fuhrmann, Manfred 340n.74, 357–8, 366–7

Gabriele, Angelo 255–6
Galen 20, 27–8, 71, 78–81, 84–6, 93, 96, 103n.99, 106, 108–11, 116–21, 126, 128–9, 133–4, 136n.214, 140, 153, 226–7, 251n.2, 252, 303–4, 306, 312–13, 340n.74, 342–3, 348n.98, 350–1, 355–6, 364–5, 369–71, 411, 415–16
Galilei, Galileo 52–3, 63–4, 86–9, 154, 172, 183–4, 193–4, 200, 253–4, 258–9, 281, 283, 333, 344–5, 401–4
Gallus, Paschalis 150
Galvani, Luigi 201, 213–17
Gascoigne, Robert Mortimer 1–2
Gassendi, Pierre 52–3, 185–6, 193–4, 200, 356n.123, 363–4, 396–8, 401, 412, 415–21, 416n.78
Gasser, Achilles Pirmin 357n.125
Gaub, Hieronymus David 97n.88
Gaun, Franz 288n.114
Gaurico, Luca 101–3, 422, 424n.100
Gauß, Carl Friedrich 37–8, 43, 322–3, 325–6
Gaza, Theodorus 83, 85–6, 89, 91–2, 98–100, 111n.148
Geminus of Rhodes 84, 110–11
Gemma, Cornelius 244–5
Gentile da Foligno 124n.178, 131–2
George of Trebizond 83, 111, 415–16
Gerard of Cremona 82–3, 358, 363–4, 369
Gerber, Traugott 176–7
Gerstner, Johann Karl Anton 265n.32
Gessner, Conrad 2n.15, 71, 128n.188, 137–42, 142*f*, 147–8, 153–4, 167, 170–1, 173, 178, 182, 219n.136, 220–1, 223–4, 226–7, 232–3, 274–5, 278–9, 315–18, 323, 327n.48, 329–36, 341–4, 347, 350–1, 356n.121, 401, 412, 415–17
Ghymm, Walther 416
Giacoletti, Giuseppe 384–5
Gilbert, William 33, 195, 198n.93, 307, 320

Gilles, Pierre 92
Giovio, Paolo 108n.127
Giuntini, Francesco 321
Gmelin, Johann Georg 175–8, 222–3, 225–6
Goens, Daniel van 278–9
Goldbach, Christian 169n.12, 172n.22, 173n.24
Gonzaga, Francesco 172
Grafton, Anthony 1
Grammatica, Francesco Vigilio 397–8, 408, 409*f*
Grape, Zacharias 279, 282
Grapheus, Cornelius 399–400
Gregory XIII, Pope 296n.144
Gregory, David 343n.86, 344–5, 360–1, 363–4
Greiling, Johann Christoph 155–6
Grischow, Augustin 215n.128
Grisogono, Federico 195, 430–1
Grospré, Robert 341, 345–6
Gross, Alan G. 209–12
Gross, Johann Georg 277
Grosseteste, Robert 291–2
Großinger, Johann Baptist 221–2
Grove, Robert 383–4, 402n.46, 403
Gruter, Jacob 88n.60
Guarino Veronese 84
Guarinonius, Hippolytus 254–5, 259–61
Guericke, Otto von 195–6, 200
Guibert, Nicolas 264
Gupalatinus, Nicolaus 89n.66
Gutenberg, Johannes 57–8

Habersack, Franz Adolph 384–5
Habicot, Nicolas 297–8
Hainzel, Johann Baptist 145
Haller, Albrecht von 18n.19, 37, 43–4, 150, 166n.6, 167, 172n.21, 174–81, 186–8, 194–5, 202, 232, 266–7, 275n.75, 365n.151
Haller, Wolfgang 92–3
Halley, Edmond 84, 86n.52, 110, 168*t*, 204, 206–7, 214–15, 401n.45, 405–6
Harchies, Josse de 138–9, 350, 385–6
Harmon, Joseph E. 209–12
Harvey, William 278–9, 293, 295–7, 311, 313–14, 383–4, 401, 402n.46, 403
Hawthorne, Nathaniel 55
Hebenstreit, Johann Baptist 403
Hebenstreit, Johann Ernst 160n.285
Heffter, Johann Carl 2n.8, 64, 151–2, 273, 275–7
Heister, Lorenz 358–9, 361
Hell, Maximilian 193–4, 416–17
Helmholtz, Hermann von 269n.52, 288n.113
Helmont, Jan Baptist van 46n.55, 69–70, 251n.3
Hemmer, Johann Jakob 192
Hermann V von Wied 256
Heron of Alexandria 84, 133–4, 143
Herschel, William 112–13, 383–4, 396–7, 401, 403
Hesiod 382–3, 390–2
Hesronita, Johannes 81n.18
Heuraet, Hendrik van 303n.161
Heurne, Johan van 357n.125, 360n.134, 364–5
Hevelius, Johannes 16n.10, 180, 336, 398n.40, 401
Hippocrates 27–8, 52, 58n.9, 80–1, 84n.33, 92–3, 106, 108–10, 135–6, 278–9, 321–2, 340–1, 350–3, 386, 408, 411, 415–16, 422–3
Hizler, Georg 423
Hofer, Johann 215–16
Hoffmann, Friedrich 321–2, 331
Hofmann, Caspar 221–2, 231
Homer 50, 77–8, 261–2, 301n.158, 382–3, 390–2
Honterus, Johannes 145, 356n.121, 357n.124, 364, 386–9
Horn, Georg 356n.123, 376n.169
Horne, Johannes van 114n.159
Horrocks, Jeremiah 180
Hottinger, Johann Heinrich 213–14, 216, 268–9, 279–80, 287
Hottinger, Salomon 268–9
Howard, Nicole 59–60
Hueber, Georg Ludwig 288n.114
Huet, Pierre-Daniel 79–80
Humboldt, Alexander von 221–2
Humphrey, Lawrence 79–80
Huswirt, Johannes 362–3, 374, 375n.166
Hutton, Charles 168*t*
Hutton, James 288–90
Huygens, Christiaan 2, 114, 166n.6, 168*t*, 278–9, 295, 401
Huygens, Constantijn 396–7
Huygens, Constantijn, the Younger 406–8

Iacopo da Cremona 102–3
Ingrassia, Giovanni Filippo 81, 111, 126
Innocent X, Pope 260
Ittig, Thomas 69–70, 267–8, 319

Jacobaeus, Matthias 399
Jacobi, Carl Gustav Jacob 38n.19, 298
Jacquier, François 114
Jacquin, Nikolaus Joseph von 220–1, 290–1
Javellus, Chrysostomus 123
Jean de Meur 358
Jefferson, Thomas 345–6
Jerome 77–80, 163, 241–2, 414n.73
Jessen, Johann von 416–17, 421, 423, 426–7
John George II, Elector of Saxony 69–70
John of Garland 133–4, 137

John of Genoa 133
Johnson, Thomas 194–5, 201
Jonston, John 137–8, 230–1, 320, 346–7
Julius Caesar 188
Juncker, Christian 152, 204
Jungermann, Ludwig 221–2, 231
Jungius, Joachim 168*t*
Junius, Hadrianus 144
Jurin, James 168*t*, 183, 214–15, 218
Jussieu, Antoine Laurent de 227–8, 229*f*

Keill, John 360–1, 363–4
Kentmann, Johannes 138–9, 228–30
Kepler, Johannes 2, 28–9, 39n.23, 56, 61–2, 86–7, 92n.72, 104n.104, 112–13, 154, 166n.6, 167–70, 168*t*, 183–4, 193–4, 241n.208, 242–8, 250, 252, 293, 295, 300–1, 303, 318, 321, 325–6, 337–8, 344–5, 363–4, 372–3, 375–6, 378–9, 381, 400n.43, 401–5, 416–17
Kies, Johann 280–1
King, Adam 389–90
Kinner, Cyprian 401
Kircher, Athanasius 23, 40, 56, 167–9, 168*t*, 182, 194–5, 300–1, 322, 326–7, 331, 334–5, 338
Kirchmaier, Georg Caspar 184
Klein, Jacob Theodor 228n.173
Krause, Johann Gottlieb 157–8
Kraz, Georg 216–17
Kronick, David A. 207, 208*t*
Kühn, Carl Gottlob 150

Lachmund, Friedrich 184n.56
LaCroix, Demetrius de 388
Laet, Johannes de 89, 111, 122–5, 321, 337–8
Lagomarsini, Girolamo 384–5
Laguna, Andrés de 342–3, 350–1
La Hire, Philippe de 242–3, 247
La Loubère, Antoine de 297n.149
Lambert, Johann Heinrich 156n.279
Lansbergen, Philips van 241n.208, 243n.219, 245–6
Lavoisier, Antoine de 31, 46–7
Ledesma, Miguel Jerónimo 92
Leeuwenhoek, Antonie van 16–17, 90, 167–9, 168*t*, 179–81, 206–7
le Febvre, François Antoine 384–5
Leibniz, Gottfried Wilhelm 36–7, 155n.269, 157n.281, 166n.6, 167, 179n.36, 214–15, 307n.166
Leo Africanus 94n.79
Leoniceno, Niccolò 92n.72, 94
Leopold, Christoph 284–8
Le Seur, Thomas 114
Lesser, Friedrich Christian 184, 189
Lhuyd, Edward 167–9, 168*t*, 185, 222–3, 228–30
Libavius, Andreas 400–1, 404
Lichtenberg, Georg Christoph 213–14
Linacre, Thomas 84, 110–11
Lindemann, Thomas 277
Linden, Johannes Antonides van der 147n.240, 150, 278–9
Linnaeus, Carl 37n.13, 56, 65–6, 112–13, 137, 167–71, 174–5, 180, 219–26, 228–30, 232–7, 247, 275–6, 278–9, 290–1, 295, 300, 353–4, 421–3, 427n.107, 428
Lipenius, Martin 150, 351–2
Lipsius, Justus 165, 404–5
Lister, Martin 168*t*
Livius Andronicus 77–8
L'Obel, Matthias de 227–8, 230, 232
Lochner, Christoph Moritz 279
Lochner, Jacob Hieronymus 277
Longueil, Gilbert de 257
Lonicer, Johannes 67n.60
Lossius, Lucas 362–3, 372–3, 375n.166
Louis III, Duke of Württemberg 376
Lubieniecki, Stanisław 180–1
Lucas, Henry 425–6
Lucian of Samosata 251
Lucretius 56, 112–13, 382–3, 386–8, 390–3, 394n.28, 395, 397–8, 401–2, 405–6
Ludwig, Christoph Gottlieb 159–60
Lusitanus, Amatus 112–13, 192–3, 196, 201
Luther, Martin 79–80

Maestlin, Michael 360–1, 363–4, 367, 375–81
Magini, Giovanni Antonio 154n.268, 167–9, 168*t*, 172
Maimonides, Moses 350–1
Major, Johann Daniel 184n.56, 185–6, 211–12, 215–17, 385–6
Manardi, Giovanni 67n.60, 180, 182
Manget, Jean Jacques 319
Manilius 52n.15, 112–13, 116–21, 117*f*, 382–5
Manincor, Anton 267n.43, 277
Manna, Pietro 185
Manuel, Chrysoloras 82–3
Marggraf, Georg 222–3
Marggraff, Michael 282
Märter, Franz Joseph 138–9, 145, 366
Marti, Hanspeter 272–3
Martini, Martino 246–7
Mattioli, Pietro Andrea 65–6, 87, 128–9, 131–2, 342–3, 346
Mauchart, Johann David 284–6
Maurolico, Francesco 241n.203, 256–7

Mazzolari, Giuseppe Maria 384–5
Melanchthon, Philip 52–3, 55–6, 242, 356n.123, 360–1, 361n.136, 363nn.144–145, 364–5, 372, 400n.43, 420–3, 425–6
Mencke, Friedrich Otto 158, 208*t*, 412, 417–18
Mercator, Gerardus 29–30, 338n.67, 416
Mercklin, Georg Abraham 150
Mercuriale, Girolamo 84–5, 108, 143, 186n.62, 338, 343, 356n.123, 376n.169
Merian, Maria Sibylla 89–90
Mersenne, Marin 166n.6, 167–9, 168*t*, 173n.24, 319
Mesmer, Franz Anton 280
Meurer, Wolfgang 137–8
Milich, Jakob 421–2, 423n.95, 425–6
Mizauld, Antoine 344n.91, 389–90, 391*f*, 397–400, 403–4
Mizler, Lorenz Christoph 115–16, 155nn.272–3, 156
Moffett, Thomas 182n.48, 230–1, 258, 264n.31
Molnár, János 384–5
Monau, Friedrich von 297–8
Mondella, Luigi 189n.67
Mondino de Luzzi 108–10, 129n.192
Montalbani, Ovidio 149
Montanus, Johannes Fabricius 182
Monte, Giovanni Battista da 108, 115–21, 126, 131–2
Monteux, Sebastien de 154n.268, 182
Morescottus, Alfonsus 341n.77
Morhof, Daniel Georg 319
Morison, Robert 220–1, 227–8, 290n.115
Morrhy, Gérard 398, 403–4
Moscati, Pietro 160
Müller, Jakob 378–9
Müller, Otto Frederik 225–6
Müller von Löwenstein, Friedrich 141, 143
Münster, Sebastian 87
Muralt, Johann von 343–4, 347
Musschenbroek, Petrus van 355–6

Napier, John 245–6, 248
Naudé, Gabriel 61, 416
Nemnich, Philipp Andreas 135–6, 140–1
Nepos, Cornelius 410–11
Neucrantz, Paul 295
Newton, Isaac 2, 3n.18, 28–33, 80, 87, 89, 92, 95, 114, 129–31, 167–9, 168*t*, 179n.36, 191n.72, 196–7, 206–7, 287n.109, 307, 318, 325n.43, 344–5, 360–1, 392n.24, 395–8, 401–2, 405–6, 415–16
Nicander 85, 108–10, 112–13, 382–3
Niccolai, Alfonso 401n.45
Nicomachus of Gerasa 24–6, 78–9, 105–6
Nicomedes 256–7
Nieremberg, Juan Eusebio 222–3, 230
Noceti, Carlo 384–5, 388–90, 392–5
Nuñez, Pedro 87–8, 111–12, 126

Oddi, Marco degli 115–16
Oddi, Oddo degli 45, 111, 115–16, 124–5
Odo Magdunensis 393
Oldenburg, Henry 99n.93, 166n.6, 167–9, 168*t*, 206–7, 292, 298
Olhoff, Johann Erich 180
Oresme 291–2
Ortelius, Abraham 167–9, 168*t*, 338–9
Osler, William 64
Oudin, François 388–9, 392n.23
Ovid 70n.71, 382–3, 395, 399–400, 403, 405–6

Pallavicino, Sforza 255n.13
Pamphilus 257
Pappelbaum, Georg Christoph 176–7
Pappus 84–6, 90, 92n.72, 106n.116, 110–11, 121–2, 124–5, 129n.192, 414
Paracelsus 31, 87–8, 135–7, 143, 258, 401
Pardies, Ignace Gaston 206–7
Pascal, Blaise 55–6, 172
Passe, Crispijn van de 220–1, 232n.184
Patrizi, Francesco 297, 320, 325–6
Paul III, Pope 296n.144
Paul of Aegina 80n.15, 84–5, 111, 113
Paulli, Simon 149, 219–20
Paullini, Christian Franz 184
Pauw, Pieter 223n.153, 357n.125, 360n.134
Peano, Giuseppe 43
Pecham, John 358
Pecquet, Jean 195–203, 199n.94, 260n.26
Peiresc, Nicolas-Claude Fabri de 167–9, 168*t*, 171
Peletier, Jacques 89, 111–12, 116–21, 216n.130, 345, 375n.166
Pena, Pierre 227–8
Pennant, Thomas 168*t*
Petermann, Daniel 282, 284–6
Petrarch 164–5, 180–1, 410–11
Petreius, Johannes 88, 319
Peucer, Caspar 182
Peuerbach, Georg von 70n.72, 94, 104n.104, 111–12, 115–21, 117*f*, 124–6, 242–3, 248–50, 340n.72, 343n.86, 344–5, 363–4, 369, 377, 417
Pflaum, Jakob 243–4
Pico della Mirandola, Giovanni 29, 324
Pictorius, Georg 350
Pindar 395

Piscator, Andreas 276–7, 281
Planer, Johann Andreas 278n.92
Plantin, Christoph 232n.184
Platner, Johann Zacharias 217
Plato 105, 141n.229, 211–12, 251–2, 258–9
Platter, Felix 371n.158
Pliny the Elder 122–3, 128–9, 128n.189, 146–7, 219, 225–6, 232–3, 316–17, 326, 330
Pliny the Younger 163, 180–1, 188
Plot, Robert 168*t*
Ploucquet, Wilhelm Gottfried 150
Plutarch 92n.72, 211–12, 251–2, 410–11
Polignac, Melchior de 131n.196, 386
Pollucius, Samuel 336
Pomata, Gianna 6
Pomponazzi, Pietro 108n.127, 115–16, 123
Pomponius Mela 29–30, 116–21, 117*f*, 127–8, 131–2
Pontano, Giovanni 111, 318, 324, 384–5
Pontanus, Johannes Isacius 168*t*
Pope, Alexander 397–8
Porphyry 412
Porzio, Lucantonio 251n.3, 253–4
Porzio, Simone 184–5
Praetorius, Martin 400–1, 403–5
Prantner, Peter 267n.43, 277
Pritzel, Georg August 149
Proclus 86n.52, 91–2, 94, 96, 106n.115, 110–11, 122n.173, 125–6, 131–2
Ptolemy 20, 28–30, 46n.57, 70n.72, 78–9, 82–5, 94, 105–6, 111, 113, 241–2, 245–7, 256–7, 315–17, 321, 324n.40, 327, 344–5, 377–9, 395, 399–400, 404, 411, 415–16
Pythagoras 303, 401

Quillet, Claude 388, 396–8
Quintiis, Camillus Eucherius de 386
Quintilian 49–50, 52
Quintus Serenus 383

Ralla, Johannes 284–6
Ramus, Petrus 358, 362–3, 369–71
Rasario, Giovanni Battista 84–5
Ratdolt, Erhard 72, 121n.170, 296n.146
Ray, John 185, 225–7, 232–3
Rechtenbach, Polycarp Michael 275–6, 284
Redi, Francesco 87, 89, 195
Regiomontanus, Johannes 59–60, 70n.72, 83, 94, 102–3, 104n.104, 169–70, 243–5, 248, 340n.72, 343n.86, 344–5, 363–4, 369, 417, 422, 425–7, 429
Reidy, Michael 209–12
Reinhold, Erasmus 111–12, 242–3, 245–6
Reisch, Georg 277, 283–4, 320, 323, 331
Renieri, Vincenzo 184–5
Reuss, Jeremias David 152, 207
Reusser, Johann Philipp 280–2
Rhazes 78–9, 91n.69, 108–10, 115–21, 126, 131–2
Rheticus, Georg Joachim 169–70, 185, 245–6, 248n.235, 257–8, 425–7
Riccioli, Giovanni Battista 317–18, 321, 324–7, 328*f*, 330–2, 334, 336–8
Roberval, Gilles Personne de 84–5, 110–13
Rodigast, Samuel 282
Rohault, Jacques 87, 99–101, 129–31
Romoli, Domenico 341
Rondelet, Guillaume 220–1, 227–8, 231, 385–6, 400–1, 402n.48, 412n.65
Rösel von Rosenhof, August Johann 89n.63, 90
Rossi, Gregorio 254n.11
Ruel, Jean 83–4, 128–9, 225–6, 321
Ruf, Jacob 89–90, 92–3
Ruland, Martin 136–9
Rümelin, Johann Heinrich 416–17
Rumpf, Georg Eberhard 90, 224–5
Ryff, Walther Hermann 345–7

Sachs von Lewenheimb, Philipp Jacob 168*t*, 185–6, 211–12, 292, 295, 327n.49
Sacrobosco, Johannes de 104n.104, 106, 108–11, 121, 123n.175, 126, 129–32, 296n.146, 315–16, 344–5, 359–61, 363–4, 377–8, 381, 398, 415–16
Saia, Nonio Marcello 300, 306
Sainte-Marthe, Scévole de 388
Saladinus de Asculo 135–6, 345–6, 372–3
Salernus, Iacobus 284–6
Salvemini, Giovanni 416
Salviani, Ippolito 220–1
Sambucus, Johannes 406–8, 413n.70
Sanches, António Nunes Ribeiro 92–3, 175–8
Sand, Christoph 99n.93, 206–7
Santangél, Luis de 183–4
Santorio, Santorio 300–3, 352n.110, 353n.113
Sapidus, Johannes 400–1, 404
Saumaise, Claude 139, 140n.226
Savastano, Francesco Eulalio 388
Savoy, Gaspare 278–9
Scaliger, Joseph 21–2, 116–21, 117*f*, 135, 154n.268, 167–9, 168*t*, 246, 293–4, 303–4, 326n.46
Scaliger, Julius Caesar 113
Scheiner, Christoph 267n.44, 281, 283–6, 293, 300, 307, 336
Schenck, Johann Georg 6n.24, 150, 292

Scheuchzer, Johann Jakob 22–3, 87, 149, 153–4, 183n.52, 194–5, 196n.86, 200, 202, 268, 279–80, 287
Schickard, Wilhelm 185–6
Schindler, Conrad 277
Schmied, Gottfried 282, 284–6
Schnabel, Johannes 279, 282
Schneider, Johann Friedemann 288n.112
Schoepf, Thomas 246–7
Schönborn, Samuel 71n.78, 340–1, 348
Schöner, Johannes 18n.15, 185, 242–3, 321, 363n.144
Schönperger, Johann Georg 267n.44
Schoock, Martin 278n.92
Schooten, Frans van 94–5, 114
Schott, Gaspar 167–9, 168*t*, 195, 319
Schreckenfuchs, Erasmus Oswald 111–12
Schreiber, Friedrich 175–7
Schröck, Lucas 71n.78
Schulze, Johann Heinrich 45–6, 192, 196
Schweling, Johann Eberhard 278–9
Schwenckfeld, Caspar 221–2
Scot, Michael 83
Scribonius, Wilhelm Adolf 367–9
Semmelweis, Ignaz Philipp 269n.52
Seneca 161n.288, 180–1
Sennert, Daniel 364–5
Servius 106–7
Severino, Marco Aurelio 92
Sfondrati, Pandolfo 294, 296n.144, 297
Sigonio, Carlo 253–4
Sigwart, Georg Friedrich 275–6
Silvaticus, Matthaeus 133–4, 134n.207, 136
Simmler, Josias 153–4, 411–12, 417
Sionita, Gabriel 81n.18
Sittard, Kornelius 256
Solinus 122–3, 127
Sommerhoff, Johann Christoph 134n.209, 135–6, 140–1
Soranus 77–8, 411, 415–16
Sorbière, Samuel 416
Spacciolo, Valerio 86n.52, 122n.173
Spener, Christian Maximilian 155n.273, 213–14
Speroni, Sperone 253–4
Spiegel, Adriaan van de 338–9
Sprat, Thomas 52–3
Sprengel, Kurt 150–1
Stahl, Georg Ernst 294, 297–8, 346
Statius 395
Stattler, Benedikt 45, 267–8, 360n.132, 366
Stay, Benedict 52–3, 384–6, 389–90, 398n.37, 401–2, 428
Stay, Christoph 52–3
Stearn, William 236–7
Steno, Nicolas 2, 33, 186n.65, 294, 296–8, 301, 399
Sterpin, John 206–7
Stifel, Michael 46–7
Stöffler, Johannes 243–4, 396–8, 401
Strabo 29–30, 84–6, 316–17
Stromeyer, Friedrich 47–8, 216–17
Struś, Józef 84–5, 98n.91
Sturm, Johann Christoph 277, 361–3, 372
Sturm, Reginald 93
Suetonius 410–11, 414–15, 426–7
Swammerdam, Jan 89–90, 98n.89, 114n.159, 167–9, 168*t*
Syllanus de Nigris 115
Szerdahélyi, György Alajos 112–13, 383–4, 396–7, 403

Tannstetter Collimitius, Georg 413
Tarillon, François 384–5
Tartaglia, Niccolò 102–3
Taruffi, Giuseppe Antonio 399, 403
Tasso, Torquato 253–4
Tazewell, William 345–6
Telesio, Bernardino 320
Themistius 123
Theon of Alexandria 83, 106n.116, 121n.170, 362n.139, 414–15
Theophrastus 32, 85, 98, 111, 122–5, 144, 219–21, 225–7, 233n.187, 291, 316–17
Thomas Aquinas 123
Thomas, Loup 384–5
Thorer, Alban 84
Thorius, Raphael 388
Thorndike, Lynn 1
Tifernas, Gregory 84, 428
Tilli, Michaele Angelo 223n.153
Toxites, Michael 136, 142
Triegler, Johann Georg 398
Trithemius, Johannes 147
Trombsdorff, Johann Nikolaus 282
Turcq, Pascasius Justus 67n.60
Turner, William 153–4, 257

Ugolino, Blasio 264
Urban VIII, Pope (Maffeo Barberini) 400, 404
Ussher, James 173n.24, 174n.30
Uterverius, Johannes Cornelius 406–8

Vadianus, Joachim 116–21, 117*f*, 127–8, 131–2, 344–5
Valentini, Michael Bernhard 88, 92
Valentini, Pietro Aloisio 364–5

Valgrisi, Vincenzo 84–5
Valla, Giorgio 110
Varenius, Bernhardus 318, 321, 330, 333–4, 337
Varolio, Constanzio 186n.62
Venusti, Antonio Maria 181–2
Vesalius, Andreas 2, 46–7, 61n.23, 63n.41, 72–3, 91n.69, 185, 187, 201n.97, 305–6, 317nn.10–11, 329, 338–9, 356n.123, 364–5, 400n.43
Vesling, Johann 365, 367
Vespucci, Amerigo 183–4, 194–5
Vettori, Benedetto 108–10, 116–21, 124
Vian, Christoph 284–8
Viète, François 293, 296–7
Virgil 56, 106–7, 331, 382–3, 386–8, 405–6, 407*f*
Virginianus, Saxirupius 404
Vitali, Girolamo 138, 243n.215, 249n.239
Vives, Juan Luis 103–4nn.102–3, 125n.180
Voigt, Gottfried 299
Vorstius, Aelius Everardus 423
Vossius, Isaac 294, 296n.144, 297–8

Wagner, Christian 176, 279
Waid, Anton 267n.43, 277
Waldung, Wolfgang 295
Wallis, John 167–9, 168*t*
Wang Shuhe 100–1
Wedel, Georg Wolfgang 22–3, 294–7, 300–2, 304, 306, 308, 371n.158
Weidler, Johann Friedrich 149
Weinhart, Ignaz 355–6, 362–3, 372, 375, 416–17, 420–3, 425–6
Welsch, Georg Hieronymus 193n.76
Westmann, Johannes 422–3
Widmann, Johannes 356n.123, 362–4, 367, 374–5
Widmann, Joseph Alphons 278–9
Wiggers, Johann Georg 411n.63
William IV, Landgrave of Hesse-Kassel 179, 182
William of Moerbeke 82–3, 102–3
Willis, Thomas 251n.3
Willughby, Francis 233
Winckler, Daniel 295
Winter, Johann 70n.72, 84n.36, 96–8, 103n.99, 111
Winther, Mathias 150–1
Winthrop, John, the Elder 168*t*
Winthrop, John, the Younger 168*t*
Wolff, Christian 277, 283–4
Woodward, John 87
Wotton, Edward 220, 226–7
Wren, Christopher 303
Wurfbain, Johann Paul 327n.49

Xylander, Wilhelm 84

Zallinger zum Thurn, Franz Seraphin von 360, 367
Zamagna, Bernardo 384–5, 389n.19
Zetzner, Lazarus 88
Zwinger, Friedrich 213n.124, 214, 320, 326, 369–71

Subject Index

Note: Tables are indicated by an italic "*t*", following the page number.

For the benefit of digital users, indexed terms that span two pages (e.g., 52–53) may, on occasion, appear on only one of those pages.

'a capite ad calcem' 138–9, 348, 383
abbreviations
  explanation of 248
  in handwriting 66
  in print 239–40, 287, 337, 348, 433
abridgements 204–5, 264, 332, 362–5
  compendia as 339–40
  'epitome' as term for 94
  of works in reviews 156
  textbooks as 359–60
  updated 344–5
academies 19–20; *see also* journals, academy
  and the vernaculars 42–3
  in Western and Southern Europe 273–4
  papers read at sessions of 419–20
  speeches given at 423
'accessus' 104–5, 115, 410–11
Act and Tripos verses 273–4, 385
adjectives, *see also* epithets
  formation of new 44–6
  in botany 238–40
advertisement of books 9–10
  frontispieces as 423–4
  in paratexts 68, 223–4, 342–3, 401
  in the colophon 68–9
  on title pages 71, 144, 244–5, 299, 315–16, 324, 326–7, 337–8, 369, 380, 395–6, 403–4
  previews as 157–8
alchemy, *see* chymistry
algebra
  evolution of 30–1
  fused with geometry 94–5
  symbolic 293
'algorithmi'
  as textbooks 347
  Latin equivalent to 'Rechenbücher' 341
  perhaps used by autodidacts 345
  style of exposition 374–5
  subtitles in 367
almanachs 241–5; *see also* 'ephemerides'
alphabetization
  determined by Latin 141
  in bibliographies 146–7, 150–2
  in dictionaries 132–4, 138
  in encyclopedias 319
  in 'historiae' 226–7
  in indexes 331
  in pharmacopoeias 348
  incomplete vs full 71
  rejection of 227
anatomy
  animal 92, 400
  human 126
  in the medical curriculum 17–18
  non-academic practitioners of 43, 297–8
  of the ear 187
  terminology of 46–8
  textbook of 358–9
anonymity 155–6, 159–60, 204–5
anthology 92, 319, 350–1, 388–9, 397–8, 435
antiquarianism 22, 166, 436
  and medicine 296–7, 338
aphorisms 351–4, 435
  astrological 84–5, 111, 344–5
  dissertations consisting of 284–6
  Galenic 350–1
  Hippocratic 106
    commentaries on 108–10
    poetic translations of 93, 385–6
  monograph consisting of 300–1
  style 56
appendix, publication as 153–4, 261, 296–7, 416–17
Arabic 34, 39
  commentary on scientific texts in 108–10
  source of terminology 35, 39–40, 46–7, 133–5
  translation from, into Latin 78–83, 85–6, 113, 340
arithmetic 24–6, 30–1; *see also* 'algorithmi'; Hindu-Arabic numerals
  teaching of elementary 341, 345, 361–3, 374
artisans
  collaborating with scholars 15–17
  involved in printing 57–8

astrology 29
degraded to pseudo-science 5, 17–18
symbols used in 46–7
table works in 244–5
astronomy 28–9
and philology 20–1
atlases in 338–9
compendia in 344–5
evolution towards celestial mechanics 32–3
monographs in 293–4
orations on 424–5
pandects in 321
praised in panegyrical poetry 399
reports in 193–4
spherical and theorical (planetary) 28–9, 363–4, 373–4, 377–8, 381
symbols used in 46–7
table works in 243–4
textbooks in 363–4, 375–81
atlas 29–30, 338–9, 433
audience of speeches 423–4*f*, 428–31
role in rhetoric 51, 55
authority
creation of 260–3
of classical authors 259, 278–9, 321–2, 378–9, 406–8
of early modern authors 209–10, 234–5
pandects associated with 325–6
autodidacticism 106–7, 340–2, 345, 355, 362–3, 377

bi- and multilingualism
in books 40, 42–3, 87, 90, 139–42, 224–5, 317–18, 335–6
in correspondence 171–2
Bible
allegorical interpretation of 428
biblical philology 50
biblical science 22–3, 108–10
Gutenberg's 57–8
Vulgate 77
binomial nomenclature, *see* terminology, technical
biology 24, 32
folk- 227
book, *see also* advertisement of books; title
collectors 61, 150–1
fairs 62–3
market 41–2, 61, 64–5, 68, 95, 293, 395–6
numbering systems in 66–7, 71, 303–4, 434
portability 223–4, 347
prices 63
privilege 62–3, 68–9
sellers 60
botany 32; *see also* pharmacology; terminology, technical
bibliographies of botanical works 149
botanical gardens 32, 222–3
long-lived Latin tradition of 43
translations of botanical works 80–1
brevity
as stylistic merit 51
of letters 163
of monographs 290
broadsheets
announcing dissertations 266–72
bloodletting calendars as 244–5
in the incunabula era 64
panegyrical poems on 397–8, 408–10
portraits-cum-epigrams as 406–8

calculus 30–1, 214–15, 280–1, 322–3
cartography 29–30; *see also* atlas; geography
maps 71–2, 217, 246–7, 433
table works in 246–7
case histories 6n.26, 192–3
catalogues 435
fair 41–2, 146–8, 155
garden 222–3
library 146–7
museum 64, 151–2, 273, 275 6
of private collections 138–9, 151–2, 222–3, 228–30, 273
of sources 146–7
sales 154, 275–6
star 241–2, 248
catechisms
compendia as 350
textbooks as 363–4, 371–3, 378–9
Catholic Church, *see also* Jesuits
against divination 29
Church Latin 35
schools run by 34–5
censorhip 63–4
chemistry, *see* chymistry
chronology 21–2
as structural principle 146, 150, 413–14
chronological tables 246
chymistry, *see also* medicine, iatrochemistry
covering alchemy and chemistry 10, 24*f*, 31
emergence of modern chemistry from 31
secretiveness 167–9, 319n.22
symbols used in 46–7
terminological confusion 319n.22
Ciceronianism 45, 79–80, 97–8
Ciceronian controversy 35
Ciceronian period 50–1

circumlocutions
  in poetic language 390–2, 405–6
  instead of technical terms 46–7, 188–9, 335–6
citation 38, 287, 336–7; *see also* intertextuality; quotation
clarity 159–60, 305–14, 333
classification, *see* taxonomy
'consilia' (medical letters) 181–2, 192–3
copper engraving 72, 200, 217, 232, 269–70
  title copper and copper title 68–9, 161
'corollaria', in dissertations 267–8, 283–4, 286
cosmography 26, 341
  textbooks of 373–4
cosmology 24
  sub- and superlunary world 13–14, 29
courts (princely) 16, 19–20, 34–5, 42–3
Creation 21–3, 279, 335
cross-reference 66–7, 216
curriculum 18, 108–10, 276–7

dead language, Latin as a 34–5
dedication 69–70
definition
  in dissertations 282, 286
  in textbooks 372–3, 379–80
  stipulative 45, 307
  definition, subdivision, and distinction 347, 366–7
derivative forms 65–6, 435
description
  graphic 188–9, 288–90, 393–4
  in botany 46–7, 56, 235–40, 263
  in classical rhetoric 197–8
  of plants in Latin 43, 236–7
  'vicarious' 56, 238
  vs diagnosis 235–6, 239
diagrams 94, 102–3, 122, 200, 247–8, 256–7, 283, 309–10, 338, 389–90, 433
  called 'theoricae' 369
  replaced by other means 333, 337–8
digests, *see* abridgements
digression 127–8, 156, 253–4, 299
  in literary theory 298–9
  mythological 390–2
'discrete' texts 8, 134–5, 173, 225, 432–3
disputation 8–9, 18
  and dialogue 251–2
  and dissertation 265–9, 273–5
  announcement on broadsheets 269–72
dissection 17–18
  didactic poem on 383
  guides to 365
  report on 196–7

Dutch 45–6, 89–90, 140, 147–8, 157
  language of correspondence 171
  translations from 90, 180, 206–7, 224, 321
  translations into 22–3, 40, 224, 341

earth sciences 33, 46–7, 317–18
edition
  bilingual 40, 77, 90, 140–1, 221–2, 224, 247, 388, 404–5
  counterfeit 61, 62n.36, 204n.101, 365n.152
  expanded 127–8, 137, 147, 388–9
  one-volume 388
  updated 150, 153–4, 359–60
elegiac couplets 93–5, 385, 396–7, 402–3, 408
ellipsis 93, 114, 121–2, 177, 433–4
'ellogium' 69–70, 284, 403–4
eloquence 50, 419, 423–4; *see also* rhetoric
emblem 385–6, 406–8
empiricism 8, 20–1, 27–8, 434
'encyclopedia', early modern meaning of 316
English
  language of modern science 9–10, 41, 43–4
  little known in early modern times 157, 206–7
  source language of Latin translations 88
Enlightenment 224, 252
  didactic poetry in the 383, 388–9
  in favour of the vernacular 41, 43–4
'ephemerides', *see also* almanachs
  astronomical/-logical 205–6, 241–5
  meteorological 43, 88, 207*t*, 214–15, 218, 245
epigrams 350, 385–6, 395–7, 404–5
  and images 400, 406–8
  paratextual 70, 256, 284, 398
  style 404
epistemology, *see also* genre, epistemic
  and the journal article 212
  and the report 190–1
  Aristotelian 14–15, 195
  scepticism 209–10
epitaph 397, 401–4
epithets
  complimentary 160
  ornamental 390–2, 405–6
  specific 234–6, 353
excursus, *see* digression
exegesis 33, 104
  exegetic vs 'quaestiones' commentaries 104–5, 115, 121, 124
experiment 14–15, 31–3
  experimental report 191, 195–8

fact, notion of 14–15, 20
faculties
  arts 17–18, 80–1, 106, 242
  medical 17–18
fictionality, notion of underdeveloped 49–50
Flood 23, 87, 279; *see also* geology
fluid text 65–6
footnotes 66–8, 213–16, 416–17, 434
  commentaries as 120*f*–1, 129–31
forgery 84–5, 110–11, 184–5
French
  as language of correspondence 171
  as lingua franca 157
  chemical terminology founded in 46–7
  translations from 88, 94–5, 224, 297–8
  translations into 22–3, 65–6, 224, 247, 264, 341, 361, 388
frontispieces 68, 423–4
  in pandects 326–7

genre
  as prototypical category 7
  definition of 4
  epistemic 6
geography 29–30; *see also* cartography
  chorography 294, 326–7
  mathematical and descriptive 148, 167–9, 246–7, 344–5, 364
geology 33, 213–14; *see also* Flood
  geological maps 217
  geological terminology 137–8
geometry 24–6, 30–1; *see also* mathematics; quadrivium
  analytic 30–1
  Euclidean/Greek 110, 121–2, 133–4, 309–10
  model of scientific method 284–6, 309
  trigonometry 28–9, 245–6, 380
German
  as language of correspondence 171, 174–5
  technical terminologies in 43, 46–7, 137–8
  translations from 87, 224
  translations into 65–6, 92, 128, 224, 247, 297–8, 350–1, 362–3, 398
grammar, Latin
  disregarded by Linnaeus 56, 235–6
  restored to classical standards 35
Greek
  ancient language of natural sciences 36
  Greek terms Latinized 45–6, 239
  in multilingual lexica 140
  little known 39, 80
  scientific terminology 39–40, 46–7, 135–6
  translations from 77–86
  word composition in 45–6

Hebrew 35, 46–7, 78–80, 140, 224–5, 287
hendecasyllables 403–4
herbals 72–3, 219–20, 224, 232, 432–3
hexameters 93, 382–3, 385, 389–90, 403, 405–6
hierarchy, textual, *see also* structure of works
  Euclidean 307
  expressed typographically 68–9, 121, 367–9, 378–9, 434
  levels of 230, 284–6, 331
Hindu-Arabic numerals 30–1, 358, 362–3, 374
'historia'
  early modern notion of 21–2, 190–1; *see also* 'observatio'
  'historia litteraria' 146–7, 150–1, 319, 373–4
history, and study of nature 21–2
Horatian metres 385, 403–4
horoscopes 29, 242–5
  collections of 21–2, 435
humanism
  and science 20–1
  ideas about education 358
  ideas about Latin 35
  medical 20–1
hyperbole 56, 418–19

illustrations 72–3, 200, 217–18, 283, 337–9, 369
  costs of 62–3
  keys to 241
  problems with 72–3, 232–3
'imitatio' 35, 50–2, 351–2
IMRaD 210, 216, 432–3
incunabula 58–9
  mainly in Latin 36
index 71, 131–2, 150, 179, 223–4, 316*t*, 330–1
  absent in monographs 296
  and marginalia 331–2
  dependent on numbering 66–7
instruments, scientific 27–8, 32–3, 217, 243; *see also* microscope; telescope
  monographs on 291–2
  paper instruments 72
intertextuality 66–7, 252–3, 302–4, 434
Italian
  as language of correspondence 171–2
  translations from 65–6, 88, 128, 180, 183–4
  translations into 87, 264, 341, 388

Jesuits
  gymnasia 18–19, 274
  mission in China 222–3
  Ratio studiorum 39, 274, 358
  style-conscious 288, 334, 337, 383

journals 203–9
'abstract' 146–7
academy 204–5
review 158–61

law 14–16, 20
layout 38, 58–9
as substitute for syntax 235–6, 240–1, 433
in incunabula 58–9
of commentaries 116–21
of dictionaries 135
of didactic poems 389
of recipes 348–9*f*
lectures
and commentaries 20, 104–5, 115
and textbooks 355–7, 364–5
announcements of 355–6, 359n.130
publication of 108, 115–16, 356n.123
students' manuscripts of 108, 376
lemmata
in commentaries 116–7*f*
in dictionaries 132–3, 138–9
libraries 61
bibliographies as virtual 146
pandects as 327n.48
Linnaean taxonomy and nomenclature 26, 43, 46–7, 65–6, 167–9, 221–2, 230, 235–6; *see also* taxonomy; terminology, technical
literacy 60–1, 164–5
literary theory 52–3
fields not covered by 54–5, 382–3
'local flora' 221–2, 231–2

manuscript 5
and print 58–9
lectures in 108, 121–2
letters in 166–78
orations in 420–1
textual corruption in 97–8, 376
maps, *see* cartography
marginalia 66–7, 86–7, 303–4, 331–2, 389, 434
mathematics 30–1; *see also* algebra; arithmetic; geometry; quadrivium
mathematical notation 71–2, 218, 337
mixed 26–7
'occasional' 293
medicine 27–8; *see also* humanism; physicians
Chinese 100–1
five parts of 27, 364–5, 373–4
iatrochemistry 27–8, 137, 258
iatromathematics 29, 345–6
iatromechanics 281, 321–2
veterinary 84, 152

metaphor 238, 326–7, 390–2, 425–6
explanatory 392, 394
to be avoided in modern articles 210
meteorology (Aristotelian) 24
astro- 29, 245
monographs in 294
vs modern 33, 43
metres, *see* elegiac couplets; hendecasyllables; hexameters; Horatian metres
microscope 90, 232, 236–7, 239, 343–4, 400
mineralogy
name of the discipline 326–7
pandects in 322
regional mineralogies 221–2
terminology of 137–8
mining 33, 322
terminology of 46–7, 137–8, 335–6

narration
first-person 8, 190–1, 194–203, 209–10, 213–14, 216–17
in rhetoric 51, 426–7
to be avoided in modern articles 190–1, 210
natural history 26
and 'historiae' 218
compendia of 346
in letters 167–9
terminological problems in 46–7
Neo-Latin 2–3, 35, 44–5
neologisms 45–6, 134–6, 307, 327, 331, 334
non-text 200, 210, 217–18, 240, 433; *see also* diagrams; illustrations; layout
nouns, *see also* style, nominal
abstract 44
noun phrases, in modern articles 210
novelty
in books 69–70, 186, 274–5, 300–2
'novus' as catchword 300–1

'observatio'
epistemological notion of 190–1; *see also* 'historia'
medical genre of 112–13, 192–3
occasionality in reports 196–7
orality 40, 42–3, 144, 265–6, 274, 412, 420; *see also* style, oral

page, folio, and column numbers 66–7
paragraph 58–9, 134–5, 215–16, 389
count 270, 284–6, 303–4, 367–9, 378–9, 428
paratext 67–8; *see also* dedication; footnotes; index; preface; table of contents; title
pathos 56, 334–5, 429–31
patronage 19, 69–70, 269–70, 274–5, 400

period, *see* Ciceronianism, Ciceronian period
persuasion 51, 55, 197–8
pharmacology 24, 27n.62, 256; *see also* botany
compendia of 345–6
pharmacopoeias 347–8
symbols used in 46–7, 433
philology 20–1, 50, 91–2, 116–21, 139, 143–4, 256–7
philosophy
Aristotelian natural 13–14, 24, 320, 341
of mathematics 125–6, 425–6
physicians 16; *see also* medicine
biographies of 413–14
municipal 19
physico-theology 21, 55–6, 188, 276–7, 429
physics
Aristotelian, *see* philosophy, Aristotelian natural
Cartesian 17–18, 32–3, 129–31, 278–9
modern 32–3
Newtonian 17–18, 32–3, 129–31, 280–1, 355–6, 363–4, 386
plagiarism 82–3, 102, 297–8
poetry, *see also* epigrams; metres
mnemotechnic advantages 350, 382
praise as function of 395
polemics 154, 184, 302–4
against alchemy 137, 404–5
avoided in reviews 156, 159–60
humanist against medieval translations 82–3
preface 70–1
printers 60; *see also* publishers
as authors of paratextual poems 398
neat layout produced by 305, 378–9
printing press 4–5, 57–8
professors
against compendia 343
as authors of dissertations 274–5
as authors of textbooks for their own courses 360, 364–5
as scientific authors 18
inaugural lectures 422
lectures published as commentaries 108
progress 15, 68–9, 292–3
effects on publications 80, 86, 126, 220–1, 292–3, 317–18, 323–5, 344–5, 358–9
'observatio' leading to 190–1
Protestantism 18–19, 21–2, 246, 274, 361n.136, 364–5
pseudo-, *see* spurious texts
psychology (Aristotelian) 24–6, 343–4, 364–5
publishers 62–3; *see also* printers
investing money in monographs 293
role in creating new text types 65
taking the initiative for publications 89, 155
wary of pandects 318

quadrivium 24–6; *see also* arithmetic; astronomy; geometry; mathematics
oration in praise of 430–1
taught at arts faculty 17–18
question, rhetorical 373, 430
questionnaire 183
quotation, *see also* citation; intertextuality
Greek 39
in dissertations 287
replaced by referencing 234–5, 434

Ramist table 227–8, 241, 259–60, 305, 330, 350, 369–71
reader
annotating books 60–1, 329, 355–6
friendliness 116–21, 303–4
reading
habits 60–1
lists 146–7, 150–1
perusal vs consultation 71, 306, 316*t*, 329
recipe 64, 348; *see also* pharmacology, pharmacopoeias; style, recipe
republic of letters 16, 144, 165, 429; *see also* scientific community
rhetoric 51–2, 390–2, 419; *see also* eloquence; pathos; persuasion; style
epistolary theory as offshoot of 163–4
'rhetoric of science' 3
textbooks of 361–2
ring composition 189, 404

scholasticism 108–10, 306, 316–17
scholastic Latin 35, 56, 77–8, 115
schools 18–19, 34–5; *see also* Jesuits, gymnasia
disputations in 265–6, 273–4
mathematics taught in 362–3
scientific community 16, 31, 43–4, 144, 429, 434–5; *see also* republic of letters
Scientific Revolution 13, 21, 57
sentence, *see also* Ciceronianism, Ciceronian period; syntax
emphasis on single 367
length 177, 209–10, 314
starting from direct object 177
'sententia' 351–3
serialization 192, 214–15, 243, 278n.92, 296–7, 413, 432–3
simile 390–2, 394
Spanish
letters in 171
translations from 224
translations into 341

'Sprachbund' 41, 47, 55n.30
spurious texts 84–5, 91–2, 128
structure of works, *see also* hierarchy, textual
  'descensus' 320
  emphasized 219–20
  Euclidean 56, 284–7, 331, 353
  standardized 54, 56, 66–7, 231, 348, 432–3
  Suetonian, of biographies 410–11, 415, 426–7
students
  as addressees of books 145, 150–1, 242, 329, 341, 355–6, 364–6, 371–2, 376–7
  as potential investigators of nature 16
  publishing professors' lectures 115–16
  transcribing professors' lectures 108
  warned against compendia 343
style
  apology for 45, 52, 216n.129
  argumentative 310–14
  doctrine of the three styles 51, 56
    'genus grande' 55–6, 196n.86, 405–6
    'genus humile' 52–3, 55–6, 287–8
  figures 3n.18, 289–90, 418–19; *see also* ellipsis; hyperbole; metaphor; question, rhetorical; simile
  higher in introductory parts 55–6, 115, 216–17, 288, 334
  keyword 160, 235–6, 239–40, 243–4, 305, 414, 433
  nominal 45–6, 235–6, 289–90, 433–4
  oral 115–16, 253
  recipe 191n.72, 247–8, 250, 374–5
summary
  in didactic poems 389–90
  in monographs 299, 305–6
  in reviews 156–8
  of letters 179
  title of articles as 215–16
syllogism 104–5, 115, 282–3, 428
synonyms
  elimination of 433
  in botany 20–1, 137, 234–5, 353
  use of, for stylistic reasons 334, 418–19, 433
syntax, *see also* Ciceronianism, Ciceronian period; sentence
  and technical terms 210, 433–4
  repetitive 250, 374
  substituted by visual features 235–6, 240, 433

table of contents 66–7
taxonomy, *see also* Linnaean taxonomy and nomenclature
  in natural history 26, 225–30, 234–5
  of scientific disciplines 23
telescope 28–9, 193–4, 293
  establishment of name 344–5
terminology, technical 132–45; *see also* Linnaean taxonomy and nomenclature
  apology for 45, 52
  binomial nomenclature 65–6, 221–2, 235–7
  botanical 235–40
  Euclidean 125–6, 131–2, 143, 284–6
theology 21, 327; *see also* physico-theology
title
  neologism as 326–7
  page 68–9, 161
    none in incunabula 58–9
    of dissertations 283–4
  rise to prominence 68–9
  sub- 66–7, 94–5, 284–6, 331, 367, 389
  types of
    'animadversiones' 104n.104
    'bibliotheca' 146
    'comment-' 103, 290–1
    'dissertatio' 265
    'enchiridion' 342
    'epitoma' 342
    'Flora' 219–20
    'icones' 219–20
    'iter' 194–5
    'manuale' 342
    'observationes' 112–13, 192–3
    'quaestiones' 123, 435
    'theatrum' 219–20, 326–7
    'thesaurus' 326–7
typeface
  gothic and roman 66, 116–21
  roman vs italic 121, 367, 378–9, 390
  size 66–9, 378–9

university 17–18, 42–3; *see also* faculties
  and commentaries 106, 108–10
  and dissertations 265–6

verbs 44–6, 238
  loss of importance 210, 240, 433–4
vernaculars, *see also* Dutch; English; French; German; Italian; Spanish
  and Latin 40
Vulgate, *see* Bible

witnesses
  in articles 209–10, 212
  in reports 196–7, 199–201
  'virtual' 197–8
women
  as authors 89–90, 247, 404–5
  as correspondents 174
  as readers 41, 60–1
  excluded from academia 16–17
  non-Latinate 40

woodcut 72–3, 128–9, 232–3, 238, 261, 338–9, 385–6
  title 68–9
world systems 28–9, 281, 363–4
  Copernican 84–5, 185–6, 258–9, 324–5, 327, 352–3, 363–4
  Eudoxan 293–4, 306
  Ptolemaic 284–6, 293–4, 324–5, 327
  Tychonian 281, 324–5, 327, 363–4

zoology
  Latin nomenclature in 43, 47, 221–2, 235–6
  monographs in 291, 295